Advances in Cryogenic Engineering
Materials

VOLUME 28

An International Cryogenic Materials Conference Publication

Advances in Cryogenic Engineering *Materials*

VOLUME 28

Edited by
R. P. Reed and A. F. Clark
National Bureau of Standards
U.S. Department of Commerce
Boulder, Colorado

PLENUM PRESS · NEW YORK and LONDON

The Library of Congress cataloged the first volume of this title as follows:

Advances in cryogenic engineering. v. 1–
New York, Cryogenic Engineering Conference; distributed
by Plenum Press, 1960–
v. illus., diagrs. 26 cm.
Vols. 1– are reprints of the Proceedings of the Cryogenic Engineering
Conference, 1954–
Editor: 1960- K. D. Timmerhaus.

1. Low temperature engineering–Congresses. I. Timmerhaus, K. D.,
ed. II. Cryogenic Engineering Conference.

TP490.A3 660.29368 57-35598

ISBN-13: 978-1-4613-3544-3 e-ISBN-13: 978-1-4613-3542-9
DOI: 10.1007/978-1-4613-3542-9

Proceedings of the Fourth International Cryogenic Materials Conference
(ICMC) held in San Diego, California, August 10–14, 1981

Library of Congress Catalog Card Number 57-35598
ISBN-13: 978-1-4613-3544-3

Softcover reprint of the hardcover 1st edition 1982
A Division of Plenum Publishing Corporation
233 Spring Street, New York, N.Y. 10013

CONTENTS

NONMETALLICS AND COMPOSITES

FLUX PINNING IN SUPERCONDUCTORS

HIGH-FIELD SUPERCONDUCTORS

A15 SUPERCONDUCTORS

MULTIPLY-CONNECTED SUPERCONDUCTORS

SUPERCONDUCTOR PROPERTIES AND MEASUREMENTS

STRAIN EFFECTS IN SUPERCONDUCTORS

SUPERCONDUCTOR PERFORMANCE

FABRICATION - SUPERCONDUCTORS

FABRICATION - STRUCTURAL ALLOYS AND COMPOSITES

INDEXES

FOREWORD

The Fourth International Cryogenic Materials Conference (ICMC) was held in San Diego, California in conjunction with the Cryogenic Engineering Conference (CEC) on August 10-14, 1981. The synergism produced by conducting the two conferences together remains very strong. In the application of cryogenic technology, materials continue to be a demanding challenge, and sometimes, an obstacle. The association of materials and cryogenic engineers increases their awareness of recent research in each other's fields and influences the course of future research.

Many contributed to the success of the 1981 conference. J. W. Morris of the University of California--Berkeley was ICMC Conference Chairman. E. N. C. Dalder of Lawrence Livermore Laboratories was ICMC Structural Program Chairman; D. C. Larbalestier of the University of Wisconsin--Madison, and D. K. Finnemore of Iowa State University were Superconducting Materials Program Chairmen. Local arrangments were expertly coordinated by R. E. Tatro of General Dynamics--San Diego. The CEC Board, especially their conference chairman, T. M. Flynn, of the National Bureau of Standards, Boulder, contributed very substantially to conference planning and implementation. All of their efforts provided the foundation of the largest CEC/ICMC ever.

We thank the Office of Naval Research and the Office of Fusion Energy and Basic Energy Sciences of the Department of Energy for providing needed financial support for the conference. Finally, we especially thank M. Stieg, who prepared the papers for the new procedures and format used in this volume.

The ICMC Board met during the conference and elected H. C. Freyhardt of the Institut für Metallphysik, University of Göttingen, West Germany and G. K. White of the National Measurement Laboratory, Sydney, Australia as new Board Members. The board also supported a joint conference with the International Cryogenic Engineering Conference in Kobe, Japan in May 1982. A separate Cryogenic Materials Series book on austenitic steels will be published from papers presented on that subject at the Kobe conference. The next (1983) joint CEC/ICMC conference will be held in Colorado Springs, Colorado.

The 1981 ICMC consisted of 93 contributed papers, 19 invited papers, 19 poster session papers, and a special workshop on low temperature codes and standards. The special workshop generated lively discussion related to the need for development of standards for low temperature material testing, specifications for low temperature materials, and codes for cryogenic structures. Two plenary talks were sponsored by ICMC: S. O. Dean of Fusion Power Associates discussed the future prospects for fusion power, and F. R. Fickett of the National Bureau of Standards reviewed low temperature materials research.

In Volume 28, a wide variety of subjects relate to superconducting materials: A15 compounds, stress-strain effects, development of in situ materials, and new developments in high-field, high-current superconductors. A number of problems were also addressed concerning the development of forced-flow conductors, internally strengthened conductors, very large conductors, the joining of superconductors, testing and evaluation, and stability of superconductors. Highlighted are the improvements of the critical parameters for practical superconductors by ternary additions and the developments necessary for the production of practical conductors for large-scale applications of superconductivity.

Development of standardized nonmetallic laminate materials, the manufacturing capability of cryogenic laminates, radiation effects, and development of basic cryogenic performance data are included. Papers describe work underway in Japan, Germany, and the United States to expand the data base on the cryogenic performance of composite laminates, unreinforced polymers, and concretes. Studies include the effects of anisotropy and biaxial stress states on the mechanical degradation of laminates at low temperatures and on the permeability of laminates to helium. Data on the ability of reinforced and unreinforced polymers to resist impact forces at cryogenic temperatures are presented for the first time. These presentations reflect an increasing sophistication in studies relating to the use of nonmetallic insulating and structural materials in fabrication of cryogenic systems.

Papers are included on the mechanical properties and fracture characteristics of austenitic steels for use at 4 K. Alloy development of high manganese austenitic steels is currently underway, primarily in Japan. It appears that these high manganese steels offer good strength, adequate toughness, and lower cost than the (Fe-Cr-Ni) austenitic stainless steels now in use in most cryogenic structures. New fracture mechanics data on nitrogen-strengthened steels are presented. There were special sessions on the distinctive problems encountered in the development of stainless steel sheaths for superconducting cables. Welding of austenitic steels continues to be a major fabrication problem in the construction of superconducting magnet structural support systems, and a number of papers address weldment properties and procedures.

The conference continues to grow, and the papers improve in quality and cover more diverse subjects, demonstrating that the cryogenic properties of materials are an essential component of an advancing technology.

R. P. Reed
A. F. Clark

BEST PAPER AWARDS

At this conference, the International Cryogenic Materials Conference Board presented its first awards, those for Best Papers from the proceedings of the 1979 ICMC held in Madison, Wisconsin. The editors selected from the proceedings, Advances in Cryogenic Engineering--Materials, Volume 26, about a dozen papers in the two categories structural and nonmetallic materials and superconducting materials. From these nominations the awards committee selected the papers below as Best Papers 1979.

AC LOSSES IN UNTWISTED SUPERCONDUCTING COMPOSITES FABRICATED BY AN *IN SITU* TECHNIQUE

A. I. Braginski and G. R. Wagner

Westinghouse R & D Center
Pittsburgh, Pennsylvania

J. Bevk

Harvard University
Cambridge, Massachusetts

MECHANICAL, ELECTRICAL, AND THERMAL CHARACTERIZATION OF G-10CR AND G-11CR GLASS-CLOTH/EPOXY LAMINATES BETWEEN ROOM TEMPERATURE AND 4 K

M. B. Kasen, G. R. MacDonald, D. H. Beekman, Jr., and R. E. Schramm

National Bureau of Standards
Boulder, Colorado

Quality is one of the major goals of the International Cryogenic Materials Conference and in all of our technical work and presentations. Papers such as these help establish ideals for our research community to follow. We thank the authors for their inspiration.

1981 INTERNATIONAL CRYOGENIC MATERIALS CONFERENCE BOARD

A. F. Clark, Chairman . . . . . . National Bureau of Standards
Boulder, Colorado, USA

R. W. Boom . . . . University of Wisconsin
Madison, Wisconsin, USA

E. W. Collings . . . . Battelle Memorial Institute
Columbus, Ohio, USA

D. Evans . . . . Rutherford Laboratory
Chilton, Didcot, England

G. Hartwig . . . . Institut für Technische Physik
Karlsruhe, FRG

T. Horiuchi . . . . Kobe Steel
Kobe, Japan

J. W. Morris, Jr. . . . . University of California
Berkeley, California, USA

R. P. Reed . . . . National Bureau of Standards
Boulder, Colorado, USA

M. Suenaga . . . . Brookhaven National Laboratory
Upton, New York, USA

K. Tachikawa . . . . National Research Institute for Metals
Ibaraki, Japan

K. A. Yushchenko . . . . E. O. Paton Institute of Electrowelding
Kiev, USSR

ACKNOWLEDGMENT

The International Cryogenic Materials Conference Board is deeply grateful for the support of the Physical Sciences Division and the Materials Division of the Office of Naval Research.

LOW-TEMPERATURE MATERIALS RESEARCH: A HISTORICAL PERSPECTIVE*

F. R. Fickett

Electromagnetic Technology Division, National Bureau of Standards, Boulder, Colorado

INTRODUCTION

The many modern large-scale applications of cryogenics, from superconducting magnets for fusion research to LNG tankers, have been supported by materials that were usually adequate for the task and that appeared usually just in time. This is, of course, not accidental. It is probably fair to say that all significant progress in cryogenic applications has, in one way or another, come about because of an advance in materials or a clever new use of existing materials. As an example, consider the Collins helium liquefier, recently described as "the single most important technical advance of the past forty years in cryogenics."[1] The basic concept of the expansion engine liquefier had been developed quite a few years earlier by Kapitza.[2] Collins's technical improvements that led to production of a successful machine were in large part the clever use of materials, including:[3] hard nitrided steel for pistons and cylinders that allowed smaller clearance and, thus, less leakage; a new design for an efficient heat exchanger using copper fins in copper nickel tubing; and keeping piston and valve rods always in tension, thus allowing use of thinner rods resulting in reduced thermal losses.

In spite of numerous similar examples, materials research is a relatively low-visibility pursuit and one that seldom commands the resources that we in the field feel are our due. Be that as it may, the investigation of the suitability of materials for use at very low temperature is an interesting enterprise with a long and reasonably honorable history, although not without its share of disasters of various magnitudes. Some of that history is presented

*Invited paper.

here along with various milestones in the development of cryogenics. No pretense is made to completeness, although some effort has gone into selecting the more significant events. The definition of cryogenic temperatures that I have chosen is that used for the first Cryogenic Engineering Conference in 1954: temperatures below -150°C.

Following the historical review, a discussion of the present status of the field is given. This is not intended to be a review of the data; several recent ones are available.[4-7] Rather, it is an overview of the general health of the field in relation to the problems of the present day. Then we take a brief look at trends, both good and bad, that seem to be developing in the materials field.

Before venturing into the past, let's take a general look at the materials used in cryogenic systems and consider some usually unwelcome, but basic, truths about their behavior at low temperatures. Nearly every system has the components shown in Table 1. In addition to the properties listed there, every ideal material should be: available with dependable, predictable, and uniform properties; inexpensive; readily obtainable in the shapes and amounts needed; easily fabricated. In addition, a prior history of successful application is always nice, but seldom available.

Table 1. Material Properties for Cryogenic Systems Components

Component	Typical Material	Desirable Properties
support and containment structure (cold)	aluminium alloy, fiberglass-epoxy, potting resin, concrete	high strength, ductility, and thermal conductivity; low total enthalpy change; easily joined
thermal standoff	fiberglass-epoxy, stainless steel	low thermal conductivity; high ductility and strength
thermal insulation	aluminized polymer, polymer foam, radiation shields	low thermal conductivity; high electrical resistivity and strength; easily installed
electrical insulation	conductor insulation, spacer blocks	high electrical resistivity and thermal conductivity

There are, of course, no surprises in this table. It is clear that a perfect material does not exist for any application, and in fact, there are few that we can even call good. In part this is due to the perversity of mother nature, but also because there has been a tendency to make do with materials that are available--materials that were actually developed for quite different purposes. The austenitic stainless steels are a classic example. Only now are steels with much lower nickel and chromium being evaluated for cryogenic applications. Of course there is the other side of the coin in which materials, such as some resins, were developed for low temperature use while ignoring the fact that all devices are at room temperature at some time in their life. The relatively good flexibility at low temperature sometimes resulted in unacceptable sagging near room temperature.

There are some basic truths regarding the effect of low temperatures on materials (Table 2). Some have their roots in basic physics; others fall into the category of "conventional wisdom." In either case there are always exceptions, but they tend to be very unusual materials. Certainly one of our goals for the future should be development of methods of circumventing these relationships. We will mention some new successes in this area later, but as an example of the many ways around a materials problem, consider relative thermal expansion. The fact that various construction materials contract differently has always plagued the low temperature scientist. The solutions that have been developed cover a wide range:

- Use ductile materials, like copper and aluminum, and let them deform;
- Make composites with matching properties like the filled epoxies;
- Build in expansion regions, like loops and bellows;
- Use high modulus materials engineered to be always in tension, such as bolts;
- Allow relative motion in at least one direction;
- Use spring loading with materials such as beryllium copper that maintain some elasticity at low temperatures;
- Let it work for you, such as introducing prestrain into composite superconductors;
- Make one material so thin that its contraction follows the base, like silvering on Dewars and certain epoxy joints.

The important thing to note in this list is that nearly every solution requires knowledge of the properties of many candidate materials. The more well-characterized materials one has at his disposal, the better will be the solution.

With the optimistic conviction that most problems can be solved if we are clever enough, let's proceed to look at how the field of cryogenic materials research has developed.

Table 2. Truths Regarding Materials at Low Temperatures

If a material has	it will also have
high thermal conductivity	high electrical conductivity
low thermal conductivity	low ductility
high electrical conductivity	low strength
low electrical conductivity	high thermal contraction
high strength	low fracture resistance
good fracture resistance	high ductility
low specific heat	low thermal stability
high reflectivity	high thermal and electrical conductivity

WHERE DID WE COME FROM?

The field of cryogenics is relatively new, having been born in the late 1800s. The branch that deals with materials is even newer, with its real start in the early 1950s. The journals that we read (Table 3) and conferences we attend (Table 4) were nearly all created within the lifetime of all but the youngest of us. It should be noted that all of the conferences listed have published proceedings that easily stand on an equal footing with the journals.

Table 3. Journals Covering Low Temperature Materials

Journal	Editor	First Published
Commmunications from the Physical Laboratory, Leiden	H. K. Onnes	1885
Progress in Low Temperature Physics	C. J. Gorter	1957
Progress in Cryogenics	K. Mendelssohn	1959
Cryogenics*	K. Mendelssohn	1960
Cryogenic Technology	W. D. English	1965
Cryogenic Engineering News	R. J. Nemmers	1965
(became CIG)	----	1969
Cryogenic Engineering, Tokyo	K. Yasukochi	1966
Materials Science and Engineering	R. Maddin	1966
Journal of Low Temperature Physics	J. G. Daunt	1969
LNG/Cryogenics	----	1973
Indian Journal of Cryogenics	A. Bose	1976

*Supplement published in 1961 covered papers from 1944-1960.

Table 4. Conferences with Sessions on Materials

Conference	First Held	
Low Temperature Physics	Cambridge, England	1946
Cryogenic Engineering	Boulder, Colorado	1954
International Institute of Refrigeration Commission 1	Eindhoven, Netherlands	1960
Magnet Technology	Stanford, California	1965
Applied Superconductivity	Upton, L.I., New York	1966
Int'l. Cryogenic Engineering	Kyoto/Tokyo, Japan	1967
Int'l. Cryogenic Materials	Kingston, Ontario	1975

The history of cryogenics from the viewpoint of a materials scientist has four fairly distinct periods. Each one is discussed only briefly here, but in that discussion I have attempted to present the important advances in materials and to indicate the forces and events motivating the developments.

The Early Years, 1850–1910

Developments in cryogenics during this period are shown in Table 5. The major accomplishments were in liquefaction of gases. The true birth of cryogenics, as defined earlier (T < -150°C), coincided with the liquefaction of oxygen in 1877. The primary motivating force during this period was a desire to understand the basic physics and chemistry of these liquids, but also the seeds of cryogenic engineering were being sown with the introduction of household refrigeration. Probably the greatest advance involving materials was the invention of the silvered vacuum flask.

Table 5. Early Developments in Cryogenic Technology

1877	L. Cailletet R. Pictet	liquid oxygen
1882	J. Violle	silvered vacuum flask
1883	S. V. Wroblewski K. S. Olszewski	liquid nitrogen; liquid hydrogen fog in 1884
1890	J. Dewar	silvered vacuum flask
1891	J. Dewar	liquid hydrogen
1892	H. K. Onnes	liquid air
1895	C. Von Linde	commercial liquid air
1904	----	commercial liquid nitrogen
1907	H. P. Cady D. F. McFarland	helium found in nitrogen gas
1908	H. K. Onnes	liquid helium

Silvering decreased the radiative heat transfer by a factor of 25, enough to allow relatively long-term storage of cryogenic liquids. Furthermore, more easily liquefied gases could now be used for precooling in the liquefaction of other gases.

Liquid helium was first made at Leiden near the end of the period. Interestingly, Leiden remained the only laboratory with this capability for fifteen years. During that time many materials measurements were made, usually on pure metals.

Commercialization of Cryogenics, 1911-1956

Some milestones of this period are presented in Table 6. They do not, however, give the flavor of the rapid commercial development that was the driving force during this time. By the end of the period massive quantities of liquid oxygen, liquid nitrogen, and liquid natural gas (LNG) were moving in commerce. Liquid hydrogen was in commercial production, but its day was yet to come. Helium recovery from natural gas had been (fortunately) started on a large scale in anticipation of an expanding air traffic of dirigibles that never materialized. The use of liquid oxygen as a rocket fuel component had its successful (and unpleasant) debut with the V2 rocket.

These were the "pure" days of superconductivity from its discovery in 1911 to the first hints of commercialization at the end of the period. The early years saw increased investigation of the basic physical properties of pure metals and some alloys. The development of a commercial liquefier made reasonable quantities of liquid helium widely available for the first time and started the days of the more engineering-oriented materials research.

The major large-scale use of materials was in the storage and shipment of cryogenic fluids. This use also provided one of the first, and most fatally dramatic, examples of what can happen when a large structure is designed with only a limited knowledge of the effect of low temperatures on materials--the Cleveland LNG explosion and fire in which 131 people died. The tank failure apparently was caused by cracking of the 3.5% Ni steel as a result of a stress configuration not previously encountered; the tank that failed was of new design. This event effectively halted the development of LNG technology for about a decade.

For the most part, the materials used during this time were the conventional low-nickel steels, copper-nickel alloys, and brasses. Some LNG storage tanks were built using a titanium-chromium-nickel steel in the USSR. Insulation was provided most often by granulated cork and balsa, although pearlite and combinations of glass fiber and foil were seeing some use toward the end of the period.

Table 6. Milestones in Cryogenics from 1911 to 1956

1911	H. K. Onnes	superconductivity
1914	G. Cabot	patent applications on LNG
1917	U.S. Bureau of Mines	helium separation from natural gas
1900s	K. Tsiolkovsky R. Goddard, H. Oberth	rocketry
1920s	C. Birdseye	attempted freezing food with liquid nitrogen
1925	Airship "Pilgrim"	first commercial helium airship
1926	R. Goddard	first rocket flight (~10 m)
1930	W. J. De Haas and J. Voogd	high-field superconductor (2 T for PbBi)
1933	W. Meissner and R. Ochsenfeld	Meissner effect
1934	P. Kapitza	expansion engine helium liquefier
1939	----	large-scale LNG storage
1940	P. Kapitza	expansion turbine for air liquefiers
1944	W. Dornberger	V2 rocket
1944	Cleveland, Ohio	LNG disaster
1947	S. Collins	commercial helium liquefier
1951	----	foil/glass fiber mat insulation
1952	National Bureau of Standards	Cryogenic Engineering Laboratory

The Space Program, 1957-1970

Table 7 presents some of the milestones of the period. The driving force for materials development was without question the space program. The early years of this period saw the development of most of the present day cryogenic materials data base.[8] Lightweight metals, especially the aluminum alloys, were tested in every way possible at temperatures down to 20 K. New plastic materials, lubricants, and insulations were developed. For the first time, sufficient data were available to allow the engineer a choice of materials for cryogenic applications. Compatibility of materials with liquid oxygen and, to a lesser extent, hydrogen became a concern. It was an oxygen compatibility problem that led to near disaster on the flight of Apollo 13. When a polymeric insulation burned an explosion was caused that led to massive failures in other systems and required a hasty return to Earth.

Table 7. Cryogenics in the Years of the Space Program

1957	A. A. Abrikosov	Type II Superconductivity
1957	J. Bardeen, L. Cooper and R. Schrieffer	BCS theory
1957	----	Sputnik
1957	----	large LNG storage facility
1958	NASA	agency created
1959	"Methane Pioneer"	liquid methane shipped across the Atlantic
early 1960s	Aerospace Industry and NASA laboratories	strong materials research program
1961	J. E. Kunzler et al.	J_c-H for Nb_3Sn
1963	D. L. Martin et al.	10-T Nb_3Sn solenoid
1964	C. Laverick	bubble chamber magnet
1965	----	stainless steel Dewars
1965	----	aluminized insulation
1965	Cambridge Accelerator	explosion and fire
1965	A. R. Kantrowitz and Z. J. J. Stekly	cryostatic stabilization of superconductors
1966	Z. J. J. Stekly et al.	MHD dipole magnet (5 MJ)
1967	----	multifilamentary NbTi
1968	J. Purcell et al.	Argonne 12-ft bubble chamber magnet (80 MJ)
1969	N. Armstrong and E. Aldrin	moon landing
1970	A. Prodell et al.	Brookhaven 7-ft bubble chamber magnet (72 MJ)
1970	Apollo 13	explosion in space

The other major disaster of the period occurred when a beryllium window in contact with liquid hydrogen in a bubble chamber failed, leading to an explosion and fire that totally destroyed the experimental hall of the Cambridge Electron Accelerator. Beryllium has some ductility in liquid nitrogen, but none at liquid hydrogen temperatures. It is suspected that machining-caused twins acted as stress risers, and a crack started in one of these regions and rapidly propagated across the window.

LNG returned to the scene with the construction of large storage facilities and the demonstration that transatlantic shipment was not only possible, but also profitable. It is easy to miss the fact that the LNG industry has been quite innovative in the development of new materials for very large-scale applications. Examples are the polymer foam insulations and containment structures of concrete and frozen earth.

Superconductivity came into its own. The phenomenon was explained theoretically to the satisfaction of nearly everyone.

The practicality of using high-field superconductors for magnets was made clear very early by the Bell Labs group. Large dc superconducting magnets were built, starting around 1965, for bubble chambers, MHD, plasma studies, and NMR work. All used heavily stabilized conductors. Several years later the multifilamentary NbTi conductors appeared and started the rapid development of the high energy physics beam-line magnets, superconducting machinery, and levitated transportation. Fields near 15 T were obtained with Nb_3Sn magnets of reasonable size about 1967.

Several new materials began to appear in cryogenic applications at about the middle of the period. Stainless steel replaced glass in most laboratory Dewars and "superinsulation" of aluminized polymer sheet started the trend away from nitrogen shielding. Composite materials, primarily fiberglass-epoxy, began to see large-scale use at low temperatures in applications ranging from filament-wound cryogenic pressure vessels to spacer blocks for the windings of large magnets.[9] Many foamed plastic insulations were developed and tested for large-scale applications with liquid hydrogen and LNG.[10] Numerous adhesives and sealants were evaluated for cryogenic applications.[11] It is an interesting commentary on the frantic pace of those days that the reviews just referenced were not compiled until well into the 1970s.

Modern Times

There is no table for this time period since almost everyone has a favorite milestone and any listing would probably just lead to trouble. Also, the next section discusses the present day materials work in some detail. It is reasonable, however, to mention some highlights.

The main driving force in this decade has been energy--its production and efficient use. The major source of funding for large cryogenic research projects in the United States shifted from NASA to the Department of Energy (and its predecessors) and the military, especially DARPA, Navy, and Air Force. Funding for LNG research reverted nearly exclusively to private industry, an industry that has shown enormous growth.

Superconductors and superconducting magnets provided many dramatic developments. Niobium titanium became the only ductile alloy superconductor in commercial production. Very early on we saw the development of the multifilamentary Nb_3Sn conductor and the introduction of numerous new and exotic superconductors of which only V_3Ga has reached the status of a commercial product. The effects of strain on superconducting composites of all types were measured and explained. Potting of magnet coils began about 1974.

The large bubble chamber magnets at CERN and NAL, the Baseball magnet at LLL, and the superconducting Levitron at Culham were all in operation by 1973. The continuing development of multifilamentary superconductors led to numerous pulsed and dc dipoles for high energy physics. Levitated transportation systems using superconducting magnets were demonstrated. The very successful Fawley motor showed the practicality of using superconductors in dc motors. The feasibility of ac machinery using superconducting rotors was also shown early in the period.

The present day status of each of these endeavors is well known and a review of modern projects is not appropriate here. Suffice it to say that the field continues to grow, although a serious problem with any of the giant systems due to come on line in the next few years could slow that growth considerably.

WHERE WE ARE NOW

One might well ask what is new at present in the field. Are we just doing more of the same on a larger scale? Is there really any need to continue materials research, given all of the data developed in the 1960s? Certainly during that time cryogenic systems reached very large proportions, but the largest structures were most often simple storage tanks or solenoidal magnets with almost no limit on the amount of copper stabilizer.

The structures with which we are now faced are not only enormous in size, they are of complex design and, in use, are subject to very large, nonsymmetric forces that often will vary with time through many cycles. Typical of these is the giant MFTF magnet that we have been seeing a lot of recently. These devices are by no means monolithic. They contain a bewildering variety of materials, all of which must work together at 4 K. As an example, consider Fig. 1, which shows the internal structure of the MFTF coil.[12] The six magnets of the Large Coil Program are of similar size and complexity and potentially subject to an even greater array of out-of-plane forces.[13] We really have no foundation of experience on which to build these fantastic devices. The production of a working fusion device is an engineering undertaking easily the equal of the space program, but it is being performed with much less funding and no fanfare.

That doesn't mean that no one is looking. These large devices will, we hope, eventually end up out in the real world to be operated by mere mortals. It is up to us to insure that they will operate safely by the use of realistic design criteria and sophisticated structural analysis. Such criteria and analysis demand the best materials data that we can supply.

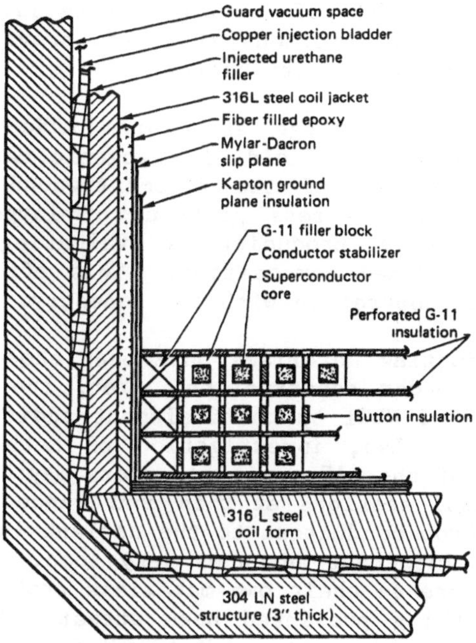

Fig. 1. Cross section of MFTF coil assembly (from Ref. 12).

Status of Materials Research

Modern structural materials work has begun to concentrate on topics that were of little concern even a few years ago. The most dramatic change is probably the concern with thick structural members and methods for joining them. Cast stainless steels are now being evaluated for low temperature applications as are structural alloy systems that do not use expensive (nickel) or strategic (chromium) materials, but depend instead on manganese to provide the requisite austenite stability. The nitrogen-modified versions of the standard stainless steel grades have shown significant increases in yield strength and, in spite of somewhat lower fracture toughness, have become the materials most often chosen for large magnet structures. And, at last, there has been an opportunity to do some relatively fundamental research on the effects of carbon and nitrogen in the austenitic stainless steels.

Research on welding, including materials, processes, and variability, is also a relative newcomer to the low temperature region. The discovery a few years ago of welds that matched base metal properties at room temperatures, but were unacceptable at 4 K, led to a program of evaluation of weldments. Basic work being done to evaluate the effect of varying ferrite and nitrogen content on cast stainless steel specimens will also provide information of value to the study of welded structures.

The development of large structures made from nonmetallic laminates has opened a new field of cryogenic materials research. The simple hoop stresses of the filament-wound tanks have been replaced by the complex stress patterns of large magnets. Most work to date has involved producing and evaluating fiberglass-epoxy materials of standardized composition. This effort has benefited greatly from close cooperation with the industry.

The superconducting materials have been chosen for all of the large magnets now under construction. Yet, the attempt to raise the critical field of NbTi by third-element additions is an active area of research. Also, investigation of methods by which the limited ductility of the A15 superconductors may be circumvented is the topic of several sessions at this conference. The enhancement of heat transfer from conductors to the coolant is a field where we may well see some important advances very soon. Large, internally cooled conductors create their own materials problems as the number of papers on sheathing presented here attest. The phenomenon of magnet training persists, continuing to show us that a superficial knowledge of the properties of the mechanical interactions in the magnet composite is not adequate.

Research for LNG systems continues with programs to evaluate the thermal and mechanical properties of very well-characterized polymer foams and concretes. The many materials in the earlier days were almost never characterized other than by type and density, and that allows an unacceptably wide range of properties for the "same" material.

Status of Materials Data

The data base for many materials is in quite good shape. The collection, compilation, and review of data on structural materials from numerous corporate reports and other obscure documents of the last twenty years has been accomplished, primarily by the staff of the now-defunct Cryogenics Division of NBS. These early data have been supplemented by two major materials research programs, both headquartered at NBS. The first, operated from 1973 to 1976 with DARPA funding, concentrated on materials related to superconducting machinery. A data handbook was published[14] that is probably the most complete available. It concentrates on a few wrought stainless steels and aluminum alloys, but data is included on other alloys and some nonmetallics. Another handbook with information on the major alloys, insulations, aggregates, and composites, plus thermodynamic data on cryogenic fluids is also available.[15]

The second program, and the only one now operating that is devoted strictly to structural materials, is the DoE-sponsored work on materials for fusion energy applications. Four reports are now in print that present data on large-section properties of various alloys; weldment studies; fatigue, crack growth, and fracture; properties of nitrogen-strengthened stainless steels and stainless steel castings; and modern developments and data on numerous nonmetallic composites. No data handbook has yet been developed, although the data should eventually find its way into the new fusion materials handbook.[16] The existence of this program should not give one cause for complacency. It is very small in relation to its constituency. It is encouraging to note that more low temperature materials research groups are now being formed, primarily in industry, but it is vitally important that their data be generally available.

The "practical" superconducting materials themselves seem to be well under control, although the modern drive for a 12-T magnet with ductile conductor may well change that. The auxiliary materials used for strength and stabilization have been reasonably well evaluated, although more specific data on copper and aluminum in various processed states would be helpful. Similarly, a compilation of properties of "other" materials, such as glues and fillers, would be extremely useful.

WHERE ARE WE GOING?

There is, of course, no real way to answer this question. Unlike the early days, the future is determined more by the vagaries of politics than by the care and cleverness with which we pursue our science. Be that as it may, it is still possible to outline some of the trends that seem to be developing.

The Good News

One of the most encouraging trends is the increase in both the number and size of mechanical properties test machines dedicated to low temperature (4-K) measurements. Before last year very few 90-kN machines were available in the United States. Now several 225-kN machines are working and at least one 450-kN machine is planned. This advance allows testing of cross sections more closely resembling those actually in use. Furthermore, the new machines can test many full-size welds, which is important for determining the effect of residual stresses.

There is a trend toward evaluating alloys in which the constituents, such as carbon and nitrogen, are varied to optimize low temperature properties. Many of these materials are now used in sufficient quantity that the mills will produce to a given composition. This is the first significant departure from the earlier approach of just collecting low temperature data on existing materials.

The growing effort to evaluate nonmetallic composites and their constituents for low temperature applications is another trend that is encouraging and has already yielded some very interesting results. Any program that investigates the tailoring of materials for specific applications is certain to pay significant dividends. Much remains to be done along these lines. A few prime areas are: anisotropic thermal and electrical conductors; high specific heat, high thermal conductivity ceramics; and laminates with controlled thermal expansion.

The development of several facilities for electrical and thermal testing of very large, high current conductors and coils is good news. Still to come is a facility that would incorporate a mechanical test capability for these same materials. The continuing drive for better superconductors is encouraging. The proposed use of sophisticated techniques of nondestructive evaluation (NDE) to analyze the performance of the large superconducting magnets will do much to enhance our understanding of the mechanisms of degradation and failure in these structures.

The Bad News

The small size of the low temperature materials research capability has resulted in specific materials choices being made before all the relevant data were available. In several instances, the ensuing data indicated that the chosen material was not the best. In at least one major program, the principal material was unacceptable and had to be changed. A particular problem arises with welds. Recent research has shown that weld quality in some stainless steels is strongly dependent on the specific manner in which the weld is made. Thus data correlating weld quality with some readily observable property must be developed, and this is a time-consuming process.

With the advent of the higher strength, nitrogen-strengthened stainless steels, allowable operating stresses have been raised for various designs. Typical practice is to use two-thirds of the yield strength at 4 K with higher stresses allowed under fault conditions. The only experience with load-controlled structures at such high stress levels is with aerospace pressure vessels--relatively simple structures. The high stress levels are often justified with fracture mechanics calculations and, thus, minimum toughness values are specified for the base metals and welds. To meet these specifications, weld materials and practices are selected to maximize toughness, a situation that most often leads to lower yield strength. The resulting mismatch in yield strength between the weld and the base metal can result in strain concentration in the weld, the region of lowest toughness.

Another important problem is the continuing lack of an effort to develop models and scaling experiments that will allow confident prediction of structure performance at low temperature from a good data base of material properties.

The evaluation of composite materials for low temperature use has shown that those materials that give the greatest increase in strength also have a large sample-to-sample variability. This effect is large enough to make the usefulness of some boron epoxies and glass epoxies doubtful at cryogenic temperatures.

ACKNOWLEDGMENTS

Most of the pleasure of preparing a paper of this sort is derived from time spent with colleagues in remembrance and wild speculation. My thanks to H. I. McHenry, D. Chelton, L. L. Sparks, and A. F. Clark for their assistance in this vein. As always, the manuscript was prepared with cheerfulness and dispatch by Ms. V. Grove.

REFERENCES

1. G. Gamota, IEEE Trans. Magn. MAG-17:19 (1981).
2. P. Kapitza, Proc. R. Soc. 147A:189 (1934).
3. H. Weinstock, Chapter 1 in: "Applications of Cryogenic Technology," R. W. Vance and H. Weinstock, eds., Tinnon-Brown, Los Angeles (1969).
4. F. R. Fickett, in: "Proceedings, Fifth International Conference on Magnet Technology," N. Sacchetti, M. Spadoni, and S. Stipcich, eds., Laboratori Nazionali del CNEN, Frascati, Italy (1975), p. 659.
5. F. R. Fickett, in: "Proceedings Sixth International Cryogenic Engineering Conference," IPC Business Press, Sussex, England (1977), p. 20.
6. F. R. Fickett, R. P. Reed, and E. N. C. Dalder, J. Nucl. Mater. 85 and 86:353 (1979).
7. F. R. Fickett and H. I. McHenry, IEEE Trans. Magn. MAG-17:2297 (1981).
8. F. R. Schwartzberg, "Cryogenic Materials Data Handbook," Martin Marietta, Denver, Colorado (1968).
9. M. B. Kasen, Cryogenics 15:327 (1975).
10. F. R. Williamson, NASA Reference Publication 1002 (1977).
11. F. R. Williamson and N. A. Olien, NASA SP-3101 (1977).
12. C. D. Henning, IEEE Trans. Magn. MAG-17:618 (1981).
13. P. N. Haubenreich, IEEE Trans. Magn. MAG-17:31 (1981).

14. "Handbook on Materials for Superconducting Machinery," MCIC-HB-04, Metals and Ceramics Information Center, Battelle Columbus Laboratories, Columbus, Ohio (1974). There are two supplements.
15. "LNG" Materials and Fluids," D. B. Mann, ed., Cryogenic Data Center, National Bureau of Standards, Boulder, Colorado (1977). There are two supplements.
16. "Materials Handbook for Fusion Energy Systems," DoE/TICC-10122 (1980).

MECHANICAL PROPERTY MEASUREMENTS AT LOW TEMPERATURES*

D. T. Read and R. L. Tobler

Fracture and Deformation Division, National Bureau of Standards, Boulder, Colorado

INTRODUCTION

Accurate material mechanical properties data at low temperatures are essential for design of economical, efficient, and safe structures for low temperature operation. Fracture toughness, tensile properties, and fatigue crack growth resistance characterize the mechanical performance of structural materials. Because construction of a sizable welded structure free from geometrical defects is a practical impossibility, adequate material toughness is needed to prevent crack growth and eventual fracture at such defects. Strength and stiffness enable a material to support its applied loads without stretching or buckling. Adequate fatigue crack growth resistance allows a part to continue to perform its function for a lifetime of load cycles. Compromises among toughness, strength, and fatigue resistance are required because no one material is best in all properties and because economy in material costs is necessary.

Previous papers[1-3] have described our earlier apparatus and techniques for obtaining mechanical property data at low temperatures. In this paper improved measurement techniques for toughness, strength, and fatigue resistance are described. Some brief remarks on the use of the data are also included.

MEASUREMENT TECHNIQUES

Fracture Toughness

Fracture toughness is the measure of a cracked material's resistance to tearing and fracture. The most direct measurements

*Invited paper.

of this property are considered to be the fracture mechanics toughness tests in which K_{Ic}, the critical stress intensity factor for Mode I fracture, or J_{Ic}, the critical J-contour integral for Mode I fracture, are measured. The stress intensity factor is better established theoretically and in practice, but it is only applicable to very large specimens or low-toughness material. The J-integral is applicable in these cases and also to relatively small specimens of high-toughness material. The J-R (resistance) curve concept allows characterization of a material by its crack propagation resistance over a range of crack extensions instead of only at the initiation of tearing.

All of these techniques employ a specimen with a fatigue-sharpened crack; the specimen is extended to failure in the low-temperature environment of interest while load and displacement values are continuously measured. Quasi-static loading is used routinely. The loading may be monotonic and continuous or it may be interrupted by periodic partial unloadings. The load and displacement values are analyzed according to various standard procedures to obtain K_{Ic}, J_{Ic}, or a J-R curve.

To date, compact specimens 25 mm x 61 mm x 64 mm have been routinely used, primarily because they offer a high toughness measurement capability for their size. But 3-point bend specimens (25 mm x 25 mm x 114 mm) are now being developed for J-integral tests at 4 K. The bend specimen is less expensive to machine.

Apparatus to cool the fracture toughness specimen to 4 K, extend it to failure in a controlled fashion, and sense and record load and displacement throughout the test are required. The apparatus at NBS is based on a commercially supplied servohydraulic testing device capable of extending a specimen under load or displacement control with loads up to 100 kN and displacements of up to 10 cm. The hydraulic actuator that supplies the load is mounted vertically at the top of the device. Tensile loads are transmitted through a load cell at room temperature, along a pull rod to the upper clevis grip (Fig. 1). The specimen is pinned to the upper and lower clevis grips (Fig. 2). The lower clevis grip is attached to a compressive load frame, which transmits the reactive load back up to the actuator mount. This re-entrant scheme allows the use of conventional Dewars without low temperature force feed-throughs. Currently, two different load frames are in service: one uses stainless steel loading columns for rigidity; the other uses fiberglass-reinforced plastic for minimal heat leak, as described previously.[2]

Load is sensed by a conventional commercially supplied load cell in the room temperature environment. Displacement is sensed by a clip gage that is mounted on knife edges spot-welded onto the specimen at the load line and retained by the spring action of the

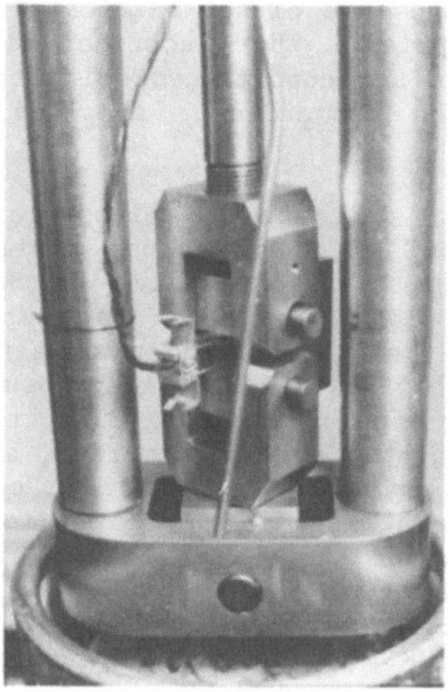

Fig. 1. Low temperature fracture toughness testing apparatus in use at NBS.

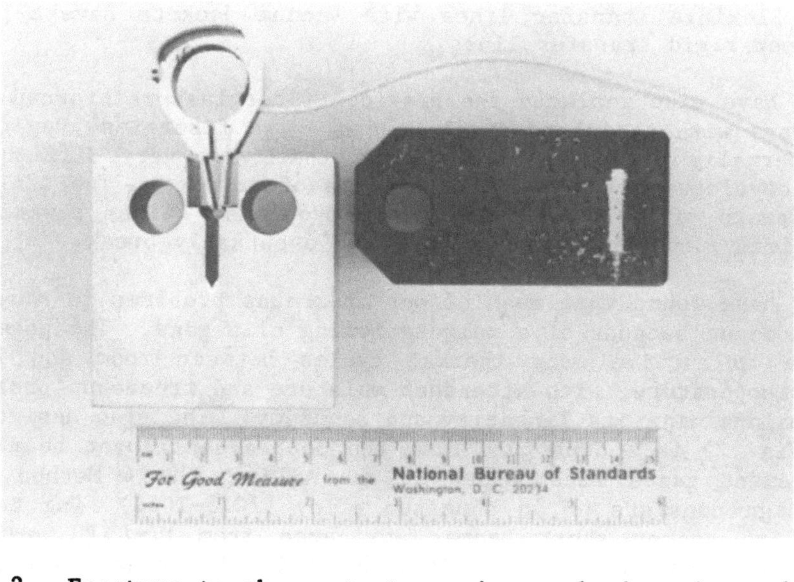

Fig. 2. Fracture toughness test specimen, clevis grip, and clip gage.

gage beams. These gages are either fabricated in-house or custom-built by a local supplier. The electrical signals from the load and displacement gages are conditioned using the conventional circuitry built into the commercially supplied mechanical testing system.

An earlier paper[2] described an elaborate 2-h cool-down procedure involving installation of copper baffles around the load frame members, sealing the cryostat lid, and vacuum-pumping of the liquid nitrogen precoolant to reduce its temperature below 76 K. However, we usually prefer the following alternative procedure, which eliminates the baffles and the time-consuming steps involving mechanical pumping. After precooling the cryostat to 76 K, the Dewar of liquid nitrogen is removed, emptied, and replaced in 30 s. Helium transfer then begins. The use of an outer Dewar of liquid nitrogen (double-Dewar arrangement) is also dispensable, especially when the test will be of short duration. This procedure reduces the time to accomplish cool-down to about 40 min, at the cost of several additional liters of liquid helium. A J-integral test may consume from 10 to 30 ℓ of helium, depending on procedural and experimental variables.

Several improvements have been made to the instrumentation. Instead of the carbon resistor liquid level indicator previously used,[2] we now have a 30-cm continuous liquid level monitor, which attaches to one of the cryostat load frame columns. This meter gives a continuous readout of the helium level in the cryostat and its relation to the specimen, rather than a simple on-off indication. Flexible transfer lines with vacuum jackets have replaced the former rigid transfer lines.

We have also replaced the previous fiberglass-reinforced plastic Dewars with stainless steel Dewars. The fiberglass Dewars are more thermally efficient,[2] but leaks that were very difficult to repair developed after one or two years of service. The stainless steel Dewars offer fewer maintenance problems. Glass Dewars were used briefly, but these are fragile and eventually break.

We have found that most of our apparatus problems in toughness testing occur because of a malfunctioning clip gage. The gage must be able to survive many thermal cycles between room and liquid helium temperature, with attendant moisture and freeze-up problems, while maintaining its linearity and accuracy. The ring design clip gage (Fig. 2) is a departure from the double-cantilever beam style displacement gages recommended in the ASTM E 399-74 Method. The new design consists of an aluminum alloy (7075-T651) ring that is slotted to accept short beams fabricated from 17-7 PH stainless steel. The rear of the ring is reduced in thickness and provides the bending member onto which four electrical resistance strain gages chosen for good low temperature performance are mounted. The linear operating range is from 2.5 to 7 mm of opening displacement (higher operating ranges are possible). Sensitivity is greater and linearity is improved, as compared to double-cantilever designs of the ASTM E 399-74 type. Conventional laboratory-quality signal conditioning circuitry is adequate to read out the displacement signal if it contains no excessive electrical noise. Noise can come from inadequate solder joints, a conductive film of debris on the gage surface, or other sources. Well constructed gages can survive millions of fatigue cycles and tens of tests at 4 K. The mechanical apparatus shown in Figs. 1 and 2 has endured hundreds of fracture toughness tests.

A recent addition to the NBS test apparatus is a minicomputer system for data acquisition, storage, and real-time analysis. The details of this system are described elsewhere.[3] The signals are applied to the input of a high-resolution analog-to-digital converter, from which their values are transmitted to a minicomputer. In the computer, load-displacement data are stored on a floppy disk. To conduct a single-specimen J-integral test, periodic unloadings are performed under manual control to obtain the instantaneous crack length by the compliance technique. This involves execution of a subprogram for extracting the specimen compliance from load and displacement data. The J-integral is also calculated from load and displacement data. The computer plots each crack length J-integral data point as soon as it is acquired. This provides the operator with guidance for taking the next data point. The loading is under control of the operator through the machine console; only the data acquisition is done by computer. This apparatus and cool-down procedure allow up to four low temperature toughness tests to be conducted in an 8-h day.

The single-specimen unloading compliance technique enables J_{Ic} to be measured with good accuracy, but gives an erroneous measurement of tearing modulus, that is, dJ/da, because of errors in the crack extension, Δa. The discrepancy between actual and compliance-measured Δa increases progressively with Δa. Since this error is less at low Δa values, the J_{Ic} measurement from the single-specimen technique is reasonably accurate (within 5% of the value measured with the multiple-specimen method), but the slope of the resistance curve is inaccurate. Comparison of results from both methods for AISI 310S at 4 K are shown in Fig. 3.

Measurement uncertainties in individual J values are considered to be 5% or less, whereas the bias due to systematic error in the inferred Δa values contributes an uncertainty of 10% or less to the J_{Ic} measurement. The cumulative uncertainty in J_{Ic} values is estimated to be ±15% or less, so the measurement uncertainty in K_{Ic} values inferred is 8% or less, since K_{Ic} is proportional to $\sqrt{J_{Ic}}$.

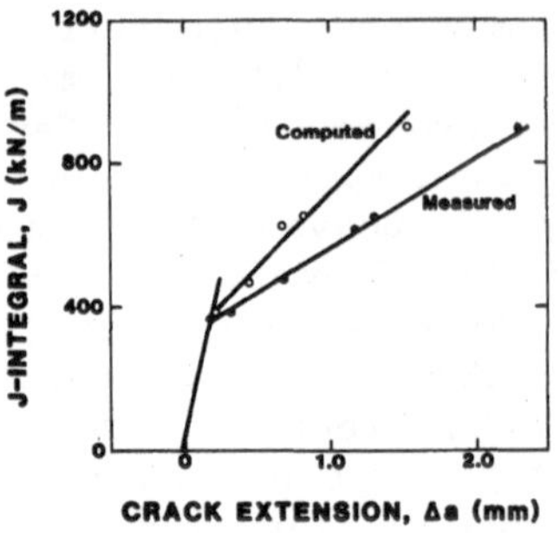

Fig. 3. Graph of J-integral versus crack extension for AISI 310S at 4 K. Crack extension values obtained from the compliance technique using one specimen (open circles) are compared with optically measured values from five specimens.

If errors in fracture toughness associated with material variability are large, the accuracy of J_{Ic} results will be further limited.

Strength

In this era of computerized fracture toughness testing, the strength must not be neglected. Tensile testing provides the material yield strength, ultimate strength, elongation to failure, and reduction of area. Elastic constants are measured to within ±0.2% at NBS by sound-propagation techniques using separate specimens and apparatus.[4]

Again, apparatus to cool tensile specimens to 4 K, extend the specimen to failure in a controlled fashion, and sense and record load and displacement is used. Screw-driven, mechanically controlled devices are in use at NBS for tensile testing (Fig. 4). These machines are more economical in initial cost and laboratory space than the hydraulic testing machines, and tensile testing does not require rapid testing machine response. Load is transmitted from the fixed upper crosshead with a built-in load cell down a pull rod to the upper specimen grip. The reactive load is transmitted from the lower grip through a compression tube to the moving crosshead of the machine. Conventional glass or stainless steel Dewars without force feed-throughs are used for immersion of the specimen in cryogen during the test. The extension of the specimen within its gage length is sensed by a cantilever beam gage whose two arms are attached to the specimen at the ends of the gage section (Fig. 5). The gage is clipped in between retaining pins, which are fixed to the specimen. When the specimen has been extended to about 10% strain, the gage simply falls away. The

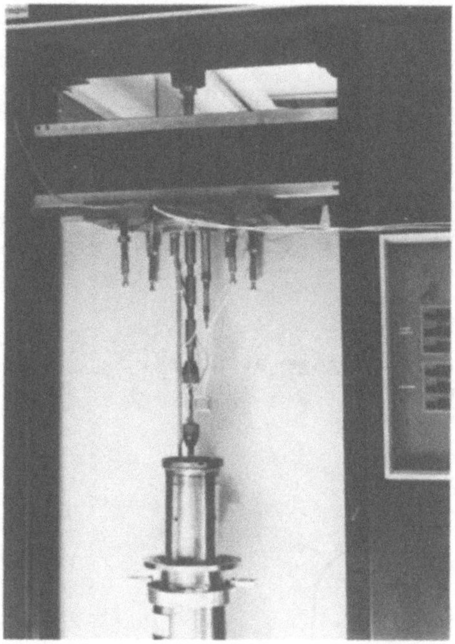

Fig. 4. Low temperature tensile testing apparatus in use at NBS.

gages are fabricated in-house or custom-made by a local supplier. Calibrations are conducted at each test temperature using a vertically mounted dial micrometer.

The yield and ultimate strengths are calculated in the usual way from the measured load and strains. Uncertainties in tensile flow strength measurements are estimated at ±2% or less, owing to uncertainties in specimen dimensions, load values as read from recorder traces, and (for yield strength) specimen strains from the extensometer. The elongation and reduction of area are obtained conventionally from specimen dimensional differences before and after testing.

Flow strength data at temperatures intermediate between the boiling points of the common cryogens are obtained by controlled thermal conduction using liquid helium or nitrogen reservoirs and heaters. The experimental arrangement shown in Fig. 6 has been used in several studies. The specimen is cooled by two cryogen reservoirs, one at the bottom of the push tube and the other mounted on the pull rod. The specimen is heated by electrical resistance heaters wound on each grip. The specimen temperature is sensed by two diode thermometers, one mounted at each end of the gage section. The current to the heaters is controlled so that the thermometers, and therefore the specimen, remain at the desired temperature (±1 K). With this apparatus temperatures from 4 to

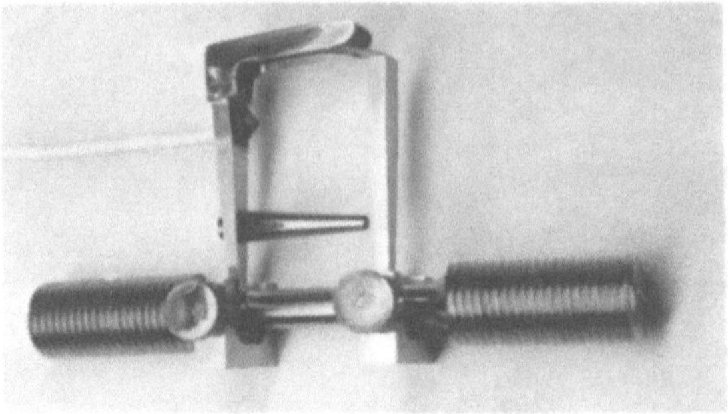

Fig. 5. Cantilever beam extension gage attached to tensile specimen.

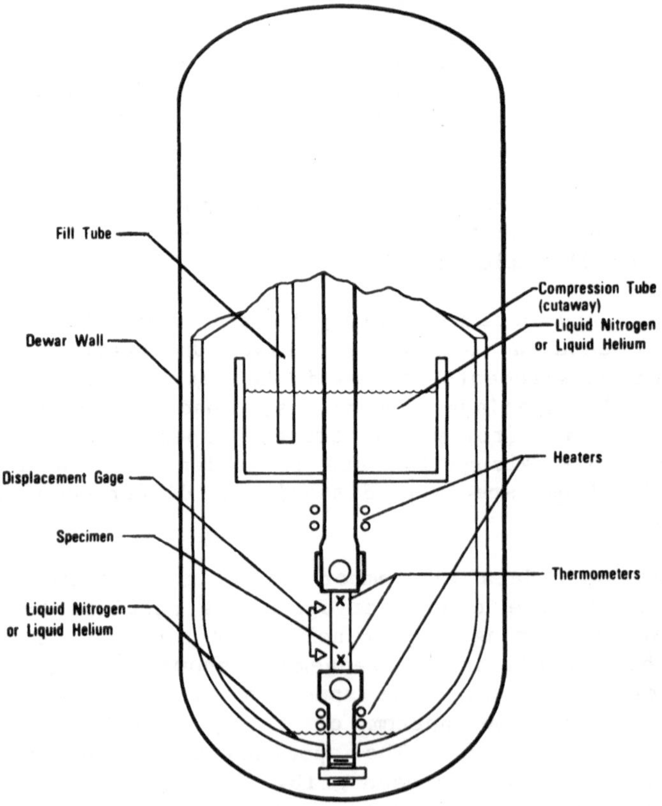

Fig. 6. Experimental arrangement for tensile testing at temperatures above cryogen boiling points.

300 K can be obtained in metal specimens. Nonmetal specimens must be surrounded by a thermal shield controlled at the desired test temperature.

In a recent series of experiments, capacitive sensing of specimen extension was used. Concentric aluminum cylinders, one mounted to each grip, form the two capacitor elements (Fig. 7). A capacitance bridge and a lock-in amplifier were used to detect specimen extension. This apparatus was needed because attachment of conventional extensometers to soft viscoelastic foams could have influenced the measured material properties. The noise level of the displacement signal was about 0.3 μm. This could be reduced by at least one order of magnitude for measurement on metals by reducing the separation between the capacitance cylinders. In this apparatus the inner capacitor tube also served as a thermal shield, allowing the temperature of the thermally nonconductive foam to be accurately controlled.[5]

Fatigue

The fatigue lifetime of a structural component is characterized by its fatigue crack growth rate (FCGR). In our laboratory the servohydraulic apparatus described earlier is also used for

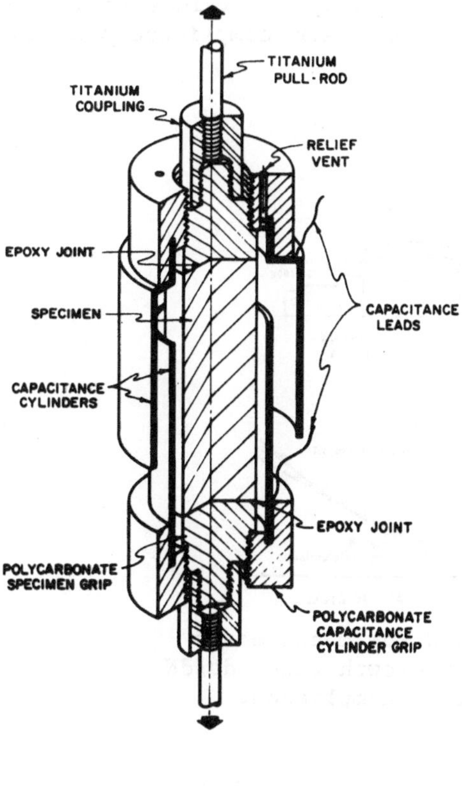

Fig. 7. Capacitance extensometer assembly for foam specimen testing.

fatigue tests. The specimens used are the same for K_{Ic} or J_{Ic} testing, but shorter initial crack sizes are specified.

The compliance method of FCGR measurement is the most suitable method for cryogenic testing. The technique is applicable at all temperatures, even if the specimen is enclosed in a cryostat and inaccessible. It is based on the correlation between compliance (specimen deflection per unit load) and crack length: compliance increases as crack length increases. This method offers advantages for testing thick specimens, because crack-front curvature is accounted for in the average crack length derived from the compliance value.

The procedure previously described is illustrated in Fig. 8. The specimen is loaded cyclically at a selected load range and at a frequency of 20 Hz. At intervals ranging from 100 to 20,000 load cycles (N), the specimen compliance is recorded and used to obtain the instantaneous crack length. The FCGR is simply the change in crack length per cycle, which is calculated as the slope of the a-versus-N curve at a given value of a. The cycle number is easily monitored by electronic or mechanical counters.

To calibrate this method, crack front striations on specimen fracture surfaces are created by changes in minimum fatigue load. After breaking open the calibration specimen, the values of a are measured, averaged, and plotted against their compliance values.

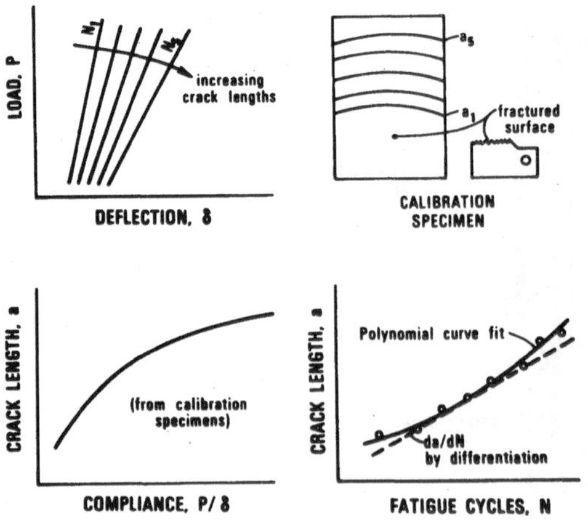

Fig. 8. Compliance method for measurement of fatigue crack growth rate, da/dN, using the specimen compliance.

The resultant curve is fit using the appropriate polynomial expression, which in turn is used to infer crack lengths from compliance data recorded during FCGR tests.

The FCGR data are commonly plotted against the cyclic applied stress intensity factor range (ΔK), which is calculated from load, crack length, and specimen dimensions. Uncertainties of measurement and material variability may be significant, and standard deviations of 20 to 40% from the average value of FCGR at a given ΔK value are typical for our results. The accuracy of FCGR data can be degraded if highly variable materials, such as welds, are tested.

USE OF TOUGHNESS AND STRENGTH DATA

A structure can tear and fracture because of inadequate toughness or inadequate strength. Toughness is required in large sections with cracks of large absolute size. Strength is a problem in smaller sections with cracks that may be small on an absolute scale but are large relative to the section size. An estimate of the size scale of interest for cracks can be made from a fracture mechanics formula for stress intensity. In a material whose yield strength is 850 MPa and whose toughness is 150 MPa$\sqrt{m}$, a crack of large absolute size would be one greater than 1 cm in length (through crack). A large section dimension would be one greater than 10 cm. In large sections, the critical stress intensity factor can be reached before the section yields, producing failure at loads "lower than expected." The greater the material toughness, the larger the crack that can be tolerated at a given load. If the material toughness is improved and permitted cracks get larger, or if the size of the section is scaled down with crack size held constant, the section containing a crack can yield before crack extension. Once a section has reached yield, it will extend at nearly constant load until the load is relieved or the section fails. The critical factor in such a case is the material strength. Higher strength holds off section yielding to a higher load. Higher toughness allows slightly more yield-load extension before fracture, but the amount is small, of the order of millimeters or less. From these considerations it is clear that inspection for large defects is necessary to insure fracture prevention in severely loaded structural members.

CONCLUSION

Techniques are described for the measurement of fracture toughness, tensile properties (including yield and ultimate strengths, elongation to failure, and reduction of area), and fatigue crack growth rates at temperatures down to 4 K. Accuracy is generally limited by specimen-to-specimen variation.

ACKNOWLEDGMENTS

J. Shepic contributed by designing and building the cryogenic clip gages. This work was supported by the U.S. Department of Energy, Office of Fusion Energy.

REFERENCES

1. R. P. Reed, A cryostat for tensile tests in the temperature range 300 to 4 K, in: "Advances in Cryogenic Engineering," Vol. 7, Plenum Press, New York (1961), pp. 448-454.
2. C. W. Fowlkes and R. L. Tobler, Fracture testing and results for a Ti-6Al-4V alloy at liquid helium temperature, Eng. Fract. Mech. 8:487-500 (1976).
3. R. L. Tobler, D. T. Read, and R. P. Reed, Strength and toughness relationship for interstitially strengthened AISI 304 stainless steels at 4 K, in: "Fracture Mechanics: Thirteenth Conference," ASTM STP 743, American Society for Testing and Materials, Philadelphia (1981), pp. 250-268.
4. H. M. Ledbetter, N. V. Frederick, and M. W. Austin, Elastic constant variability in stainless steel 304, J. Appl. Phys. 51:305-309 (1980).
5. J. M. Arvidson, "Capacitive Technique for High-Sensitivity Low Temperature Extension Measurement," National Bureau of Standards, Boulder, Colorado, to be published.

RECENT DEVELOPMENTS IN FILAMENTARY COMPOUND SUPERCONDUCTORS*

K. Tachikawa

National Research Institute for Metals, Ibaraki, Japan

INTRODUCTION

The filamentary compound superconductors are indispensable for generating magnetic fields over 10 T at 4.2 K. The composite process and the <u>in situ</u> process have been developed to fabricate mechanically brittle high-field superconducting compounds into filamentary conductors. In this article, recent developments of filamentary compound superconductors mainly achieved at the National Research Institute for Metals (NRIM) in Japan are described.

COMPOSITE PROCESSING OF A15 COMPOUNDS

The fabrication of multifilamentary Nb_3Sn and V_3Ga wires is based on the composite process in which a composite of niobium or vanadium cores in a Cu-Sn or Cu-Ga solid-solution alloy matrix is fabricated and then heat-treated.[1,2] The copper acts as a mother metal for tin or gallium and, moreover, enhances the formation of the Nb_3Sn or V_3Ga layer through diffusion reaction.[3] A15 V_3Si and V_3Ge are also formed by the composite process using a vanadium core and a Cu-Si or Cu-Ge alloy matrix.[4,5] However, a second compound phase richer in silicon or germanium is formed together with A15 V_3Si and V_3Ge. The typical formation temperatures of Nb_3Sn, V_3Ga, V_3Si, and V_3Ge in the composite process are about 750°, 650°, 800°, and 800°C, respectively.

The first stranded-type multifilamentary V_3Ga conductor was commercially produced by Furukawa Electric Company.[6] This conductor, reinforced by a central tungsten wire, showed an overall

*Invited paper.

critical current density of 3.3×10^4 A/cm^2 at 15 T. A small 10-T magnet, wound after the reaction, was constructed in 1975 using this conductor. The multifilamentary V_3Ga conductor is very stable under a rapid field change; the critical current of small coils of this conductor shows no degradation even at an excitation rate of over 20 T/s. In 1979, a 13-T high-stability magnet was made of the prereacted, multifilamentary V_3Ga tape.[8] Meanwhile, industry-scale production of multifilamentary Nb_3Sn conductors of large current-carrying capacities is proceeding in several countries using advanced fabrication techniques.[8-11] Large-scale applications of these conductors to magnetic fusion reactors[12] and high energy accelerators[13] are in progress.

EFFECTS OF THIRD-ELEMENT ADDITIONS TO COMPOSITE-PROCESSED A15 COMPOUNDS

Transition Temperature and Upper Critical Field

Among the composite-processed A15 compounds, the most drastic change of the transition temperature, T_c, and the upper critical field, H_{c2}, is found when aluminum is added to the vanadium core of the V_3Ge composite conductor;[14] the T_c and the H_{c2} (4.2 K) of V_3Ge are increased from 7 K and 2 T to 12.5 K and 18 T by the aluminum addition, as demonstrated in Fig. 1. The H_{c2} is more significantly increased than T_c, possibly owing to the increase of the normal state resistivity, ρ_n, of V_3Ge by the partial substitution of aluminum for germanium.

The H_{c2} of the composite-processed Nb_3Sn is increased by about 4 T through the titanium addition to the niobium core[15,16] and by about 7 T through the simultaneous addition of hafnium to

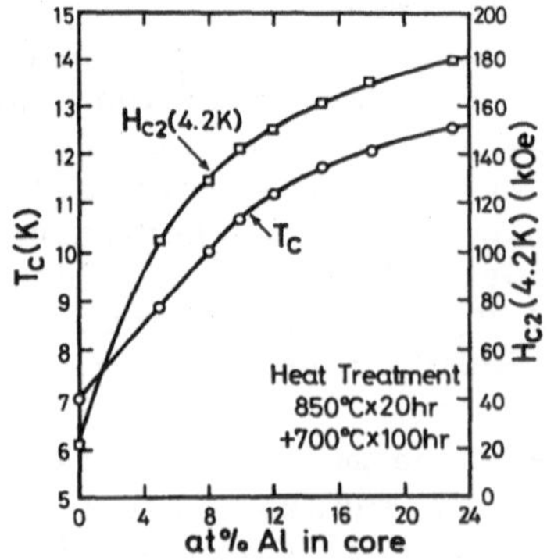

Fig. 1. T_c and H_{c2}(4.2 K) of V-Al/Cu-9 at.% Ge composite conductors as a function of aluminum concentration in vanadium core.

the core and gallium to the matrix.[17,18] These simultaneous additions realize an H_{c2} of 26 T at 4.2 K and raise T_c by about 0.9 K. The H_{c2} of the composite-processed V_3Ga is increased by about 2 T through the simultaneous addition of gallium to the vanadium core and magnesium to the matrix.[19] The increase in ρ_n by the third element addition may not contribute much to the enhancement in H_{c2} of V_3Ga since its H_{c2} is strongly limited by the Pauli paramagnetic effect.

Microstructures

The addition of IVa elements, such as Ti, Zr, and Hf, to the niobium core produces fine Nb_3Sn grains in the composite-processed Nb_3Sn.[20] The addition of small amounts of magnesium to the Cu-Sn[21] or Cu-Ga matrix[19] also makes Nb_3Sn or V_3Ga grains significantly finer. The IVa element addition to the niobium core and the magnesium addition to the matrix prevent columnar growth of Nb_3Sn or V_3Ga grains. Figure 2 shows the effect of the magnesium addition on the grain structure of the composite-processed Nb_3Sn.[21]

The addition of 2 at.% of IVa elements to the niobium core increases the Nb_3Sn growth rate 2-3 times over that of the pure niobium core conductor.[20] The addition of small amounts of magnesium to the Cu-Sn or Cu-Ga matrix also significantly increases the formation rate of Nb_3Sn[21] or V_3Ga.[19] The grain refinement of Nb_3Sn or V_3Ga caused by these additions may account for the increased growth rate in terms of the enhanced grain boundary diffusion of tin or gallium. With respect to the enhancement of the V_3Ga formation rate, the addition of aluminum to the Cu-Ga matrix is as if the gallium concentration in the matrix were increased;[22]

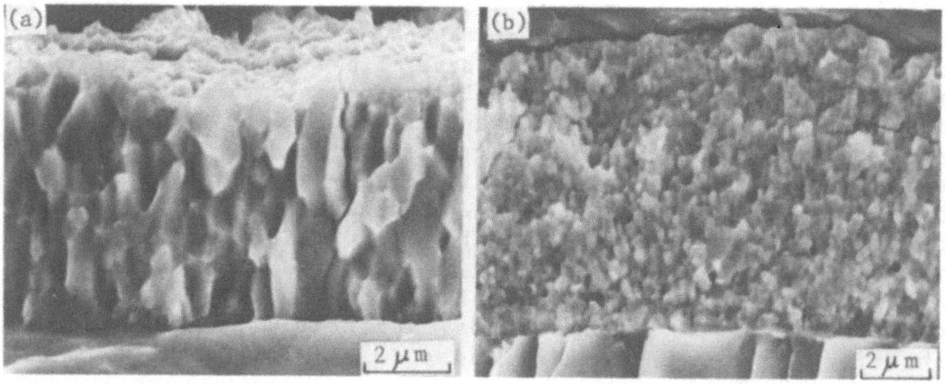

Fig. 2. Scanning electron micrographs of Nb_3Sn grain structures formed at 800°C for 20 h: (a) Nb/Cu-7Sn; (b) Nb/Cu-7Sn-0.5Mg. Arrows indicate niobium substrate.

the addition of zinc to the Cu-Sn matrix is as if the tin concentration in the matrix were increased.[23] Thus, the aluminum or zinc addition can reduce the amount of gallium or tin needed in the matrix.

X-ray microanalysis on the IVa element addition to the niobium core revealed that the amount of titanium incorporated in the Nb_3Sn layer is considerably larger than that of zirconium or hafnium.[20] This may be due to the difference of atomic radius among titanium, zirconium, and hafnium; the atomic radius of titanium is slightly smaller than that of niobium, whereas those of zirconium and hafnium are considerably larger than that of niobium. The titanium added to the Cu-Sn matrix is more readily incorporated into the Nb_3Sn layer than that added to the niobium core, probably because the diffusion of titanium in the Cu-Sn matrix is much faster than that in the niobium core.[24] The lattice parameter of the Nb_3Sn decreases monotonically with increasing titanium concentration in the niobium core, indicating that the titanium is dissolved in the Nb_3Sn layer.[20] The gallium substituted for the tin in the matrix is also partly incorporated into the Nb_3Sn layer.[20]

Critical Current Densities in High Magnetic Fields

The superconducting properties of the composite-processed Nb_3Sn are insufficient for generating magnetic fields higher than 12 T, since its critical current density, J_c, falls rapidly in these fields. The addition of titanium to the niobium core or to the Cu-Sn matrix raises J_c in high fields, and a J_c (for Nb_3Sn layer) of over 1×10^5 A/cm^2 is obtained at 15 T.[16] The optimal amount of titanium addition to the core is about 2 at.%. The overall J_c is more significantly increased since the Nb_3Sn layer thickness is also increased by the titanium addition. The hafnium addition to the niobium core shifts the J_c versus H curve to higher fields without changing its shape.[17] The gallium addition to the Cu-Sn matrix changes the shape of the J_c versus H curve of the composite-processed Nb_3Sn tape from convex downwards to convex upwards in high fields like the curve for V_3Ga.[17] The simultaneous addition of hafnium to the core and gallium to the matrix is most effective in improving the high-field performance of Nb_3Sn, and a J_c of over 1×10^5 A/cm^2 is obtained at 18 T.[18] For the composite-processed V_3Ga, the addition of gallium to the vanadium core has been reported to be effective in enhancing J_c.[25] The simultaneous addition of gallium to the core and a small amount of magnesium to the Cu-Ga matrix is most effective in enhancing J_c in high fields, and a J_c of over 1×10^5 A/cm^2 is obtained at 20 T.[19]

The composite-processed multifilamentary Nb_3Sn conductors with improved high-field performances are being developed in industry. Figure 3(a) shows the cross section of a multifilamentary Nb_3Sn tape with Nb-2 at.% Ti cores produced by Furukawa Electric Company, and Fig. 3(b) shows that of a multifilamentary Nb_3Sn wire with Nb-5 at.% Hf cores and Cu-5 at.% Sn-4 at.% Ga matrix produced by Hitachi Cable Company. Figure 4 shows J_c (core + bronze + barrier) versus H curves of the Nb_3Sn conductors with these additions. The overall J_c of the Nb_3Sn wire with hafnium and gallium addition is about 2×10^4 A/cm^2 at 16 T, which is nearly one order of magnitude larger than that of the pure Nb_3Sn wire. Improvements demonstrated in Fig. 4 make it feasible to generate magnetic fields over 15 T at 4.2 K by the composite-processed multifilamentary Nb_3Sn. The J_c of the multifilamentary tape with titanium addition shown in Fig. 4 is slightly lower than that of the multifilamentary wire, maybe owing to the anisotropy in J_c.

Stress Effect

The addition of hafnium to the core and gallium to the matrix improves ε_{irr} of the composite-processed Nb_3Sn, the strain where the irreversible degradation in critical current, I_c, takes place. This improvement in ε_{irr} becomes significant with increasing volume fraction of unreacted core and with decreasing Nb_3Sn layer thickness. An ε_{irr} of nearly 1% has been obtained for the Nb-5Hf/Cu-5Sn-4Ga conductor.[18] Scanning electron microscopy indicates that the irreversible degradation in I_c is initiated by a local

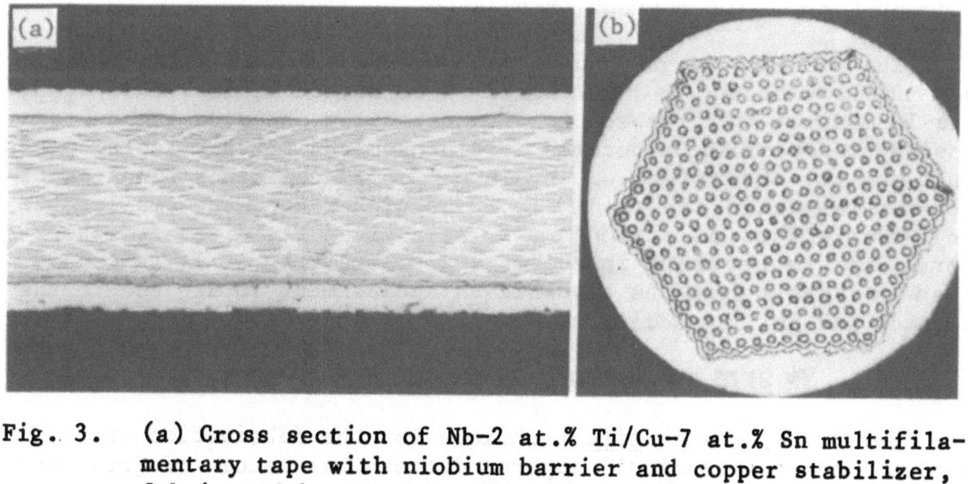

Fig. 3. (a) Cross section of Nb-2 at.% Ti/Cu-7 at.% Sn multifilamentary tape with niobium barrier and copper stabilizer, fabricated by Furukawa Electric Company. Total thickness: 0.20 mm, width: 5 mm; core number: 2440; (b) Cross section of Nb-5 at.% Hf/Cu-5 at.% Sn-4 at.% Ga multifilamentary wire with niobium barrier and copper stabilizer fabricated by Hitachi Cable Company. Diameter: 0.70 mm; core number: 331.

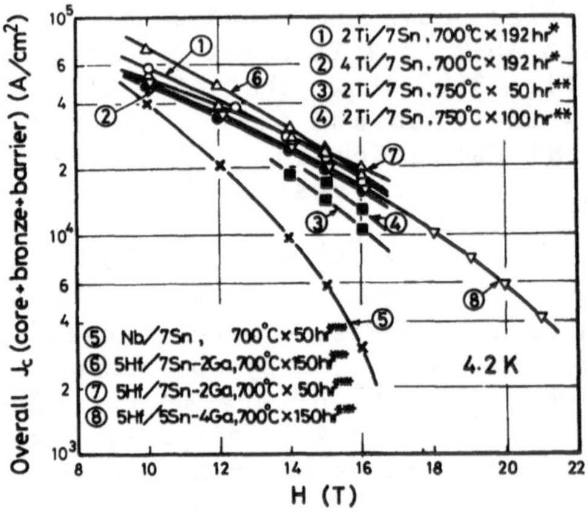

Fig. 4. Magnetic field dependence of overall J_c (core + bronze + barrier) of multifilamentary Nb_3Sn conductors with improved high-field performances.
* 160-core wire; bronze ratio: 3.0; fabricated by NRIM.
** 2440-core tape; bronze ratio: 1.7; by Furukawa.
*** 331-core wire; bronze ratio: 2.5; by Hitachi.

deformation in the core and matrix. The addition of hafnium and gallium may strengthen the core and the matrix after the reaction and prevent the outbreak of the local deformation. This might account for the improvement in ε_{irr}.

The hafnium and gallium addition also improves the strain sensitivity of I_c in the region where strain is less than ε_{irr}; the I_c maximum with respect to strain usually observed in the composite-processed Nb_3Sn becomes less pronounced by this addition.[26] In general, the effect of strain on I_c is a distinct function of reduced magnetic field, H/H_{c2}. The higher H_{c2}, the smaller the relative effect of strain on I_c at a given field level. Therefore, the improvement in the strain sensitivity of I_c is considered to result from the significant enhancement in H_{c2} due to the hafnium and gallium addition.

IN SITU PROCESSED FILAMENTARY A15 CONDUCTORS

Recently, the *in situ* technique for the fabrication of filamentary composites of A15 superconductors has become of great interest as a possible alternative to the conventional composite process. This process consists of casting a two-phase material and subsequent cold working and heat treatment, thus producing an array of aligned and discrete superconducting filaments in a ductile matrix. A large number of studies have demonstrated that the *in situ* formed Nb_3Sn superconductors have a J_c comparable to that of conventional conductors, with the advantage of superior mechanical properties.[27-29] Recent investigations on the in situ process of the Nb_3Sn conductor have concentrated on its scale-up.[30]

In contrast to rather extensive efforts to develop the in situ Nb_3Sn composite, only a few attempts to produce V_3Ga composites by the in situ approach have been reported.[31,32] A systematic study of the in situ process of the V_3Ga composite superconductor has been carried out at NRIM in Japan, including attempts toward large-scale production.[33] In the case of Cu-V alloys, the large miscibility gap in the liquid region makes it difficult to produce a homogeneous ingot by the conventional casting method. It is found, however, that the continuous casting method using an arc melting furnace is the most feasible method to produce a large Cu-V alloy ingot with sufficient homogeneity.[34] Figure 5 shows one of the processes for the preparation of long-length in situ V_3Ga composite superconductors. It involves (1) continuous arc melting and casting of a Cu-V alloy ingot 40 mm in diameter and about 150 mm in length using a consumable electrode, (2) rolling and drawing the ingot into a wire or tape of long length, (3) continuous gallium coating on the wire or tape, and finally, (4) heat treatment to form V_3Ga filaments.

Typical results of overall J_c versus applied fields up to 21.5 T are shown in Fig. 6. The J_c of the commercial V_3Ga multifilamentary wire fabricated by the conventional composite process[6] and that of the in situ processed Nb_3Sn wire[27] are also plotted for comparison. The high-field performance of the in situ processed V_3Ga is considerably superior to that of the in situ processed Nb_3Sn, analogous with the results on the conventional composite-processed conductors. A large overall J_c of about 1 x 10^5 A/cm^2 at 4.2 K and 17 T has been obtained for the (Cu-40 at.% V) + 18 at.% Ga composite 0.2 mm in diameter.

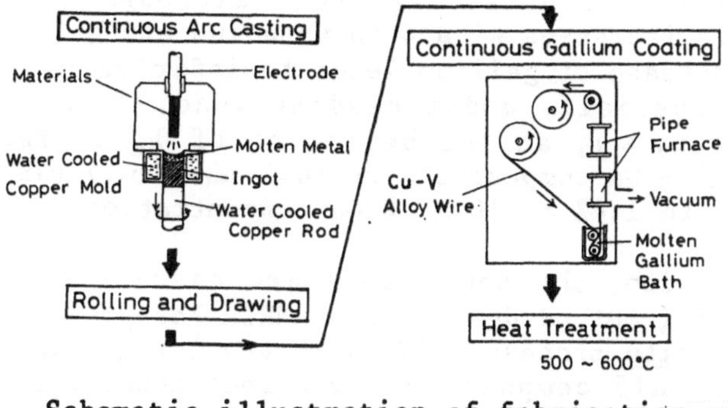

Fig. 5. Schematic illustration of fabrication process of long-length in situ V_3Ga wire.

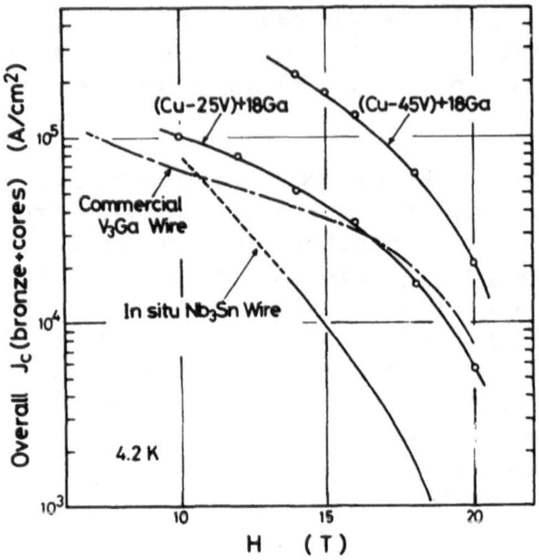

Fig. 6. Typical magnetic field dependence of overall J_c of in situ V_3Ga wire; data on commercial V_3Ga multifilamentary wire[6] and in situ Nb_3Sn wire[27] are shown for comparison.

The mechanical behaviour of the in situ V_3Ga composite is strongly dependent on its alloy composition. Its ultimate tensile strength ranges from 60 to 80 kg/mm^2, increasing with the volume fraction of the vanadium filaments. These strength values are significantly higher than that of the commercial V_3Ga multifilamentary wire, probably owing to the smaller spacing of the filaments, which acts as a barrier to the motion of dislocations in the matrix.[33] The in situ A15 composite conductors show somewhat different behaviour of I_c degradation under tensile or bending stress from that of the composite-processed A15 conductors. The I_c maximum does not clearly appear, and when the load is removed, the in situ V_3Ga composite shows a complete recovery of I_c even after the degradation has started. Although J_c of the in situ V_3Ga composite increases with both vanadium and gallium concentrations, the increase of gallium leads to inferior mechanical load tolerance, especially under bending load.[34] For example, the bending strain, ε_d, at the beginning of I_c degradation of the (Cu-35 at.% V) + Ga composite wire (0.3-mm diam.) varies from 0.6% at 18 at.% Ga to 1.4% at 10 at.% Ga concentration.

In conclusion, the continuous arc casting plus continuous gallium coating method developed at NRIM has proved that scaling-up of the in situ process of V_3Ga is promising. In the design of these in situ A15 composites for magnet conductor use, a proper choice of the alloy composition is necessary, depending on which is of major interest, current-carrying capacity or mechanical tolerance. Furthermore, additional optimization with regard to stability and ac losses will be required for applying the in situ composites to magnet operation in time-varying magnetic fields.

MULTIFILAMENTARY C15 V_2(Hf,Zr) WIRES

The V_2Hf-based C15 crystal-type compounds have gained much interest as a high-field superconductor for magnetic fusion reactor use owing to the promising combination of high H_{c2} (exceeding 20 T at 4.2 K) and high tolerance to neutron irradiation.[35-37] The superconducting properties of 7-core V_2(Hf,Zr) wires fabricated by a composite process using a V-1 at.% Hf alloy matrix and Zr-25 at.% Hf alloy cores have been reported.[38] Recently, V_2(Hf,Zr) wires with 133 cores have been fabricated. In this case, 19 V-1 at.% Hf cores were inserted into the holes drilled in a Zr-35 at.% Hf matrix. The resulting composite was cold-drawn with intermediate annealing at 750°C into a wire and then cut into 7 pieces. These pieces were encased in a Zr-Hf alloy pipe and cold-drawn into a 133-core composite wire. The diameters of individual cores after drawing were 10-20 μm. Finally, the wire was reacted at temperatures between 850° and 1,000°C to form the C15 filaments. The intermediate annealing at 750° C was best for matching the workability of the Zr-35 at.% Hf alloy to that of the V-1 at.% Hf alloy. The 133-core wire with a V-1 at.% Hf matrix and Zr-35 at.% Hf cores is also being successfully fabricated by a similar process.

The formation rate of V_2(Hf,Zr) layers in the V-1 at.% Hf(core)/Zr-35 at.% Hf(matrix) composite wires with single and 133 cores is seen in Fig. 7. The V_2(Hf,Zr) layer thickness is almost independent of the number of cores and does not saturate with heat-treatment time until each core is completely reacted with the matrix. In this composite, all the elements contained in the matrix and the cores contribute to form V_2(Hf,Zr) layers, and the compositions of the matrix and the cores are essentially unchanged with the C15 layer thickness. In the composite-processed A15 wire, on the contrary, the A15 layer thickness tends to saturate since the tin or gallium concentration in the matrix decreases rapidly with the A15 layer thickness. The multifilamentary C15 wire with a small matrix/core area ratio may show larger overall J_c than the multifilamentary A15 wire with the same ratio, owing to a thicker layer in the former wire. In other words, the volume fraction of the residual matrix after the reaction may be smaller in the C15 wire than in the A15 wire.

The I_c and J_c versus magnetic field curves for V-1 at.% Hf/Zr-35 at.% Hf wires with 1, 7, and 133 cores are shown in Fig. 8. The J_c increases slightly with increasing number of cores. In fields below 12 T, a higher J_c is obtained for the C15 layer formed below 925°C; in fields above 12 T, however, a higher J_c is obtained for that above 925°C. Heat treatment at higher temperatures slightly increases H_{c2} and, therefore, J_c in high fields. A J_c(4.2 K) of 1 x 10^5 A/cm^2 is obtained at 12 T for the 133-core wire. The T_c and the H_{c2} of this wire are 9.4 K and 20.5 T, respectively. Measurement of the multifilamentary wire composed of V-1 at.% Hf matrix and Zr-35 at.% Hf cores is also in progress, and comparable superconducting properties have been obtained so far.

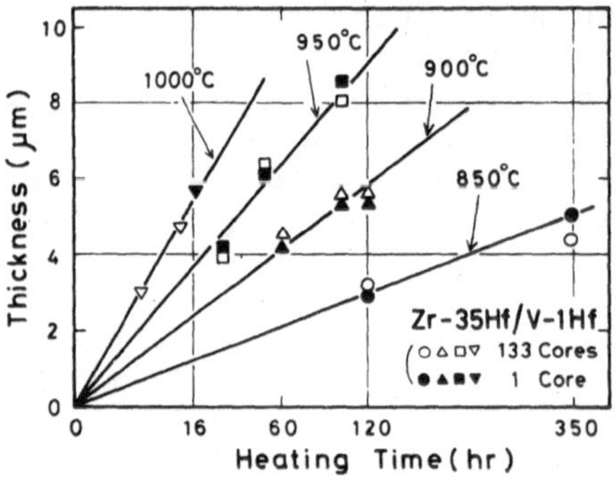

Fig. 7. Thickness of V_2(Hf,Zr) layers in composite-processed V-1 at.% Hf/Zr-35 at.% Hf wires with single and 133 cores as a function of reaction time, reaction temperatures being between 850° and 1,000°C.

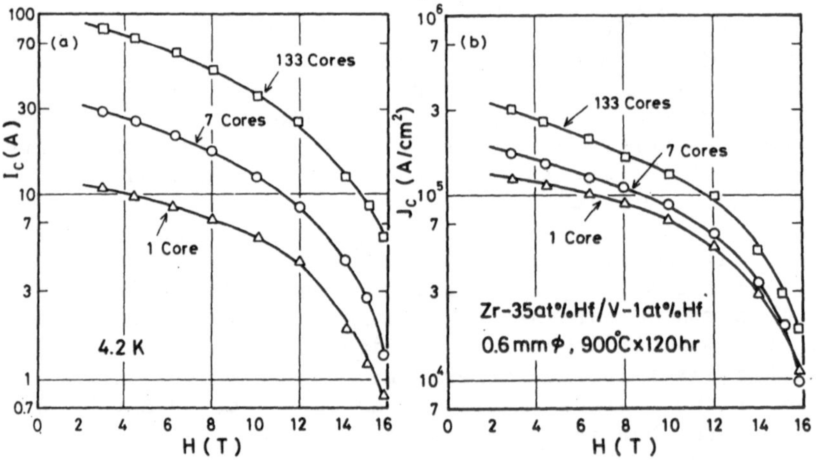

Fig. 8. Magnetic field dependence of I_c and J_c of composite-processed V-1 at.% Hf/Zr-35 at.% Hf wires with single, 7, and 113 cores, reacted at 900°C for 120 h.

REFERENCES

1. A. R. Kaufman and J. J. Pickett, Bull. Am. Phys. Soc. 15:833 (1970).
2. K. Tachikawa, "Proc. ICEC-3" Iliffe Sci. Tech. Pub., Surrey, (1970) p. 339.
3. K. Tachikawa and Y. Tanaka, Jap. J. Appl. Phys. 6:782 (1967).
4. K. Tachikawa, Y. Yoshida, and L. Rinderer, J. Mater. Sci. 7:1154 (1972).
5. K. Tachikawa, R. J. Burt, and K. T. Hartwig, J. Appl. Phys. 48:3623 (1977).
6. Y. Furuto, T. Suzuki, K. Tachikawa, and Y. Iwasa, Appl. Phys. Lett. 24:34 (1974).
7. K. Itoh and K. Tachikawa, Appl. Phys. Lett. 26:67 (1975).
8. Y. Furuto, Y. Tanaka, S. Meguro, T. Suzuki, and I. Inoue, in: "Filamentary A15 Superconductors," M. Suenaga and A. F. Clark eds., Plenum Press, New York (1980), p. 115.
9. S. Shimamoto, K. Yasukochi, T. Ando, N. Tada, K. Aihara, and Y. Hotta, IEEE Trans. Magn. MAG-15:406 (1979).
10. E. Gregory, E. Adam, W. Marancik, P. Sanger, and C. Spencer, in: "Filamentary A15 Superconductors," M. Suenaga and A. F. Clark eds., Plenum Press, New York (1980), p. 47.
11. H. Hillman, H. Pfister, E. Springer, M. Wilhelm, and K. Wohlleben, in: "Filamentary A15 Superconductors," M. Suenaga and A. F. Clark, eds., Plenum Press, New York (1980), p. 17.
12. P. N. Haubenreich, J. N. Luton, and P. B. Thompson, IEEE Trans. Magn. MAG-15:520 (1979).
13. W. B. Sampson, S. Kiss, K. E. Robin, and A. D. McInturff, IEEE Trans. Magn. MAG-15:117 (1979).
14. K. Tachikawa, H. Sekine, and K. Togano, IEEE Trans. Magn. MAG-15:762 (1979).
15. M. Suenaga, W. B. Sampson, and T. S. Luhman, IEEE Trans. Magn. MAG-17:646 (1981).
16. K. Tachikawa, T. Asano, and T. Takeuchi, to be published in Appl. Phys. Lett.
17. H. Sekine and K. Tachikawa, Appl. Phys. Lett. 35:472 (1979).
18. H. Sekine, T. Takeuchi, and K. Tachikawa, IEEE Trans. Magn. MAG-17:383 (1981).
19. K. Tachikawa, Y. Tanaka, Y. Yoshida, T. Asano, and Y. Iwasa, IEEE Trans. Magn. MAG-15:391 (1979).
20. T. Takeuchi, T. Asano, Y. Iijima, and K. Tachikawa, to be published in Cryogenics.
21. K. Togano, T. Asano, and K. Tachikawa, J. Less-Common. Met. 68:15 (1979).
22. Y. Yoshida, K. Tachikawa, and Y. Iwasa, Appl. Phys. Lett. 27:632 (1975).
23. H. Wada, M. Kimura, and K. Tachikawa, J. Mater. Sci. 13:1943 (1978).

24. K. Tachikawa, T. Takeuchi, T. Asano, Y. Iijima, and H. Sekine, Effects of the IVa element additions on composite-processed Nb_3Sn, in: "Advances in Cryogenic Engineering--Materials," Vol. 28, Plenum Press, New York (1982), p. 389.
25. D. G. Howe and L. S. Weinman, IEEE Trans. Magn. MAG-11:251 (1975).
26. J. W. Ekin, H. Sekine, and K. Tachikawa, J. Appl. Phys. 52:6252 (1981).
27. R. Roberge, S. Foner, E. J. McNiff, Jr., B. B. Schwartz, and J. L. Fihey, Appl. Phys. Lett. 34:111 (1979).
28. J. D. Verhoeven, F. A. Schmidt, E. D. Gibson, J. E. Ostenson, and D. K. Finnemore, Appl. Phys. Lett. 35:555 (1979).
29. J. P. Harbison and J. Bevk, J. Appl. Phys. 49:6031 (1978).
30. J. D. Verhoeven, F. A. Schmidt, E. D. Gibson, J. J. Sue, J. E. Ostenson, and D. K. Finnemore, IEEE Trans. Magn. MAG-17:251 (1981).
31. J. Bevk, F. Habbal, C. J. Lobb, and J. P. Harbison, Appl. Phys. Lett. 35:93 (1979).
32. J. L. Fihey, R. Roberge, S. Foner, E. J. McNiff, Jr., and B. B. Schwartz, in: Advances in Cryogenic Engineering--Materials," Vol. 26, Plenum Press, New York (1980), p. 343.
33. K. Togano, H. Kumakura, and K. Tachikawa, IEEE Trans. Magn. MAG-17:985 (1981).
34. H. Kumakura, K. Togano, and K. Tachikawa, High-field critical current and mechanical properties of in situ processed V_3Ga superconductors, in: "Advances in Cryogenic Engineering--Materials," Vol. 28, Plenum Press, New York (1982), p. 515.
35. K. Inoue and K. Tachikawa, IEEE Pub. No. 72 CHO 682-5-TABSC, IEEE, New York (1972), p. 415.
36. K. Inoue and K. Tachikawa, J. Jap. Inst. Met. 39:1265 (1975); ibid, p. 1274.
37. B. S. Brown, J. W. Hafstrom, and T. E. Klippert, J. Appl. Phys. 48:1759 (1977).
38. K. Inoue, T. Kuroda, and K. Tachikawa, IEEE Trans. Magn. MAG-15:635 (1979).

POWDER METALLURGY PROCESSES—A REVIEW OF THE STATUS AND PROMISE*†

S. Foner

Francis Bitter National Magnet Laboratory‡ and Plasma Fusion Center
Massachusetts Institute of Technology, Cambridge, Massachusetts

INTRODUCTION

The first demonstration of a superconducting material with a high critical current at high magnetic fields was reported in 1961 by Kunzler et al.[1] for Nb_3Sn. It was produced by a powder process. This development was the first promise of superconducting materials for high field magnet applications. Improved superconducting materials have been developed over the last two decades so that the promise has been partially fulfilled. Although Kunzler's materials employed a powder process, practical fabrication technologies have favored other fabrication techniques for A15 materials. Recently several powder processing approaches have been developed as alternatives to the now more conventional processes. This review gives source references and brief descriptions of the infiltration, ECN, and hot and cold powder processes. It also outlines some advantages and disadvantages of these processes and indicates the status of their development. In order to assess a process as a viable industrial alternative to existing technologies, factors such as cost of initial materials, fabrication costs, and performance of the final wires are required. This assessment is difficult to make until industrial transfer occurs. However, it is possible to make promises based on apparent advantages over existing processes. This review follows historical precedence.

*Invited paper.
†Work supported by the U.S. Department of Energy.
‡Supported by the National Science Foundation.

INFILTRATION PROCESS

A recent review of the infiltration process at Lawrence Berkeley Laboratory was given by Pickus et al.[2] A flexible Nb_3Sn tape was fabricated by roll-compacting 270 mesh, hydride-dehydride, Nb powder. It was sintered at 2250°C for 3 min to produce an interconnected porous structure. This tape was infiltrated with molten Sn at 850°C for 1 min to completely fill the pores. The tape was then cold rolled, and a diffusion reaction carried out for 3 min at 950 to 975°C. The filaments were typically 30 x 5 μm and a $J_c = 10^5$ A/cm^2 at 10 T was reported.[2]

The procedure for Nb_3Sn wire used -200 + 400 mesh Nb powder, isostatically cold-compacted at 30 ksi, vacuum sintered at 2250°C from 30 to 60 min for a 3/16-in diam. rod, then immersed in a tin bath (350-400°C) for infiltration. A Monel cladding was then applied and the resultant bar was rolled and wire-drawn to produce a wire with a reduction ratio of 4000. Heat treatment at 950°C for 2 to 3 min produced a Nb_3Sn wire. A $J_c = 4 \times 10^4$ A/cm^2 at 16 T was measured for the core of the Nb_3Sn wire.

Prealloyed powder compacts without infiltration were fabricated, but the results were not promising.[2] However, infiltration of ternary A15 materials using sintered Nb powder as above, but infiltrants of Al-Ge or Al-Si eutectics were successful in producing Nb-Ge-Al and Nb-Al-Si wires. The problems associated with this approach are discussed in Ref. 2.

Attempts to transfer the infiltration technology to industry were made. At present this approach for industrial processing has been discontinued.

ECN POWDER PROCESS

The group at the Netherlands Energy Research Foundation, ECN, Petten, The Netherlands has developed powder processes for Nb_3Sn, V_3Ga and V_3Si.[3,4,5] All the processes start with prealloyed powders $\leq$ 10 μm in diam. placed in a tube of Nb or V that is surrounded by a pure Cu tube. The powders, $NbSn_2$, V_2Ga_5, VSi_2, are brittle, but since the final size of the powder compact fiber is generally larger than the particle size, this may not be critical. For V_3Si production, VSi_2 powder is put in a V tube with added Cu as a catalyst. Cold reduction of the tube (or a bundle of tubes) results in tubes of V with a 40-μm central powder region. The reaction of the powder in the tube results in relatively thick layers (up to 10 μm) V_3Si and the VSi_2 can be fully reacted. The reaction between 800 and 1000°C is $V + VSi_2 + Cu \rightarrow V_5Si_3 \rightarrow V_3Si$. A similar procedure is used for V_3Ga. Reaction takes place between 550 to 650°C, given by $V + V_2Ga_5 + Cu \rightarrow V_6Ga_5 \rightarrow V_3Ga$, again with an intermediate product (V_6Ga_5) that can be converted completely to

V_3Ga. For both wires no Kirkendall voids were found[5] and J_c decreased with increasing A15 layer thickness. Prolonged and complete reaction produced high T_c material.

The most recent work[3] concentrated on stabilized Nb_3Sn development and discussed production technology. The reaction for Nb_3Sn is Nb + $NbSn_2$ → Nb_6Sn_5 → Nb_3Sn. Reaction temperatures ranged from 625 to 700°C.[2] The Nb tube is surrounded by a pure Cu tube and can be bundled. Reduction in cross section is done by swaging, wire drawing, rolling, or extrusion. Single-core rods with a circular Nb tube filled with $NbSn_2$ powder surrounded by a Cu tube having a circular hole and hexagonal outer geometry were used for fabrication of multiple filament wires. The Sn is fully reacted with the Nb tube. Because there is a relatively large amount of Sn available compared with the bronze process, thick Nb_3Sn layers (up to 7 µm) can be produced with a reduced number of filaments. Adiabatic stability imposes an upper limit of about 30-µm diam. on the inner core of the filament.[3] Material with filament cores from 30 to 75 µm for fully reacted materials showed no dependence of core size on J_c. Fully reacted wires with overall J_c values of 1.5×10^4 A/cm^2 at 14 T were reported. The J_c for the 19-filament wire shows no prestress (an immediate reduction in J_c with strain is observed).

The authors note that conventional wire drawing would be limited to units of about 20 kg. Extrusion was attempted as a larger scale technology. Initial tests resulted in perforations of the Nb tube. Modifications of the powder specifications are expected to permit hydrostatic extrusion of a 150-kg billet of multifilamentary Nb_3Sn. A cooperative program is being carried out by ECN with the Metals Research Institute, TNO, Apeldoorn, and Holec Wire b.v., Nijmegen, both of The Netherlands.

HOT POWDER PROCESS

Hot powder metallurgy processing has been developed at the University of Göttingen. Early work with hot extrusions of Nb + Cu powders did not give good results because of solution hardening of the Nb by interstitial oxygen.[7,8,9] A method of high temperature processing was demonstrated by Bormann and Freyhardt,[10] who incorporated a getter (e.g., Al, Mg, Zr, Hf, or Ca powder) in the compact. Hot extrusion between 950 and 1050°C results in reduction of the Nb, producing a low microhardness. Large areal reductions of the billet are feasible with high extrusion temperatures (above 900°C) and the resultant compact can be deformed easily by cold rolling or wire drawing without intermediate annealing or bundling. External Sn plating, diffusion, and reaction between 500 to 600°C yielded materials with high J_c at high field. The Cu-30 wt.% Nb gave overall critical currents of 10^4 A/cm^2 at 16 T. A similar approach[10,11] with powdered Cu and V for Cu-30 wt.% V + Ga gave

$J_c = 2$ to 4×10^4 A/cm^2 at 16 T with reaction temperatures at 500 to 600°C. As expected, performance of the V_3Ga is superior to Nb_3Sn at high fields.

Large green compacts (~100 kg) can be produced by this hot processing. So far billets from 2 to 8 kg have been produced and scale-up does not appear to pose problems. Final filament sizes can be varied by choice of initial powder size, and, as for the in situ process, complete reaction is achieved for fine filaments, resulting in high values of J_c. Furthermore, additives of In, Ga, and Ta can be incorporated. Commercial processing is being explored in cooperation with Siemens and VAC/Hanau.

COLD POWDER PROCESSING

The success of in situ processing for Cu-Nb-Sn and Cu-V-Ga[12] led to the cold powder processing approach at MIT as an alternative. Flükiger et al.[13] demonstrated that cold processing was feasible starting with 40-μm hydride-dehydride Nb and Cu powder. Compaction of the powder in a Be-Cu tube, swaging, and wire drawing followed by external Sn plating, diffusion, and reaction at 650°C yielded submicron filaments of Nb_3Sn with high J_c values at high field. Both the J_c and strain tolerance for such Cu-Nb-Sn materials comparable to the best in situ materials.[13,14,15,16]

External diffusion of Sn into the Cu-Nb places a limit on the wire size to less than 0.5-mm diam. As an alternative approach, Sn powder was introduced with the Cu and Nb powder in the billet. The resulting wire has no size limitation because of the short internal diffusion paths. Cold processing produced multifilamentary wires in a "one-step" process also with high J_c and good strain tolerance.

The powder processing approach permits great flexibility. The filament size can be controlled by the starting powder size and reduction ratio, as long as the powders are ductile. Cold processing avoids metallurgical problems of reaction between powders prior to the final heat treatment. The laboratory scale fabrication permits inexpensive tests of various processes with small quantities of material prior to scale-up. Examples of this flexibility were demonstrated[17] by cold powder processing with Sn introduced in prealloyed Cu-Sn powders for a "one-step" process, Cu-NbTa-Sn produced with prealloyed NbTa powders resulting in increased values of J_c at high fields, hot hydrostatic extrusion with 40-μm Nb powder and an Al getter[10] at low temperature, hot hydrostatic extrusion of large 150- to 600-μm Nb powders with no getter, and fabrication of Nb-Al multifilamentary composites.[18,19] All wires showed high J_c and high strain tolerance. Further developments are summarized in another paper.[20]

The development of the Nb-Al wires is unique to the cold powder process; both hot processing and in situ processing are not feasible because reaction of the Nb and Al occurs at a relatively low temperature.

Recently Hong et al.[21] at Lawrence Berkeley Laboratory have used this cold powder process to produce Nb-Al materials. They have employed two-stage heat treatments and have examined the microstructure with TEM and STEM and the interfacial reactions in this material. Jorda et al.[22] reported briefly on powder processed NbAl with a $T_c \sim 12$ K, well below that obtained by the groups at MIT or LBL.

SUMMARY

Advantages of ductile powder processes are that they are quite versatile and permit easy fabrication of fiber sizes in the submicron range with a minimum number of processing steps. There is a diversity of possible approaches, but if hot processing is used for Nb_3Sn or V_3Ga, external diffusion of Sn or Ga is required (applied internally or externally) so that strands are restricted to less than 0.5-mm diam. If cold processing is used then there are possibilities of including Sn in the Cu-Nb composite from the start. Also, Nb_3Al can be fabricated with no additions. Brittle powders can also be used if the fiber size is not restricted to very small dimensions.

The J_c of powder processed materials is comparable to the best in situ or continuous fiber processed materials, and the strain tolerance is very good for the ultrafine fiber materials. However, as losses are reported to be large at low fields. For dc applications this is not a problem, but for ac applications this problem should be solved.

My assessment is that the hot and cold powder processes both are promising for practical conductor development. Furthermore, ac losses can be minimized in various ways for the powder and in situ processed materials. Whether the economics will dictate acceptance in industry is currently being evaluated. Certainly for less conventional materials, e.g., Nb_3Al, there is a possibility of superior performance as the processing is developed further.[23] Evaluation of the cost, ease of manufacture and adaptability of the powder process to conventional manufacturing technology will determine whether this approach is accepted. It should be noted that as the conventional processes for Nb_3Sn are further developed, various problems continue to arise after twenty years of development. As the powder process is improved, it should be continually reevaluated; it is very likely that the strain tolerance and other performance chracteristics of the alternative processed materials may be the deciding factors for high field applications.

I have not attempted to compare the powder processes to other competing technologies. The closest competitor for ultrafine fiber materials is the in situ process. The performance of the hot and cold powder processes is comparable to the best in situ materials. Conventional continuous fiber technologies are also being developed for ultrafine fibers and the J_c and strain tolerance is comparable to the best powder and in situ materials. Another approach, the modified jellyroll, is not a powder process, but shows promise. All must now be examined as possible commercial alternatives to conventional processes.

ACKNOWLEDGMENTS

I wish to thank Prof. H. Freyhardt for forwarding extensive references and comments on the hot process. I also wish to thank M. Hong for discussions concerning the powder process work at Lawrence Berkeley Laboratory.

REFERENCES

1. J. E. Kunzler, E. Buehler, F. S. L. Hsu, and J. H. Wernick, Phys. Rev. Lett. 6:89(1961).
2. M. R. Pickus, J. T. Holthuis, and M. Rosen, in: "Filamentary A15 Superconductors," M. Suenaga and A. F. Clark, eds., Plenum Press, New York (1980), p. 331 and references cited therein.
3. J. D. Elen, J. W. Schinkel, A. C. A. van Wees, C. A. M. van Beijnen, E. M. Hornsveld, T. Stahlie, H. J. Veringa, and A. Verkaik, IEEE Trans. Magn. MAG-17:1002 (1981).
4. C. A. M. van Beijnen and J. D. Elen, IEEE Trans. Magn. MAG-15:87 (1979).
5. J. D. Elen, C. A. M. van Beijnen, and C. A. M. van der Klein, IEEE Trans. Magn. MAG-13:470 (1977).
6. C. A. M. van Beijnen and J. D. Elen, IEEE Trans. Magn. MAG-11:243 (1975).
7. L. Schultz, H. C. Freyhardt, R. Bormann, and B. L. Mordike, in: "Proceedings of LT14," Vol. 2, North Holland, Amsterdam (1975), p. 59.
8. R. Bormann, L. Schultz, H. C. Freyhardt, and B. L. Mordike, Z. Metall. 70:467 (1979).
9. R. Bormann, L. Schultz, and H. C. Freyhardt, Appl. Phys. Lett. 32:79 (1978).
10. R. Bormann and H. C. Freyhardt, Appl. Phys. Lett. 35:944 (1980).
11. R. Bormann and H. C. Freyhardt, IEEE Trans. Magn. MAG-17:270 (1981).
12. R. Roberge and S. Foner, in: "Filamentary A15 Superconductors," M. Suenaga and A. F. Clark, eds., Plenum Press, New York (1980), p. 241.

13. R. Flükiger, R. Akihama, S. Foner, E. J. McNiff, Jr., and B. B. Schwartz, Appl. Phys. Lett. 34:473 (1979).
14. R. Flükiger, S. Foner, E. J. McNiff, Jr., and B. B. Schwartz, Appl. Phys. Lett. 35:810 (1979).
15. R. Flükiger, R. Akihama, S. Foner, E. J. McNiff, Jr., and B. B. Schwartz, Appl. Phys. Lett. 34:763 (1979); 35:430, (1979).
16. R. Flükiger, R. Akihama, S. Foner, E. J. McNiff, Jr., and B. B. Schwartz, in: "Advances in Cryogenic Engineering - Materials," Vol. 26, Plenum Press, New York (1980), p. 377.
17. R. J. Murphy, R. Akihama, S. Foner, and B. B. Schwartz, IEEE Trans. Magn. MAG-17:989 (1981).
18. R. Akihama, R. J. Murphy, and S. Foner, Appl. Phys. Lett. 37:1107 (1980).
19. R. Akihama, R. J. Murphy, and S. Foner, IEEE Trans. Magn. MAG-17:274 (1981).
20. S. Foner, S. Pourrahimi, J. Otubo, C. L. M. Thieme, H. Zhang, R. J. Murphy, R. Akihama, B. B. Schwartz, and W. Marancik, Powder metallurgy processing of Nb-Al and Cu-Nb-Sn and scale-up, paper presented at the International Cryogenic Materials Conference, San Diego, California (Aug. 10-14, 1981).
21. M. Hong, Lawrence Berkeley Laboratory, Berkeley, California, private communication.
22. J. L. Jorda, R. Flükiger, A. Junod, and J. Müller, IEEE Trans. Magn. MAG-17:557 (1981).
23. J. Bevk, in: "Conference on Rapidly Quenched Metals III," Vol. 2, B. Cantor, ed., Metal Society of London (1979), p. 17.

DEFORMATION OF METASTABLE AUSTENITIC STEELS AT LOW TEMPERATURES

R. P. Reed and R. L. Tobler

Fracture and Deformation Division, National Bureau of Standards, Boulder, Colorado

INTRODUCTION

The reaction of a solid to continuously increasing applied tensile load is portrayed using a stress-versus-strain curve. From stress-strain curves the Young's modulus, yield strength, ultimate tensile strength, percent elongation, and work hardening characteristics can be obtained. Typically, following the elastic deformation region, a metal or alloy work hardens at a decreasing rate until localized specimen necking initiates and the ultimate strength of the specimen has been reached.

The stress-strain characteristics at low temperatures of commercial grades of polycrystalline austenitic stainless steels containing about 18 wt.% Cr and 8 wt.% Ni are not conventional. The anomalous behavior is thought to be caused by the martensitic transformation of face-centered cubic austenite to body-centered cubic (α') and hexagonal close-packed phases. This paper suggests that the plastic deformation of metastable austenites is composed of three stages and uses a series of Fe-18Cr-8Ni-1-6Mn, 0.1-0.2N alloys, tested at 4 K, to demonstrate the usefulness of this characterization.

EXPERIMENTAL PROCEDURES

The tensile tests were conducted in liquid H using a strain rate of 0.02 min^{-1}. Strain was recorded using a clamp-on strain-gage extensometer. Strains larger than about 0.10 were computed using load-time graphs of each test. Other experimental details, alloy compositions, and alloy hardness and grain size are discussed by Tobler and Reed.[1,2]

TENSILE STRESS-STRAIN CURVES

Typical stress-strain curves of metastable polycrystalline austenitic stainless steels at temperatures between 4 and 295 K are illustrated in Fig. 1 for an Fe-18Cr-8Ni alloy. Compositions refer to weight percent. Alloys of this type remain ductile at low temperatures, exhibit considerable work hardening, have a region of low work hardening at 195, 76, and 4 K, and have discontinuous yielding at 4 K. The stress-strain curves for a series of Fe-18Cr-8Ni-1-6Mn with 0.1 and 0.2N are presented in Figs. 2 and 3, respectively. For purposes of clarity, the discontinuous yielding serrations have been deleted; instead, only the maximum flow stress values are plotted and the load drops are omitted. The presence of an easy glide region is very apparent, followed by a region of high work hardening.

DISCUSSION

For characterization and analysis, the stress-strain curves of such metastable polycrystalline austenitic alloys can be considered to have three stages of plastic deformation. These are referred to as stages I, II, and III. The stress-strain curve in Fig. 4 identifies these stages and (following single-crystal nomenclature) assigns stress and strain symbols for the onset of the three stages. Stage I represents the first stage of plastic deformation, beginning at the elastic limit and ending at the onset of easy glide (usually about $\varepsilon = 0.02$). The yield strength is included in

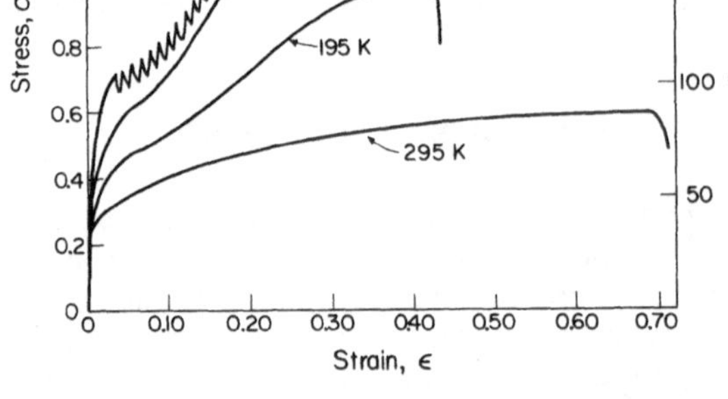

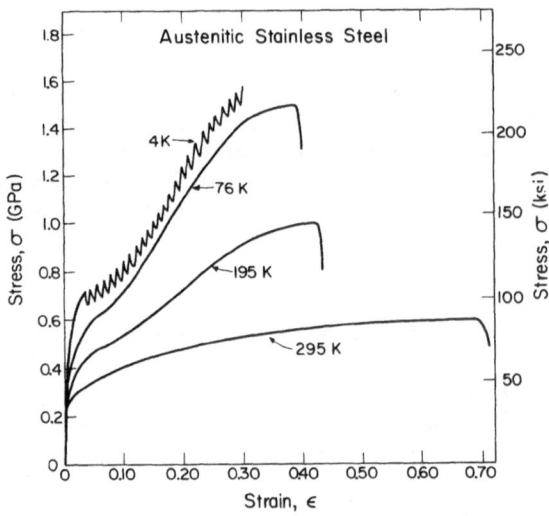

Fig. 1. Typical stress versus strain curves (engineering) for polycrystalline Fe-18Cr-8Ni austenitic stainless steel.

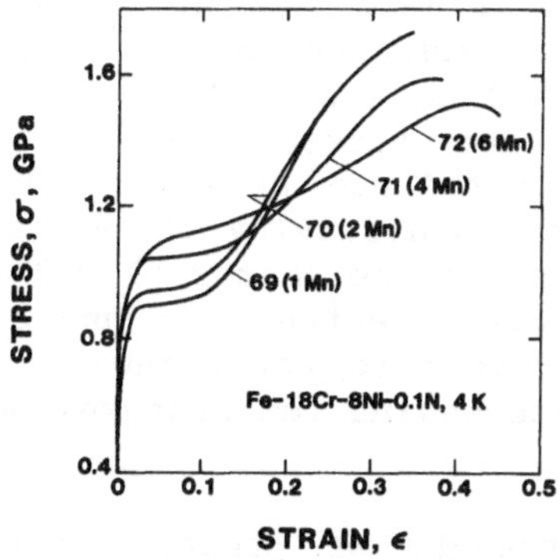

Fig. 2. Stress-strain curves at 4 K for series of Fe-18Cr-8Ni-0.1N alloys with 1-6 wt.% Mn.

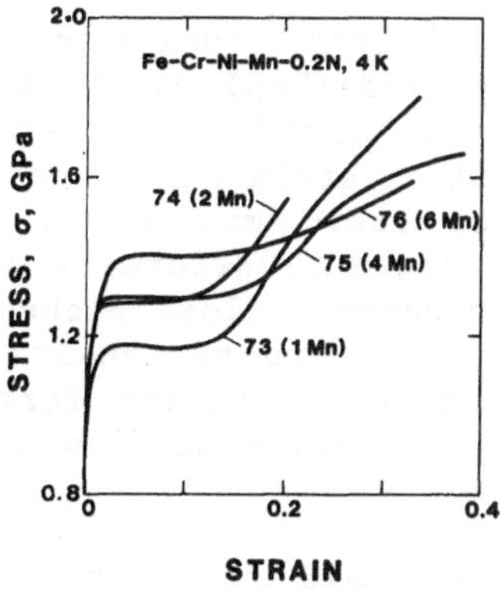

Fig. 3. Stress-strain curves at 4 K for series of Fe-18Cr-8Ni-0.2N alloys with 1-6 wt.% Mn.

this stage. Stage II is the easy glide region. Stage III represents the region of high work hardening.

Since stable austenitic steels do not exhibit the stages of the stress-strain curves discussed above, it is likely that these stages are associated with martensitic transformation. The remainder of this paper concerns correlations between the stresses and strains of the three stages and the parameters, such as composition, that affect austenite stability or represent martensite concentration.

Stage I

In our study, sufficiently accurate measurements of ε_I and σ_I were not obtained of the onset of stage I. It is likely that only very small amounts (< 1%) of martensite form in this stage. Suzuki

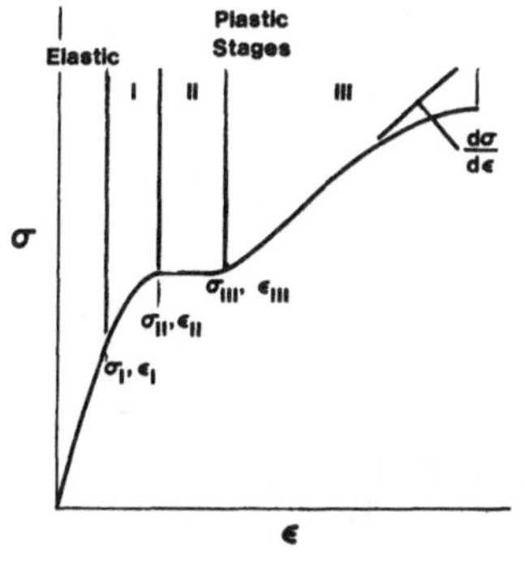

Fig. 4. Identification of three stages of plastic deformation in austenitic stainless steels at low temperatures.

et al.,[3] from studies on an 18Cr-9Ni steel, suggest hexagonal close-packed (hcp) martensite formation. Olsen and Azrin,[4] using a TRIP steel (Fe-9Cr-8Ni-4Mo-2Si) that is less stable than a normal austenitic stainless steel, suggest that stress-assisted body-centered cubic martensite formations contribute substantially to early austenite deformation. Within the sensitivity of our x-ray (> 2% hexagonal close-packed, > 2% α') and magnetometer ($\geq$ 0.2% α') measurements, no martensite formation was detected. In any case, the initial martensite formation is thought to contribute to the observed decrease of flow strength at decreasing temperatures below about 200 K.[3,4,5]

The yield strength, σ_y, (flow strength at plastic strain = 0.002) is within stage I. As shown in Fig. 5, σ_y is linearly dependent on Mn content, increasing with increasing Mn. The linear dependence matches that of σ_{II}, the stress at which stage II begins. The increase of σ_y with Mn is considerable, about 33 MPa per wt.% Mn, but not as high as the increase of σ_y from the addition of N. From Fig. 5, it is shown that an addition of 0.1 wt.% N increases σ_y about 50%. Note that the dependence of σ_y on wt.% Mn is independent of N content. The strength increases from N and Mn additions may arise from two sources: (1) increased austenite stability,[6,7] which results in less martensitic transformation and, consequently, less work hardening and (2) solid-solution strengthening; approximately 0.1 wt.% N is equivalent to 8 wt.% Mn in contribution to strengthening.

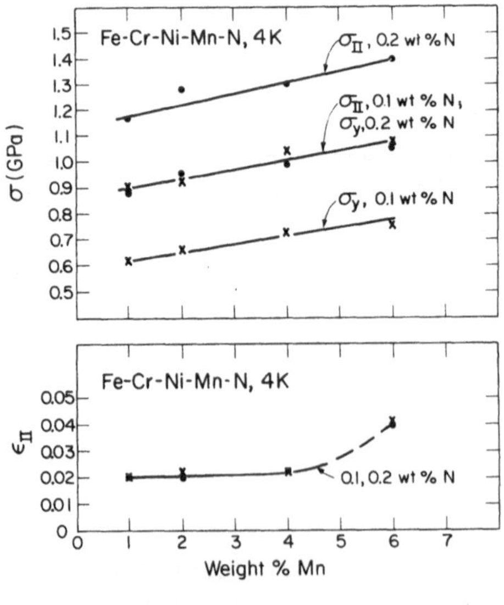

Fig. 5. Tensile yield strength and stage II σ_{II} and ε_{II} at 4 K as a function of wt.% Mn.

Stage II

The stress to initiate stage II (σ_{II}) linearly increases with Mn content, but ε_{II} is essentially independent of Mn until 6% Mn is added (Fig. 5); at this concentration ε_{II} is larger. The N addition also has a much stronger influence on σ_{II} than Mn. The strain, ε_{II}, is independent of N content. In this stage, there is little work hardening. Work softening has been reported,[8] and this stage has been termed "easy glide."[9] Reed and Guntner,[9] noting the easy glide region and measuring (using x-rays) the increase of hcp martensite in the region, suggested that stage II was caused by hcp formation in slip bands. Suzuki et al. observed bcc formation at slip-band intersections using transmission electron microscopy and proposed that these martensite laths act as "windows," promoting dislocation mobility at slip-band intersections.

Stage III

The transition to the region of high work-hardening rates, stage III, is denoted by a strain, ε_{III}, and a stress, σ_{III}, which are plotted in Fig. 6. The strain, ε_{III}, has greater scatter and apparently mildly increases with increasing Mn. The stress, σ_{III}, also increases mildly and linearly with increasing Mn content. In contrast to stage II, the dependence of ε_{III} and σ_{III} on Mn appears to be conditioned by the N content. Stage III begins at lower stresses and larger strains for 0.1 wt.% N than for 0.2 wt.% N.

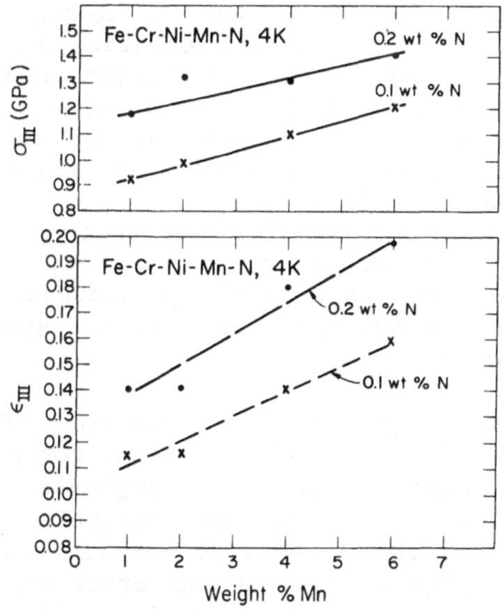

Fig. 6. Stage III σ_{III} and ε_{III} at 4 K as a function of wt.% Mn.

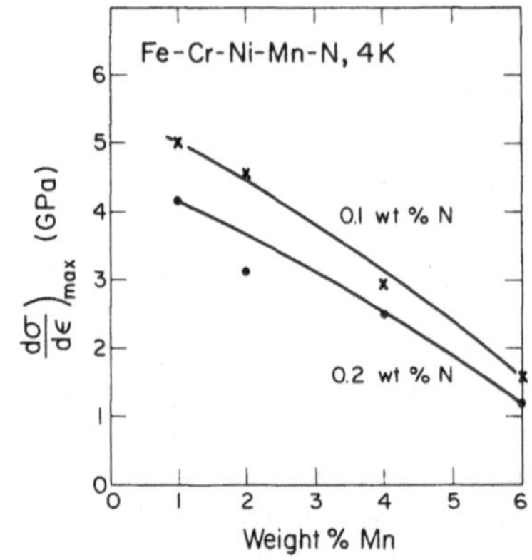

Fig. 7. Rate of work hardening versus wt.% Mn for alloy series at 4 K.

The rate of work hardening, $d\sigma/d\varepsilon$, decreases with increasing Mn, as shown in Fig. 7. The addition of N also lowers the rate of work hardening and slightly affects the dependence of Mn content; it contributes to the reduction of $d\sigma/d\varepsilon$ about ten times more strongly (on the basis of wt.% addition) than Mn.

In Fig. 4 of the Tobler and Reed paper in this volume,[1] the "normalized" percent α' martensite (divided by the percent uniform elongation within the measurement length) is plotted as a function of composition [Mn + Ni + 10 (C+N)]. Determined from these data and Fig. 7, the dependence of the work-hardening rate in stage III on percent α' martensite per unit elongation is shown in Fig. 8. Clearly, there is a strong linear correlation, and the rate of work hardening can be regarded as a function of austenite stability with respect to α' formation as determined by composition. This dependence may imply that α' formation in stage III is distinct from stage II (and possibly stage I) formation at slip-band intersections. A second possiblity is that sizable α' formation produces sufficient accommodation deformation, resulting from the associated $\gamma \rightarrow \alpha'$ volume expansion, to increase the defect concentration (and work hardening) more than the specimen strain produced by transformation.

The fracture toughness variation with composition is difficult to assess (see Ref. 1). Typically, at 4 K for a series of stable and metastable austenitic alloys, the toughness [K_{Ic}(J)] varies as the reciprocal of the σ_y.[1] Since the stability of the alloys has been demonstrated to affect the stress-strain characteristics, one may expect that the plastic zone at the crack tip of the specimens may also be affected. The extent of easy glide in stage II may influence the shape of the zone of crack-tip plasticity; more easy

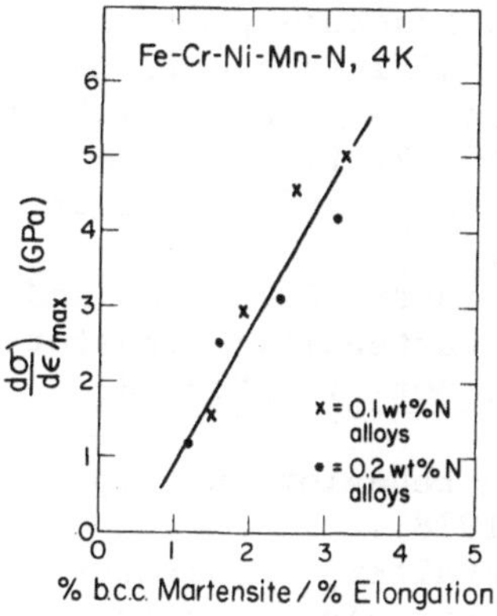

Fig. 8. Rate of work hardening versus "normalized" martensite function (wt.% α' martensite divided by % elongation) for alloy series at 4 K.

glide may result in more planar and local extension of the plastic zone ahead of the crack tip. Such net-section yielding phenomena would be expected to result in lower effective fracture toughness. Indeed, a linear relationship is obtained if one assumes that $K_{Ic}(J) \leq a(\varepsilon_{III} - \varepsilon_{II}) + b(1/\sigma_y)$ where the proportionality constants a and b are related by a ≈ 0.1 b with the units of $K_{Ic}(J)$ in $MPa\sqrt{m}$ and σ_y in MPa.

SUMMARY

We have suggested that stress-strain curves of metastable austenitic stainless steels have three stages of plastic deformation. From stress-strain curves for a series of Fe-18CR-8Ni alloys with varying Mn (1-6 wt.%) and N (0.1, 0.2 wt.%) contents, the parameters of stress and strain associated with the onset of the three stages have been demonstrated to correlate very well with alloy content. Alloy content affects austenite stability. Correlations of thermodynamic stability, elastic-plastic fracture toughness, and solid-solution strengthening with the parameters of the three stages will be explored in future papers.

ACKNOWLEDGMENTS

This work was supported by the Office of Fusion Energy, Department of Energy and by the National Science Foundation.

REFERENCES

1. R. L. Tobler and R. P. Reed, Tensile and fracture properties of manganese-modified AISI 304 type stainless steel, in: "Advances in Cryogenic Engineering - Materials," Vol. 28, Plenum Press, New York (1982), p. 83.

2. R. L. Tobler and R. P. Reed, Tensile and fracture properties of manganese-modified AISI 304 type stainless steel, in: "Materials Studies for Magnetic Fusion Energy Applications at Low Temperatures IV," NBSIR 81-1645, R. P. Reed and N. J. Simon, eds., National Bureau of Standards, Boulder, Colorado (1981), pp. 77-100.
3. T. Suzuki, H. Kujima, K. Suzuki, T. Hashimoto, S. Koite, and M. Ichihara, Plastic deformation and martensitic transformation in an iron-base alloy, Scripta Met. 10:353-358 (1976).
4. G. B. Olson and M. Azrin, Transformation behavior in trip steels, Metall. Trans. A 9:713-721 (1978).
5. R. L. Tobler, R. P. Reed, and D. S. Burkhalter, Temperature dependence of yielding in austenitic stainless steels, in: "Materials Studies for Magnetic Fusion Energy Applications at Low Temperatures III," NBSIR 80-1627, R. P. Reed, ed., National Bureau of Standards, Boulder, Colorado (1980), pp. 51-78.
6. G. H. Eichelman and F. C. Hull, The effect of composition on the temperature of spontaneous transformation of austenite to martensite in 18-8-type stainless steel, Trans. Am. Soc. Met. 45:77-104 (1953).
7. I. Williams, R. G. Williams, and R. C. Capellero, Stability of austenitic stainless steels between 4 K and 373 K, in: "Proceedings of the Sixth International Cryogenic Engineering Conference," IPC Science and Technology Press, Guildford, Surrey, England (1976), pp. 337-341.
8. R. P. Reed and C. J. Guntner, Stress-induced martensitic transformations in 18Cr-8Ni steel, Trans. AIME 230:1713-1720 (1964).
9. C. J. Guntner and R. P. Reed, The effect of experimental variables including the martensitic transformation on the low-temperature mechanical properties of austenitic stainless steels, Trans. Am. Soc. Met. 55:399-419 (1962).

DISCONTINUOUS DEFORMATION MODES OF A NITROGEN-STABILIZED AUSTENITIC STEEL

B. Obst and D. Pattanayak

Institut für Technische Physik, Kernforschungszentrum Karlsruhe Karlsruhe, Federal Republic of Germany

INTRODUCTION

The instability of plastic flow at low temperatures appears to be quite a general phenomenon of many metallic materials. Several theories have been advanced to account for this effect: strain-induced phase transformations[1] and twins,[2] avalanche-like barrier crossing by dislocation pile-ups and their sudden multiplication under the action of the load,[3] and thermal instability of deformation resulting from a large temperature dependence of the flow stress coupled with low specific heat.[4-6] This last "thermal model" is commonly given as the cause for discontinuous yielding. However, the detailed mechanism has not been definitely explained. Particularly, the considerable localized plastic flow that must have occurred before the temperature rises[7] can by no means be ruled out, and the available experimental results are quite contradictory.

So, at the present time, it appears that no consensus exists as to the meaning of the elementary step of discontinuous deformation at very low temperatures. Possibly it is the result of various processes occurring almost simultaneously throughout the entire specimen.

The present paper is concerned with a metastable AISI 316 LN-type austenite steel. The transformation behaviour and the mechanical properties were thoroughly investigated. On the basis of these results, an attempt is made to elucidate the physical nature of the discontinuous jerky flow in the low-temperature plastic strain.

Table 1. Chemical Composition of the Steel Used

Element	C	N	Ni	Mn	Si	Mo	Cr	P	S	Fe
wt.%	0.032	0.16	13.70	1.26	0.41	2.68	16.70	0.016	0.016	bal.

MATERIALS AND EXPERIMENTAL METHODS

The chemical composition of the alloy investigated is given in Table 1 (the German Mat. No. is 1.4429). Tensile specimens of 20-mm gauge length and 5-mm diameter were cut from 60-mm-thick plates, as supplied by the manufacturer (Krupp, Essen). After machining, the samples were cleaned and sealed in an Ar-filled quartz tube. Then they were homogenized at 1050°C for 90 min. Thereafter, the tube was removed from the furnace, the Ar atmosphere was broken, and the specimens were quenched in ice water. This treatment ensured a recrystallized and texture-free state.

The mechanical properties were studied by isothermal tensile tests using an Instron-type hard beam machine in the temperature range 4 to 80 K. This is, approximately, the temperature range M_s to M_d. Deformation rate was usually fixed at 0.3 mm/min. Stress and strain values were measured in a temperature-variable flow cryostat by means of a quartz load cell and an Instron clip-type strain gauge extensometer, respectively.

After straining the specimens to fracture, they were characterized at room temperature. The intensity of magnetization was measured because it provides a convenient means of estimating the volume fraction of α'-martensite, f, formed during the tensile test. Microstructures were studied by optical metallography, transmission electron micrographs, and by x-ray diffraction using Cr-K_α radiation.

EXPERIMENTAL RESULTS

Figure 1 shows stress-strain diagrams in the range 6 to 36 K. The work-hardening rates were remarkably low for all strains and they showed no significant dependence on the test temperature. At liquid helium temperature, load drops were observed right from the beginning of plastic flow, and most of the deformation appeared to take place by jerky extension. At somewhat higher temperatures, the onset of discontinuities was preceded by a smooth plastic deformation. As can be seen in Fig. 2a, this behaviour seems to be independent of the temperature of prestraining, T_d. Once serrations had begun, they occurred repeatedly until ultimate rupture. Their magnitude increased with increasing strain and was roughly the same for all T_d's, provided the total elongation of the

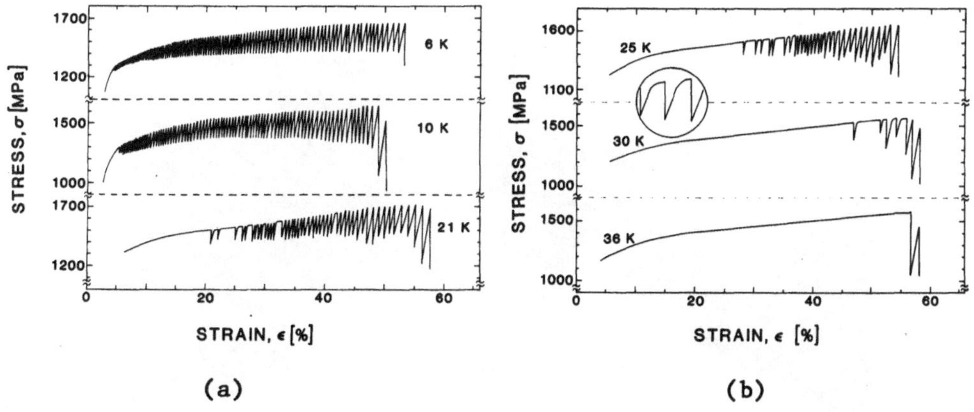

Fig. 1. Variation of stress with strain for various testing temperatures.

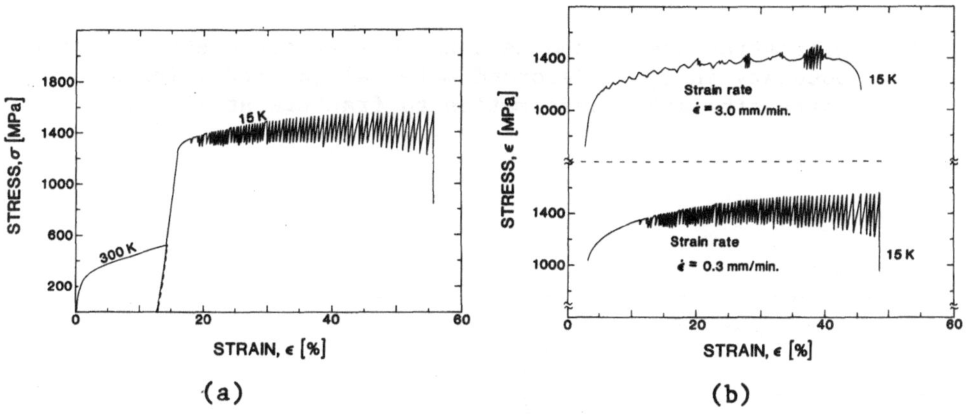

Fig. 2. Influence of prestrain at 300 K on the onset of serrations (a) and serrations at different strain rates (b).

specimens was the same. The abrupt load drops in the stress-strain curves were accompanied by local necks and clearly audible clicks. After each abrupt drop, which was accompanied by sudden local elongation, further extension was elastic until the original load was reached and plastic flow resumed.

When the rate of deformation was increased, the continuous flow at the beginning of deformation became very small and the stress-strain curve displayed irregular undulation (Fig. 2b). A higher deformation rate is equivalent to a lower deformation temperature as far as the discontinuous character of the deformation is concerned.

(a) (b)

Fig. 3. Microstructure of the metastable austenitic steel: twin boundary in the undeformed material (a) and twinned martensite after deformation to fracture at 15 K (b).

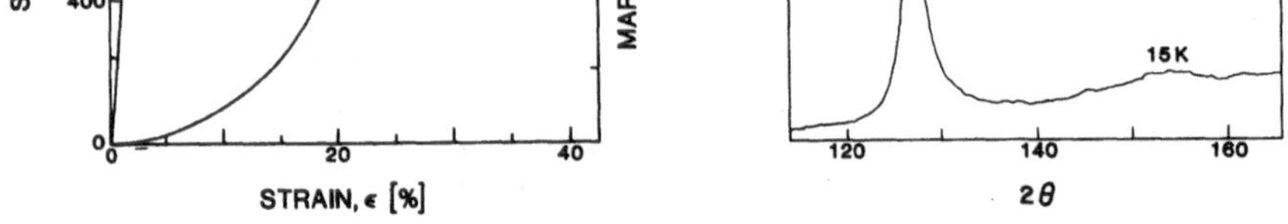

(a) (b)

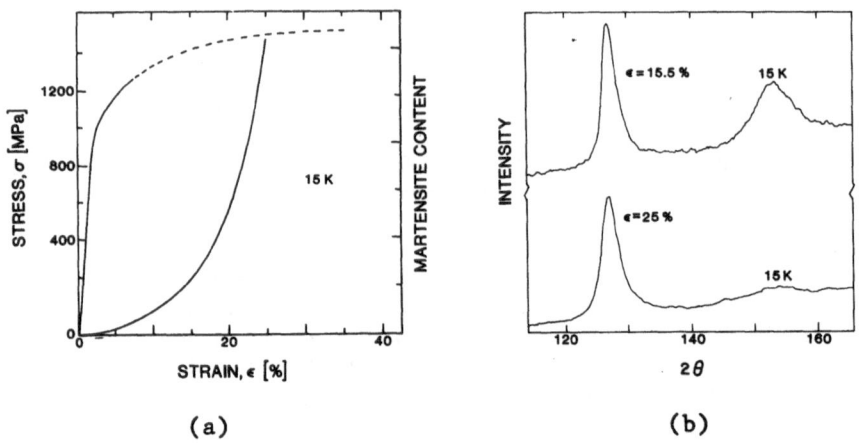

Fig. 4. Stress-strain curve at 15 K and amount of martensite induced during this deformation process (a); γ:α' x-ray diffraction peaks (γ:220, α':211) at two different strains (b).

In this alloy, no α'-martensite formed on cooling to 4.2 K, i.e., this steel in the softened condition remained fully austenitic. It must be considered stable (Fig. 3a). However, substantial transformation from the austenite phase, γ, to the ferromagnetic martensite, α', was observed under uniaxial deformation (Fig. 3b). The degree of transformation clearly correlates with the extent of deformation (Fig. 4). It was also affected by T_d.

Figure 5 shows the variation of α'-martensite content in specimens as a function of straining temperature. X-ray diffraction studies indicated that the strain-induced martensite was not uniformly distributed. An increased α'-formation was found in the necking region--in accordance with the results shown in Fig. 4. It is, however, evident that the average martensite volume fraction vs. temperature was maximum at about 40 K. The total elongation was also found to be maximum at 40 K, whereas the corresponding values at LHe and room temperature were approximately equal.

DISCUSSION

That the austenite γ (fcc, paramagnetic) does not transform on cooling[8] (i.e., $M_s \approx 0$ K), but does transform on deformation below a critical temperature ($M_d > M_s$) to two types of martensite [α' (bcc, ferromagnetic) and ε (hcp, paramagnetic)] is an important property of this type of steel. The α'-phase formed at low temperature remains stable on warming up to room temperature,[9] thus facilitating the magnetic measurements (Fig. 5). Furthermore, the intrinsic stacking fault energy (SFE) in this alloy is low at room temperature (~20 erg/cm^2) and decreases rapidly on cooling. This

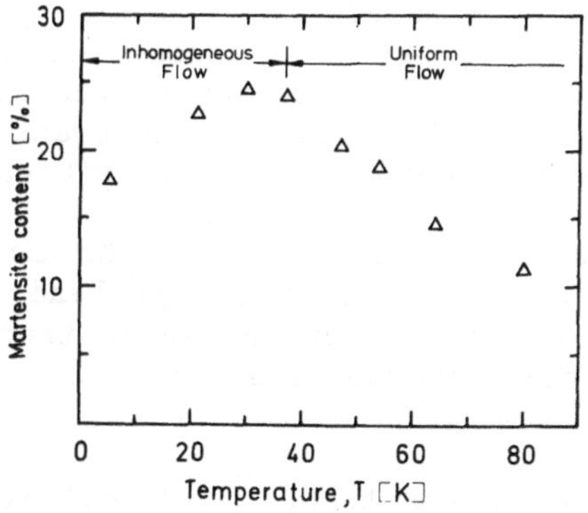

Fig. 5. Content of strain-induced α'-martensite (vol.%) vs. test temperature (specimen deformed to fracture).

low SFE can be directly related to the formation of many stacking faults (SFs) during low temperature deformation and insures that the ε-phase, which is achieved by propagation of intrinsic SFs, is stable relative to the austenite.[10-12]

For the alloy investigated, the deformation of austenite at low temperatures takes place heterogeneously in localized bands. In these bands, the deformation can proceed by formation of ε-platelets, twinning in the parent phase, or gliding of dissociated dislocations. Intersecting bands are the most commonly reported strain-induced nucleation sites for α'martensite.[13-17]

Consequently, the process relating to the genesis of α'-embryos has been rationalized in terms of an alternative deformation mode of the material, using a dislocation model. It is construed to take place more readily during plastic deformation than under applied stress after deformation (if the conditions for thermodynamic stability permit it), and the moving defects thus created are considered to play a very important part in the transformation process. This agrees well with earlier results: dynamic stresses promote the transformation to a much greater extent than static stresses.[13]

In alloys where the austenite SFE is low, the intersection event may be pictured as a moving array of appropriate partial dislocations of the parent lattice, which is impeded by a stationary packet of SFs (not necessarily ε).[11-13] High stress concentrations may develop in front of the pile-ups such as to induce the nucleation process.

If a "spreading" of the entering partials over a number of successive $\{111\}_\alpha$ planes is accomplished during the intersection event concurrently with a shuffling of atomic planes in the intersected volume, then the passage of these dislocations through the intersection can occur by the formation of bcc martensite.[12,18] In this shear and shuffling mechanism, the movement of the transformation dislocations is considered to be the critical stage.[19]

To put the result in other words, the active slip dislocations, blocked at an intersecting (slip) band, first nucleate an α'-martensite along the intersection, which then lets the pile-ups penetrate through the intersecting volume. (For an excellent review, we refer the reader to Ref. 12.)

It is well established that, first and foremost, nucleation controls martensitic reactions and not the subsequent growth, which has a very low activation energy. Thus, the volume fraction of martensite is expected to increase gradually with plastic strain; this is clearly confirmed by our experiments (Fig. 4).

In the situation illustrated above, where partial dislocations pile-up at the intersection of two deformation bands, the specimen is in a highly nonequilibrium mechanical state. Instabilities are a natural consequence of this state. So, if for some reason, the α'-nucleation process does not take place (e.g., when the SFE decreases towards zero at temperatures close to the M_s temperature, the increasing dissociation of dislocations immobilizes them and limits their ability to serve as nucleation sites), the high stress concentrations are not released and the probability, then, for a catastrophic breakaway of the pile-ups is greatly increased. This may appear as a sudden load drop on a macroscopic scale.

We emphasize that, according to our model, as described above, any rise in temperature that has been frequently reported in the literature is a consequence of the heat liberated in such adiabatic conditions and that it is not a prerequisite for the instability of plastic flow at low temperatures.

Finally, in the range of inhomogeneous plastic flow, i.e., below about 40 K, the total strain, ε_{tot}, for the material deformed in tension is given by (Fig. 5):

$$\varepsilon_{tot} = (\varepsilon_{uni} + \varepsilon_{disc}) + \Delta(\varepsilon_{uni}, T)$$

$$\text{with } \Delta \to 0 \text{ for } \varepsilon_{uni} \to 0$$

ε_{uni} and ε_{disc} represent the uniform and discontinuous strain parts of "conventional" glide, respectively, whereas Δ is due to the enhanced elongation by the martensitic transformation upon deformation (TRIP effect). This Δ-elongation, in turn, and therefore the volume fraction of α', are found to be interrelated again to ε_{uni} (Figs. 1 and 5).

In the range of uniform flow ($T \geq 40$ K), the increase in α' with decreasing temperature is due to the chemical driving force, which adds to the low SFE-dislocation structure. There are no indications that further transformations would not occur below 40 K if deformation were to be carried out continuously.

Finally, a general result that follows for the austenite steel investigated is that, depending on the defects created in the course of deformation at very low temperatures, a catastrophic breakaway of piled-up dislocations may occur and compete with the plastic-deformation-induced transformation--concurrently with the plastic deformation modes of the parent phase. Thus, the two effects, serrations and martensite formation, are complementary so that the dislocation structure most favourable for the one mode of deformation (a combination of leading partial dislocations to produce sessile dislocations) is detrimental to the other mode of deformation (gliding of slip dislocations through an intersecting deformation band by "appropriate spreading" of the partials and shuffling of atomic planes).

CONCLUDING REMARKS

Mechanical properties of a metastable austenitic steel (DIN 1.4429 ~ AISI 316 LN) have been studied at low temperatures. The characteristic features of the stress-strain curve and possible interrelations with the nucleation process of the ferromagnetic α'-phase are discussed. The most important results obtained are summarized as follows:

1. In the M_s to M_d temperature range, at about T_p = 40 K, the total amount of austenite transformed to α'-martensite during deformation shows a marked peak (Fig. 5). A direct relationship to TRIP characteristics is also found.

2. With decreasing probability of α'-nucleation in the temperature range below T_p, an increasing number of serrations in the stress-strain curves is observed (Figs. 1 and 5).

3. Introducing the usual slip bands while deforming at room temperature will not significantly change the onset of serrations at low temperatures (Fig. 2a). This agrees well with the statement given in Ref. 13 that the formation of α'-martensite is closely related to slip bands rather than ε bands.

4. The strain-induced formation of α'-martensite (bcc or bct) in γ (fcc) austenite has been interpreted by adopting the Bogers-Burgers double shear mechanism.[18] Using the extension thereof by Olson and Cohen[12] to embrace dislocation motion, a change in the relation of plastic flow and transformation at the peak temperature (Fig. 5) results immediately; this follows if the stacking fault energy is low.

5. Thus, lattic transformation upon deformation and catastrophic breakaway of dislocations piled-up at the interface of deformation (slip) bands are thought of as two competing deformation modes of the material. The strong influence of temperature on the onset of these modes is attributed to the strong variation of the stacking fault energy with temperature.

6. The characteristics of low-temperature plastic strain are by no means compatible with any "heating hypothesis."[4,5] To take some examples: the large uniform flow preceding the beginning of serrations (Fig. 1), the influence of prestrain at 300 K on the inhomogeneous flow at 15 K (Fig. 2a), and the interrelation between the amplitudes of the load drops with the absolute strain value, irrespective of the test temperature.

A hypothesis of the kind that relies on the small magnitude of the specific heat at low temperatures is completely ruled out by similar experimental findings obtained in low-alloy steel at rather high temperatures.[20]

ACKNOWLEDGMENTS

The authors are grateful to K. Dürr for conducting the stress-strain measurements at low temperatures. They appreciate the stimulating interest and encouragement offered throughout this work by Prof. H. Wühl, and they also wish to thank Dr. U. Wolfstieg for his support in x-ray diffraction and Dr. Förster and the Institut for making available the use of their facilities.

REFERENCES

1. G. V. Uzhik, Izvest. Akad. Nauk SSSR, Otdel. Tek. Nauk 1:57 (1955).
2. T. H. Blewitt, R. R. Coltman, and J. K. Redman, "Defects in Crystalline Solids," Phys. Soc., London (1955).
3. A. Seeger, "Dislocations and Mechanical Properties of Crystals," Wiley, New York (1957).
4. Z. S. Basinski, Proc. Roy. Soc. A240:229 (1957).
5. Z. S. Basinski, Aust. J. Phys 13:354 (1960).
6. E. T. Wessel, Trans. Am. Soc. Met. 49:149 (1957).
7. P. Haasen, Trans. AIME 212:42 (1958).
8. D. C. Larbalestier and H. W. King, Cryogenics 13:160 (1973).
9. R. P. Reed, Acta Metall. 10:865 (1962).
10. J. A. Venables, Philos. Mag. 7:35 (1962).
11. F. Lecroisey and A. Pineau, Metall. Trans. 3:387 (1972).
12. G. B. Olson and M. Cohen, J. Less-Common Met. 28:107 (1972).
13. R. Lagneborg, Acta. Metall. 12:823 (1964).
14. T. Suzuki, H. Kojima, K. Suzuki, T. Hashimoto, and Ichihara. Acta. Metall. 25:1151 (1977).
15. G. B. Olson and M. Azrin, Metall. Trans. 9A:713 (1978).
16. J. W. Brooks, M. H. Loretto, and R. E. Smallman, Acta. Metall. 27:1829 (1979).
17. A. Sato, H. Kasuga, and T. Mori, Acta. Metall. 28:1223 (1980).
18. A. J. Bogers and W. G. Burgers, Acta. Metall. 12:225 (1964).
19. W. G. Burgers and J. A. Klosterman, Acta. Metall. 13:1005 (1970).
20. N. Niikura, M. Yamada, J. Tanaka, and H. Ichinose, in: "New Aspects of Martensitic Transformation," Proc. JIMIS-1, Kobe (May 10-12, 1976), p. 321.

THE INFLUENCE OF MAGNETIC STRUCTURE ON TEMPERATURE DEPENDENCE OF THE YIELD STRENGTH OF 20Cr-16Ni-6Mn STEEL AT LOW TEMPERATURES

K. A. Yushchenko

E. O. Paton Institute of Electrowelding, Ukr.SSR Academy of Sciences, Kiev, USSR

B. I. Verkin, V. Ya. Ilichev, and I. N. Klimenko

Physical and Technical Institute of Low Temperatures, Ukr.SSR Academy of Sciences Kharkov, USSR

INTRODUCTION

Investigations of the temperature dependence of the yield strength of austenitic stainless steels containing 18% Cr and 8 to 25% Ni have shown that their yield strengths had an anomalous behavior in the low-temperature range.[1] It was assumed that this effect is of a magnetic nature. Subsequent investigations of the magnetic state of these alloys by means of magnetic susceptibility and gamma-resonance spectroscopy[2,3] revealed that their magnetic state depended on nickel content.

Thus, according to the magnetic susceptibility data,[2] an increase of the nickel content from 8 to 25% in Fe-Cr-Ni alloys changes the magnetic state of those steels at low temperatures from the preferentially antiferromagnetic to the preferentially ferromagnetic state. And it does not mean at all that the Fe-Cr-Ni steels are magnetically homogeneous below the phase transition temperatures. Both the antiferromagnetic and ferromagnetic orders have short-range characteristics. The magnetic state of these systems is characterized by the coexistence of regions with preferentially positive or negative bonds between the central iron atom and its nearest neighbors. Investigation of the Fe-Cr-Ni steels by gamma-resonance spectroscopy[3] showed that long-range magnetic order is not established up to 30 K. The processes of magnetic moment relaxation in regions with the short-range order play an essential part in the magnetic state determination of those alloys. The nature of magnetic moment relaxation depends on nickel concentration.

CURRENT STUDIES

Figure 1 shows the temperature dependence of magnetic susceptibility of 20Cr-16Ni-6Mn steel. At room temperature, $\chi = 2.1 \times 10^{-5}$ cm^3/g. Temperature dependence of magnetic susceptibility has a well-developed maximum, which implies a magnetic ordering in the structure of this steel. One may assume that the magnetic structure of the 20Cr-16Ni-6Mn steel changes with decreasing temperature similar to that of the Fe-Cr-Ni alloys of different nickel content studied earlier. It seems that preferentially antiferromagnetic ordering of the cluster type exists to about 40 K with a few ferromagnetic clusters, and then a sharp increase of the ferromagnetic phase takes place indicated by the essential increase of magnetic susceptibility.

Therefore, it was interesting to investigate the temperature dependence of yield strength of the stable 20Cr-16Ni-6Mn stainless steel, which differs from the steels studied earlier by the content of magnetically ordering elements, such as chromium and manganese; its nickel content is within the concentrations studied before.

The temperature dependence of the yield strength, $\sigma_{0.2}$, of 20Cr-16Ni-6Mn steel is shown in Fig. 2 (curve ●). A sharp drop in yield stress is seen to occur at $50 \text{ K} < T < 60 \text{ K}$. The value of flow stress, $\Delta\sigma_{0.2}$, changes about 7% from its value at T = 60 K. A further decrease in temperature from 50 to 4.2 K results in a monotonic increase of flow stress.

Figure 3 shows the nickel content dependence of the temperature of the anomalous drop in $\sigma_{0.2}$ for stainless alloys studied in Ref. 1 and for 20Cr-16Ni-6Mn steel. The temperature of anomalous

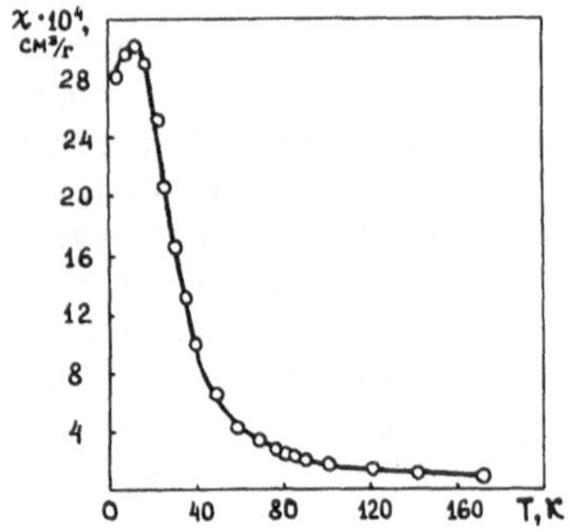

Fig. 1. Temperature dependence of the magnetic susceptibility of 20Cr-16Ni-6Mn steel.

drop of yield strength in 20CR-16Ni-6Mn agrees well with the results of Ref. 1. This indicates that nickel plays a major role in affecting the mechanism of anomalous drop of yield strength in the stainless Fe-Cr-Ni steels at low temperatures.

Assuming that the anomalous drop of yield strength is due to the peculiarities of magnetic structural change in stainless steels, one may expect the applied magnetic field to have some influence on the temperature and/or the value of anomaly (curve o) shows the temperature dependence of yield strength in 20Cr-16Ni-6Mn steel in an applied magnetic field of 20 kOe. The magnetic field application is seen to result in a monotonic increase of yield strength temperature dependence in the temperature range of the anomaly, which is observed at H = 0.

The drop of yield strength seems to be due to the unpinning of dislocations, which gain some additional energy. It is also obvious that the additional energy appears as a result of changes in magnetic structure of the material at definite temperatures.

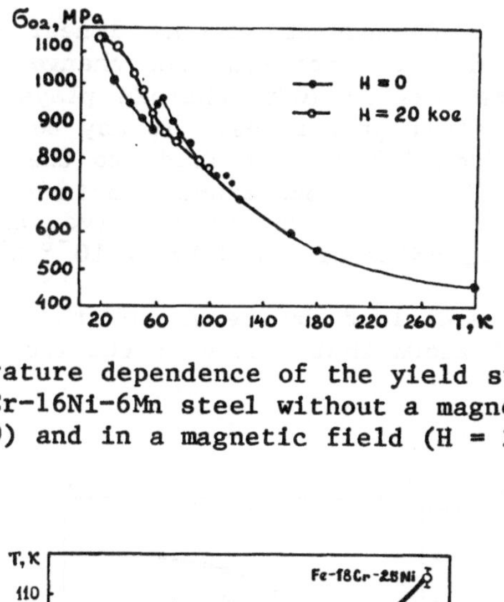

Fig. 2. Temperature dependence of the yield strength ($\sigma_{0.2}$) of 20Cr-16Ni-6Mn steel without a magnetic field (H = 0) and in a magnetic field (H = 20 kOe).

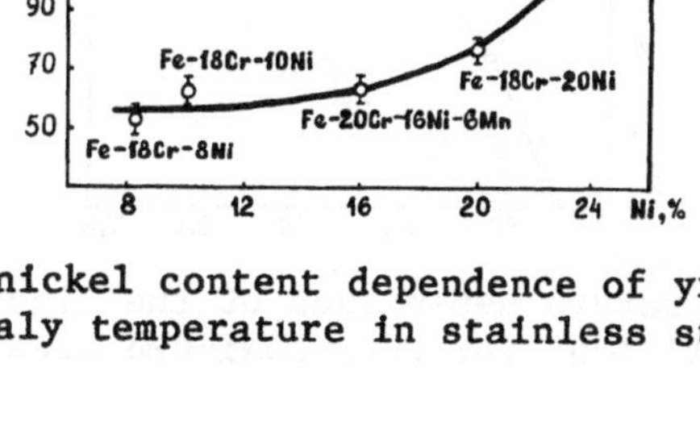

Fig. 3. The nickel content dependence of yield strength anomaly temperature in stainless steels.

From the papers devoted to gamma-resonance spectroscopy,[3] it follows that at definite temperatures in Fe-Cr-Ni alloys, ferromagnetic clusters appear in which relaxation of the magnetic moments takes place at certain frequencies, apparently in the range $1.4 \times 10^8 s^{-1}$ to $2 \times 10^7 s^{-1}$.

At the same time, if a dislocation is considered to be an oscillating segment, the frequency of its oscillation can be determined from the papers on internal friction. It proved to be $10^8 s^{-1}$. Thus, one may assume that in Fe-Cr-Ni alloys there are two interacting oscillating systems, each with its own frequency. Those frequencies seem to coincide under certain conditions (the resonance phenomenon), which could lead to a sharp increase in amplitude of dislocation segment oscillation and facilitate the unpinning of dislocations from obstacles.

Other explanations of the anomalous drop in Fe-Cr-Ni alloys at certain temperatures are also possible.

Application of an external magnetic field essentially decreases the relaxation frequency of cluster magnetic moments, and the effect of softening shifts to higher temperatures.

The temperature dependence of the linear expansion coefficient, α (Fig. 4), may be proof that the change of magnetic structure at low temperatures in 20Cr-16Ni-6Mn plays an essential part and affects its physical properties. It may be seen that the temperature decrease from 300 to 26 K leads to the monotomic decrease of α down to 0. If the temperature continues to decrease, the linear expansion coefficient becomes negative, and in the range of T = 15 K, the minimum value of α (-0.197×10^{-6} K^{-1}) is observed.

This can be explained by means of the model proposed by Weiss.[4] It is well known that γ-iron of the fcc structure may have

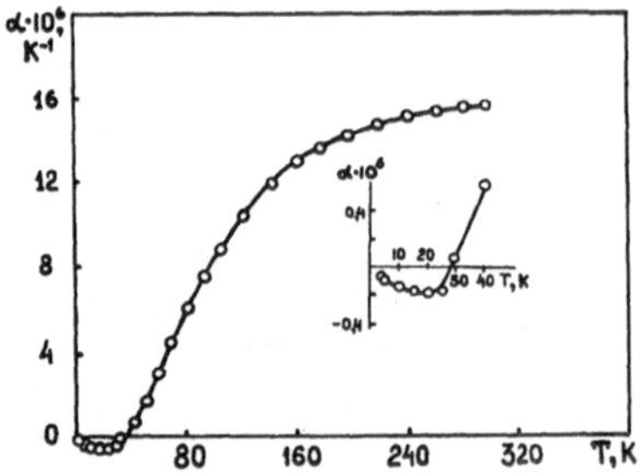

Fig. 4. Temperature dependence of the linear expansion coefficient for 20Cr-16Ni-6Mn steel.

two electron configurations (γ_1 and γ_2),[5] which differ in lattice parameter (a γ_1 = 3.54 Å; a γ_2 = 3.64 Å). In each of them, there is an established magnetic order: antiferromagnetic for the γ_1 state and ferromagnetic for the γ_2 state. For example, the probability of the coexistence of γ-iron atoms both in ferromagnetic and in antiferromagnetic states is realized in a number of iron-nickel alloys of the Invar type, and this may be the cause of anomalies in the temperature dependences of Young's modulus and linear expansion coefficient.[4] The existence of antiferromagnetically and ferromagnetically ordered regions in stainless steels enables the $\gamma_1 \rightarrow \gamma_2$ transitions because alloying γ-iron with nickel makes the γ_1 configuration unstable.[4] These transitions, in their turn, are accompanied by a lattice parameter increase and may give rise to the observed anomaly of the linear expansion coefficient in 20Cr-16Ni-6Mn steel.

Thus, the cause of the anomalous behavior of the linear expansion coefficient in 20Cr-16Ni-6Mn steel at $T < 30$ K and of the anomalous drop of Young's modulus, E, in the low-temperature range for stainless alloys seems to be the complex mixed magnetic state of stainless steels containing a large percent (> 30%) of iron. Hence, the magnetic state of austenite may be assumed to influence the processes of dislocation motion as well and, consequently, the mechanical properties of these materials.

CONCLUSIONS

The temperature of the initiation of a decrease in flow strength with a decrease in temperature is related linearly to nickel concentration. The anomaly temperature fell onto the straight line of the nickel content dependence of anomaly temperatures for stainless steels containing 8 to 25% nickel concentrations.

1. An anomalous drop of yield strength in 20Cr-16Ni-6Mn steel was found at $T < 60$ K.

2. The application of an external magnetic field of 20 kOe led to the monotonic increase in temperature dependence of yield strength in the temperature range of anomaly that was observed at $H = 0$.

3. An anomaly of the linear expansion coefficient in 20Cr-16Ni-6Mn steel was found at low temperatures.

REFERENCES

1. V. Ya. Ilichev, M. M. Medvedev, I. A. Shapovalov, and I. N. Klimenko, Fiz. Met. Metalloved. 44(1):199 (1977).

2. V. Ya. Ilichev, I. N. Klimenko, and E. N. Khatz'ko, Fiz. Nizk. Temp. 3:370 (1978).
3. I. N. Klimenko, V. Romanov, and V. Ilichev, Cryogenics 19(4) (1979).
4. R. Weiss, Proc. Phys. Soc. 82(526):281 (1963).
5. Z. Kaufman, E. Claugherty, and R. Weiss, Acta Metall. 11:323 (1963).

MECHANICAL EVALUATION OF NITROGEN-STRENGTHENED STAINLESS STEELS AT 4 K

Y. Takahashi, K. Yoshida, M. Shimada, and E. Tada

Japan Atomic Energy Research Institute, Ibaraki, Japan

R. Miura

Japan Steel Works, Ltd., Muroran, Japan

S. Shimamoto

Japan Atomic Energy Research Institute, Ibaraki, Japan

INTRODUCTION

Because the structural materials of large superconducting magnets for fusion machines suffer high magnetic stress at 4 K, the knowledge of their mechanical properties at cryogenic temperature is highly desirable. Austenitic stainless steels have good properties for use in superconducting fusion magnets, but their relatively low yield stress at 4 K is serious drawback. Though some studies[1,2] show this shortcoming can be overcome by adding nitrogen and carbon, there is as yet no reliable data base regarding the structural properties of the improved steels. During the design of the Japanese LCT coil, it became clear that the yield stress of the structural steel must be greater than 700 MPa at 4 K. For this reason, JAERI began to evaluate the mechanical properties of nitrogen-strengthened stainless steels at 4 K.

This paper describes the results of the mechanical evaluation of the stainless steels JIS SUS304 and SUS316 doped with various levels of nitrogen and carbon and also reports briefly the results of verification tests of the structural steel SUS304LN used in the Japanese LCT coil.

Table 1. Chemical Compositions of SUS304 Grade (wt%)

C	Si	Mn	P	S	Ni	Cr	Mo	N
0.016	0.35	1.44	0.020	0.011	9.92	18.46	-	0.056
0.015	0.59	1.13	0.019	0.009	9.78	18.47	-	0.095
0.018	0.44	1.13	0.018	0.009	10.30	18.06	-	0.123
0.015	0.53	1.07	0.020	0.009	10.10	18.05	-	0.147
0.036	0.54	1.19	0.020	0.009	10.02	17.42	-	0.052
0.039	0.54	1.14	0.019	0.008	9.81	18.89	-	0.097
0.041	0.47	1.15	0.020	0.013	9.78	18.33	-	0.134
0.039	0.58	1.23	0.020	0.008	9.99	18.35	-	0.174
0.071	0.52	1.18	0.021	0.009	9.96	18.49	-	0.061
0.070	0.56	1.19	0.020	0.010	10.00	18.31	-	0.090
0.074	0.56	1.18	0.020	0.009	10.03	18.14	-	0.138
0.077	0.57	1.19	0.020	0.010	10.01	18.20	-	0.178

Table 2. Chemical Compositions of SUS316 Grade (wt%)

C	Si	Mn	P	S	Ni	Cr	Mo	N
0.017	0.57	1.13	0.018	0.009	12.10	16.69	2.49	0.056
0.018	0.56	1.13	0.019	0.009	11.80	16.48	2.46	0.088
0.016	0.45	1.18	0.021	0.008	11.78	16.37	2.38	0.122
0.014	0.56	1.18	0.019	0.009	11.91	16.27	2.50	0.166
0.039	0.53	1.13	0.018	0.009	12.00	16.62	2.46	0.057
0.041	0.54	1.20	0.019	0.010	11.87	16.21	2.48	0.088
0.039	0.52	1.24	0.018	0.008	11.91	16.42	2.36	0.120
0.045	0.58	1.19	0.018	0.008	12.02	16.38	2.49	0.157
0.074	0.53	1.17	0.018	0.009	11.69	16.31	2.48	0.047
0.076	0.42	1.16	0.020	0.009	11.89	16.64	2.45	0.084
0.085	0.50	1.22	0.019	0.009	11.65	16.23	2.44	0.121
0.077	0.52	1.31	0.021	0.008	11.77	16.52	2.50	0.154

MATERIALS AND SPECIMENS

Materials

The steels studied were SUS304 and SUS316 with contents of carbon from 0.016 wt% to 0.077 wt% and nitrogen from 0.05 wt% to 0.18 wt%. The chemical compositions are listed in Tables 1 and 2, respectively. Twelve plates of each were produced from twenty-four 50-kg induction-melted heats. The ingots were then heated and soaked at 1150°C for 3 h and then forged to 55-mm-thickness plates. They were cut into 220-mm-long pieces, reheated and soaked at 1150°C for 3 h, hot-rolled to 22-mm-thickness plates, and finally solution-treated at 1040°C for 2 h. Their measured average grain diameters ranged from 110 μm to 180 μm.

The plate SUS304LN [produced by the electroslag remelting method (ESR)] and the welding rod modified D316 (coated arc welding rod) used in the Japanese LCT coil case were also examined in verification tests. Their chemical compositions are listed in Table 3.

Table 3. Chemical Compositions of the Structural Steel of the Japanese LCT Coil (wt%)

	C	Si	Mn	P	S	Ni	Cr	Mo	N
(a)	0.017	0.52	1.05	0.023	0.008	10.45	18.84	-	0.149
(b)	0.037	0.38	5.26	0.009	0.004	16.38	19.28	2.85	0.036

Specimens

Round tensile specimens oriented transverse to the rolling direction were used. Their gage length was 35 mm and the diameter of gage section was 7 mm according to JIS Z 2201.

Charpy V-notched (CVN) specimens were bars 55 mm in length with a square cross section having sides of 10 mm according to JIS Z 2022.

EXPERIMENTAL PROCEDURES

Tensile Test

Tensile tests at 4 K were carried out in a special tensile apparatus, TURRET DISC, reported by one of the authors,[3] which made it possible to test up to ten specimens within a load limit of ten

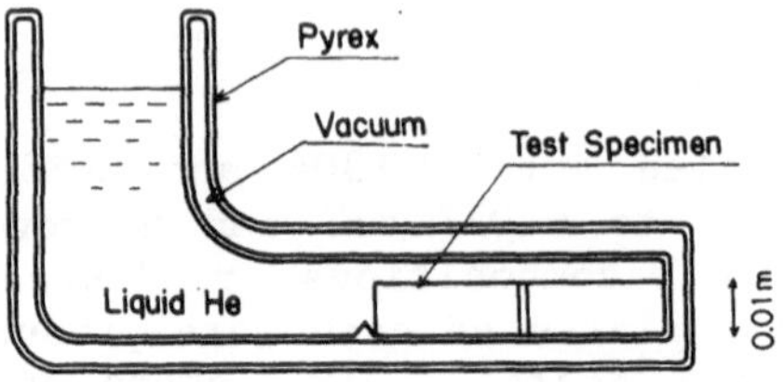

Fig. 1. Glass Dewar used for Charpy test at 4 K.

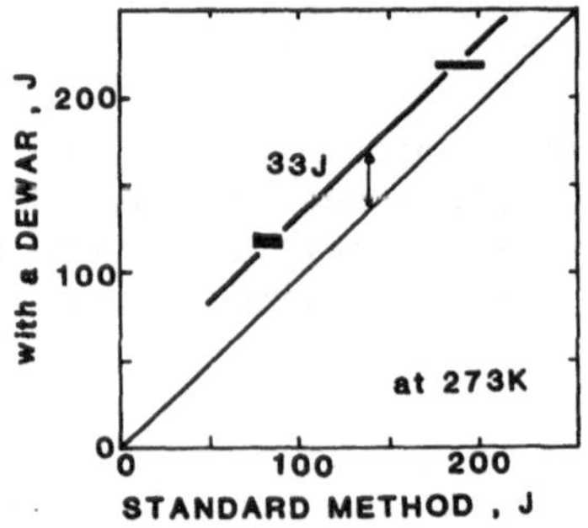

Fig. 2. Comparison of the impact energy obtained by the standard procedure and by using the Dewar.

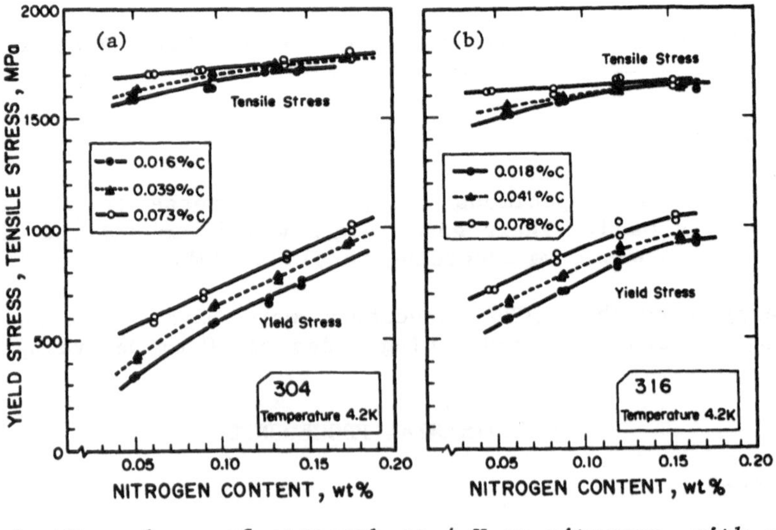

Fig. 3. Dependence of strength at 4 K on nitrogen, with carbon level as a parameter. (a) SUS304; (b) SUS316.

tons in one experimental cycle. Yield stresses (0.2% offset stress) were obtained quite precisely because strain was measured by a strain gage attached directly to the specimens. A strain rate of 7×10^{-5} s^{-1} was applied until yielding occurred; then the rate was changed to 1×10^{-3} s^{-1} to failure. Yield stress, tensile stress, elongation, and reduction of area were measured.

Charpy Tests

Charpy tests at 4 K were conducted by a different technique from the standard one. A specimen was inserted in a small glass Dewar designed for the Charpy test as shown in Fig. 1. It took about 4 min for the liquid helium in the Dewar to vaporize after the Dewar was removed from the liquid helium bath. Thus the specimen temperature of 4 K was insured for the duration of the test. Because the whole system of specimen and liquid-filled Dewar was struck by a hammer, the impact energy obtained had to be calibrated beforehand. The calibration was made at 273 K (as shown in Fig. 2) using standard specimens whose impact energies were well known.

RESULTS AND DISCUSSION

Tensile Properties

Every test at 4 K produced a serrated stress-strain curve. The tensile properties of the steels tested are plotted in Figs. 3-5. In Fig. 3, the dependence of the yield stress and the tensile stress of SUS304 and SUS316 on nitrogen, with carbon levels as a parameter, is shown. Figure 4 shows the dependence of the yield stress on the combined nitrogen and carbon content. The yield stresses of SUS304 and SUS316 increase almost linearly with nitrogen and carbon contents. The yield stress of SUS316 is higher than that of SUS304 by 118 MPa in the case of the same nitrogen and carbon content, and there is a departure from the linear trend for nitrogen and carbon contents of 0.11 wt% or less. This departure is probably due to austenite instability, as Tobler and Reed have pointed out.[2]

Multiple regression analysis indicates that nitrogen is more effective in raising yield stress than carbon and that the yield stresses of both steels can be expressed as follows:

$$\text{Y.S.[MPa]} = 135 + 2893[\%C] + 3707[\%N] \quad \text{for SUS304}$$

$$\text{Y.S.[MPa]} = 390 + 2491[\%C] + 2913[\%N] \quad \text{for SUS316}$$

The tensile stress of both steels increases gradually with carbon and nitrogen contents. The tensile stress of SUS304 is higher than that of SUS316 by about 100 MPa, opposite to the case of the yield stress. Figure 5 shows that elongation of both steels increases

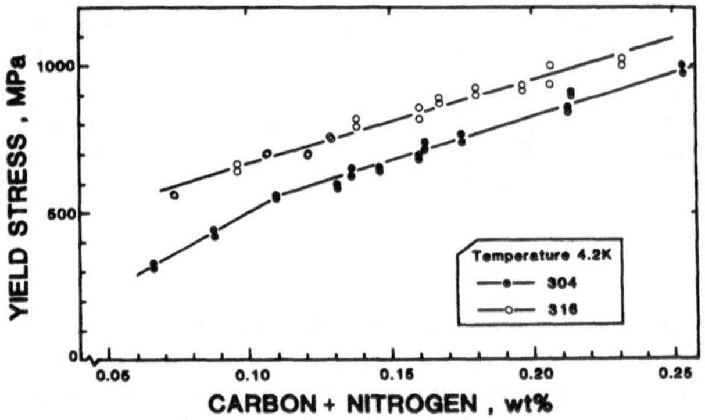

Fig. 4. Dependence of yield stress at 4 K on combined carbon and nitrogen level.

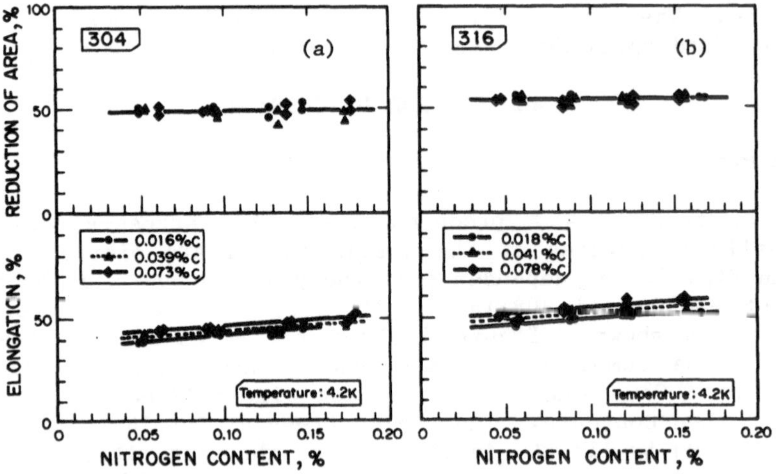

Fig. 5. Dependence of ductility at 4 K on nitrogen, with carbon level as a parameter. (a) SUS304; (b) SUS316.

very slightly with nitrogen and carbon contents, and reduction of area of both steels does not depend on carbon or nitrogen content. It became clear that addition of nitrogen and carbon had no bad effects on the ductility of these steels at 4 K.

Toughness

The CVN impact energy of SUS304 at 4 K is plotted as a function of nitrogen content in Fig. 6. The impact energy decreases gradually with nitrogen and carbon contents but it remains at the relatively high level of 90 J. The SUS304LN of the LCT coil case was found to have an extremely high impact energy, more than 270 J at 4 K. On the other hand, steel having the same chemical composition as that of SUS304LN has an impact energy around 100 J. This difference means that the steel produced by the ESR method is very tough, probably because it contains very small amounts of impurities and inclusions. A linear relation was found between impact energy and lateral expansion at 4 K, as shown in Fig. 7.

Materials of the Japanese LCT Coil Case

Fig. 8-(a) shows the mechanical properties of the SUS304LN of the LCT coil case as a function of temperature. J_{Ic} was also measured using 25.4-mm-thick ASTM E339 compact specimens by the R-curve method at 4 K, 77 K, and RT, but the J_{Ic} obtained at RT was invalid. As one can see in Fig. 8-(a), the steel has an excellent combination of strength and toughness. Fig. 8-(b) shows the results of the deposited metal used in the Japanese LCT coil. Though

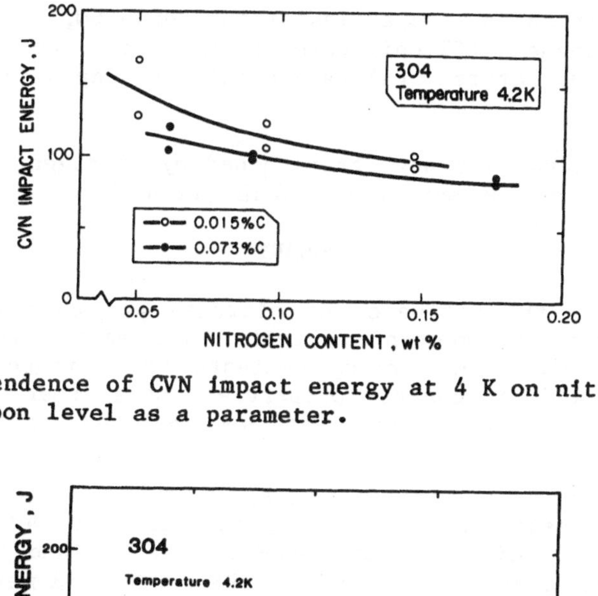

Fig. 6. Dependence of CVN impact energy at 4 K on nitrogen, with carbon level as a parameter.

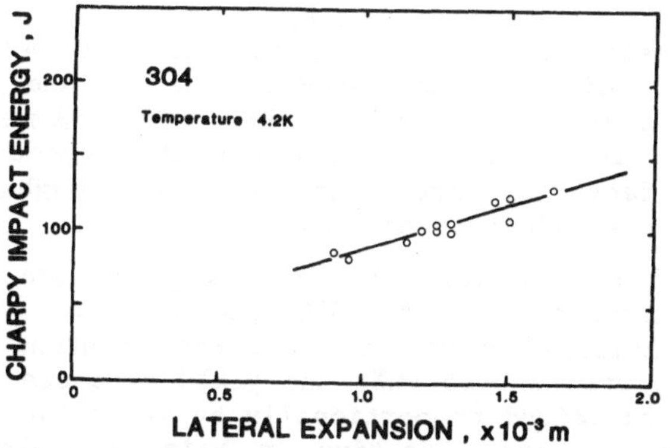

Fig. 7. CVN impact energy of SUS304 plotted against lateral expansion at 4 K.

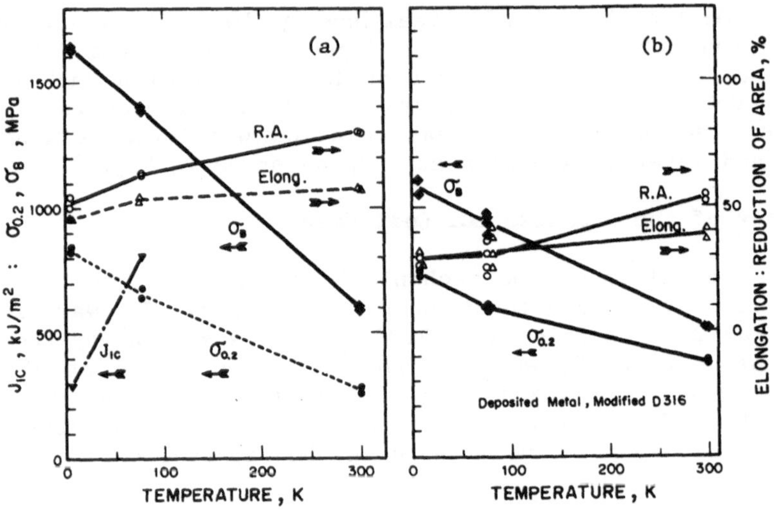

Fig. 8. Mechanical properties of the SUS304LN, used in the Japanese LCT coil case, plotted as a function of temperature. (a) base metal; (b) deposited metal.

the deposited metal was not strengthened by nitrogen, it proved to have a yield stress of more than 700 MPa at 4 K.

SUMMARY

To clarify the influence of nitrogen and carbon on the mechanical properties, mechanical tests have been carried out for SUS304 and SUS316 with various contents of nitrogen and carbon. The tensile and toughness properties have been investigated broadly. The results of these tests and analyses can be summarized as follows:

1. According to the tensile tests, Charpy tests, and fracture toughness tests, the structural material and weldment of the Japanese LCT coil case have enough strength and toughness for the planned test of this coil at ORNL. The ESR method, used in the Japanese LCT coil case, enhances the impact toughness of SUS304 by a factor of three over the impact toughness of SUS304 produced by the induction-melted method.

2. Nitrogen and carbon have the following influence on the mechanical characteristics of SUS304 and SUS316:
   a. As the combined nitrogen and carbon content is increased from 0.1 wt% to 0.25 wt%, the yield stresses of SUS304 and SUS316 are raised proportionally by about 1.8 times.
   b. Nitrogen is more effective in raising yield stress than carbon.

c. Addition of nitrogen and carbon has no negative effect on ductility.
d. As the nitrogen and carbon contents increase, the impact energy decreases gradually.

ACKNOWLEDGMENTS

The authors wish to thank L. Dresner and R. Kensley for their valuable comments.

REFERENCES

1. D. T. Read and R. P. Reed, Fracture and strength of selected austenitic stainless steels at cryogenic temperatures, in: "Materials Studies for Magnetic Fusion Energy Applications at Low Temperatures--II," F. R. Fickett and R. P. Reed, eds., National Bureau of Standards, Boulder, CO (1979).
2. R. L. Tobler and R. P. Reed, Interstitial carbon and nitrogen effects on the tensile and fracture parameters of AISI 304 stainless steel, in: "Materials Studies for Magnetic Fusion Energy Applications at Low Temperatures--III," R. P. Reed, ed., National Bureau of Standards, Boulder, CO (1980).
3. T. Horiuchi, M. Shimada, T. Fukutsuka, and S. Tokuda, Design and construction of an apparatus for testing materials at cryogenic temperature, in: "Proceedings of the Fifth International Cryogenic Engineering Conference," K. Mendelssohn, ed., Mid-Country Press, London (1974).

TENSILE AND FRACTURE PROPERTIES OF MANGANESE-MODIFIED AISI 304 TYPE STAINLESS STEEL

R. L. Tobler and R. P. Reed

Fracture and Deformation Division, National Bureau of Standards, Boulder, Colorado

INTRODUCTION

The AISI-300 series stainless steels are austenitic Fe-Cr-Ni alloys offering relatively low strength but excellent cryogenic ductility and toughness. Recently, nitrogen-strengthened grades, such as AISI 304 N or AISI 304 LN, have attracted attention as possible substitutes for applications demanding higher strength. Nitrogen is a relatively inexpensive strengthener, and it stabilizes the austenitic structure with respect to martensitic phase transformation. A disadvantage of the nitrogen-strengthened grades is that they are more difficult to fabricate. To overcome this disadvantage, manganese additions (greater than the normal 1 or 2 wt.%) have been recommended. The potential advantages of manganese-modified type AISI 304 LN steels are: (1) consistent fabricability, (2) consistent weldability, and (3) reliable supply in all product forms.[1] Other studies[2-4] have reported potential advantages with manganese additions. It was the purpose of this study to investigate the effects of manganese on the 4-K mechanical properties of nitrogen-strengthened AISI 304 L type alloys. Therefore, a series of steels containing 1-6% Mn and 0.1-0.2% N were prepared, tested, and compared with previous results for Fe-18Cr-10Ni stainless steels.[5,6]

MATERIALS

Ten stainless steel plates were obtained from a commercial steel manufacturer. These steels had chemical compositions as listed in Table 1. Heats 69, 70, 73, and 74 have compositions corresponding to AISI 304 LN; the other heats are modifications. Grain size ranged randomly from 49 to 57 μm in diameter and hardness from R_B 82 to 94. No precipitation at grain boundaries was

Table 1. Compositions of Normal (≤ 2% Mn) and Modified (> 2% Mn) Type 304 LN Stainless Steel.

Heat No.	Weight Percent C	Mn	P	S	Si	Cr	Ni	Mo	Cu	N
69	0.025	1.02	0.022	0.013	0.53	18.19	8.68	0.32	0.30	0.11
70	0.028	2.00	0.021	0.013	0.63	18.10	8.66	0.31	0.28	0.11
71	0.025	3.85	0.023	0.013	0.59	18.22	8.70	0.31	0.29	0.11
72	0.021	5.81	0.023	0.014	0.64	18.06	8.62	0.33	0.30	0.11
73	0.025	1.19	0.023	0.015	0.60	19.36	7.81	0.32	0.28	0.19
74	0.022	2.01	0.023	0.013	0.64	19.35	7.89	0.32	0.30	0.19
75	0.024	3.85	0.022	0.014	0.65	19.25	7.84	0.31	0.31	0.21
76	0.033	5.79	0.024	0.014	0.61	19.48	7.83	0.31	0.31	0.21
77	0.026	4.03	0.023	0.014	0.62	18.71	8.22	0.31	0.30	0.15
78	0.031	5.80	0.024	0.014	0.62	18.42	8.29	0.31	0.30	0.15

observed from light microscopy examination of the microstructures for all heats.

All steels were produced from ingots (12.7 cm x 12.7 cm x 17.8 cm) that were hot-rolled at 1450 K to plates approximately 25-mm thick. The plates were annealed at 1340 K for 1/2 h, then water quenched and acid pickled. Grain boundary segregation was suspected in the lower manganese alloys. Solution treatments at 1560 K for 1 h with a water quench were made on heats 69 and 78. Following this anneal, compact fracture specimens were machined.

EXPERIMENTAL PROCEDURE

Tensile tests were conducted using equipment, procedures, and specimen geometry previously described.[5] The specimens were machined in the transverse orientation. The load-vs-deflection curves were recorded using the outputs from a commercial load cell and strain-gage extensometer. The crosshead rate was 8.5 μms^{-1}.

Single-specimen J-integral tests were used to measure fracture toughness. The technique incorporates a computer to enable digital data acquisition and J-resistance curve plotting during the tests, as described previously.[5,6] The compact fracture toughness specimen was 24.6-mm thick and 50.8-mm wide. The notch was machined in the TL orientation, as defined in ASTM E-399-81.[7]

The fracture specimens were fatigue cracked at 76 K to a crack length, a, between 27.25 and 31.44 mm (a/W ≃ 0.535 to 0.620). The maximum load used in fatigue precracking was 22 kN. The specimens were then loaded in displacement control. As the load increased, partial unloadings (10%) were performed periodically. Using a crack-length-vs-compliance correlation, the crack length at each unloading was inferred and used to obtain the crack extension increment (Δa) from the difference of the initial precrack length. The J value at each unloading point was calculated using the expression: $J = \lambda A/Bb$, where λ is the Merkle-Corten factor, A is the area under the load-vs-deflection curve at the unloading point, and b is the specimen ligament ($b = W - a$).

Scatter in the J-resistance curves led to uncertainties of ±15% in the values of J_{Ic}.

The J_{Ic} values were converted to K_{Ic} estimates, denoted $K_{Ic}(J)$, using a Young's modulus value of 206.8 GPa, a Poisson's ratio of 0.30, and the standard formula relating the elastic J to K_{Ic}.[5]

Magnetic measurements, using a bar magnet torsion balance, were used to estimate the percent bcc (ferromagnetic) martensite that formed during deformation. A calibration had previously been made relating data obtained from the torsion balance device to volume percent bcc martensite.[8] The measurements were all made within the gage length, in the uniform elongation section.

X-ray techniques previously described,[8] (measurement and computation of integrated peak intensities of fcc, bcc, and hcp structures) were used to estimate the amount of bcc martensite on the fracture surfaçe of compact tension specimens. The area of the specimens examined was restricted to near the precrack tip to ensure some relationship to the plastic-zone at initiation of crack growth. In all x-ray measurements, no hcp phase was ever detected on the crack surface.

Hardness measurements were made using a diamond pyramid indentor and a 10-g load. Hardness traverses on annealed specimens with polished surfaces were conducted to attempt to detect segregation effects at grain boundaries.

RESULTS AND DISCUSSION

Tensile Properties

The tensile properties of the steels tested in this study are plotted in Figs. 1 through 3 and tabulated in Ref. 5. In these figures, comparison is made with trend lines from a previous study.[6] As shown in Fig. 1, the yield strength (σ_y) data for the present Fe-18.06-19.5Cr-7.81-8.70Ni steels (open symbols) agree well with previous data trends for Fe-18Cr-10Ni steels (solid lines). The 295-K data confirm a previous conclusion: the yield strength is a linear function of C + N content[6] for the composition range studied. The 4-K data also confirm the existence of a kink

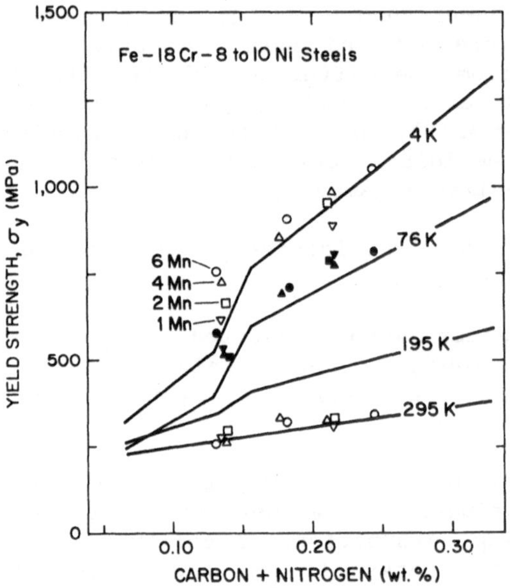

Fig. 1. Effect of carbon plus nitrogen content on yield strength of stainless steels at various temperatures.

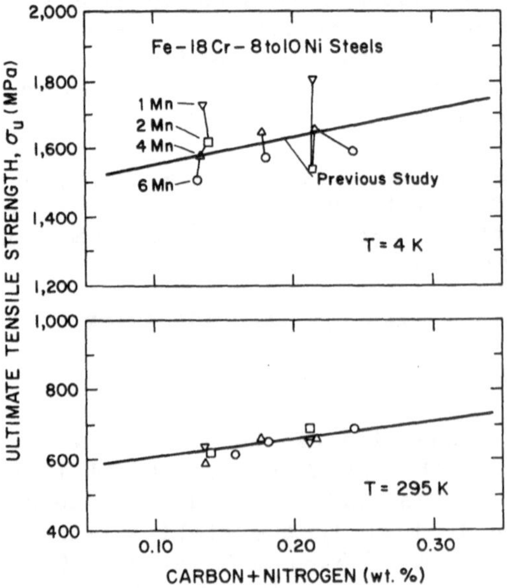

Fig. 2. Effect of carbon plus nitrogen interstitials on ultimate strength at two test temperatures.

in the σ_y-vs-C + N plot at C + N contents between 0.127 and 0.157 wt.%. In our previous study,[6] this kink was attributed to the structural instability of the austenitic phase at lower interstitial contents. In the present study at 4 K there is a clear trend of increasing σ_y values with higher manganese contents at C + N contents from 0.131 to 0.138 wt.%. Apparently, the kink at 4 K is removed by the manganese additions, which have a stabilizing effect with respect to the martensitic transformation at this critical interstitial content.

The ultimate tensile strength results of the present study, shown in Fig. 2, are also in general agreement with expected trends. The data at room temperature fall on the same trend line that was previously reported for the Fe-18Cr-10Ni stainless steels. The 4-K data show a minor variation, probably owing to the austenite stabilizing effect of manganese modifications. As more manganese is added at a given interstitial concentration level, the austenite is progressively more stable during tests at 4 K; as a result, less martensite is formed during testing and ultimate tensile strength is reduced as manganese concentration is increased to 6%.

Figure 3 shows the spread of the elongation and reduction of area measurements for the ten steels. Not all of the ductility values are typical of commercial stainless steels. In particular,

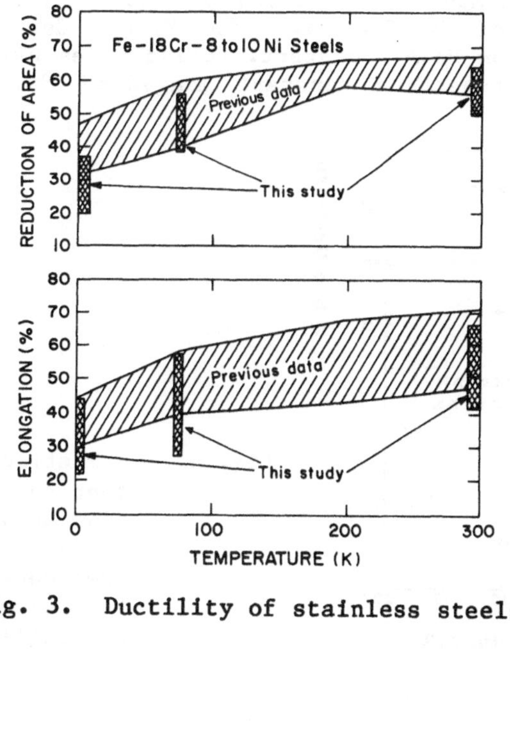

Fig. 3. Ductility of stainless steels.

the reduction of area values for the developmental steels containing 1-4% manganese are lower than the values previously reported for Fe-18Cr-10Ni steels[6] or other commercially available AISI 304 LN steels. Only the 6% manganese-modified steels possess ductility values at the expected levels.

Magnetic measurements made on the fractured tensile specimens were converted to percent martensite, and "normalized" by dividing by the strain corresponding to the uniform elongation where the measurements were made. This procedure assumes a linear dependence of martensite formation with strain. The data are plotted versus Mn + Ni + 10(C + N) concentration in weight percent in Fig. 4. At higher solute concentrations, the amount of martensite per unit strain is reduced. A ratio of the relative alloying contributions was chosen to provide a good fit to the data, and the result follows approximately the prediction of the dependence of M_d temperature (the temperature above which no deformation-induced martensite is possible) found by Williams et al.[9] for compression of Fe-Cr-Ni alloys at low temperatures. However, from Fig. 4, it appears that the dependence on nitrogen content is not linear with Mn + Ni content. Also, the dependence of bcc martensite at compact specimen fracture surfaces is not linear with solute alloying concentration.

Fracture Toughness

The J_{Ic} results obtained for one specimen of each of the ten different steel compositions tested are tabulated[5] and the $K_{Ic}(J)$ values are plotted in Fig. 5. The $K_{Ic}(J)$ data follow regular trends depending on Mn + N content. Solid lines have been drawn to connect data representing constant manganese contents; for each

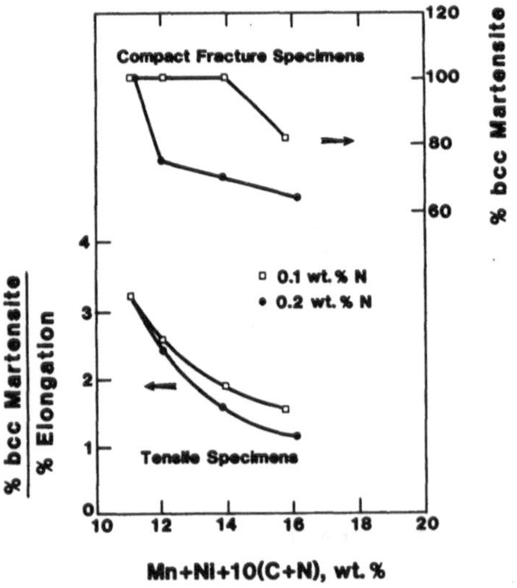

Fig. 4. The effect of alloy content on martensite formation at 4 K at the crack tip in compact tensile specimens and within uniform elongation section of tensile specimens.

value of manganese content, the 0.1-N composition is represented by the symbol at the lowest yield strength, whereas the 0.2-N composition is represented by the higher yield strength, and 0.15-N is intermediate. From Fig. 5 it is clear that higher nitrogen at a given manganese content lowers the $K_{Ic}(J)$ value. It is also clear that manganese has a favorable effect: it raises both the strength and the toughness of these developmental alloys.

The major observation is that the 6%-Mn-modified steels have acceptable fracture toughness, whereas the modifications containing 1-4% Mn display inferior properties. The 1 and 2% Mn steels are particularly poor, having estimated K_{Ic} values as low as 117 $MPa \cdot m^{1/2}$. Their load-deflection fracture toughness test records showed a nearly linear-elastic behavior with regions of instability (serrations). This evaluation is based on comparison with the previous results for a series of nine Fe-18Cr-10Ni stainless steels containing variable C + N contents. As shown in Fig. 5, the 6% Mn-modified steels exhibit an inverse $K_{Ic}(J)$-vs.-σ_y relationship at 4 K and, within the measurement error indicated by the error bars shown in the figure, the inverse-linear trend falls along the lower bound of the scatter band of data for the Fe-18Cr-10Ni steels.

The cause of the unexpected low $K_{Ic}(J)$ values of the 1-4% Mn grades is not clear, and we have examined several possibilities. The nominal compositional variations between the 1-2% Mn steels of

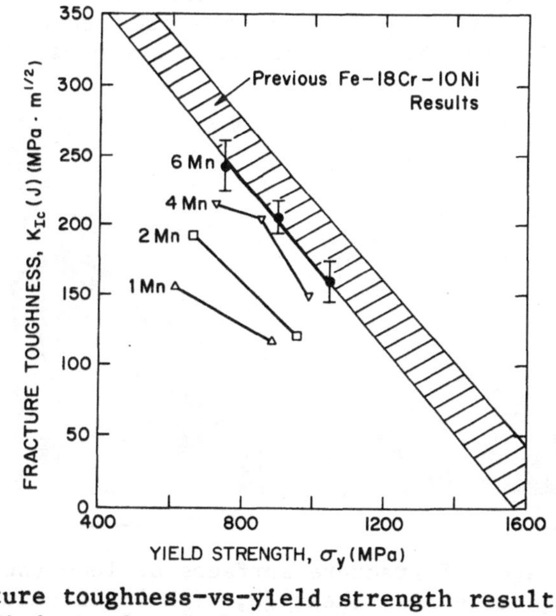

Fig. 5. Fracture toughness-vs-yield strength results for Mn-N modified steels of the present study compared with previous results for AISI 304 type stainless steels.

this study and the Fe-18Cr-10Ni-1.5Mn steels of the previous study include a 1.25-2.5% difference in nickel content, which may be significant. Metallurgical factors, such as interstitial segregation, may also contribute to the low fracture toughness of the 1-4% Mn steels of the present study. However, light microscopy revealed no evidence of sensitization or other obvious microstructural differences. Microhardness traverses across grain boundaries and centers failed to detect any hardness variations.

Evidence of an abnormal condition in the 1, 2, and 4% Mn steels was obtained by scanning electron microscopy performed on the fracture surfaces of the compact specimens subsequent to J_{Ic} testing at 4 K. Figure 6 presents typical results. Specimens containing 4% Mn or less exhibited unusual features consisting of isolated facets (brittle fracture) surrounded by a matrix of numerous very fine dimples. In contrast, the three heats containing 6% Mn (heats 72, 76, and 78) all showed a more completely dimpled fracture mode, with relatively large dimples; this type of fracture was also observed in the Fe-18Cr-10Ni steels of our previous study, which had good fracture toughness. The only apparent distinction between thermal-mechanical treatments of the two alloy series was the temperature of hot-working of the billet. The hot-rolling temperature of the initial 304 LN alloy series was 1560 K (2350°F) and the rolling temperature of the 1-6 wt.% Mn alloy series was 1450 K (2150°F). In addition, plates from heats 69 and 78 were re-solution-annealed at a higher temperature, machined into compact tensile specimens, and tested at 4 K. The values of $K_{Ic}(J)$ (heat 69-133 $MPa{\cdot}m^{1/2}$; heat 78-172 $MPa{\cdot}m^{1/2}$) were about 15% lower than those reported in Fig. 5. This decrease was presumably associated

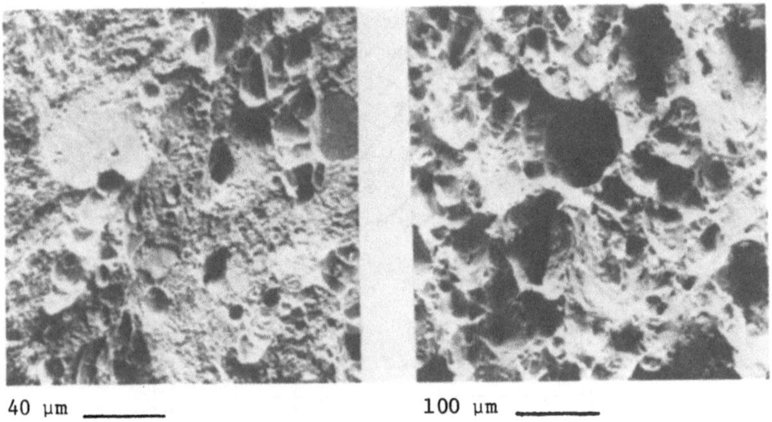

Fig. 6. Micrographs of fracture surfaces of low- (heat 69, left) and high-toughness (heat 72, right) austenitic stainless steels.

with the increased grain size. In any case, the re-solution treatment failed to eliminate the problem of low toughness.

X-ray diffractometer measurement data of the amount of martensitic transformation phases on the fracture surface near the crack tip of compact tensile specimens tested at 4 K are also shown in Fig. 4. As the alloy stability increases, the amount of transformation during early crack formation is reduced. However, in all heats and tests, the amount of transformation was extensive (>64%), indicating large amounts of local plastic deformation in the region of the crack tip. We have also detected large amounts of transformation in the tough Fe-18Cr-10Ni austenitic steels.[6] In Fig. 4 notice that a tough alloy (heat 72) exhibited more crack tip martensite (81%) than two of the alloys (heats 74, 75) that exhibited marginal toughness. Therefore, it is difficult to argue that martensitic transformation, per se, results in reduced toughness. The rate of transformation per unit strain and the mechanisms of transformation must be considered. A possibility remains that the martensitic transformation, which influences the stress-strain characteristics, produces a modified plastic zone. This modified zone in turn may influence J-integral measurements.

CONCLUSION

The data of this study show that 6% Mn-modified AISI 304 LN stainless steel can be produced having conventional tensile and fracture toughness properties equivalent to standard AISI 304 LN stainless steel when tested at 4 K. However, the 1-4% Mn heats tested in this study displayed some unusual, apparently brittle fracture features in association with relatively low fracture toughness. The cause of this behavior has not yet been established, but the problem deserves a full investigation.

ACKNOWLEDGMENTS

The stainless steel plates were produced by Armco, under the direction of R. H. Espy of Armco's Research and Technology Division. E. L. Brown, now at the Colorado School of Mines, contributed metallographic assistance, and D. Baxter of NBS conducted many of the tensile and fracture toughness tests. The work was sponsored by the Department of Energy, Office of Fusion Energy.

REFERENCES

1. R. H. Espy, "Stainless Steel Alloy Development," presented at NBS/DoE Workshop on Materials at Low Temperatures, Vail, Colorado, October 16-18, 1979.

2. T. Kato, M. Fujikura, T. Soh, and K. Ishida, Effects of alloying elements on the austenite stability and mechanical properties of austenitic stainless cast steels at cryogenic temperatures, Denki Seiko 48(3):179-190 (1977).
3. J. Hochmann, Le role des additions de manganese dans les aciers inoxydables austenitiques, Mater. Tech., Spec. Issue 65: 69-87 (1977).
4. A. B. Meade and R. Oppenheim, Tieftemperatureigen schaftor stickstofflegierter austenitische Chrom-Nickel-(Molybdan) sowie Chrom-Mangan-Nickel-Stahle und deren Eignung als kaltzahe Stahle, DEW-Techn. Ber. 12:13-19 (1972).
5. R. L. Tobler and R. P. Reed, Tensile and fracture properties of manganese-modified Fe-18Cr-8Ni type austenitic stainless steels, in: "Materials Studies for Magnetic Fusion Energy Applications at Low Temperatures - IV," NBSIR 81-1645, R. P. Reed and N. J. Simon, eds., National Bureau of Standards, Boulder, Colorado, (1980), pp. 77-100.
6. R. L. Tobler, D. T. Read, and R. P. Reed, Strength and toughness relationship for interstitially strengthened AISI 304 stainless steels at 4 K, in: "Materials Studies for Magnetic Fusion Energy Applications at Low Temperatures - IV," NBSIR 81-1645, R. P. Reed and N. J. Simon, eds., National Bureau of Standards, Boulder, Colorado, (1981), pp. 37-76.
7. Standard method of test for plane strain fracture toughness of metallic materials (Designation E399-81), in: "1981 Annual Book of ASTM Standards," Part 10, American Society for Testing and Materials, Philadelphia (1981), pp. 588-618.
8. R. P. Reed and C. J. Guntner, Stress-induced martensitic transformations in 18Cr-8Ni steel, Trans. AIME 230: 1713-1720 (1964).
9. I. Williams, R. G. Williams, and R. C. Capellaro, Stability of austenitic stainless steels between 4 K and 373 K, in: "Proceedings of the Sixth International Cryogenic Engineering Conference, IPC Science and Technology Press, Guildford, Surrey, England (1976), pp. 337-341.

MECHANICAL PROPERTIES OF HIGH MANGANESE STEELS AT CRYOGENIC TEMPERATURES

T. Horiuchi, R. Ogawa, and M. Shimada

Asada Research Laboratory, Kobe Steel, Ltd., Kobe, Japan

S. Tone, M. Yamaga, and Y. Kasamatsu

Kakogawa Works, Kobe Steel, Ltd., Kakogawa, Japan

INTRODUCTION

Structural steels for fusion magnet applications at any ambient temperatures have hitherto been austenitic stainless steels, such as SUS 304L and 316L. Although these steels have good fracture toughness even at liquid helium temperature, they also have several drawbacks: low yield strength and instability of the austenitic phase against deformation.[1]

Recently, we have developed a new, nonmagnetic steel plate (NONMAGNE 30) for the structural material of "JT-60" which is under construction at JAERI. This alloy is stabilized and strengthened with mainly C and Mn and has excellent mechanical and magnetic properties at room temperatures. It has a yield strength of more than 1 GPa at 4 K, but its toughness at 4 K is not sufficient.

The purpose of this study was to evaluate the cryogenic mechanical properties of two kinds of high Mn steels (N-strengthened high Mn stainless steel and high C, high Mn steel) in an attempt to develop a new cryogenic steel with a higher yield strength than 304LN or 316LN accompanied with stable austenitic phase.

EXPERIMENTAL PROCEDURE

Alloys were prepared by vacuum-melting 10-kg and 90-kg ingots and air-induction melting a 90-kg ingot. The ingots were forged and hot-rolled to 15-mm and 20-mm thickness plates. All plates

Table 1. Chemical Compositions of High Mn Stainless Steels

	C	Si	Mn	Ni	Cr	P	S	N	Nieq.
A	0.030	0.11	15.5	4.00	12.98	0.004	0.0090	0.140	17
B	0.035	0.10	16.2	3.92	14.67	0.001	0.0093	0.193	19
C	0.036	0.10	18.3	3.98	11.93	0.001	0.0072	0.204	20
D	0.032	0.11	18.4	4.92	13.80	0.001	0.0078	0.233	22

Table 2. Chemical Compositions of High Mn Stainless Steels

	C	Si	Mn	Ni	Cr	P	S	N	Nieq.
E	0.031	0.11	18.64	5.06	16.11	0.001	<0.01	0.261	23.1
F	0.037	0.12	18.09	5.08	18.10	0.001	<0.01	0.243	22.5
G	0.032	0.10	18.18	7.12	16.19	<0.001	<0.01	0.251	24.7
H	0.033	0.10	18.16	3.11	16.30	0.001	<0.01	0.231	20.1
I	0.034	0.10	18.16	1.07	16.53	0.001	<0.01	0.241	18.4
J	0.031	0.10	15.01	5.08	16.23	0.001	<0.01	0.237	20.6
K	0.032	0.11	20.00	5.10	15.94	<0.001	<0.01	0.246	23.4
L	0.033	0.10	22.00	5.12	15.94	<0.001	<0.01	0.245	24.5
M	0.037	0.10	28.60	5.12	15.98	<0.001	<0.01	0.247	27.9

Nieq.=Ni% + 30 x C% + 30 x N% + 0.5 x Mn%

Table 3. Chemical Compositions of High C - High Mn Steels

	C	Si	Mn	P	S	Cu	Ni	Cr	N
N	0.58	0.25	14.57	0.004	0.006	-	1.85	2.11	0.012
O	0.60	0.33	18.13	0.003	0.005	-	2.11	1.96	0.025
P	0.63	0.29	20.28	0.003	0.005	-	2.01	1.89	0.022
Q	0.61	0.32	24.00	0.004	0.005	-	2.01	2.01	0.030
R	0.60	0.38	17.96	0.003	0.006	-	-	-	0.022
S	0.59	0.33	17.95	0.004	0.006	-	3.05	-	0.023
T	0.62	0.32	17.82	0.004	0.004	-	6.01	-	0.021
U	0.60	0.41	17.32	0.005	0.006	0.98	3.09	-	0.026
V	0.63	0.33	18.20	0.004	0.005	3.01	2.92	-	0.013
W	0.63	0.34	18.10	0.003	0.006	3.13	3.04	1.91	0.014
X	0.61	0.30	18.20	0.003	0.005	2.96	3.04	4.91	0.027

were then solution-treated at 1050°C or 1100°C for 60 min, followed by quenching into water. Chemical compositions of steels are given in Tables 1-3. Several kinds of austenitic stainless steels, SUS 304L, 316L, 304LN, 316LN, 310S, AISI 205, Nitronic 33 (N33), and 40 (N40) were prepared as reference materials.

Tensile specimens were cut from plate centers at depths of 1/4 and 3/4 thickness. The tensile tests were conducted at room temperature, 77 K, and 4 K by using the tarlet disk-type tensile cryostat.[2] Toughness of each plate was evaluated by Charpy impact tests at RT and 77 K and also tensile tests of fatigue notched specimens at RT, 77 K, and 4 K. The fracture toughness of these fatigue notched specimens was determined by Rice's method, and the magnetic permeabilities at the fractured surface of the tensile and Charpy impact specimens were measured by a low permeability meter.

RESULTS

Nitrogen-Strengthened High Mn Stainless Steels

The yield strength and elongation of typical commercial austenitic stainless steels, SUS 304L, 316L, 304LN, 316LN, 310S, AISI 205, A286, Nitronic 33 and 40, are given in Fig. 1. All

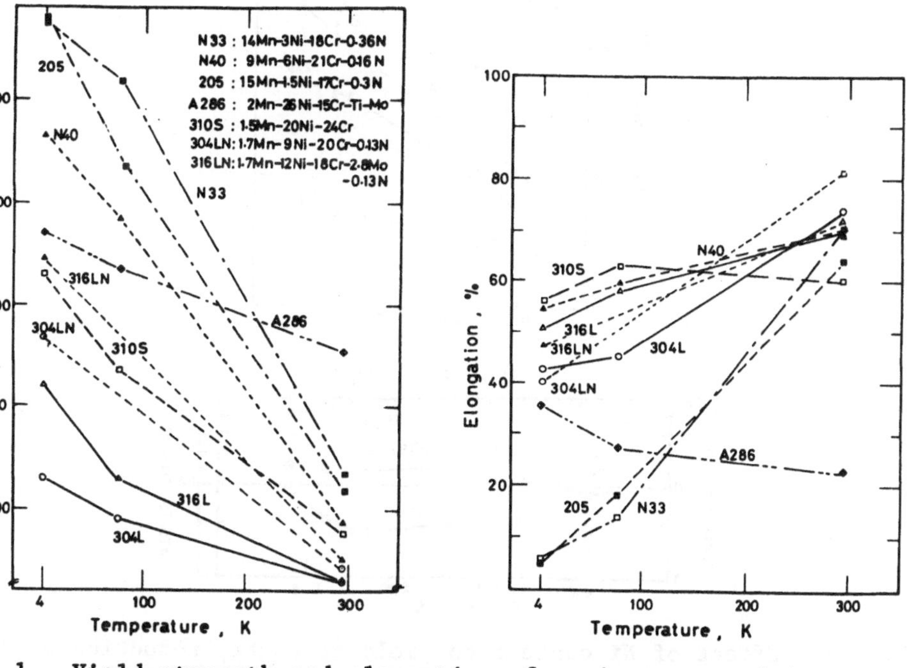

Fig. 1. Yield strength and elongation of various austenitic stainless steels at temperatures between 4 K and RT.

alloys were as solution treated except A286 which was solution treated and aged. Alloys 304LN and 316LN, which contain 0.13% N, have yield strengths about 200 MPa higher than those of 304L and 316L without deterioration of ductility. High Mn stainless steels, AISI 205, Nitronic 33 and 40 have shown high yield strengths of more than 1 GPa at 4 K. However, AISI 205 and Nitronic 33, which contain more than 0.3% N, have shown poor ductility at 4 K and 77 K.

Taking these results into consideration, investigations have been made on the effect of Mn, Ni, Cr, and N content on the cryogenic mechanical properties and the stability of N-strengthened high Mn stainless steels based on AISI 205 or N33 type steels.

Figure 2 shows the effect of N content on the yield strength of 15%-18% Mn alloys (A-D in Table 1). A yield strength of more than 1 GPa can be achieved by N content of more than 0.2%.

The dependency of yield strength on the C+N content of high Mn stainless steels is similar to that of 304LN. The notch yield

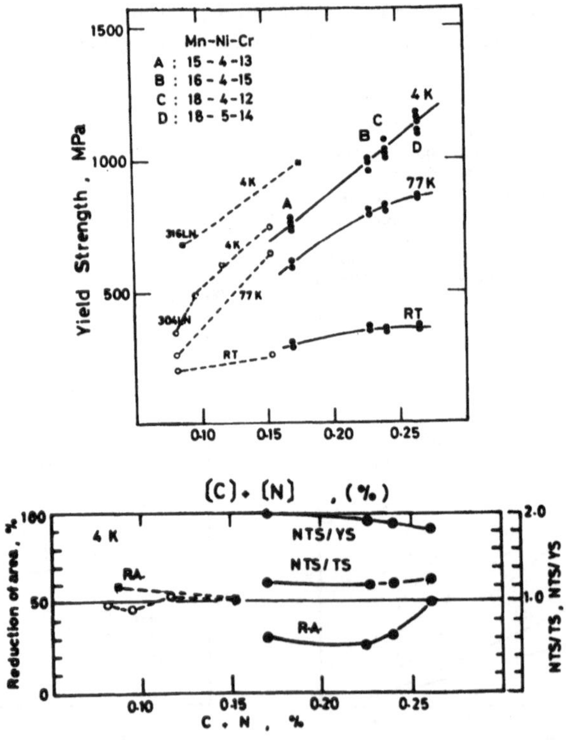

Fig. 2. Effect of Ni content on yield strength, reduction of area, and notch tensile properties of high Mn stainless steels.

ratio and notch tensile ratio indicate a low notch sensitivity of these steels at 4 K. The 18Mn-5Ni-14Cr steel (D in Table 1) of high Ni equivalence shows good ductility at 4 K. This improvement is attributed to an increase in the stability of austenite.[3,4] The effect of Mn, Ni, and Cr contents on the tensile properties of high Mn stainless steels that have constant C+N content ($\approx$0.28%) are given in Fig. 3 (a)-(c). The magnetic permeabilities measured at the fractured surfaces of the specimens tested at 4 K are also given in Fig. 3. Over the range 15 to 28% Mn, the strength and ductility at 4 K increase slightly [Fig. 3(a)]. The austenite phase is stabilized as Mn contents increase from 15% to 18% α', and ε' martensite measured on the fractured specimen decreases with increase in Mn content. The formation of α' martensite by deformation at 4 K is suppressed in the steels containing more than 18% Mn. The ductility of 18% Mn alloy at 4 K increases remarkably with increase in Ni content, but slows down over 5% Ni, as shown in Fig. 3(b). Over the range 14% to 18% Cr, the ductility of 18% Mn alloy decreases slightly. In 18% Cr alloy, δ-ferrite was observed in the as-cooled condition.

High C, High Mn Steels

A newly developed high C, high Mn steel plate (NONMAGNE 30) comprises 0.6% C, 14% Mn, 2% Ni, and 2% Cr and is an economical austenitic steel, which has excellent mechanical and magnetic properties at room temperatures. However, at cryogenic temperatures, this steel shows insufficient toughness in spite of high yield strength of more than 1 GPa at 4 K. The effects of Mn, Cu, Ni, and Cr contents on the mechanical and magnetic properties of high C, high Mn steels were studied at cryogenic temperatures on the basis of the chemical composition of NONMAGNE 30.

The effect of Mn and Ni contents on the yield strength at 4 K and Charpy impact value at 77 K are given in Fig. 4. The increase in Mn contents enhanced Charpy impact value quite sharply, but slightly affected the yield strength at 4 K, and on the other hand, the Cu addition was effective in increasing the toughness. The Cr addition also enhanced the yield strength, but deteriorated the toughness.

Considering these results from a practical point of view, 0.6% C-18% Mn-3% Cu-3% Ni steel was selected for cryogenic use; the cryogenic mechanical properties of this steel plate of 20-mm thickness are given in Fig. 5. The yield strength of this plate is almost the same as that of NONMAGNE 30 at the temperatures between RT and 4 K, but the ductility and the fracture toughness are remarkably improved. The fracture toughness is nearly equal to that of SUS 316L.

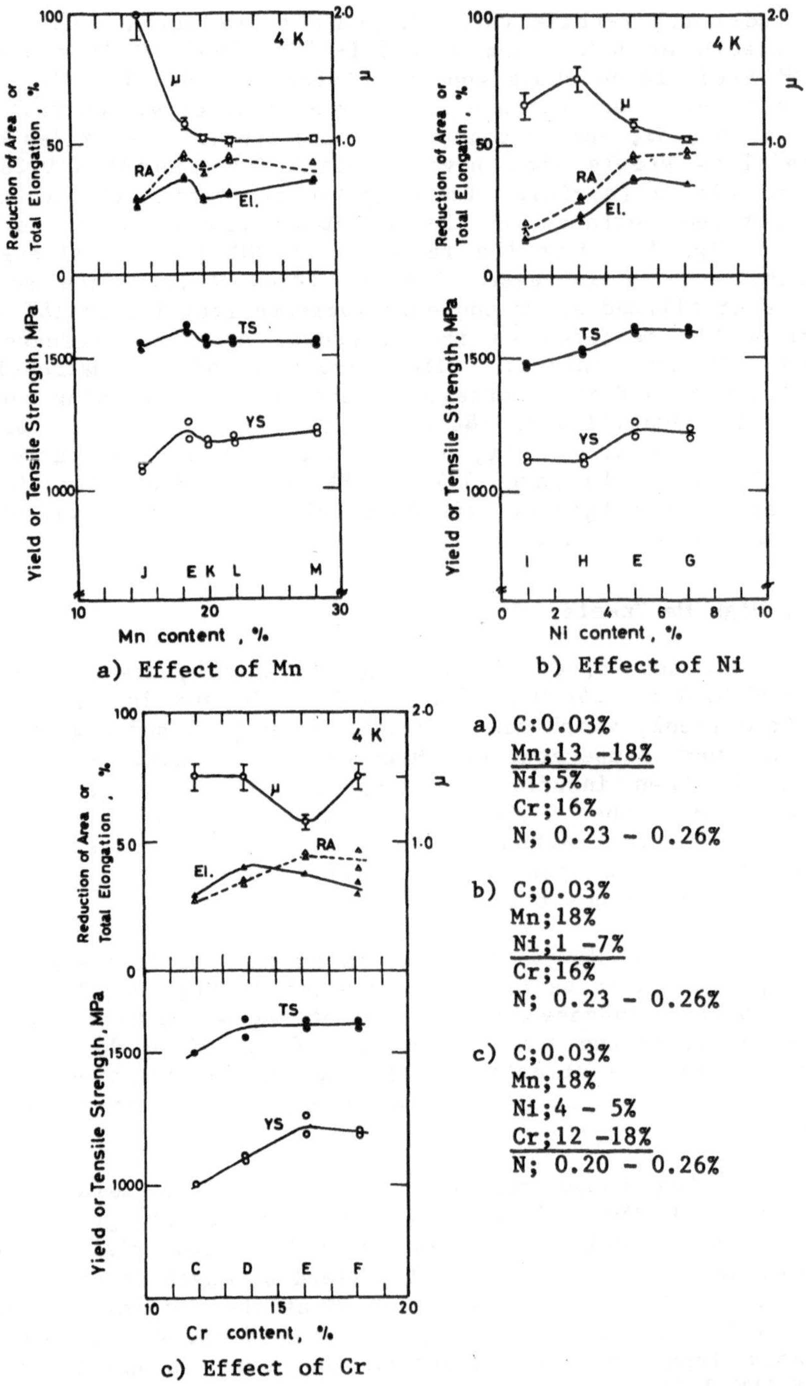

Fig. 3. Effect of Mn, Ni, and Cr contents on tensile properties of high Mn stainless steels at 4 K. μ: Magnetic permeability.

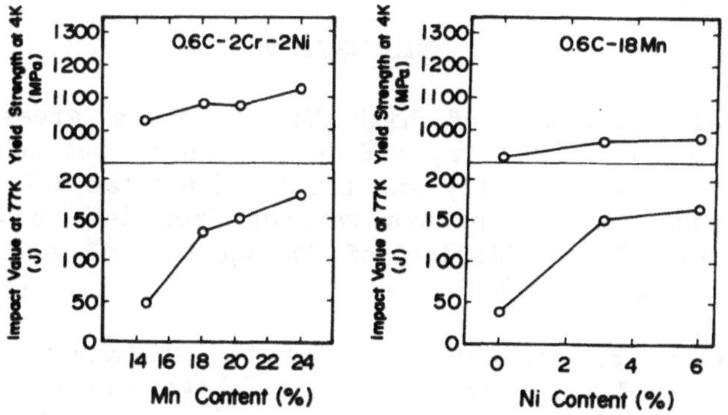

a) Effect of Mn. b) Effect of Ni.

Fig. 4. Effect of Mn and Ni content on yield strength (4 K) and Charpy impact value (77 K) of high C, high Mn steels.

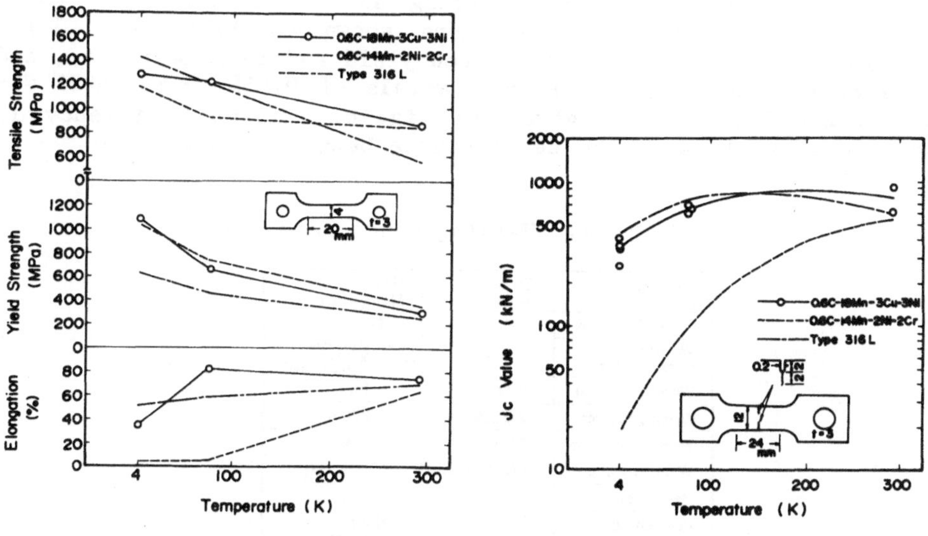

a) Tensile properties. b) Fracture toughness.

Fig. 5. Tensile properties and fracture toughness of 0.6C-18Mn-3Cu-3Ni.

The austenite phase of this plate was stabilized and no martensitic products were observed at the fractured surface of specimens tested at 4 K.

DISCUSSION

The yield strength of high Mn stainless steel at 4 K was slightly affected by Mn, Cr, and Ni contents, but affected mainly by N content. The ductility and the toughness at 4 K were improved by the increase in Mn contents ranging from 14% to 22% and in Ni contents under 5%. Addition of Mn up to 28% deteriorates the Charpy impact value at 77 K.

Chromium is an essential element for stainless steels, but it has been reported that for high Cr (>13%) and high Mn contents (>20%) in Fe-Mn-Cr system, δ-ferrite is retained in the as-cooled condition.[5] Over the range 13%-18% Cr of 18% Mn alloy that contains 3-5% Ni and 0.25% N, δ-ferrite was observed in only the 18% Cr alloy. Therefore, optimum ranges of Mn, Ni, and Cr contents of N-strengthened high Mn alloys are 28% > Mn $\geq$ 18%, Ni $\geq$ 5%, and 18% > Cr $\geq$ 13%.

Two types of high Mn steel plates (0.03% C-18% Mn-5% Ni-14% Cr-0.23% N steel and 0.6% C-18% Mn-3% Cu-3% Ni steel) were welded by Electron Beam (EB) welding and Metal Inert Gas (MIG) welding with the welding conditions shown in Table 4. Joint efficiency of EB welded joints, notch yield and tensile ratios, and the fracture toughness of EB weld metals of 0.03% C-18% Mn-5% Ni-14% Cr-0.23% N steel are given in Fig. 6. Joint efficiency is nearly equal to 100% at the temperatures between RT and 4 K.

Table 4. Welding Conditions

Welding Method	Electron Beam	Gas Metal - Arc
Plate Thickness	20mm	20mm
Welding Consumables	——	DW-14M (1.6mm φ) Ar + 20% CO_2 (25 ℓ/min)
Welding Position	Horizontal	Flat
Welding Current	120 mA	270 A
Welding Voltage	50 kV	28 V
Welding Speed	60 cm/min	23 cm/min
Heat Input	6 kJ/cm	19.7 kJ/cm
Preheating	None	None
Interpass Temperature	——	Lower than 150°C
Number of Passes	1	6
Groove Preparation (unit : mm)	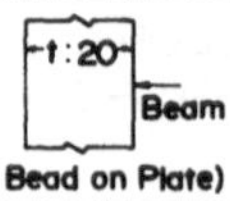 (Bead on Plate)	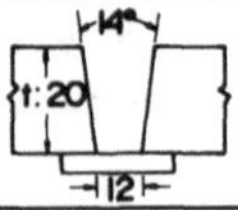

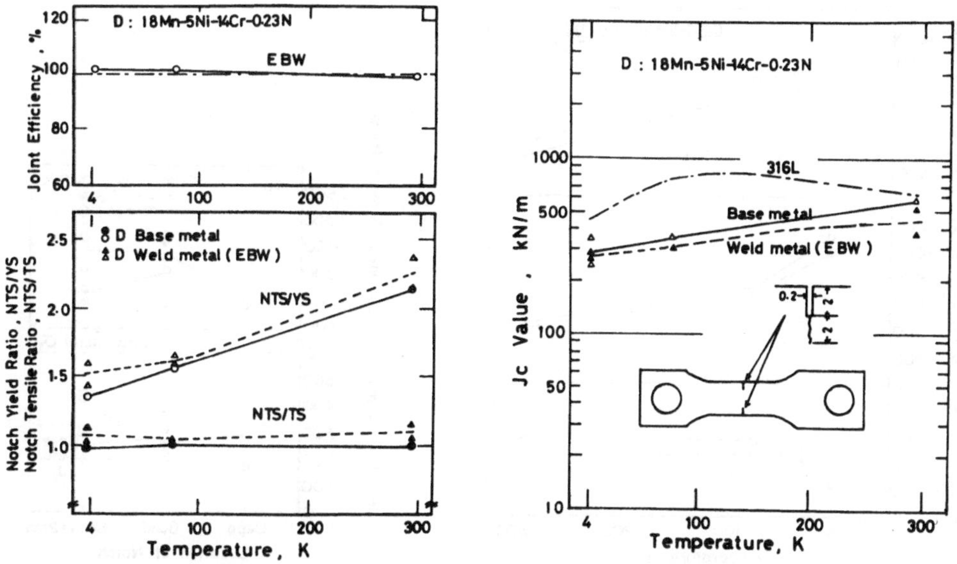

a) Joint efficiency and notch tensile properties.

b) Fracture toughness.

Fig. 6. Mechanical properties of base metal, EB welded joints, and weld metals of 18Mn-5Ni-14Cr-0.23N plate.

Fracture toughness and notch yield and tensile ratios of weld metals show values equal to those of the base metals. Mechanical properties of EB and MIG welded joints and the fracture toughness of each location of welded joints of 0.6% C-18% Mn-3% Cu-3% Ni steel are given in Fig. 7. The EB and MIG welded joints show good ductility and toughness at cryogenic temperatures, although the research in the welding consumables for MIG welding was insufficient.

SUMMARY

The mechanical properties of two types of high Mn steels (N-strengthened high Mn stainless steel and high C, high Mn steel) for cryogenic use were evaluated at temperatures between RT and 4 K.

1. The high Mn stainless steels are strengthened with N addition of more than 0.2% to meet the yield strength of more than 1 GPa at 4 K.
2. Optimum ranges of Mn, Ni, and Cr contents in the N-strengthened high Mn stainless steels ($\geq$0.25% N) showing good ductility and toughness at 4 K are 28% > Mn $\geq$ 18%, 18% > Cr $\geq$ 13%, and Ni $\geq$ 5%. The austenite phase is stabilized in this range of each element.

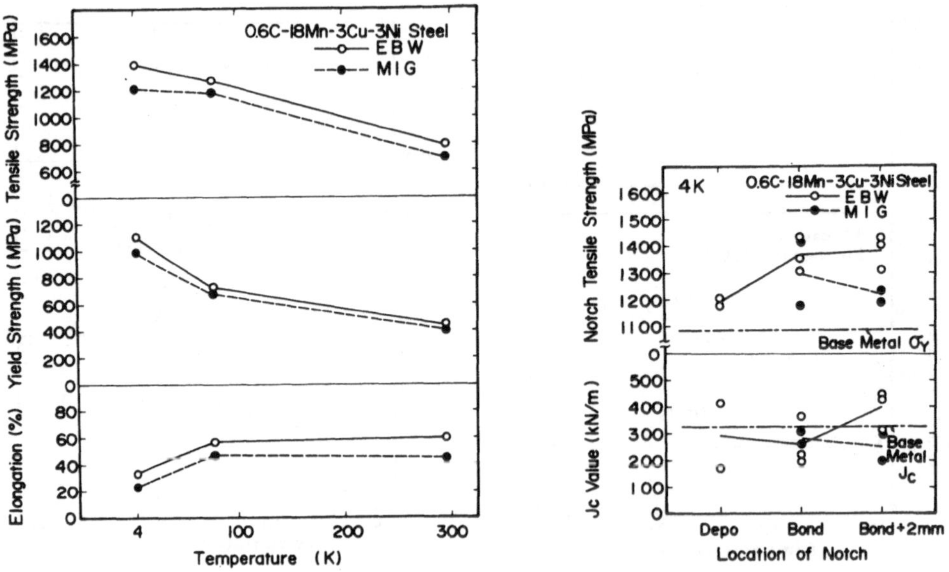

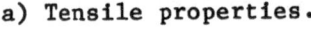

a) Tensile properties.

b) Fracture toughness of each position of welded joints.

Fig. 7. Mechanical properties of EB and MIG welded joints of 0.6C-18Mn-3Cu-3N steel.

3. Fracture toughness of high C, high M steels (0.6% C) increases with increase in the Mn, Ni, and Cu contents. Addition of Cr enhances the yield strength, but deteriorates the fracture toughness.
4. The 0.6% C-18% Mn-3% Cu-3% Ni steel has excellent ductility of the austenite phase against plastic deformation at 4 K.
5. Nitrogen-strengthened high Mn stainless steels of 0.3% C-18% Mn-5% Ni-14% Cr-0.23% N and high C, high Mn steels of 0.6% C-18% Mn-3% Cu-3% Ni have high yield strength, excellent ductility and toughness at 4 K, and show sound EB and MIG welded joints.

REFERENCES

1. Y. Hirai, H. Sonoi, R. Ogawa, and T. Horiuchi, in: "Proceedings of the 1st JIM Symposium on New Aspects of Martensitic Transformation," (1976), p. 423.
2. T. Horiuchi, M. Shimada, T. Fukutsuka, and S. Tokuda, in: "Proceedings of ICEC5," IPC Science and Technology Press, Guildford, Surrey, England (1974), p. 465.

3. S. K. Hwang and J. W. Morris, Jr., in: "Advances in Cryogenic Engineering," Vol. 24, Plenum Press, New York (1975), p. 137.
4. J. W. Morris, Jr., S. K. Hwang, K. A. Yushchenko, V. I. Belotzerkovetz, and O. G. Kvasnevskii, in: "Advances in Cryogenic Engineering," Vol. 24, Plenum Press, New York (1975), p. 91.
5. T. Kato, S. Fukui, M. Fujikura, and K. Ishida, Trans. Iron Steel Inst. Jpn. 16:673 (1976).

HIGH-ALLOY MANGANESE-ALUMINUM STEELS FOR CRYOGENIC APPLICATIONS

J. Charles, A. Lutts, and A. Berghezan

Université Catholique de Louvain, Louvain-la-Neuve, Belgium

INTRODUCTION

In most of the recently developed austenitic steels for cryogenic applications, nickel is replaced partially or totally by simultaneous additions of manganese and nitrogen. Such alloys also contain a sufficient amount of chromium (12-18%)* to render them stainless while the nitrogen additions both strengthen the alloy and stabilize the austenite.

Our present study is based upon the two following considerations: replace all the nickel by manganese, and chromium by moderate additions of aluminum, in order to develop a new and more economical alloy for cryogenic applications.

From the work of Schumann[1-3] and the more recent results of Holden, Bolton, and Petty,[4] only those binary Fe-Mn alloys containing 28-50% manganese possess stable austenites when cooled to room temperature. For manganese contents between 15 and 28% there exists a two-phase region--austenite + hexagonal ε-martensite--whereas for still lower manganese contents (< 15%), the austenite transforms partially into tetragonal α'-martensite responsible for the sharp decrease in toughness at low temperature.

In this work we have studied the mechanical properties in tension and resilience at 300 and 78 K of both Fe-(20-40)Mn alloys with and without 5% Al additions. We restricted our study to maximum aluminum additions of 5%, since, according to Schmatz,[5-6] higher aluminum contents impair the stability of the austenite.

*All compositions used in this paper are in wt.%.

Since the alloys do not contain any strong carbide-forming elements, we also studied the influence of increasing carbon content on these mechanical properties and the stability of the austenitic phase in the alloy Fe-23Mn-5Al.

EXPERIMENTAL PROCEDURE

The alloys were prepared in a sealed induction furnace from ARMCO iron, electrolytic manganese powder, and refined aluminum (99.99%).

The sealed melting chamber was evacuated to a pressure of 10^{-1} torr to dry and degas the charge. When the temperature of the latter attained approximately 500°C (930°F), the partial vacuum was replaced by an argon protective atmosphere (static pressure of about 1 bar). The liquid metal was then cast under the same argon atmosphere into a water-cooled copper mould. The resulting ingots weighed 1.2 kg (2.6 lbs) or 2.4 kg (5.2 lbs), depending upon the size of the mould employed.

Carbon additions were introduced using a master alloy Fe-13Mn-5C prepared under the same conditions described above.

The ingots were given a thermomechanical treatment consisting of deformation by rolling to a 33% reduction in thickness, followed by a homogenization treatment of 48 h at 1150°C (2100°F) to break up the large-grained, cored, as-cast structure. This first treatment was followed by a second thermomechanical treatment to refine the grain, which consisted of a new plastic deformation of 33% reduction in thickness by cold-rolling, followed by a recrystallization heat-treatment of about 25 min at 925°C (1700°F).

When ε-martensite was present in the structure, cold-rolling was replaced by hot-rolling at 900°C (1670°F).

Specimens for tension tests, having threaded ends (L_0 = 25 mm, Φ = 5 mm), as well as the Charpy KCV impact specimens, were then machined. In the case of those alloys exhibiting a low yield strength of a metastable austenite, a heat treatment of about 15 min at 900°C (1670°F) was performed on the machined specimens sealed in quartz tubes under an argon atmosphere before testing.

Tensile tests were performed on an Instron machine using a speed of 5 mm/min. Those tests performed at room temperature used an extensometer to determine the yield strength after 0.2 and 1% elongation. No extensometer was used in the 78-K tests, performed by immersing the specimen assembly and specimen grips in a bath of liquid nitrogen.

Optical microscopic examination was performed after mechanical polishing and chemical etching using a solution of 5% Nital for binary Fe-Mn alloys and a solution containing 87% H_2O, 10% HNO_3, and 3% HF for the alloys containing 5% Al. Those alloys exhibiting metastable austenite that could transform into ε-martensite during mechanical polishing were electrolytically polished in a solution composed of 10% $HClO_4$ and 90% $C_6H_{14}O_2$ before chemical etching.

The proportion of the different phases present in these alloys was obtained from quantitative x-ray diffraction analyses.

BINARY Fe-(20-40)Mn ALLOYS

Structures

The increasing stability of the austenitic phase with higher manganese contents for the Fe-Mn alloys cooled to room temperature is illustrated in the optical micrographs, Fig. 1. In those alloys containing more than 28% manganese, a fine-grained twinned structure, characteristic of the face-centered cubic phase, can be observed in Fig. 1c. When the manganese content was less than 28%, the austenite partially decomposed into ε-martensite. This is clearly visible in Fig. 1a and b where the ε-martensite laths appear in large numbers inside the original austenitic grains.

Mechanical Properties

Figure 2 shows the variation in hardness (HV_{50}), toughness (KCV), yield strength ($E_{0.2}$,E_1), and tensile strength (TS), as well as the total elongation in traction (A) for these binary alloys containing 20 to 40% manganese at 300 and 78 K.

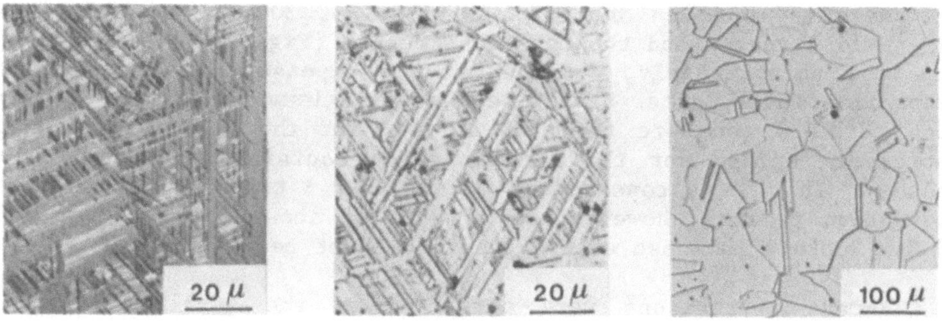

(a) (b) (c)

Fig. 1. Optical micrographs of Fe-Mn alloys: (a) Fe-20Mn (γ + ε); (b) Fe-24Mn (γ + ε); (c) Fe-29Mn (γ). Etching: Nital 5%.

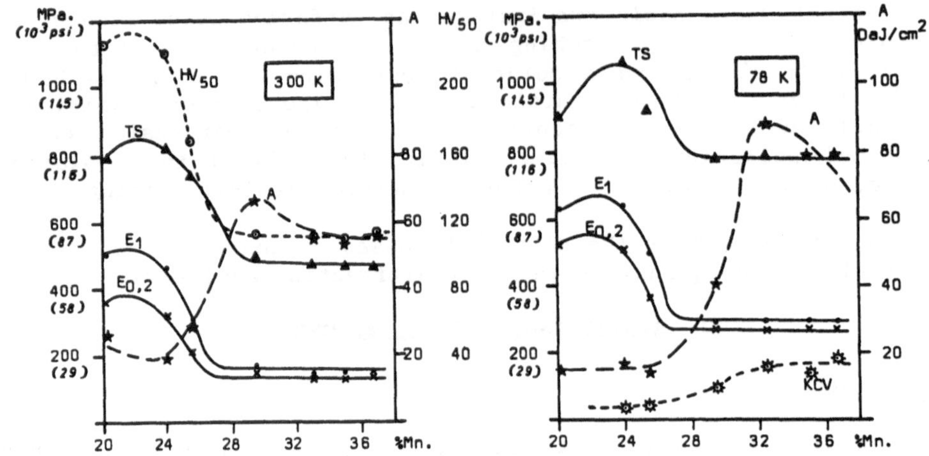

Fig. 2. Mechanical properties of Fe-(20-40)Mn alloys at 300 K and 78 K. TS = tensile strength; E_1 and $E_{0.1}$ = yield strength; A = % elongation; HV_{50} = Vickers hardness (50 kg); KCV = Charpy V-notch value.

The rapid increase in hardness (HV_{50}) observed (Fig. 2) when the manganese content was less than 28% is a consequence of the partial transformation of the austenite into ε-martensite, as revealed by our study of the microstructure of the Fe-(20-40)Mn alloys. The hardness increase appears to be proportional to the amount of ε-martensite, since it increased as the manganese content decreased. The hardness increase resulting from the $\gamma \rightarrow \varepsilon$ transformation in these alloys (HV_{50} = 220) was, however, considerably less than that due to the $\gamma \rightarrow \alpha'$ transformation in steels ($HV_{50} \simeq$ 700).

We also found an important increase at 300 K and at 78 K in the yield strength and the tensile strength (Fig. 2). This hardening was, unfortunately, obtained at the expense of a decrease in elongation which, after having reached a maximum (60% total elongation at 300 K and more than 80% at 78 K for the alloy Fe-32% Mn), decreased rapidly for the Fe-24Mn alloy containing both γ and ε phases. This is a consequence of the $\gamma \rightarrow \varepsilon$ transformation. The elongation remains, however, at a value of about 15% at 300 K and 78 K. Rupture in these alloys occurs without necking.

X-ray diffractions were performed on the Fe-32Mn alloy in three different conditions: annealed, tested to rupture at both room temperature and at 78 K. They show an important transformation of austenite into ε-martensite induced by plastic deformation. These x-ray diffraction results are confirmed by the micrograph shown in Fig. 3.

The maximum elongation in tension can be explained by the $\gamma \rightarrow \varepsilon$ transformation during plastic deformation also observed in Fe-Ni-C steels (TRIP steels). The shift of the maximum elongation (28Mn at 300 K and 32Mn at 78 K) confirmed the increasing instability of the austenite as the testing temperature of the alloys decreased. We should note the low yield strength in these binary Fe-Mn austenites: 130 MPa at 300 K and 220 MPa at 78 K, which was, fortunately, compensated by a large reserve of ductility.

The toughness of the Fe-(32-36)Mn austenitic alloys at 78 K was especially large: ≈160 J/cm^2 (see Fig. 2). This represents an important advantage of the face-centered cubic structures, presenting a good toughness at low temperature, which amply justifies their use for cryogenic applications. As soon as ε-martensite was present, a large decrease in toughness was observed. This value remained, however, approximately 30-40 J/cm^2 in the case of the Fe-(24-25)Mn alloy, which exhibited a mixed, γ + ε, microstructure.

Conclusions

The mechanical properties of the annealed, binary Fe-Mn alloys exhibit restricted interest for cryogenic applications; when the manganese content is less than 28%, the presence of ε-martensite is responsible for a hardening effect and a simultaneous decrease in ductility, whereas, on the basis of its mechanical properties, the Fe-36Mn alloy with its stable austenitic phase presents all the requirements for cryogenic applications, except that its yield strength is too low (130 MPa).

Fe-(20-40)Mn-5Al ALLOYS

Structures

The as-cast Fe-(22-36)Mn-5Al alloys exhibited a mixed γ + ε structure associated with an important coring. After homogenization, however, they presented a stable austenitic structure, except the Fe-22Mn-5Al alloy, where several small isolated regions of ferrite remained. No traces of ε-martensite were visible (Fig. 4a).

Mechanical Properties

Figure 5 shows the hardness and toughness curves as well as the tensile properties (yield strength, tensile strength, and elongation) of the Fe-(22-40)Mn-5Al alloys.

No rapid variation of hardness as a function of manganese was observed, which confirms the absence of ε-martensite in all these alloys due to the presence of 5% Al. The slight increase in hardness observed in the case of the Fe-22Mn-5Al alloy can probably be traced to the small ferrite grains that inhibit an exaggerated

Fig. 3. Scanning electron micrograph of Fe-32Mn alloy after being tested to rupture at 78 K. Austenite transforms into ε-martensite.

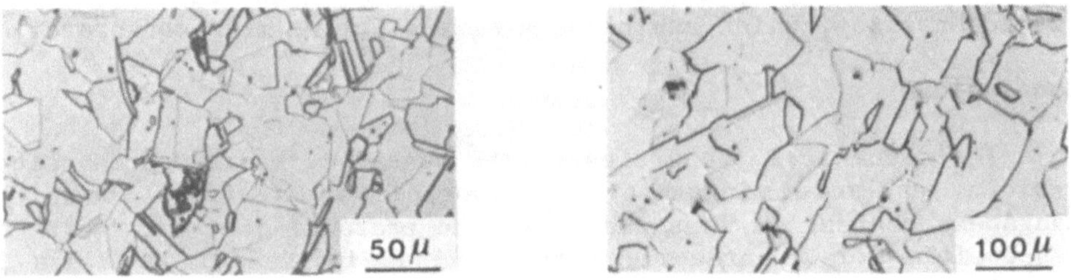

(a) (b)

Fig. 4. Optical micrographs of annealed alloys: (a) Fe-22Mn-5Al (γ + α); (b) Fe-34Mn-5Al (γ).

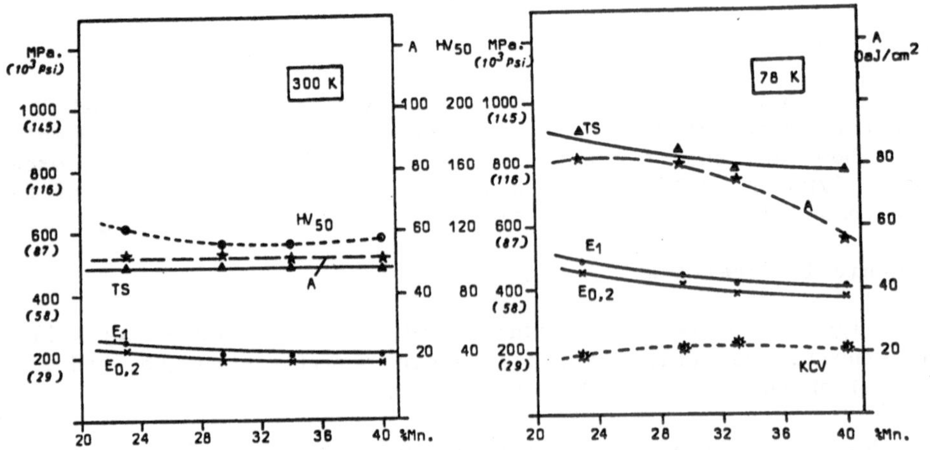

Fig. 5. Mechanical properties of the Fe-(20-40)Mn-5Al alloys at 300 K and 78 K. TS = tensile strength; $E_{0.2}$ and E_1 = yield strength (offset method); A = elongation; HV_{50} = Vickers hardness (50 kg); KCV ≐ Charpy V-notch value.

grain growth during the homogenization heat treatment. The tensile properties are in good agreement with the microstructure in that only a slight increase in the yield strength and the tensile strength was observed as the manganese content decreased from 40 to 22%. On the other hand, the elongation and the toughness were especially high in all these alloys. This can be explained by the suppression of the martensitic transformation in the alloys of lower manganese content due to the presence of 5% Al.

We attribute the 70-MPa increase of the yield strength at 1% elongation to the substitutional solid solution hardening effect of 5% Al.

Finally, we observed in all these alloys a very high Charpy V-notch value (210 J/cm^2), which guarantees their dependability for cryogenic applications.

Conclusions

The addition of 5% Al to these binary Fe-(20-40)Mn alloys greatly enhances the stability of the austenitic phase by suppressing the $\gamma \rightarrow \varepsilon$ transformation. The quantity of aluminum must, however, be restricted to 5%, especially in the case of low-manganese alloys in order to avoid the formation of ferrite. Finally, the most surprising result is the simultaneous solid-solution hardening effect of aluminum when added to the binary Fe-Mn alloys and the increase in both toughness and elongation.

Fe-23Mn-5Al-(0-1.0)C ALLOYS

Structures

Additions of increasing amounts of carbon to the Fe-23Mn-5Al base alloy had a powerful austenitic stabilizing action, as has already been shown.[7,8] The proportion of ferrite in the as-cast, cored structure decreased with increasing carbon additions. The as-cast Fe-24Mn-5Al alloy containing 0.9 C was also totally austenitic (Fig. 6). Even with such high carbon additions, the formation and/or precipitation of a carbide phase was not detected.

Mechanical Properties

Figure 7 shows the changes in the tensile mechanical properties as well as the toughness and hardness determined as a function of carbon content of the alloys Fe-23Mn-5Al-(0-0.9)C. Carbon additions had the expected hardening effect: the hardness, yield strength, and tensile strength all increased in a manner proportional to the carbon content. Elongation was only moderately influenced at room temperature. At liquid nitrogen temperature, however, a sharp decrease in ductility was observed when the carbon

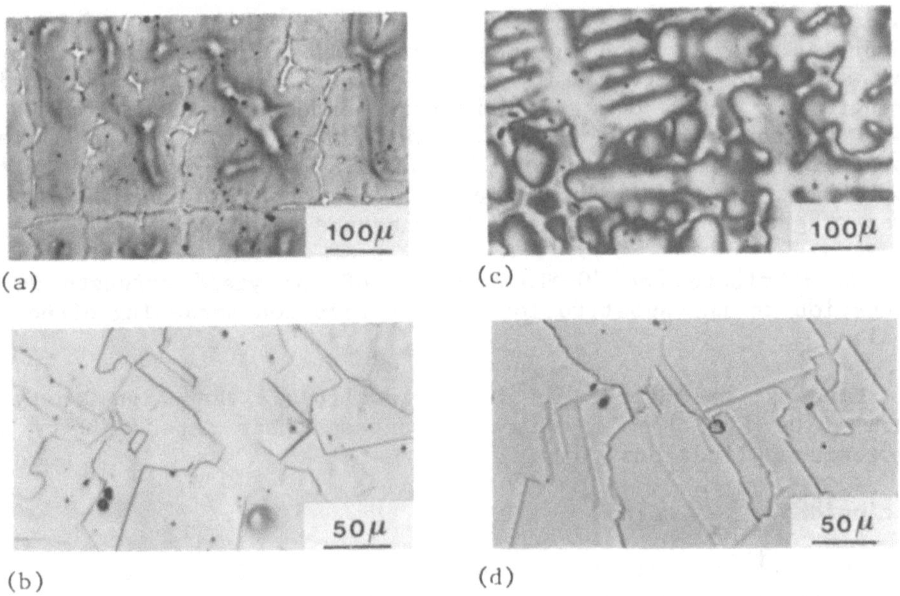

Fig. 6. Optical micrographs showing the interdendritic segregation in Fe-Mn-Al alloys, the austenitic stabilizing effect of carbon, and the absence of carbide phase precipitations, even with 0.9C in both as-cast and annealed alloys.
(a) Fe-23Mn-5Al-0.2C as-cast structure ($\alpha + \gamma$);
(b) Fe-23Mn-5Al-0.2C annealed structure (γ);
(c) Fe-24Mn-5Al-0.9C as-cast structure (γ);
(d) Fe-24Mn-5Al-0.9C annealed structure (γ).

content was greater than 0.5%. However, both the ductility and toughness remained at a safe level (A $\simeq$ 40% and KCV = 100 J/cm^2) (Fig. 8).

Conclusions

This large reserve of ductility represents an important guarantee for dependable service under cryogenic conditions and can also be exploited by a predeformation, as for example, the Fe-22 Mn-5Al-0.25C alloy leading to an important increase in the mechanical properties (yield strength, tensile strength).

Let us stress here that hardening, which is proportional to the carbon content, was obtained without a marked decrease in ductility. It can be explained by the total absence of a precipitated carbide phase. As a result, all the carbon added entered into interstitial solid-solution, in contrast to what was observed in the nickel-chromium austenites (AISI 304). We attribute this

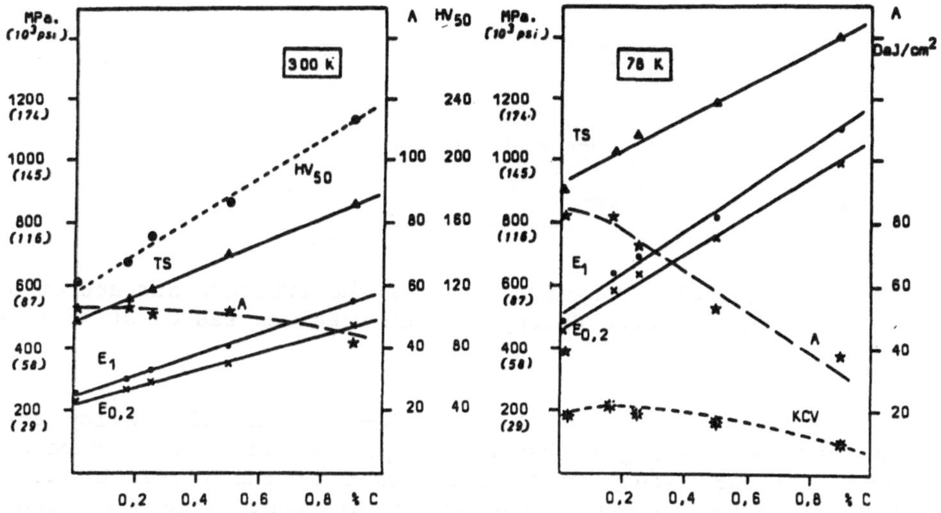

Fig. 7. Mechanical properties of the Fe-23Mn-5Al-(0-1)C alloys at 300 K and 78 K. TS = tensile strength; $E_{0.2}$ and E_1 = yield strength (offset method); A = total elongation; HV_{50} = Vickers hardness (50 kg); KCV = Charpy V-notch value.

remarkable behavior to the absence of any strong carbide-forming elements in these Fe-Mn-Al austenites.

CONCLUSIONS

We have shown that the stabilization of the austenitic phase in binary Fe-Mn alloys during plastic deformation in tension at 78 K requires manganese contents greater than 32%. Thus, the Fe-(32-36)Mn alloys, because of their large reserve of ductility at 78 K, are of interest for cryogenic applications. However, their yield strengths at room temperature (130 MPa) are quite low.

Additions of 5% Al to the binary Fe-(22-40)Mn alloys increase the stability of the austenite, since the Fe-23Mn-5Al alloy conserves a stable austenitic structure even after plastic deformation in tension at 78 K. The $\gamma \rightarrow \varepsilon$ transformation, which causes low toughness in the binary Fe-(20-30)Mn alloys, is suppressed.

Finally, since this Fe-23Mn-5Al alloy is characterized by a large reserve of ductility, especially at 78 K, we studied the variation of its mechanical properties with increasing carbon additions. This study illustrated the powerful austenitic stabilizing effect of this element as well as its solid-solution hardening effect. Hardening is proportional to the content added without an important decrease in ductility.

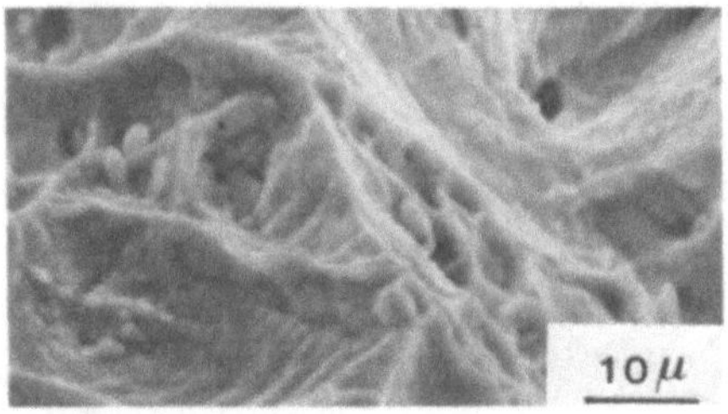

Fig. 8. Scanning electron micrograph of the fracture surface of a Fė-23Mn-5Al-0.2C Charpy V-notch specimen tested at 78 K.

We therefore emphasize that the Fe-24Mn-5Al-0.25C alloy possesses very interesting mechanical properties at room temperature and 78 K. Future study will be welcomed in the fields of weldability and corrosion resistance, since this austenitic alloy shows considerable promise for cryogenic applications, such as liquid gas pipelines.

ACKNOWLEDGMENTS

This research was made possible by the financial support of the Manganese Centre, Paris. The authors wish to thank Mr. P. L. Dancoisne, Director of the Manganese Centre, for his continuous interest in this study.

REFERENCES

1. H. Schumann, Arch. Eisenhuettenw. 38(8):647 (1967).
2. H. Schumann, Arch. Eisenhuettenw. 38(9):743 (1967).
3. H. Schumann, Neue Hütte Jg. 14(9):542 (1969).
4. A. Holden, J. D. Bolton, and E. R. Petty, J. Iron Steel Inst. Lond. 209:721 (1971).
5. D. J. Schmatz, Trans. Metall. Soc. AIME 15:112 (1959).
6. D. J. Schmatz, Trans. Am. Soc. Met. 52:898 (1960).
7. J. Charles and A. Berghezan, Cryogenics 21:278 (1981).
8. J. Charles, A. Berghezan, A. Lutts, and P. L. Dancoisne, Met. Prog. 120:71 (1981).

PROPERTIES OF LOW-CARBON 25Mn-5Cr-1Ni AUSTENITIC STEEL FOR CRYOGENIC USE

H. Yoshimura

Hikari Works, Nippon Steel Corporation, Tokyo, Japan

H. Masumoto

Yawata Works, Nippon Steel Corporation, Tokyo, Japan

T. Inoue

Fundamental Research Laboratory, Nippon Steel Corporation, Tokyo, Japan

FUNDAMENTAL INVESTIGATIONS OF OPTIMUM ALLOY SYSTEM

The Object of the Development

Owing to remarkable advances in cryogenic techniques, industrial applications of superconducting magnets are now under development in such fields as nuclear fusion reactor research, MHD electric power generation, and the magnetic levitation railway. The properties required for the structural materials of such equipment are: (1) high strength and toughness at the temperature of liquid He, (2) nonmagnetism, (3) little thermal conductivity and a small expansion coefficient, (4) microstructural stability during cold working and temperature changes, and (5) good weldability and machinability. One steel that closely meets those requirements is 300-series austenitic stainless steel. The features of austentitic steel are derived from its face-centered cubic lattice structure, which is substantially nonmagnetic and ductile. But this stainless steel is not necessarily the most suitable steel for cryogenic use because of its high cost, low strength, microstructural instability during cold working and temperature drops, poor machinability, and the scarcity of Ni resources. This may also be supported by the fact that this particular steel was originally developed for corrosion-resistant or heat-resistant use. On the other hand, medium-C, high-Mn austenitic steel is well known as a nonmagnetic steel, but it has an instable microstructure at low temperature.

Considering the above-mentioned situation, the authors tried to develop a new nonmagnetic Mn-Cr austenitic steel for cryogenic use. In this paper we describe the details of the development of this new steel and its properties.

Investigation of Fundamental Alloy System

At first, the combinations of alloying elements that stabilize austenite at the temperature of liquid Ni or during cold working were investigated on steels experimentally melted in a vacuum. The Ca content was kept between 0.01 and 0.15% to improve notch toughness and reduce the thermal expansion coefficient. The Mn content was varied from 5 to 35%, and in addition, Cr up to 15% and Ni up to 1% were alloyed moderately to stabilize the austenite. Ingots weighing 7-100 kg were heated at 1250°C and rolled into plates 13-mm thick, then solution treated by heating at 1050°C for 1 h followed by water quenching. Afterwards the microstructure, magnetic permeability (μ), average thermal expansion coefficient (α) between 77 K and room temperature (RT), and mechanical properties were investigated.

Figure 1 illustrates the structural stability, plotted on a Mn-Cr diagram, against cold working up to 25% or temperature down to 77 K. At RT, austenite remains stable in the region higher in Mn and/or Cr than the borderline that connects the points of 10Mn-10Cr and 20Mn-0Cr. In the region to the left of this boundary, austenite transforms partially into α'-martensite and exhibits ferromagnetism. This boundary is shifted towards the high Mn side by cold working or by lower temperatures. Cold working of 25% or cooling down to 77 K changes the boundary into the line between about 20Mn-10Cr and 25Mn-0Cr. The Mn acts more effectively than Cr in stabilizing the austenite. As long as the microstructure was completely austenite, μ did not exceed 1.002. With an increase in Mn content, α decreased and reached a minimum of 7×10^{-6}/K at 25% Mn. This value is about half that of stainless steel.

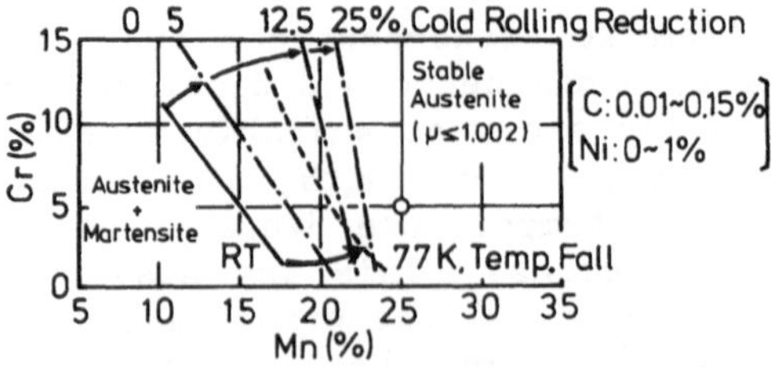

Fig. 1. Structural stability of low-C Mn-Cr steels.

In the region of stable austenite, 0.2% offset proof stress (PS) was about 200 MPa, tensile strength (TS) ranged from 500 to 600 MPa at RT, and the absorbed energy of a 2-mm Charpy V-notched impact test specimen (vE) was about 100 J at 77 K. The addition of 1% Ni to 25Mn-5Cr steel raised the vE to 150 J and increased the stability of the austenite. On the contrary, with the formation of α'-martensite in another region, strength was raised and vE was lowered progressively. On the basis of the microstructural stability, μ, α, and notch toughness, an alloy system of 0.15C-25Mn-5Cr-1Ni was selected as the base composition.[1]

Increase in Strength by Microalloying

The steel of the above-mentioned base composition possessed a PS of about 200 MPa at RT, which seemed to be rather low for structural steel. Therefore, as the second step, an attempt was made to increase the strength by microalloying, or by the addition of small quantities of alloying elements. Alloying elements Nb, V, and Ti up to 0.1% were added separately to the steel of the base composition with N up to 0.2%. Test procedures were the same as those in the former section.

For increasing the PS of the solution-treated steel, the addition of Nb with N was most effective. With this increase in Nb and N content, the PS increased remarkably by solution hardening of N and by grain refinement due to precipitated Nb nitrides, but the effects of Nb and N reached saturation at 0.1% and 0.15%, respectively. The V could increase the PS, but was less effective for TS than Nb. The effect of Ti fell between those of Nb and V.

The relations between the austenite grain diameter, d, and PS or vE are shown in Fig. 2. Obviously, PS increases with the refining of grain size, and the following Petch's equation holds true:

$$PS(MPa) = 73.5 + 27.0\ d(mm)^{-1/2}$$

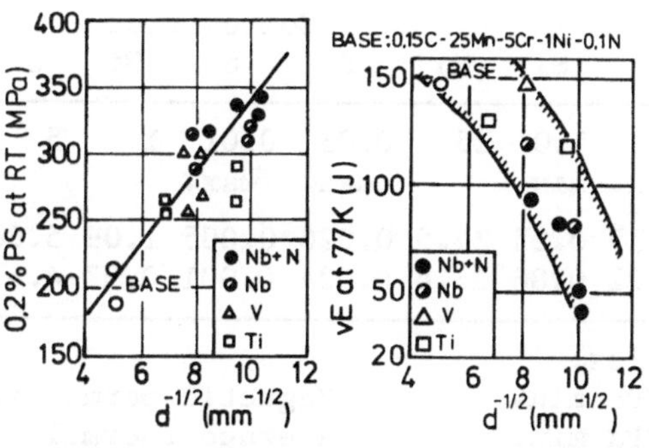

Fig. 2. Effects of grain diameter on proof stress and toughness.

The value of TS was also found by adding 392 MPa to this equation. Unlike that of ferritic steel, the vE decreases with the refining of grain size; however, comparison of PS with vE in Fig. 2 indicates that microalloying will make the vE above 50 J at 77 K compatible with the PS above 300 MPa at RT. Moreover, experimental steel of base composition with 0.05% Nb exhibited a vE of 123 J at 27 K.[2] Consequently it was decided to add 0.1% Nb and 0.1% N to the base composition.[1,3] This modification caused no significant change in α and μ. Welding materials for this steel were also investigated at the same time.

The decrease in vE with grain refinement is explained as follows: In austenitic steel, which always fractures in a ductile manner, the grain boundary acts merely as a lattice defect and operates to increase dislocation density. Therefore, for finer grain sizes, more work hardening is required and less energy is absorbed in the impact test before the specimen fractures.[1]

PROPERTIES OF THE PRODUCTS MANUFACTURED ON A COMMERCIAL SCALE

Manufacturing Procedures and Testing Methods

On the basis of the experimental results mentioned above, the optimum chemical composition and the objective properties of the newly developed low-C high-Mn austenitic steel (25Mn steel) were established. Subsequently, several types of products were produced of this steel in two heats during a trial commercial-scale operation. No particular problem arose in the process. The optimum composition and ladle analysis of two trial heats and objective properties are listed in Table 1. Steel M1 is lower in Ca and higher in Ni content than steel M2. Material properties were

Table 1. Optimum Composition and Ladle Analysis of Two Trial Steels (%) and the Objective Properties

	Composition									
	C	Si	Mn	P	S	Ni	Cr	Nb	Al	N
Optimum	0.3 max.	1.0 max.	25	0.03 max.	0.01 max.	1	5	0.01	0.03	0.01
Steel M1	0.15	0.25	24.5	0.026	0.005	1.09	5.08	0.05	0.01	0.087
Steel M2	0.22	0.86	25.6	0.024	0.001	0.97	4.67	0.04	0.01	0.040

Objective properties:
PS at RT: 294 MPa min.
TS at RT: 686 MPa min.
vE at 77 K: 49 J min.
Magnetic permeability: 1.02 max.
Average thermal expansion coefficient: 7×10^{-6}/K

investigated mainly on as-rolled or solution-treated heavy plates 13- to 80-mm thick at RT, 77 K, and 4 K or so. Test items were similar to those for steel plate for welded structures.

Strength and Toughness at Room Temperature and 77 K

Tensile test data and notch toughness at RT and 77 K are listed in Table 2. The as-rolled steel exhibits a PS above 350 MPa and a TS above 700 MPa at RT and vE above 50 J at 77 K. These values well exceed the target values. These propertiess of as-rolled steel can be varied by control of the finishing rolling temperature. Solution treatment lowered the PS to about 300 MPa and improved the vE in all products. In Fig. 3 are shown the results of the COD test (BS 5762). The critical COD of 25Mn steels is excellent. Even the values of as-rolled plates are about 3 times

Table 2. Tensile Test Data and Toughness of Two Trial Steels

Steel	Heat* treat.	Temp. (K)	t† (mm)	PS (MPa)	TS (MPa)	El‡ (%)	vE(J)§ L	vE(J)§ T
M1	R	RT	13	475	767	58	148	111
			40	388	737	56	165	177
	R	77	13	851	1439	35	75	52
			40	775	1277	20	81	74
	ST	RT	13	352	693	69	187	139
			40	311	712	61	209	186
	ST	77	13	648	1266	33	97	78
			40	---	---	--	104	103
M2	R	RT	27	326	716	62	187	137
			80	423	790	51	---	---
	R	77	27	626	1233	34	72	64
			80	---	---	--	131	135
	ST	RT	27	315	705	67	202	154
			40	276	720	64	240	216
			80	287	709	66	229	224
	ST	77	27	643	1262	35	133	132
			40	565	1278	46	159	137
			80	---	---	--	147	141

* Heat treatment, R: as-rolled, ST: solution treated
† Plate thickness
‡ Elongation
§ L: longitudinal, T: transverse

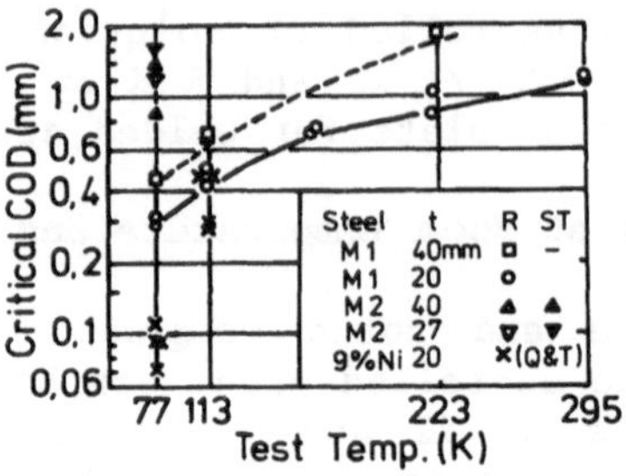

Fig. 3. Critical COD of 25Mn steels and 9% Ni steel.

better than those of 9% Ni steel specimen of the same thickness at 77 K. The NRL drop-weight test at 77 K showed no crack and the NDT temperature lies far below 77 K. In the duplex ESSO test at 77 K and a stress of 243 or 373 MPa, brittle fracture was arrested at the weld metal between the 25Mn steel plate and the crack initiating plate.

Figure 4 exhibits strength and toughness after heating at 550-1050°C. Heating at 550-850°C has little effect on strength, but vE at 77 K deteriorates a fair amount. This reduction in absorbed energy is caused by embrittlement of grain boundaries owing to the precipitation of P and Cr carbides in grain boundaries. The nose of this precipitation lies at about 600°C. The carbides precipitated at this temperature range are not ferromagnetic and μ is hardly affected. Conversely, by heating at 950°C and above, ductility is improved due to the solution of precipitates into matrix, and PS decreases to about 300 MPa due to grain coarsening. This is simply solution treatment. Notch toughness after welding thermal cycles is illustrated in Fig. 5. At 77 K, vE exceeds 100 J in the as-welded state, but is lowered to one half by annealing at 600°C. Consequently, annealing or heating at about 600°C, or slow cooling in this temperature range should be avoided for 25Mn steel.

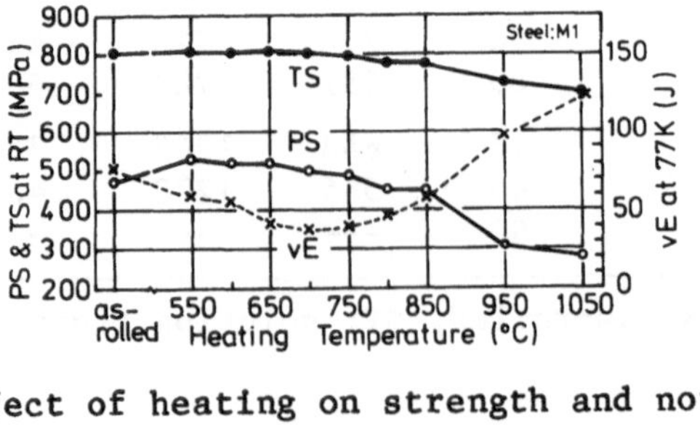

Fig. 4. Effect of heating on strength and notch toughness.

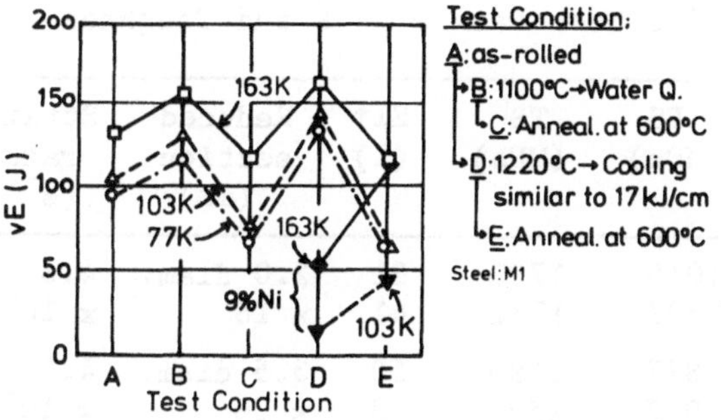

Fig. 5. Change of toughness by welding thermal cycles.

Welding material of a chemical compound similar to that of the base metal with several percent of Mo was also developed. Submerged-arc welded joints using this material at a heat input of 40 kJ/cm exhibited a TS of 677 MPa and an elongation of 37% at RT; the vE at 77 K exceeds 60 J throughout this weld joint in the as-welded state.

All the test results stated proved that this 25Mn steel is fully ductile at 77 K and superior to 9% Ni steel in toughness even in the as-rolled or as-welded state. Therefore, 25Mn steel will be able to substitute for 9% Ni steel as a cryogenic steel, if the design stress or codes are appropriately modified for this steel.

Strength and Toughness at Cryogenic Temperatures

Tensile data and notch toughness of solution-treated steel at cryogenic temperatures are shown in Table 3. The change in these properties against temperature is shown in Fig. 6, including data of AISI 304 for drawing comparison. Both PS and TS increase remarkably when the temperature falls. At 77 K, PS and TS of 25Mn steel exceed 550 and 1250 MPa and at 4 K exceed 850 and 1450 MPa, respectively. At 4 K, TS is nearly equal to, and PS is slightly lower than, those values of nitrogen-strengthened stainless steels.[4] The higher strength of steel M1 than that of steel M2 is attributed to the higher Ni content of M1. This indicates that the increase in N content is effective in raising strength at cryogenic temperatures also in high-Mn steels. The TS of 304 at 4 K is on a level with that of 25Mn steel, but the PS is about one half of that of 25Mn steel. Toughness decreases with the fall in temperature, but the vE of 25Mn steel at about 25 K still remains above 100 J. Cold working of 10% or subsequent aging at 250°C caused little decrease in vE (Table 4). All Charpy impact specimens of 25Mn steel fractured in a fully ductile manner.

Table 3. Tensile Test Data at 4 K and Toughness at about 20 K.

Steel	t* (mm)	PS (MPa)	TS (MPa)	El† (%)	Reduced section (mm)	Strain rate (s^{-1})	vE (J)	Test temp. (K)
M1	13	1015	1798	31	2.0 diam.	4.6	123	27
304‡	60	527	1782	41	x 18	x 10^{-4}	196	17
M2	40	877	1495	50	3.5 diam.	4.2		
	80	907	1554	48	x 20	x 10^{-4}		

* Plate thickness
† Elongation
‡ 0.06C-0.9Si-1.01Mn-0.030P-0.006S-9.1Ni-18.4Cr-0.019N

The strength and ductility of weld joints and weld metal composition are listed in Table 4. The PS of weld metal is slightly greater than, the TS is less than, and the vE values of the weld metal and HAZ are equivalent to those of the base metal.

The load-displacement curves of the tensile test at 4 K are shown in Fig. 7. The load of 25Mn steel increased proportionally to the displacement until yielding and after that increased rather slowly with serrations. The 304 steel yielded at comparatively low load, and subsequently, at increased load with serrations till fracture at nearly the same stress as that of 25Mn steel. The higher yield stress of 25Mn steel is due to solution hardening by N and C, fine grain size, and low stacking-fault energy. The serration after yielding may be caused mainly by formation of ε-martensite and partially by twinning and detwinning due to adiabatic deformation.[5] The 304 steel is different from the 25Mn steel in

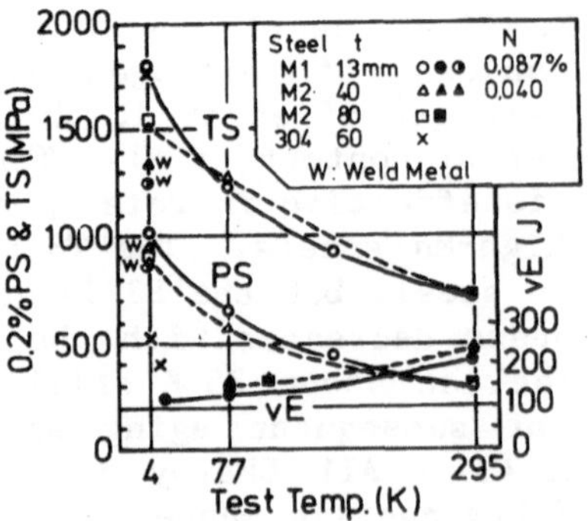

Fig. 6. Change of strength and toughness with temperature.

Table 4. Tensile Test Data at 4 K and Toughness at about 23 K of Weld Joints and Weld Metal Composition

Weld Metal	PS (MPa)	TS (MPa)	El (%)	vE of M1 (J, subsize specimen, 5-mm thick)			
M1	870	1258	31	Weld metal	44, 31	Base metal	50
M2	961	1338	34	Fusion line	45	-10% strained	41
				HAZ	50, 26	- aged at 250°C	42

Composition of Weld Metal (%)

W.M.	C	Si	Mn	P	S	Ni	Cr	Mo	N	Heat Input ($kJ \cdot cm^{-1}$)
M1	0.14	0.18	25.9	0.002	0.014	2.00	4.9	2.9	0.004	20
M2	0.18	0.57	24.4	0.006	0.002	2.91	6.3	1.4	0.024	11-16

the large amount of work hardening that results from the transformation of austenite into α'-martensite by cold working.

Other Mechanical and Physical Properties

Machinability. Machinability was decided by the tool life to complete failure or tool wear in drilling or turning. When a high-speed tool steel was used, the machinability of the 25Mn steel was slightly inferior to that of 304 steel, but the machinability of the 25Mn steel was improved greatly by use of a carbide tool. For this 25Mn steel, carbide tools are recommended.

Fatigue property. Rotating bending fatigue strength at RT was about 40% of TS (e.g., σ_w = 314 MPa to TS = 737 MPa). This strength ratio is nearly equal to that of other high-Mn steels or austenitic stainless steels. The fatigue crack propagation rate at RT is well expressed by Paris's equation: for instance, for steel M1 as-rolled

$$da/dn(mm/cycle) = 3.50 \times 10^{-9} \Delta K(MPa)^{3.08}$$

Thermal expansion coefficient. The average α between 77 K and RT was 7.22×10^{-6}/K. This value corresponds to the target and is about half of that of stainless steel.

Magnetic permeability and microstructural stability. Magnetic permeability was measured by the method of ASTM A 342-64. Ferromagnetic phase was detected by a "ferrite indicator" and by the relative change of magnetization in a magnetic field of 1 T. It was confirmed that if the α'-martensite detected by the ferrite

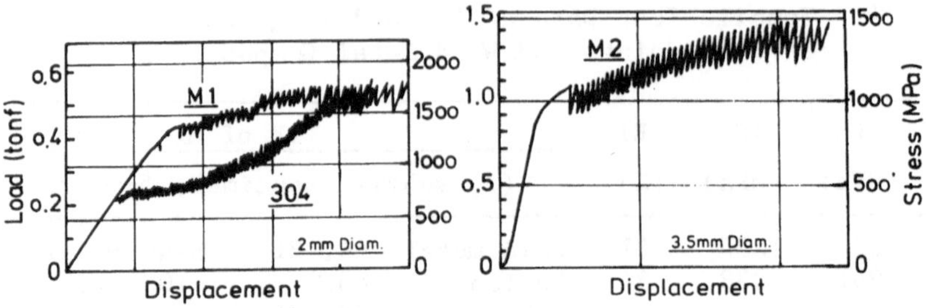

Fig. 7. Load-displacement curves at 4 K.

indicator was 0.2% and 1.5%, μ was increased to 1.27 and 1.42, respectively. Therefore this machine is useful as a very sensitive and convenient detector of ferromagnetic phase.[3] The 25Mn steel exhibited a μ always below 1.005 and no ferromagnetic phase after each treatment such as hot rolling, heat treatment, welding, and cold working. Even on the fracture surface of the test specimens after tensile or Charpy impact tests, no ferromagnetic phase was detected except for very small indications on the fracture surface at 4 K.

CONCLUSION

1. The authors developed a nonmagnetic, austenitic structural steel for cryogenic use. Investigations on low-C, high Mn-Cr steels yielded a low-C 25Mn-5Cr-1Ni-Nb-N steel, which has high strength and toughness at cryogenic temperatures.

2. The plates of this steel manufactured on a commercial scale were superior to 9% Ni steel in ductility at 77 K even without heat treatment, and at 4 K they were fully nonmagnetic and possessed a proof strength above 850 MPa, a tensile strength above 1450 MPa, and excellent toughness. Other properties were suitable for structural materials.

3. All data obtained indicate that this low-C, high-Mn steel can be used as a structural material for low temperature use or as a nonmagnetic material for cryogenic use.

ACKNOWLEDGMENTS

The mechanical tests at liquid He temperature were carried out at the Research Institute for Applied Mechanics of Kyushu University and the National Research Institute for Metals. The authors are grateful to these institutes for their assistance. The authors also wish to thank Prof. K. Kitajima of Kyushu University and Prof. I. Tamura of Kyoto University for their very helpful suggestions.

REFERENCES

1. H. Yoshimura, N. Yamada, H. Yada, H. Honma, and T. Ito, Microstructures, low temperature toughness and thermal expansion coefficients of high manganese-chromium austenitic steels, Trans. Iron Steel Inst. Jpn. 16:98 (1976).
2. H. Yoshimura, T. Shimizu, H. Yada, and K. Kitajima, Toughness of high manganese-chromium austenitic steel at liquid helium temperature, Trans. Iron Steel Inst. Jpn. 20:187 (1980).
3. H. Yoshimura, T. Shimizu, and N. Yamada, Microstructure, magnetic permeability and electric resistivity of high manganese-chromium-nickel steel, Tetsu-to-Hagané 65:1434 (1979), in Japanese.
4. D. T. Read and R. P. Reed, Toughness, fatigue crack growth, and tensile properties of three nitrogen-strengthened stainless steels at cryogenic temperatures, in: "Metal Science of Stainless Steels," AIME, New York (1979), pp. 92-121.
5. H. Yoshimura, T. Shimizu, and K. Kitajima, Tensile and impact properties of 25Mn-5Cr-1Ni austenitic steel at liquid helium temperature, submitted to Trans. Iron Steel Jpn.

CRACK INITIATION AND ARREST CHARACTERISTICS OF 9%Ni STEELS WITH VARIOUS CHARPY V-NOTCH ENERGY VALUES

Y. Nakano, S. Suzuki, and A. Kamada

Research Laboratories, Kawasaki Steel Corporation, Chiba, Japan

K. Hirose

Chiba Works, Kawasaki, Steel Corporation, Chiba, Japan

INTRODUCTION

Nine-percent Ni steel has been used to construct LNG tanks. Since the brittle fracture of an LNG tank results in catastrophic damage, high fracture toughness is required of the 9%Ni steel.

As a means of quality control, 2-mm V-notch Charpy test values, such as the impact energy and the lateral expansion, are used. To assure the eligibility of 9%Ni steel against brittle fracture, the Charpy test parameters must be related to fracture mechanics parameters, which include crack opening displacement (COD) and J_{Ic} for crack initiation and dJ/da for ductile crack propagation and crack arrest characteristics.

The present paper describes the relation between the Charpy energy value and the crack initiation, propagation, and arrest characteristics of 9%Ni steels.

EXPERIMENTAL PROCEDURES

Materials

The materials tested were quenched and tempered (QT), double normalized and tempered (NNT), and as-rolled 9%Ni steel plates. All steels except one were 30-mm thick. The heat treatment of the steels was performed in accordance with ASTM Standard A553 and

Table 1. Chemical Composition (wt.%) of 9%Ni Steels

Steel	C	Si	Mn	P	S	Ni
A	0.051	0.26	0.63	0.0029	0.0008	9.10
B	0.043	0.27	0.60	0.0020	0.0013	9.14
C	0.064	0.25	0.58	0.0020	0.0023	9.06
D	0.052	0.25	0.62	0.0050	0.0050	9.13
E	0.061	0.25	0.61	0.0032	0.0013	9.17
F	0.051	0.21	0.59	0.0040	0.0010	9.16

Table 2. Mechanical Properties of 9%Ni Steels

Steel	Heat Treatment	Direction	Tensile Test at RT			Charpy Test at 77 K		
			YP (MPa)	TS (MPa)	El. (%)	vE (J)	SA (%)	LE (mm)
A	QT	T - L	659	709	23.6	250	100	2.27
B	QT	T - L	662	708	21.0	224	100	2.17
C	QT	T - L	675	744	24.0	200	100	1.99
D	QT	T - L	678	744	22.0	130	98	1.53
D	QT	T - L	656	710	23.0	124	100	1.67
D	NNT	L - T	586	741	26.8	118	60	1.48
D	NNT	T - L	587	742	25.6	52	43	0.79
D	As rolled	T - L	717	833	13.2	5	0	0.21
E	QT	T - L	690	733	21.7	78	77	0.97
F	NNT	T - L	563	730	24.2	96	55	1.22

Table 3. Chemical Composition (wt.%) of Weld Metal Produced by TIG Welding

C	Si	Mn	P	S	Ni	Cr	Mo	W
0.04	0.24	0.26	0.003	0.002	52.1	12.72	13.60	2.26

Table 4. Mechanical Properties of Weld Metal Produced by TIG Welding

Tensile Test at RT			Charpy Test at 77 K		
YP (MPa)	TS (MPa)	El. (%)	vE (J)	SA (%)	LE (mm)
441	770	35	71	100	1.28

A353. The Charpy impact energy of the steels was controlled by the sulphur content, testing direction, and heat treatment.

The chemical composition and mechanical properties of the steels tested are listed in Tables 1 and 2, respectively.

Some of the steels were welded by TIG welding at a heat input of about 40 kJ/cm. The chemical composition and the mechanical properties of a typical weld metal are listed in Tables 3 and 4, respectively.

Fracture Toughness Test

The three-point bend test and center-notched wide plate tension test were conducted to study the crack initiation and ductile crack propagation characteristics; the hybrid ESSO test was performed for the determination of crack arrest characteristics. All tests were conducted at 103 K.

The specimen geometries are shown in Fig. 1. The three-point bend test was carried out on the steel plates and a weld metal, whereas the center-notched wide plate tension test was performed on the specimens that had through-thickness notches along the fusion lines (WM/HAZ = 50/50) of the weld joints. Each hybrid ESSO specimen consisted of a starter section and a test section that was welded to the starter section with the welding consumable for 785-MPa high-tensile-strength steel plates. The crack starter and the welding consumable used had such low toughness that they were expected to absorb little energy during the crack propagation. For a hybrid ESSO test of a weld joint, the specimen configuration was designed so that a brittle crack propagates to meet the fusion line of the weld joint.

EXPERIMENTAL RESULTS AND DISCUSSION

Three-Point Bend Test

Figure 2 shows the plot of COD at 103 K vs. the Charpy impact energy at 77 K, vE(77K). The number attached to each datum point indicates the sulphur content in the unit of ppm, and L-T and T-L indicate the testing directions. In an L-T specimen, for example, the testing direction that was perpendicular to the notch was parallel to the rolling direction and the direction parallel to the notch was in the transverse direction, as defined by ASTM Standard E399. The COD values, δ_c and δ_{Pmax}, were calculated according to British Standard BS 5762 from the clip gage displacements, V_c and V_{Pmax}, corresponding to the brittle fracture and the maximum load if no brittle fracture took place, respectively.

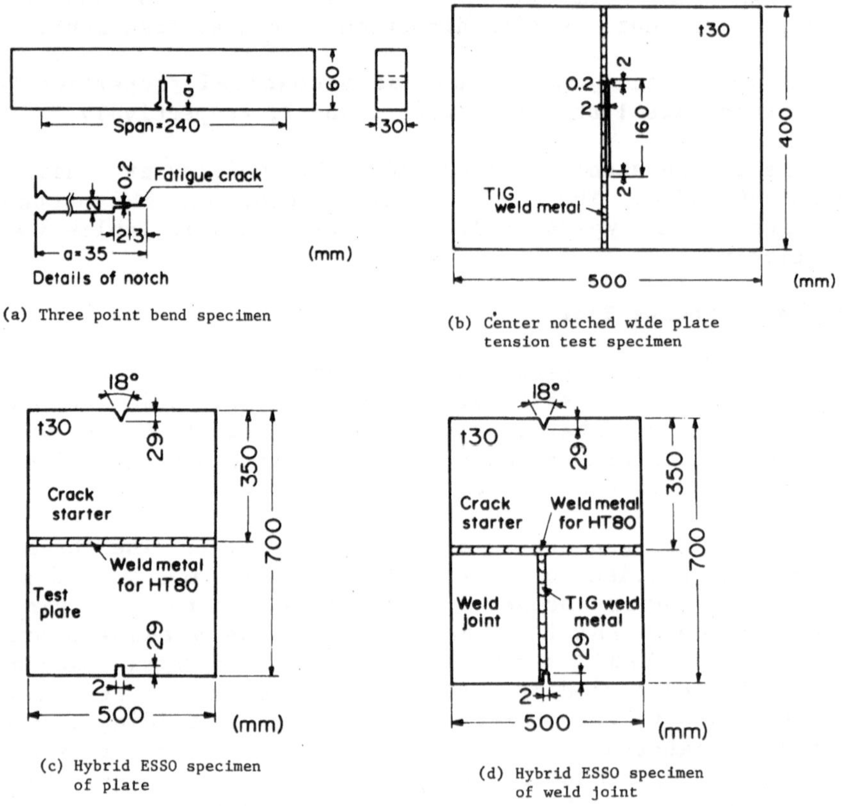

Fig. 1. Specimen geometries.

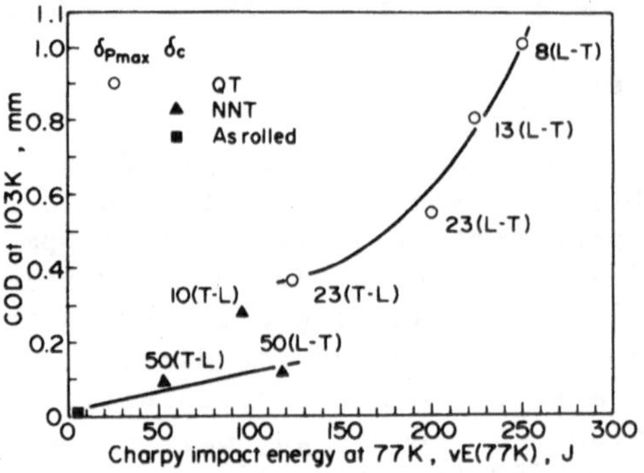

Fig. 2. Plot of COD at 103 K vs. Charpy impact energy at 77 K.

The QT steels whose vE(77K) was greater than 124 J did not exhibit brittle fracture. The COD at the maximum load, δ_{Pmax}, increased with vE(77K). The reduction of sulphur content increased δ_{Pmax}. When the sulphur content was the same, the testing direction affected COD.

The NNT and as-rolled steels exhibited brittle fracture and gave smaller COD values than the QT steels, though the COD of the NNT steel with the sulphur content of 10 ppm was much larger than those of 50 ppm. This fact is attributed to the microstructures. The microstructure of the NNT steels mainly consisted of upper bainite, whereas that of the QT steels consisted of lower bainite and tempered martensite. Such a difference in microstructures seems to contribute to brittle fracture significantly, but not much to ductile fracture.

The Charpy test specimen surface of the NNT steels exhibited both ductile and brittle fracture, whereas that of the QT steels exhibited ductile fracture alone. The largest value of the impact energy is attained by ductile fracture. The COD, on the other hand, was determined at the brittle crack initiation or the maximum load, whichever came first. If a brittle crack initiates, therefore, the COD value becomes small, as observed in the NNT steels. These facts explain the result that the QT and NNT steels, which had almost same impact energy, showed a great difference in COD.

Figure 3 shows the plot of J vs. crack extension for the QT steels and the weld metal produced by an ordinary TIG welding. The J-integral was calculated by the following equation:

$$J = 2U / B (W - a) \qquad (1)$$

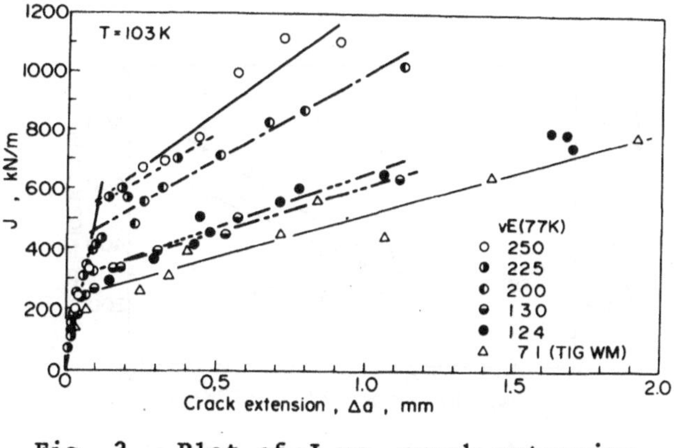

Fig. 3. Plot of J vs. crack extension.

where U, B, W, and a are the energy absorbed by a specimen at a given value of displacement, the specimen thickness, the specimen width, and the notch length, respectively.

A single blunting line can represent the experimentally determined relation between J (kN/m) and the crack extension (= stretched zone width), Δa (mm), for all the QT steels and the weld metal. It can be described by the following equation:

$$J = 5296\ \Delta a \tag{2}$$

The coefficient of Eq. (2) is much greater than $2\sigma_f$ ($\simeq$2000 MPa) proposed by ASTM E24 Committee, where σ_f is the flow stress as calculated by the average of the yield and tensile strengths.

The intersection point between the blunting line and each R-curve approximated by a straight line gives the ductile crack initiation resistance, J_{Ic}. The J_{Ic} value of the weld metal was smaller than those of the 9%Ni steels. The slope, dJ/da, of a straight-line approximation to the R curve gives the resistance to ductile crack extension.

Figure 4 shows the plots of J_{Ic} and dJ/da vs. vE(77K). The figure indicates that the increase in Charpy impact energy improves the resistance to ductile crack initiation and propagation. In the figure, J_{Ic} and dJ/da of the weld metal are shown by arrows. They correspond to those of a 9%Ni steel with vE(77K) of about 105 J.

Center-Notched Wide Plate Tension Test

The center-notched wide plate tension test of the weld joints of QT and NNT 9%Ni steels showed that cracks that initiated from the tip of a through-thickness notch located along the fusion line

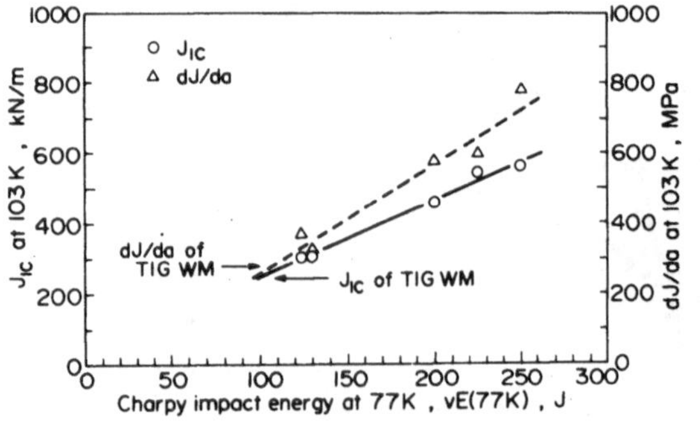

Fig. 4. Plots of J_{Ic} and dJ/da at 103 K vs. Charpy impact energy at 77 K.

of the weld joint penetrated into the weld metal and led the specimen to a final ductile fracture. This is due to the softness of the weld metal compared to that of the plate. The maximum net stress and COD at the maximum load of each specimen were greater than 830 MPa and 3.0 mm, respectively, and no pop-in was observed.

Hybrid ESSO Test

Table 5 lists the hybrid ESSO test results. As observed in the table, all the QT steels, including a 22.5-mm-thick steel whose vE(77K) was 78 J, arrested the brittle crack that had run 350 mm in the starter section at a stress as large as 490 MPa. The load to which each ESSO specimen was subjected appears to remain constant while the brittle crack propagates as far as a half of the specimen width.[1] The static-stress intensity factor at the tip of the arrested crack, therefore, is about 580 MPa$\sqrt{m}$ according to Japan Welding Engineering Society Standard, WES 3003. This value of arrest toughness is great enough to arrest a brittle crack having propagated about 4 m at the design stress of 164 MPa. The NNT steels, on the other hand, did not arrest the brittle crack even when the applied stress was as small as 196 MPa. It is

Table 5. Hybrid ESSO Test Results

Steel	Thickness (mm)	Heat Treatment	vE(77K) (J)	Stress (MPa)	Go/No Go
A	30	QT	250	294 392 490	No Go No Go No Go
B	30	QT	224	196 294 392	No Go No Go No Go
C	30	QT	200	196 294 392	No Go No Go No Go
D	30	QT	124	228 294 392 490	No Go No Go No Go No Go
D	30	NNT	118	196 294	Go Go
D	30	NNT	52	196 392	Go Go
E	22.5	QT	78	392 490	No Go No Go
F	30	NNT	96	196 392	Go Go

concluded, therefore, that the QT steels have superior brittle crack arrest characteristics compared with the NNT steels.

The hybrid ESSO test of TIG weld joints of QT [vE(77K) = 124 and 225 J] and NNT [vE(77K) = 52 J] steels showed that the brittle crack did not propagate in the heat affected zone. A brittle crack propagated in the NNT steel plate and a ductile crack propagated in the weld metal of the weld joints of the QT steels. These results and those of the center-notched, wide plate tension test suggest that the brittle crack does not propagate in the heat affected zone.

Figure 5 shows an example of the fracture appearance around the welded portion of a hybrid ESSO specimen of a QT steel. The test section exhibits a diamond-shaped, flat fracture surface

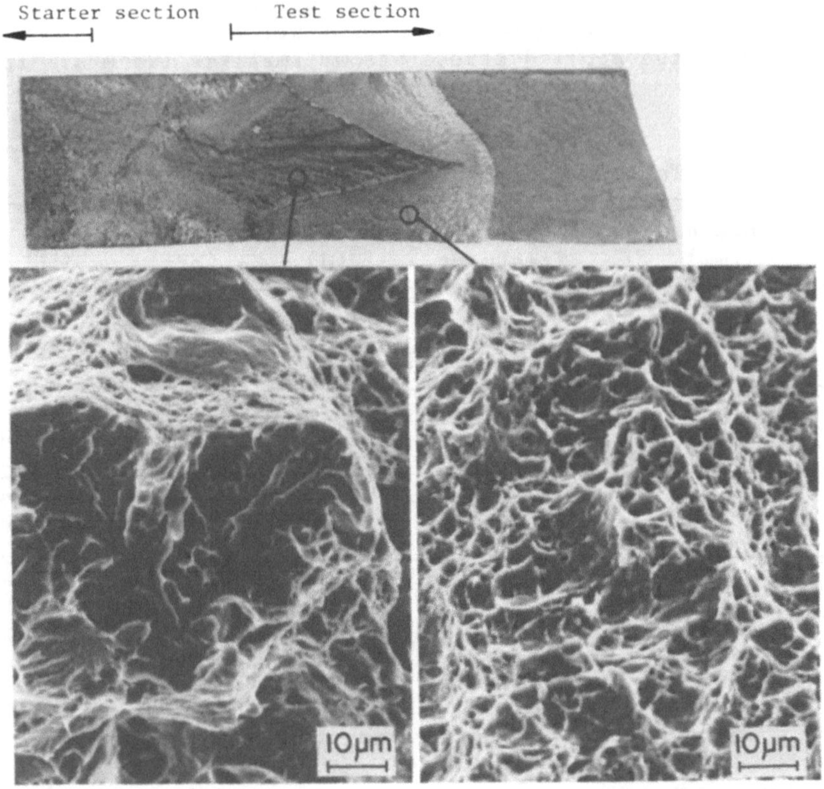

(a) Flat fracture surface (b) Shear fracture surface

Fig. 5. An example of the fracture appearance of a hybrid ESSO specimen of a QT 9%Ni steel.

whose scanning electron micrograph [Fig. 5(a)] showed cleavage facets surrounded by tiny dimples and a shear fracture surface constituted by dimples [Fig. 5(b)].

Figure 6 plots the flat and shear crack length vs. vE(77K) for the 30-mm-thick QT steels. The figure shows that the shear crack length decreased strongly as vE(77K) increased, whereas the flat crack length was not affected significantly by vE(77K). It is concluded, therefore, that the increase in the Charpy impact energy improves the resistance to the ductile crack propagation. This fact is in good agreement with the static test results.

CONCLUSIONS

The crack initiation, propagation, and arrest characteristics of 9%Ni steels at 103 K were studied with respect to the 2-mm V-notch Charpy impact energy at 77 K. The main results obtained are as follows:

1. The increase in the Charpy impact energy, which was mainly attained by reducing the sulphur content, improved the resistance to the ductile crack initiation and propagation for the QT steels.

2. The QT steels had larger COD values than the NNT and as-rolled ones even if the Charpy impact energy was the same. This fact was attributed to the difference in microstructures.

3. The resistance to ductile crack initiation, J_{Ic}, and propagation, dJ/da, of the weld metal produced by TIG welding at a heat input of about 40 kJ/cm corresponded to those of a QT steel with vE(77K) of about 105 J.

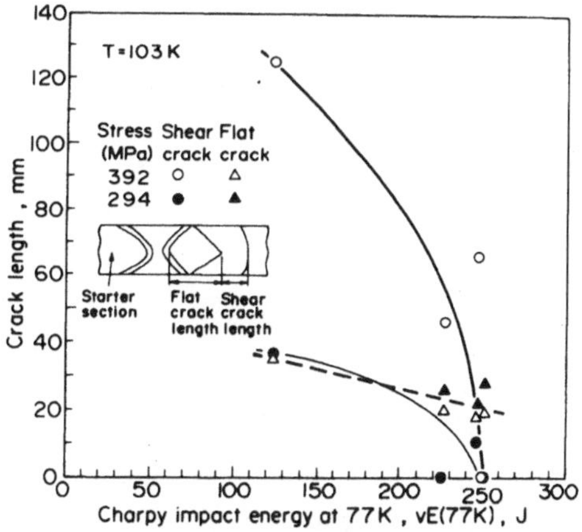

Fig. 6. Plots of flat and shear crack length vs. Charpy impact energy at 77 K for QT 9%Ni steels.

4. The center-notched wide plate tension test and the hybrid ESSO test of TIG weld joints suggested that the brittle crack did not propagate in the heat affected zone.

5. All QT steels tested, including a 22.5-mm-thick one whose vE(77K) was 78 J, arrested the brittle crack that had run 350 mm in the starter section at a stress as large as 490 MPa.

REFERENCE

1. Y. Nakano and M. Tanaka, Crack arrest toughness of structural steels evaluated by compact test, J. Iron Steel Inst. Jap. 67(7):979 (1981).

THE USE OF LOW-TEMPERATURE STRENGTHENING OF STEELS IN WELDED CRYOGENIC STRUCTURAL CODES

K. A. Yushchenko and S. A. Voronin

E. O. Paton Institute of Electrowelding, Ukr.SSR Academy of Sciences, Kiev, USSR

The ultimate strength and yield strength of the majority of steels and alloys at cryogenic temperatures increase with a decrease in temperature. This is characteristic mainly of the widely used austenitic, ferritic, and martensitic steels. The design of vessels for storage of cryogenic liquids is erroneously based on strength characteristics of metal at room temperature. Sometimes faulty design is the result of improper test methods, but more often it is the result of insufficient data on the capacity of real structures, their strength margin, and safety. During the last few years, investigations in various countries in the field of cryogenic materials have made the application of the low-temperature strength in design more reasonable. By increasing the design values of allowable stresses, fewer materials and cryogenic liquids are needed in service, and, therefore, expenses are reduced.

In structural design standards for strength, the allowable stresses [σ] are defined from the accepted values of the safety factor (n) and the strength levels of the material (conventional tensile yield strength, $\sigma_{0.2}$, $\sigma_{1.0}$, and ultimate tensile strength, σ_{UTS}):

$$[\sigma] = \frac{\sigma}{n} \qquad (1)$$

For statically loaded pressure vessels, the following dependences are used:

$$[\sigma] = \frac{\sigma_{1.0}^{293K}}{n_{YS}} \quad \text{or} \quad [\sigma] = \frac{\sigma_{UTS}^{293K}}{n_{UTS}} \qquad (2)$$

where $\sigma_{1.0}^{293K}$ is the yield strength at 293 K, determined at a relative residual deformation equal to 1%; σ_{UTS}^{293K} is the strength limit at 293 K; n_{YS} is the safety factor, equal to 1.5 and based on yield strength; n_{UTS} is the safety factor, equal to 2.4 and based on ultimate tensile strength.

The smaller of the two values calculated from equations (2) is put into the design calculations. When the temperature decreases, the strength increases. Therefore, the value of allowable stresses should also increase, taking into account that in most cryogenic applications loading of the material takes place at low temperatures. It is expedient to calculate the allowable stresses in equations (2) from the values of $\sigma_{1.0}$ and σ_{UTS} at operational temperatures. It is possible under the following conditions:

1. The brittle transition temperature at all loading conditions for a given steel must be lower than the operational temperature.

2. Failure must be tough and the fracture surface must be tough or slightly quasi-cleavaged. Also, during temperature changes, there must be no sharp decrease in strength properties and in fracture characteristics caused by phase transformations.

A weldment, owing to the presence of the weld, is characterized by inhomogeneity of its properties and is the weakest place of the structure. The strength properties of a cast metal similar in composition to the parent metal are lower. Availability of stress concentrators in the form of reinforcement or penetration, poor fusion, or lack of penetration may visibly change the properties of weld and weldment in general. Therefore, it is necessary to introduce a coefficient into the equations for weldments that takes into account the weldment strength decrease. In this case, the equations of allowable stresses (2) become:

$$[\sigma^T] = \frac{\sigma^T_{1.0}}{n_{YS}} \phi^T \quad \text{or} \quad [\sigma^T] = \frac{\sigma^T_{UTS}}{n_{UTS}} \phi^T \tag{3}$$

where T is the predetermined operating temperature of the structure, and ϕ^T is the equal strength coefficient of a weldment at low temperature, T, numerically equal to the relationship of the weldment ultimate strength to the base metal strength. The ϕ^T values are determined from test data.

Let us analyze the temperature and stress concentration influence on the strength of some steels and their weldments: 03Kh12N10MT belongs to the martensitic class; 12Kh18N10T,

Table 1. Chemical Composition of Steels

Steel Grade	Elements Content, wt.%									
	C	Si	Mn	Cr	Ni	N	S	P	Mo	Ti
03Kh12N10MT	0.025	0.1	0.1	12.0	9.6	0.03	0.007	0.007	0.65	0.15
03Kh13AG19	0.027	0.21	18.5	12.6	0.7	0.15	0.007	0.028	-	-
12Kh18N10T	0.12	0.75	1.3	18.0	10.5	-	0.011	0.010	-	0.4
03Kh20N16AG6	0.025	0.4	6.0	20	16	0.20	0.004	0.009	-	-

03Kh13AG19, and 03Kh20N16AG6 belong to the austenitic class (the first two are metastable).

Chemical composition of steels is given in Table 1. Welding of steels was submerged arc with a heat input equal to 7000 kcal/cm. The metal thickness was 12 and 20 mm. Specimens cut from the base metal and from the weldment were tested. Specimens were circularly notched and unnotched, flat with reinforcement and without reinforcement, and double-notched along the weld and fusion line. The stress concentration coefficient, K_t, for notched specimens was 3 to 12; for the double-sided weld reinforcement, 1.4 to 2.

As temperature decreases, steels show various sensitivities to stress concentration. Figures 1 through 3 are diagrams of changing steel strength in the presence of stress concentration. The stress concentration coefficient, K_t, is 6 to 12 depending on temperature. The stable austenitic steel 03Kh20N16AG6 displays an insignificant sensitivity to stress concentrations as temperature

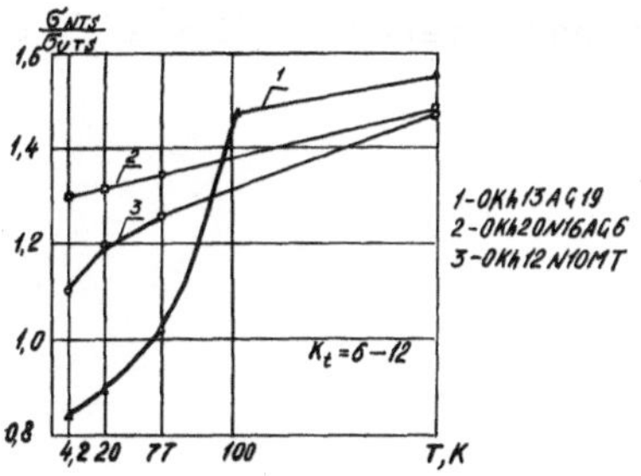

Fig. 1. Temperature sensitivity of steels to a notch (tensile measurements).

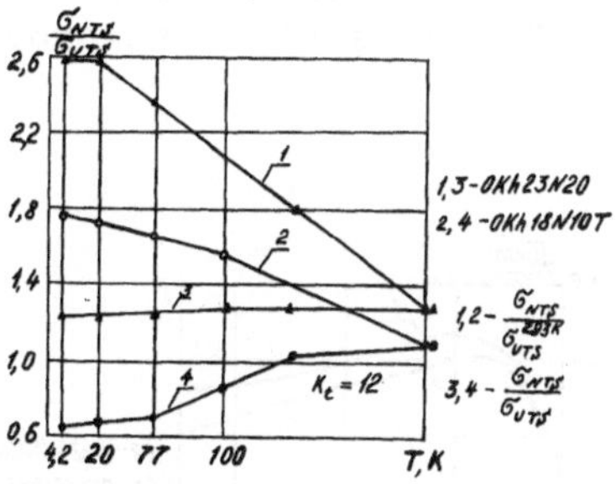

Fig. 2. Temperature sensitivity of steels to a notch (tensile measurements).

decreases. For the martensitic steel 03Kh22N10MT, $\sigma_{NTS}/\sigma_{UTS}$ decreases more sharply as temperature decreases (σ_{NTS} is notch tensile strength). There is no sharp strength decrease for specimens of these steels with concentrators when the stress concentration coefficient, K_t, is changed from 1.5 to 12 in the low-temperature range.

As temperature decreases, the metastable steels 12Kh18N10T and 03Kh13AG19 display a significant sensitivity to stress concentration that is caused by martensite formation during the deformation process.

Structures with deformation martensite display a significant sensitivity to stress concentrations at low temperatures. In particular, the presence of 80% cold deformation induced martensite in 12Kh18N10T steel leads to the net strength decrease to 650 MPa at $K_t \geq 6$. Martensite forms at relatively small amounts of deformation. In the stress concentration area, such deformation levels may be reached at much lower general deformations that precede specimen fracture.

Thus, we may suppose that in metastable steels the fracture process starts in the area of the stress concentration by the intensive transformation at a relatively small amount of deformation. The amount of transformation in the area of the concentrator is connected with the stress concentration coefficient, the level of applied stresses, and the chemical composition of the metal. The fracture stress for metastable steels at low temperatures depends upon the stress concentration coefficient value. Such dependence of the fracture stress on K_t is shown in Fig. 2. For circular specimens of 12Kh18N10T steel, a sharp strength decrease is observed when K_t changes from 1.4 to 12. For flat specimens, a decrease of $\sigma_{NTS}/\sigma_{UTS}$ takes place at smaller values of K_t. The test data show that at any stress concentrator, the strength decreases to the yield strength across the net section. The influence of stress concentration is the same on weldments made with wires of composition analogous to the base metal. Figure 3 shows

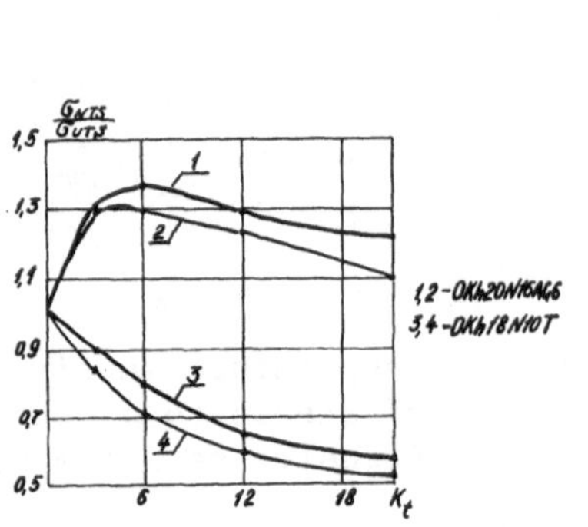

Fig. 3. Sensitivity of steels (1,3) and weld (2,4) to stress concentration (tensile measurements).

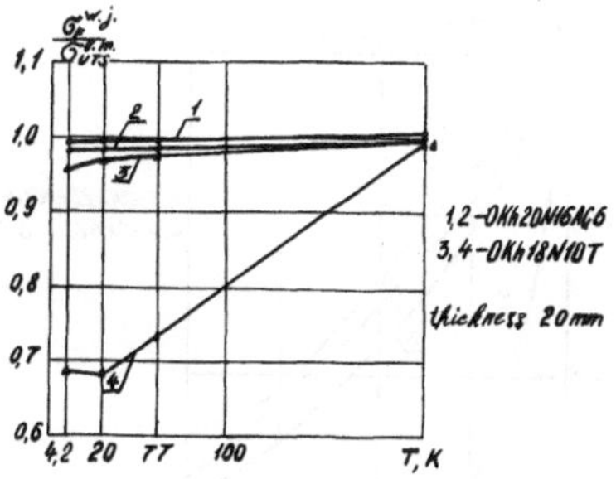

Fig. 4. The temperature influence on equal tensile strength coefficients of weldment with reinforcement (2,4) and without it (1,3).

the influence of temperature on the weld strength at values of K_t of 6 to 12. The influence of weld reinforcement on weldment strength with K_t = 1.8 is shown in Fig. 4.

A similar picture is observed while testing weldments where the stress concentrator in the form of poor penetration is placed in the weld metal (Figs. 5 and 6). Figure 7 shows the influence of the defective area on the relationship of the ultimate strengths of specimens with and without defects ($K_\alpha = \sigma_F^{w.def.} / \sigma_{UTS}^{w.}$). Here, the amount of strength decrease evidently depends on the stability of the weld metal structure. In the welds with stable austenite, the dependence of large stresses on the relatively defective area is close to linear, whereas the unstable systems undergo a sharp strength decrease at small defect sizes (up to 5%) and then change proportionally to the decrease of the working area of specimen (Fig. 8).

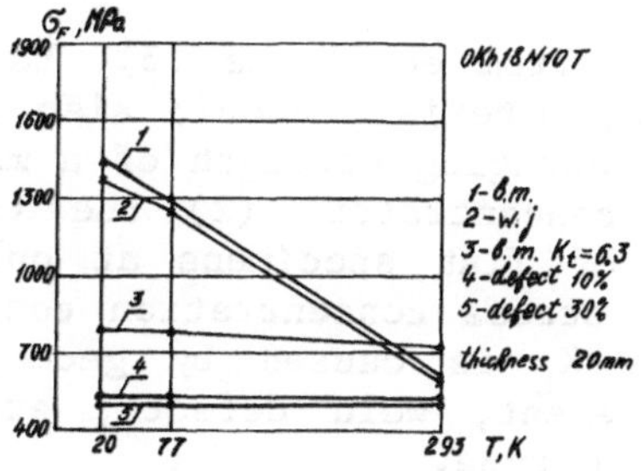

Fig. 5. The steel and weldment strength.

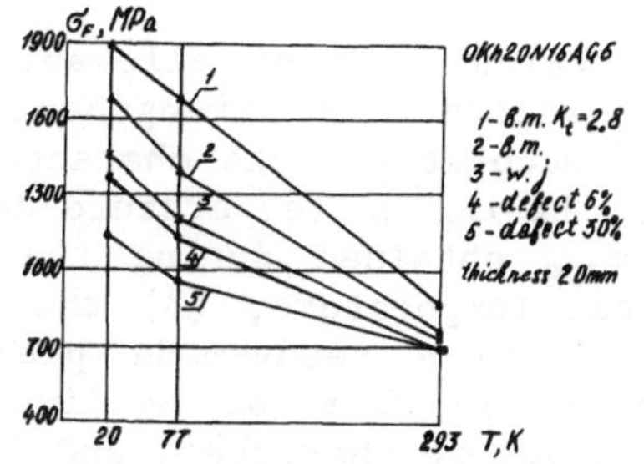

Fig. 6. The steel and weldment strength.

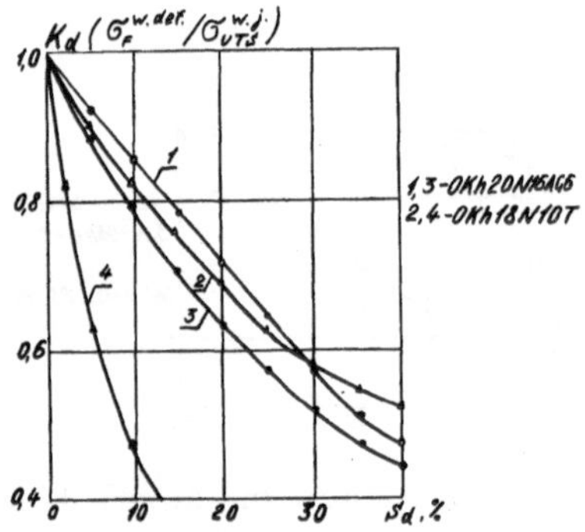

Fig. 7. The influence of the defective area on the defect coefficient of various welds.

Biaxial stress influences upon the behavior of steels in the presence of stress concentrators were investigated.

The hemispheres, 1.5-mm thick with a diameter of 420 mm, were loaded with internal pressure at 77 and 20 K. A correlation of the metal behavior at uniaxial and biaxial loadings was obtained. The amount of strength decrease of metastable steels in the presence of stress concentrators was very evident and rose above 50%.

From the given data it follows that various materials manifest different sensitivities to stress concentrators at low temperatures. Analysis of the behavior of steels and their weldments at low temperatures resulted in a schematic generalization, shown in Fig. 9. Here the possibility of changing (1) the ultimate strength of a specimen by varying the stress concentrators, σ_F^K, and (2) the strength properties of a metal by varying the temperature is shown. For materials that transform to a brittle state, curve 3 is typical; those that are not sensitive to stress concentration correspond to curve 1; and those of little sensitivity, to curve 2.

Calculation of allowable stresses from equations (3), taking into account the strengthening for any material, should also take into account (1) the character of the changing strength of a material having a temperature-dependent concentrator, (2) the value σ_F^K/σ_{UTS} obtained during the testing of flat specimens at operational temperature, (3) the value of stress concentration coefficients in a ready-made piece (where K_t is caused by geometric changes in form or welding reinforcement, weld defects, etc.), (4) fracture character, and (5) fracture type.

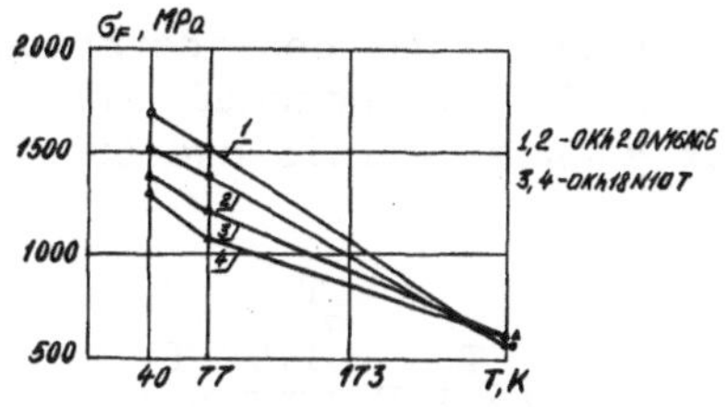

Fig. 8. Temperature dependence of the strength of hemispheres made of base metal (1,3) and of welded hemispheres (2,4).

For stable austenitic materials of the 03Kh20N16AG6 type, calculations can be made by substituting into equations (3) the values $\sigma^T_{1.0}$ and σ^T_{UTS} at operational temperature. For the steels sensitive to stress concentration at decreasing temperature, it is necessary to take into account the data concerning the decrease of strength relations, σ^K_F/σ_{UTS}, obtained from flat-specimen tests. On such steels, the low-temperature strengthening cannot be completely used. For the metastable steels of 03Kh13AG19 and 12Kh18N10T types, only $\sigma^T_{1.0}$ and $\sigma^{K,T}_F$ may be put into the equations, taking into account the fact that the fracture in the presence of a stress concentrator cannot take place at stress levels lower than $\sigma^T_{1.0}$ across the net section. In addition, a necessary condition for the application of steels at a given temperature is the ability for fracture following considerable plastic deformation. Taking into account that $\sigma^T_{1.0}$ and $\sigma^{K,T}_F$ may have similar values, but $\sigma^{K,T}_F > \sigma^T_{1.0}$

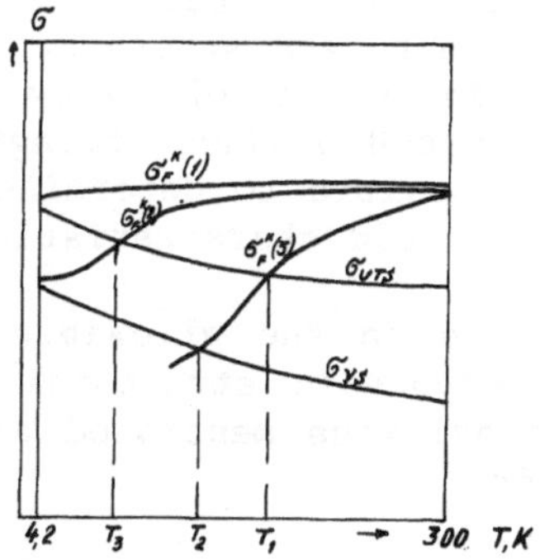

Fig. 9. Diagram of the application of metal ultimate strength with the concentrator (σ^K_F) and strength (σ_{UTS}) depending on temperature.

and $\sigma_F^{K,T} \rightarrow \sigma_{1.0}^T$, and taking into consideration their low-temperature strengthening, it is recommended that the calculation of allowable stresses for such materials use the following criterion:

$$[\sigma^T] = \frac{\sigma_{1.0}^T}{n_{UTS}} \phi^T$$

From this equation follows the conclusion: the level of allowable stresses for low-temperature materials rises with the increase of the yield strength. Thus, increasing the yield strength of a metal (which for the austenitic steels is achieved by means of alloying the metal with nitrogen) is an effective way of decreasing the amount of material in cryogenic structures.

Both at normal and at low temperatures the weld strength may be inferior to that of the base metal, even if their chemical composition is the same. But the welds of the austenitic steels and some high-strength alloys are as strong as the base metal because they exceed the minimum strength level for the base metal specified by technical documents.

For welding of cold-resistant steels, wires assuring a decreased or increased strength may be used. At cryogenic temperatures, the equal strength factor with respect to the base metal may change from 0.6 to 1.0.

Thus, for structural metals obtained from manufacturing plants, it was necessary to introduce into the standards and specifications requirements regulating the minimum characteristics of their cryogenic yield strength and strength. Statistical data processing concerning the influence of alloying elements and heat treatment on the temperature-dependent strength properties of steel and weld metal makes it possible to determine the minimum level of the ultimate strength and yield characteristics.

Table 2 presents data on the allowable stresses that may be used in calculation of cryogenic structural material requirements proceeding from the propositions mentioned above. Recommendations are given for sound welds.

The advantage of nitrogen-strengthened steels over ordinary Cr-Ni steels (higher yield strength) and the advantage of stable austenitic steels over the metastable steels are quite evident, especially from the analysis of ultimate strength characteristics at cryogenic temperatures.

Table 2. Allowable Stresses in Calculation of Welded Structures

Steel Grade and Weld Wires	T, K	Minimum Calculation Values, MPa	MPa	Equal Strength Coef. of Weldments	Allowable Stress, MPa	Specific Strength, MPa/g/cm^3
0Kh18N10T,	293	210	540	1.0	140	17
W-04Kh19N9	77	320	1400	0.6*	220	
	20	440	1580	0.6*	290	
	4.2	410	1520	0.6*	270	34
03Kh20N16AG6,	293	370	650	1	250	32
W-01Kh19N15G6M2AV2	77	820	1230	1	510	
	20	930	1440	1	600	
	4.2	980	1480	1	616	78
03Kh13AG19,	293	370	650	1	250	32
W-0Kh15N9G6AM	77	760	1340	0.7*	390	50
03Kh12N10MT	293	800	1000	0.8	415	53
	77	1000	1400	0.8	460	
	20	1200	1600	0.8	530	68

*Weldment with weld reinforcement.

CONCLUSIONS

1. The low-temperature strengthening of metal may be used in the design of welded cryogenic structures.

2. It is necessary to know the temperature dependence of the metal properties when the stress concentrators, such as notches or reinforcements in weldments, are present. Steels having austenitic metastable structure decrease the strength of weldments.

3. Manufacturers should guarantee the minimum yield and ultimate strengths of their materials at cryogenic temperatures.

4. Calculation of the low-temperature allowable stresses based on characteristics of the material ultimate strength derived from the test data of smooth specimens may lead to erroneous results.

5. Increases in the level of allowable stresses in welded low-temperature steels must be attained by increasing the yield strength.

A REVIEW OF THE EFFECTS OF IONISING RADIATION ON PLASTIC MATERIALS AT LOW TEMPERATURES*

D. Evans and J. T. Morgan

Rutherford and Appleton Laboratory, Chilton, Didcot, Oxon, England

INTRODUCTION

Materials based on organic polymers have many applications where they are subjected to very low temperatures. The properties responsible for their selection are generally low thermal or electrical conductivity combined with low mass and, in the case of polymer matrix composites, high strength and modulus. Most existing materials have been optimised for service at room temperature or above and are not necessarily suitable for cryogenic service. A number of epoxide resin systems having improved resistance to thermal cracking at low temperatures have been developed,[1,2] and these should be considered for use in cryogenic applications.

One important role of polymeric materials is in the construction of superconducting magnets, where they may be used for conductor insulation, impregnation and consolidation of windings, and structural components such as supports or clamps.

Large-scale applications of superconducting magnet technology are envisaged for accelerators and associated beam lines and for fusion reactors; in such applications the materials will be exposed to ionizing radiation in addition to very low temperatures. In general, organic materials are likely to be the most radiation sensitive of the construction materials and may, consequently, determine the life span of the magnet.

*Invited paper.

INTERACTION OF RADIATION WITH PLASTIC MATERIALS AT LOW TEMPERATURE

High energy radiation, on its passage through a plastic material, loses energy by interaction with the electrons and nuclei of the polymer molecules. Each particle or photon of the radiation possesses sufficient energy to separate an orbital electron from its nucleus or to break the bond between adjacent atoms. The basic mechanisms, therefore, are:

Ionization
Excitation
Displacement of a nucleus
Scattering and emission of secondary radiation.

Much of the energy absorbed will eventually be degraded and appear as heat.

Ionization and excitation are by far the more important effects as far as plastic materials are concerned. These processes render molecules unstable, producing reactive fragments which then react with each other or with the basic polymer molecules and in so doing considerably change their chemical structure.

The reactions ensuing from the initial ionization or excitation occur as a sequence of events about which very little is known, but for plastic materials certain generalisations may be made:

1. The most likely reaction is the removal of hydrogen atoms, those attached to aliphatic systems being more vulnerable than those attached to aromatic parts of the molecule. The ability of aromatic nuclei to delocalize electrons is thus seen as a stabilizing influence on the reactivity of the resultant species. This is in contrast to UV-initiated oxidative degradation to which aromatic systems are prone (e.g., aliphatic resin systems are more resistant to weathering than aromatic systems).

A hydrogen atom that is removed may:

a. combine with a similar species to form a hydrogen molecule;

b. abstract a hydrogen atom from an adjacent group to form a hydrogen molecule and a double bond;

c. abstract a hydrogen atom from an adjacent molecule to form a hydrogen molecule and a crosslink.

2. Scission of other covalent bonds will lead to a reduction in molecular weight of the polymer and production of low molecular weight fragments.

3. It is the fate of the species of radicals produced in 1 and 2 that determines the manner in which radiation affects the polymer. It may result in depolymerisation (e.g., polymethyl methacrylate), cross-linking (e.g., polyethylene), or chain scission (e.g., PTFE).

4. Gaseous degradation products are also formed as a result of these reactions.

At very low temperatures, the ionization and excitation produced by incident radiation would not be expected to differ significantly from that produced at room temperature; neither would the abstraction of hydrogen atoms. This is illustrated in Figure 1, which shows the evolution of hydrogen from polyethylene at various temperatures.[3] A discontinuity is seen only at the glass transition temperature, the point at which movement of the polymer chain segments becomes severely restricted. This restriction of the movement of polymer chains is responsible for the differences produced by irradiation at low temperature.

Any reaction that requires molecular movement, for example the reaction between adjacent molecules, is less likely to occur at low temperature; it is, however, likely that some reactivity will be preserved so that reaction occurs as molecular mobility increases on warming. A reduced yield of cross-linkages has been observed for irradiations at low temperatures.[4] This may also explain the differences observed[5-7] in the mechanical properties of some materials when tested with and without an intermediate warming to ambient conditions. Chemical reactions would also explain anomalies

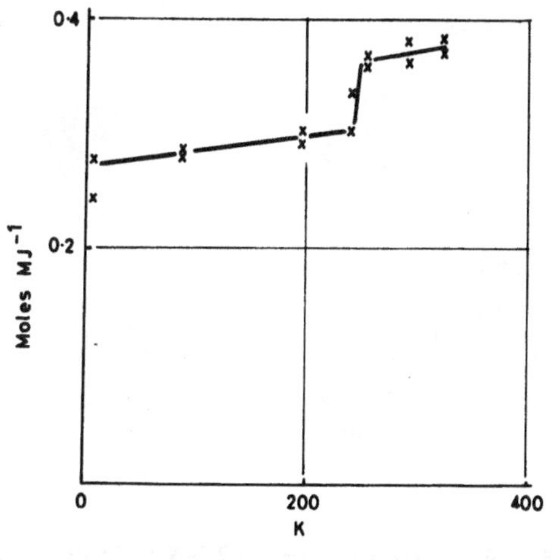

Fig. 1. Hydrogen evolution from polyethylene irradiated at various temperatures.[3]

in specific heat measurements during warming after low temperature irradiation; both exothermic and endothermic effects are observed.[5] Reproducible colour changes have been observed during warming after low temperature irradiation.[5]

Gas evolution during irradiation is of relatively minor importance at normal or elevated temperatures because normal diffusion processes are sufficient to cope with its removal. For low temperature irradiations, however, the release of gases during warming becomes of major importance since it is unlikely that even careful control of the rate of temperature rise will eliminate the sudden release of gas.

This effect has been observed to cause a foam-like structure,[8] severe cracking,[9] and delamination in composite materials.[9,10] (Figure 2.)

One further effect of the deposition of energy by radiation is that localized heating may occur, especially in materials having low thermal conductivity; this may require additional cooling capacity[11] and, in extreme cases, cause problems due to differential thermal expansions in materials that are essentially brittle at low temperature.

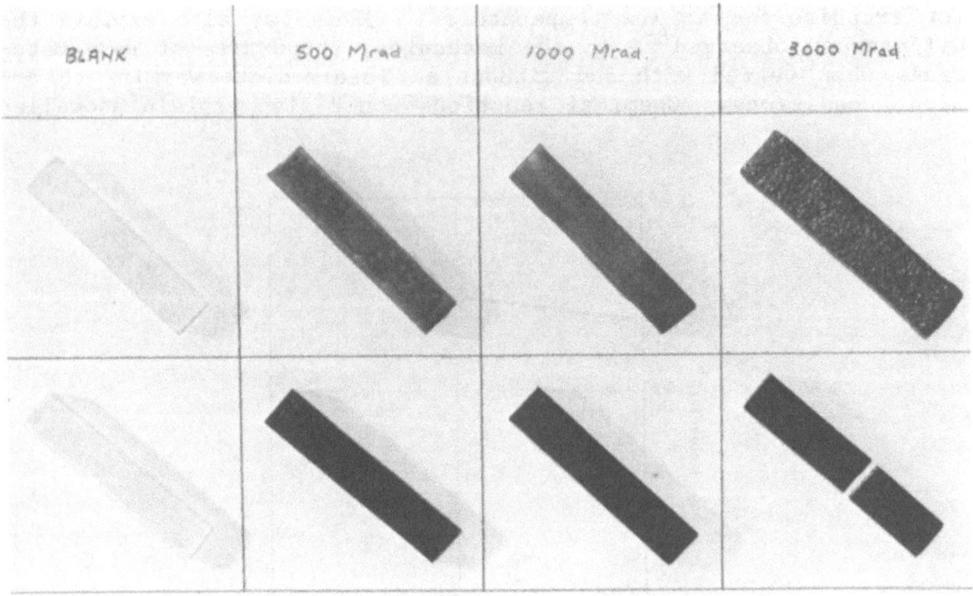

Fig. 2. Effect of low temperature (77-K) irradiation and subsequent warming to room temperature on two epoxide resin systems.

Finally, it should be stated that consideration of individual effects in a material does not necessarily mean that the performance of a component subjected to irradiation at low temperature can be reliably predicted. Changes may be caused or enhanced by a combination of effects (e.g., is there an influence of stress on radiation damage to insulators?).

RADIATION UNITS AND THE CONCEPT OF ℓG

Currently the Gray (Gy) is used as a unit of radiation to quantify the amount of energy deposited in a plastic material. In conventional energy terms, one Gray corresponds to an absorption of 1 J/kg by the plastic material under consideration. Leaving aside irradiation with neutrons, the effect of which varies markedly with material and neutron energy, the following conversion factors and definitions apply:

1 Gy ≡ 100 rads
1 rad ≡ 100 ergs absorbed/g of material
1 Gy ≡ 6.24×10^{18} eV/kg
1 Gy ≡ 10^{-3} W•s/g

The Concept of ℓG

A single number that provides sufficient information about the properties of a material in an ionizing radiation environment would enable an engineer to make a realistic choice between competing materials. Whilst it is clear that such an all-embracing hope could never be realised in one exhaustive and immutable unit, we believe that trends may be adequately indicated by a simple number. Our conceptual unit is ℓG, which we define as the logarithm of the radiation dose (in Gray) to reduce a particular mechanical property by 25%.

E.g., If the flexural strength of a plastic material is reduced by 25% after a radiation dose of 2×10^6 Gray, then the ℓG becomes 6.30 ($\log_{10}$ of 2×10^6).

We believe that a statement such as ". . . the material has an ℓG of 6.30 . . ." will be sufficiently informative for many purposes and will reduce the complex subject of radiation chemistry to a problem of manageable proportions for the busy engineer and designer.

NEUTRON IRRADIATION

Fast Neutrons

This is the major form of radiation present in nuclear reactors. Neutrons deposit energy in polymer molecules mainly by collision with atomic nuclei. The main effect of fast neutron irradiation is the interaction of neutrons with hydrogen atoms (hydrogen atoms are usually present in the greatest number in a polymer molecule) to produce fast protons, which result in intense local ionization and excitation. Therefore, for a given energy absorption, changes resulting from neutron irradiation may not, necessarily, be the same as for electron or γ-radiation.

Slow Neutrons

Most elements have a larger capture cross section for slow neutrons than for fast neutrons, and in many cases, the nucleus resulting from the capture of a neutron is unstable. The resulting transformation is accompanied by the emission of a γ-photon, an electron, or a proton, which cause ionization and excitation within the material. The probability of this capture and transformation

Materials	Curve	Property	Irradiation Temperature (K)	Test Temperature (K)	Initial Value (MPa)	Dose*@ Irrad. Temp (Gy)	ℓG
Polystyrene	A	Tensile strength	77	77	51	$>2.0 \times 10^6$	>6.30
PVC(Polyester plasticised)	B	Tensile strength	77	77	67	$>1.3 \times 10^6$	>6.10
Polymethyl methacrylate (Plexiglas)	C	Flexural strength	77	77	213	2.2×10^6	6.34
Polyethylene	D	Tensile strength	77	77	129	1.8×10^6	6.25
PVC(unplasticised)	E	Tensile strength	77	77	109	$>2.0 \times 10^6$	>6.30
Polymethylmethacrylate	F	Flexural strength	4.2	295	133	1.0×10^5	5.00
PVC(unplasticised)	G	Flexural strength	77	77	152	$>5.0 \times 10^6$	>6.70
Polyethylene	H	Flexural strength	77	77	232	1.3×10^7	7.11

* Dose to reduce given property by 25%.

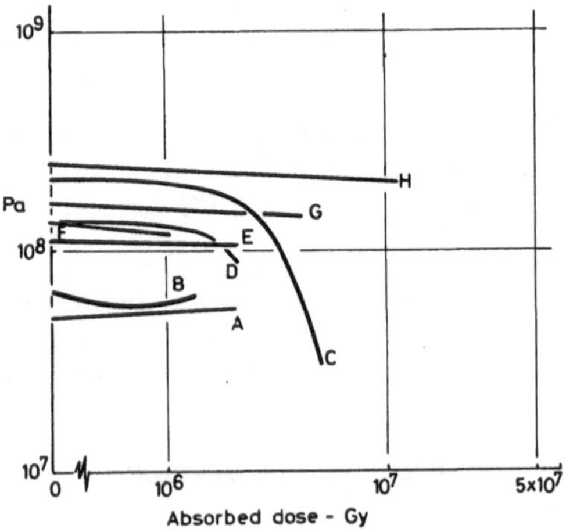

Fig. 3. Thermoplastic materials. Mechanical properties as a function of low temperature radiation dose.

process depends strongly on the elements present; it is low for hydrogen but much higher when say chlorine or nitrogen atoms are present. It therefore follows that if polyethylene and PVC were exposed for the same time to thermal neutrons, the PVC would be subjected to a much higher radiation dose than the polyethylene.

DISCUSSION

From a practical point of view, the most significant effects induced in polymeric materials by ionizing radiation relate to changes in mechanical properties. Mechanical property data of many of these materials as a function of low temperature radiation dose are given in Figures 3-8.

Materials	Curve	Property	Irradiation Temperature (K)	Test Temperature (K)	Initial Value (MPa)	Dose*@ Irrad. Temp (Gy)	ℓG
Polycarbonate	A	Tensile strength	77	77	153	$1.5x10^6$	6.18
ABS	B	Tensile strength	77	77	100	$>5.0x10^6$	>6.70
Polyimide(15% MoS_2)	C	Tensile strength	20	20	102	$>3.0x10^6$	>6.50
Polyphenylene Oxide	D	Flexural strength	77	77	152	$>1.0x10^7$	>7.00
ABS	E	Flexural strength	77	77	167	$1.1x10^7$	7.04
Nylon 6	F	Flexural strength	77	77	304	$7.8x10^5$	5.89
Polyethylene terephthalate	G	Flexural strength	77	77	315	$2.0x10^6$	6.30
Polyimide(Vespel)	H	Tensile strength	77	77	202	$>5.0x10^6$	>6.70
Nylon 6	I	Tensile strength	77	77	163	$7.8x10^5$	5.89

* Dose required to reduce given property by 25%.

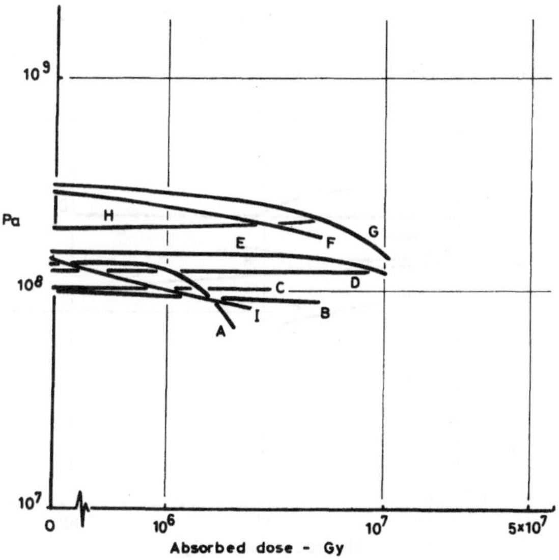

Fig. 4. Thermoplastic materials (continued). Mechanical properties as a function of low temperature radiation dose.

It is possible to measure changes of several orders of magnitude in the electrical conductivity of plastics during irradiation, but in general, this is of little practical significance.[6] For most plastic materials, dielectric strength remains unchanged in the region where mechanical properties are not seriously changed, and it is therefore reasonable to base the serviceability of these materials on mechanical criteria.

The state of the material after warming to room temperature may have important consequences. Polyethylene adequately illustrates this point by exhibiting a fair resistance to radiation when mechanical properties are measured immediately following the low temperature radiation. However, even after small doses when left in air for a period of time, mechanical properties, whether measured at room or low temperature, may indicate that serious embrittlement has occurred due to post irradiation oxidation. Also, if a material is irradiated cold and retained in a cryogenic environment between periods of irradiation, thus building to a significant

Materials	Curve	Property	Irradiation Temperature (K)	Test Temperature (K)	Initial Value (MPa)	Dose*@ Irrad. Temp (Gy)	ℓG
Polyimide(Kapton)	A	Tensile strength	77	77	245	$>1.3 \times 10^6$	>6.11
Polyimide(Kapton)	B	Tensile strength	20	20	325	1.3×10^6	6.11
Polyethylene terephthalate	C	Tensile strength	77	77	260	3.8×10^5	5.58
Polyimide(Kapton)	D	Tensile strength	5	77	320	$>9.0 \times 10^6$	>6.95
Polyphenylene terephthalamide (Nomex 77)	E	Tensile strength	5	77	186	2.4×10^6	6.38
Fluorocarbon (Polyvinylidine fluoride	F	Tensile strength	21	20	140	4.0×10^5	5.60

* Dose required to reduce given property by 25%.

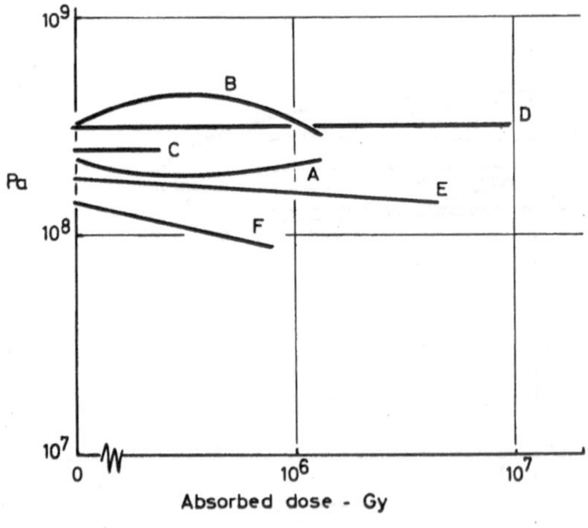

Fig. 5. Plastics, films, and fibres. Mechanical properties as a function of low temperature radiation dose.

total integrated radiation dose, serious damage may occur on eventual warming to room temperature. This is illustrated in Figure 2, which shows two different epoxide resin systems after irradiation at low temperature and warming to room temperature. The presence of low molecular weight (gaseous) products, released suddenly on warming has caused serious mechanical damage. One material, which is tough and semiflexible at room temperature, has undergone considerable volumetric expansion and become 'foam-like' in character. The second material, a brittle epoxide system at room temperature, has not changed dimensionally but the release of gaseous products has resulted in physical failure of the material.

Materials	Curve	Property	Irradiation Temperature (K)	Test Temperature (K)	Initial Value (MPa)	Dose*@ Irrad. Temp (Gy)	ℓG
PTFE/glass	A	Tensile strength	77	77	52	$>1.0x10^6$	>6.00
PTFE/glass	B	Tensile strength	20	77	63	$3.0x10^5$	5.47
PTFEbonded glasscloth-Armalon	C	Tensile strength	20	77	580	$5.5x10^5$	5.74
Polyester/glass	D	Tensile strength	20	20	275	$>1.3x10^6$	>6.11
Silicone	E	Tensile strength	77	77	358	$>8.0x10^6$	>6.90
Phenolic resin/glass	F	Tensile strength	20	20	455	$>5.0x10^6$	>6.70
Melamine/glass	G	Tensile strength	21	21	786	$>2.2x10^6$	>6.30
NEMA G10 CR	H	Flexural strength	77	4.9	862	$>1.0x10^7$	>7.00
Phenolic resin/asbestos	I	Tensile strength	20	20	93	$>5.0x10^6$	>6.70
Stycast 2850	J	Flexural strength	77	77	119	$>1.0x10^7$	>7.00
Stycast 2850	K	Flexural strength	77	4.9	254	$1.0x10^7$	7.00
Epon 828/Z/Silica	L	Flexural strength	77	4.9	225	$>1.0x10^6$	>6.00
Phenolic resin/glass	M	Tensile strength	20	20	290	$>5.0x10^6$	>6.70
Phenolic resin/linen	N	Tensile strength	20	20	80	$>5.0x10^6$	>6.70

* Dose required to reduce given property by 25%.

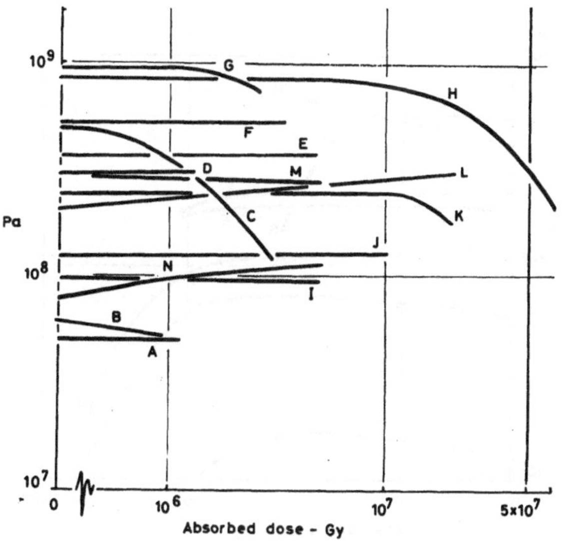

Fig. 6. Composite materials. Mechanical properties as a function of low temperature radiation dose.

MATERIALS AND APPLICATIONS

Fluoroplastics

This group of materials[5,7,12-16] has poor resistance to radiation at room temperature, and the same is true at low temperature. Their toughness at low temperatures combined with their low friction characteristics makes them attractive for certain applications, but they should not be considered satisfactory for radiation levels exceeding 10^5 Gy (Figure 9).

Thermoplastics

Of the thermoplastics,[3,7-9,14-18] polyphenylene oxide (PPO) and polyimide are considered to be satisfactory for use in a low

Materials	Curve	Property	Irradiation Temperature (K)	Test Temperature (K)	Initial Value (MPa)	Dose*@ Irrad. Temp (Gy)	ℓG
MY740/Texaco D400	A	Flexural strength	77	295	14.2	9.7×10^6	6.99
Rutherford No. 53	B	Flexural strength	77	295	112	1.2×10^7	7.11
MY740/CY208/HY906	C	Flexural strength	77	295	60	1.6×10^7	7.23
Araldite D/HY956	D	Tensile strength	77	77	122	$>1.0 \times 10^6$	>6.00
Epikote 828/DDM	E	Flexural strength	77	77	172	$>1.0 \times 10^7$	>7.00
Epikote 191/ DDM	F	Flexural strength	77	77	211	$>1.0 \times 10^7$	>7.00
Epikote 828/ Polyamide	G	Tensile strength	5	77	510	3.0×10^6	6.40
Epikote 828/ K61 B	H	Tensile strength	5	77	390	$>3.0 \times 10^6$	>6.40

* Dose required to reduce given property by 25%.

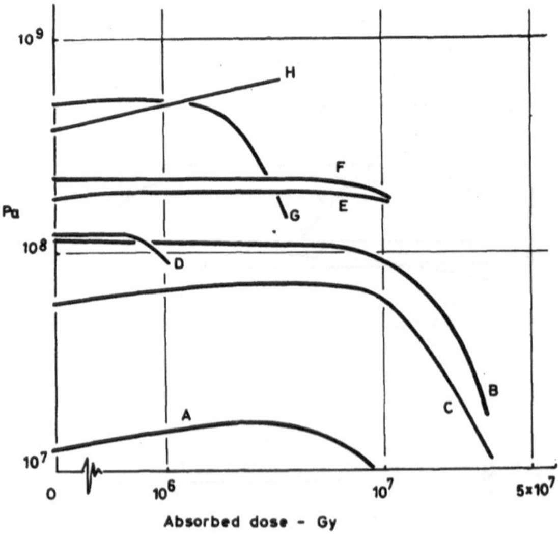

Fig. 7. Unfilled epoxide resins. Mechanical properties as a function of low temperature radiation dose.

temperature environment for radiation levels exceeding 10^7 Gy. Polyethylene also has fairly good radiation resistance at low temperatures and is used, for example as wire insulation.

Polypropylene and polymethyl methacrylate, in common with other polymers possessing a tertiary or quaternary substituted carbon atom in the polymer chain, have relatively poor resistance to radiation.

Polyethylene terephthalate is used at low temperatures in the form of superinsulation, but it should be remembered that it has only moderate radiation resistance and its use above 10^7 Gy is not advised (Figure 9).

Materials	Curve	Property	Irradiation Temperature (K)	Test Temperature (K)	Initial Value (MPa)	Dose*@ Irrad. Temp (Gy)	ℓG
Rutherford No. 71	A	Flexural strength	77	295	94	>1.0x10^7	>7.00
DGEBA/Adiprene L 100/Epon 871/ MOCA	B	Flexural strength	77	295			
MY740/HY951	C	Flexural strength	77	295	129	2.2x10^7	7.36
MY740/HY906	D	Flexural strength	77	295	115	4.1x10^7	7.62
Epikote 171/MNA/BDMA	E	Flexural strength	77	77	106	>1.0x10^7	>7.00
Epikote 828/DDM/Epikote 154	F	Flexural strength	77	77	167	>7.9x10^6	>6.90
MY745/HY905/DY063/EPN 1139	G	Flexural strength	77	295	204	>1.0x10^7	>7.00
Rutherford No. 79	H	Flexural strength	77	295	22.6	4.0x10^6	6.60

* Dose required to reduce given property by 25%.

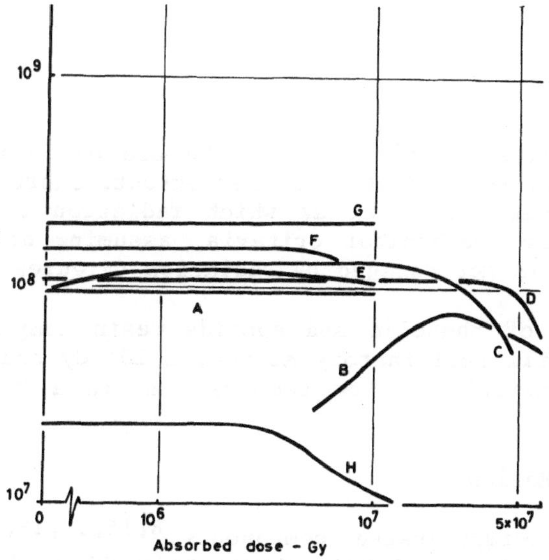

Fig. 8. Unfilled epoxide resins (continued). Mechanical properties as a function of low temperature radiation dose.

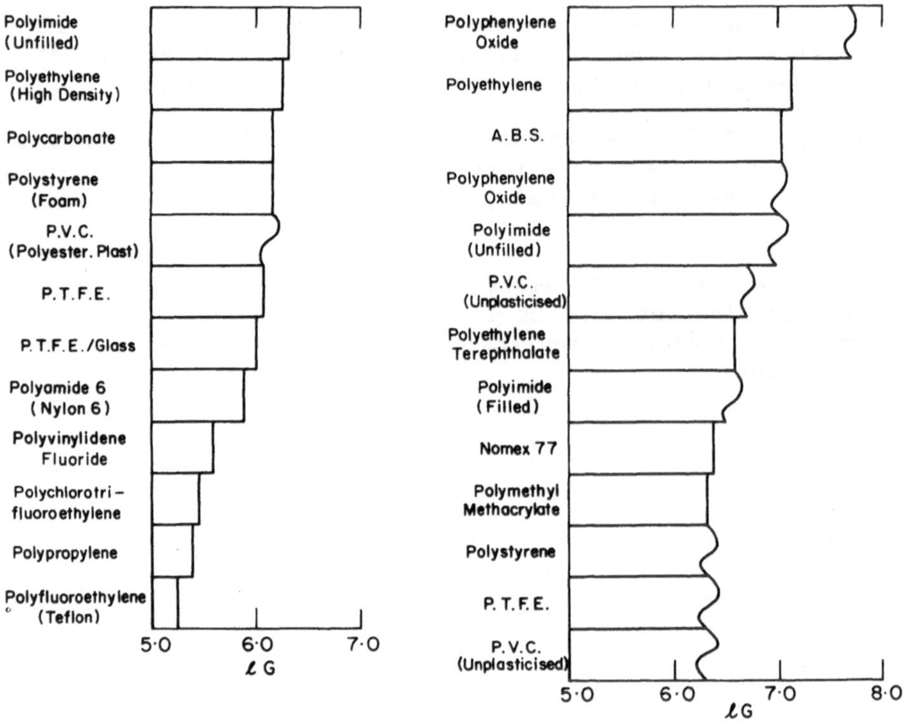

Fig. 9. Effect of low temperature irradiation on thermoplastic materials.

Composites

In composites,[7,9,12–14,19,20] the use of reinforcement in the form of long fibres and, to a lesser extent, short fibres and inert fillers improves the level at which radiation damage becomes observable through mechanical criteria, assuming of course that the reinforcement is not damaged by radiation (Figures 10 and 11).

A number of phenolic and epoxide resin composites with glass fibres are still satisfactory above 2 x 10^7 Gy and may find use as structural materials at low temperatures in a radiation environment.

Impregnating Resins

For resin-impregnated components,[8,15,17,18,21,22] it is not always possible to completely eliminate small resin-rich areas. It is, therefore, desirable to use resin systems that retain some

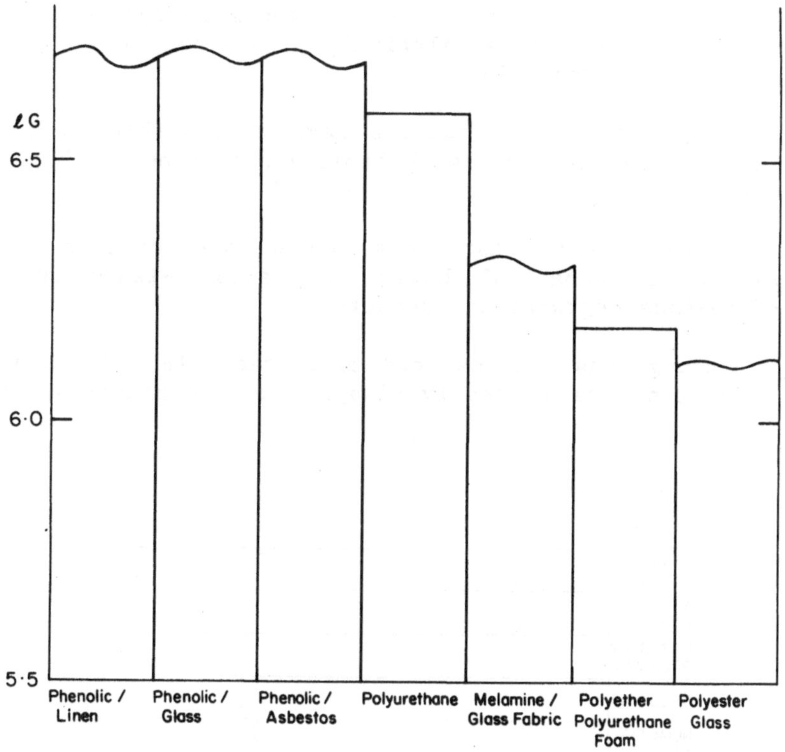

Fig. 10. Effect of low temperature irradiation on thermosetting plastics.

ductility at temperatures below ambient (room), in order to minimize cracking of the impregnant owing to differential thermal contraction.

A number of unfilled epoxide resin systems have been shown to possess radiation stability to $>10^7$ Gy, but most of these materials are brittle in nature even at room temperature and therefore are not regarded as 'crack resistant' for cryogenic use. Such impregnating resins that have been identified have been shown to possess only moderate resistance to ionizing radiation, particularly when extensibility is used as the criteria for assessing radiation resistance. These systems, based on aliphatic amine curing agents, show a reduction in the elongation at break after irradiation, although the breaking stress stays approximately constant.

If we assess the available literature on impregnating resins, it is possible to make certain generalisations:

1. Acid anhydride hardeners, in general, result in brittle systems after curing with a relatively high level of gas evolution on irradiation (see Table 1).

2. Aliphatic amine hardeners make it possible to formulate 'tough' epoxide systems, but with limited resistance to high energy radiation.

3. Aromatic amine hardeners confer the greatest radiation stability, on a mechanical basis, and also release the lowest levels of gaseous degradation products.

4. Although cast resins tend to increase in modulus on irradiation, the converse is nearly always true for glass-fabric-based composites.

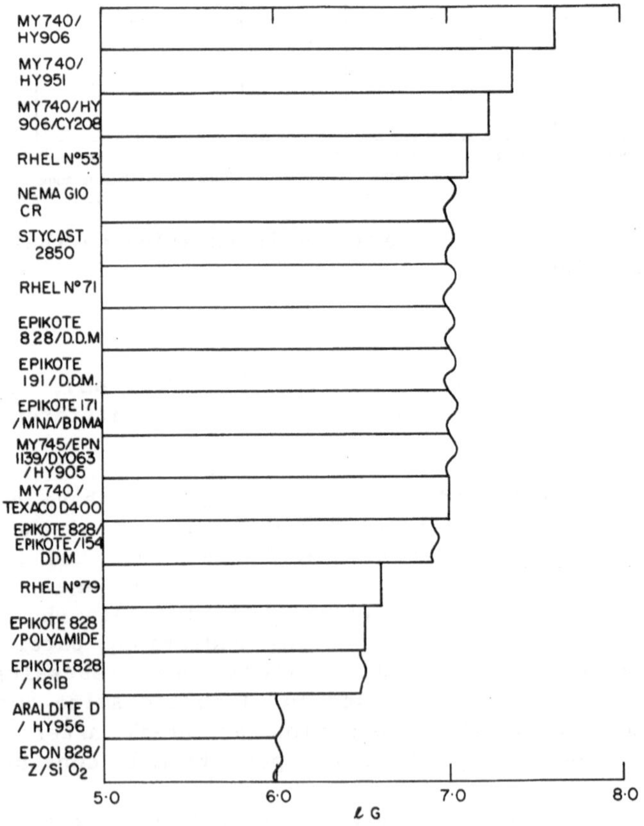

Fig. 11. Effect of low temperature irradiation on epoxide resins.

Table 1. Gas Evolution from DGEBA* Epoxide Resin Cured with Various Hardeners[22]

	Gas Evolved ($cm^3/g/Gy$)	Composition (% volume)				
		H_2	CO_2	N_2/CO	CH_4	C_2H_6
Aromatic amine	2.5×10^{-7}	77.1	0.8	20.9	0.9	0.3
Aliphatic amine	4.3×10^{-7}	88.4	-	10.0	1.3	0.3
Acid anhydride	7.3×10^{-7}	19.9	56.9	23.2	-	-

*DGEBA = Diglycidyl ether of bisphenol A.

Consequently, we have experimented with formulations based on aromatic amine hardeners in an attempt to produce radiation-stable, thermally shock-resistant impregnating materials. In Figure 12, the relative response of these systems to low temperature irradiation is shown. Two sysems, Rutherford Formulations 53 and 71, are based on aliphatic amines and have achieved a measure of acceptance as thermally shock-resistant materials. The new system, Rutherford Formulation 220, achieves the same standard of resistance to thermal cracking but offers improved radiation resistance.

CONCLUSIONS

The available literature on low temperature radiation effects in plastic materials, although fairly extensive, is surrounded with trade names, a profusion of units, and formulating and test variables. For instance, the radiation stability of a material may be judged on mechanical criteria, but these will vary significantly if the material was 'annealed' during the radiation period or if the total dose was achieved without intermediate warming. Therefore, the assessment of radiation damage should be carried out on specimens that simulate service conditions, including irradiation under load, if stress is part of the in-service specification.

In general terms, low temperature irradiation produces less mechanical damage than the equivalent dose at room temperature, but factors such as stress and outgassing may be significant, and with some materials, particularly thermoplastics, postirradiation oxidation at room temperature may be an important additional factor. There must remain some doubt as to how much of the improvement in radiation stability at low temperature can be attributed to the oxygen-free environment.

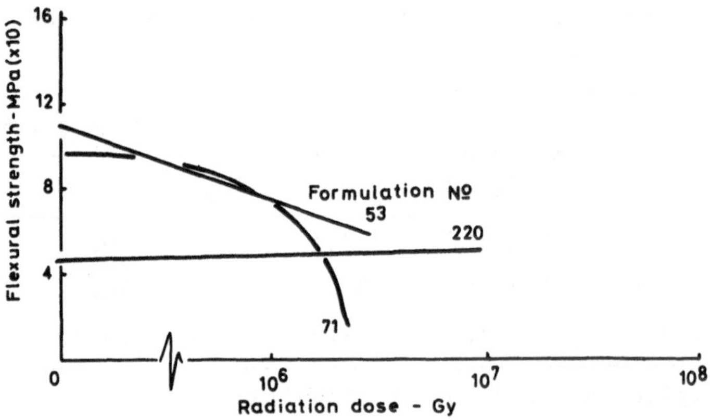

Fig. 12. Effect of radiation on thermally shock-resistant epoxide resins.

Low temperature radiation stability is not necessarily related to thermal stability. PTFE is one of the most thermally stable polymers known but offers little resistance to damage by high energy radiation. It is also true that some materials, such as polymethyl methacrylate, that offer resistance to UV radiation have little resistance to ionizing radiation.

Although 20 K is not believed to be an anomalous temperature, results have been reported on specimens irradiated in liquid hydrogen that suggest some interaction with the cryogen may be occurring.

REFERENCES

1. D. Evans and J. T. Morgan, in: "Fundamentals and Applications of Nonmetallic Materials at Low Temperatures," G. Hartwig and D. Evans, eds., Plenum Press, New York, in press.
2. L. M. Soffer and R. Molho, in: "Cryogenic Properties of Polymers," T. T. Serafini and J. L. Koenig, eds., Marcel Dekker, New York (1968).
3. J. T. Morgan et al., Internal Report No. RPP/E13, Rutherford Laboratory, Chilton, Didcot, England (1969).
4. R. M. Black, Nature 178:303-306 (1956).
5. M. H. Van de Voorde, IEEE Trans. Nucl. Sci. 20:693 (1973).
6. C. J. Long, R. H. Kernohan, and R. R. Coltman, Jr., in: "Nonmetallic Materials and Composites at Low Temperatures," A. F. Clark, R. P. Reed, and G. Hartwig, eds., Plenum Press, New York (1979), p. 141-153.
7. C. W. Funk and C. E. Dixon, Trans. Am. Nucl. Soc. 9:406 (1966).

8. D. Evans, J. T. Morgan, and G. B. Stapleton, Internal Report No. RHEL/R220, Rutherford Laboratory, Chilton, Didcot, England (1971).
9. E. E. Kerlin and E. T. Smith, Biennial Report, 1962-1964, Contract NAS8-2450. Microfiche FZK-188-2 (May 1964).
10. M. H. Van de Voorde, in: "Proceedings ICEC 4," IPC Science and Technology Press, Guildford, Surrey, England (1972).
11. H. Ullmaier, in: "Proceedings of the International Conference on Radiation Effects & Tritium Technology for Fusion Reactors II" (1976).
12. E. E. Kerlin and E. T. Smith, Biennial Report, 1964-1966, Contract NAS8-2450. Microfiche F2K 290 (1966).
13. H. Brechna, SLAC-PUB-469 (August 1968).
14. E. C. McKannan and R. L. Gause, J. Spacecraft, 2(4):558-564 (1965).
15. M. H. Van de Voorde, CERN Report 77-03 (Chapter IV), CERN, Geneva, Switzerland.
16. M. H. Van de Voorde, IEEE Trans. Nucl. Sci. 18(3):784-785 (1971).
17. S. Takamura and T. Kato, in: "Nonmetallic Materials and Composites at Low Temperatures," A. F. Clark, R. P. Reed, and G. Hartwig, eds., Plenum Press, New York (1979), pp. 155-163.
18. M. H. Van de Voorde, USA Particle Accelerator Conference, Chicago (March 1971).
19. R. M. Kernohan, C. J. Long, and R. R. Coltman, J. Nucl. Mater. Part 85-86(A), 79:379-383 (1979).
20. W. Weleff, in: "Advances in Cryogenic Engineering," Vol. 11, Plenum Press, New York (1965), pp. 486-491.
21. M. H. Van de Voorde, Report No. ISR-MA/75-38, CERN, Geneva, Switzerland.
22. J. T. Morgan et al., Rutherford Laboratory Internal Report No. RHEL/R196, (1970).

References Consulted but not Cited.

23. S. Nishijima, S. Ueta, and T. Okada, Cryogenics 21(5):312-313 (1981).
24. W. Weleff et al., Final Report GTR16, Aerojet General Corp., Sacramento, California (1966).
25. B. S. Brown, J. Nucl. Mater. 97:1-14 (1981).
26. D. Evans et al. RPG Symposium, Plastics & Rubber Institute, Bristol (February 1976).
27. G. Hill et al., IEEE Trans. Nucl. Sci. 18(3):761-763 (1971).
28. S. Nishijima and T. Okada, Cryogenics 18(4):215-219 (1978).
29. S. Takamura and T. Kato, Cryogenics 20:441-444 (1980).
30. H. Brechna and W. Maurer, in: "Proceedings 8th International Conference on High Energy Accelerators," CERN, Geneva, Switzerland (1971) (IUPAP).

31. S. Nishijima and T. Okada, Cryog. Eng. 12(6):224-229 (1977).
32. T. Kato and S. Takamura, Cryog. Eng. 14(4):178-183 (1979).
33. M. B. Kasen, Cryogenics, 15(6):327-349 (1975).
34. J. R. Coombe and R. P. Shogan, Cryog. Ind. Gases, 5(2):21-25 (1970).

MECHANICAL PERFORMANCE OF GRAPHITE- AND ARAMID-REINFORCED COMPOSITES AT CRYOGENIC TEMPERATURES

M. B. Kasen

Fracture and Deformation Division, National Bureau of Standards, Boulder, Colorado

INTRODUCTION

Advanced structural composites reinforced with boron, graphite, or aramid fibers have unique mechanical, thermal, and electrical properties that make them attractive alternatives for metals in many cryogenic applications. A most promising application is the structure of the central solenoid of the poloidal field system in Tokamak magnetic fusion energy (MFE) devices.[1] Depending on the flux rise time of a particular design, energy loss due to generation of eddy currents in metallic structures might place an excessive burden on the refrigeration system. Here, a replacement of the metallic structure with advanced composites could suppress eddy currents while providing strength and stiffness equal to that of steel. The low thermal conductivity of graphite-reinforced epoxy laminates in the 77-4 K range combined with high strength and modulus suggests the possibility of fabricating more efficient thermal isolation supports than heretofore available.

This paper reports the effect of cryogenic temperatures on two types of graphite-reinforced epoxy-matrix laminates, one fabricated with fibers of intermediate strength and intermediate modulus and the other fabricated with a fiber of lower strength but high modulus. Results of studies on a composite laminate reinforced with an aramid fiber are also presented and discussed. These laminates were fabricated with a conventional, fully reacted, commercial TGDDM-DDS resin system representative of those used in aerospace construction.

The effect of cryogenic temperatures on the mechanical performance of boron-reinforced epoxy has been discussed in a previous publication.[2] This reference also reports data obtained with epoxy matrix laminates reinforced with high-strength graphite fiber and with glass. Unfortunately, subsequent work[3] has shown that the highly flexibilized epoxy matrix used in fabricating these latter materials was somewhat detrimental to their performance. These prior data are summarized in the present paper for comparative purposes.

MATERIALS AND TEST METHODS

All materials were fabricated by aerospace firms to industrial quality standards with void contents <1.2%. The program evaluated uniaxial longitudinal strength and modulus (6 ply), uniaxial transverse strength and moduli (15 ply), uniaxial longitudinal and transverse compressive strength (30 ply), and in-plane shear strength and modulus (±45°, 10 ply). Young's moduli, Poisson's ratios, and the coefficients of variation (standard deviation expressed as a percentage of the average value) were determined by stress cycling specimens several times to 15-50% of the ultimate strength, depending on orientation. Transverse Poisson's ratios, ν_{21}, were calculated from the relationship

$$\nu_{21} = \nu_{12}\ (E_{22}/E_{11})$$

where E_{22} was obtained from uniaxial transverse specimens and ν_{12} and E_{11} were obtained from uniaxial longitudinal specimens. Young's modulus in tension was determined from strain gages bonded to the specimens. Compressive Young's modulus was not measured; however, data obtained from dynamic modulus studies suggest that differences between tensile and compressive moduli are very small in uniaxially reinforced laminates.[4]

Three or four specimens were tested for each condition at 295 K, 76 K, and 4 K. Test procedures were the same as for earlier work.[3]

RESULTS AND DISCUSSION

The ability of the high-modulus graphite/epoxy laminates to retain their integrity upon cooling to 4 K was an initial concern because of internal stress created by the combined negative coefficient of thermal contraction of the fiber and the positive coefficient of the matrix. This concern was not warranted in the uniaxial laminates used in this study. Repeated thermal cycling of the 30-ply high-modulus laminate from 296 K to 4 K failed to produce detectable cracking either in the bulk of the laminate or adjacent to voids. Crossply laminates may sustain thermal shock damage, but this was not investigated.

Table 1. Temperature Dependence of the Static Mechanical Properties of High-Modulus Graphite/Epoxy Laminates (Average Values)

Temperature	Young's Modulus			Poisson's Ratio		Proportional Limit			Ultimate Strengh			Ultimate Strain	
(K)	(GPa)	(10^6psi)	CV(%)*		CV(%)*	(MPa)	(10^3psi)	CV(%)*	(MPa)	(10^3psi)	CV(%)*	(%)	CV(%)*
						Tensile: Longitudinal (0°)							
295	323	46.8	3.7	0.454	5.1		Linear		726	105	13.2	0.22	14.1
76	328	47.6	1.0	0.374	20.0		to		717	104	9.1	0.21	9.7
4	326	47.2	1.6	0.385	12.5		failure		716	104	6.3	0.22	5.3
						Tensile: Transverse (90°)							
295	6.70	0.97	1.5	0.009	6.2		Linear		22.2	3.22	7.5	0.34	7.5
76	8.70	1.26	5.6	0.010	6.0		to		19.9	2.89	4.3	0.23	17.8
4	8.94	1.30	1.2	0.011	5.4		failure		16.7	2.41	15.2	0.19	17.2
						Tensile: Crossply (± 45°)							
295	20.6	2.99	2.2	0.917	1.5	54.4	7.89	8.9	89.2	12.9	1.9	0.50	7.0
76	25.0	3.62	2.1	0.904	3.2	35.5	5.15	2.1	72.3	10.5	1.0	0.46	3.3
4	25.7	3.73	1.7	0.900	3.0	36.1	5.24	8.2	72.5	10.5	3.3	0.43	6.8
						Compressive: Longitudinal (0°)							
295			5.9						431	62.5	18.4	0.17	13.3
76			5.9			378	53.8	16.5	642	93.1	4.4	0.22	10.3
4			6.2			304	44.1	10.0	694	101	10.5	0.25	10.6
						Compressive: Transverse (90°)							
295			12.6				Linear		157	22.8	6.2	2.53	6.0
76			1.7				to		149	21.7	8.1	1.51	6.0
4			3.1				failure		150	21.8	14.9	1.62	20.5

In-Plane Shear

Temperature	Shear Modulus			Shear Strength		
(K)	(GPa)	(10^6psi)	CV(%)*	(MPa)	(10^3psi)	CV(%)*
295	5.17	0.750	2.64	44.6	6.47	1.91
76	6.55	0.950	1.36	36.2	5.24	1.12
4	6.88	0.998	2.41	36.3	5.26	3.14

*CV = coefficient of variation among individual specimens.

Table 2. Temperature Dependence of the Static Mechanical Properties of Medium-Modulus Graphite/Epoxy Laminates (Average Values)

Temperature	Young's Modulus			Poisson's Ratio		Proportional Limit			Ultimate Strength			Ultimate Strain	
(K)	(GPa)	(10^6psi)	CV(%)*		CV(%)*	(MPa)	(10^3psi)	CV(%)*	(MPa)	(10^3psi)	CV(%)*	(%)	CV(%)*
Tensile: Longitudinal (0°)													
295	186	27.0	3.9	0.345	14.7				1187	172	6.8	0.60	9.6
76	192	27.9	7.7	0.376	16.4				1127	163	10.8	0.58	4.4
4	189	27.5	3.9	0.380	2.6				1045	152	7.2	0.51	----
Tensile: Transverse (90°)													
295	8.62	1.25	4.1	0.016	5.1				27.8	4.03	10.9	0.33	11.4
76	12.0	1.73	5.1	0.024	5.5				27.6	4.00	3.4	0.24	6.3
4	12.1	1.76	3.9	0.025	2.4				23.3	3.38	12.8	0.20	15.4
Tensile: Crossply ($\pm 45^o$)													
295	17.1	2.48	3.7	0.823	1.8	49.0	7.11	4.7	113	16.4	1.1	1.0	17.8
76	22.3	3.24	4.1	0.737	3.3	40.2	5.83	6.0	91.4	13.3	3.6	0.56	19.9
4	23.0	3.34	4.6	0.726	9.0	37.3	5.34	18.7	88.9	12.9	2.1	0.62	42.5
Compressive: Longitudinal (0°)													
295									813	118	7.9		
76									878	127	11.0		
4									801	116	11.2		
Compressive: Transverse (90°)													
295									117	16.9	20.0		
76									132	19.2	19.6		
4									166	24.0	12.4		

In-Plane Shear

Temperature	Shear Modulus			Shear Strength		
(K)	(GPa)	(10^6psi)	CV(%)*	(MPa)	(10^3psi)	CV(%)*
295	4.67	0.678	2.9	56.5	8.20	1.1
76	6.41	0.929	4.6	45.8	6.63	3.6
4	6.47	0.939	5.0	44.5	6.45	2.1

*CV = coefficient of variation among individual specimens.

Table 3. Temperature Dependence of the Mechanical Properties of Aramid/Epoxy Laminates (Average Values)

Temperature	Young's Modulus			Poisson s Ratio		Ultimate Strength			Ultimate Strain	
(K)	(GPa)	(10^6psi)	CV(%)*		CV(%)*	(MPa)	(10^3psi)	CV(%)*	(%)	CV(%)*
				Tensile: Longitudinal (0°)						
295	71.4	10.4	13.9	0.440	7.0	1132	164	18.7	1.58	4.3
76	99.4	14.4	5.8	0.315	7.6	1154	167	1.7	1.07	5.2
4	99.4	14.4	13.2	0.376	22.0	1142	166	5.5	1.13	6.7
				Tensile: Transverse (90°)						
295	2.51	0.36	19.6	0.016	22.8	4.17	0.605	--	0.16	--
76	3.59	0.52	--	0.011	--	3.63	0.527	--	0.12	--
4	4.56	0.66	--	0.017	--	5.56	0.806	--	0.14	--
				Tensile: Crossply (±45°)						
295	6.89	1.00	4.4	0.776	2.1	68.7	9.97	2.1	1.52	11.4
76	8.72	1.26	1.6	0.731	4.0	59.1	8.57	1.0	1.01	1.1
4	9.83	1.43	7.4	0.834	7.2	62.1	9.01	5.6	0.85	15.1
				Compressive: Longitudinal (0°)						
295						225	32.7	11.8		
76						290	42.1	15.5		
4						366	53.0	8.7		
				Compressive: Transverse (90°)						
295						84.4	12.3	11.3		
76						109.7	15.9	18.4		
4						117.2	17.0	25.0		

In-Plane Shear

Temperature	Shear Modulus			Shear Strength		
(K)	(GPa)	(10^6 psi)	CV(%)*	(MPa)	(10^3psi)	CV(%)*
295	1.90	0.276	2.2	34.3	4.97	2.1
76	2.56	0.371	2.1	29.6	4.29	1.0
4	2.67	0.387	6.3	31.2	4.52	5.3

*CV= coefficient of variation among individual specimens

Results of the high- and medium-modulus graphite/epoxy test program and the aramid/epoxy test program appear in Tables 1 through 3. These data are compared with data obtained on other materials tested in earlier programs in Figs. 1 through 6.

Graphite/Epoxy Laminates

The data in Tables 1 and 2 indicate that the uniaxial strength properties of the high- and medium-modulus graphite fiber-reinforced laminates either remain relatively unaffected or improve on cooling to cryogenic temperatures. A possible exception is transverse tensile strength, which appears to suffer some decrease. These data are at variance with earlier data[5] that indicated a systematic trend toward reduced uniaxial tensile strength on cooling. The temperature dependence of in-plane shear is quite similar for these two laminate types, with a substantial increase occurring between 25 K and 76 K.

A substantial decrease in ±45° tensile strength is noted for these materials between 295 K and 76 K, with little added decline on further cooling. A converse temperature dependence of Young's modulus is observed. This reflects the added significance of the matrix and matrix-fiber interface properties in off-axis loading as compared with the uniaxial case. Cooling embrittles the matrix by blocking relaxation mechanisms, greatly increasing the notch sensitivity of the epoxy. This is exacerbated by residual stress created at the matrix-fiber interface and is due to differences in thermal contraction; the effect is maximized in the ±45° orientation where the interface is in shear.

Aramid Epoxy Laminate

The data in Table 3 suggest that cooling an aramid/epoxy laminate to cryogenic temperatures has little effect on the uniaxial longitudinal and transverse tensile strength, whereas a substantial increase may develop in the compressive strength. Cooling increases the uniaxial longitudinal modulus, but not to the extent reported in earlier literature.[5] Cooling also increases the uniaxial transverse and in-plane shear modulus while having little effect on in-plane shear strength.

Comparison with Other Composite Types

Figures 1 through 6 provide a comparison of the cryogenic performance of the graphite and aramid laminates tested in the current program with those tested previously. In interpreting these data, the reader must be aware that the performance of the high-strength graphite laminate and the glass-reinforced laminate fabricated with the highly flexibilized Resin 2 epoxy matrix are not representative of the performance of laminates fabricated with a fully reacted

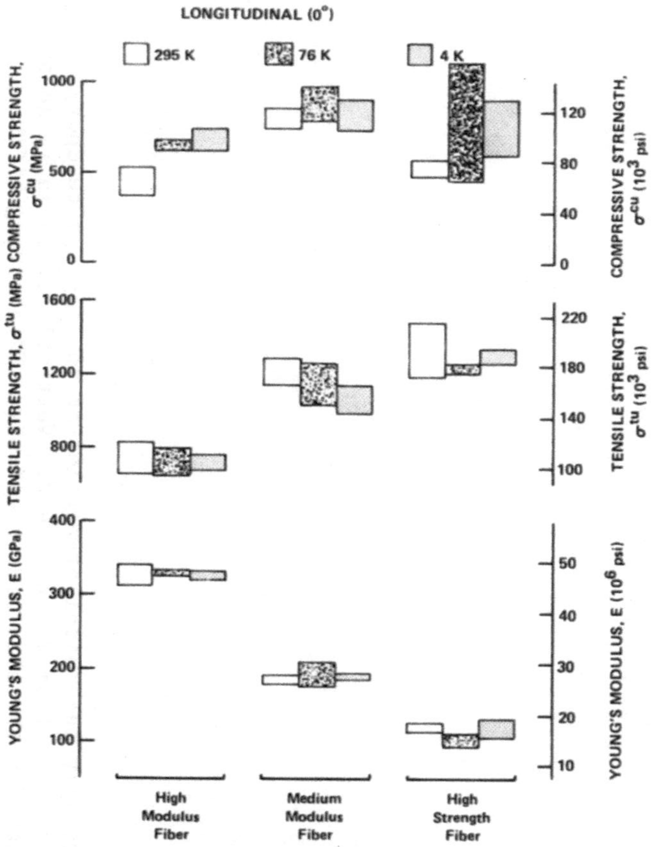

Fig. 1. Longitudinal mechanical properties of three graphite/epoxy laminates from 295 K to 4 K. Vertical spread is range of data for three to five specimens. High strength fiber laminate fabricated with a highly flexibilized matrix.

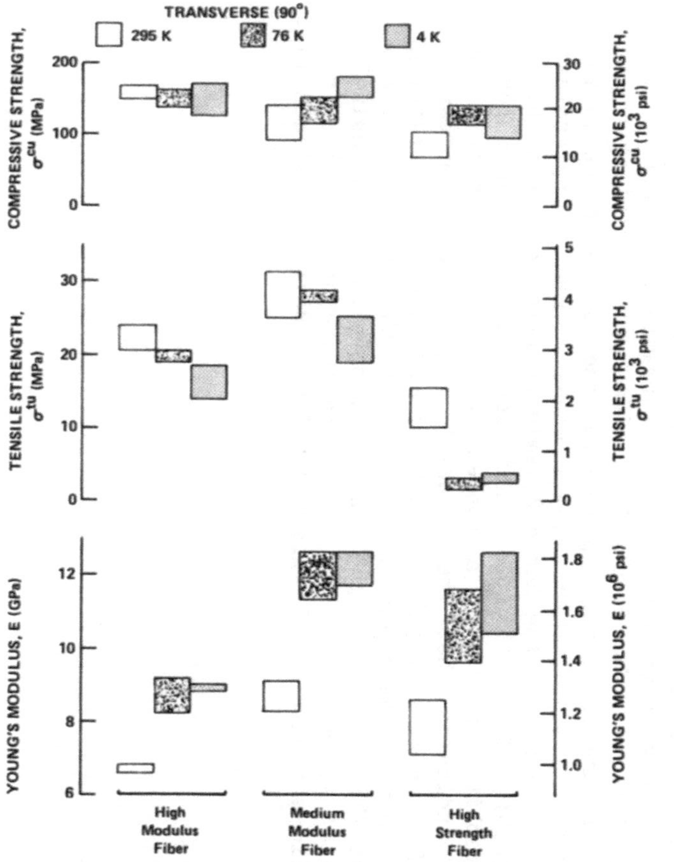

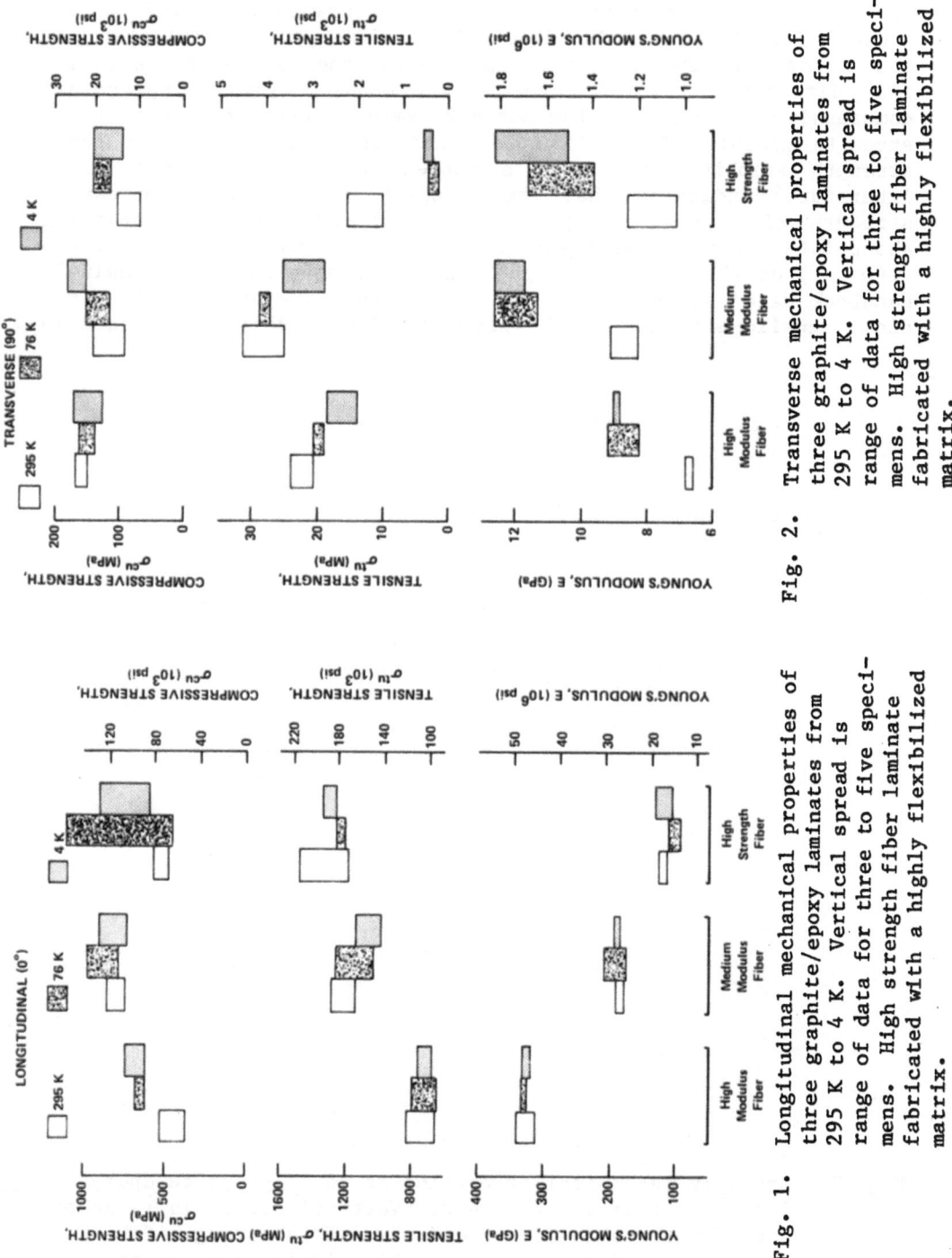

Fig. 2. Transverse mechanical properties of three graphite/epoxy laminates from 295 K to 4 K. Vertical spread is range of data for three to five specimens. High strength fiber laminate fabricated with a highly flexibilized matrix.

epoxy system. Available evidence indicates that both the room-temperature and cryogenic-temperature performance is reduced by the Resin 2 formulation. The deleterious effect of flexibilizing is most evident in the compressive properties and in properties in the ±45° direction. For example, Fig. 1 shows a lower longitudinal compressive strength for the Resin 2 laminate reinforced with high-strength graphite fiber than for laminates reinforced with lower strength graphite fibers in a conventional epoxy matrix. The literature[6] indicates that the uniaxial longitudinal compressive strength of a high-strength graphite fiber laminate containing about 0.62 fiber volume fraction should be on the order of 965 MPa (140 ksi) at 295 K, or 75% higher than that developed with Resin 2. Handbook data[7] suggest that the room-temperature uniaxial longitudinal tensile strength of the glass-reinforced Resin 2 laminate in

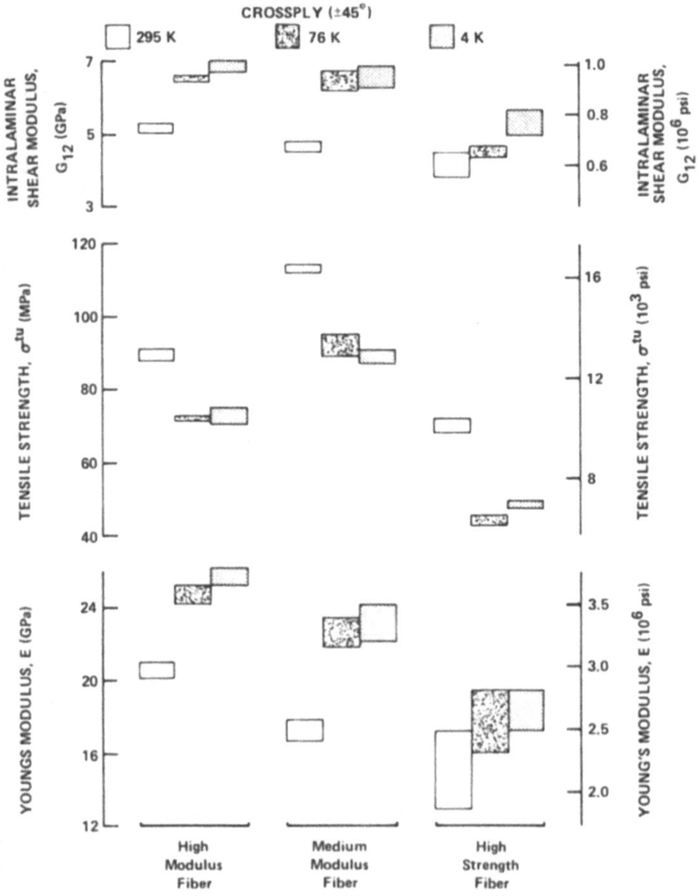

Fig. 3. Crossply mechanical properties of three graphite/epoxy laminates from 295 K to 4 K. Vertical spread is range of data for three to five specimens. High strength fiber laminate fabricated with a highly flexibilized matrix.

Fig. 4 is about 30% lower than that expected from a laminate of similar glass content in a fully reacted epoxy matrix. Cooling increased the tensile strength of this laminate, but only to a level of the conventional laminate at room temperature. A similar degradation is noted for the transverse tensile and compressive properties. The poor room-temperature performance of the Resin 2 matrix laminates is due to low resin strength, whereas the poor cryogenic performance is attributed to the high level of residual stress created by the large thermal contraction of the flexibilized epoxy compared with that of conventional unflexibilized systems.

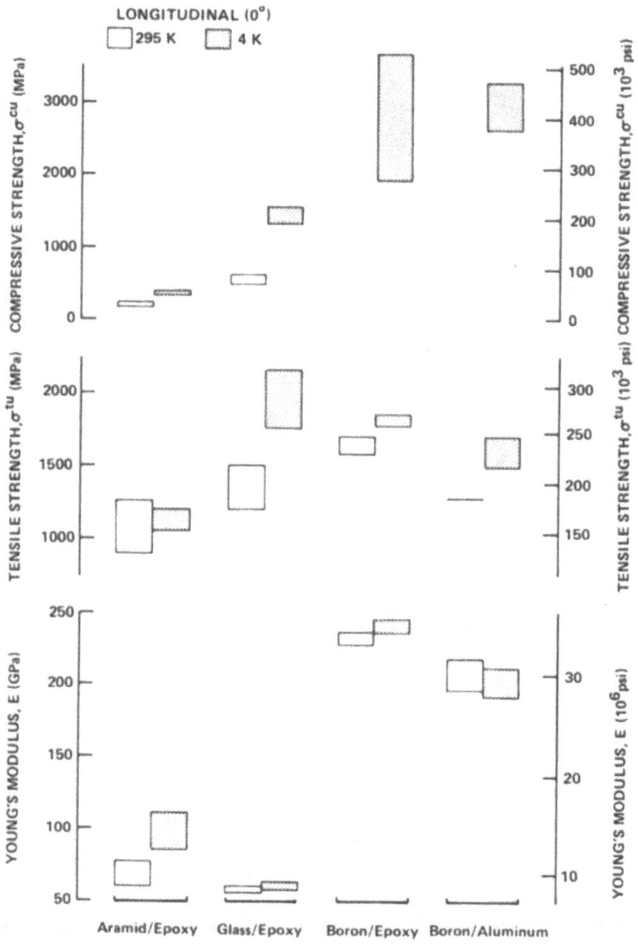

Fig. 4. Longitudinal mechanical properties of three aramid, glass, and boron laminates at 295 K and 4 K. Vertical spread is range of data for three to five specimens. Glass/epoxy laminate fabricated with a highly flexibilized matrix.

These data suggest that uniaxial longitudinal tensile strength of the aramid-fiber-reinforced laminate about matches that of the medium-modulus graphite laminate, approximately the middle point of the graphite group. The aramid-reinforced laminate provided a uniaxial longitudinal Young's modulus between that of glass and the lowest modulus graphite. The primary deficiencies of the aramid laminate are low compressive strength, low transverse strength, and low shear strength.

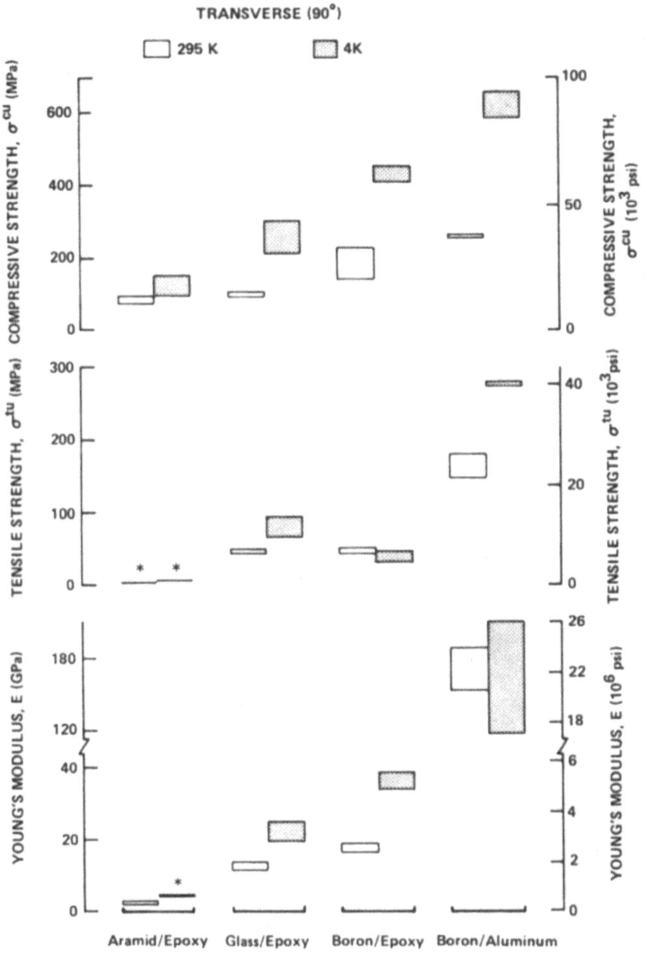

Fig. 5. Transverse mechanical properties of aramid, glass, and boron laminates at 295 K and 4 K. Vertical spread is range of data for three to five specimens except as noted. Glass/epoxy laminate fabricated with a highly flexibilized matrix. *Only one specimen.

The data suggest that the boron-reinforced laminates provide the best combination of strength and modulus, with particularly outstanding compressive strength performance. Glass reinforcement provides high strength in the fiber direction; however, the modulus is low. Graphite fiber laminates offer a compromise: high modulus at low strength, low modulus at high strength, or intermediate properties trading one parameter against the other. Transverse properties are low, suggesting that crossply elements should be incorporated into any functional laminate structures using these materials. Aramid fiber laminates are best used to carry tensile loads.

A recent review of the effect of cryogenic temperatures on laminate performance[3] suggests that the major problem in composite

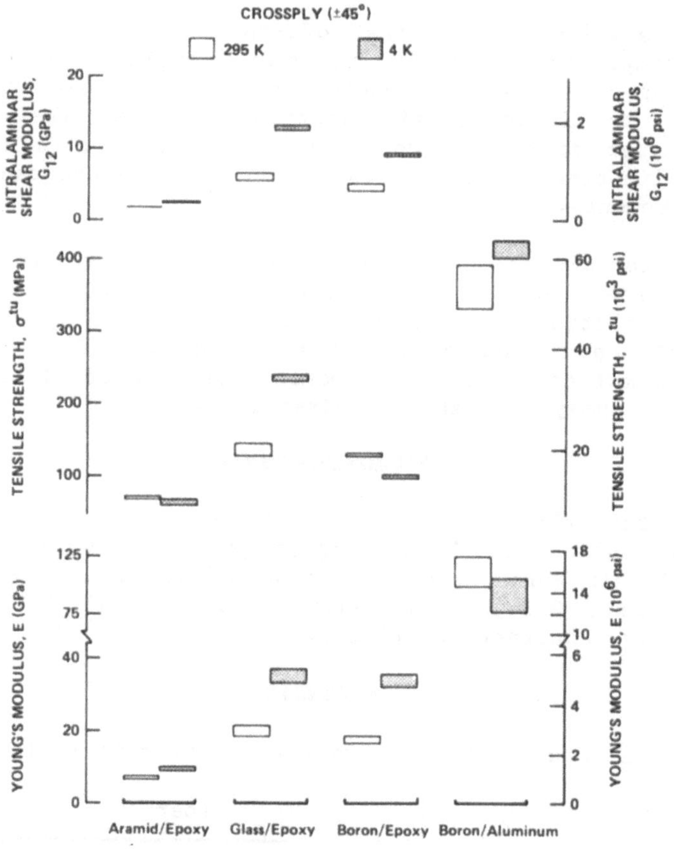

Fig. 6. Crossply mechanical properties of aramid, glass, and boron laminates at 295 K and 4 K. Vertical spread is range of data for three to five specimens. Glass/epoxy laminate fabricated with a highly flexibilized matrix.

material performance at cryogenic temperatures is the high notch sensitivity of the polymeric matrix. This limits glass-reinforced laminates to a useful stress of 20 to 30% of maximum stress if substantial damage accumulation in the matrix cannot be tolerated. High notch sensitivity, related to the absence of operable relaxation mechanisms in the polymer, also causes the laminates to become very sensitive to the presence of small flaws--in fracture mechanics terms, the critical crack size becomes extremely small. This substantially increases the variability in properties at low temperatures, in particular for compression, shear, and off-axis properties. Resin systems providing increased cryogenic toughness combined with low thermal contraction are under development; however, none are yet proven.

CONCLUSIONS

These studies confirm that off-the-shelf laminates developed for room-temperature service are viable engineering materials at 4 K. Averages of static mechanical and elastic properties improve on cooling, but it is becoming increasingly clear that embrittling of the polymer matrix and the consequent decrease in strain capability is limiting the extent to which effective use can be made of such improvements.

The materials characterization work was restricted to mechanical properties. Data on the effect of cryogenic temperatures on thermal contraction and specific heat of boron epoxy, boron aluminum graphite epoxy, and glass epoxy are found in Ref. 8. Information on thermal conductivity of boron epoxy, boron aluminum, and on two graphite epoxy laminates is given in Ref. 9.

ACKNOWLEDGMENTS

The author thanks Mr. Raymond Schramm and Mr. Ralph Tobler for their contributions to the testing program. The work was sponsored by the U.S. Air Force Aero Propulsion Laboratory, Wright-Patterson Air Force Base, Ohio, under Contract No. MIPR FY 1455 79000 603, Dr. C. Oberly, contracting officer.

APPENDIX

Laminates studied in this work are commercially identified as:

	Fiber	Matrix
High-modulus graphite	GY-70	934
Medium-modulus	HM-S	934
Aramid	Kevlar 49	934
High-strength graphite	AS	Resin 2

	Fiber	Matrix
Glass	S	Resin 2
Boron/epoxy	5.6 mil	2387
Boron/aluminum	5.6 mil	6061

The selection of trade name products reflected a desire to establish the static mechanical cryogenic performance of conventional laminates as a baseline for comparison with the performance of experimental laminates. Selection of these systems was not based on the assumption of their superior performance in comparison with that of other systems nor does it imply endorsement or approval of such products by NBS.

REFERENCES

1. C. J. Long, "Application of Advanced Composites in Tokamak Magnet Systems," ORNL/TM-6047, Oak Ridge National Laboratory, Oak Ridge, Tennessee (1977).
2. R. E. Schramm and M. B. Kasen, Cryogenic mechanical properties of boron-, graphite-, and glass-reinforced composites, Mater. Sci. 30:197 (1977).
3. M. B. Kasen, Cryogenic properties of filamentary-reinforced composites: An update, Cryogenics 21:323 (1981).
4. H. M. Ledbetter, Dynamic elastic modulus and internal friction in fibrous composites, in: "Nonmetallic Materials and Composites at Low Temperatures," A. F. Clark, R. P. Reed, and G. Hartwig, eds., Plenum Press, New York (1979), p. 267.
5. M. B. Kasen, Mechanical and thermal properties of filamentary-reinforced structural composites at cryogenic temperatures--2: Advanced composites, Cryogenics 15:701 (1975).
6. "Advanced Composite Design Guide: IV-Materials," Wright-Patterson Air Force Base, Ohio (1973).
7. "Plastics for Aerospace Vehicles; Part 1. Reinforced Plastics," MIL-HDBK-17A, Department of Defense, Washington, D.C. (1971).
8. F. J. Jelenik and E. W. Collings, Low-temperature thermal expansion and specific heat properties of structural materials, in: "Materials Research for Superconducting Machinery," Vol. VI, R. P. Reed, H. M. Ledbetter, and E. C. Van Reuth, eds., ADA 036919, National Bureau of Standards, Boulder, Colorado (1976).
9. J. G. Hust, Thermal conductivity, in: "Materials Research for Superconducting Machinery," Vol. VI, R. P. Reed, H. M. Ledbetter, and E. C. Van Reuth, eds., ADA 036919, National Bureau of Standards, Boulder, Colorado (1976).

REINFORCED POLYMERS AT LOW TEMPERATURES*

G. Hartwig

Institut für Technische Physik, Kernforschungszentrum Karlsruhe
Karlsruhe, Federal Republic of Germany

INTRODUCTION AND SURVEY

Because of their low electrical and thermal conductivities, fibre-reinforced composites are a necessary supplement to metals in low-temperature technology. Their high specific strength and excellent fatigue behaviour favours a substitution of metals by composites in many applications. But problems like their high brittleness and their low interlaminar shear strength must not be overlooked. A comparison of elementary properties of usual fibres is given in Table 1.

In Fig. 1 typical stress-strain diagrams are plotted for the important fibre types and an epoxy resin. Their relatively low fracture strain should be noted; the 3% fracture strain for fibreglass is the highest value. The fracture strain of the epoxy resin is only of the order of 2% at 4.2 K, from which the typical residual strain of 1.2% from cooling down to LHe must be subtracted. (The contraction of the fibres can be neglected in this rough estimation.) At higher temperatures the fracture strain of a polymer is larger, and part of the residual stress is compensated by viscoelastic flow. But at low temperatures, polymers are known to be brittle. So the polymeric matrix will be damaged and the interlaminar shear strength reduced, when use is made of the maximum strength, for example, of fibreglass composites ($\varepsilon \approx 3\%$). This is true for static and dynamic loads. For carbon fibres the fracture strain is smaller than and similar to that of the matrix including the residual strain. General features of common fibre composites are summarized in Table 2. The low stiffness of glass-fibre composites, the high thermal insulation capability of carbon-fibre composites, the rather high negative thermal expansion, and the very poor workability of Kevlar composites should be noted.

---

*Invited paper.

Table 1. Fibres for Reinforcement

FIBRE GLASS	- tough in all directions. - low stiffness. - high strength. - high fracture strain. - strong bond with resins. - low thermal contraction
CARBON FIBRES	- brittle; especially perpendicular to the fibre direction. - high stiffness or strength. - low fracture strain. - medium bond with resins. - very low thermal contraction. - carrier material for screens
KEVLAR FIBRES	- very brittle; especially perpendicular to the fibre direction (at 4.2 K) - tendency for defibration - medium fracture strain. - medium bond with resins. - negative thermal contraction (in fibre direction).

For technical applications, e.g., supporting parts, characteristic ratios like specific strength (strength per weight) or thermal conductivity per strength are decisive. The weight and volume of a plant or machinery is controlled by the first ratio; the second ratio determines the cooling costs. The orders of magnitude of the most important ratios are summarized in Table 3. For comparison, the respective values of a low temperature austenitic steel are included (see Ref. 1). To allow a meaningful comparison,

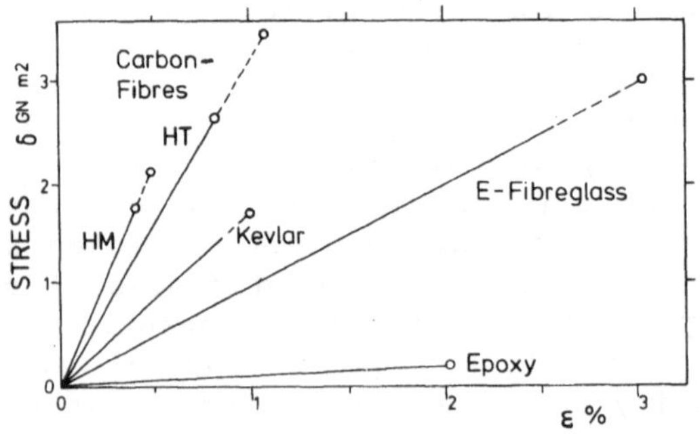

Fig. 1. Stress-strain diagrams of fibres and of a matrix at 4.2 K.

Table 2. General Features of Fibre Composites*

Materials Unidirectional Quasi-isotropic	Stiffness	Strength	Fatigue	Thermal Insulation	Electrical Insulation	Thermal Contraction ΔL/L(293–4.2 K)	Workability
Fibre-glass/ epoxy	0	++	+	+	++	$\gtrsim 1 \times 10^{-3}$	++
Carbon-fibre/ epoxy (high tensile)	+	++	++	++	+	$\lesssim 0.6 \times 10^{-3}$	++
Carbon-fibre/ epoxy (high-modulus)	++	+	++	++	+	$\lesssim 0.6 \times 10^{-3}$	++
Kevlar-fibre/ epoxy	+	+/++	+	+/++	++	-0.2×10^{-3} (unidirectional negative)	-
Whisker/epoxy	-	0/+		+	+	$\lesssim$ 0.8%	+

* ++ = excellent; + = good; 0 = poor; - = very poor.

a separation is made into conditions of one-, two- and three-dimensional loads. It is obvious that for one-dimensional load conditions all fibre composites considered here are much more favourable than steel. All these composites have a specific strength and a thermal conductivity per strength greater by one order of magnitude or more than steel. Generally, the best ratios apply to carbon-fibre composites, followed by Kevlar and fibreglass composites.

For two-dimensional load conditions, the situation is much more balanced, but most of the characteristic ratios, at least of

Table 3. Characteristic Ratios at 4.2 K

MATERIALS	E/ρ [J/gr]	σ/ρ [J/gr]	λ/E [m^3/sec·K]	λ/σ_T [m^3/sec·K]	λ/σ_c [m^3/sec·K]
unidirectional $_{\parallel}$ (0°)	LOAD IN ONE DIRECTION				
Fibre-glass/Epoxy	4×10^4 1)	12×10^2	12×10^{-13}	$4. \times 10^{-11}$	$8. \times 10^{-11}$ 2)
Carbon-fibre/Epoxy (high tensile)	10 "	15 "	2. "	1.5 "	≤1.5 "
Carbon-fibre/Epoxy (high modulus)	15 "	9 "	1.5 "	2.5 "	≤2.5 "
Kevlar-fibre/Epoxy	10 " 1)	12 "	5. "	4.5 "	≤4.5 "
quasi-isotropic* (0°,90,±45)	LOAD IN TWO DIRECTIONS				
Fibre-glass/Epoxy	1.5×10^4 3)	$4. \times 10^2$	$20. \times 10^{-13}$	$10. \times 10^{-11}$	
Carbon-fibre/Epoxy (high tensile)	3. "	4. "	6. "	5. "	
Carbon-fibre/Epoxy (high modulus)	4. "	(2.5)*) 1.5 "	5. "	(9)*) 13. "	
Kevlar fibre/Epoxy	3 "	3.5 "			
isotropic	LOAD IN THREE DIRECTIONS				
FE-Ni-Whisker/Epoxy (22 Vol%)	1.8×10^4	$\times 10^2$	$30. \times 10^{-13}$	$\times 10^{-11}$	$\times 10^{-11}$
Steel (316 LN)	3. " 4)	1.5 "	14. "	25. "	25. "

*) hybrid composite

E: Young's modulus; ρ: density; σ: fracture strength; λ: thermal conductivity carbon- and Kevlar-fibre composites, are remarkably greater than those for steel. In case of high loads acting on all three dimensions, metals like steel are unmatched. But many cases might be covered by high loads acting in two directions and one low load in the other direction; this can be met by quasi-isotropic composites. Three-dimensional fibre composites are feasible but very expensive. One possibility is a whisker/epoxy composite. The rod-shaped whisker crystals have a very low ratio of thickness to length and thus allow a maximum shear transmission by the matrix. Therefore, a higher strength than for the unfilled matrix (200 MPa at 4.2 K) should be possible. The main problem is to achieve a high filling factor without holes in the matrix. In our preliminary test a 22 vol.% filling with Fe-Ni whiskers gave no real improvement of strength. Only the stiffness was increased by a factor of two: $E \approx 22$ GPa. By means of pressure or ultrasonic treatment, a filler content of 35-40 vol.% should be possible.

Omitting the condition of high loads in three dimensions, it is obvious that carbon-fibre composites as a whole possess the most attractive properties. In addition, there is one line of development that is becoming more and more important: carbon-fibres can be used as excellent and mechanically very strong carriers of screens, e.g., of superconducting materials. A very recent approach consists in the treatment of carbon-fibres, e.g., with copper, which converts the fibres into good electrical and thermal conductors, even at low temperatures (Union Carbide Corporation).

For future designs it will be feasible to construct superconducting magnets out of the same basic material with modified components: superconducting carbon-fibres, normal conducting carbon-fibres, and insulating carbon-fibres for supports and armatures. This "magnet composite" would consist of a mechanically very strong and resistant base material with nearly equal thermomechanical properties. With respect to this goal, the most important mechanical and thermal low temperature properties of the unmodified carbon-fibre composites are treated in more detail.

CARBON-FIBRE COMPOSITES

Materials

With respect to price and properties, the following fibre types were selected:

High-tensile fibres T 300 (Torayca; Japan); strength: 3.9 GPa.
High-modulus fibres M 40 A (Torayca; Japan); modulus: 370 GPa.

For the matrix, two epoxy resins were chosen:

A flexibilized system: Cy 221/Hy 979.
A rigid system: Ly 556/Hy 917 (both Ciba Geigy).

The filling factor was 60 vol.%.

For one- and two-dimensional load conditions, the fibre arrangement is: (a) unidirectional and (b) quasi-isotropic in one plane.

A hybrid system containing glass laminates as crack stoppers between the layers of carbon fibres was investigated. Fibre arrangement and contents were: quasi-isotropic; 20 vol.% glass laminate; 40 vol.% carbon fibres.

PROPERTIES

As regards most of the mechanical properties, there is no significant difference between 77 K and 4.2 K. Therefore, only results obtained at 77 K are presented. The thermal properties were determined down to 4.2 K.

Mechanical Stiffness and Strength

The high stiffness and unidirectional composites with M 40 A fibres and the high strength with T 300 fibres can be seen from Table 4. The stiffness of unidirectional and quasi-isotropic composites obeys the rules of stress analysis, whereas the strength of quasi-isotropic materials is too low for M 40 A composites. One reason is the interaction of neighbouring layers. Under load in the 90° ply or even more in the 45° ply, a crack is formed that propagates to the 0° load-bearing layer. According to the poor transverse strength of carbon-fibres, a small stress concentration at the crack tip is sufficient for cutting this load-bearing ply.

Table 4. Carbon Fibre/Epoxy

Materials	Young's modulus E [GPa]	60 Vol %; 77 K		Fracture strain ϵ_T %	Fatigue endurance limit σ/σ_T	Poisson's ratio
		Tensile strength σ_T [GPa]	Interlam. shear str. τ [GPa]			
High tensile T 300 (0°)	140	2.2	105	1.5	80 - 85 %	0.32
quasi-isotropic (0°, 90°, ± 45°)	53	0.5	-	1.1	65-70 %	0.31
High modulus M40A (0°)	240	1.3	76	0.6		0.23
quasi-isotropic (0°, 90°, ± 45°)		0.19	-	0.4	65 %	
Hybrid composite 20 Vol % fibre glass 40 Vol % carbon-fibres M40A (0°, 90°, ± 45°)	62	0.33	-	0.54		0.3

Epoxy resin: Cy 221/Hy 979 (Ciba-Geigy)

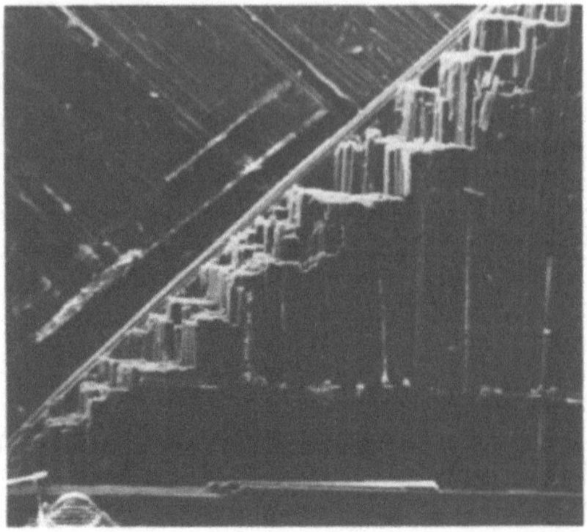

Fig. 2. Fracture area of a quasi-isotropic carbon-fibre composite (M 40 A).

Figure 2 demonstrates the failure of the load-bearing 0° ply by a 45° and a 90° ply. The load-bearing fibres are cut stepwise into planes. By contrast, the fracture areas of unidirectional composites have a very defibred shape with fibre pull-out (see Fig. 3). This effect can be avoided by interlayers of tough glass laminates as crack stoppers. They somewhat reduce the stiffness and strength of the total composite, but the interaction between layers is reduced. The significantly improved values of this hybrid composite are listed in Table 4.

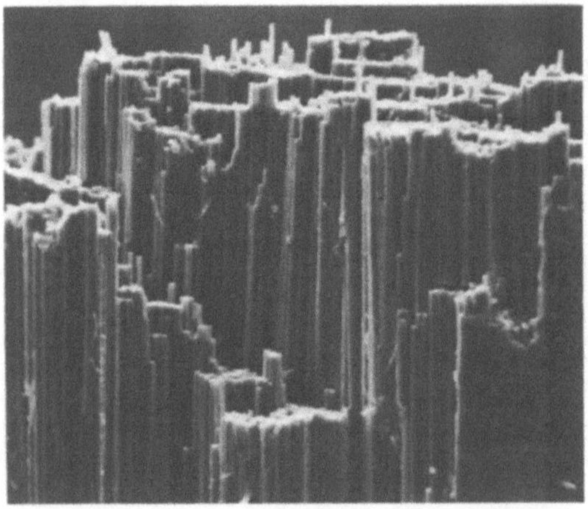

Fig. 3. Fracture area of a unidirectional carbon-fibre composite.

Interlaminar Shear Strength

The interlaminar shear strength was measured by bending thin plates with the following dimensions: thickness, d, = 2 mm; width, b, = 10 mm; length, l, = 16 mm. From the relation $\sigma = 0.75\ K/(b\ d)$ the interlaminar shear strength, σ, was determined via the fracture force, K. At higher temperatures this property naturally is remarkably dependent on temperature and is influenced by the matrix. But at low temperatures ($\lesssim$ 77 K) there is no temperature dependence. Nevertheless, composites with T 300 fibres show remarkably higher values than those with M 40 A fibres.

Fracture Strain, ε_T, and Poisson's Ratio, μ

The values of ε_T range from 0.6% (M 40 A) to 1.5% (T 300), both unidirectional. This is covered by the fracture strain of the matrix after subtraction of the residual strain. The fracture strain of the matrix is 2.6% at 77 K; the residual strain: $\Delta L/L$ (293-77 K) $\approx$ -1.1%. Thus, a free strain of $\Delta\varepsilon \approx 1.5\%$ is available, which is sufficient for both types of composites considered. For quasi-isotropic composites, the situation is more critical since multiaxial stresses are present.

Poisson's ratio, μ, shows no temperature dependence at low temperatures, but it differs for M 40 A and T 300 fibre composites.

Fatigue Behaviour

The fatigue endurance limit is of special interest. The measurements were performed in the threshold-tension mode with sinusoidal load. To this purpose, samples with load ratios, σ/σ_T, that survived more than 10^7 load cycles were considered. For unidirectional composites, a fatigue endurance life of $\sigma/\sigma_T \approx$ 80-85% was found. For comparison: the fatigue endurance life of the epoxy matrix was measured to be 40%.[6] This means 40% of the free matrix strain $\Delta\varepsilon \approx 1.5\%$, which is 0.6%. This covers the 80% load ratio of M 40 A composites, but for T 300 fibre composites a matrix failure must be envisaged. This could be detected by measuring the interlaminar shear strength of precycled samples.

For quasi-isotropic composites with T 300 fibres, a fatigue endurance life of $\sigma/\sigma_T \approx$ 65-70% was found. So the excellent fatigue behaviour of unidirectional (0°) and quasi-isotropic composites can be demonstrated.

Thermal Contraction

The integral thermal contraction $\Delta L/L$ (293-4.2 K) is summarized in Table 5 for the composites under consideration. As typical examples, the temperature dependence of $\Delta L/L$ and α, the

Table 5. Thermal Properties

MATERIALS	Thermal contraction $-\Delta L/L$ (293 - 4.2 K)	Thermal conductivity λ [W/cmK] 7 K	293 K
M40A/epoxy " (0°) ∥	2.15×10^{-4}	5.8×10^{-4}	$30. \times 10^{-2}$
(0°,90°,±45°)		3.4 "	12. "
" (90°)⊥	79.8 "	3.3 "	1.8 "
T300/epoxy " (0°) ∥	2.74×10^{-4}	4.3×10^{-4}	5.8×10^{-2}
(0°,90°,±45°)	8.46 "	3.4 "	3. "
" (90°)⊥	66.2 "	3.5 "	0.6 "
		error ≈ 5 %	
		range: factor 1.7	factor 50
Epoxy-resin Cy221/Hy979		6.9×10^{-4}	23×10^{-4}

Composite: $\frac{\lambda\ (7\ K)}{\lambda (293 K)} = 2.0 \times 10^{-3}$; fibre: $\frac{\lambda\ (7\ K)}{\lambda (293 K)} = 3.3 \times 10^{-3}$

coefficient of thermal expansion, are plotted in Fig. 4. The existence of a shoulder in $\Delta L/L$ for quasi-isotropic composites should be mentioned, which leads to a drop in α.

Thermal Conductivity

The thermal conductivity, λ, of carbon-fibre composites that differ in the type and arrangement of fibres is summarized in Table 5. The huge decrease of λ is due, to a large degree, to the freezing-in of the electronic part of the thermal conductivity. At room temperature, the values of λ differ by a factor of 50, whereas at 7 K they only differ by a factor of 1.7 (see the curves of Fig. 5). It is interesting to note that below 14 K the composites exhibit a lower thermal conductivity than the expoxy matrix. There are several reasons and effects responsible for this diminishing range at low temperatures:

Diminishing range of unidirectional $\lambda_{\parallel}$ and $\lambda_{\perp}$ values at low temperatures. At room temperatures, there is, according to the

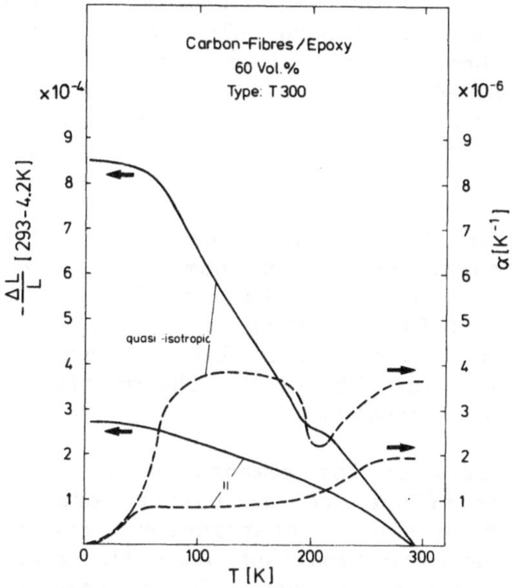

Fig. 4. The integral thermal contraction and the coefficient of thermal expansion versus temperature for a unidirectional and a quasi-isotropic composite.

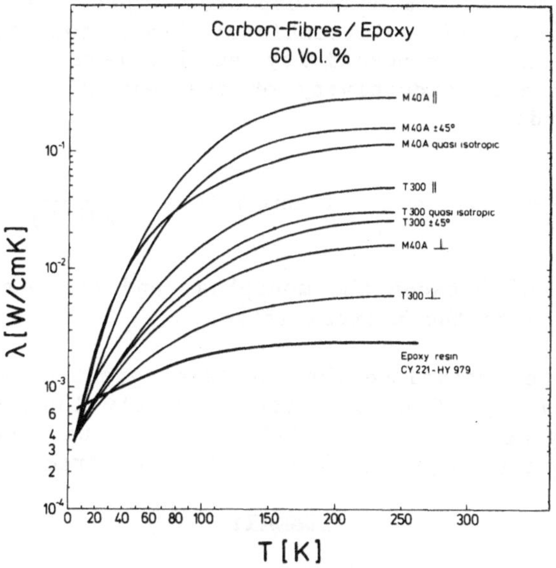

Fig. 5. Thermal conductivity of composites and of the epoxy resin matrix versus temperature.

heat conductivity tensor of carbon fibres, a higher conductivity in the fibre direction than perpendicular to it. At low temperatures, this difference has largely disappeared, since only long-wave phonons are activated which propagate almost uniformly within and perpendicular to the graphite planes. The highly energetic, short-wave phonons at room temperatures are bound to propagate mainly in the graphite planes.

Convergence of the λ values from T 300 and M 40 A fibre composites at low temperatures. At room temperature, the fibre conductivity is much higher than that of the matrix. The different conductivities of different fibres are dominant. But at low temperatures the fibre conductivity becomes comparable to or lower than that of the matrix. Therefore, the matrix, which was the same for all composites, dominates.

Boundary effects at low temperatures. At low temperatures, the so-called size effect and the Kapitza effect become dominant, if the boundaries within a heterogeneous system are very large. The size effect becomes important when the mean phonon range is larger than the sample dimensions. In the case of composites, this relates to the fibre thickness or the matrix spacing in between.

The Kapitza effect results from boundary scattering by phonon mismatch of differently refractive materials. Both effects increase with decreasing fibre thickness. For both types of fibres considered, the fibre thickness was equal and had a value of 7 µm. Measurements were performed on unpotted fibres and compared with conductivity values of unidirectional fibre composites. It was found that the fibre conductivity at low temperatures is higher than the normalized conductivity of the composite. The following ratios were found:

$$\text{fibres: } \frac{\lambda(7\ \text{K})}{\lambda(293\ \text{K})} \approx 3.3 \times 10^{-3}; \quad \text{composite: } \frac{\lambda(7\ \text{K})}{\lambda(293\ \text{K})} \approx 2 \times 10^{-3}.$$

The factor of 1.65 between the monophase and the biphase system is mainly attributed to the Kapitza effect.

It should be emphasized that it is of great advantage when the thermal conductivity of a composite is nearly independent of fibre type and arrangement. Then optimization can be made with respect to other parameters, e.g., load or thermal contraction.

SUMMARY

Carbon-fibre composites exhibit an extremely high mechanical stiffness or strength. Even quasi-isotropic composites show a specific strength or stiffness that is comparable to or higher than that for steel. The thermal conductivity per strength and stiffness of carbon-fibre composites is much lower than that for steel. Thus, at very low temperatures, carbon-fibre composites have a much higher insulation capability than steel. The thermal conductivity of the carbon-fibre composites considered is rather independent of the fibre arrangement below 10 K. Thus, the fibre configuration can be optimized with respect to other parameters. The thermal contraction is very small in the fibre direction; even for quasi-isotropic composites, the value of the integral thermal contraction is: $\Delta L/L(293-4.2\ K) \lesssim x\ 10^{-4}$.

The mechanical properties were improved by a hybrid system. A combination of strong carbon-fibres and tough glass-fibres results in a better composite.

ACKNOWLEDGMENTS

The excellent work of A. Kuhn, P. Raber, B. Stark, and B. Vogeley is appreciated. The carbon-fibre composites were manufactured by Messerschmitt, Bölkow, Blohm (MBB-München-Ottobrunn). The work was supported in part by the Bundesministerium für Forschung und Technologie, Bad Godesberg.

REFERENCES

1. A. Nyilas and H. Krauth, Use of heavy section austenitic welds for 4-K service, in: "Advances in Cryogenic Engineering--Materials," Vol. 28, Plenum Press, New York (1982), p. 853.
2. M. B. Kasen, Adv. Compos. Cryog. 15:327 (1975).
3. E. L. Stone, L. O. El-Marazki, and W. C. Young, Compressive fatigue tests on a unidirectional glass/polyester composite at cryogenic temperatures, in: "Nonmetallic Materials and Composites at Low Temperatures," A. F. Clark, R. P. Reed, and G. Hartwig, eds., Plenum Press, New York (1978), p. 283.
4. M. B. Kasen, G. R. McDonald, D. H. Beckman, Jr., and R. E. Schramm, Mechanical, electrical, and thermal characterization of G-10CR and G-11CR glass-cloth/epoxy laminates between room temperature and 4 K, in: "Advances in Cryogenic Engineering--Materials," Vol. 28, Plenum Press, New York (1980), p. 235.
5. W. Weiß, Low temperature properties of carbon-fibre reinforced epoxies, in: "Fundamentals and Applications of Nonmetallic Materials at Low Temperatures," D. Evans and G. Hartwig, eds., Plenum Press, New York (1982).
6. W. Weiß, Fatigue behaviour of epoxy resins at low temperatures, Prog. Colloid Polym. Sci. 64:68 (1978).

DEGRADATION OF FIBER-REINFORCED COMPOSITE MATERIALS AT CRYOGENIC TEMPERATURES, PART I—UNIAXIAL TENSILE AND PURE TORSIONAL FATIGUE

S. S. Wang and E. S.-M. Chim

Department of Theoretical and Applied Mechanics
University of Illinois, Urbana, Illinois

INTRODUCTION

Recently, a considerable amount of interest has been generated in cryogenic engineering applications of fiber-reinforced composites.[1-3] The ability of composite materials to withstand cyclic stresses at cryogenic temperatures is of crucial importance in their suitability for such applications. Since this is a new development in the field of composite material technology, very limited information concerning this critical material issue is currently available. In this research, uniaxial tensile and pure torsional fatigue studies were conducted on the NEMA/ASTM G-10 composite laminates in a cryogenic environment.

Depending on the applications, cryogenic fatigue failure is considered to occur if any one of the following exists: (1) fracture by losing load-bearing capacity (structural failure), (2) excessive stiffness degradation (functional failure), and (3) excessive energy dissipation leading to heat generation and loss of superconductivity (functional failure). In this paper, attention is focused on the second problem area in which the change of material properties during cyclic loading is considered. The major objective is to examine the stiffness degradation and associated damage mechanisms in the G-10 composite subjected to tensile and torsional cyclic stresses.

Details on cryogenic fatigue fracture and energy dissipation in the composite can be found in Ref. 4. Results obtained in the study give basic information on fatigue resistance design and life

Table 1. Mechanical Properties (Initial, Monotonic) of G-10 Composite Laminate

Property		Temperature (K) 295	77	4
Young's Modulus, E (GPa)	(warp)	2799	3130	3592
	(fill)	2241	2703	2910
Poisson's Ratio, ν	(warp)	0.175	0.19	0.21
	(fill)	0.14	0.18	0.21
Tensile Strength, σ_{UTS} (MPa)	(warp)	429	825	862
	(fill)	257	459	496
Tensile Fracture, Strain, ε_f (%)	(warp)	1.89	3.54	3.67
	(fill)	1.55	2.53	2.70
Shear Strength, τ_{UTS} (MPa)	(warp)	57	135	
	(fill)		73	79

prediction for the cryogenic composite. They also provide a reference for studying more complex multiaxial fatigue failure of the material.[5]

MATERIAL AND SPECIMEN PREPARATION

Test specimens were made of NEMA/ASTM G-10 grade composite laminates, which are epoxy resin reinforced with E-glass woven fabrics. The reinforcing glass fabrics are unbalanced in the composite. The fiber content in the fill direction is about 75% (by volume) of that in the warp direction. Basic mechanical properties of the composite laminate are given in Table 1.

Two types of test specimens were prepared: flat coupon and tubular specimens. Flat coupon specimens were made in accordance with ASTM specification D638-77a.[6] The finished dimensions were 10.16 cm x 28.58 cm with a gage section of 5.08 cm x 9.53 cm, and an average thickness of the coupon was 0.28 cm. Two sets of flat coupon specimens were used: one with fibers of fill and warp directions running parallel and perpendicular to the loading axis, respectively [denoted as (0/90) or cross-laminated composite] and the other with fibers running 45° and -45° off the loading axis [denoted as (±45) composite]. Surfaces of the specimens remained as molded during preparation. Tubular specimens were trimmed mechanically from laminated tubes in accordance with a modified ASTM specification to ensure a uniform stress distribution and final fracture in the gage section.[7] The specimen had a finished total length of 61 cm, an inner diameter of 5.08 cm, and a gage length of 10.16 cm with wall thickness of 0.25 cm. The fill fibers were along the axial direction of the tube.

TEST PROCEDURE

Fatigue tests of the flat coupon specimens were conducted in a closed loop servohydraulic cyclic fatigue test system under constant-amplitude loading of frequency, f, 1 Hz. The minimum-to-maximum stress ratio, R, was approximately 0.01 in all tests. Test specimens were submerged completely in liquid nitrogen in a specially designed cryostat throughout the test. An extensometer specially designed for fatigue experiments at cryogenic temperatures was used to measure axial deformation. Both loading and deformation were monitored.

Fatigue tests of the tubular specimens were carried out in a MTS multiaxial fatigue test system. The test procedure and conditions were the same as described previously. Both uniaxial tensile and pure torsional fatigue tests for the tubular specimens were performed. A cryogenic tension-torsion extensometer was designed to measure simultaneously both axial and torsional deformations in the tubular specimens. A detailed description of the test system is given in Ref. 5.

STIFFNESS DEGRADATION DURING FATIGUE

Uniaxial Tensile Fatigue of Cross-Laminated Composite

The stiffness change of the composite during fatigue has been one of the most important problems in the fatigue failure of the material. For the flat-coupon (0/90) composite, changes of Young's modulus in the fill direction associated with different cyclic loading levels during fatigue are shown in Fig. 1. The results clearly indicate that the stiffness of the composite decreases monotonically with loading cycles and that the rate of stiffness reduction increases continuously during fatigue. At a given number of cycles, increasing the range of cyclic stress, $\Delta\sigma$, increases both the magnitude and the rate of stiffness degradation. These phenomena are found to be consistent with the so-called wear-out process in the common fatigue failure of fiber-reinforced composites[11] in which damage cumulation occurs uniformly in the specimen and progressively deteriorates the stiffness of the material. Final fatigue fracture of the composite occurs at approximately the same level of stiffness degradation irrespective of the fatigue stress level and number of load cycles. The final stiffness retention is about 85% of the initial tensile modulus.

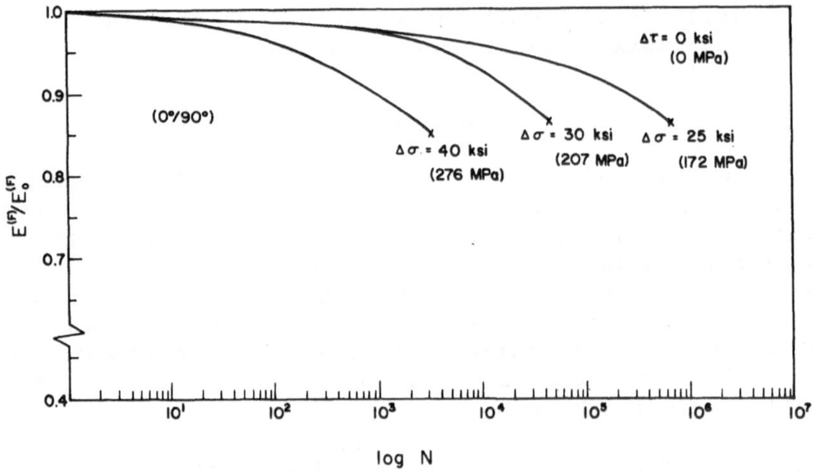

Fig. 1. Degradation of composite modulus $E^{(F)}$ in fill direction of G-10 laminate with 0/90 fiber orientation during uniaxial tensile fatigue [flat coupon specimen, R ≃ 0.01, f = 1 Hz, T = 77 K (-321°F)].

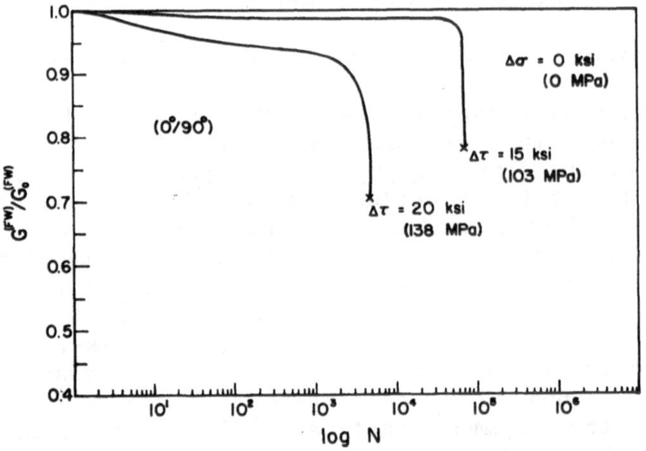

Fig. 2. Degradation of composite shear modulus $G^{(FW)}$ of G-10 laminate with 0/90 fiber orientation under pure torsional cyclic stresses, Δτ [tubular specimen, R ≃ 0.01, f = 1 Hz, T = 77 K (-321°F)].

Pure Torsional Fatigue of Cross-Laminated Composite

Degradation of torsional stiffness of (0/90) composite is shown in Fig. 2. The torsional stiffness remains relatively constant after the first few hundred cycles; very little stiffness reduction is observed until just before the final fracture when a drastic shear stiffness drop occurs. A threshold state of torsional fatigue damage seems to exist before the catastrophic fracture. The threshold level appears to be stress dependent. An increase in the applied shear stress lowers the threshold level. The final torsional stiffness retention is about 70% of its initial value.

Uniaxial Tensile Fatigue of (±45) Composite

In the (±45) composite, degradation phenomena similar to the (0/90) case were observed and are shown in Fig. 3. Due to the unfavorable fiber orientations, the rate of stiffness change in the (±45) case during fatigue is higher, and the stiffness retention at fracture is lower than that in the (0/90) case. The final stiffness retention, which is about 70% of the initial tensile modulus, is again approximately the same for the (±45) composite irrespective of the cyclic stresses applied. The macroscopic property degradation illustrated in Figs. 1 and 3 clearly reveals the significant cumulation of massive microscopic damages, such as microfracture at the crossovers, matrix cracking, and interface debonding in the composite during the wear-out process of fatigue. The matrix-dominated fatigue degradation leads to a functional failure before cyclic structural fracture if the stiffness reduction becomes larger than the range acceptable in the design and analysis.

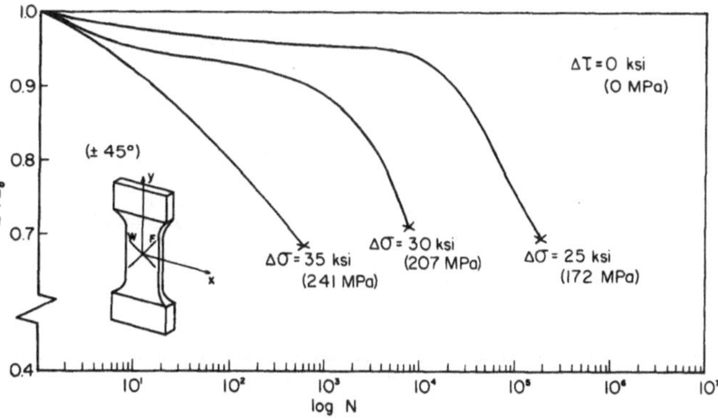

Fig. 3. Degradation of composite modulus $E^{(Y)}$ along loading direction in uniaxial tensile fatigue of G-10 composite laminate with ±45 fiber orientation [flat coupon specimen, $R \approx 0.01$, $f = 1$ Hz, $T = 77$ K (-321°F)].

The final fatigue fracture in the composite is thought to be governed by the life of the strong, load-bearing glass fibers and, thus, less sensitive to the matrix-dominated damage.

Effects of Surface Finishing

It is well known that the surface condition significantly influences fatigue failure. In Fig. 4, stiffness degradations of a flat coupon with virgin surfaces and of a tubular specimen with machined surfaces under the same level of axial cyclic stress were compared. The tubular specimen clearly shows a larger amount and higher rate of stiffness reduction, which are definitely attributed to the differences in surface conditions of the two types of specimen. The machining process introduced a considerable amount of microscopic damage on the surface of the tubular specimen. Cooling of the composite down to cryogenic temperature induced additional thermal stresses, which aggravated the situation even more. Thus, the tubular specimen started out with a much higher amount of defects than the flat coupon and experienced more severe degradation during fatigue. Despite the great difference in stiffness reduction, the fatigue lives for fracture are about the same for both cases. This indicates further that fatigue stiffness degradation is a matrix-dominated damage mechanism, whereas fatigue fracture is mainly associated with fiber-controlled failure.

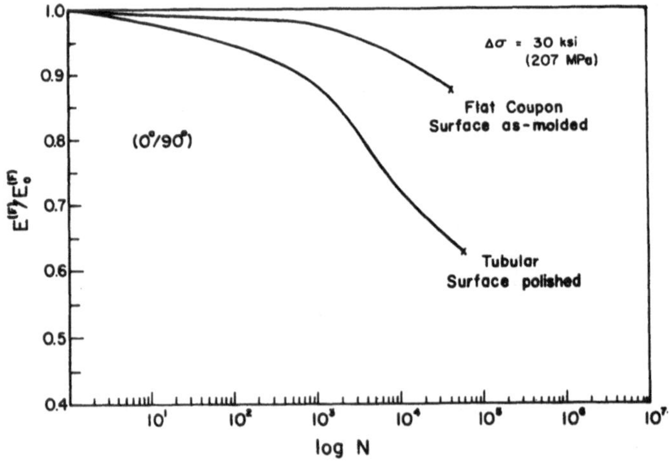

Fig. 4. Effect of surface condition on modulus degradation in G-10 composite with 0/90 fiber orientation under uniaxial tensile fatigue [R ≃ 0.01, f = 1 Hz, Δσ = 207 MPa, T = 77 K (-321°F)].

FATIGUE DAMAGE MECHANISMS

Fatigue damage mechanisms in composites are generally very complex due to their heterogeneous microstructure. A fractographic study of test specimens was conducted to deduce information of damage mechanisms of fatigue in the G-10 composite. The fracture surface appearance of a tubular composite that failed in uniaxial tensile fatigue is shown in Fig. 5(a). Fracture is clearly defined in a plane normal to the loading direction. The fracture surface was further examined under a scanning electron microscope. Little fiber pull-out along the loading direction was observed and transverse fibers lay close to the fracture surface. A typical broken-fiber bundle is seen in Fig. 6(a). The fibers generally failed in a brittle manner with a smooth and clear-cut fracture surface. Most of the fibers in the bundle fractured at approximately the same level very close to the fracture surface. Pull-out lengths of the fibers were generally small. A resin-rich area of the fracture surface, which had a glassy appearance, is shown in Fig. 6(b). Cracks normal to the loading direction dominate the failure of this area. Characteristics of a typical fiber-matrix interface are shown in Fig. 6(c). Debonding of fibers from the matrix was evident in almost every interface region, especially at the ends of fiber bundles. Some of the interface cracks extended into the matrix. Delamination between adjacent plies was not evident in the uniaxial tensile fatigue. The fracture surface of a tubular specimen that failed in pure torsional fatigue is shown in Fig. 5(b). The failure mode was completely different from that of a tensile fatigue case. Fracture followed a spiral path with massive delamination and subsequent buckling of fabric layers. SEM fractographs

(a) (b)

Fig. 5. Fracture surface of G-10 composite laminates (a) under uniaxial tensile fatigue [$\Delta\sigma$ = 207 MPa, R ≃ 0.01, f = 1 Hz, T = 77 K (-321°F)], (b) under pure torsional fatigue [$\Delta\tau$ = 103 MPa, R ≃ 0.01, f = 1 Hz, T = 77 K (-321°F)]. (Reduced 19%.)

showed microscopic fracture surface characteristics unique to torsional fatigue failure with extensive fiber/matrix interface and matrix cracking as well as microbuckling of fibers. The pull-out lengths of fibers were relatively long, as shown in Fig. 7(a). The matrix surface was much rougher [Figs. 7(b) and 7(c)] than the surface in the tensile fatigue case.

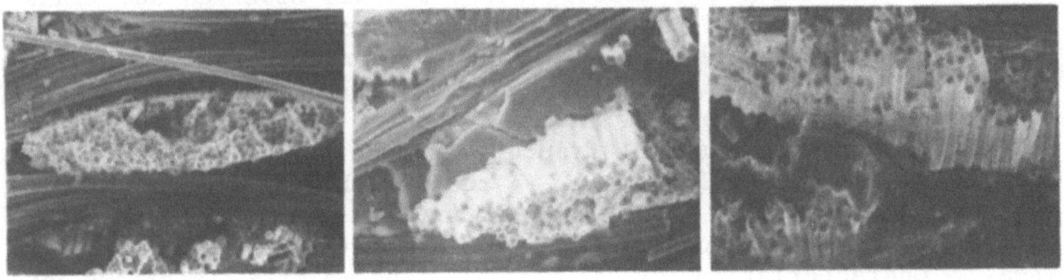

(a) (b) (c)

Fig. 6. SEM fracture surface characteristics of glass-fabric-reinforced G-10 composite under uniaxial tensile fatigue [$\Delta\sigma$ = 207 MPa, R $\simeq$ 0.01, f = 1 Hz; T = 77 K (-321°F) (a) fiber bundle fracture (X100), (b) matrix cracking (X150), (c) interface debonding (X150). (Reduced 19%).

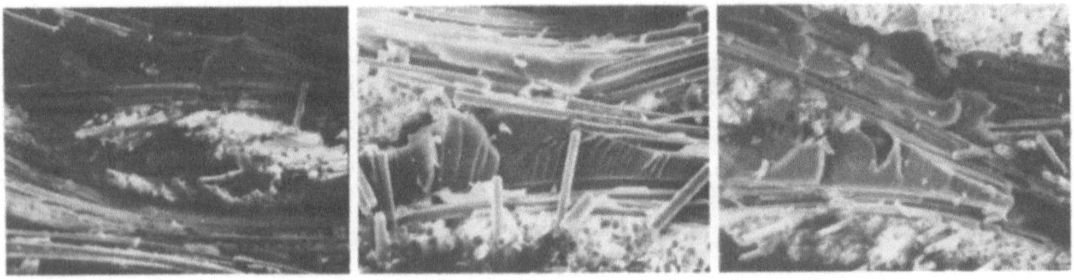

(a) (b) (c)

Fig. 7. SEM fracture surface characteristics of G-10 composite under uniaxial tensile fatigue [$\Delta\sigma$ = 103 MPa, R $\simeq$ 0.01, f = 1 Hz; T = 77 K (-321°F) (a) fiber bundle fracture (X100), (b) matrix-rich region (X150), (c) interface debonding (X100). (Reduced 19%).

CONCLUSIONS

Fatigue degradation of G-10 grade, glass-fabric-reinforced composite laminates subjected to cyclic, uniaxial tensile and pure torsional loading at cryogenic temperatures has been studied. Based on the results obtained, the following conclusions may be reached:

1. Fatigue degradation of the composite laminate at cryogenic temperatures can be clearly described by the change of material stiffness.

2. Stiffness reduction in all cases decreases monotonically with the number of loading cycles.

3. The amount and rate of stiffness degradation depend on the applied cyclic stress range.

4. The axial stiffness reduction in the (±45) composite under uniaxial tensile fatigue is more severe and the rate of degradation is more rapid than that in the (0/90) laminate under the same loading mode.

5. In pure torsional fatigue, the steady-state stiffness reduction covers most parts of the fatigue life, and the final stage of catastrophic change of material stiffness occurs at a threshold level depending upon the level of the applied cyclic load.

6. The surface condition is of great importance in controlling the fatigue damage initiation, growth, and cumulation. Stiffness degradation in a laminate with machined surfaces is more severe than that in a composite with virgin surfaces.

7. Fatigue damage mechanisms are related to loading modes and microstructures. For the (0/90) composite under tensile fatigue, damage is controlled by fracture of load-bearing fiber bundles, first at crossovers and then elsewhere. For the (±45) composite under tensile fatigue and the (0/90) laminate under pure torsional fatigue, matrix-dominated damage by interface debonding, matrix cracking, and fiber and fabric buckling prevails throughout the material.

REFERENCES

1. M. B. Kasen, Composite laminate applications in magnetic systems, in: "Proceedings of the Second International Conference on Composite Materials," (ICCM-II), B. Noton, R. Signorelli, K. Street, L. Phillips, eds., The Metallurgical Society of AIME, New York (1978).

2. A. F. Clark, R. P. Reed, and G. Hartwig, eds., "Nonmetallic Materials and Composites at Low Temperatures," Plenum Press, New York (1979).
3. R. P. Reed, ed., "Materials Studies for Magnetic Fusion Energy Applications at Low Temperatures-III," Technical Report NBSIR 80-1627, National Bureau of Standards, Boulder, Colorado (1980).
4. S. S. Wang, E. S.-M. Chim, D. F. Socie, J. V. Gauchel, and J. L. Olinger, Tensile and torsional fatigue of fiber reinforced composites at cryogenic temperatures, submitted to J. Eng. Mater. Technol., Trans. ASME (1981).
5. S. S. Wang, E. S.-M. Chim, and D. F. Socie, Biaxial fatigue of fiber reinforced composites at cryogenic temperature, Part I: Fatigue fracture life and mechanisms; Part II: Stiffness degradation and energy dissipation, submitted to J. Eng. Mater. Technol., Trans. ASME (1981).
6. Plastic - general test methods, nomenclature, in: "Annual Book of ASTM Standards, Part 35," American Society for Testing and Materials, Philadelphia (1980).
7. R. R. Rizzo and A. A. Vicario, A finite element analysis for stress distribution in gripped tubular specimens, in: "Composite Materials: Testing and Design (Second Conference)," ASTM STP 497, American Society for Testing and Materials, Philadelphia (1972).
8. W. W. Stinchcomb and K. L. Reifsnider, Fatigue damage mechanisms in composite materials: A review, in: "Fatigue Mechanisms," ASTM STP 675, American Society for Testing and Materials, Philadelphia (1979).

DEGRADATION OF FIBER-REINFORCED COMPOSITE MATERIALS AT CRYOGENIC TEMPERATURES, PART II—MULTIAXIAL FATIGUE

S. S. Wang and E. S.-M. Chim

Department of Theoretical and Applied Mechanics
University of Illinois, Urbana, Illinois

INTRODUCTION

In an associated paper,[1] cryogenic fatigue fracture of glass-fiber-reinforced composites subjected to in-phase, biaxial cyclic stresses has been studied. Both the fatigue fracture life and associated failure mechanisms in the composites have been examined. It has been discussed[2] that in many cryogenic engineering applications of fiber-reinforced composites, the consideration of fatigue fracture alone is not adequate for a reliable design and that other parameters measuring progressive fatigue damage in the material may be needed also. Since many cryogenic composite devices and structures, which are usually under a complex pulsating state of electromagnetic and mechanical stresses, are designed on the basis of stiffness and/or stability considerations, naturally, degradation of stiffness or modulus of the composite material subjected to a complex state of cyclic stress is of critical importance. Furthermore, stringent requirements of maintaining a proper cryogenic temperature level to achieve optimum performance (for example, superconductivity) led to the consideration of hysteresis energy dissipation or internal damping in the material during cyclic loading.

In this research, two important subjects were studied: degradation of stiffness and hysteresis energy dissipation in the G-10 grade cryogenic composite[2] under biaxial tension-torsion fatigue at liquid nitrogen temperature. Owing to space limitations, only results of the stiffness degradation are presented in this paper. The study of hysteresis energy dissipation can be found in Ref. 1. The objective of this study was to investigate the progressive fatigue damage of the material through quantitative measurements of parameters that characterize the physical processes of fatigue degradation and that are related also to the performance of the composite under a complex cyclic stress state. From a broader consideration of various failure parameters, a better understanding of the multiaxial fatigue degradation behaviour of the material may be achieved. More complete fatigue failure criteria may be established for the life prediction and reliable fatigue-resistant design of cryogenic composite materials and structures.

MATERIALS, MULTIAXIAL FATIGUE TEST SYSTEM, AND PROCEDURE

The test specimens were made of NEMA/ASTM G-10 composite laminates commonly used in cryogenic engineering applications. The microstructure and basic mechanical properties of the composite laminates were reported in Ref. 2. Standard laminated composite tubes of 0.64-cm (0.25-in) wall thickness and 244-cm (8-ft) length were trimmed to the geometric configuration shown in Ref. 1. A computer-controlled, closed-loop, servohydraulic test system developed in the Materials Engineering Research Laboratory at the University of Illinois had been used for the biaxial tension-torsion fatigue testings. Specially designed cryogenic extensometers were constructed to measure both the axial and torsional deformations simultaneously.[1] All experiments were performed in a liquid nitrogen environment with T = 77 K (-321°F). A detailed description of the materials and specimen preparation, the fatigue test system, and the experimental procedure can be found in Refs. 1 and 2.

STIFFNESS DEGRADATION UNDER BIAXIAL FATIGUE

As a quantity that is physically well defined and experimentally measurable, the stiffness degradation is of significant importance and primary interest here because it relates microscopic damage to macroscopic performance of the composite under fatigue loading. The concept of using the stiffness change as a measure of microscopic damage in fiber-reinforced composites has been noted by many researchers, for example, Refs. 3-6. Broutman and Sahu[3] were among the first to provide a quantitative correlation between the change of modulus and the development of microcracks in a glass-epoxy composite during fatigue. The notion of employing the stiffness change as a definition of fatigue failure of composite materials was first proposed by Salkind.[7] Degradation of fiber-reinforced composites in a severe thermal environment, such as cryogenic temperatures, has just received the needed attention recently.[8,9] The property change of a composite material in a cryogenic environment is a new area of current development in composite materials technology and requires a considerable amount of effort for future cryogenic engineering applications.[10,11]

Owing to the complexities involved in both testing and theories, fatigue degradation of fiber-reinforced composites subjected to a loading condition other than the conventional uniaxial or flexural fatigue has not been studied extensively. In fact, fatigue degradation and fracture of fiber-reinforced composites under multiaxial loading are areas of very much current research. Francis et al.[12] and Owen and Rice[13] have tested graphite-epoxy and glass-polyester composites under combinations of tension, torsion, and internal pressure. Fracture envelopes have been constructed for several cases with different states of biaxial cyclic stress. To the authors' knowledge, there is no information available in the literature on the subject of progressive degradation of stiffness of fiber composites under biaxial cyclic fatigue, not mentioning additional complexities introduced by the cryogenic environment.

Tensorial Nature of Stiffness Degradation

In the form of classical lamination theory,[14] the stiffness of a fiber-reinforced composite laminate under combined tension-torsion fatigue at a cryogenic temperature may be related to the deformation and loading by the following:[14]

$$\underset{\sim}{N} = \underset{\sim}{A}\, \underset{\sim}{\varepsilon}_o + \underset{\sim}{B}\, \underset{\sim}{\kappa} - \underset{\sim}{N}^T, \tag{1}$$

where $\underset{\sim}{N}$ and $\underset{\sim}{N}^T$ are forces resulting from mechanical and thermal loadings; $\underset{\sim}{\varepsilon}_o$ and $\underset{\sim}{\kappa}$ are the midplane strains and curvatures, and $\underset{\sim}{A}$ and $\underset{\sim}{B}$ are extensional and coupling stiffness matrices. For the G-10 composite with fill and warp directions parallel and perpendicular to the axial loading, $\underset{\sim}{B}$ vanishes identically for the orthotropic system, and $\underset{\sim}{A}$ and $\underset{\sim}{N}^T$ may be expressed explicitly by

$$\underset{\sim}{A} = \underset{\sim}{Q}\, t \tag{2}$$

$$\underset{\sim}{N}^T = \underset{\sim}{Q}\, \underset{\sim}{\alpha}\, \Delta T\, t, \tag{3}$$

where t, $\underset{\sim}{\alpha}$, and ΔT are the laminate thickness, thermal expansion coefficients, and the temperature change, respectively. The $\underset{\sim}{Q}$ matrix is the laminate stiffness matrix with its components in a plane stress condition given by:[14]

$$Q_{11} = E^{(F)}/[1 - \nu^{(FW)}\, \nu^{(WF)}], \tag{4}$$

$$Q_{22} = E^{(W)}/[1 - \nu^{(FW)}\, \nu^{(WF)}], \tag{5}$$

$$Q_{12} = Q_{21} = \nu^{(WF)}\, Q_{11} = \nu^{(FW)}\, Q_{22}, \tag{6}$$

$$Q_{66} = G^{(FW)}, \tag{7}$$

where the superscripts F and W refer to the fill and warp direction, respectively. In general, the product $\nu^{(FW)}\, \nu^{(WF)}$ is orders of magnitude smaller than other terms in the above equations and may be negligible in actual numerical computation. Therefore, the stiffness component, Q_{11}, for example, may be approximated by $E^{(F)}$, and Q_{22}, by $E^{(W)}$.

The composite laminate stiffness, $\tilde{Q}$, in Eqs. 2-7 is noted to be an average of the contribution of each identical, unbalanced fabric-epoxy layer within the laminate. It is important to note that during the course of cyclic fatigue, the stiffness of the composite, $\tilde{A}$ and $\tilde{Q}$, changes with the cumulation of the microscopic damage. In other words, the stiffness of the composite is a function of the number of fatigue loading cycles. It is also noted that in order to describe fully the fatigue degradation, the tensorial nature of the stiffness matrix of the composite must be considered, i.e., the change of all of the components of $\tilde{Q}$ [or the four independent elastic constants $E^{(F)}$, $E^{(W)}$, $G^{(FW)}$, and $\nu^{(FW)}$]. The necessity of simultaneous consideration of all of the four independent elastic constants during fatigue is one of the most unique features in the evaluation of fatigue damage of fiber composites. As will be shown later, the situation is especially true in the case of fiber composites under multiaxial cyclic loading in which the degradation of stiffness is much more severe than that in a unidirectional case.

The determination of the changes of all of the four independent elastic constants during biaxial cryogenic fatigue is a formidable task. Owing to space limitations, the two most critical ones, $E^{(F)}$ and $G^{(FW)}$, are presented in the paper for illustrative purposes for the G-10 composite under tension-torsion fatigue at liquid nitrogen temperature.

Effects of Cyclic Torsional Stress

Degradation of stiffness (in terms of modulus changes) as a function of the cyclic torsional stress level is examined first. The change of Young's modulus along the fill direction, $E^{(F)}/E_o^{(F)}$, where $E_o^{(F)}$ is the initial fill modulus, is shown in Fig. 1 for the G-10 composite under cyclic fatigue with a constant axial stress amplitude, $\Delta\sigma$, of 172 MPa (25 ksi) and various torsional stress amplitudes, $\Delta\tau$'s. For comparative purposes, degradation of $E^{(F)}$ under uniaxial tensile fatigue of the same loading amplitude is also given. Much more significant stiffness degradation of the composite is observed in a biaxial cyclic stress state. The three stages of degradation in uniaxial fatigue are observed also in the present biaxial fatigue. The initial drop of the fill modulus (stage I), due to significant damage on the material in first few cycles, is not clearly shown. However, the steady-state decrease in modulus (stage II) associated with propagation of initial flaws and continuous initiation of new flaws in the material, is quite obvious. A shorter duration of the steady-state degradation of the Young's modulus is found in the biaxially fatigued composite with a

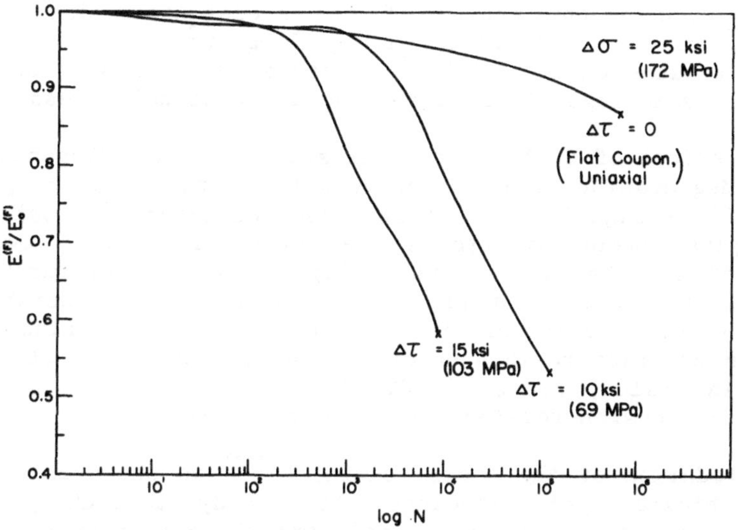

Fig. 1. Influence of cyclic torsional stress on fill-modulus $E^{(F)}$ degradation in biaxial cryogenic fatigue.

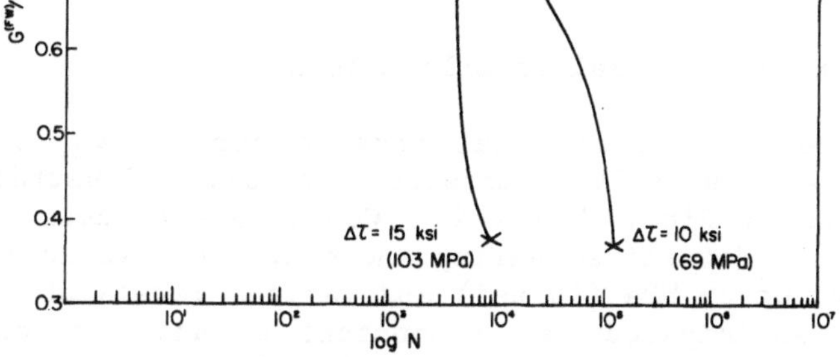

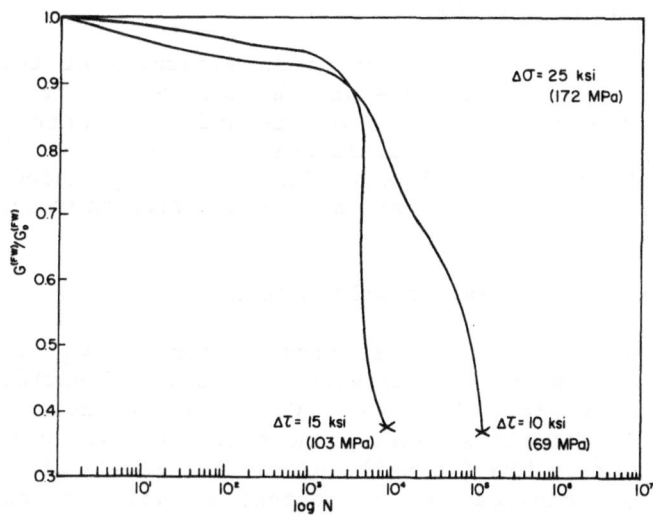

Fig. 2. Influence of cyclic torsional stress on shear-modulus $G^{(FW)}$ degradation in biaxial cryogenic fatigue.

higher cyclic torsional stress. For the present case, a small increase in torsional cyclic stress from 69 MPa (10 ksi) to 103 MPa (15 ksi) reduces the course of the steady-state degradation by almost an order of magnitude, i.e., from 10^3 to 10^2 cycles. The rate of modulus change, dE/dN, in stage II, however, is approximately the same regardless of the cyclic torsional stress level.

The effect of cyclic torsional stress on the cryogenic biaxial fatigue degradation is most appreciable in the final stage of the cyclic life (stage III) where stiffness deteriorates most rapidly. At any given number of cycles the stiffness under a lower $\Delta\tau$ is clearly higher than that under a higher $\Delta\tau$. It is further noted that the stiffness in a biaxial situation is considerably lower than that in a uniaxial one. For example, the final modulus reduction at fracture in a uniaxial tensile fatigue is about 15% of its initial value, but a 45-50% reduction of $E^{(F)}$ is observed in the biaxial tension-torsion cryogenic fatigue.

Degradation of the shear modulus $G^{(FW)}$ is shown in Fig. 2 for the same biaxial cyclic loadings. The steady-state change of shear modulus is found to extend to approximately one order of magnitude longer than that of $E^{(F)}$. For instance, the stage II degradation of $G^{(FW)}$ for the case of $\Delta\tau$ = 69 MPa (10 ksi) and $\Delta\sigma$ = 172 MPa (25 ksi) ends at about 10^4 cycles, whereas this occurs at about 10^3 cycles for $E^{(F)}$ in Fig. 1. The final stage of shear modulus degradation becomes much more rapid and the rate of the stiffness change, dG/dN, becomes much steeper as the cyclic torsional stress is increased in the biaxial fatigue.

The more significant reduction in stiffness of the composite under biaxial cryogenic fatigue with higher cyclic torsional stress mainly results from the severe matrix and fiber/matrix interface cracking and the interlayer delamination introduced by the torsional stress. These additional extensive microdamage modes appear all over the composite and are confirmed by electron microscopic examinations.[1]

Influences of Cyclic Axial Tensile Stress

Influences of cyclic axial tensile stress, $\Delta\sigma$, on stiffness degradation of the G-10 composite are best illustrated by the results shown in Figs. 3 and 4. Continuous changes of $E^{(F)}$ and $G^{(FW)}$ have been determined during the course of cyclic loading with a constant $\Delta\tau$ of 69 MPa (10 ksi) and various levels of cyclic axial stress. As anticipated, at a constant $\Delta\tau$ and a given number of loading cycles, increasing the amplitude of cyclic axial stress clearly increases both the magnitude and rate of stiffness degradation. When the axial stress amplitude is low, e.g., $\Delta\sigma$ = 172 MPa (25 ksi), both the stage II and stage III degradation of the material properties can be seen clearly from the figures and the

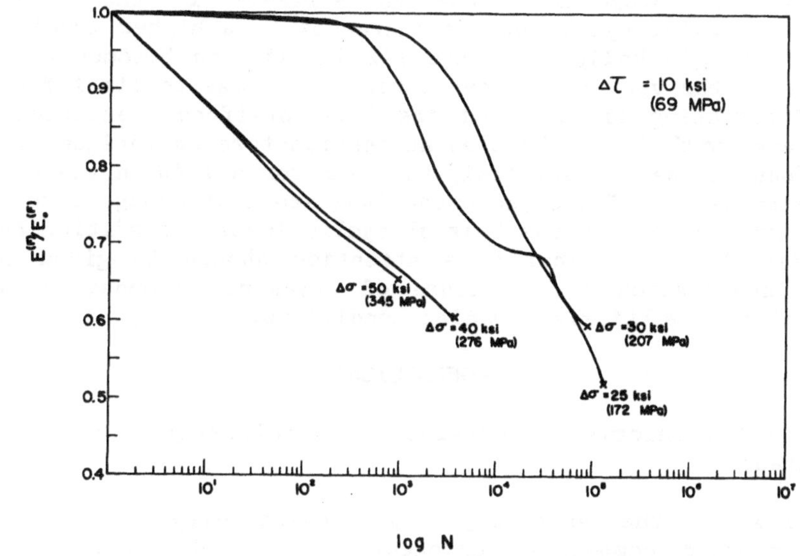

Fig. 3. Influence of cyclic axial stress on fill-modulus $E^{(F)}$ degradation in biaxial cryogenic fatigue.

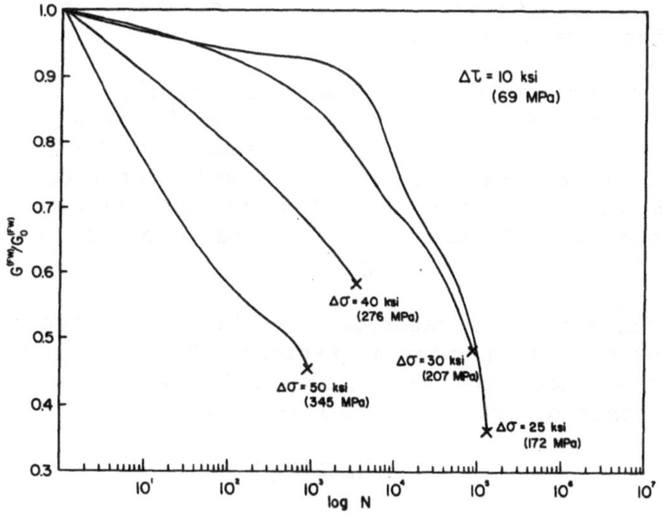

Fig. 4. Influence of cyclic axial stress on shear-modulus $G^{(FW)}$ degradation in biaxial cryogenic fatigue.

duration of the stage II degradation decreases with the increase in $\Delta\sigma$. As the axial cyclic stress increases to a higher level, e.g., $\Delta\sigma$ = 276 MPa (40 ksi), the stage III degradation becomes unobservable due to the very rapid development of damage to final fracture. It is interesting to note that the final stiffness retention seems to increase with $\Delta\sigma$ in the present tension-torsion fatigue, whereas it is found to be approximately the same for all $\Delta\sigma$ in the uniaxial tensile fatigue.[2] These phenomena indicate that damage development is much more severe in the biaxial cyclic loading condition than in the uniaxial case. Thus more attention should be given to the fatigue degradation and fracture of cryogenic composites in the more realistic multiaxial loading conditions.

CONCLUSIONS

From the information obtained, the following conclusions may be reached:

1. Owing to the anistropy and inhomogeneity of the fiber-reinforced composite, cryogenic fatigue degradation in the material is inherently tensorial in nature, i.e., tensor modulus changes must be considered instead of one single component.

2. In a multiaxial cryogenic fatigue loading condition, the magnitude and rate of stiffness degradation are found to be much more significant than those in a uniaxial fatigue case. In the fatigue-resistant design of cryogenic composite components and structures under complex stresses, these should be taken into consideration.

3. Three stages of modulus reduction during uniaxial fatigue appear also in the present biaxial cryogenic fatigue. Although stage I is not clearly shown in the present study, stage II (steady-state) and III (rapid deterioration) may be defined for each individual loading condition. In general, much more rapid deterioration of shear modulus in stage III is observed than that of Young's modulus.

4. The final stiffness retention in biaxial cryogenic fatigue is much lower than in uniaxial fatigue. The magnitude and cyclic life associated with the final stiffness depend on the cyclic stress state of the multiaxial loading.

ACKNOWLEDGMENTS

The authors wish to express their sincere gratitude to Drs. J. L. Olinger and J. V. Gauchel of Owens-Corning Fiberglas Corporation (OCF) for their encouragement and fruitful discussion and to Prof. D. F. Socie of the University of Illinois for his assistance in conducting the tests during the course of this study. The support by OCF through Contract DML9815 to the University of Illinois is also gratefully acknowledged.

REFERENCES

1. S. S. Wang, E. S.-M. Chim, and D. F. Socie, Biaxial fatigue of fiber-reinforced composites at cryogenic temperatures, Part I: Fatigue fracture life and mechanisms; Part II: Stiffness degradation and energy dissipation, submitted to J. Eng. Mater. Technol., Trans. ASME (1981).
2. S. S. Wang, E. S.-M. Chim, D. F. Socie, J. V. Gauchel, and J. L. Olinger, Tensile and torsional fatigue of fiber reinforced composites at cryogenic temperatures, submitted to J. Eng. Mater. Technol., Trans. ASME (1981).
3. L. J. Broutman and S. Sahu, Progressive damage of a glass fiber reinforced plastic during fatigue, in: "Proceedings of the 24th SPI Annual Technical Conference," The Society of Plastic Industry, Washington, D.C. (1969).
4. T. K. O'Brien and K. L. Reifsnider, Fatigue damage: Stiffness/strength comparisons for composite materials, J. Test. Eval. 5:384 (1977).
5. W. W. Stinchomb and K. L. Reifsnider, Fatigue damage mechanisms in composite materials: A review, in: "Fatigue Mechanisms," ASTM STP 675, American Society for Testing and Materials, Philadelphia (1979).
6. T. Gottesman, Z. Hashin, and M. A. Brull, "Reduction of elastic moduli of fiber composites by fatigue crack accumulation," Technical Report No. 4, Contract No. N00014-78-0544, Office of Naval Research, Arlington, Virginia (1980).
7. M. S. Salkind, Fatigue of composites, in: "Composite Materials: Testing and Design (Second Conference)," ASTM STP 497, American Society for Testing and Materials, Philadelphia (1972).
8. M. B. Kasen, R. E. Schramm, and D. T. Read, Fatigue of composites at cryogenic temperatures, in: "Fatigue of Filamentary Composite Materials," ASTM STP 636, American Society for Testing and Materials, Philadelphia (1977).
9. R. L. Tobler and D. T. Read, Fatigue resistance of a uniaxial S-glass/epoxy composite at room and liquid helium temperatures, J. Compos. Mater. 10:32 (1976).
10. M. B. Kasen, Composite laminate applications in magnetic fusion energy superconducting magnet systems, in: "Proceedings of the Second International Conference on Composite Materials (ICCM-II)," B. Noton, R. Signorelli, K. Street, and L. Phillips, eds., The Metallurgical Society of AIME, New York (1978).

11. R. P. Reed, ed., "Materials Studies for Magnetic Fusion Energy Applications at Low Temperatures - III," Technical Report NBSIR80-1627, National Bureau of Standards, Boulder, Colorado (1980).
12. P. H. Francis, D. E. Walrath, D. F. Sims, and D. N. Weed, Biaxial fatigue loading of notched composites, J. Compos. Mater. 11:488 (1977).
13. M. J. Owen and D. J. Rice, Biaxial strength behavior of glass reinforced polyester resins, submitted to "Composite Materials: Testing and Design (Sixth Conference)," ASTM STP XXX," American Society for Testing and Materials, Philadelphia (1981).
14. R. M. Jones, "Mechanics of Composite Materials," Scripta (McGraw-Hill), Washington, D.C. (1975).
15. R. W. Hertzberg and J. A. Manson, "Fatigue of Engineering Plastics," Academic Press, New York (1980).

CHARACTERIZATION OF GLASS-REINFORCED COMPOSITES FOR CRYOGENIC APPLICATIONS*

J. V. Gauchel, J. L. Olinger, and D. C. Lupton

Research and Development Division, Owens-Corning Fiberglas Corporation, Granville, Ohio

INTRODUCTION

The reaction of a composite system to a service load is a complex function of the physical and mechanical properties of its components (matrix and reinforcement); the relative directionality of the applied loads versus the geometry of the reinforcement (anisotropy effect); the magnitude of the applied loads, and the rate at which they are introduced (viscoelasticity effect). Although all these parameters may not be important in a particular design with a selected reinforcement and simple loading, understanding that these effects exist and being prepared to include them in design criteria is critical for developing cost-effective high performance composite parts. As composite systems are introduced as secondary and primary structures for cryogenic applications, characterization of the interaction of stress and low temperatures on the performance of these systems is also necessary.

Evaluation of a composite material for cryogenic applications requires investigating both the structural and functional failure characteristics of that system and relating these characteristics to the service requirements of the particular application being considered. Good initial physical and mechanical properties at cryogenic temperatures are not sufficient to guarantee long-term performance. The effect of dynamic thermal and mechanical stresses on the integrity of a system must be considered in order to determine acceptable design criteria.

*Invited paper.

Owens-Corning has investigated the effect of reinforcement form on the response of glass-reinforced composite laminates to dynamic stresses. Laminates have been exposed to cyclic thermal shock, tensile fatigue, and creep loadings at cryogenic temperatures. During the course of the experiments, the relative containment performance of these systems has been monitored. The results of these tests indicate that a random reinforcement gives the best overall compromise of mechanical and containment properties.

This paper presents the results of the above investigation. Emphasis is placed on relating mechanical evaluation and manufacturing technology to potential design criteria for composite systems used in cryogenic applications.

REINFORCEMENT EFFECTS

Glass reinforcements come in many shapes, sizes, and forms. The proper form of reinforcement for a given application depends on the functional performance criteria for a particular application, the size and shape of the part, and how the part is to be manufactured. For an application such as a small cryogenic liquid containment vessel, composites reinforced with several fiber forms, such as fabric, woven roving, continuous-strand mat or chopped glass, alone or in combination, could be used to manufacture the part. Structurally, the tank could be designed so that each composite system could carry the static loads. However, in order to perform its function as a containment vessel, the part must be capable of holding a liquefied gas--both initially and after repeated thermal shock (i.e., filling). The effect of repeated thermal shock on the containment properties of the composite can be measured by using an apparatus designed by Owens-Corning. These tests effectively rank reinforcements with respect to functional performance.

Several composites manufactured with common forms of glass reinforcement were subjected to fully restrained cyclic thermal shock. The relative containment efficiency of these materials was determined by measuring the number of cycles required to induce a detectable flow of gaseous helium through the sample at -300°F. The cycle consisted of dropping liquid nitrogen on the top surface of a 19-in diameter, 1/8-in-thick flat disc that was restrained at its circumference by insulated steel plates. Liquid nitrogen was added until the back surface temperature of the disc reached -300°F. A radiant heater was then engaged, and the back side temperature returned to 75°F. Care was taken so that the top surface of the laminate never exceeded 130°F. The procedure was then repeated at an average rate of one cycle every 10 to 15 min. At 100-cycle increments, the dome over the top of the sample was pressurized with gaseous helium to 40 psig while the sample was maintained at -300°F. If no flow was detected after 5 min, the pressure was released and the test continued.

Table 1 details the cycles to leakage for laminates reinforced with woven roving, continuous-strand mat, woven E&S glass fabrics, and chopped E&S glass. All samples were prepared using the same resin system and similar cure cycles. The results indicate that glass fabric and chopped glass rovings are the most effective reinforcements under these test conditions.

Although fiber form is an important parameter, it is not the only determining material parameter. As mentioned above, the laminates tested were manufactured with the same resin system. This system was selected after a careful screening of available systems using as a criterion resistance to induced cracking under restrained thermal shock. During the course of the work, a modification was made to the resin system that reduced the residual thermal stresses caused during the curing cycle. The resulting effect on the laminate performance is also shown in Table 1. (This modified resin system will be referred to during the remainder of the paper as Cryoglas™.)

Table 2 gives data generated at NBS in Boulder on the mechanical performance at 4 K of the Cryoglas™ resin relative to several other standard systems. Cryoglas™ exhibits good strength at temperature with excellent strain to failure.

ELECTRICAL AND MECHANICAL CHARACTERIZATION

Having established the fiber form and resin formulation, a series of experiments were performed that characterized the mechanical and electrical performances of laminates manufactured from combinations of these materials. Since one of the major

Table 1. Effect of Reinforced Form on Cycles to Leakage of Glass-Reinforced Epoxy Resin Composites

Reinforcement Form	Cycles to Leakage
Woven Roving	100
Continuous-Strand Mat	600
Woven E&S Glass Fabrics	1500
Chopped E&S Glass	1800
Modified Matrix Chopped E-Glass Mat	2500

applications for composite systems in the cryogenic industry is as an insulator for superconductors, it was necessary to assess the electrical performance of a typical resin-fiber combination. ASTM Standard electrical tests were performed on the chopped-strand-mat-reinforced Cryoglas™ resin. NEMA grade G-10 laminates manufactured by Synthane Taylor were used as a control. The results of these experiments are shown in Table 3. The Cryoglas™ mat system performs similar to the G-10 system in every area except water absorption. Considering the availability of liquid water at cryogenic temperatures, this should not be a problem.

Table 2. Four-Point Bend Tests of Epoxy Resin Samples at -452°F (4 K)

Epoxy Type	Spec. No.	Young's Modulus (GPa)	Young's Modulus (10^6 psi)	Ultimate Strength (MPa)	Ultimate Strength (10^3 psi)	Ultimate Strain (%)
Araldite 6004 plus Lindride 16 and 12	A1	8.16	1.18	143	20.8	1.8
	A2	7.30	1.06	128	18.6	1.8
	A3	7.91	1.15	119	17.3	1.6
	Avg	7.79	1.13	130	18.9	1.7
IGC	B3	8.77	1.27	195	28.3	2.4
	B4	9.28	1.35	202	29.3	2.3
	Avg	9.03	1.31	199	28.8	2.4
OCF Cryoglas™ Resin	1	9.00	1.31	208	30.1	2.3
	2	8.50	1.23	206	29.9	2.3
	3	8.91	1.29	199	28.8	2.2
	4	8.99	1.30	172	24.9	1.9
	5	8.63	1.25	173	25.1	2.0
	Avg	8.81	1.28	192	27.8	2.2

Table 3. Room Temperature Electrical Properties of OCF Cryoglas™ and Synthane Taylor G-10 Woven Cloth/Epoxy Laminate

		Test Method	Cryoglas™ Mat Laminate	G-10 Laminate
Arc Resistance	Time, Sec	ASTM D495-73		
	Average		131.4	123.6
	Minimum		128.5	122.8
Dielectric Constant	@ 60 Hz	ASTM D150-74	5.33	5.71
	@ 1,000,000 Hz		4.62	4.90
Dissipative Factor	@ 60 Hz		0.014	0.011
	@ 1,000,000 Hz		0.019	0.017
Loss Index	@ 60 Hz		0.074	0.061
	@ 1,000,000 Hz		0.086	0.082
Breakdown Voltage		ASTM D229-72		
	Average		62,750	60,250
	Minimum		49,000	54,000
	Maximum		69,000	69,000
Liquid Absorption	Avg %	ASTM D570-63	0.14	0.06

Table 4. Material Properties Summary of 181 Style Fiberglass Fabric-Reinforced Cryoglas™ Resin

Property	Symbol	Unit	Test Temperature					
			80 F (300 K)			-320 F (78 K)		
Direction			0	45	90	0	45	90
Tensile Strength	σ_{ult}	PSI x 10^3	40.3	18.2	39.7	90.2	58.2	90.4
Tensile Modulus	E	PSI x 10^6	2.68	0.85	2.79	4.03	2.75	4.07
Poisson's Ratio	γ	Ratio	0.21	0.98	0.14	-	-	-
Shear Strength	σ_{shear}	PSI x 10^3	-	-	-	-	-	-
Flexural Strength	σ_{ult}	PSI x 10^3	-	-	-	-	-	-
Flexural Modulus	E	PSI x 10^6	-	-	-	-	-	-
Change in Temperature	ΔT	Fahrenheit	80 F to -320 F					
Direction			0		45		90	
Thermal Coefficient of Expansion	α	in./in. °Fx10^6	8.10		8.44		7.76	

Table 5. Material Properties Summary of Cryoglas™ Chopped Mat Cryogenic FRP Laminate

Property	Symbol	Units	Test Temperature (F)					
			80	75	-100	-105	-270	-320
Tensile Strength	σ_{ult}	PSI x 10^3	-	30.1	49.2	-	61.2	-
Tensile Modulus	E	PSI x 10^6	-	1.16	-	-	2.44	-
Compressive Strength	$\sigma_{edgewise}$ $\sigma_{flatwise}$	PSI x 10^3 PSI x 103	- -	14.7 45.0	-	-	72.6 103.0	-
Shear Strength	σ_{shear}	PSI x 10^3	-	3.9	-	-	10.2	-
Flexural Strength	σ_{ult}	PSI x 10^3	-	18.9	-	-	-	-
Poisson's Ratio	γ	Ratio	-	0.39	-	-	0.31	-
Specific Heat *	C_p	Btu/Lb F	0.27	-	-	-	0.11	-
Coefficient of Thermal Expansion	α	In./In. Fx10^{-6}	75 F to -105 F		-105 F to -320 F		75 F to -320 F	
			12.07		9.18		10.58	
Density *	p	Lbs/Ft^3	101					

* Values are approximate, as data shown is for a laminate of the same resin and glass content, but a continuous strand mat reinforcement

Table 6. Material Properties Summary of Hybrid Cryogenic FRP Laminate.*

Property	Symbol	Unit	Test Temperature					
			80 F			-320 F		
Direction			0	45	90	0	45	90
Tensile Strength	σ_{ult}	PSI x 10^3	31.3	18.2	29.7	58.7	42.1	62.0
Tensile Modulus	E	PSI x 10^6	1.77	1.13	1.68	2.65	2.44	2.78
Poisson's Ratio	γ	Ratio	0.26	0.57	0.22	-	-	-
Shear Strength	σ_{shear}	PSI x 10^3	-	-	-	-	-	-
Flexural Strength	σ_{ult}	PSI x 10^3	41.0	29.8	39.7	-	-	-
Flexural Modulus	E	PSI x 10^6	21.2	0.94	2.09	-	-	-
Change in Temperature	ΔT	Fahrenheit	80 F to -320 F					
Direction			0		45		90	
Thermal Coefficient of Expansion	α	In./In. Fx10^6	8.72		10.05		8.67	

* Construction: 2 plies 181 fabric/2 plies (3) oz chopped fiber mat/2 plies of 181 fabric all impregnated with Cryoglas™ resin

Tables 4, 5, and 6 give the static mechanical and physical properties versus temperature for Cryoglas™ resin reinforced with fabric, chopped-strand mat, and a hybrid of fabric and chopped-strand mat. Data in the warp, fill, and 45° direction for the fabric and hybrid system are also presented. Several general trends can be seen throughout the data. Both tensile strength and modulus double as the temperature is reduced from room temperature to liquid-nitrogen temperature. The exceptions are the data for a fabric-reinforced system loaded at 45° to the warp. Here, modulus and strength tripled. This enhanced temperature dependence may be attributed to the large dependency of the 45° specimen on resin shear strength. The performance and temperature variations of the Cryoglas™ systems are quite similar to those of the G-10 material currently being used in cryogenic applications.[1]

TIME-DEPENDENT MECHANICAL PERFORMANCE

Although initial static properties are important in designing cryogenic structures, the effect of long-term static and cyclic loads on cryogenic performance must also be factored into any practical design, either by safety factor or by actual evaluation. Data on the time-dependent load capability has been developed for Cryoglas™ mat-reinforced systems. Table 7 depicts the effect of

Table 7. Creep-Stress Rupture Data on G-10 and Cryoglas™ Mat Laminates.

Material Designation	Test Temperature	Percent of Ultimate Tensile Strength	Tensile Stress (psi)	Strain at Failure (In./In.)	Elapsed Time to Failure (Hours)
Synthane Taylor G-10	23 C	80	34,272	0.0113	36.6
		70	31,416	0.0105	12.6
		50	28,560	0.0094	1,056. *
OCF Cryoglas™ Mat (BP-17)	23 C	80	19,232	0.0165	3.2
		75	18,030	0.0243	4.6
		70	16,828	-	1,633.5 *
	-270 F	90	65,718	-	<10 sec-14.5 hrs
		80	58,416	-	14.2-361.
		75	54,765	-	>2,000 *
		68	49,377	-	850.5 *
		60	43,568	-	134.4 *

* Test terminated - specimen did not fail.

Note: Specimens approximately 0.1 inch thick x 1.0 inch wide.

long-term static load on the load-bearing capability of Cryoglas™ material at room temperature and at -270°F. Run-out values of approximately 70% of static ultimate strength are achievable with the mat reinforcement. By comparison, fabric-reinforced G-10 supports about 50% of its ultimate tensile strength long-term. If the stress is cyclic, further reduction in load-bearing performance is shown. For a mat-reinforced system subjected to a cyclic load at 1 Hz, run-out occurs at a stress level of approximately 40% of ultimate tensile strength with an r value of 0.05. The r value is defined as the minimum stress applied divided by the maximum stress applied. At constant maximum stress level, the smaller the r value, the greater the amplitude of the oscillatory component of the stress. Figure 1 shows the typical results for Cryoglas™ mat laminates fatigued at various r values over maximum stress level range of 40 to 90% of ultimate tensile strength. Although the scatter level is high, the general trends indicate that as the r value is increased, the cycles to failure increase. Conversely, as the percentage of the cyclic component of stress increases, the cycles to failure at constant maximum stress level decrease.

A further description of the effect of cyclic stresses can be seen in Figure 2, where the S-versus-N curves for G-10 and Cryoglas™ mat laminates at -270°F and r = 0.05 are plotted. In the region of 10^3 to 10^5 cycles to failure, the stress level required to cause failure is the same for both systems--this, despite the fact that G-10 has an ultimate tensile strength of nearly double

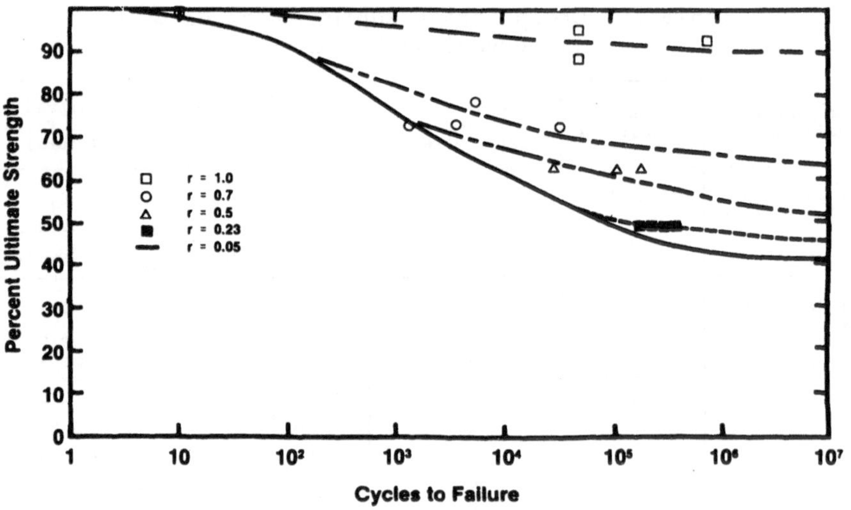

Fig. 1. Effect of r value on the fatigue life of Cryoglas™ mat laminates at -270°F.

that of the mat-reinforced system. Data on fabric loaded in the 45° direction in the 10^3 to 10^5 cycles-to-failure range fall about 10% below the curves established for the fabric in the warp direction. Thus, for any application requiring long-term performance under cyclic loading, fabric and mat systems would have the same design limits.

Functional failure can also be stress dependent. During the course of the fatigue testing on the Cryoglas™ mat-reinforced laminates and G-10 fabric laminates, data on hysteresis, modulus decay, and leakage were established. In general, modulus decay and damping increase dramatically prior to actual failure in a laminate. For uniaxially stressed G-10 laminates, failure occurred at a modulus decay level of approximately 15%--independent of the stress level applied. Large damping changes occurred at approximately 60% of the cycles to ultimate failure. Both modulus decay and damping increase are functions of the state of stress. A detailed description of these effects may be found in papers by Wang et al.[2-4] Use of these phenomena as a measure of degree of damage prior to failure in laminates has been previously hypothesized by Di Benedetto et al.[5]

The effect of stress level on leakage is also interesting. As the stress level approached the run-out stress level, the cycles required to induce leakage approached the cycles to failure. Thus, containment vessels designed using dynamic mechanical ultimates do not need a further safety factor to account for leakage failures.

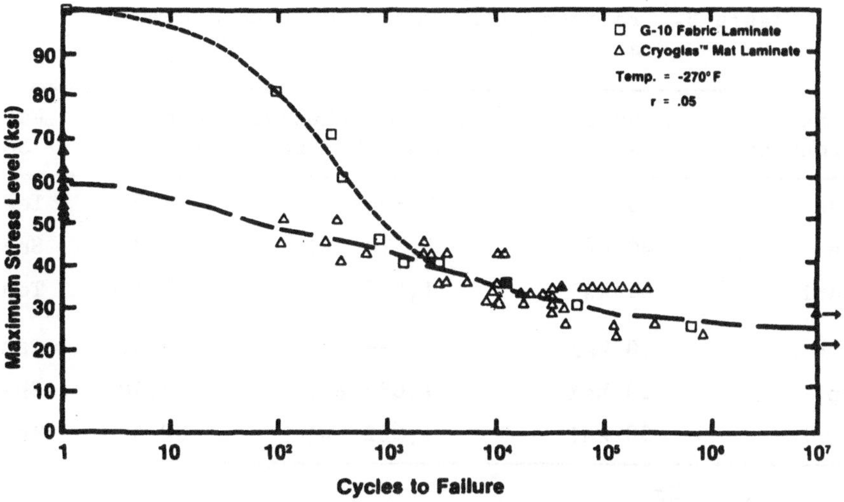

Fig. 2. S-N curve for tensile fatigue of cryogenic grade laminates. The solid symbols indicate laminates that had been soaked one year in LNG.

STRUCTURAL CONSIDERATIONS

Although test specimen evaluation is important for developing design data, the proof of a materials system is to make something from it and have it work in the field. To build structures from composite systems, it is either necessary to mold them in one piece or join them by some form of coupling. Adhesive bonding is the fastening method of choice. Table 8 shows the effect of bond design on the cryogenic fatigue performance of Cryoglas™ mat laminates bonded with Ren DA 509-13 polyurethane adhesive. Proper design of the joint forces the failure to occur in the adherend at nearly the same cycles to failure as the parent material. This means that, given proper consideration to joint design, large bonded structures may be assembled from components without affecting the functional performance of the system.

Owens-Corning has likewise looked at seamless molded parts as potential structures for cryogenic applications. One such prototype is shown in Figure 3. This 22-gal all chopped-fiberglass-reinforced tank was molded in one piece by a Preform Bladder/Resin Transfer technique. Modifications of this technique have allowed Owens-Corning to attach a 12-in long, 3-in diameter neck integrally during the molding operation. Thus, a one-piece seamless prototype inner vessel for a liquefied-gas Dewar can now be molded. Evaluation of the vacuum and outgassing performance of these structures is currently underway.

Table 8. Fatigue* Performance of Bonded Samples† at -270°F

Joint Conformation	Adherend Stress Level	Applied Adhesive Shear Stress	Cycles to Fail	Failure Mode
None	40 ksi	----	2,500	Tensile
Lap	40 ksi	1,393 psi	800	Shear
Scarf	40 ksi	1,417 psi	1,100	Tensile
None	30 ksi	----	95,000	Tensile
Lap	30 ksi	1,085 psi	5,480	Shear
Scarf	30 ksi	1,062 psi	90,400	Tensile

*Cycle rate: 1 Hz.
†Adherend: Cryoglas™ Mat Reinforced Laminate; Adhesive: REN DA 509-13.

Fig. 3. Prototype vessel for liquefied gas containment.

SUMMARY

Performance of glass-reinforced composites for cryogenic applications have been shown to be a complex function of materials, loading, time, and temperature dependence. Evaluation of materials systems for specific applications requires matching the design criteria to the in-service conditions of that part. Dynamic mechanical evaluation through fatigue and thermal shock testing provides design criteria that should be conservative for both structural and functional performance for most composite cryogenic structural applications. Fiber and resin selection based on ultimate static properties may be misleading when dealing with real world structures having loads applied over long times.

REFERENCES

1. "LNG Materials and Fluids Handbook," 1st Edition, D. Mann, ed., National Bureau of Standards, Boulder, Colorado (1977).
2. S. S. Wang, E. S-M. Chim, D. F. Socie, J. L. Olinger, and J. V. Gauchel, Tensile and torsional fatigue of fiber reinforced composites at cryogenic temperatures, submitted to J. Eng. Mater. Technol., Trans. ASME.
3. S. S. Wang, E. S-M. Chim, and D. F. Socie, Biaxial fatigue of fiber reinforced composites: Part I: Fatigue fracture life and mechanisms, submitted to J. Eng. Mater. Technol., Trans. ASME.
4. S. S. Wang, E. S-M. Chim, and D. F. Socie, Biaxial fatigue of fiber reinforced composites: Part II: Stiffness degradation and energy dissipation, submitted to J. Eng. Mater. Technol., Trans. ASME.
5. A. T. Di Benedetto, J. V. Gauchel, R. L. Thomas, and J. W. Barlow, Nondestructive determination of fatigue crack damage in composites using vibration tests, J. Mater., JMSLA, 7(2):211-215 (1972).

MECHANICAL PROPERTIES OF AN INSULATOR FOR THE JAPANESE LCT COIL

K. Koizumi, K. Yoshida, E. Tada, M. Nishi, and S. Shimamoto

Japan Atomic Energy Research Institute, Ibaraki, Japan

M. Nagai, K. Kadotani, and N. Tada

Hitachi, Ltd., Ibaraki, Japan

INTRODUCTION

Superconducting fusion magnets operate in high electromagnetic forces at liquid helium temperature. Electrical insulators in a winding must sustain high compressive forces as well as high voltage. Far fewer studies have been made on mechanical properties than on electrical properties of insulation design. Therefore, we have investigated mechanical and electrical characteristics of several kinds of insulation materials at 4 K in order to select insulators for the Japanese LCT coil.[1]

This paper describes the experimental results of tensile, compressive, bending, bond shear, and thermal contraction properties as well as breakdown voltages on several insulation materials from 4 K to 300 K.

DESIGN CRITERIA OF THE INSULATOR

Because the LCT coil is operated under large magnetic forces and at high current at liquid helium temperature, the insulation materials should have not only adequate electrical properties to avoid electrical shorts between pies and turns and between the winding and the helium vessel, but also sufficient mechanical strength to transmit the magnetic force from turn to turn, pie to pie, and from the winding to the helium vessel.

Also, when the coil is cooled down to 4 K, the gaps between the insulator and the winding and between the insulator and the helium vessel must be zero in order to transmit the magnetic force from the winding to the helium vessel. Thus, the thermal contraction of the insulator must match that of the conductor as closely as possible.

According to stress, electrical, and thermal analysis, the insulators used for the LCT coil should satisfy the following conditions:

1. The insulator between pies and between the winding and the helium vessel:
   a. Compressive strength of more than 400 MPa at 4 K.
   b. Thermal contraction the same as that of the copper.
   c. Breakdown voltage of more than 3,000 V at 4 K.
2. The insulator between turns:
   a. Compressive strength of more than 400 MPa at 4 K.
   b. Bond shear strength of more than 400 MPa at 4 K.
   c. Bonding material to hold the insulator between turns and pies should not decrease the cooling surface by flowing over the corner of the conductor.

TEST RESULTS AND DISCUSSION

Nine samples were selected as the first candidate materials for the insulator of the LCT coil, and the following preliminary tests were carried out at room temperature and 77 K: (1) tensile, (2) bending, (3) compression, (4) breakdown voltage, (5) thermal contraction, and (6) shear.

Table 1. Insulation Materials Studied for the Japanese LCT Coil

Sample symbol	Insulator sample	Usage
I-1	Glass-cloth tape(TE-2513) impregnated with epoxy resin(KE-5111) and cured at R.T. for 48 hours.	Insulator between turns
I-7	Epoxy resin(CEMEDIN-1500) was sandwithed with cured epoxy glass-cloth tape(SGT-F7036) and cured again at R.T. for 48 hours.	Insulator between turns
S-2	Epoxy glass-cloth laminates F.R.P. plate (VL-E200)	Insulator between pies,and between the winding and the helium vessel
L-1	Glass-cloth sheet(WE-35D100BS) impregnated with polyester resin (PS-518A-20) and cured at R.T. for 24 hours.	Insulator between the winding and the helium vessl

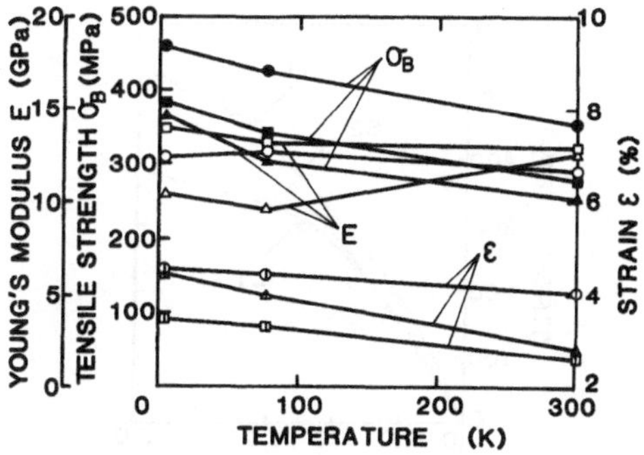

Fig. 1. Tensile properties of various insulation materials; σ_B: ● I1, ■ S1, ▲L1; E: ○ I1, □ S2, △L1; ε: ⦶ I1, ⊞ S2, ⧊L1.

On the basis of these test results, the four samples shown in Table 1 were selected as candidate materials for the insulator. Tensile tests, compression tests, cyclic loading tests, and breakdown voltage test were carried out at 4 K.

Mechanical Properties

Tensile test results. Ultimate strength, Young's modulus, and strain to failure in the warp direction obtained by tensile tests at 300 K, 77 K, and 4 K are plotted in Fig. 1. Three test specimens, which have a 40-mm gauge length, were tested at each test temperature. Figure 1 shows that the ultimate strength generally increases with decreasing temperature. The average ultimate strengths of insulator samples I1, S2, and L1 are 457, 383, and 367 MPa at 4 K, respectively; these values are more than 30% higher than their respective room temperature values. The strain to failure also increases with decreasing temperature, and the strain of all samples is around 4% at 4 K. The Young's modulus of sample L1 decreases with decreasing temperature because sample L1 uses a polyester resin instead of epoxy. Strain-stress curves of various insulators at 4 K are shown in Fig. 2. All samples show serrated yielding at 4 K owing to microscopic fracture of the resin.

Compressive test results. The compressive strength in the normal direction was measured at 300 K and 77 K. Three test specimens, 10-mm square x 3.0-mm thick, were used for all test samples. With 400-MPa compressive stress limited by the test capacity, materials listed here did not fracture at 4 K. To evaluate the compressive strength with cyclic loading, a cyclic compressive stress of 196 MPa was applied to the samples up to 10 cycles at 300 K and 77 K. After cyclic loading, the compressive strength was measured at room temperature. The stress of 196 MPa corresponds to the maximum stress in the insulation materials when

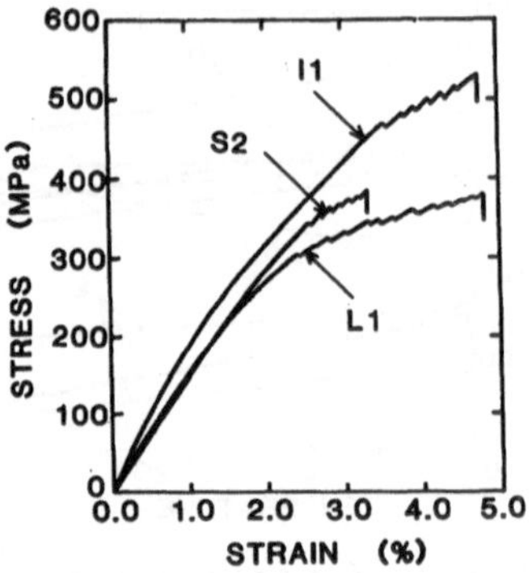

Fig. 2. Stress-strain curves of various insulation materials.

Table 2. Compressive Strength of Insulation Materials at 300 K and 77 K

Sample Symbol	Test Temperature (K)	Initial Compressive Strength (MPa)	Compressive Strength at 300 K after Cyclic Load: 196 MPa x 10^4 cycle at 300 K (MPa)	Compressive Strength at 300 K after Cyclic Load: 196 MPa x 10^4 cycle at 77 K (MPa)
I1	300	331.6 ±36.0	324.3 ±34.3	325.0 ±29.4
	77	753.0 ±9.0	---------	---------
S2	300	584.7 ±30.0	588.0 ±49.0	609.2 ±41.7
	77	1032.3 ±6.0	---------	---------
L1	300	476.9 ±17.0	393.6 ±51.5	457.3 ±20.0
	77	901.6 ±10.0	---------	---------

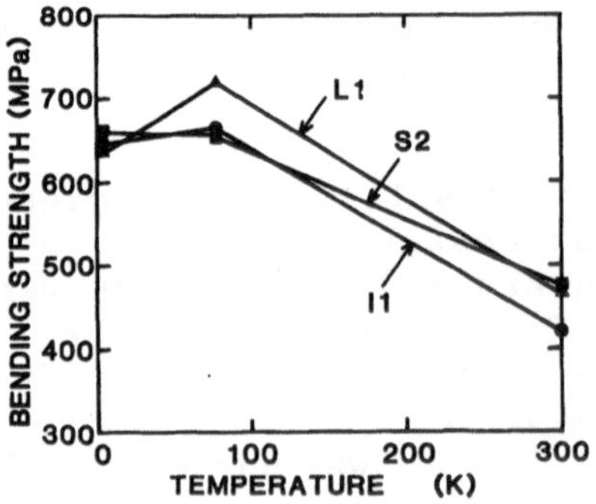

Fig. 3. Bending strength of various insulators.

the LCT coil is operated in the normal condition. Average values of three test specimens are summarized in Table 2. The compressive strength at 77 K is about twice that at room temperature. Cyclic loading at 300 K and 77 K did not significantly affect the compressive strength of any of the samples.

Bending test results. The bending strength in the warp direction was measured as a function of temperature. Three test pieces of each sample, 40-mm long and 7-mm wide, were cut from 1.0-mm sheets. Figure 3 shows the test results obtained by the three-point bend tests. The bending strength of all samples increases with decreasing temperature down to 77 K, but the strength at 4 K is almost the same as that of 77 K. The bending strength of all samples is 640 ±70 MPa at 4 K.

Bond shear test results. Bond shear strengths of insulators I1 and I7, bonded between chemically oxidized flat copper surfaces, were measured by the lap shear method at 300 K and 77 K. Test pieces, 30-mm long and 12-mm wide, were cut from 0.5-mm sheets and bonded with epoxy resin. For comparison, the strength of insulator I1 bonded between untreated copper surfaces was also measured. Three test specimens were used for each measurement. The results are summarized in Table 3. The bond shear strength depends on surface treatment of the conductor. Compared with the ordinary copper surface, the chemically oxidized surface improves the bond shear strength by 93% at 300 K and 26% at 77 K.

The results show that sample I1 has more than sufficient bond shear strength to serve as insulator between turns. However, when winding with the I1 material as turn-to-turn insulator, the winding speed had to be kept low because of the need to heat-treat the epoxy. Furthermore, the epoxy resin damaged the specially roughened conductor surface by flowing beyond the corner of the conductor. Therefore, various resin materials were selected and tested to avoid such flowing. Finally, the insulation material

Table 3. Bond Shear Strength of Insulation Materials Between Turns

Sample Symbol	Test Temperature (K)	Copper Surface (MPa)	Oxidized Copper Surface (MPa)
I1	300	409.3 ±38.3	789.5 ±190.6
	77	630.8 ±147.0	792.8 ±104.8
I7	300	----------	694.2 ±140.0
	77	----------	797.1 ±88.4

I7, which has almost the same bond shear strength at 77 K as I1, was used for the Japanese LCT coil.

The conductor of the Japanese LCT coil has a specially roughened surface to increase the heat flux. The conductor surface touching the pie-to-pie insulators is filled with polyester compound to make a flat surface for good contact between the conductor and insulator. The effect of thermal cycling on the shear strength of this bond was measured between 77 K and 300 K. The polyester compound and sample S2 were bonded between conductors and then thermally cycled up to 30 times. The results are shown in Fig. 4. The bond shear strength decreases with increasing number of cycles up to 10 cycles, but thereafter the bond shear strength does not change. The residual bond shear strength is around 2.4 MPa after 10 cycles.

Thermal contraction properties. The rigidity of winding depends on thermal contraction of insulators. The thermal con traction of all samples was measured between 300 K and 77 K with a quartz dilatometer. Table 4 shows the test results. Since a substantial amount of insulator S2 is used between pies and between the winding and the helium vessel, the thermal contraction of this material has an important effect on rigidity of winding. Sample S2 has twice the thermal contraction of copper in the direction normal to the lamination; this value is approximately 14% smaller than that of G-10CR measured at NBS.[2] A gap due to the difference of the thermal contraction between insulator and conductor at 4 K can be avoided by prestress with shrinkage of the coil case.

Electrical Properties

Electrical properties of insulation materials were measured in 4-K liquid helium, in 4-K gas helium, in liquid nitrogen, and in 300-K air. Test results are shown in Table 5.

Insulator samples I1 and S2 did not break down in any test condition, but creep discharge was observed in all test cases. The creep discharge voltage of insulator samples I1 and S2 are 5.9 kV ac and 10.5 kV ac, respectively, in liquid helium. Breakdown voltage of insulator sample L1 is 14.0 kV ac in liquid helium.

The test result indicates that the insulation materials have enough margin for a quench voltage of 1.0 kV.

CONCLUSION

To select the insulation materials used for the Japanese LCT coil, the mechanical, electrical, and thermal properties of

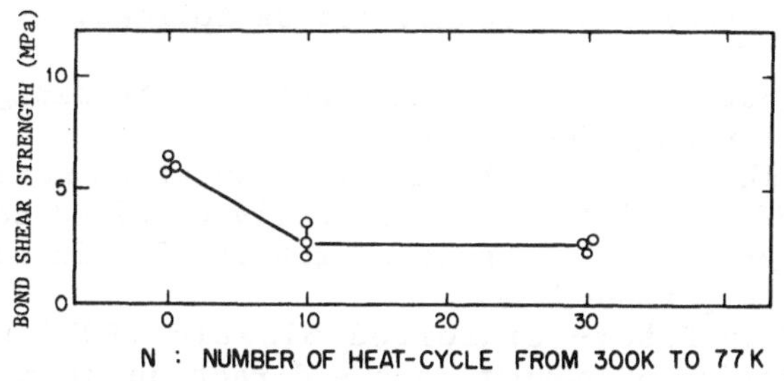

Fig. 4. Bond shear strength of pie-to-pie insulator.

Table 4. Thermal Contraction of Insulation Materials

Sample Symbol	Thermal Contraction (%) α//1 warp	α//2 fill	α//1 normal	β* total
I7	0.26	0.26	0.75	1.27
L1	0.32	0.32	0.83	1.47
S2	0.24	0.24	0.55	1.03
Copper	0.26	0.26	0.26	0.78

* β ≑ α//1 + α//2 + α1

Table 5. Electrical Properties of Insulator Used for the Japanese LCT Coil

Insulator Sample	Insulation Media	Thickness (mm)	Creep Discharge Voltage (kV ac)	Breakdown Voltage (kV ac)
I1	Liq.He (4 K)	0.70	5.90 kV rms	-----
	Gas He (4 K)	0.70	3.10	-----
	Liq.N_2 (77 K)	0.68	12.30	-----
	Air (300 K)	0.68	1.80	-----
S2	Liq.He (4 K)	1.10	10.70	-----
	Gas He (4 K)	1.10	6.36	-----
	Liq.N_2 (77 K)	1.11	14.50	-----
	Air (300 K)	1.11	3.10	-----
L1	Liq.He (4 K)	1.02	-----	14.00
	Gas He (4 K)	----	-----	-----
	Liq.N_2 (77 K)	0.93	-----	21.75
	Air (300 K)	1.02	20.00	-----

various insulation materials were measured at 300 K, 77 K, and 4 K. The test results are summarized as follows:

1. All insulation materials satisfy the design criteria of compressive strength, 400 MPa at 4 K.
2. Bond shear strength of the insulation materials between turns is 797 MPa at 77 K. The chemically oxidized copper surface improves the bond shear strength by 26% at 77 K in comparison with ordinary flat copper surface.
3. Voidless fiber-reinforced plastic (FRP) has twice the thermal contraction of copper in the direction normal to lamination.
4. Breakdown voltage of all insulation materials is higher than the design criteria.

From the test results, the following materials were selected:

1. Epoxy impregnated glass-cloth tape (SGT-F7036) as insulator between turns and pies.
2. Voidless FRP (VL-E200) as insulator between pies and between the winding and the helium vessel.
3. Laminated glass cloth sheets with polyester (WE-35D100s) as insulator between the winding and the helium vessel.

ACKNOWLEDGMENTS

The authors would like to thank Drs. S. Mori, Y. Iso, and Y. Obata for their continuing encouragement on this project. The authors wish to thank L. Dresner and R. Kensley for their valuable comments.

REFERENCES

1. S. Shimamoto, T. Ando, T. Hiyama, H. Tsuji, Y. Takahashi, E. Tada, M. Nishi, K. Yoshida, K. Okuno, K. Koizumi, T. Kato, K. Yasukochi, R. S. Kensley, K. Oka, M. Shimada, Y. Sanada, Y. Ibaraki, Evolution of the Japanese Test Coil Work for the Large Coil Task, IEEE Trans. Magn. MAG-17:1734 (1981).
2. M. B. Kasen, G. R. MacDonald, D. H. Beekman, and R. E. Schramm, Mechanical, electrical, and thermal characterization of G-10CR and G-11CR glass-cloth/epoxy laminates between room temperature 4 K, in: "Advances in Cryogenic Engineering--Materials," Vol. 26, Plenum Press, New York (1980), p. 235.

THE MANUFACTURE AND PROPERTIES OF RADIATION RESISTANT LAMINATES*

J. R. Benzinger

Spaulding Fibre Company, Inc., Tonawanda, New York

INTRODUCTION

Many studies have been made of the effects of radiation on polymers and insulators, but only a few on high pressure thermosetting laminates.[1] There was virtually no published information on the simultaneous effects of radiation and low temperature until the recent work at Oak Ridge National Laboratory (ORNL) by Coltman et al.[2] on the Spauldite® SCR® grade laminates, some particle-filled epoxies, Nomex® paper, and Kapton® films.

One conclusion of their report was that at 1×10^{10} rads, the flexural strength of the particle-filled epoxies was at "end of life," while the glass-cloth-filled epoxies retained usable strength. However, the compressive strength of all of the materials was reduced to possibly barely useful levels, and loss in voltage breakdown strength of the laminates suggested design adjustments. Previous results showed that at 2×10^8 rads neither electrical nor mechanical properties underwent significant change, but the strengths of the glass fabric epoxies were greatly reduced at 2.4×10^9 rads.[3]

In some locations of the magnet, such as adjacent to the neutron beam injector, Santaro et al.[4] and recently Engholm[5] concluded that higher doses would require additional shielding or redesign at considerable expense. It was suggested as an alternate approach that it might be more prudent to improve the radiation resistance of the insulation or search for more radiation resistant materials.

---

*Invited paper.

Inorganic insulators, like ceramics, are much more resistant to radiation, but tend to be very brittle. This rules them out for most fusion reactor applications since superconducting magnets are reported by Reed et al.[6] to be subjected to very large contractural and magnetic field-induced forces. Consequently, critical applications in these magnets will require large-scale use of tough, high-pressure industrial thermosetting laminates. The choice of material, according to Kasen et al.,[7] is based on considerations other than toughness, such as availability, economics, machinability, high strength-to-weight ratios, and low electrical and thermal conductivities.

SFC PROGRAM

As part of the ongoing cryogenic radiation research program on laminates at the Spaulding Fibre Company (SFC), several experimental variants of G-10CR and G-11CR were investigated with a view toward improving radiation resistance, mechanical strength, and cost. A literature search revealed improved radiation stability by incorporating particle fillers in epoxy resin castings and moldings. Adaptation to laminates was expected to provide similar improvement, at least to the resin part of the structure. The particle filler selected for experimentation was quartz (pure SiO_2), which was reported as the least damaged of several inorganics tested to a fast neutron fluence of 10^{21} n/cm^2.

Polyimides are known to perform well at high temperatures and have been reported as having good resistance to radiation. For this reason, a polyimide of the aromatic diamine bismaleimide family (Kerimid 601) was selected for the polyimide variant.

To improve mechanical strength properties and radiation effects, a satin-style fabric woven from S-2® glass was used to reinforce one of the polyimide variants. An E-glass fiber mat was used as the reinforcement on two of the variants for applications where requirements may not be as severe, but where cost is the major factor.

Test Results

The bar graph in Figure 1 compares the compressive strength of the GEM-CR (continuous-strand glass-epoxy mat) variant with a NEMA GPO-1 (glass mat polyester) laminate tested at ambient (293 K) and liquid nitrogen (77 K) temperatures. The GEM-CR variant, like G-10CR, increases considerably in strength at cryogenic temperatures. Only a slight increase in strength is noted for the polyester. The graph in Figure 2 is similar to Figure 1 in all respects except that it compares interlaminar shear (unsupported) as a function of temperature. Tests were conducted in the laboratories at MIT.

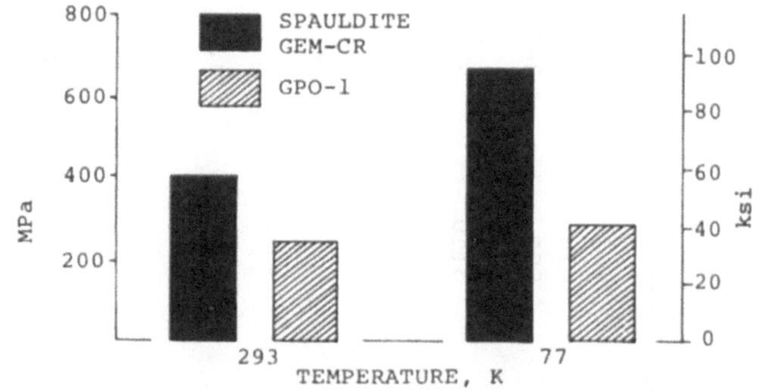

Fig. 1. Compressive strength edgewise as a function of temperature.

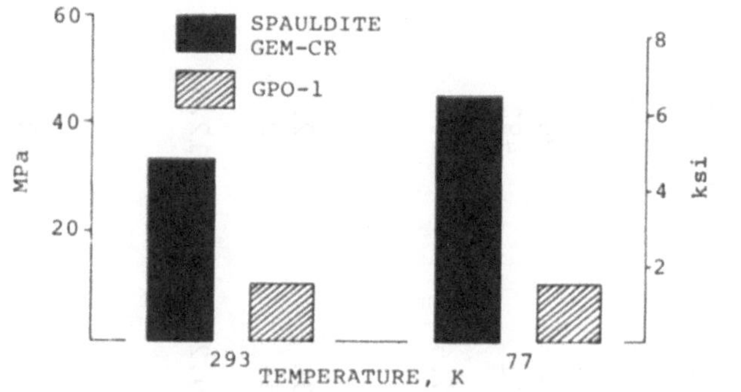

Fig. 2. Interlaminar shear strength (unsupported) as a function of temperature.

Shown in Table 1 are the mechanical strength properties for some SCR variants tested to ASTM standards at SFC. Surprisingly the GEM-CR mat variant with the G-10CR epoxy resin had equivalent or better properties than the control material. The GEM-CRF variant containing 24% quartz showed a significant reduction in properties. However, the strength properties of the quartz-filled variant, G-10CRF, except for interlaminar shear, were as good or better than the G-10CR control.[8]

Polyimide resins are hard and tend to be a little brittle, as evidenced by the lower interlaminar shear and bonding strengths of the polyimide variants, when compared with G-10CR. However, bonding strengths are better than G-11CR and exceed the minimum bonding strength requirements for NEMA glass-epoxy (G-10, G-11), glass-phenolic (G-3), and glass-melamine (G-5, G-9) laminates. Compressive strengths of the polyimides were good, with the S-2 glass variant significantly better than any of the laminates

Table 1. Typical Mechanical Strength Properties of Some SCR Variants*

Mechanical Strength Properties at 295 K of a 12.7-mm† Specimen		Glass Mat		Glass Fabric			
		Epoxy		Epoxy		Polyimide	
		Unfilled GEM-CR	Filled GEM-CRF	Filled G-10CRF	Control G-10CR	Spaulrad™ E	S
Compressive, MPa† Edgewise	Warp	370	331	459	414	441	538
	Fill	370	331	369	331	414	496
Normal		580	331	493	483	572	862
Interlaminar Shear, MPa† Guillotine Method A	Warp	46.9	27.6	38.6	48.3	31.0	27.6
Flexural, Normal, MPa†	Warp	492	214	521	496	-	-
	Fill	492	214	432	379	-	-
Bonding, kN† 10-mm ball		12.9	10.2	12.5	12.9	9.3	9.8
Physical Properties							
Percent Resin		29	29	23	33	27	26
Density, g/cc		1.93	1.93	2.06	1.85	2.00	1.98

* Spaulding Fibre Company, Inc. laboratory values based on the average test data of pilot plant laminates.

† 12.7 mm = 0.5 in, 1 MPa = 145 psi, 1 kN = 224.8 lbs.

tested. Specimens that were tested parallel to laminations tended to fracture by the delamination mode rather than by the diagonal break usually observed with G-10CR.

Listed in Table 2 are some additional test data by SFC for the polyimide variants made with S-2 and E-glass. The mechanical strengths of the S-2 glass variant are significantly better than the values for the E-glass polyimide. On the other hand, note that the E-glass variant has better electrical properties. Its resistance to dielectric breakdown parallel to laminations and after a two-day soak in warm water is remarkably better than that of the S-2 glass variant.

MIT-ITR PROGRAM

The program at the Massachusetts Institute of Technology (MIT) Plasma Fusion Center, sponsored by DoE, is part of an investigation of material properties for an Ignition Test Reactor (ITR), which is being designed to study the physics of fusion ignition.[9] The reactor magnet consists of large flat plates of metal separated by thin insulators. The insulator must survive 10,000 cycles to 77 K of 140 MPa pulsed pressure, 8.4 MPa interlaminar shear, and a lifetime radiation fluence of 10^{20} n/cm^2. Because of a large bank of test data, the commercial glass-epoxy laminates, G-10CR and G-11CR, were chosen as likely candidates for evaluation.

Experimental Laminates

Early in the program it was decided to include also some of the SCR grade variants and some experimental laminates prepared from materials expected to be more resistant to radiation damage. Criteria for selection were that the materials be generic and commercially available.

Test Results

Initial compressive fatigue testing was conducted at MIT on unirradiated specimens of glass-silicone, glass-epoxy, mica-glass, and glass-polyimide. All of the insulators, with the exception of mica-glass, survived a compressive stress of 310 MPa after 60,000 cycles.

Subsequently glass-silicone and glass-epoxy were irradiated in the Advanced Test Reactor at the Idaho National Engineering Laboratory (INEL) to a dose of 3.8 x 10^{11} rads. Compressive testing at a level exceeding ITR requirements indicated survival of the glass-epoxy and failure of the glass-silicone. Six laminates, including the S-glass polyimide variant, were then

Table 2. Additional Typical Properties of the Polyimide Variants

295 K Properties		Spaulrad E-Glass	Laminates S-2 Glass
Mechanical Properties 3.8-mm thick specimen			
Flexural, Normal, MPa	Warp	510	807
E-300/523 K	Warp	346	-
	Fill	420	745
Compressive, Edgewise, MPa	Warp	333	471
	Fill	316	430
Compressive, Normal, MPa		572	893
Izod Impact, Edgewise,	Warp	0.93	1.63
Notched, kJ/m	Fill	0.48	1.29
Tensile, MPa	Warp	372	600
	Fill	262	540
Electrical Properties 0.8-mm specimen			
Dielectric Breakdown, s/s kV		100+	95
Condition D-48/323 K, s/s kV		90	15
Permittivity at 1 MHz		4.8	5.0
Dissipation Factor at 1 MHz		0.012	0.016
Arc Resistance, s		183	185

1 MPa = 145 psi, 1 kJ/m = 18.7 ft.lbs./in, 1 mm = 0.0394 in.

irradiated in the MIT Reactor to a dose of 2.3 x 10^{10} rads. All survived compressive strength fatigue tests exceeding ITR requirements.

Ambient test results both at MIT and INEL suggest that the commercial SCR grade laminates, as well as the polyimides, will withstand the ITR environment. To confirm this assumption, future plans at MIT are to conduct combined shear and compression strength tests during irradiation down to 77 K.

NBS PROGRAM

The National Bureau of Standards (NBS) at Boulder, Colorado recently initiated a program to screen the mechanical performance of the glass fabric polyimide variant at cryogenic temperatures to compare the results with the standard grades G-10CR and G-11CR.

Test Results

Listed in Table 3 are some preliminary test data generated by NBS on the Spaulrad polyimide at 295 K, 76 K, and 4 K. Included in the table for comparison are the original NBS test data on the SCR pilot plant laminates. The test results appear to indicate that increases in mechanical strengths at low temperatures are significantly less for the polyimide than for the epoxies with one exception: For Young's modulus of elasticity in tensile, the polyimide increases in strength more like the epoxies.

ORNL-MFE PROGRAM

The program at ORNL is an ongoing special materials program for the Office of Magnetic Fusion Energy (MFE). Its purpose is to study irradiation effects on organic insulators used in the superconducting coils and magnets of fusion reactors. Aware of the reported superior radiation resistance of polyimides over epoxies, ORNL initiated a study of the mechanical properties of polyimides irradiated at various doses in liquid helium. Two laminates, Spauldite (Spaulrad) and Kerimid, reinforced with E-glass woven fabric were tested along with Vespel®, an unfilled polyimide.

Irradiation Conditions

The highest dose at which the specimens were irradiated was 10^{10} rads of gamma rays, 3.1×10^{21} n/m^2 of thermal-neutron fluence and a fast-neutron fluence of 8.7×10^{20} $n/m^2 > 0.2$ MeV. Irradiation time at 4.9 K in liquid helium was 189 h. Methods and conditions were the same as those reported in previous studies of epoxies at ORNL. See Reference 2 for details of the 10_B fission dose.

Test Results

Shown in Figure 3 are graphs of the mechanical strength test results by Coltman and Klabunde.[10] The flexural strength of the unirradiated Spaulrad tested 25% stronger than Kerimid at 300 K and 40% stronger at 77 K. After irradiation at 10^{10} rads, Spaulrad lost 38% of its original flexural strength at 77 K. Nevertheless, throughout the entire dose range, it still remained

Table 3. Mechanical Screening of Polyimide Versus the SCR Epoxies (Pilot Plant)*

Temperature, K	Young's Modulus, E, GPa†		Poisson's Ratio, ν		Tensile Strength, MPa†		Compressive Strength, MPa†		Shear Strength, MPa† Guillotine	
	Warp	Fill	Warp	Fill	Warp	Fill	Warp	Fill	Warp	Fill
G-10CR										
295	28.0	22.4	0.150	0.144	415	257	375	283	42.3	
76	33.7	27.0	0.190	0.183	825	459	834	557	61.3	72.9
4	35.9	29.1	0.211	0.210	862	496	862	598	72.6	78.8
G-11CR										
295	32.0	25.5	0.157	0.146	469	329	396	315	40.6	
76	37.3	31.1	0.223	0.214	827	580	804	594	56.5	56.6
4	39.4	32.9	0.212	0.215	872	553	730	632	56.2	57.0
Polyimide Variant										
295	32.7	28.4	0.167	0.151	385	265	355	316	36.9	36.5
76	35.7	32.4	0.210	0.187	633	379	447	466	44.5	45.9
4	39.4	34.7	0.244	0.189	638	407	555			

* Specimen thickness = 3.8 mm (0.15 in).
† 1 MPa = 145 psi, 1 GPa = 1.45 x 10^5 psi.

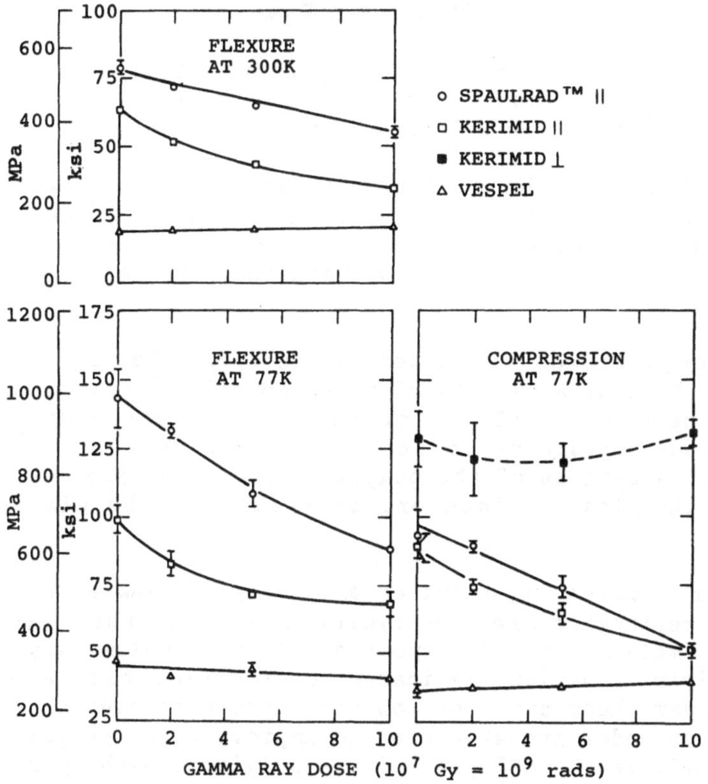

Fig. 3. Results of fracture strength tests on pure and glass-fabric-filled (gff) polyimides after irradiation at 4.9 K followed by warmup to 307 K. Each datum point shows the average value of three tests, and error bars indicate average deviation. Points without error bars indicate data scatter was too small to show. See text for accompanying 10_B fission dose.[10]

25-40% stronger than Kerimid and 2-3 times stronger than Vespel. After a dose of 2.4 x 10^9 rads, the better polyimide laminate remains about 5 times stronger than the best epoxy (G-11CR) for tests at 77 K.

For compressive strength tested on edge or parallel to laminations at 77 K, there was a loss in strength of about 40% after irradiation for the glass-polyimide laminates. However, there was no significant loss in compressive strengths in the normal or perpendicular direction after irradiation. These test data tend to be in agreement with results reported earlier by MIT for compressive strength or irradiated laminates tested in the same direction.

FUTURE PROGRAMS

Plans for the future are to continue studies of the irradiation effects on the other SCR grade variants. The next most likely candidate is the S-2 glass variant of the polyimide.

Nearly all of the radiation effect studies to date, because of a lack of facilities, have been limited to gamma rays. In future programs, the behavior of G-10CR, G-11CR, and Spaulrad will also be studied after irradiation with the fluence from a neutron source.

Although it was concluded in one study that an amine silane tested the best of a variety of glass coupling agents, an update study is needed. Analysis of the most recent ORNL radiation effects study appears to indicate that the degradations in the mechanical strengths of the polyimide laminates are due to either damage of the glass surface or, more likely, the glass-polymer interface.

Other epoxies reported as promising candidates for improved radiation resistance are the cycloaliphatics, the highly functional aromatics, and the epoxy novolacs. Preliminary screening of the polyimide variant indicates lower values than epoxies for interlaminar shear and most low-temperature mechanical strengths. One approach under investigation to improve these properties, and cost as well, is the alloying of the polyimide with epoxies.

CONCLUSION

High pressure industrial laminates can be modified to make them more resistant to radiation. In addition, variants of the SCR grade laminates made with S-2 glass provide higher mechanical strength and less residual radioactivity than E-glass laminates. The good results, obtained with some of the lower cost variants made with glass mat fibers and quartz particle fillers, afford the potential for future cost savings.

Owing to a lack of funding, only the polyimide variant, Spaulrad, was tested for radiation damage at low temperatures. Preliminary mechanical screening of this polyimide indicated lower shear and bonding strengths at 295 K and less increase in strength at low temperatures when compared with those of G-10CR and G-11CR. However, after irradiation the polyimide variant was less damaged and remained 5 to 10 times stronger than the epoxies at 77 K. Current pricing of the polyimide is about 4 to 6 times more than epoxies.

Although more testing needs to be done, the polyimide variant, notwithstanding some limitations, appears to offer a practical alternate solution to the insulator problem in fusion reactors. The SCR grades, being less expensive, will continue to be used at lower dose locations.

ACKNOWLEDGMENTS

The author wishes gratefully to thank H. Becker and E. Erez of MIT for their contribution of test data and information. He is especially grateful to M. Kasen and R. Schramm of NBS for use of the test data in Table 3 and to R. Coltman of ORNL for permission to reproduce the graphs of Figure 3 in advance of publication. The author also wishes to thank L. Moskal and D. Sieracki for the preparation and testing of the laminates.

REFERENCES

1. M. G. Young, "Radiation Effects on Laminated Plastic Materials," UD-NEMA Facility (April 1965).
2. R. R. Coltman, C. E. Klabunde, R. H. Kernohan, and C. J. Long, "Radiation Effects on Organic Insulators for Superconducting Magnets, Annual Progress Report for Period Ending September 30, 1979," ORNL/TM-7077, Oak Ridge National Laboratory, Oak Ridge, Tennessee (1979).
3. R. H. Kernohan, R. R. Coltman, Jr., and C. J. Long, "Radiation Effects on Organic Insulators for Superconducting Magnets," ORNL/TM-6708, Oak Ridge National Laboratory, Oak Ridge, Tennessee (1979).
4. R. T. Santaro, J. S. Tang, R. G. Alsmiller, Jr. and J. M. Barnes, Nucl. Technol. 37:65 (1978).
5. B. A. Engholm, "Preliminary Radiation Criteria and Nuclear Analysis for ETF," Fourth ANS Topical Meeting on the Technology of Controlled Nuclear Fusion (October 1980).
6. R. P. Reed, F. R. Fickett, M. B. Kasen, and H. I. McHenry, in: "Materials Studies For Magnetic Fusion Energy Applications at Low Temperature - 1," Report No. NBSIR 78-884, National Bureau of Standards, Boulder, Colorado (1978), p. 243.
7. M. B. Kasen, G. R. MacDonald, D. H. Beekman, and R. E. Schramm, Mechanical, electrical, and thermal characterim zation of G-10CR and G-11CR glass-cloth/epoxy laminates between room temperature and 4 K, in: "Advances In Cryogenic Engineering - Materials," Vol. 26, Plenum Press, New York (1980), pp. 235-244.
8. J. R. Benzinger, Manufacturing capabilities of CR-grade laminates, in: "Advances In Cryogenic Engineering - Materials," Vol. 26, Plenum Press, New York (1980), pp. 252-258.

9. E. A. Erez and H. Becker, Radiation damage in thin sheet fiberglass insulators, in: "Fundamentals and Applications of Nonmetallic Materials And Composites At Low Temperatures," Plenum Press, New York (1982).
10. R. R. Coltman, Jr. and C. E. Klabunde, "Mechanical Strength of Low-Temperature-Irradiated Polyimides: A Five-to-Tenfold Improvement in Dose Resistance over Epoxies," Second Topical Meeting on Fusion Reactor Materials, Seattle, Washington (August 1981).

MECHANICAL AND THERMAL PROPERTIES OF GLASS-FIBER-REINFORCED COMPOSITES AT CRYOGENIC TEMPERATURES

A. Khalil and K. S. Han

University of Wisconsin, Madison, Wisconsin

INTRODUCTION

Because of the large forces encountered in large superconducting energy storage magnets, forces have to be transmitted through high strength struts to the room temperature structure. Besides strength, the material of these struts has to have a low thermal conductivity in order to minimize the refrigeration requirement for the energy storage system. The candidate materials that possess high strength-to-weight ratios and low thermal conductivities are fiberglass-epoxy and polyester composites. However, the cost of the selected material is also an important factor because of the large amount required in large superconducting magnet energy systems (SMES). Therefore, the choice of a strut material has to be based on a careful balance between the different factors involved. Fiberglass-reinforced epoxy composites have been successfully employed in different applications in fusion and magnetohydrodynamic (MHD) programs and several studies were carried out to investigate the mechanical and thermal properties of these composites at low temperatures.[1-3] On the other hand, fiberglass-reinforced polyester composites are cheaper.

In this study, more extensive investigations were carried out on the fiberglass-epoxy composites (G-10CR), polyester fiberglass (Extren), and glass-fiber-wound epoxy composite tubes (GFW AT-1008). The mechanical properties investigated were compressive strength, elastic modulus, fatigue, and fracture behavior at 300 K and 77 K. Thermal conductivity, heat diffusivity, and specific heat were also measured in the temperature range of 4-300 K. It is generally recognized that processing variables in fabricating these composites can have some effects on the mechanical and thermal properties.

MECHANICAL PROPERTIES

Experiments

A compression and tension apparatus for operation at room and cryogenic temperatures was attached to a 4.44×10^5 N (100,000 lbf) closed-loop electrohydraulic test machine manufactured by MTS, which can be operated in either a load or displacement control mode. Loads were measured with a built-in load cell, and strain gauges were used to monitor displacements. Load and displacement were plotted on an x-y recorder.

The compressive strength and fatigue specimens were: (1) tubes--2.54-cm o.d., 0.32-cm thick, and 5.08-cm long, (2) round rods--1.57-cm diameter and 5.11-cm long, and (3) rectangular rods--3.05-cm long, 1.02-cm wide and 1.02-cm thick. To prevent premature failures due to end splitting and brooming and also to ensure good vertical alignment of the specimens, end caps were made from 5.08-cm diameter, 1.27-cm thick discs of tool steel. A round or rectangular hole just large enough for the specimens was machined 0.64-cm deep. For tube testing, the inside was also supported with a steel rod of 1.91-cm diameter, which was screwed to the end cap. Based on the compressive strength, fatigue tests were conducted using the same specimens at peak stress levels of 0.9, 0.8, 0.7, and 0.6 of ultimate strengths. The minimum stress level was maintained constant (12.9 MPa) to simulate charge-discharge behavior of a superconducting energy storage magnet. The fatigue cycle frequency used was 1 Hz to reduce the frequency effect on the fatigue limit. At higher frequencies, the internal temperature rises in the specimens, which might cause a deterioration of the fatigue life of composites. All tests were conducted at 300 K and 77 K. No measurements were conducted at 4 K because the changes in mechanical properties between 4 K and 77 K are reported to be negligible.[1]

Compressive Strength and Modulus

Table 1 summarizes the compressive strength and modulus data of the tested materials at 300 and 77 K. The composite properties strongly depend on the fiber orientation, fiber aspect ratio, and curing procedure. The Extren sample (pultruded glass-fiber-reinforced polyester composite) had a lower modulus and strength than the G-10 sample. Extren has 44 wt.% glass fibers; most of them are short glass fibers. Only a small amount of long glass fibers are used to stimulate the pultrusion process. The short glass-fiber mat has a lot of potential crack propagating sites, such as pores and fiber ends, that reduce its mechanical properties. The mat portion alone had even lower properties (10.3 GPa modulus and 82.7 MPa tensile stress) than Extren material at 300 K. However, the cheaper price of Extren makes it competitive with the other

Table 1. Elastic Modulus, Compressive Strength, and Thermal Conductivity of the Tested Samples

Sample	T, K	E, GPa	σ, MPa	k, W/mK	σ/k
G-10 tube	300	17.2	179.2	0.890	201.3
	77	34.5	448.2	0.300	1494
	5			0.055	
G-10 tube, pressure cured	300	31	413.7	1.102	375.4
	77	56.0	848.0	0.330	2569.7
	5			0.090	
Wound tube	300	39.3	358.5	1.029	348.4
	77	78.3	606.7	0.317	1913.9
	5			0.093	
Extren tube	300	46.5	234.4	0.961	243.9
	77	62.4	717.0	0.380	1886.8
	5			0.070	
Extren sheet	300	17.2	144.8	0.680	212.9
	77	31.7	262	0.290	903.4
	5			0.041	

materials. G-10 rod, which is a unidirectional glass-fiber-reinforced epoxy composite processed by compression molding, had a high modulus and strength because of its fiber orientation and fewer pores owing to the compression molding process. Various sizes of G-10CR tubes, cured without pressure, had a 17.2 GPa modulus and 179.1 MPa compressive strength at 300 K. However, the same material cured under pressure, had a 31 GPa modulus and 413.7 MPa compressive strength at 300 K. Epoxy resin could penetrate through the glass-fiber cloth under pressure resulting in fewer pores in the composite. Also it has a higher glass-fiber content than regular G-10CR tube. One other material tested was glass-fiber-wound tube, which had a 39.3 GPa modulus and 358.5 MPa ultimate strength. The higher modulus is due to the higher glass content (80 wt.%). However, the ultimate strength is lower due to its fiber orientation.

Fatigue

Extren plate, G-10CR rod, and G-10 tube were cycled under compressive load at various stress levels. Although there was a scattering in fatigue life, it was observed that the ratio of peak and ultimate stresses fell on a straight line against its fatigue life regardless of material and temperature, as shown in Fig. 1. Fatigue life can be estimated using this result. The actual strut system undergoes 18,250 compressive fatigue cycles for 50 years because of the charge-discharge operation in a storage magnet. Peak stress of about 60% of ultimate stress gives this fatigue life. However, because of the scattering of fatigue life and safety factor, it is necessary to design such a strut system using 50% of its ultimate stress.

Failure Behavior

The failure behavior of the composites during compression testing and fatigue testing was the same. The failure in Extren plate initiated by delamination along the long fibers at the early stage of testing. It gave a large hysteresis loop energy during fatigue tests. This delamination was followed by shear failure through the specimen leading to a catastrophic failure. The G-10 rod failed in a shear mode in a short time leading to crushing. There was no sign of crack initiation or propagation. The failure surface was tilted approximately 45° along the fiber direction, as shown in Fig. 2.

Tubes failed near the end cap region and the cracks passed around the tube. This failure was due to the localizations of transverse shear and moment.[4] Because of end caps around the thin tube, shear force and moment were developed at the tube ends, and these effects were localized near the end caps. This caused a failure near the end caps. Testing without end caps gave the same modulus but a slightly lower ultimate strength because of the premature failure by crushing at tube ends. The fiber-wound tube

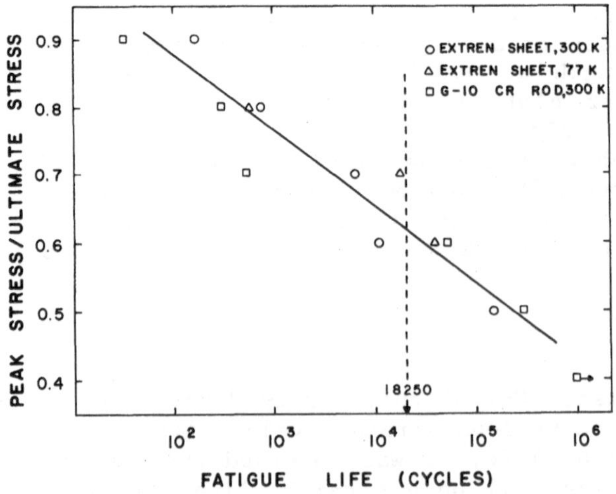

Fig. 1. Fatigue life as a function of stress level for different samples.

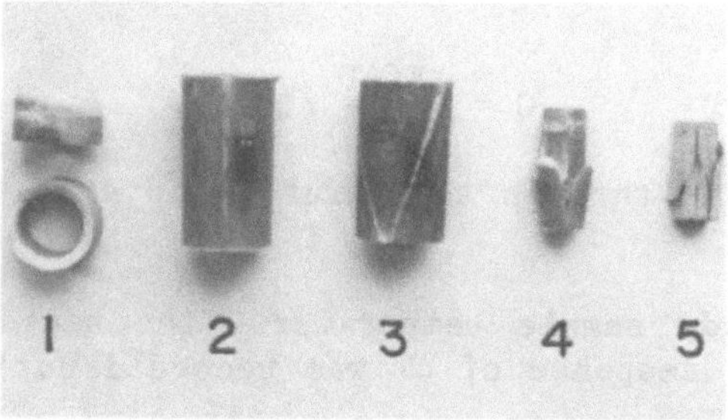

Fig. 2. Typical specimens broken by compression test; 1. G-10 tube, 2. Molded G-10 tube, 3. Wound tube, 4. G-10CR rod, 5. Extren sheet.

failure passed through the glass fibers in a V-notch shape. The failure studies showed that the direction and length of glass fibers determined the failure behavior of composites.

THERMAL PROPERTIES

Experimental Rig and Procedure

Thermal conductivity, heat diffusivity, and specific heat of the composite samples were measured at different temperatures in the range of 4 K to 300 K using the continuous-flow helium cryostat with a temperature-controlled heat sink. The tested samples were clamped to the heat sink with a sample holder. To avoid the problem of heat leaks through the sample clamping mechanism used in most conventional rigs, two identical samples were used. The samples were separated by a thin copper heater plate and clamped firmly between two flanges by a threaded copper rod to maintain a good interface thermal contact. The top flange was attached to the temperature-controlled heat sink of the continuous-flow helium cryostat and shielded with two radiation shields in vacuum. The sample temperature was controlled within ±50 mK in the temperature range of 4 K to 300 K. The temperature drop across the sample was monitored by a differential Chromel vs. Au-Fe 0.07% thermocouple, and the heat sink temperature was measured with a silicon diode. All the sample wiring was thermally anchored to the heat sink, and correction was made for the conduction through the thermocouple and heater leads.

The differential method was used for determining the thermal conductivity (k) by applying a small temperature differential (0.5 to 1.5 K) across the sample and calculating k as follows:

$$k = \frac{Q\ell}{A\Delta T} \tag{1}$$

or

$$Q = \frac{Ff\Delta T}{\ell} / (\sigma/k) \tag{2}$$

i.e., the heat leak through the strut is inversely proportional to σ/k.

At each steady sample temperature, the heater was turned on, and the transient response of ΔT was recorded until a steady state was reached.

The sample specific heat and heat diffusivity were obtained by substituting the time-dependent ΔT into the solution of the heat diffusion equation.[5]

The thermal conductivity of the tested composite depends mainly on glass-fiber content and orientation, resin type, and sample temperature. The same is true for specific heat, except that it might not be sensitive to fiber orientation with respect to heat flow direction.

The experimental error in measuring k or c did not exceed 5 to 15% in the considered temperature range. The tested samples were 1.7-cm-long tubes of 2.54-cm o.d. and 0.32-cm thickness, and sheet samples were 3.5-cm long, 2-cm wide, and 1.27-cm thick.

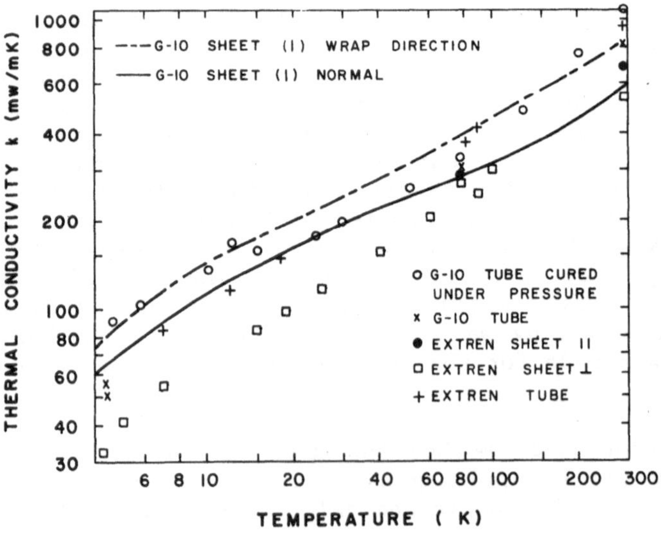

Fig. 3. Thermal conductivity of Extren in comparison with published data[1] on G-10 sheets.

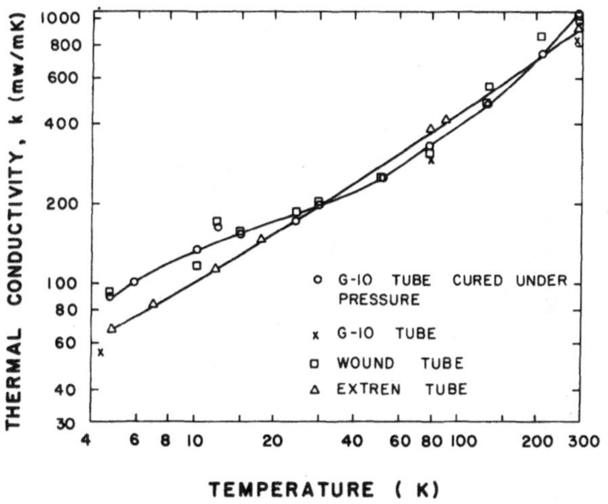

Fig. 4. Thermal conductivity as a function of temperature for different tube samples.

Results and Discussion

Figure 3 shows the thermal conductivity data on G-10 tubes and Extren tubes and sheets in comparison with the published data[1] on G-10 sheets in both the normal and wrap directions. The thermal conductivity values of the G-10 tube cured under pressure were close to those of G-10 the sheet in the wrap direction. The thermal conductivity values of the regular G-10 tube were up to 35% less than those of the pressure-cured G-10 tube at 5 K, and up to 20% less at room temperature because the regular G-10 tube is less dense and contains less glass. The σ/k values were still higher for the pressure-cured tubes, as shown in Table 1.

The thermal conductivity values of the fiberglass polyester sheet samples were up to 50% smaller than those of the G-10 sheet in the direction normal to fibers at liquid helium temperature because they contain less glass and also they contain some voids due to the pultrusion process. The same trend is also seen in the wrap direction at temperatures lower than 77 K.

Figure 4 shows a comparison of four different tube samples. The thermal conductivity change with temperature of the fiberglass-filament-wound tube was close to that of the pressure-cured G-10 tube. Table 1 shows that the pressure-cured G-10 tube sample possessed the highest σ/k over the entire temperature range. To investigate the effect of fatigue on the thermal conductivity, a G-10 tubular sample was cycled up to 43,000 cycles at 50% of ultimate stress. The observed decrease in k was very small and within the experimental error limits.

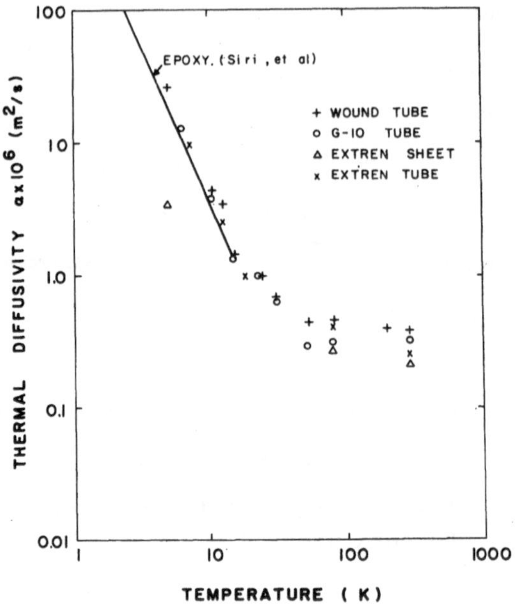

Fig. 5. Thermal diffusivity of tested samples as a function of temperature.

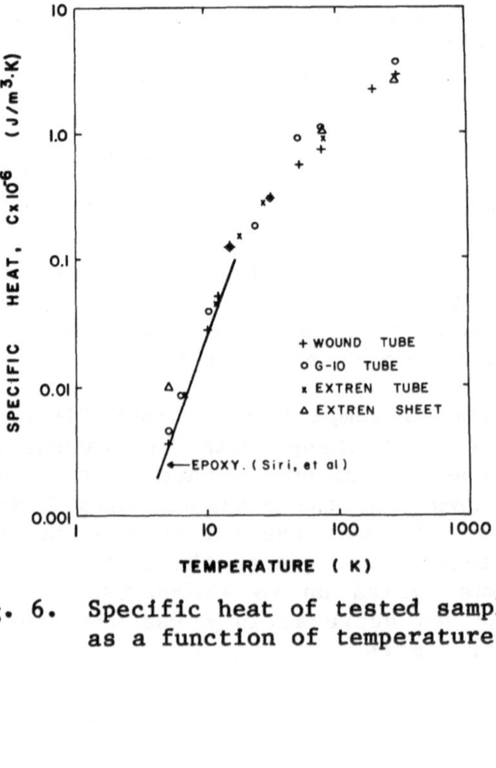

Fig. 6. Specific heat of tested samples as a function of temperature.

The thermal diffusivity and specific heat of the different samples are shown in Figs. 5 and 6 as a function of temperature. The data are plotted in comparison with data of Siri et al.[6] for the epoxy resin from 2 K to 15 K. The thermal diffusivity and specific heat of G-10, filament-wound, and Extren tubes followed a power law trend close to that of the epoxy in this temperature range. Above 20 K the slope of α decreases continuously until it almost saturates at room temperature. Figure 6 also shows that the specific heat of polyester fiberglass sheet sample is more than double that of the glass-fiber filament-wound and G-10 tubes at liquid helium temperature.

ACKNOWLEDGMENTS

This work was supported by the Department of Energy and the Wisconsin Electric Utilities Research Foundation.

NOTATION

A = sample cross sectional area = Ff/σ, m^2
c = specific heat, J/m^3
E = elastic modulus, GPa
F = force applied to the strut, N
f = factor of safety
k = thermal conductivity, mW/mK
ℓ = sample length in the heat flow direction
Q = heater power, W
T = temperature, K
x = distance in the heat flow direction, m

Greek Symbols

α = thermal diffusivity, m^2/s
τ = time, s
ΔT = temperature difference, K
σ = ultimate compressive stress, MPa

REFERENCES

1. M. B. Kasen, G. R. MacDonald, D. H. Beekman, Jr., and R. E. Schramm, Mechanical, electrical, and thermal characterization of G-10CR and G-11CR glass-cloth/epoxy laminates between room temperature and 4 K, in: "Advances in Cryogenic Engineering--Materials," Vol. 26, Plenum Press, New York (1980), p. 235.
2. J. R. Benzinger, Properties of cryogenic grade laminates, paper presented at the Electrical and Electronic Insulation Conference, Boston, Massachusetts (Oct. 1979).

3. M. de F. F. Pinheiro, D. J. Radcliffe, and H. M. Rosenberg, The thermal expansion and thermal and electrical conductivity of carbon and glass fiber/epoxy-resin composites from 2 to 290 K, in: "Proceedings of ICEC-7," IPC Science and Technology Press, Guildford, Surrey, England (1978), p. 494.
4. K. S. Han, R. A. Peterson, and R. E. Rowlands, Compressive tests of composite tubes at 300 K and 77 K, in: "Advances in Cryogenic Engineering--Materials," Vol. 28, Plenum Press, New York (1982), p. 253.
5. W. M. Rosenow and J. P. Hartnett, "Handbook of Heat Transfer," McGraw-Hill, New York (1973), pp. 3-83.
6. A. Siri, G. Sissa, R. Vaccarone, P. Fernandez, and C. Salvo, Low temperature measurements of thermal diffusivity in composite epoxies, in: "Proceedings of ICEC-7," IPC Science and Technology Press, Guildford, Surrey, England (1978), p. 499.

COMPRESSIVE TESTS OF COMPOSITE TUBES AT 300 K AND 77 K

K. S. Han, R. A. Peterson, and R. E. Rowlands

University of Wisconsin, Madison, Wisconsin

INTRODUCTION

Glass-fiber composites, with their high strength and stiffness-to-weight ratios and low thermal conductivity, are suitable as a strut material in a superconductive energy storage magnet. An A-frame-type strut system was previously designed and built at the University of Wisconsin-Madison using glass-fiber reinforced composite plates.[1] That strut system has been extensively investigated and has proved to be satisfactory. However, based on strength, thermal properties, and material and refrigeration costs, we have subsequently designed a tube-type strut system that is superior to A-frame-type composite plate structures for withstanding the large magnet stresses of 200 MPa that occur in a superconducting magnet of 5000 MWh capacity.[2] The tube-type composite strut arrangement is shown in Figure 1. The mechanical requirements for such a composite tubular strut are a compressive strength exceeding 200 MPa (30 ksi) and a modulus of at least 20 GPa (3×10^6 psi).

Although considerable theoretical and experimental design information is available for loaded homogeneous isotropic tubes, such is not the case for anisotropic tubes. This dearth of data is especially true at cryogenic temperatures. Engineering data on composite tubes are needed under these conditions to design a strut to carry the magnet forces from a 1.8-K superconducting magnet in an underground energy storage system to 300-K bedrock walls. The present study includes experimental compressive stress and strain measurements of composite tubes at 300 K and 77 K, a failure analysis, and a temperature distribution study along the tube length.

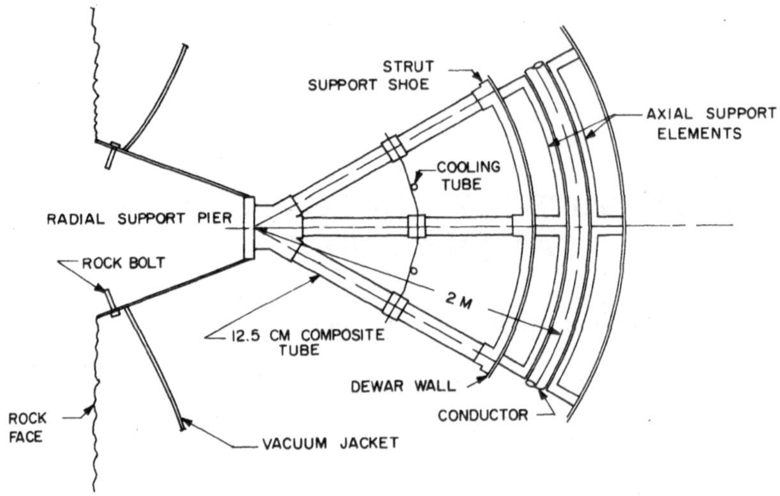

Fig. 1. Schematic of tube-type compressive structure to carry magnetic forces from magnet to rock.

EXPERIMENTS

The material used was G-10 glass-fiber-cloth reinforced epoxy composite tubes manufactured by Spaulding Fiber Co. End caps were used to prevent premature failure due to splitting or brooming behavior of the tube ends under compressive load. For a 9.53-cm tube, a 12.7-cm-diameter by 2.54-cm-thick tool steel block was machined to provide a hole 9.53 cm in diameter and 0.51 cm deep. An inner core of a 8.89-cm-diameter and 0.51-cm-thick tool steel block was attached concentrically to the main end cap to support the inside of the tube. No attempt was made to bond the tubes to the end caps. The end caps were machined with approximately 0.013-cm radial clearance inside and outside the tubes. This end-cap arrangement provides essentially a fixed-end condition.

The hydraulic compression testing machine has a capacity of 2.66×10^5 N (60,000 lbf). Strains were measured using strain gages and a digital strain indicator, and the load was indicated by the compression-machine dial. To simulate the behavior of an actual strut in the proposed storage magnet, the top end of the composite tube was kept at room temperature (300 K), while the bottom 5 cm was maintained at liquid nitrogen temperature (77 K). The Dewar was a container of aluminum alloy covered with polymeric foam. The level of the liquid nitrogen in this bottom container was maintained constant during a test. Several foil strain gages (model nos. EA-06-250BF-350 and SK-09-125AD-350) and temperature sensors (model no. ETG-50B) from Micro Measurements Group were bonded along the outside surface of the tube. The temperature distribution was stabilized within 15 min of bringing the ends of the tube to their prescribed conditions. The tubes were loaded to failure in 1.1 x 10^4 N (2500 lbf) increments once the temperature equilibrium had been established. Strains were recorded at each load increment, and the longitudinal Young's modulus was computed from the loads and measured strains.

COMPRESSIVE TESTS AT 300 K

For the compression testing of the 9.53-cm-o.d. by 8.89-cm-i.d. G-10CR tubes, a length of 81.3 cm were used. Two different strain-gage configurations were used. One configuration involved three strain gages at the center of the sample, located at equal intervals around the circumference to check for possible bending of the sample due to misalignment or buckling. The other configuration used five strain gages along the sample length to examine localized deformation under compression.

Tests with the three-gage configuration showed that only uniform strains take place. The testing facilities, end caps, and tube end conditions are therefore believed to be adequate to give pure compressive loading during tests. The sample dimensions, 9.53-cm-o.d. by 8.89-cm-i.d. by 81.3-cm-long, preclude buckling failures. The modulus obtained from these data was 20 GPa.

Uniform strains were also obtained for the tests with the five-gage configuration, and the modulus from these data was 20.7 GPa. The ultimate compressive strength of the 9.53-o.d. sample was 206.7 MPa. To check the effect of length on compressive strength, 11.43-cm-long samples with 9.53-cm-o.d. and 8.89-cm-i.d. were also tested. An identical compressive strength of 206.7 MPa confirmed that the 81.3-cm samples were failing in pure compression. All the samples tested failed near the end-cap region at 300 K.

FAILURE ANALYSIS

An investigation of the end-cap region shows why failure occurred at this location. Because the end caps tended to constrain the radially expanding tube, which was subjected to longitudinal compression, edge-bending moments and transverse shear forces developed at the ends of the tube. In the case of edge-loaded semi-infinite cylindrical shells, the bending moment and shear force at any distance, x, along the tube length are expressed by[3]

$$M_x = M_o C(\lambda x) + Q_o B(\lambda x)/\lambda \tag{1}$$

$$Q_x = 2M_o\lambda\ B(\lambda x) + Q_o D(\lambda x) \tag{2}$$

where

$$A(\alpha) = e^{-\alpha} \cos \alpha$$
$$B(\alpha) = e^{-\alpha} \sin \alpha$$
$$C(\alpha) = A(\alpha) + B(\alpha)$$
$$D(\alpha) = A(\alpha) - B(\alpha)$$
$$\alpha = \lambda x$$
$$\lambda = [3(1-\nu^2)/R^2t^2]^{1/4}$$

If the end caps have infinite longitudinal stiffness and thus do not permit rotation of the cylinder edges, the edge bending moment, M_o, and transverse shear force, Q_o, at the juncture of the tube with the end cap are as follows:[3]

$$M_o = -\nu \, \theta_H(\lambda x) \, N_x/2\lambda^2 R \qquad (3)$$

$$Q_o = \nu \, \theta_M(\lambda x) \, N_x/\lambda R \qquad (4)$$

where

ν = Poisson's ratio, taken to be 0.2 for G-10CR
N_x = $P/2\pi R$, the applied load per unit of edge length
$\theta_H(\alpha)$ = $(\sinh \alpha - \sin \alpha)/(\sinh \alpha + \sin \alpha)$
$\theta_M(\alpha)$ = $(\cosh \alpha - \cos \alpha)/(\sinh \alpha + \sin \alpha)$
R = mean radius of tube
t = thickness of tube wall

As an example of this failure analysis, we have calculated the edge moments and shear forces for a 9.53-cm-o.d. by 8.89-cm-i.d. by 81.3-cm-long G-10CR sample. The values were M_o = 122.3 N (27.55 lbf) and Q_o = 264 N/cm (151 lbf/in) at the end of the sample.

The bending moments and shear forces, M_x and Q_x, along the length of the same sample, 9.53-cm-o.d. by 8.89-cm-i.d. by 81.3-cm-long G-10CR, were calculated (see Figs. 2 and 3). Owing to these localized shear and moment effects on the sample, the region near the end cap was subjected to higher stresses, and failure therefore occurred near an end fixture.

COMPRESSIVE TESTS AT 77 K

Figure 4 shows the stress-strain data from the eight strain gages bonded to a G-10CR tube subjected to uniaxial compression. The tube was 9.53-cm-o.d. by 8.89-cm-i.d. by 81.3-cm-long. Gage location is indicated in the figure. Strains were quite uniform at the room temperature locations and gave an elastic modulus of 20.7 GPa. The gages at the bottom end, which were immersed in liquid nitrogen during testing, gave a modulus of 28.3 GPa. The cylinders failed at the top end, which was maintained at room temperature, with an audible noise. Failure at this location was not

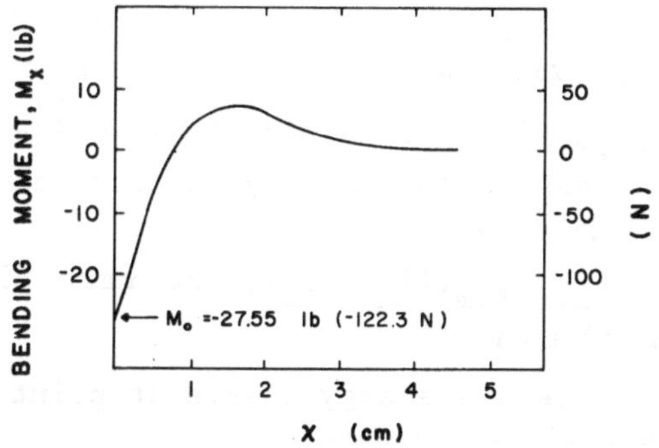

Fig. 2. Variation of bending moment, M_x, along the tube as a function of the distance, x, from an end plate.

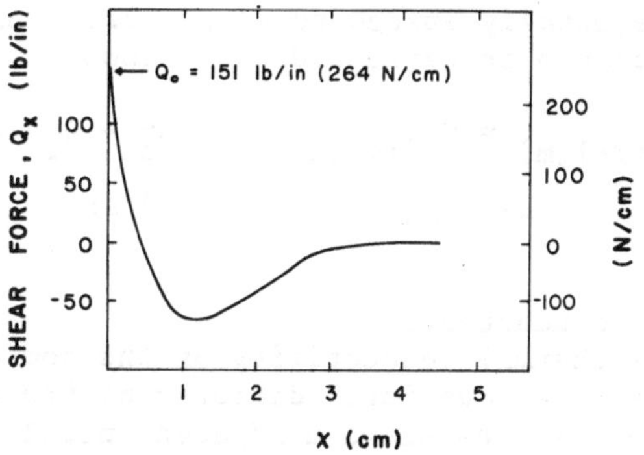

Fig. 3. Variation of shear force, Q_x, along the tube as a function of the distance, x, from an end plate.

surprising in that the material is weaker at room temperature (top) than at cryogenic temperature (bottom). The failure strength from the tests was 206.7 MPa.

TEMPERATURE DISTRIBUTION

The temperature distribution along a composite tube is predicted numerically. It is assumed that the temperature is constant across a transverse plane at any arbitrary distance from an end of the tube and that the air inside the tube behaves exactly the same as the outside air. For a steady state, the net heat flow into a control region is zero, i.e.,[4]

$$q_{c,(m-1)\to m} + q_{a,\infty\to m} + q_{c,(m+1)\to m} = Q_m = 0 \tag{5}$$

where

$q_{c,(m-1)\rightarrow m} = K_{k(m-1,m)}(T_{m-1}-T_m)$, the rate of heat flow by conduction from nodal point (m-1) to point m.

$q_{a,\infty-m} = K_{a(\infty,m)}(T_\infty-T_m)$, the rate of heat flow by convection from the surrounding gas to the surface of the subvolume of nodal point m.

$q_{c,(m+1)\rightarrow m} = K_{k(m+1,m)}(T_{m+1}-T_m)$, the rate of heat conduction from (m+1) to m.

Q_m = Rate of internal energy change at point m, which must be zero in a steady state.

Because of the low thermal conductivity, k, of the composite material, it is assumed that the temperature at x = 5.08 cm from the liquid nitrogen level is 300 K. This was confirmed by experiments. The end 5.08 cm of the tube was subdivided into eight sections with nine equally spaced nodal points. For the cylindrical tube, the conductance is expressed as follows:

$$K_{c(m-1,m)} = K_{c(m+1,m)} = k\pi(D_o^2 - D_i^2)/4\Delta x \tag{6}$$

$$K_{a(\infty,m)} = h\pi(D_o + D_i)\,\Delta x \tag{7}$$

where

K = conductance
k = thermal conductivity of the composite tube
D_o and D_i = outer and inner diameter of the tube
Δx = distance between adjacent nodal points along the tube
h = composite's unit surface conductance.

With these expressions, Eq. 5 can be written

$$T_{m-1} + T_{m+1} + [4h\Delta x^2/k(D_o-D_i)]T_\infty - 2[2+4h\Delta X^2/k(D_o-D_i)]T_m =$$

$$Q_m/[k\pi(D_o^2-D_i^2)/4\Delta X] = Q_m' \tag{8}$$

The temperatures were predicted at the seven inside nodal points using equation 8, relaxation techniques, and the known temperatures at x = 0 (77 K) and at x = 5.08 cm (300 K). Thermal conductivity values of 7.62 $mWm^{-1}K^{-1}$ at 77 K and 20.31 $mWm^{-1}K^{-1}$ at 300 K were used and a linear distribution of conductivity was assumed between these values.[5] A value of h = 7.2 $mWm^{-1}K^{-1}$ was used.[6] The calculated temperature distribution for a G-10CR tube with 9.53-cm-o.d. by 8.89-cm-i.d. by 8.13-cm-long is shown in Fig. 5 together with the experimental results. The experimental and calculated results are in good agreement. The highly localized cryogenic effect on the cylinder was due to the low thermal conductivity of the composite materials.

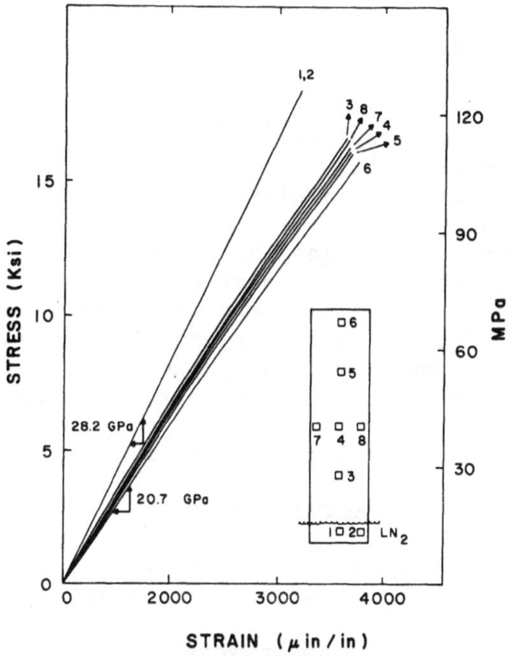

Fig. 4. Measured stress-strain at various locations of a composite tube subjected to longitudinal compression. Top of sample maintained at 300 K while bottom at 77 K.

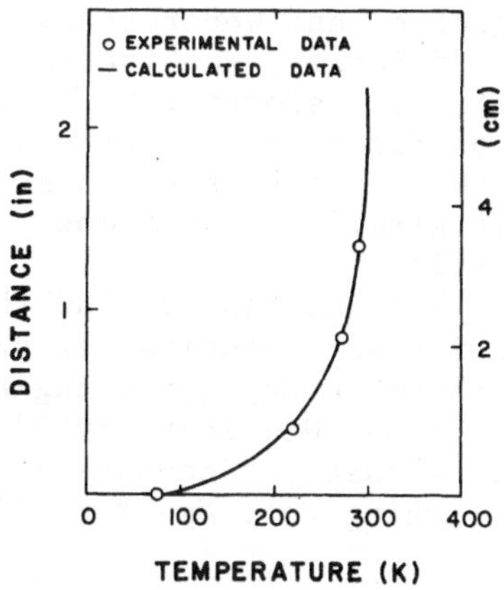

Fig. 5. Temperature distribution along the tube as a function of the distance from an end plate.

CONCLUSIONS

The following conclusions can be made:

1. The composite tubes failed under compression in the region of the steel end cap owing to the additional stresses resulting from the restraint.

2. The 81.3-cm-long samples were long enough to isolate the effect of each end.

3. The range of tube lengths (11.43 to 81.3 cm) all yielded the same measured mechanical properties.

4. The modulus and ultimate compressive strength of the tested composite materials were larger at liquid nitrogen temperatures than at room temperature, as expected.

5. Tubes with a small portion of the length immersed in liquid nitrogen failed at the room temperature end because of the relatively weaker mechanical properties at room temperature.

ACKNOWLEDGMENTS

This work was supported by the Department of Energy and the Wisconsin Electric Utilities Research Foundation.

REFERENCES

1. R. W. Boom, "Wisconsin Superconductive Energy Storage Project, Vol. 1, Vol. 2, 1977 Annual Report and Vol. 4, University of Wisconsin, Madison, Wisconsin.
2. S. Van Sciver, K. Han, K. Hartwig, G. McIntosh, and W. Young, Conductor and structural component studies for diurnal energy storage, in: "1980 Annual Contractor's Review of Mechanical, Magnetic, and Underground Energy Storage," U.S. Department of Energy, Washington, D.C. (1980).
3. P. Seide, "Small Elastic Deformations of Thin Shells," Noordhoff International Publ., Netherlands (1975).
4. G. M. Dusinberre, "Heat Transfer Calculations by Finite Differences," International Textbook Company, Scranton, Pennsylvania (1961).
5. A. Khalil and K. S. Han, Mechanical and thermal properties of glass-fiber-reinforced composites at cryogenic temperatures, in: "Advances in Cryogenic Engineering--Materials," Vol. 28, Plenum Press, New York (1982), p. 243.
6. W. Rohsenow and J. Hartnett, "Handbook of Heat Transfer," McGraw-Hill Publishing Co., New York (1973).

IMPACT TESTS OF REINFORCED PLASTICS AT LOW TEMPERATURES*

S. Nishijima, M. Takeno, and T. Okada

The Institute of Scientific and Industrial Research, Osaka University, Suita, Osaka, Japan

S. Namba

Junior College of Engineering, University of Osaka Prefecture, Osaka, Japan

INTRODUCTION

The applied forces to the component materials of a superconducting magnet turn to varied and complex form accompanied with increase in magnet size and/or pulsative mode of operation. These materials are required to withstand the varied forces and show sufficient performance at cryogenic temperatures. It is important, therefore, to elucidate the behavior of each magnet component against the various forces at low temperatures.

In this series of work, impact forces have been studied on reinforced plastics, which are expected to have excellent static mechanical properties. Materials, especially plastics,[1-3] are expected to turn hard and brittle when they are loaded at the higher strain rate and the lower temperatures. The mechanical behavior of reinforced plastics against impact forces at cryogenic temperatures has not been studied.

The Charpy impact test and drop weight test were conducted and the absorbed energy was measured. Static flexural tests were also conducted to compare the impact and static response of the material used in the present investigation.

* This work is supported by Grant in Aid for Fusion Research No. 56055028, Ministry of Education in Japan.

EXPERIMENTS

Impact Tests

Two types of impact tests, the Charpy and drop weight tests, were conducted. Charpy tests were performed at RT (room temperature) and LNT (liquid nitrogen temperature); drop weight tests at RT, LNT, and LHeT (liquid helium temperature). In both tests the impact velocity was approximately 2.9 m/s. In the drop weight test, the specimen is sprayed with liquid helium. After the weight is cooled down to LNT, it is dropped on the specimen. The temperature rise of the specimen during the striking was detected by a thermocouple (AuFe-Chromel), which is attached to the back surface of the specimen. The weight was dropped from a height that corresponds to the breaking energy measured by preliminary experiments. If the specimen broke, the potential energy of the weight was decreased by a certain value; if the specimen did not break, the potential energy of the weight was increased. The energy that broke the specimen with 50% probability and its standard deviation were calculated.

Specimens

The samples were epoxy resin[4] (named A), glass-cloth-reinforced epoxy (B), carbon-cloth-reinforced epoxy (C), commercial glass-cloth-reinforced epoxy (D), and commercial glass-mat-reinforced epoxy (E). Samples B and C were made by the hand layup

Table 1. Specifications of Specimen

Specimen	Method of Manufacture	Test	Weight Fraction of Reinforcement (%)
A Epoxy		Charpy	0
B Glass-cloth-reinforced epoxy	hand layup	Charpy	48
C Carbon-cloth-reinforced epoxy	hand layup	Charpy	56
D Commercial glass-cloth-reinforced epoxy	dry press	Charpy, Drop Weight, Flexural	62
E Commercial glass-mat-reinforced epoxy	dry press	Charpy, Drop Weight	60

method using the same epoxy as sample A. All cloths for reinforcement were fabricated in plain weave. The specifications of each sample are given in Table 1.

The shape of the specimen in the Charpy impact test was 80 mm x 10 mm x 4 mm with a 1-mm-deep V-shaped notch (Type I). This shape was used for samples A, B, and C. The other, used for samples D and E, was 100 mm x 10 mm x 4 mm with a 6-mm-deep V-shaped notch (Type II) shown in Fig. 1. The error in cutting out the specimens was within 5%. Testing samples B and C were notched in the form of flatwise and cut out with the off-axis angle of 0° to 45° every 15° to clarify the anisotropy.[5-7] For samples D and E, both flatwise and edgewise samples were prepared; they were cut out with the off-axis angle of 0° to 90° every 15°, that is, a total of 14 types of specimens were made (see Fig. 2).

In the drop weight test, samples D and E were used. The specimens were disc type with a 60-mm diameter and 1-mm thickness. Figure 3 represents the setting of the specimen and the dimension of the weight. The specimen is set between two holders that are tightened with six bolts. The flexural test is carried out with the same sample and same supporting system as the Charpy test. Each datum point represents the average of at least five samples in the Charpy and flexural tests.

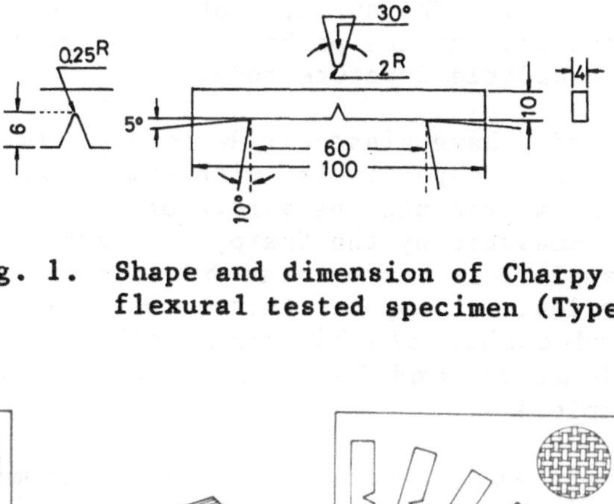

Fig. 1. Shape and dimension of Charpy and flexural tested specimen (Type II.)

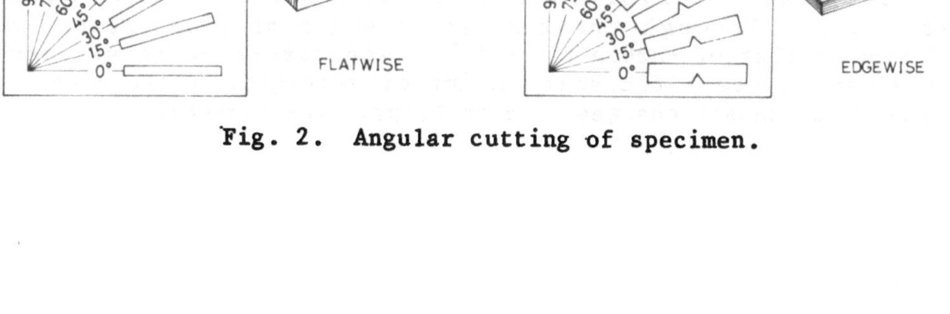

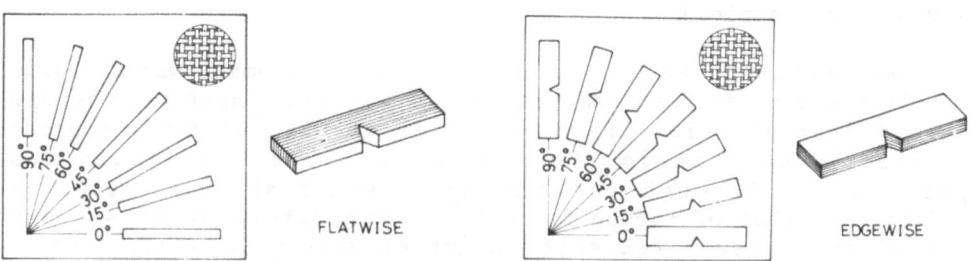

Fig. 2. Angular cutting of specimen.

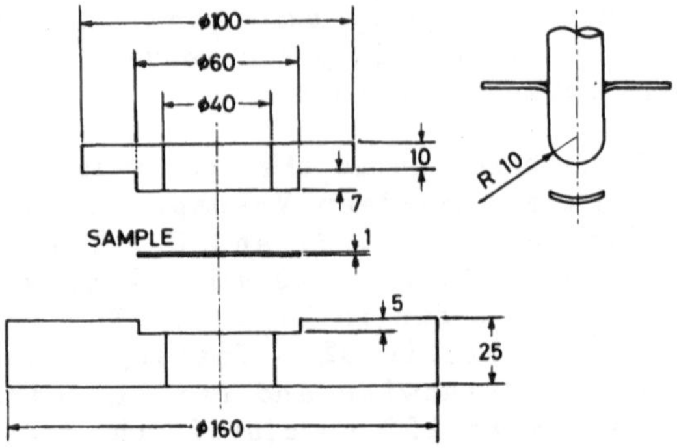

Fig. 3 Supporting system of specimen and weight of drop weight test.

RESULTS AND DISCUSSION

The Charpy impact strengths of sample A (epoxy) were 1.33 ± 0.51 and 0.60 ± 0.37 $kgf \cdot cm/cm^2$ at RT and LNT, respectively. Decreasing the temperature from RT to LNT caused a decrease in the impact strength of approximately one-half.

Most of the surface area of the specimen examined at RT showed a rough and large, fibrous pattern; the front view showed the warped path of crack. Consequently large fracture toughness is represented at RT. The surfaces of the specimen tested at LNT were flat and smooth, and the paths of crack growth were straight, illustrating the brittle fracture mode.

Samples B and C have glass-cloth reinforcement and carbon-cloth reinforcement, respectively, with same epoxy matrix used for sample A. Figure 4 presents the directional angle dependences of impact strength measured by the Charpy test for samples B and C at RT and LNT. Sample B is represented by circles and sample C is shown by triangles. It appears that both kinds of reinforcements (glass cloth and carbon cloth) remarkably improved the impact strength both at RT and LNT, in comparison with the impact strength of sample A.

Although the weight fraction of reinforcement was approximately the same in both samples (B and C), the impact behaviors were different. The impact strength of GRP (glass-reinforced plastics) was superior to that of CRP (carbon-reinforced plastics), although the tensile strength of glass fiber is inferior to that of carbon fiber.[7-8] Temperature decrease does not always cause a decrease in impact strength; in the case of sample C, it causes changes of anisotropy. The directional angle

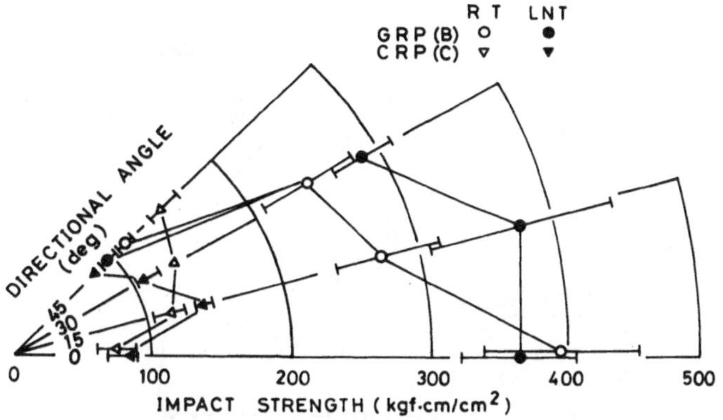

Fig. 4 Effects of directional angle on impact strength of sample B (GRP) and C (CRP) at RT and LNT.

dependences of impact strength are thought to show the symmetrical behavior with the 45° direction because both reinforcements are plain cloths. This was confirmed by sample D.

The off-axis angle dependence of impact strength was observed on sample D at RT and LNT in both edgewise and flatwise orientations. The differences between edgewise and flatwise are clarified. The results of the Charpy impact test on sample D are shown in Fig. 5. In every case, the off-axis angle dependences show symmetrical behavior with the 45° of directional angle. This is attributed to the plain-cloth reinforcement.

The impact strength of the flatwise sample was larger than that of the edgewise one, as it was reported at RT.[9] In the case of flatwise specimens at RT, the impact strength increased with directional angle and reached a maximum at the directional angle of 45°. When it was tested at LNT, the impact strength was a maximum at a directional angle of 0° or 90° and a minimum at 45°. Temperature decrease does not always mean a decrease of impact strength, as samples B and C demonstrated. In the case of an edgewise specimen, the directional angle dependence at RT coincides with that at LNT.

Flexural tests were also performed on a flatwise sample at RT, LNT, and LHeT; the absorbed energy required to break the specimen was calculated as the area under the load-displacement curves.

Figure 6 shows the load-displacement curves with the 0° directional angle of sample D obtained by flexural tests at RT, LNT, and LHeT at the deformation speed of 10 mm/min. The elastic

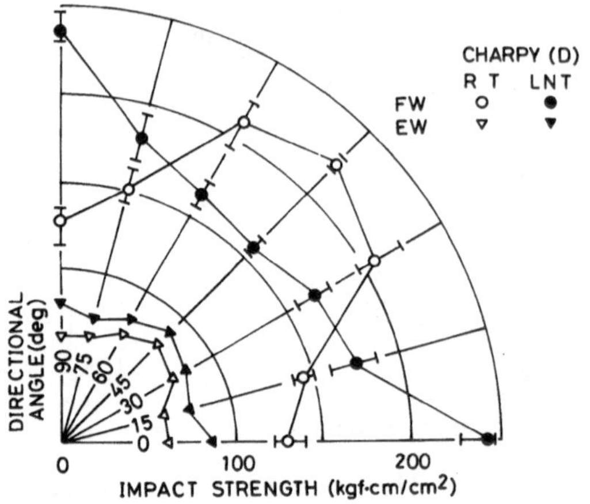

Fig. 5. Differences of impact strength between flatwise and edgewise specimens of sample D (commercial glass-reinforced epoxy) at RT and LNT.

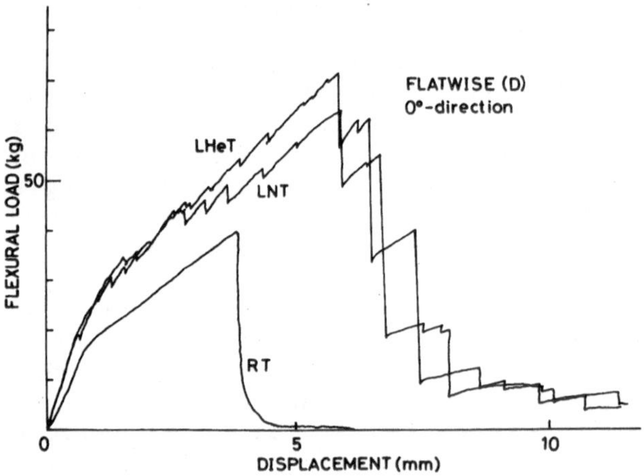

Fig. 6. Load-displacement curves of sample D with 0° directional angle in flexural test at RT, LNT, and LHeT.

modulus, maximum load, and breaking displacement of the sample tested at RT was smaller than that at LNT or LHeT. The load-displacement curve obtained at LHeT was similar to that at LNT. Many sudden load decreases were observed.

The absorbed energy required to break the specimen was calculated as the area under the load-displacement curve. The off-axis angle dependence of absorbed energy is represented in

Fig. 7 and appears to agree well with that of the Charpy impact strength. The results obtained at LHeT were similar to those at LNT.

In Fig. 8 the absorbed energy of flatwise specimens obtained by a static flexural test is compared with that by a Charpy test. The absorbed energy measured by the Charpy test was approximately 10 kgf·cm larger in every direction than that measured by the flexural test. Since Charpy impact strength (or absorbed energy) includes the kinetic energy of sample, the net absorbed energy was smaller than Charpy impact strength.[10] Hence, this evaluated kinetic energy is larger than 10 kgf·cm. In the case of LNT, the static absorbed energy agreed well with that of Charpy impact test. This means the absorbed energy by the Charpy impact test at LNT was smaller than that by flexural test. This can be detected in the case of edgewise specimens and suggests that less energy is absorbed when reinforced plastics are loaded at high speed and low temperatures.

Figure 9 represents the Charpy impact strength of sample E. Sample E was reinforced by chopped strand mat and hence did not show anisotropy. The impact strength of the edgewise sample was also smaller than that of flatwise sample. Sample E was not affected by decrease of temperature.

Drop weight tests were made on sample D. The number of samples tested were 25 at RT, 20 at LNT, and 9 at LHeT. Figure 10

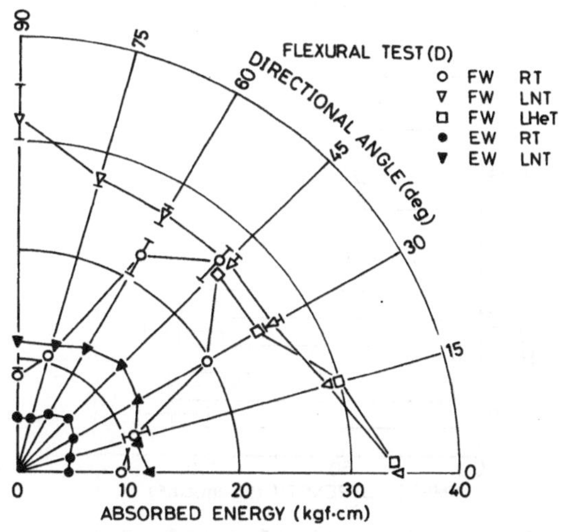

Fig. 7. Absorbed energy change obtained by flexural tests accompanied angle in both flatwise and edgewise sample D at RT, LNT, and LHeT.

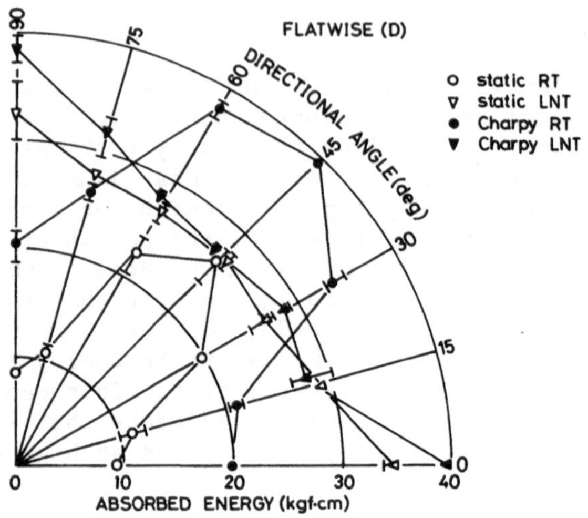

Fig. 8. Comparison of absorbed energy obtained by Charpy and flexural tests at RT and LNT for flatwise sample D.

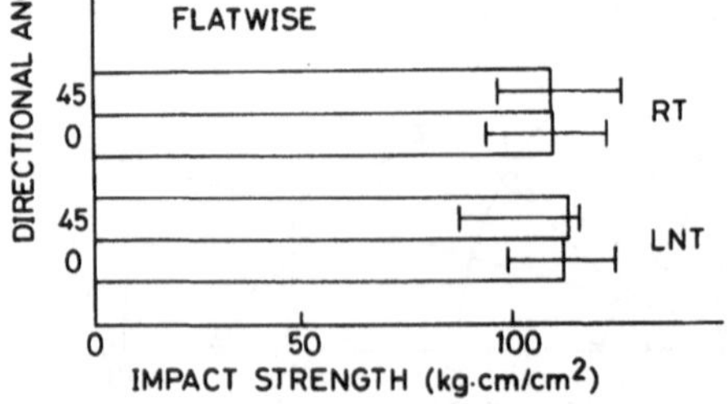

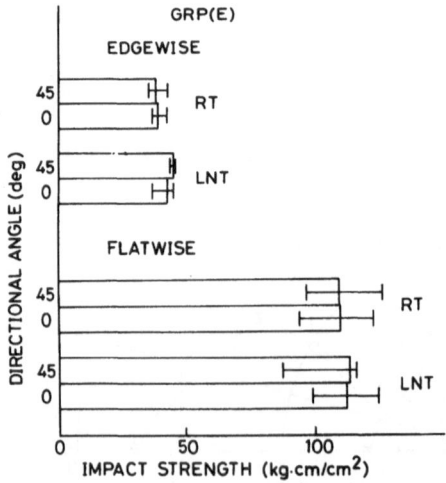

Fig. 9. Impact strength of sample E (commercial glass-mat-reinforced epoxy) at RT and LNT.

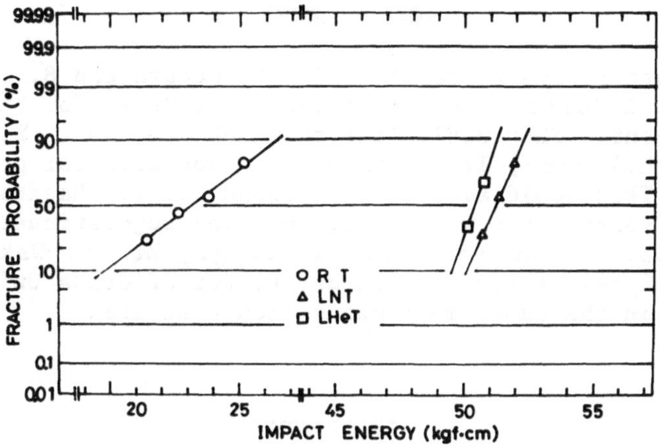

Fig. 10. Results of drop weight test at RT, LNT, and LHeT represented on probability paper.

presents the results of drop weight test at three temperatures. These results are plotted on a probability paper. When a distribution obeys a normal distribution, it is represented as a line on this paper. The steeper the slope of a line is, the smaller the standard deviation. The energy of 50% probability of fracture was 23.0 $\pm$2.3 kgf·cm at RT, 51.0 $\pm$0.66 kgf·cm at LNT, and 50.7 $\pm$0.58 kgf·cm at LHeT. The fracture energy increased when the temperature decreased from RT to LNT and showed little decrease when the temperature changed from LNT to LHeT. The degree of scattering decreased with temperature.

CONCLUSIONS

Charpy and drop weight tests were conducted on reinforced plastics at RT, LNT, and LHeT. The following conclusions were drawn:

1. The Charpy impact strength of resin alone decreases with temperature from RT to LNT, yet that of reinforced plastic does not always decrease with decreasing temperature.
2. The directional dependence of impact strength changes with temperature, reinforcement, and method of manufacture.
3. The absorbed energy loaded at high speed could be smaller than that at low speed at low temperatures and hence it would be important to examine the impact test at cryogenic temperatures.
4. The results of the drop weight tests agree well with those of Charpy tests and also show that temperature decreases do not always decrease impact strength.

ACKNOWLEDGMENTS

The authors are grateful to Dr. S. Uemura and S. Kobayashi of Government Industrial Research Institute Osaka for their help in the experiment. They would like to thank President Y. Kobayashi of Kobayashikigata, Inc. and Mr. N. Tsukamoto for preparing the specimen. They would also like to thank Dr. T. Hagihara of Osaka Kyoiku University for various stimulating suggestions. They wish to thank Dr. J. Yamamoto, Mr. T. Tsuji, Mr. Y. Wakisaka, and Mr. H. Makiyama of Low Temperature Center of Osaka University for their help in the experiment using liquid helium.

REFERENCES

1. S. Nishijima and T.Okada, Cryogenics 20(2):86 (1980).
2. K. H. Sayer and B. Harris, J. Compos. Mater. 7:129 (1973).
3. E. J. Mcquillen and L. W. Gause, J. Compos. Mater. 10:79 (1976).
4. K. Okazaki, M. Niguchi, and K. Shibayama, Kohbunshi Ronbunshu (in Japanese) 34(3):187 (1977).
5. J. M. Lifshitz, J. Compos. Mater. 10:92 (1976).
6. D. F. Sims, J. Compos. Mater. 7:124 (1973).
7. "Composite Material Engineering," T. Hayashi, ed., Maruzen (1977).
8. S. Kumazawa, J. Soc. Mater. Sci. Jpn. 21(229):893 (1972).
9. S. Uemura and Y. Masuda, Bull. JSME 738-1:45 (1973).
10. S. Uemura, Trans. JSME 44(282):1830 (1978).

CURRENT STATUS OF STANDARDIZED NONMETALLIC CRYOGENIC LAMINATES*

M. B. Kasen and R. E. Schramm

Fracture and Deformation Division, National Bureau of Standards, Boulder, Colorado

INTRODUCTION

National Bureau of Standards personnel are continuing to work with magnet fabricators and with representatives of the U.S. laminating industry to establish standard grades of nonmetallic insulator and structural support materials for use in critical parts of magnetic fusion energy (MFE) cryogenic systems. Standardized materials are required for several reasons: (1) excessive variability in the 4-K mechanical properties among nonmetallic laminates purchased to present industrial generic designations, (2) the impracticality of establishing performance specifications for retention of properties after neutron and gamma irradiation at 4 K, and (3) lack of knowledge of the significant elements in a laminate that determine such performance.

The objective of the program is to assist designers in materials selection by establishing commercial sources of standard materials for which a cryogenic data base has been established. Achievement of this objective will simplify communication between designer and laminate producer, enabling a balance to be achieved between performance and cost, and laying a sound foundation for systematic materials development in response to demonstrated need.

This is, of course, an open-ended task--there will always be a desire for better material performance at lower cost and there will always be avenues for improvement. But the program recognizes that designers require assistance in materials selection for current designs as well as improved materials for future designs. The overall approach is illustrated by the flow sheet of Fig. 1. We

*Invited paper.

start by assessing current materials needs and by consulting with industry on existing laminate systems best capable of meeting these needs. A common specification for a high-grade laminate of the defined type is then established by the laminating industry. At this point, the standard laminates differ from those of a generic designation primarily in the restrictions placed on the components used in their manufacture and on the manufacturing procedure. These restrictions are intended to be sufficiently stringent to insure commercially available products whose cryogenic performance can be meaningfully characterized and whose performance variability is minimal. Deficiencies that are subsequently revealed through experience or additional testing can then be systematically addressed since they are associated with defined materials systems.

This paper reviews progress in this effort during the past several years, discusses the present activity, and offers suggestions for future work.

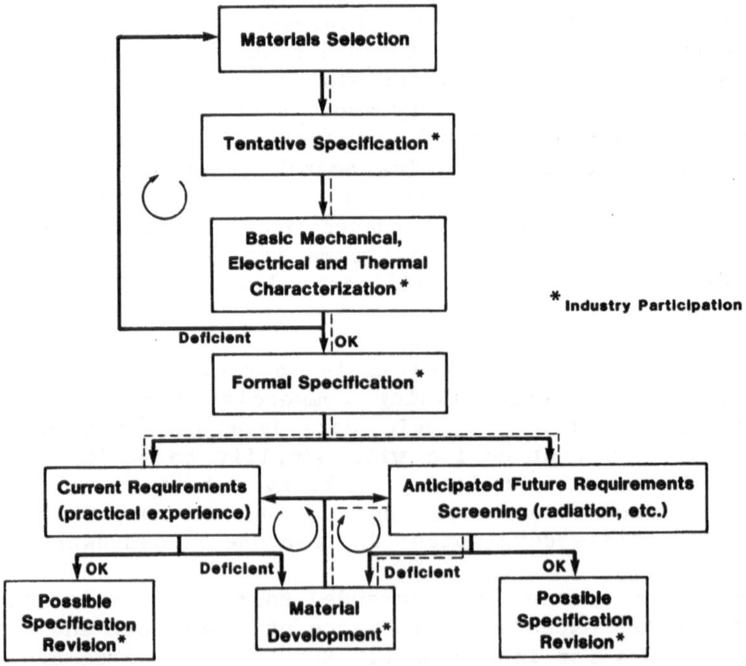

Fig. 1. Flow chart depicting a proposed approach to meeting immediate needs of nonmetallic laminate development for the cryogenic industry. Dashed lines indicate the present state of development of the G-10CR and G-11CR grades of industrial laminates.

PROGRESS TO DATE

A survey of proposed MFE designs[1] revealed that high-pressure glass-cloth-reinforced epoxy laminates of the NEMA G-10 or G-11 type were likely to be widely used for electrical insulation and structural support for superconducting magnet windings. It was known, however, that the cryogenic mechanical properties of such material produced by different manufacturers could differ as much as 40% at 4 K. This stimulated a series of discussions with representatives of the laminating industry, culminating in the establishment of tentative specifications for controlled grades of G-10CR and G-11CR. These materials are currently being manufactured by four U.S. firms and have been well received by the cryogenic industry. The effect of cryogenic temperatures on the mechanical, electrical, and thermal performance of these materials has been reported in the literature,[2] and characterization work is continuing at NBS and other laboratories. The specifications are currently undergoing revision to provide the fabricators with maximum discretion in fabrication while maintaining minimum product variability.

As anticipated, the G-10CR and G-11CR products have served as baseline materials for the study of several variants. An early variant replaced the conventional boron-containing type E glass reinforcement with a boron-free type in expectation of improved radiation resistance; however, no substantial improvement was observed.[3] Another variant, tentatively designated G-10CR-L, replaced the conventional glass reinforcement with a lightweight type having less thickness per layer for the same volume fraction of glass. This variant is useful in fabricating parts with small radii of curvature and thin sections where it is desirable to maximize the number of layers of reinforcement. Cryogenic tests have thus far indicated performance very similar to that of the original CR grade except for some decrease in compressive strength. It is probable that this variant will become a standard grade.

Radiation studies conducted at Oak Ridge National Laboratory[3] have indicated substantial degradation in mechanical properties of G-10CR and G-11CR after exposure to neutron and gamma irradiation at 4 K under fluences at levels expected in the superconducting magnets of working MFE systems. These results led to the investigation of a G-10CR variant in which the epoxy matrix was replaced by a polyimide resin. Subsequent tests at Oak Ridge National Laboratory[4] indicated a much improved radiation resistance for this variant, suggesting that it might be a candidate for insulators and structural supports in MFE magnets.

Unfortunately, the polyimide variant, tentatively designated PG-10CR, has several disadvantages of potential significance. One is high cost--six to eight times that of G-10CR. Another is relatively low cryogenic mechanical strength compared with that of G-10CR. As illustrated in Figs. 2 and 3, the polyimide variant develops significantly lower tensile, compressive, and interlaminar shear strengths at all temperatures, with the difference increasing notably at cryogenic temperatures. The 30-40% lower interlaminar shear strength of the polyimide variant is potentially the most significant deficit, because many nonmetallic magnet components are loaded in shear. Young's modulus is the only property in which the polyimide variant excells.

POSSIBILITIES FOR THE FUTURE

In the near term, it will be necessary to continue the development of standard laminate grades for use in radiation environments. Results described above suggest that radiation resistance and cost may go hand in hand, in which case it would be desirable to provide designers with a series of laminate types so that material selection could be made on a cost-versus-performance basis. Ideally, the series would include laminates reinforced with glass mat as well as glass fabric and material that could be fabricated by low- as well as high-pressure molding techniques. Discussions are being held to determine the feasibility of such development.

Because so little is known of the factors affecting degradation of laminate propeties due to irradiation at 4 K, the achievement of this goal will require close cooperation between researchers in industry and several scientific disciplines. A successful program must work with extremely well-characterized materials and

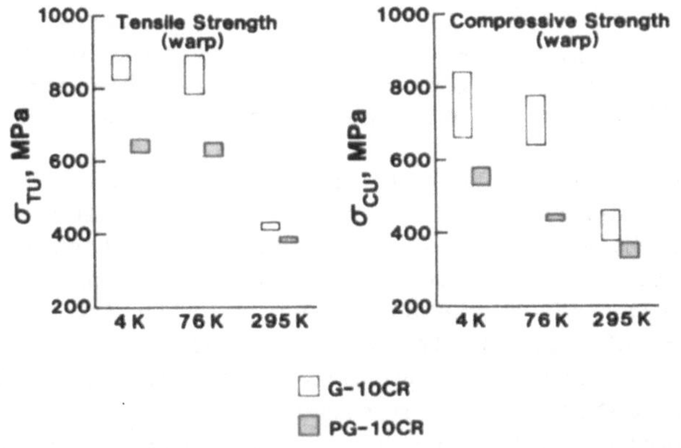

Fig. 2. Comparison of the effect of cryogenic temperature on the tensile and compressive strength of G-10CR and the polyimide-matrix variant tentatively designated PG-10CR. Preliminary data for two PG-10CR specimens of each condition tested edgewise (warp direction).

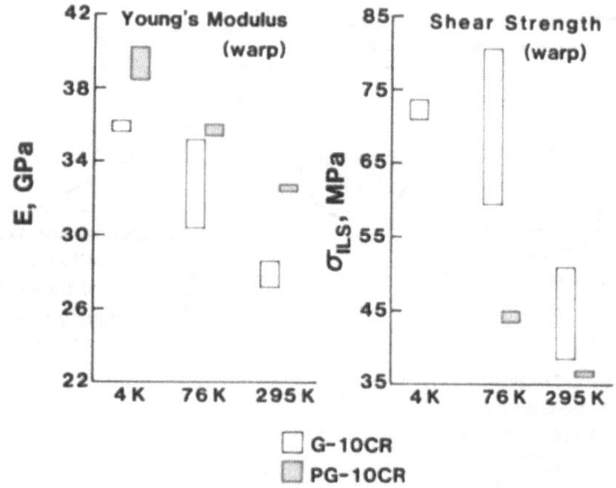

Fig. 3. Comparison of the effect of cryogenic temperature on Young's modulus and interlaminar shear strength (guillotine method) of G-10CR and the polyimide-matrix variant tentatively designated PG-10CR. Preliminary data for two PG-10CR specimens of each condition tested edgewise (warp direction).

would probably require an iterative research cycle of the type illustrated in Fig. 4. As the cryogenic radiation resistance of a polymer is likely to be a strong function of the exact molecular structure,[5] an essential part of the cycle is the correlation of molecular level damage with radiation level and property change. Only in this way will the required scientific basis be established.

The need for standardization largely reflects the absence of a nonmetallic materials classification system in which material performance is associated with a coding system. In the long run, it will be necessary to develop a system for composite laminates that is the equivalent of the systems developed for metals by AISI, the Aluminum Association, and other similar voluntary standard bodies. Although composites are indisputedly more complex than metals, there seems no _a priori_ reason why such an approach would not be feasible for composite laminates, recognizing that the significant engineering properties of a laminate at any temperature are largely defined by a few parameters. For example, the intrinsic mechanical, elastic, thermal, and electrical properties of a high quality laminate should be defined within relatively narrow limits by a numerical classification system that establishes the type and flexibility of the matrix, the type and configuration of the reinforcement, and the reinforcement volume fraction. At the very least, it should be possible to establish meaningful lower bounds

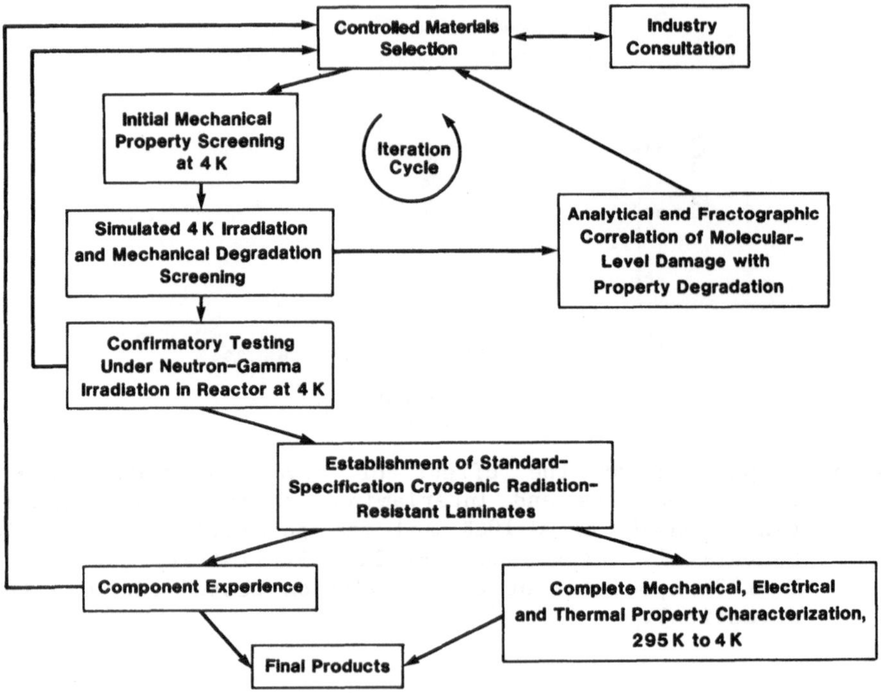

Fig. 4. Proposed flow chart for development of radiation-resistant nonmetallic laminate insulators.

on such properties. The basic elements of such a coding system have been considered in a previous publication.[6]

CONCLUSIONS

The author has reviewed ongoing efforts to provide a commercial supply of standardized, controlled-performance laminates required to meet the short-term needs of the cryogenic industry for critical components in MFE superconducting magnet design and construction. The approach provides a basis for systematic development of material variants and for new materials as required to meet the needs of future designs. The approach recognizes that development of nonmetallic insulators and structural supports in magnets subjected to intense irradiation at 4 K will require a much better understanding of the factors affecting material performance in this environment and that this will be achieved only by an interdisciplinary approach.

The lack of terminology allowing a designer to specify nonmetallic laminate materials by a performance criteria is a continuing problem. The ongoing standardization effort is designed to meet immediate technical needs; however, a long-range effort to establish an accepted standards notation system for nonmetallics comparable to that used in the metals industry is worthy of consideration.

ACKNOWLEDGMENTS

This work was supported by the Office of Magnet Fusion Energy, United States Department of Energy.

REFERENCES

1. R. P. Reed, F. R. Fickett, M. B. Kasen, and H. I. McHenry, "MFE Low Temperature Materials Program: A Survey," National Bureau of Standards, Institute for Basic Standards, Boulder, Colorado (1977).
2. M. B. Kasen, G. R. MacDonald, D. H. Beekman, and R. E. Schramm, in: Mechanical, electrical, and thermal characterization of G-10CR and G-11CR glass-cloth/epoxy laminates between room temperature and 4 K, in: "Advances in Cryogenic Engineering - Materials," Vol. 26, Plenum Press, New York (1980), p. 235.
3. R. R. Coltman, Jr., C. E. Klabunde, R. H. Kernohan, and C. J. Long, "Radiation Effects on Organic Insulators for Superconducting Magnets," ORNL/TM-7077, Oak Ridge National Laboratory, Oak Ridge, Tennessee (1979).
4. R. R. Coltman, Jr., C. E. Klabunde, and C. J. Long, "The Effect of a 100 MGy (10^{10} Rads) Gamma-Ray Dose at 5 K on the Strength of Polyimide Insulators," Report SSD No. 81-10, Oak Ridge National Laboratory, Oak Ridge, Tennessee (1981).
5. B. S. Brown, Radiation effects in superconducting fusion-magnet materials, J. Nucl. Mater. 98:1 (1981).
6. M. B. Kasen, Standardizing nonmetallic composite materials for cryogenic applications, in: "Fundamentals and Applications of Nonmetallic Materials at Low Temperatures," G. Hartwig and D. Evans, eds., Plenum Press, New York (1981) (in press).

CONCRETE: COMPRESSIVE PROPERTIES, FLEXURAL AND SPLITTING STRENGTHS, AND K_{Ic} AT LOW TEMPERATURES

J. M. Arvidson

Fracture and Deformation Division, National Bureau of Standards, Boulder, Colorado

INTRODUCTION

One important low temperature application for concrete is its use as a structural material for containment vessels for the storage of liquefied natural gas (LNG) at 111 K.[1] Other applications of prestressed concrete include floating platforms, underwater structures, floating storage vessels, barges, sailboats, and fishing boats.[2,3] Such structural applications require knowing the material properties of concrete at low temperatures.

The work described in this paper provides material property data on two selected concrete mixes that were tested at room (295 K), dry-ice and alcohol (195 K), liquid nitrogen (76 K), and liquid helium (4 K) temperatures. The properties reported are compressive Young's modulus; yield strength (at 0.2% offset); maximum, ultimate, flexural, and splitting strengths; K_{Ic} and elastic and plastic elongations. Material properties such as these, particularly at cryogenic temperatures, are of interest to many and are not currently found in the literature.

MATERIALS CHARACTERIZATION

The physical properties of two concrete mortars (designated mix 1 and mix 2) were studied in this program. Mix 1 was made by a commercial cement company and mix 2 was made at the National Bureau of Standards. Both mixes utilized portland type I cement and a vinsol resin air entraining agent. The aggregate used in mix 1 is commonly known as Elgin sand and that used in mix 2 is commonly referred to as Clear Creek sand. The chemical makeup and sieve analysis of each type of aggregate are shown in Tables 1 and 2.

Table 1. Primary Components of Aggregate Composition by Percentage of Particle Count Remaining on Standard Sieves

Mix	Component	Standard Sieve Number		
		8 (2.36 mm)	16 (1.19 mm)	30 (0.60 mm)
1	Carbonate	80.3	79.8	64.3
	Chert	8.9	11.0	-
	Granite	3.6	2.4	-
	Basalts	3.4	2.0	1.2
	Quartzite	3.0	4.3	-
	Quartz-Chalcedony	-	-	26.5
	Feldspar	-	-	7.6
2	Granite	67.4	66.4	33.2
	Metamorphics	7.2	3.7	1.5
	Basalts	6.8	2.7	1.8
	Quartz	12.4	17.2	44.9
	Feldspar	5.7	7.2	15.9

Table 2. Sieve Analysis of the Aggregates Used in Mix 1 and Mix 2

Standard Sieve Number	Standard Sieve Size (mm)	Percent Passing Mix 1 (Elgin)	Percent Passing Mix 2 (Clear Creek)
4	4.75	100	99
8	2.36	78	80
16	1.18	63	59
30	0.60	46	29
50	0.30	10	7
100	0.15	0	0

Both mortars were mixed using the following ratios: water/cement = 0.5, aggregate/cement = 3.4. The air content of the hardened mortars was 11.0 and 17.7 vol.%, respectively, and the evaporable moisture contents were 8.1 and 10.8 wt.% relative to the oven dry weight. The mixing procedure described in ASTM C192 was utilized for both mixes.

Mix 1 was designed to produce a mortar with a compressive strength of approximately 17.2 MPa (2500 psi). To maintain this strength until testing, the 10-day-old specimens were placed in a freezer held at 0°C. The hydration process, which causes concretes to increase in strength as they age, is slowed considerably at this temperature. Prior to the 0°C aging, the specimens were aged in covered glass molds for 24 h at 23°C and then removed from the molds and placed in saturated lime water at 23°C for 7 days. Upon removal from the lime water bath, they were wrapped in plastic and aluminum foil for shipment.

Mix 2 was made in two batches separated by 8 months. Mix design and conditioning for both batches were identical. The mortar was formed in airtight plastic molds and aged in the mold for 2 days at 23°C. After removal from the molds, the specimens were immersed in a saturated lime water bath at 23°C. They remained in the bath until tested. The age of the specimens at the time of testing was from 79 to 215 days.

SPECIMENS

The solid-core specimen geometry was chosen for the compression and splitting strength tests; specimens measured 5.1 cm in diameter and 9.9-cm long, as recommended by ASTM C469-65, C496-71, and C873-77T. The flexural strength test specimens were 5.3 cm x 1.8 cm x 4.6 cm per ASTM C78-75.

The ends of all specimens to be tested in compression were ground parallel and perpendicular to the longitudinal axis to within $\pm 6.3 \times 10^{-4}$ cm. All mix 1 samples were completely saturated with lime-water solution and then kept at 0°C until the time for testing. All mix 2 samples were stored immersed in a lime-water solution for 100% moisture content and removed just prior to testing.

TEST PROCEDURE

Tests were performed using a commercial 1.0 MN (2.2×10^5 lbf) servohydraulic static/fatigue test machine. According to the manufacturer, the load weighing system accuracy is ±0.5% of indicated load or ±0.25% of load range in use, whichever is greater, on all load ranges. The crosshead speed used for all tests was 0.005 cm/min and is accurate to ±0.1% of set speed at all loads and speeds. Compression specimens were instrumented with specially designed, diametrically opposed cryogenic strain-gaged extensometers. This method for sensing strain worked very well at all temperatures, including total immersion in liquid nitrogen and liquid helium. Prior to each test, particular care was taken to insure good coaxial alignment of all test components (e.g., actuator ram, specimen, Dewar) to promote uniaxial loading.

Tests were conducted at room temperature (295 K) in air, in dry ice and alcohol (195 K), in liquid nitrogen (76 K), and in liquid helium (4 K). Specimen failure modes indicated a well-aligned system with a minimum of bending (i.e., uniform, conically shaped fractures emanating from the specimens ends to center).

Table 3. Effect of Temperature on Compressive Young's Modulus and Compressive Strength for Moist Concrete Mix No. 1.

Test Temperature, K	Compressive Young's Modulus, GPa	Compressive Young's Modulus, 10^{-6} psi	Compressive Strength σ Maximum*, MPa	σ Maximum*, 10^{-3} psi	σ Ultimate†, MPa	σ Ultimate†, 10^{-3} psi
295	15.9	2.30	-	-	23.0	3.34
	8.6	1.25	-	-	21.8	3.16
	22.4	3.25	-	-	23.4	3.40
	$\bar{x}$ = 15.6	2.27			22.7	3.30
195	27.2	3.95	-	-	-	-
	-	-	56.5	8.20	53.8	7.81
	38.1	5.53	49.1	7.12	47.6	6.90
	-	-	57.4	8.32	56.9	8.25
	$\bar{x}$ = 32.7	4.74	54.3	7.88	52.8	7.65
76	14.3	2.07	-	-	71.1	10.31
	31.0	4.49	-	-	107.8	15.64
	17.7	2.57	-	-	90.9	13.18
	22.8	3.30	-	-	87.9	12.75
	$\bar{x}$ = 21.5	3.11	-	-	89.4	12.97
4	23.0	3.34	-	-	-	-

* σMaximum: maximum stress the specimen reached during the test.
† σUltimate: stress at failure.

Table 4. Effect of Temperature on Compressive Properties for Moist Concrete Mix No. 2

Test Temperature, K	Compressive Young's Modulus, GPa	Compressive Young's Modulus, 10^{-6} psi	Proportional Limit, MPa	Proportional Limit, 10^{-3} psi	Ultimate Compressive Strength, MPa	Ultimate Compressive Strength, 10^{-3} psi
295	15.25	2.21	14.45	2.10	31.23	4.53
	19.80	2.80	15.65	2.27	28.40	4.12
	$\bar{x}$ = 17.28	2.51	15.05	2.19	29.82	4.32
195	30.65	4.45	33.04	4.79	89.19	12.94
	33.78	4.90	28.87	4.19	93.22	13.52
	$\bar{x}$ = 32.22	4.68	30.96	4.49	91.21	13.23
76	38.59	5.60	-	-	120.9	17.53
	44.71	6.48	-	-	102.5	14.87
	$\bar{x}$ = 41.65	6.04			111.7	16.20

Some testing was carried out to determine plane strain fracture toughness of mix 2. Short rod specimens with a single chevron in the configuration proposed by Barker and Baratta[4] were prepared. The diameter was 5.1 cm and the length was 10.2 cm. Testing at all temperatures was carried out in a newly designed fixture that operates like a pinch-type clothespin to load the chevron-notched specimen in tension.

This fixture has a lever-arm ratio of 2:1 and can be fitted into most universal testing machines. The plane strain fracture toughness, $K_{Ic}SR$, was calculated using the following equation proposed by Barker and Baratta[4] and modified by Beech and Ingraffea:[5]

$$K_{Ic}SR = AP_c/B^{3/2}$$

where A is a dimensionless calibration constant ~22.5 ±15% (dependent on specimen geometry), P_c is the load at crack instability, and B is the specimen diameter.

All specimens but one fractured across the chevron; one fractured in the body of the specimen.

RESULTS

It is interesting to note the compressive Young's modulus for mix 1 increased in value from room temperature to 195 K, and then at 76 and 4 K it decreased by approximately 30% (see Table 3). The compressive strength, however, increased as the test temperature decreased. Also, only at 195 K did mix 1 show a maximum compressive strength before reaching its ultimate.

The compressive modulus, proportional limit, and ultimate compressive strength for mix 2 all increased in value as the test temperature decreased (see Table 4). As shown in Table 5, the splitting tensile strength increased from 295 to 195 K and then remained the same down to 76 K. The modulus of rupture (flexural strength), however, shows a peak value at 195 K.

The plane strain fracture toughness, $K_{Ic}SR$, for short rod specimens of mix 2 are tabulated vs. temperature in Table 6. The room temperature results on moist mix 2 concrete are comparable to the fracture toughness reported[6] for two kinds of rock:

Type of Rock	$K_{Ic}SR$, MPa$\sqrt{m}$	psi$\sqrt{in}$
Limestone	0.99	900
Granite	2.48	2250

Table 5. Effect of Temperature on Splitting and Flexural Strength for Moist Concrete Mix No. 2

Test Temperature, K	Splitting Tensile Strength,*		Modulus of Rupture,†	
	MPa	10^{-3} psi	MPa	10^{-3} psi
295	2.32	0.34	5.25	0.76
	2.64	0.38	5.25	0.76
	$\bar{x}$ = 2.48	0.36	5.72	0.83
			$\bar{x}$ = 5.41	0.78
195	6.62	0.96	13.44	1.95
	9.69	1.41	13.88	2.01
	$\bar{x}$ = 8.15	1.19	$\bar{x}$ = 13.66	1.98
76	5.96	0.86	9.26	1.34
	8.54	1.24	11.91	1.73
	$\bar{x}$ = 7.25	1.05	$\bar{x}$ = 10.59	1.54

*ASTM C496-71 Splitting tensile strenth = T = $2P/\pi \ell d$
where: T = splitting tensile strength, MPa (or psi)
P = maximum applied load, N (or lbf)
l = length, mm (or in)
d = diameter, mm (or in)

†ASTM C78-75 modulus of rupture = R = $3P\ell/2bd^2$
where R = modulus of rupture, MPa (or psi)
p = maximum applied load, N (or lbf)
l = span length, mm (or in)
b = average width of specimen, mm (or in)
d = average depth of specimen, mm (or in)

Table 6. Fracture Toughness of Short Rod Moist Concrete Specimens*

Temperature, K	$K_{Ic}SR$, MPa$\sqrt{m}$	(psi$\sqrt{in}$)	P_c,kg	(lb)
295	0.88	(800)	41.0	(90.28)
	0.95	(860)	41.8	(92.16)
	0.88	(800)	39.3	(86.54)
195	2.26	(2,060)	101.0	(222.56)
	2.24	(2,040)	99.9	(220.30)
	2.04	(1,860)	91.8	(202.32)
	2.20	(2,000)†	97.9	(215.80)
76	1.91	(1,740)	85.1	(187.70)

*All test specimens were $5.08^{+0.00}_{-0.05}$ cm in diameter.

†Specimen fractured in body away from chevron.

The results in Table 6 show that there was a substantial increase in $K_{Ic}SR$ as temperature was lowered initially, but no additional increases as temperature was lowered further. Splitting strength showed this same trend with decreasing temperature. It is likely that the pronounced increase in toughness with decreasing temperature is related to changes in the internal structure of the moist concrete specimens as the water entrained in the pores freezes and expands and then contracts with further cooling below the freezing temperature. Some discussion of possible changes has been published by Monfore and Lentz.[7]

The plane strain fracture toughness of moist mix 2 concrete seems rather low when compared with the fracture toughness of structured steels, i.e., 54.9 MPa$\sqrt{m}$ (50 ksi$\sqrt{in}$). This comparison suggests that moist mix 2 concrete may not be a suitable structural material under tensile loading. Nevertheless, the low cost and widespread availability of such concrete may favor its application in compressive-loading situations.

DISCUSSION

Mix 1 specimens were conditioned at 1°C prior to testing. Upon cooling, the differential in thermal expansion between water and concrete (one positive, the other negative) can create the possibility of microcracking.[6] The coefficient of thermal contraction varies from 6.0 x 10^{-6} per degree K for lightweight aggregate concrete to 8.2 x 10^{-6} per degree K for 7.0-bag sand and gravel concrete.[7] The lower values of Young's modulus, proportional limit, and strength for mix 1 as compared with those of mix 2 could be attributed to the preconditioning at 0°C. The process of cracking in concrete while under compressive loads has been analyzed by several experimentalists. Higgins and Bailey indicate a proposed failure mechanism for this phenomenon.[8,9]

A greater degree of scatter exists in the results for mix 1, which is attributed to material variability. With mix 2 special care was taken during the mixing and molding process to insure a more homogeneous specimen.

The compressive ultimate strength for both mixes showed an increase with decreasing temperature. This trend is comparable to that of steels.

ACKNOWLEDGMENTS

The author is grateful to Larry L. Sparks, Thermophysical Properties Division, NBS, and Bruce W. Christ and Edward Steketee, Fracture and Deformation Division, NBS, for their valuable assistance and suggestions in formulating this test program. This work was sponsored by the Maritime Administration.

REFERENCES

1. B. C. Geriwick, "Design and Construction of Prestressed Concrete Vessels," Paper Number OTC 1886, presented at the Fifth Annual Offshore Technology Conference, Houston, Texas, (April 29-May 2, 1973) (mailing address: Offshore Technology Conference, 6200 North Central Expressway, Dallas, Texas 75206).
2. A. L. Dinsenbacher and F. E. Braver, Material development, design, construction, and evaluation of a ferro-cement planning boat, Mar. Technol. 277 (1974).
3. B. C. Geriwick, State of the art report--durability of concrete structures under water, presented at the International Association Colloquium, Liege, Belgium (June 5, 1975).
4. L. Barker and Barrata, Comparison of fracture toughness measurements by the short rod and ASTM standard method of test for plane strain fracture toughness of metallic materials (E 399-78), J. Test. Eval. 8(3):97 (1980).
5. J. F. Beech and A. R. Ingraffea, Three dimensional finite element calibration of the short-rod specimen, accepted for publication in the Int. J. Fract. (Oct. 1980).
6. S. Mindess, Application of fracture mechanics to cement and concrete, presented at the 81st Annual Meeting of the American Ceramic Society, Cincinnati, Ohio, Paper 1-S II-79 (1979).
7. G. E. Monfore and A. E. Lentz, Physical properties of concrete at very low temperatures, J. PCA Res. Dev. Lab. 4(2):33 (1962).
8. D. D. Higgins and J. E. Bailey, A microstructural investigation of the failure behavior of cement paste, in: "Hydraulic Cement Pastes: Their Structure and Properties," Cement and Concrete Association, London (1976), pp. 283-296.
9. A. R. Ingraffea and H. Y. Ko, Determination of the fracture parameters for rock, paper to appear in: "Proceedings of the International Symposium on Absorbed Specific Energy," Budapest, Hungary (Sept. 1980).

APPENDIX A

The following is a list of ASTM specifications that have been used in part or whole to perform the tests outlined in this report.

1. ASTM C469-65 (Reapproved 1975) "Static Modulus of Elasticity and Poisson's Ratio of Concrete in Compression."
2. ASTM C39-72 "Compressive Strength of Cylindrical Concrete Specimens."
3. ASTM C78-75 "Flexural Strength of Concrete (using simple beam width third-point loading)."

4. ASTM C496-71 "Splitting Tensile Strength of Cylindrical Concrete Specimens."
5. ASTM C192-76 "Making and Cracking Concrete Test Specimens in the Laboratory."
6. ASTM C138-77 "Unit Weight; Yield, and Air Content (Gravimetric) of Concrete."
7. ASTM C670-77 "Preparing Precision Statements for Test Methods for Construction Materials."
8. ASTM C802-77 "Conducting an Interlaboratory Test Program to Determine the Precision of Test Methods for Construction Materials."
9. ASTM C873-77T "Compressive Strength of Concrete Cylinders Cast in Place in Cylindrical Molds."
10. ASTM C192 "Making and Curing Concrete Test Specimens in the Laboratory."

TENSILE, COMPRESSIVE, AND SHEAR PROPERTIES OF POLYURETHANE FOAM AT LOW TEMPERATURES*

J. M. Arvidson and L. L. Sparks

National Bureau of Standards, Boulder, Colorado

INTRODUCTION

Polyurethane foams are used in many cryogenic structural as well as insulating applications. Knowledge of material performance in extreme environments (e.g., at LNG temperature, 111 K) is essential for efficient design.

The tensile, compressive, and shear properties of polyurethane foam, having a density of 32 kg/m^3, were determined using a unique new test fixture developed specifically for this program. This fixture provides the capability of determining the above material properties at any temperature from ambient (295 K) to near 4 K, in a cold helium gas atmosphere. In addition, tests can be conducted equally as well in liquids, such as helium (4 K), nitrogen (76 K), dry ice and alcohol (195 K), and others. The fixture was also modified to perform tests on materials at the above temperatures while under static pressures ranging from atmospheric to approximately 0.3 MPa (30 psi).

The second, and unique, feature of this apparatus is a specimen strain sensing technique.[1] The problem was to develop a strain extensometer and adapt it to these soft viscoelastic materials in a way that does not influence the material's properties during a test. The result was the development of concentric, overlapping cylinder capacitance strain extensometer that works well in cryogenic environments, is not attached directly to the material being tested, and is highly accurate and linear over large ranges.[2,3] For our foam specimen geometry (5.1-cm diam. x 10.2 cm), the capacitance extensometer system had a linearity

* Work supported by the Gas Research Institute.

range in excess of 2.5 cm (1.0 in). On the basis of specific requirements for sensitivity and extent of linearity, the experimentalist can design for the degree of optimization.[4] As long as the capacitance extensometer is situated in a stable fluid (i.e., single phase--no boiling) the result is a very good low noise output signal. The device will work quite satisfactorily in any gas as well. The original calibration of the system, for example, could have been performed at room temperature in air. To conduct a test in any other media (e.g., liquid nitrogen), the original calibration need only be corrected for the change in dielectric constant.[5]

MATERIAL CHARACTERIZATION

The material tested in this study is a nominal 32 kg/m^3 polyurethane foam (PU) designated as GM30. This amorphous, organic polymer is a thermosetting foam. Our supply of this material was obtained from the NBS Office of Standard Reference Materials (OSRM), Washington, D.C. The OSRM distributed this and other expanded plastics for the Products Research Committee.[6] These materials were commercially produced and designated as General Materials.

Our bulk supply of GM30 was in the form of a 0.1 x 1.22 x 1.83 m slab. The physical tests were conducted on specimens taken from the center. The orientation of the elongated cell axis for the material used in the physical properties tests was determined optically. The ratios of cell height to cell width for the principal orthogonal planes of the physical test specimens were: $x/y = 1.33 \pm 0.24$, $z/x = 1.02 \pm 0.18$, $z/y = 1.40 \pm 0.26$. The uncertainties given represent estimates of one standard deviation.

The chemical formula of the polyurethane resin is $C_{10.47}H_{11.99}O_{2.38}N$. The cellular gas content, although of secondary importance to the properties being reported here, was found by mass spectrography to be: CCl_3F, 71.2%; N_2, 6.1%; O_2, 21.8%; A, CO_2, CO, and F, 1%. The cellular gas pressure was approximately 66 kPa (0.65 atm). The apparent density of the resin-gas composite was found to be 32.4 ± 0.5 kg/m^3 at ambient temperature and humidity. Conditioning at 23°C and 50% relative humidity caused an estimated 2.7% volume increase and a resulting density of 31.7 kg/m^3.

SPECIMENS

The tensile specimens were rods 9.9 cm long and 2.9 cm in diameter. For the determination of ultimate strength, a reduced section specimen was used. The gage length was approximately 5.1 cm and 1.9 cm in diameter. All tensile specimens were epoxied to threaded polycarbonate grips. The reduced section geometry forced fracture to occur within the gage length, thereby eliminating the effect of biaxial stresses at the grip ends. All other mechanical properties, such as Young's modulus, proportional limit, and yield stress, were derived using the rod geometry.

The compression specimens were 2.54-cm long and 2.9 cm in diameter.

Shear specimens were approximately 1.9 cm x 2.54 cm x 0.4 cm. These specimens were epoxied to flat plates and each plate was attached to a pull-rod. An aluminum cylinder was slipped over the specimen and plates to help maintain alignment during testing and to minimize induced torque.

TEST PROCEDURE

Most tests were conducted at 295, 111, and 76 K. One test was conducted at 45 K. Except where noted, all tests were conducted in the vapor above liquid helium, the desired temperature being maintained using a heater/controller. To minimize thermal shock to the specimen, the liquid helium was transferred at a very slow rate ($\sim$0.2 ℓ min^{-1}). The tensile and compressive properties include: Young's modulus, proportional limit, yield strength (at 0.2% offset), tensile and compressive strengths, and elongation (elastic and plastic). No ASTM shear strength test method is available for soft cellular materials such as these foams. After several methods were carefully considered,[7] a version of the guillotine type shear test was selected. Many specimen geometries (thicknesses and widths) were tested to find the best combination that most consistently failed in shear. At least three tests were conducted at each temperature, and in some cases, more tests were run to determine material variability. Prior to testing, all specimens were "conditioned" in an environmental chamber for not less than 4 days at 23°C and 50% relative humidity. Each specimen was tested shortly after removal from the environmental chamber. All tests were conducted on a conventional compression-tension test machine at a strain rate of 5×10^{-3} min^{-1}. (Varying the rate from 5×10^{-4} to 5×10^{-2} min^{-1} had no measurable effect on results.)

RESULTS

The results are presented in Tables 1 through 3 and in Figs. 1 through 7. The error bars on the figures show the data spread from replicate tests. Scatter is typically higher for compression and shear tests than for tensile tests, since the former tests are more sensitive to problems such as misalignment. Care must be exercised in interpreting orientation effects: two orientations (L and T) are designated, but there is an irregular cell structure, as noted in the discussion.

Table 1. Summary (Average Values) of Tensile Test Results for Polyurethane Foam

Material Property	Specimen Orientation*	295 K	111 K	76 K	45 K
Young's Modulus, MPa (psi)	L	13.61 (1,980)	24.41 (3,540)	26.37 (3,820)	26.95 (3,910)
	T	7.26 (1,050)	12.81 (1,860)	14.27 (2,070)	-
Proportional Limit, MPa (psi)	L	0.166 (23.91)	-	-	-
	T	0.077 (11.17)	-	-	-
Yield Strength (at 0.2% Offset), MPa (psi)	L	0.265 (38.46)	-	-	-
	T	0.130 (18.93)	-	-	-
Ultimate Tensile Strength, MPa(psi)	L	0.458 (66.43)	0.491 (71.23)	0.490 (71.08)	-
	T	0.365 (52.85)	0.368 (53.43)	0.386 (55.93)	-
Elongation, %					-
Elastic	L	3.1	2.0	2.0	-
Plastic		2.5	0	0	-
Total		5.6	2.0	2.0	-
Elastic	T	5.0	2.9	2.7	-
Plastic		10.6	0	0	-
Total		15.6	2.9	2.7	-

*L: longitudinal; T: transverse

Table 2. Summary (Average Values) of Compression Test Results for Polyurethane Foam.

Material Property	Specimen Orientation	295 K	111 K	76 K
Young's Modulus, MPa (psi)	L	8.48 (1,230)	14.88 (2,160)	13.26 (1,920)
	T	6.03 (880)	6.69 (970)	6.40 (930)
Proportional Limit, MPa (psi)	L	0.067 (9.70)	0.079 (11.47)	0.152 (22.03)
	T	0.073 (10.63)	0.150 (21.75)	0.118 (17.14)
Yield Strength (at 0.2% Offset), MPa (psi)	L	0.114 (16.57)	0.146 (21.15)	0.204 (29.60)
	T	0.109 (15.80)	0.179 (26.04)	0.172 (24.98)
Maximum Compressive Strength, MPa (psi)	L	0.287 (41.62)	0.311 (45.12)	0.338 (49.02)
	T	0.218 (31.60)	0.247 (35.91)	0.227 (32.93)
Elongation, %				
Elastic	L	3.3	1.9	2.5
Plastic		4.4	12.4	9.4
Total		7.7	14.3	11.9
Elastic	T	3.0	3.4	3.3
Plastic		5.4	10.5	15.2
Total		8.4	13.9	18.5

Table 3. Summary (Average Values) of Shear Test Results

Test Temperature, K	Specimen Orientation	Shear Strength, MPa	psi
295	L	0.217	31.47
	T	0.217	31.43
111	L	0.266	38.62
	T	0.219	31.74
76	L	0.242	35.14
	T	0.225	32.65
76*	L	0.241	34.95
	T	0.216	31.39

*Test conducted in LN_2.

As shown in the figures, Young's modulus, proportional limit, yield strength, ultimate strength, and shear strength all increased as temperature was lowered from 295 to 76 K, with the longitudinal specimens offering relative superiority to the transverse. Tensile elongations (both elastic and plastic) decreased, and the foam became completely brittle at 111 K and 76 K. Figure 6 indicates a significant deformation capability in compression at low temperatures, but the "plastic" strain is accomplished by a collapse of the cell structure (the polyurethane material is actually incapable of exhibiting true plastic deformation at the cryogenic temperatures).

DISCUSSION

In many respects the polyurethane foam behavior is typical of polymers in general. Expected behavior for a temperature reduction from 295 K to 76 K includes large mechanical property changes (complete loss of ductility, doubling of E_T). Certain other properties of the foam (e.g., longitudinal shear strength, longitudinal E_c) appeared to exhibit a maximum with higher values at 111 K than at 295 or 76 K. This may not be true material behavior, but an apparent effect influenced by data scatter and orientation irregularities.

Sectioning after testing revealed that cell orientation varied somewhat from the orientation of the mold axes. It is generally thought that foam specimens from the center of a large billet have uniform cells oriented with respect to the rise direction, but this is not true in the case of foams formed on a continuous or bun-line production facility. The cell orientation

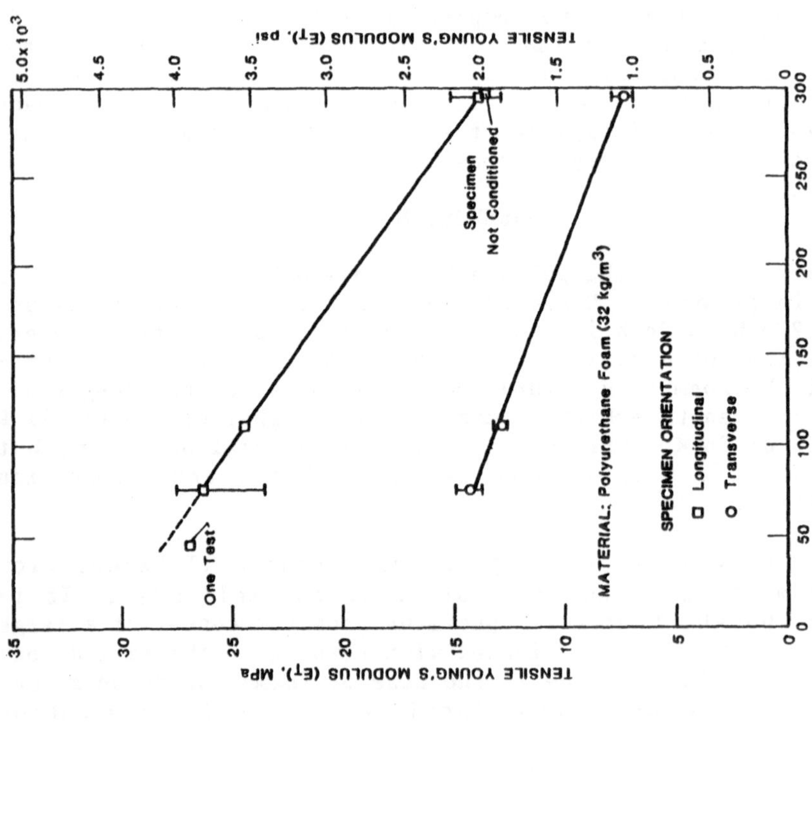

Fig. 1. Tensile Young's modulus vs. temperature.

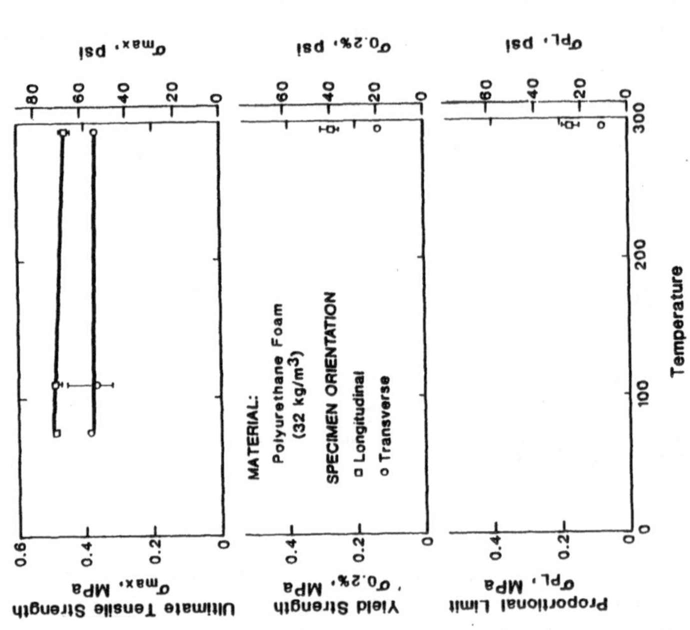

Fig. 2. Tensile test results: proportional limit, yield strength, and ultimate tensile strength vs. temperature.

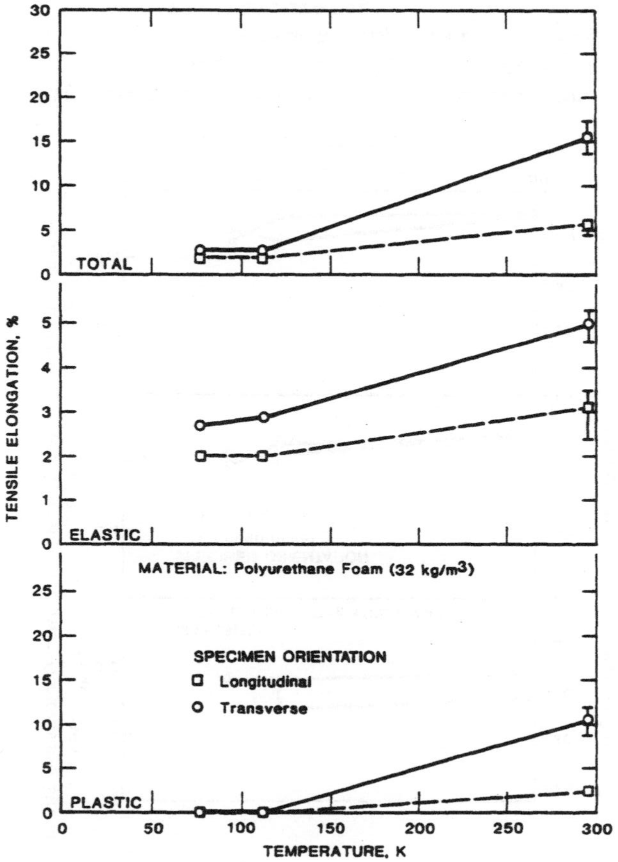

Fig. 3. Tensile strain vs. temperature.

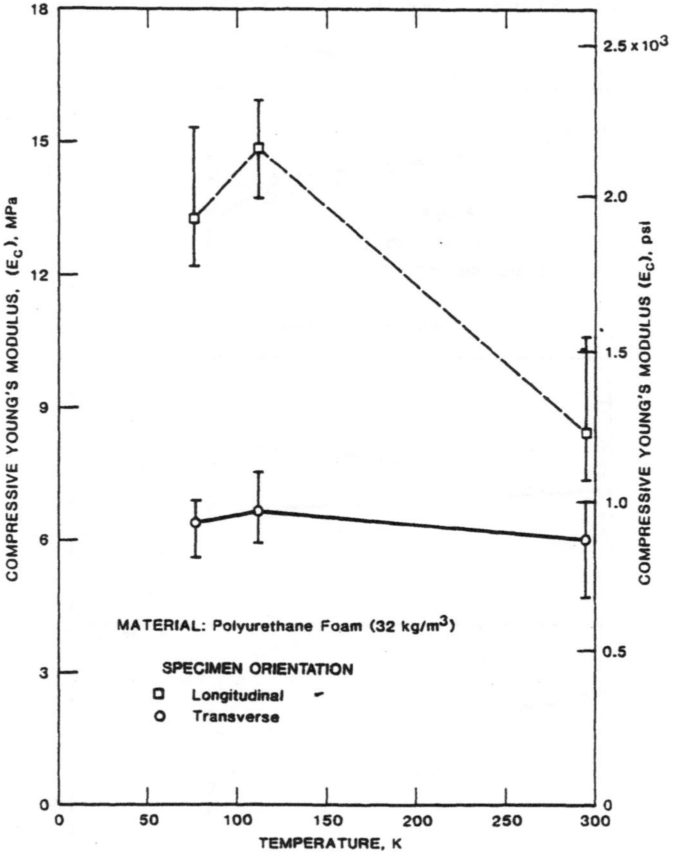

Fig. 4. Compressive Young's modulus vs. temperature.

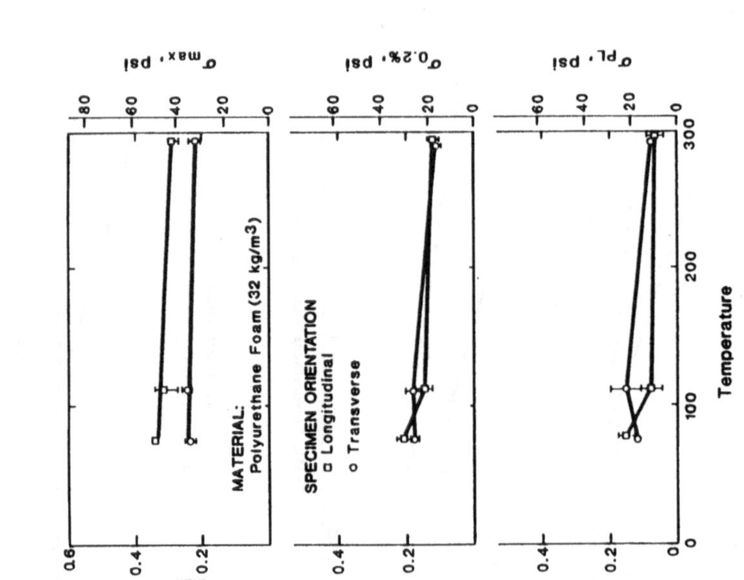

Fig. 5. Compression test results: proportional limit, yield strength, and maximum compressive strength vs. temperature.

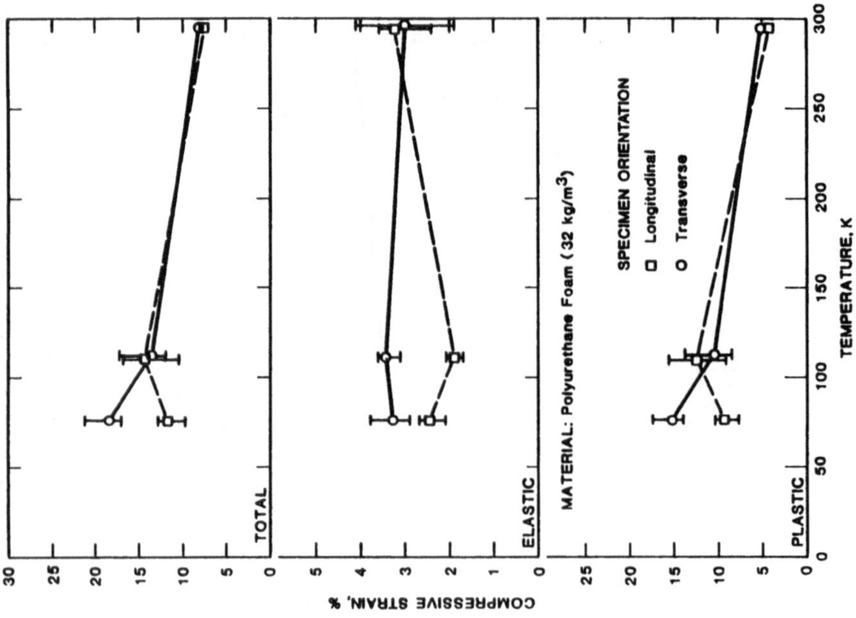

Fig. 6. Compressive strain vs. temperature.

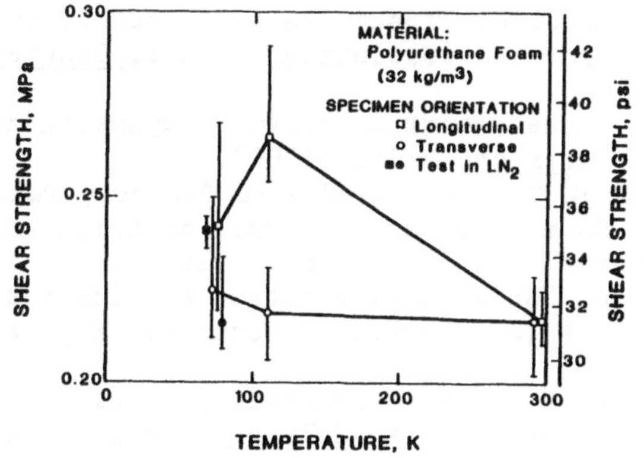

Fig. 7. Shear Strength vs. temperature.

reflects the movement of the foaming resin. Cells in the x-z plane of the material reported here have nearly a 45° inclination to the x-axis, whereas in the y-z plane there is no inclination from the z-axis. Future tests will include measurement of physical properties perpendicular and parallel to the cell axes, rather than the mold axes as reported in this study.

Specimens tested in liquid nitrogen appeared to have slightly lower shear strength values than those tested at the same temperature in cold helium gas. Apparently, this is an indication of environmental sensitivity. Nitrogen at low temperatures is deleterious to a number of other polymeric materials. Thermoplastics have exhibited crazing, or reduced fracture strengths, or both when tensile tested in cold liquid or gaseous nitrogen, as compared with vacuum or helium environments.[8] The environmental effect must be taken into account in design applications, but it is not known whether LNG represents an aggressive or benign environment.

REFERENCES

1. R. P. Reed, J. M. Arvidson, and R. L. Durcholz, Tensile properties of polyurethane and polystyrene foams from 76 to 300 K, in: "Advances in Cryogenic Engineering," Vol. 18, Plenum Press, New York (1973), pp. 184-193.
2. J. M. Roberts, R. B. Herring, and D. E. Hartman, The use of capacitance gauge sensors to make precision mechanical property measurements, in: "Materials Technology," American Society for Mechanical Engineers, New York (1968), pp. 87-96.
3. "High-Temperature Capacitive Strain Measurement System," NASA Tech. Brief B75-10069, NASA (1975).

4. P. C. F. Woldendale, Capacitive displacement transducers with high accuracy and resolution, J. Sci. Instrum. (J. Phys. E) 1:817 (1968).
5. G. R. White, Measurement of thermal expansion at low temperatures, Cryogenics, 2:151 (1961).
6. "Materials Bank Compendium of Fire Property Data," Products Research Committee, J. W. Lyons, chairman, National Bureau of Standards, Washington, D.C. (1980).
7. W. G. Jurevic, "Structural Plastics Applications Handbook Supplement 1 Test Methods," Technical Report AFML-TR-67-332 (1969).
8. A. Hiltner and E. Baer, Mechanical properties of polymers at cryogenic temperatures, Polymer 15:805 (1974).

CRYOGENIC TESTS OF GLASS-EPOXY-BASED ELECTRICAL INSULATION*

J. D. Taylor, P. S. Martin,† M. Pripstein, and M. A. Green

Lawrence Berkeley Laboratory, University of California, Berkeley, California

INTRODUCTION

A thin, superconducting solenoid for the Time Projection Chamber (TPC) experiment at PEP was constructed at Lawrence Berkeley Laboratory (LBL) in 1979[1] and tested in 1980. A failure of the ground plane insulation damaged the coil to the point that it required rebuilding. An extensive study of this failure[2] indicated that an iron chip embedded in the bore tube had penetrated the insulation. Before rebuilding the coil, the insulation system was investigated with the goal of determining the most reliable techniques and materials for withstanding high voltages in the coil package.

The experience with the TPC coil and its prototypes indicates that glass cloth vacuum-impregnated with epoxy is an excellent material for cryogenic applications from the mechanical standpoint.[3] Further, since the LBL assembly shop had extensive experience with the epoxy formulation used in the coil, there was reluctance to change that component. Therefore, the investigation concentrated on different types of glass cloth and on composites containing glass cloth.

*This work was supported by the Director, Office of Energy Research, Office of High Energy and Nuclear Physics, Division of High Energy Physics of the U.S. Department of Energy under Contract No. W-7405-ENG-48.

†Present address: Fermi National Accelerator Laboratory, Batavia, Illinois.

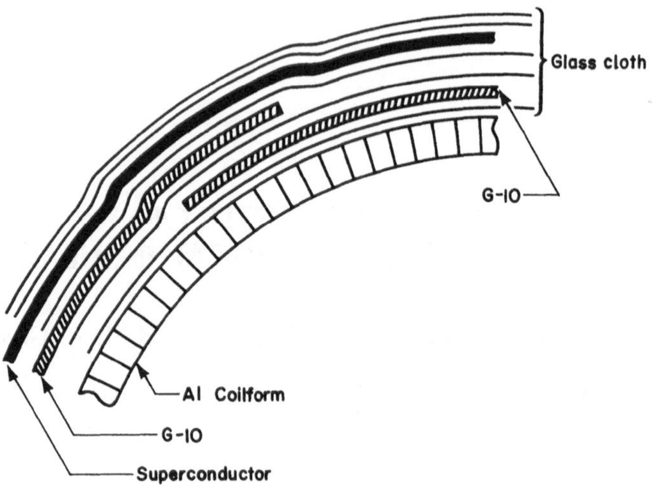

Fig. 1. Cross-sectional view of a typical sample, showing "sandwich" construction.

PROCEDURE

The samples of insulation were wrapped on an aluminum tube 8 cm in diameter and 30-cm long. Formvar-coated rectangular wire (0.9 mm x 3.6 mm) was wound spirally and closely packed over the insulation. Another layer of glass cloth was wrapped on top of the wire before epoxy impregnation. Figure 1 shows the cross section of a typical sample. The samples were vacuum cast and cured using the epoxy resin system described in Ref. 3. Figure 2 shows a sample cylinder after casting but before thermal cycling.

The wire used in making the samples was superconductor from the same lot used to wind the TPC magnet in 1979. This wire is rectangular with bare dimensions of 3.6 mm x 0.9 mm. It was insulated with Formvar, 0.05-mm thick. When the superconductor was tested at 750 V from a wire brush covering a one-centimeter length, the insulation failed at a nearly continuous rate. Almost all of the Formvar insulation failures occurred because the Formvar was not continuous on the corners. (A sample of wire was tested for breakdown through the Formvar at the center of a flat face. The 0.05-mm layer of Formvar broke down at 15 kV. The Formvar itself is good insulation, provided it is continuous.)[4]

Since the goal of the experiment was to learn how the insulation system as a whole worked, we decided to use previously insulated wire. Two samples were wound with uninsulated wire. Almost all of the insulation system failures, when dissected, were found to have occurred at the edge of the wire. Since the insulation is

Fig. 2. Typical sample after casting, before thermal cycle.

very bad on the corners of the wire, the insulation values measured should probably be considered to be due to the insulation components other than Formvar.

The samples were subjected to severe thermal shock by dipping in liquid nitrogen (LN) until boiling stopped and then in warm water. This thermal cycle was repeated 25 times. After thermal cycling, the samples were tested at voltages incremented to an upper limit of 25 kV. The sample was held for 60 s at each voltage.

TEST RESULTS

Twenty-three samples were tested: Tables 1 and 2 give a summary of these samples and their performance. The samples fell into two categories: Samples were made either with glass cloth alone between conductor and ground plane or with a sandwich of additional insulation material and glass cloth separating conductor and ground plane. The additional insulation material was intended to act as a chip barrier as well as to ensure continuous insulation in the event of imperfect epoxy filling when potting. All sandwich samples were capable of holding off 25 kV. Many of the sandwiches were holding off more than 40 kV/mm (1 kV/0.001 in) at 25 kV, with the Nomex sample holding off in excess of 80 kV/mm. A test was performed on an area of FR-4, a fire-retardant glass epoxy laminate

Table 1. Sandwiched Insulation Samples (All Withstanding 25 kV)

Sample 1	Sandwich material*	Glass cloth Thickness (inches)	Glass cloth Width (inches)	Overall nom. thickness (inches)	Voltage stress at 25 kV (kV/mm)
4	0.016 FR-4	0.010 x 1 0.007 x 2	12.0 0.75	0.040	25
6	0.016 FR-4	0.030 x 1 0.007 x 2	12.0 0.75	0.060	16
7	0.005 Kapton	0.010 x 1 0.007 x 2	12.0 0.75	0.029	34
8	0.16 FR-4	0.010 x 1 0.007 x 2	12.0 0.75	0.040	25
15	0.016 FR-4	0.010 x 1 0.007 x 2	1.5 0.75	0.040	25†
17	0.16 DMD‡	0.003 x 2	1.5	0.022	45
18	0.007 mica glass§	0.003 x 2	1.5	0.013	76
19	0.16 FR-4	0.003 x 2	1.5	0.025	39
20	0.0055 Nomex	0.003 x 2	1.5	0.0155	86
25	0.007 G-10	0.003 x 3	0.75	0.16	62

*FR-4, G-10, and Kapton were sandblasted; DMD, mica glass, and Nomex were not sandblasted.
†Not thermally cycled.
‡DMD is a Dacron-Mylar-Dacron laminate.
§Mica glass is also known at "Isomica," produced by US Samica Corp.

similar to NEMA G-10 that was 20 cm x 20 cm. The 0.016-in-thick piece of FR-4 held off 25 kV even before adding it to a sandwich of additional glass fabric and epoxy.

The glass fabric samples were wrapped using narrow tape spiral wrapped around the sample form. This was done in one of two techniques:

1. Half-lap: the trailing edge of the tape on a given turn would fall near the center of the tape of the previous turn and fall near the leading edge of the tape two turns previous while not overlapping it.

2. Butt lap: the trailing edge of the tape on a given turn would fall near the leading edge of the previous turn but not overlap it.

Two major trends developed in the tests of the glass fabric (nonsandwich) samples: (1) The finest weave samples held the highest voltage gradients, around 40 kV/mm, and (2) when samples broke down, the fault often was at the edge of the glass fabric tape. This would represent an area where there was one less layer than nominal of glass fabric, but the overall thickness of the laminate would not be changed since the void was filled with epoxy.

When the samples were thermally cycled, a very limited amount of cracking was observed. This cracking occurred in areas where unfilled epoxy had built up in layers of a thickness greater than 0.8 mm. In areas where the epoxy was well filled with glass, there was no cracking. If the glass cloth was not treated for use with epoxy (Volan treatment), a cloudy appearance developed in the laminate after only a few cool downs. The cloudy appearance is caused by failure of the bond at the boundary between glass and epoxy. This did not seem to affect performance of the insulation electrically, but it probably affects the strength of the composite. It is felt that the special glass treatment will result in a stronger composite structure over the long term.

After the thermal cycling and electrical breakdown tests, the samples were dissected. The strength of the bond between various materials was observed as well as any signs of delamination. The DMD was found to have separated at the commercially joined boundaries between the Dacron and Mylar. The mica glass was found to be easily parted at the mica surfaces; this is a property of mica rather than an adhesive failure. The bond between the treated glass and sandblasted NEMA G-10 or FR-4 is very strong, comparable to that of NEMA G-10 itself. The bond of untreated glass cloth seemed to be slightly less strong, followed by Kapton and Nomex.

Table 2. Glass Cloth Alone Insulation Samples

Sample number	Glass thickness*	Tape width (in)	Filling picks (per in)	Warp ends (per in)	Breakdown voltage† (kV)	Voltage stress at breakdown (kV/mm)
3	0.007 x 3	0.75	31	41	22.5	42
5	0.007 x 3	0.75	26	48	20	37
9	0.010 x 3	1.5	15-1/2	17-1/2	22.5	30
10	0.015 x 2	1.5	14	26-1/2	10‡	13‡
11	0.010 x 2	3.0	17	18	7.5§	15§
12	0.022 x 2	3.0	15	27-1/2	25	22
13	0.014 x 2	3.0	14-1/2	16-1/2	10	14
14	0.010 x 2	3.0	17	18	17‖	33‖
16	0.007 x 3	0.75	31	41	25#	47#
21	0.007 x 2	3.0	30	38	16.5	46
22	0.010 x 2	1.5	15-1/2	17-1/2	6	12
23	0.007 x 3	0.75	31	41	10**	19**
24	0.003 x 3	0.75	40	64	15	66

* x 2 = half lap; x 3 = half lap + buttlap.

† Formvar insultion was bad at wire edges (see text). Samples 16 + 22 used uninsulated wire.

‡ A special fabric with glass fibers woven across the tape (filling picks) and Dacron fibers running along the length of the tape (warp ends).

§ Failed at 7.5 kV after cycling; sample had held 10 kV before thermal cycling; with fault removed it went to 14 kV.

‖ Not thermally cycled.

Did not fail at 25 kV.

**Failed at 10 kV before cycling; with fault removed it held 25 kV after cycling.

CONCLUSIONS

On samples 11 and 23, the initial breakdown voltage seemed unusually low. When the conductor was peeled back in the failed region, the remainder of the sample went to much higher voltages. This would indicate that local imperfections in the materials are likely to be the cause of a failure in a large casting, even though the material may have good characteristics when tested in small batches. By using the sandwich of nonporous insulator between layers of glass cloth fabric, the chance of failures of this sort should be markedly reduced. The nonporous insulation should be checked for pinhole faults before assembly. Since this layer also represents a barrier to chip penetration, the sandwich fabrication is our first choice insulation method. We prefer glass-epoxy such as FR-4 or NEMA G-10 since we have had previous experience with them in similar potted coils. Materials such as Kapton or Nomex look promising for future development.

Glass fabric provides a path for epoxy flow to all the nonporous insulators; it should be a fine woven fabric for maximum insulation value. Additional epoxy flow channels are also provided in the TPC magnet design, owing to the large size of the coil.

ACKNOWLEDGMENTS

The authors wish to acknowledge the work of Al Barone who was instrumental in epoxy casting and who, along with Philippe Eberhard, Ron Ross, and Andrew DuBois, participated in useful discussions on design.

REFERENCES

1. M. A. Green et al., "Progress on the Superconducting Magnet for the Time Projection Chamber Experiment (TPC) at PEP," Report LBL-11007, Lawrence Berkeley Laboratory, Berkeley, California (May 1980).
2. M. A. Green et al., Ground plane insulation failure in the first TPC superconducting coil, presented at the International Conference on Magnet Technology, Karlsruhe, West Germany (30 March-3 April 1981).
3. M. A. Green, D. E. Coyle, P. B. Miller, and W. F. Wenzel, Vacuum impregnation with epoxy of large superconducting magnet structures, in: "Nonmetallic Materials and Composites at Low Temperatures," Plenum Press, New York (1979), p. 409.
4. J. Billan, Selection of the insulation of superconducting wires for d.c. magnet, "Proceedings of the International Symposium on High Voltage Insulation for Low Temperature Application," CERN-ISR-LTD/76-13, Wroclaw, Poland (13-17 September 1976).

FLUX PINNING IN HIGH-CURRENT-CARRYING SUPERCONDUCTORS*

E. J. Kramer

Department of Materials Science and Engineering and the Materials Science Center
Cornell University, Ithaca, New York

INTRODUCTION

For most magnetic applications of superconducting wire materials (e.g., superconducting power machines, tokamak coils, etc.), it is desirable for these wires to carry high current densities in high magnetic fields, H. It is well known, however, that an upper limit, the critical current density, J_c, exists, which depends strongly on the metallurgical microstructure of the wire as well as on H and temperature. The basic mechanism underlying the microstructure sensitivity of J_c is known in general terms to be the pinning of the flux line lattice, FLL, of the mixed state by defects in the crystal lattice of the superconductor.[1-5] There continues to be considerable controversy, however, about two questions: 1) What is the nature of the elementary interaction pinning force, f_p, between the FLL and a single lattice defect? and 2) How should these elementary interactions be summed to find the volume pinning force, F_p, which is the fundamental quantity that determines J_c since $F_p = J_c \times B$? In what follows, recent evidence bearing on these questions is reviewed, with special attention given to applications of these ideas to high-field high-current superconductors, such as niobium-titanium alloys and the A15 compounds.

SPECIFIC PINNING FORCES

One can identify many types of metallurgical defects that might act as pinning centers. These include point defects, such as vacancies or interstitials; larger, roughly spherical or disk-shaped defects, such as voids and precipitate particles; line

*Invited paper.

defects, such as dislocations; and planar defects, such as grain boundaries, dislocation cell walls, phase interfaces, and surfaces.

A naive and simple first cut at the problem of estimating the elementary interaction forces of various defects is to define a specific pinning force, Q, which may be derived from the experimentally measured F_p's. For point defects and other small equiaxed defects such as voids, i.e, "point" pinning centers, Q will be defined as

$$Q = F_p/\rho$$

where ρ is the defect density. For line defects, Q_L is defined as

$$Q_L = F_p/\rho_L$$

where ρ_L is the length of defect line per unit volume. For planar defects, Q_P is defined as

$$Q_P = F_p/S_v$$

where S_v is the surface area of planar defect per unit volume. Table 1 shows the Q's for various types of defects in niobium (and in one case a niobium-hafnium alloy[6]) at a reduced magnetic induction, $b = B/B_{c2}$, of 0.5. It is apparent that there is an appreciable range in Q in the various categories, depending on the type of defect. For example, voids are by far the strongest point pins and surfaces are stronger pinning centers than grain boundaries.

Table 1. Specific Pinning Forces of Defects in Niobium at b = 0.5

	Defect Type	Specific Pinning Force, Q	Remarks	Ref.
"POINTS"	Frenkel pair	5×10^{-18} N		5
	Cascades	3×10^{-16} N		5
	Voids	3×10^{-14} N	Volume = 5×10^{-25} m^3 Diameter = 10 nm	5
"LINES"	Precipitates (normal conducting)	2×10^{-16} N	Volume = 5×10^{-25} m^3	6
	Dislocation loops	2×10^{-15} N	Diameter = 10 nm	5
	Straight dislocations	2×10^{-7} N		7
	Grain boundaries	100 N/m^2	Resistivity Γ = 97 ratio	8
"PLANES"	Surface	3000 N/m^2		9

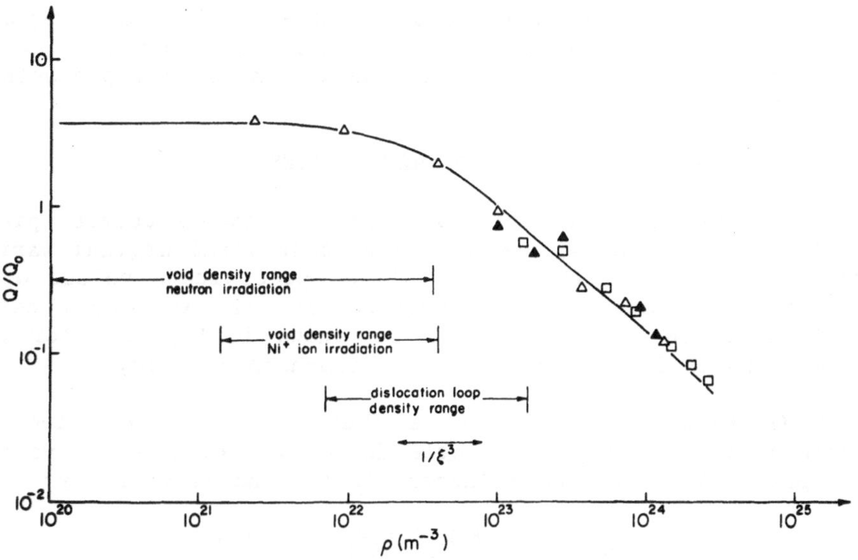

Fig. 1. The ratio of the specific pinning force, Q ($= F_p/\rho$), to the Q ($= Q_0$) at $\rho = 10^{23}$ m^{-3} plotted vs defect density. F_p is determined at a constant reduced induction b = 0.55.

Now let us examine some of the implicit assumptions underlying the concept of a specific pinning force more closely. Is Q really independent of point pin density, ρ, or planar specific area, S_v? I think the answer in both cases is yes, as long as ρ or S_v is not too large. Figure 1 shows data on Q vs ρ for cascades and Frenkel pairs in niobium.[5] For $\rho \ll \xi^{-3}$ where ξ is the coherence length. The specific pinning force is independent of ρ, a conclusion that is confirmed by other experiments.[1,5,10] Similar conclusions on Q_P for grain boundary pinning may be drawn from most experiments on F_P as a function of grain size in the A15 compounds[11,12] despite some evidence to the contrary.[13]

Above $\rho = \xi^{-3}$, Q falls off with ρ approximately as 1/ρ or even somewhat faster.[5] The reason is not hard to see. The superconductor cannot "resolve" individual defects much closer together than a coherence length. Although the fluctuation in the number of pins in neighboring elements of volume ξ^3 goes as $\sqrt{\rho\xi^3}$, the individual elementary interaction of these pins must decrease at least as $\rho^{-1/3}$, the linear distance between defects,[5] leading to a Q that decreases with ρ. Graphic illustrations of the high pin density represented by the limit $\rho = \xi^{-3}$ are two electron micrographs (Fig. 2) of arrays of bubbles in niobium produced by neon-ion bombardment of niobium foils at high temperatures (courtesy of Dr. Volker Graeger, Institut für Metallphysik, Göttingen). The micrograph on the left corresponds to $\rho \simeq 4 \times 10^{22}\ m^{-3}$, approximately $1/\xi^3$ for niobium; that on the right corresponds to $\rho \simeq 2.5 \times 10^{23}\ m^{-3}$, well above $1/\xi^3$. The middle schematic represents the scale of the FLL.

VOIDS AND PRECIPITATES

What metallurgical variables affect the specific pinning forces? For point pins, the most important metallurgical variable is defect size, i.e., the small void or precipitate volume or the dislocation loop area. Increasing defect size for small defects markedly increases Q. An example is shown in Figure 3 for voids and precipitates in niobium and a niobium-hafnium alloy.

It is instructive to compare these Q's with the elementary interaction forces, f_p, for single defects. For a small void, f_p is due primarily to the core interaction[2,4] and is given by[5]

$$f_p = 0.44\ \xi^2 k_0^{\ 3}\ \mu_0 H_c^{\ 2}(1 - b)V \tag{1}$$

where $k_0 = 2\pi/a_0$, H_c is the thermodynamic critical field, and V is the void volume. Figure 4 shows the comparison between Q and f_p

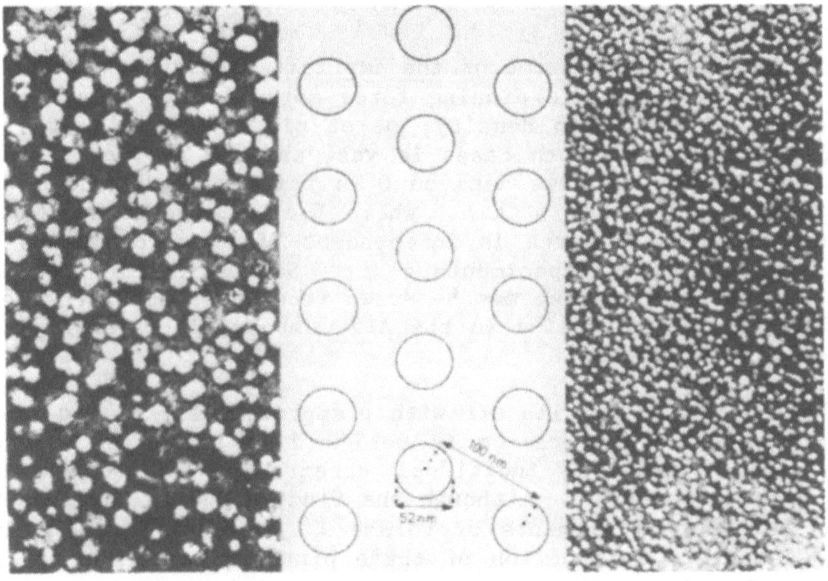

Fig. 2. Void (neon bubble) arrays in neon-ion-bombarded niobium. The left micrograph corresponds to a void density $\rho = 4 \times 10^{22}\ m^{-3}$, and the right corresponds to $\rho = 2.5 \times 10^{23}\ m^{-3}$. The center shows the scale of the FLL. (From Ref. 14).

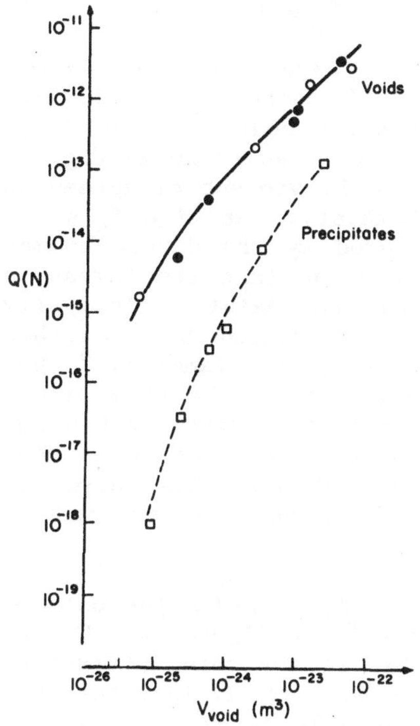

Fig. 3. The specific pinning force, Q, vs the volume of voids in niobium or precipitates in a niobium-hafnium alloy[6] (at b = 0.6).

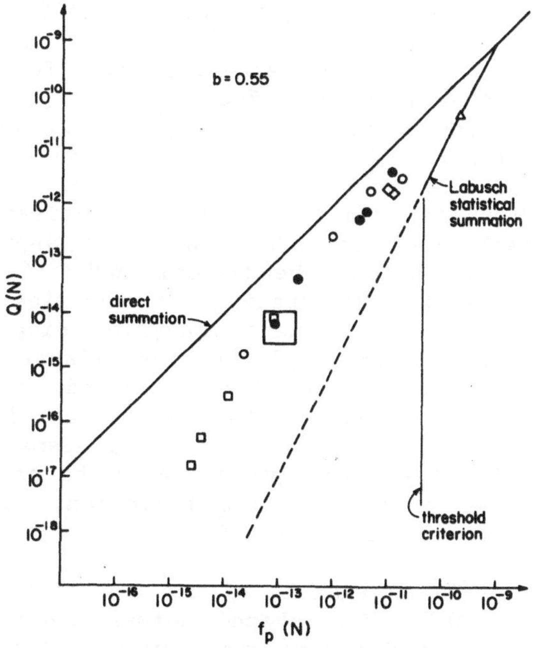

Fig. 4. The specific pinning force Q, vs the elementary interaction force, f_p, for point pins in niobium at b = 0.55 and T = 4.2 K. The circles represent data on voids whereas the squares represent data on dislocation loops (after Ref. 7).

for voids and dislocations loops. (The f_p for the loops comes from the interaction of strain field of the FLL with the stress field of the loop and is proportional to loop area.)[15]

The specific pinning forces is expected to be less than f_p because of the summation problem. The rigidity of the FLL should prevent most of the pins, if spaced randomly, from exerting their maximum f_p's simultaneously. Two extreme summation laws are the direct summation[2,16]

$$Q = f_p \tag{2}$$

and the so-called statistical theories,[17] which give in the latest version[18]

$$Q = f_p(f_p - f_t)/(f_p + f_t); \quad f_p > f_t \tag{3}$$

where f_t is a threshold pinning force that depends on the elastic constants of the FLL [$f_t \simeq a_0^2(C_{44}C_{66})^{1/2}$ where a_0 is the FLL lattice parameter and C_{44} and C_{66} are elastic moduli of the FLL]. Below the threshold Q is supposed to be zero. As shown in Fig. 4, the data fall on a single curve, which is in violent disagreement with the predictions of the statistical theory. At high f_p's, in fact, the data are much better approximated by the direct summation. Recent theoretical work[19-21] has shown that the threshold force of the statistical theories does not exist, principally because mean square flux-line displacements caused by all other pins at a given pin increase linearly with sample diameter.[19] Unfortunately, the theory derived to replace Eq. 3 is in scarcely better agreement with the data and in addition predicts that Q increases linearly with ρ. It seems prudent at the moment to consider the curve in Fig. 4 to be an empirical summation curve for point pins. Similar summation curves have been constructed at other reduced fields.[22]

As an example of the use of this curve, consider the data on the thin, disk-shaped normal precipitates[6] in Figure 3. It had been thought that these precipitates pin via the core interaction given by Eq. 1. From the summation curve however, it is clear that the f_p of a precipitate must be 0.1 to 0.01 of that of a void of the same volume, with the discrepancy increasing for smaller (thinner) precipitates. The ratio, in fact, is approximately given by $(t/\xi_0)^2$ where t is the precipitate thickness and ξ_0 is the BCS coherence length. These results are in good agreement with a theory of the proximity effect correction to f_p for thin precipitates.[23] It is worth noting that although the proximity effect makes small precipitates ineffective, flux pinning centers in the niobium alloys, normal precipitates in the A15 compounds, in which ξ_0 is thought to be $\simeq$ 5 nm, can be much smaller and still be effective pinning centers.

GRAIN BOUNDARIES AND CELL WALLS

Although some pinning centers in high-field, high-current superconductors have been identified as precipiates, pinning by planar defects, such as dislocation cell walls (niobium-titanium) and grain boundaries (A15 compounds), seems to be a more general feature. Until recently, pinning by such defects was poorly understood principally because the mechanism for f_p was uncertain. For example, there are at least three possible interaction mechanisms between a grain boundary and the FLL. These are:

1. The interactions between grain boundary strain fields and the stress fields of the FLL.[24]

2. The interaction between the grain boundary and the FLL due to the anisotropy of the upper critical field H_{c2} (crystalline anisotropy interaction).

3. The interaction between the grain boundary and the FLL due to the electron scattering from the boundary (electron scattering interaction).

The stress field interaction has recently been shown to vanish for high-angle grain boundaries in which any dislocation structure is very closely spaced.[8,25] The magnitude of the crystal anisotropy contribution has been recently estimated to be[26,27]

$$(\hat{f}_p)_{CA} = \frac{(1 - b)}{3} \mu_0 H_c^2 \frac{\delta H_c}{H_{c2}} \qquad (4)$$

where $\delta H_{c2}/H_{c2}$ is the relative change in H_{c2} across the boundary. (The elementary interaction force per unit area for a planar defect is denoted by $\hat{f}_p$). Since the magnitude of the H_{c2} anisotropy is decreased as the impurity content of the materials is increased,[28] one expects $(\hat{f}_p)_{CA}$ to decrease with the impurity parameter $\alpha \equiv 0.88\ \xi_0/\ell$, where ℓ is the electron mean free path.

The most important theoretical advance, however has been a paper by Zerweck[29] in which he developed a model for the electron scattering from the grain boundary. A similar, but improved, model was developed later by Yetter.[30] The assumptions of the latter treatment are as follows:

1. Electrons are scattered with a probability, P, by the boundary. Previous experimental and theoretical results indicate that strong electron scattering occurs by dislocation cores that make up the boundary, and thus for random high angle boundaries (closely spaced cores in the boundary), P is close to 1.[31]

2. Using this fact, the electron mean free path is computed at various points from the grain boundary using the simple geometrical averaging model illustrated in Figure 5.

3. Assuming a local relation between the Ginzburg-Landau parameter, κ, and ℓ, the increase in κ is computed at various positions from the boundary. Typical κ profiles are also shown in Figure 5. For high purity materials, the increase in κ is small but extends far from the boundary, whereas for lower purity materials, the κ charge is larger and more localized. At high magnetic induction, b, the resulting $(\hat{f}_p)_{ES}$ is estimated[30] to be

$$(\hat{f}_p)_{ES} = \frac{\sqrt{2\pi}}{3} (1 - b)\ \mu_0 H_c^2\ g_1 \frac{\Delta\tilde{\kappa}}{\kappa} (g_1) \qquad (5)$$

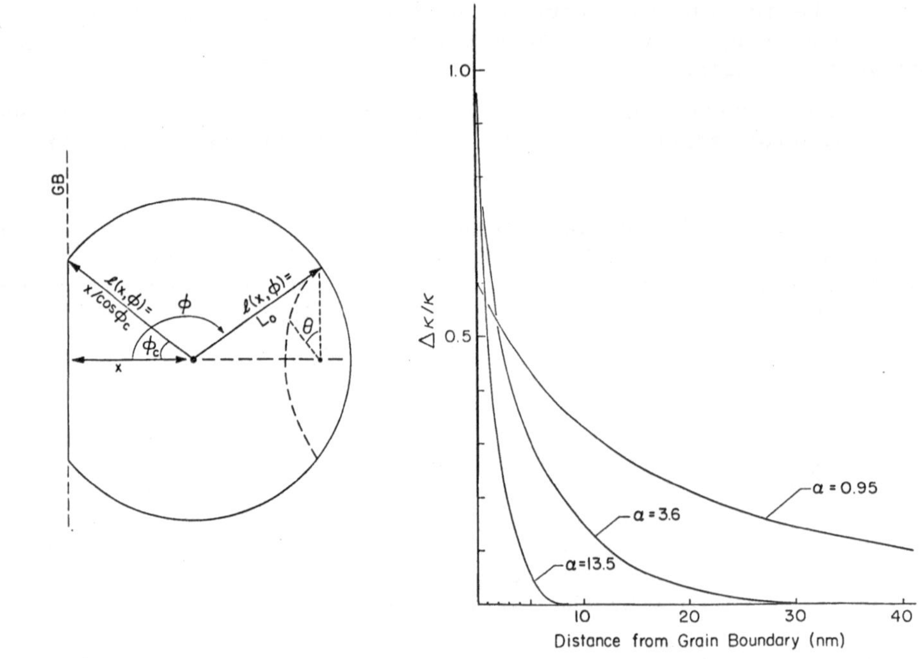

(a) (b)

Fig. 5. (a) The geometry of the calculation of the local mean free path at a given position from the grain boundary; (b) the relative change in Ginzburg-Landau parameter, κ, as a function of distance from the grain boundary for several values of α, the impurity parameter (after Ref. 30).

where g_1 is the shortest reciprocal lattice vector of the FLL and $\widetilde{\Delta\kappa}/\kappa\ (g_1)$ is the one-dimensional Fourier transform of the relative change in κ as a function of distance from the grain boundary. Since the change in κ occurs over an electron mean free path length from the boundary and since the magnitude of the κ change itself is dependent on ℓ, $\widetilde{\Delta\kappa}/\kappa$ and thus $(\hat{f}_p)_{ES}$ are strong functions of α, increasing rapidly at low α. The Zerweck model predicts a maximum at $\alpha \approx 1$, and a subsequent slow decrease for higher α's, whereas a revised version of the Yetter model puts the maximum at a much higher $\alpha \simeq 10$.

Figure 5 also provides a useful physical insight into why electron scattering is not an important pinning mechanism for point or line pins. The decrease in ℓ occurs primarily from the truncation by the plane of a spherical region with radius equal to ℓ.

Since point or line pins do not cast appreciable "shadows," they do not perturb either ℓ or κ locally, and thus do not give rise to an appreciable electron-scattering interaction.

The structure of a dislocation cell wall produced by severe plastic deformation of metals and alloys usually consists of a very high density of positive and negative edge dislocations. Its electron-scattering properties, therefore, should be not much different from a high-angle grain boundary (i.e., $P \simeq 1$). The idea from the Zerweck model that an isolated dislocation ("line defect") should produce a negligible local change in κ, whereas a dense planar array of such dislocations should produce a substantial local change in κ, may well account for the early observation that only small increases in flux pinning were produced by plastic deformation until plastic strains at which the cell wall structure begins to form were exceeded.[32]

Another interesting prediction of this model is that, in contradiction to assumptions in the literature, equilibrium (monolayer) segregation of solute atoms to the boundary should produce little if any change in $(\hat{f}_p)_{ES}$. If the probability of scattering, P, is already close to 1, adding solute atoms to the plane of the boundary can do little to enhance it. Yetter,[31] in experiments where he allowed lead or thallium to diffuse down the boundaries in the columnar grain structure of a lead-bismuth film, has confirmed this prediction in that appreciable changes in grain boundary pinning only occur when solutes have time to diffuse distances $\simeq\xi$ from the boundary.

The experimental tests of these ideas have been aided significantly by the development of techniques to measure flux pinning from single grain boundaries in bicrystals, thus eliminating the summation problem. The elegant experiments of Das Gupta et al.[8] on niobium bicrystals used the fact that $(\hat{f}_p)$ for any planar defect is strongly decreased when the FLL makes a significant angle to the plane of the defect. By measuring the anisotropy of F_p with respect to the magnetic field direction, they were able to isolate the Q_{GB} from single, symmetric, high-angle tilt boundaries. Such boundaries should have negligible $(\hat{f}_p)_{CA}$ or $(\hat{f}_p)_{SF}$ contributions. The results gave Q_{GB}'s on the order of 100 N/m^2, rather strong pinning, which must be attributed to the electron-scattering mechanism. Lunnon et al.[26] developed an alternative method of preparing bicrystals by cutting them from very large grained niobium sheet. Grain boundaries with particular character could be selected, which is not possible for bicrystals grown from the melt.[8] Grain boundary pinning peaks were observed in more than ten such bicrystals. A check of the α dependence of $(\hat{f}_p)_{ES}$ was possible since data on a symmetric tilt bicrystal with a resistivity ratio, Γ, of 14 could be compared with Das Gupta et al.'s results on a $\Gamma = 97$ bicrystal.

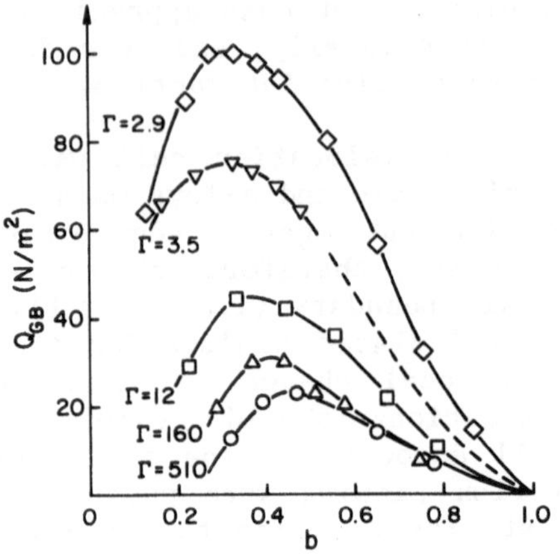

Fig. 6. The specific pinning force per unit area of grain boundary roughly parallel to H vs reduced magnetic induction in niobium foils of various resistivity ratio, Γ. The grain size in each foil is about 50 μm.

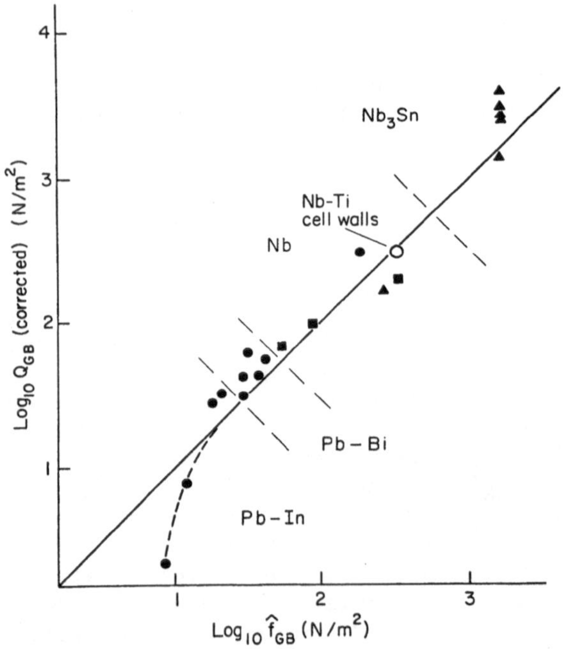

Fig. 7. The specific pinning force per unit area of grain boundary plotted against the elementary interaction force computed for the electron scattering mechanism (after Ref. 30). At b = 0.3.

The $(\hat{f}_p)_{ES}$ values inferred for the lower Γ bicrystal are about 5 times higher than the higher Γ one, in rough agreement with the predictions of the theories.

This experimental finding is qualitatively confirmed by recent measurements of Q_{GB} in 5-μm-thick niobium foils of different purities.[33] These foils were recrystallized to the same final columnar grain structure, with column axis perpendicular to the foil (grain size for each is d ~50 μm) and purified by outgassing at high temperatures. The grain boundary pinning is isolated by monitoring the height of the peak in J_c observed when H is normal to the foil. Figure 6 shows the result that Q_{GB} at low reduced inductions increases strongly with impurity parameter. There is no trace of the peak in $(\hat{f}_p)_{ES}$ at $\alpha \approx 1$ (which corresponds to $\Gamma \approx 12$) predicted by the initial theories[29,30] but the data are in rough agreement with the revised Yetter model. The magnitude of Q_{GB} predicted for various materials is quite close to that measured. Figure 7, taken from Yetter's thesis[30] shows experimental measurements of Q_{GB} plotted against the computed $(\hat{f}_p)_{ES}$. Since for equiaxed grains (triangles) and for columnar grains or cells (circles) only ~1/3 and ~1/2 respectively of the grain boundaries can be even close to parallel with H, the original Q data for these classes were corrected by multiplying by factors of 3 and 2, respectively. The agreement is probably too good, considering the theoretical uncertainties.

Given the strong trend in Q_{GB} with purity shown in Figure 6, one might conclude that $(\hat{f}_p)_{CA}$ is negligible compared with $(\hat{f}_p)_{ES}$ and a theoretical comparison of Eqs. 4 and 5 confirms that this should be so above α's of 0.1. Experimentally, however, a comparison of a bicrystal with close to the maximum CA contribution with one with no CA contribution at $\alpha \simeq 1$ reveals that the electron scattering and crystal anisotropy contributions are roughly equal.[26] Although the CA contribution for an average grain boundary in a random texture polycrystal is only approximately 37% of the maximum contribution, one would expect it to show up in the thin foil results at high Γ. It may well be that the recrystallization texture of the foils also acts to reduce the CA contribution.

The stress field interaction appears to be negligible even for very low-angle boundaries in agreement with theoretical predictions.[34] [Since the dislocation cores in such boundaries would be widely spaced the $(\hat{f}_p)_{ES}$ is also negligible.] No Q_{GB} could be detected from several bicrystals with misorientations less than 10° and Γ's between 2 and 14. No Q_{GB} could even be found in niobium single crystals ($\Gamma \simeq 2$) that had been deformed by bending and polygonized by annealing to produce an array of parallel low-angle boundaries where each crystal contained at least 25 parallel low-angle dislocation walls.

The importance of the defect structure within a high-angle boundary has not yet been quantitatively assessed, but it would seem likely that special boundaries such as high-coincidence-site boundaries (e.g., twin boundaries) would scatter electrons less strongly and have a smaller $(\hat{f}_p)_{ES}$ since the "good fit" at such boundaries would reduce the density of defects in the boundary. In support of such a role, a comparison of a niobium bicrystal boundary, which had nearly a symmetric twin orientation, with a symmetric random high-angle boundary (neither boundary has a CA contribution) reveals that the inferred $(\hat{f}_p)_{ES}$ from the twin is about half that of the random boundary.[34]

Given these facts, it seems likely that the electron-scattering mechanism is the most important grain boundary and cell-wall-pinning mechanism in alloy superconductors (e.g., niobium-titanium). Since A15 compounds are often in the clean limit, the H_{c2} anisotropy may play a more important role there.[34] From the evidence in niobium, it would appear desirable to attempt to produce "dirty" (A15) compounds (without reducing T_c and H_c). Not only would this increase $(\hat{f}_p)_{ES}$ for the grain boundaries, but it would also increase F_p at high fields by increasing H_{c2} (decreasing b).

THE SUMMATION PROBLEM AT HIGH REDUCED INDUCTION

Fifteen years ago Fietz and Webb[35] discovered a remarkable property of F_p. If F_p was plotted vs b, the reduced magnetic induction, as a function of temperature, T, the curves could be scaled to superimpose by multiplying by a factor of $[H_{c2}(T)]^m$, which implies that

$$F_p(b,T) = g(b)\, H_{c2}{}^m(T) \qquad (6)$$

where g(b) in this so-called scaling law is a function only of b. Measurements on many other systems have confirmed that most superconductors obey the scaling law at least approximately. One might expect that if nature were kind, g(b) would be a universal function for all defects or at least a universal function for all defects with the same elementary interaction mechanism. That this expectation is not the case can be seen for grain boundaries in Figure 6 and for voids in Figure 8 (from the Ph.D thesis of Graeger).[14] In these respective cases, the boundary area per unit volume and the void density are constant; the only microstructural variables that change are background mean free path in the case of grain boundaries and diameter in the case of voids. The shift in the shape of these curves cannot be blamed on a shift in the f_p vs b or $(\hat{f}_p)_{ES}$ vs b curves. The f_p for voids simply scales with the void volume (Eq. 1) while the slight shift in $[(\hat{f}_p(b)]_{ES}$ for grain boundaries is in the wrong direction (the maximum in $[(\hat{f}_p(b)]_{ES}$ shifts to

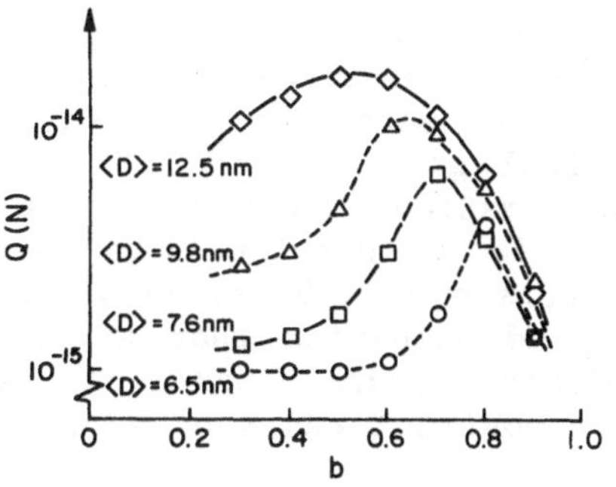

Fig. 8. The specific pinning force versus reduced magnetic induction for voids with different average diameters, <D>. The void density in each case is approximately $6 \times 10^{22}\ m^{-3}$.

slightly higher b with increasing α). The change in shape of the Q vs b curves must be due to a change in the summation condition as the values of f_p of the voids or $(\hat{f}_p)_{ES}$ of the grain boundaries are increased. Certain qualitative features become clear. At high reduced magnetic inductions well above the peak in Q vs b, Q is rather insensitive to f_p or $(\hat{f}_p)$. At the peak and well below it Q increases strongly with f_p. For example, at b = 0.8 the range in Q_{GB} is about a factor of 4, whereas it is at least a factor of 10 at b = 0.3. The change in sensitivity for voids is even more pronounced, ranging from only a factor of 2 at b = 0.8 to a factor of 10 at b = 0.3. Similar shifts in g(b) may be observed in all the high-field, high-current superconductors[3,36] where the changes that occur in microstructure are not as well established. Microstructures with large pinning usually have a peak in the $F_p(b)$ curve at low b; those with weak pinning have a smaller, sharper peak in the $F_p(b)$ curve at higher b. These latter microstructures exhibit a peak in the J_c vs B curve just below B_{c2}, the so-called peak effect.[37] These systematic changes in g(b) with microstructure seem to indicate a shift in mechanism of summation above and below the peak in g(b). One idea[2] is that as the FLL softens [C_{66} decreases as $(1 - b)^2$ as $b \to 1$] full "synchronization" of pinning occurs above the peak and Q increases to the direct summation line, $Q = f_p$. This mechanism may be ruled out for the voids in Figure 8, since at high b the ratio Q/f_p, which might be termed the summation efficiency (f_p is computed from Eq. 1), is largest for the smallest voids (smallest f_p's) and <u>decreases</u> in all cases as $b \to 1$!

In the second model, it is assumed that the FLL can shear plastically around and between the strongest pins.[36] This model predicts the saturation observed in g(b) and given reasonable agreement with the high-field dependence of F_p on b in many superconductors.[3] It also accounts for the disappearance of FLL history effects at b's above the peak[6,38,39] as well as an abrupt change in the character of the voltage current characteristic[40] at b's just above the peak. At the same time, it has received substantial criticism on theoretical grounds.[41,42] Since the detailed mechanism(s) of plastic deformation of the FLL is (are) not easily accessible to experiment, it is difficult to assess the validity of these objections (but, see Ref. 5, p. 11). At the moment however, it is the only proposed mechanism that gives the insensitivity of F_p at high b's to microstructure. Certainly experiments with other well-characterized microstructures are needed that probe the summation problem at high b.

CONCLUSIONS

1. The important microstructural variables for point pins, such as voids and precipitates, are pin volume, V, and density, ρ, with F_p at low b's increasing linearly with ρ and somewhat more strongly with V.

2. For planar pins, such as grain boundaries and cell walls, electron scattering is thought to be the most important pinning mechanism. At low b, the $(\hat{f}_p)_{ES}$ and Q_{GB} increase strongly with α, the impurity parameter. Grain boundary character and structure also have effects on grain boundary flux pinning. This strong dependence of the electron scattering mechanism on sample purity has important implications for flux pinning by grain boundaries in the A15 compounds, since they are now known to be relatively clean superconductors. Selected additions of impurity atoms (without depressing T_c) are predicted to be beneficial in two ways, by increasing $(\hat{f}_p)_{ES}$ directly and by increasing H_{c2}, thus increasing the flux pinning at high fields.

3. The summation problem remains basically unsolved. At low b, an empirical summation curve may be used for point pins. At high b, the pinning becomes insensitive to the metallurgical microstructure. The FLL plastic shear model can account for most of these high b observations, but theoretical questions as to its validity remain.

ACKNOWLEDGMENTS

The financial support of this work by the Air Force Office of Scientific Research is greatly appreciated. The experimental effort at Cornell has also benefited substantially through the use of the facilities of the Materials Science Center which is funded by NSF. I also acknowledge the important contributions of my coworkers and students at Cornell, Dr. W. Yetter, Dr. M. Lunnon, and Mr. D. A. Thomas.

REFERENCES

1. J. D. Livingston and H. Schadler, Prog. Mater. Sci. 12:183 (1964).
2. A. M. Campbell and J. E. Evetts, Adv. Phys. 21:199 (1972).
3. E. J. Kramer, J. Electron. Mater. 4:839 (1975).
4. H. Ullmaier, "Springer Tracts in Modern Physics," 76 (1975).
5. E. J. Kramer, J. Nucl. Mater. 72:5 (1978).
6. C. C. Koch, A. Das Gupta, D. M. Kroeger, and J. O. Scarborough, Philos. Mag. B40:361 (1979).
7. E. J. Kramer, J. Appl. Phys. 49:742 (1978).
8. A. Das Gupta, C. C. Koch, D. M. Kroeger, and Y. T. Chou, Philos. Mag. B38:369 (1978).
9. A. Das Gupta and E. J. Kramer, Philos. Mag. 26:779 (1972).
10. G. Antesberger and H. A. Ullmaier, Philos. Mag. 29:1101 (1974).
11. J. D. Livingston, Phys. Status Solidi 44:295 (1977).
12. R. M. Scanlan, W. A. Fietz, and E. F. Koch, J. Appl. Phys. 46:2244 (1975).
13. A. W. West and R. D. Rawlings, J. Mater. Sci. 12:1862 (1977).
14. V. Graeger, Ph.D. dissertation, Universität Göttingen, FRG (1980).
15. E. J. Kramer, Philos. Mag. 33:331 (1976).
16. D. Dew-Hughes, Philos. Mag. 30:293 (1974).
17. R. Labusch, Cryst. Lattice Defects 1:1 (1969).
18. A. M. Campbell, Philos. Mag. B37:149 (1978).
19. A. I. Larkin and Yu. N. Ovchinnikov, J. Low Temp. Phys. 34:409 (1979).
20. E. H. Brandt, Phys. Lett. 77A:484 (1980).
21. T. Matsushita and K. Yamafuji, J. Phys. Soc. Jpn. 47:1427 (1980).
22. E. J. Kramer, J. Appl. Phys. 49:742 (1978).
23. E. J. Kramer and H. C. Freyhardt, J. Appl. Phys. 51:4930 (1980).
24. C. S. Pande and M. Suenaga, Appl. Phys. Lett. 29:443 (1976).
25. M. E. Lunnon and E. J. Kramer, unpublished.
26. M. E. Lunnon, D. A. Thomas, and E. J. Kramer, Philos. Mag., in press.
27. E. J. Kramer and G. S. Knapp, J. Appl. Phys. 46:4595 (1975).
28. E. Seidl and H. W. Weber, in: "Anisotropy Effects in Superconductors," H. W. Weber, ed., Plenum Press, New York (1977), p. 57.
29. G. Zerweck, J. Low Temp. Phys. 42:1 (1981).
30. W. Yetter, Ph.D. thesis, Cornell University, Ithaca, New York (1980).
31. R. A. Brown, J. Phys. F.-Met. Phys. 7:1477 (1977).
32. A. V. Narlikhar and D. Dew-Hughes, Phys. Status Solidi 6:383 (1964).
33. D. A. Thomas, Ph.D. thesis, Cornell University, Ithaca, New York (1982).

34. K. Togano and K. Tachikawa, J. Appl. Phys. 50:3495 (1979).
35. W. A. Fietz and W. W. Webb, Phys. Rev. 178:657 (1969).
36. E. J. Kramer, J. Appl. Phys. 44:1360 (1973).
37. E. J. Kramer, in: "Proceedings of the International Discussion Meeting on Flux Pinning on Superconductors," P. Haasen and H. C. Freyhardt, eds., Akademie der Wissenshaften, Göttingen, FRG (1975), p. 240.
38. M. Steingart, A. Putz, and E. J. Kramer, J. Appl. Phys. 44:5580 (1973).
39. R. W. Rollins, H. Kupfer, and W. Gey, J. Appl. Phys. 45:5392 (1974).
40. B. D. Lauterwasser and E. J. Kramer, Phys. Lett. 53A:410 (1975).
41. A. M. Campbell, in: "Proceedings of the International Discussion Meeting on Flux Pinning on Superconductors," P. Haasen and H. C. Freyhardt, eds., Akademie der Wissenshaften, Göttingen, FRG (1975), p. 47.
42. T. Matsushita, T. Tanaka, and K. Yamafuji, J. Phys. Soc. Jpn. 46:756 (1979).

FLUX PINNING FOR *IN SITU* PREPARED SUPERCONDUCTING COMPOSITES

J. J. Sue, D. K. Finnemore, J. E. Ostenson, E. D. Gibson, and J. D. Verhoeven

Ames Laboratory—USDoE, Iowa State University, Ames, Iowa

INTRODUCTION

In situ superconducting composites[1-3] differ from bronze process[4,5] Nb_3Sn-Cu multifilamentary conductors in two major features. For in situ conductors, the filaments are discontinuous and have much smaller dimensions. This means that there must be a substantial amount of filament-filament coupling through the Cu and that the Cu-Nb_3Sn interface area is comparable to the grain boundary area. The filaments for in situ wire normally have a lacy ribbon shape, 1 grain thick and about 8 to 12 grains wide. This is in contrast to the rather thick (~1 μm) Nb_3Sn layers found in bronze process material.

In spite of the rather large differences in filament size and morphology, the flux pinning forces for in situ wires are very similar to bronze process material. Both materials show an increase in critical current, J_c, as the grain size decreases and indeed the results are quantitatively similar.[6] In addition, the dependence of J_c on both magnetic field, H, and strain, ε, is rather similar for the two types of materials. The purpose of the work reported here is to use the magnetic field dependence of J_c in conjunction with a detailed study of grain morphology via transmission electron micrographs to determine the major mechanisms that control J_c and contribute to flux pinning in these materials. Because the filaments are only one grain wide and have a great deal of Cu-Nb_3Sn surface area, it is important to know if these Cu-Nb_3Sn boundaries are effective pinning sites. Because the filaments are discontinuous, it is important to know whether the filament-to-filament crossover links are the factor controlling J_c.

RESULTS AND DISCUSSION

From the magnetic field dependence of J_c, the pinning force, $F_p = J_c \times B$, is found to be a universal function of the reduced field, $h = H/H_{c2}$, of the form

$$F_p = Ah^{1/2}(1-h)^2 \tag{1}$$

where A is a constant. The fit is rather good over the entire range from h = 0.1 to h = 1.0, as shown in Fig. 1. The maximum in F_p vs. h always occurs at h = 0.2. This is just the magnetic field dependence expected for surface pinning with a core interaction.[7] For this type of pinning the constant A is related to the effective pinning area per unit volume, S_v, by the relation

$$A = \frac{\mu_o S_v H_{c2}^2}{4\kappa_1^2} \tag{2}$$

where μ_o is the permeability of free space and κ_1 is the Ginsburg-Landau parameter. For these samples κ_1 typically is 35, and S_v is found to be about $4 \times 10^3 cm^{-1}$.

There are several reasons to believe that surface pinning[7] rather than shear of the flux line lattice[8] controls J_c for these

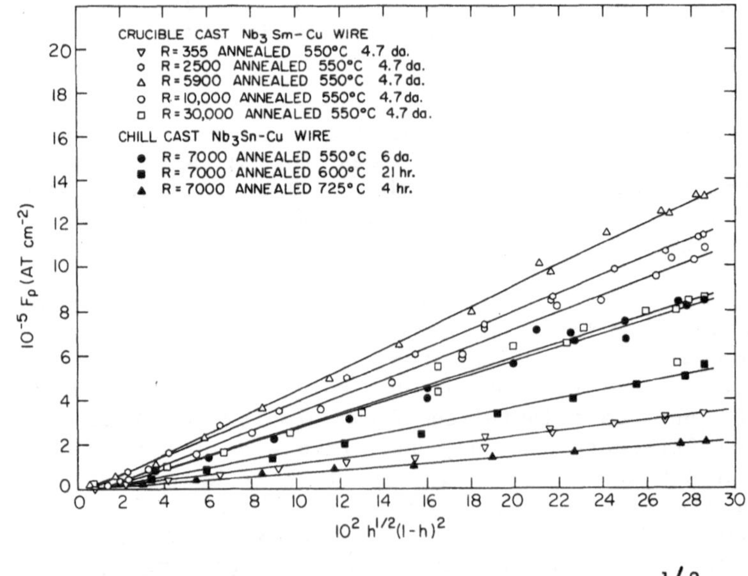

Fig. 1. Plot to show the linearity of F_p vs $h^{1/2}(1-h)^2$ for a wide variety of conditions.

samples. If J_c were controlled by shear of the flux line lattice rather than surface pinning, then[8,9]

$$A = 0.56\ H_{c2}^{5/2}\ \kappa_1^{-2}[1-a_o(\rho^{1/2})]^{-2} \tag{3}$$

where a_o is the flux line lattice spacing and ρ is the pinning site density. Because a_o depends on H, A also would depend on H, which is contrary to experiment. This is graphically shown in Fig. 1 where F_p is plotted against $h^{1/2}(1-h)^2$. To show the relative magnitude of the $[1-a_o(\rho)^{1/2}]^2$ term, we have chosen ρ to be the grain diameter of 800 Å and have plotted F_p vs $[1-a_o(\rho^{1/2})]^{-2}$ and F_p vs $h^{1/2}(1-h)^2$ in Fig. 2. The data clearly are not linear for any reasonable value of ρ. A second reason to believe that surface pinning and not shear forces control J_c is that the value of κ_1 derived by fitting the data to Eq. (3) is too small by a factor of 3 or 4. That is, $\kappa_1 = H_{c2}/\sqrt{2}\ H_c = 35$ whereas κ_1 derived from Eq. (3) is about 8. This was previously observed for bronze process wire by Suenaga and Welch.[9] This simply means that the flux line lattice pinning at the grain boundaries breaks down long before the magnetic field is high enough for shear to occur. The fact that the Nb_3Sn J_c data are linear on a so-called Kramer plot[8] of $J_c^{1/2}H^{1/4}$ vs H does not mean that shear forces control J_c at high field. Both core pinning on surfaces and shear have the $h^{1/2}(1-h)^2$ behavior at high field. From the behavior at low field and from the magnitude of J_c, one can conclude that core pinning on surfaces is the dominant mechanism.

There are at least two different kinds of surface that might contribute to the effective pinning area determined from the J_c data, the Nb_3Sn-Cu interfaces and the Nb_3Sn-Nb_3Sn grain boundaries.

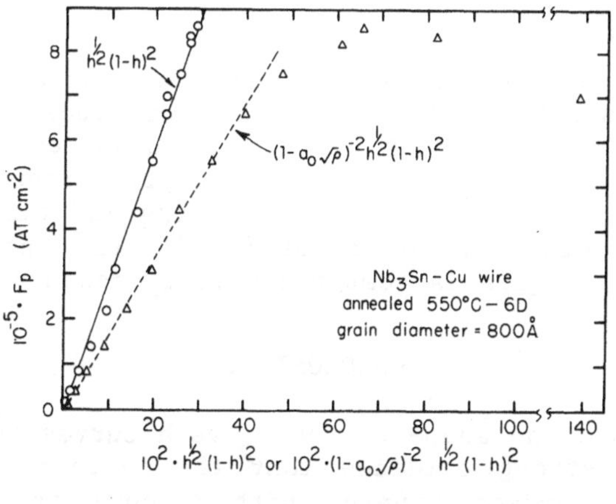

Fig. 2. A comparison of the surface pinning function, F_p vs $h^{1/2}(1-h)^2$, and shear pinning function, F_p vs $h^{1/2}(1-h)^2[1-a_o(\rho)^{1/2}]^{-2}$.

For _in situ_ materials these two areas are comparable because the filaments generally are only one grain thick. This is in contrast with bronze process materials where the Nb_3Sn is 1 to 3 μm thick. To determine the magnitude of these areas, a series of transmission electron microscope photographs was taken, and the Nb_3Sn-Nb_3Sn grain boundary area, S_{GB}, was determined from the average of several hundred grains.

Before conversion to Nb_3Sn, the Nb filaments have a flat ribbon shape having a width-to-thickness aspect ratio in the range of 10 to 40. After conversion to Nb_3Sn, some of this character is retained in that the filaments are consistently one grain thick and eight to ten grains wide. The grains are distinctly anisotropic and tend to be about 1.7 times longer along the wire axis than in the other two directions. A typical grain is 1200 Å along the filament, 900 Å in the width direction, and 700 Å in the thickness direction. There is a substantial amount of coarsening, especially along the filament, and there is a tendency for the filaments to break up into long chains of grains. This network of grains has a lacy appearance.

A typical sample has about 5×10^{14} grains per cm^3 and a total Cu-Nb_3Sn surface area of 1.5×10^5 cm^{-1}. Of course, only a small fraction of this boundary is oriented properly to pin the vortex lattice. If one envisions an array of vortices perpendicular to the filament intersecting a long chain of grains parallel to the filament, it is only the intersection of the 40 Å diameter core of the vortex with the edge of a 700 to 900 Å diameter grain that is effective. This means typically that a strip 120 Å along B by 1200 Å along the filament constitutes the pinning area of each Cu-Nb_3Sn boundary. This then gives about 7×10^3 cm^{-1} available for core pinning on these boundaries. This is comparable to S_v measured from J_c.

In addition to the Cu-Nb_3Sn boundaries, there are many Nb_3Sn-Nb_3Sn grain boundaries that have projected areas parallel to the pinning force. About half of the grain boundaries are perpendicular to filament axis and of this area the projection parallel to F_p is about half. This gives 5×10^{14} grains/cm^3 times 900 Å times 700 Å times 1/2 times 1/2 or about 8×10^3 cm^{-1} pinning area per unit volume. This again is comparable to S_v measured from J_c.

CONCLUSIONS

The magnitude and shape of the F_p vs h curves for _in situ_ composites provide strong evidence that the dominant mechanism controlling J_c is surface pinning with a core interaction. The effective pinning area per unit volume determined from J_c data is found to be $S_v = 4 \times 10^3/cm^{-1}$. The data are not consistent with models in which J_c is controlled by shear in the flux line lattice. From an analysis of transmission electron microscope photographs of extracted filaments, one finds an area of about 7×10^3 cm^{-1} available from Cu-Nb_3Sn boundaries and a comparable figure for Nb_3Sn-Nb_3Sn grain boundaries. Since these areas are not fully effective, this disagreement of only a factor of 2 in the pinning area is rather good agreement with theory. Measurements of tape samples in which the filaments are oriented will be needed to determine the relative contribution of these two boundaries to the pinning.

ACKNOWLEDGMENTS

John Clem has made important contributions to this work. This work was done at the Ames Laboratory, Iowa State University, Ames, Iowa, operated for the U.S. Department of Energy by I.S.U. under contract No. W-7405-Eng-82. The research was supported by the Director of Energy Research, Office of Basic Energy Sciences, WPAS-KC-02-01. Part of the work was performed while three of the authors (J.J.S., D.K.F., and J.E.O.) were Guest Scientists at the Francis Bitter National Magnet Laboratory, which is supported at M.I.T. by the National Science Foundation.

REFERENCES

1. J. E. Ostenson, D. K. Finnemore, J. D. Verhoeven, and E. D. Gibson., Appl. Phys. Lett. 37:662 (1980).
2. J. Bevk and J. P. Harbison, J. Mater. Sci. 14:1457 (1979).
3. R. Roberge, S. Foner, E. J. McNiff, Jr., B. B. Schwartz, and J. L. Fihey, Appl. Phys. Lett. 34:111 (1979).
4. M. Suenaga and W. B. Sampson, Appl. Phys. Lett. 18:584 (1971).
5. M. Suenaga and W. B. Sampson, Appl. Phys. Lett. 20:443 (1972).
6. D. K. Finnemore, J. D. Verhoeven, E. D. Gibson, and J. E. Ostenson, Preparation and properties of in situ prepared filamentary Nb_3Sn superconducting wire, in: "Filamentary A15 Superconductors," M. Suenaga and A. F. Clark, eds., Plenum Press, New York (1980), p. 259.
7. D. Dew-Hughes, Philos. Mag. 30:293 (1974).
8. E. J. Kramer, J. Appl. Phys. 44:1360 (1973).
9. M. Suenaga and D. O. Welch, Flux pinning in bronze-processed Nb_3Sn wires, in: "Filamentary A15 Superconductors," M. Suenaga and A. F. Clark, eds., Plenum Press, New York (1980), p. 131.
10. A. DasGupta, C. C. Koch, and D. M. Kroeger, Philos. Mag. B 38(4):367 (1978).

NEUTRON IRRADIATION OF NbTi WITH DIFFERENT FLUX PINNING STRUCTURES

H. W. Weber, F. Nardai, and C. Schwinghammer

Atominstitut der Österreichischen Universitäten, Vienna, Austria

R. K. Maix

Department for Magnets and Superconductors, BBC Brown Boveri & Cie., Zurich, Switzerland

INTRODUCTION

A variety of experiments on neutron irradiation effects in the alloy superconductor NbTi has been reported in the literature so far (for a review of this subject see, for example, reference 1). The data refer to materials of different composition, different thermomechanical treatments, different irradiation temperatures, and different neutron energies. Despite this heterogeneity of parameters, the general response of NbTi to neutron irradiation has become quite clear: whereas the primary superconductive parameters, like the transition temperature, T_c, remain almost unaffected, the critical current densities, J_c, and, hence, the bulk pinning forces, P_v, vary significantly because of changes of the metallurgical microstructure. This latter effect is, however, strongly dependent on the metallurgical starting conditions and the temperature maintained during irradiation. Accordingly, the experimental results published so far and referring to different, but in general not very well characterized, materials and to different irradiation temperatures show considerable variations with respect to the J_c dependence on neutron fluence, but have in common that the overall effects up to neutron fluences of 10^{22} - 10^{23} m^{-2} ($E > 1$ MeV) do not exceed about 20%.

To investigate the influences of metallurgical starting conditions and irradiation temperature in a more systematic way, neutron irradiation experiments on a variety of differently prepared NbTi superconductors were made first at the ambient pool water and second at liquid nitrogen temperatures. The results on flux pinning in these materials prior to irradiation and their changes upon irradiation at different temperatures are presented in the following sections.

SAMPLE CHARACTERIZATION

In the course of a previous extensive study on the metallurgical optimization of NbTi,[2] a large number of stabilized single core conductors varying in their Ti content (42, 49, and 54 wt% Ti), heat treatment, and final cold working conditions were prepared under conditions that are applicable to commercial conductor production. A detailed survey of the material parameters[3] and the experimental equipment used to determine the critical current densities[4] has been published recently; some data, which refer to the samples discussed in the present paper, are summarized in Table 1.

To demonstrate our main conclusions on the flux pinning mechanisms in NbTi, a summary of data obtained prior to neutron irradiation is shown in Fig. 1, where the maxima of the bulk pinning curves, P_{vmax}, and the magnetic fields, B_{max}, (where these maxima occur) are plotted versus the metallurgical variation parameters for all three Ti concentrations.

It will be noted that the maxima of the pinning forces increase drastically with increasing annealing temperature in Nb-42 wt% Ti, whereas the corresponding effect is small in the other materials. On the contrary, final cold work leads to an increase of the pinning forces up to a factor of 2 in the Ti-rich

Table 1. Summary of Material Parameters

Sample Number	Wt% Ti	Annealing Temp., °C	Final Cold Work, %	Cu:NbTi	Remarks
1	42	320	0	1:1	lowest J_c conductor
19	42	400	71	1:1	optimum conductor of the 42 wt% series
32	49	400	91	1:1	optimum conductor of the 49 wt% series
33	54	400	91	1:1	optimum conductor of the 54 wt% series
34	49	multiple		2.4:1	multifilamentary conductor S-48-3
35	49	multiple		2.4:1	as above, but with a short final annealing treatment

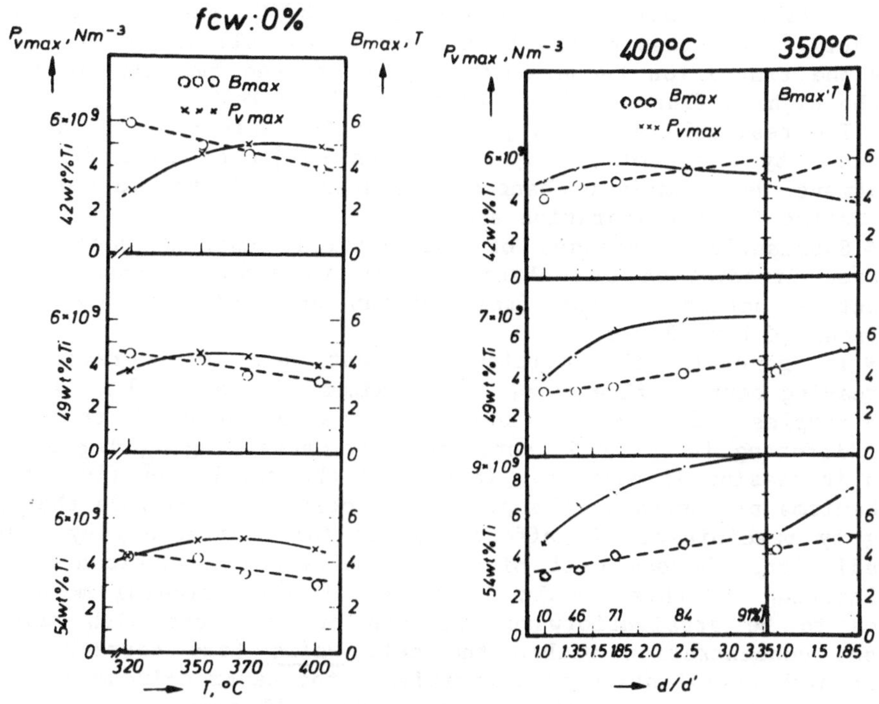

Fig. 1. Maxima of the bulk pinning curves, P_{vmax}, and corresponding field values, B_{max}, versus the metallurgical variation parameters (annealing temperature, final cold working); d/d' = ratio of diameters before and after cold working.

materials, whereas the response of Nb-42 wt% Ti is small and depends even on the initial annealing treatment.

Considering previous electron-microscopic investigations of the cell-size dependence on annealing temperature and final cold work as well as on the formation and growth of normal conducting α precipitates, we arrive at the following conclusions:[3] Flux pinning in Nb-42 wt% Ti (and presumably in all materials with Ti concentrations of less than 45 wt%) is caused mainly by dislocation cells, which act through the Δκ effect, and by dislocations. On the contrary, flux pinning in the Ti-rich alloys is caused predominantly by precipitate pinning with minor contributions of the other mechanisms.

SUMMARY OF RESULTS OBTAINED FOR NEUTRON IRRADIATION AT AMBIENT TEMPERATURE

A detailed discussion of data pertaining to irradiation of all the samples of different composition and thermomechanical treatment at the ambient pool water temperature (~80°C) has been published recently.[3] The results may be summarized as follows:

• The transition temperature, T_c, decreases by only ~0.15 K up to the highest fluence of 9 x 10^{22} m^{-2} (E > 1 MeV).

• The residual resistivity ratios, which were measured at 10 K, decrease by a factor of ~2 in samples subjected to a final heat treatment, but remain unchanged in samples subjected to cold working as the final preparation step.

• Substantial increases of the critical current density upon neutron irradiation are observed only in samples whose original defect structure is far from optimum and only at low neutron fluences (< 1 x 10^{22} m^{-2}).

• In general, the critical current density decreases with increasing neutron fluence up to a maximum of ~25% at 9 x 10^{22} m^{-2}.

• Samples subjected to identical thermomechanical treatments, but differing in their Ti content, show decreasing J_c degradations with increasing Ti content. In view of the results on flux pinning mechanisms presented in the previous section, we conclude that pinning by precipitates is affected least (or unaffected) by neutron irradiation, whereas cell pinning is affected significantly. An explanation of this result in terms of flux pinning mechanisms seems to be straightforward: because of the very high original defect concentration within the cell walls (and correspondingly their high normal state resistivities), the neutron-induced defects will increase the resistivity within the cell cores in a much more significant way. As a consequence, the difference of Ginzburg-Landau parameters will decrease and weaken the flux pinning effect. On the other hand, a redissolution of precipitates is very unlikely at the temperatures and fluences involved.

• From a comparison of the J_c dependence on neutron fluence for the "high-J_c" superconductors, we note that the irradiation-induced degradation of J_c is, in general, smaller than the improvements of J_c, which can be achieved by an optimized treatment prior to irradiation. Hence, from a practical point of view, optimized conductors are preferred.

RESULTS OF LOW TEMPERATURE IRRADIATION

Low temperature neutron irradiation leads to a quite different radiation-induced defect structure because of the lack of thermally activated defect disintegration. To take this aspect into account, the program was extended to irradiation at 77 K, although on a considerably reduced number of conductors (see Table 1). Two sets of these samples are irradiated under identical conditions: one is transferred into the measuring cryostat without warming up; the other one is warmed up to room temperature for 30 h after each irradiation step and prior to the J_c measurement.

A summary of data on the changes of critical current density with neutron fluence is shown in Fig. 2. The solid lines refer to

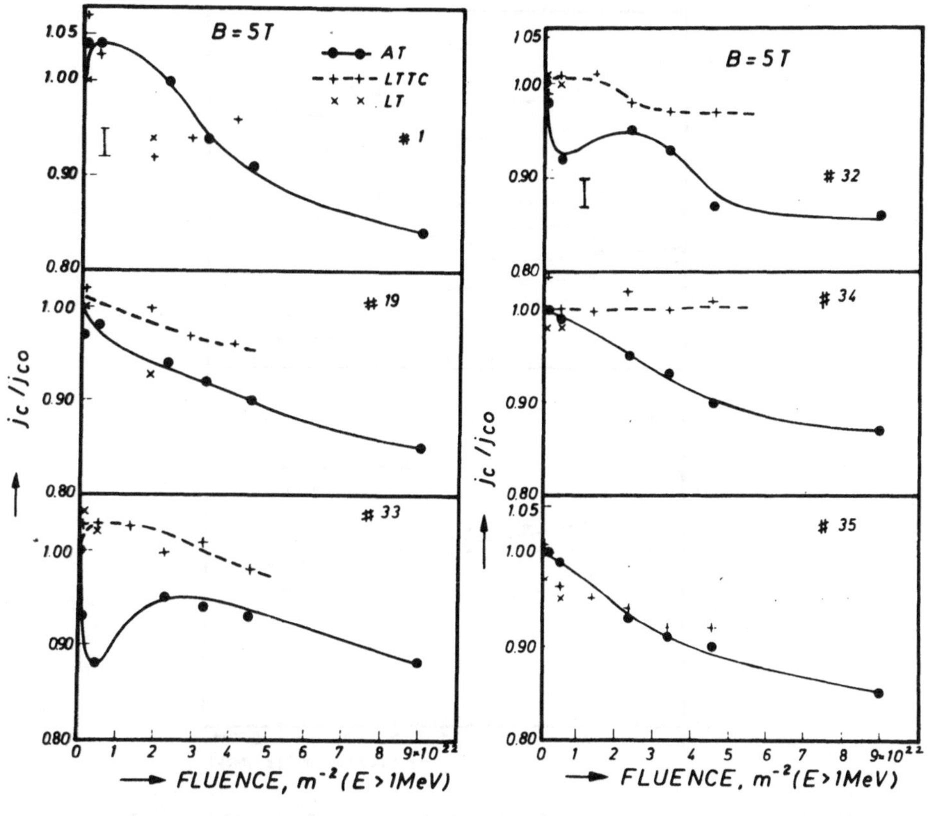

Fig. 2. Changes of critical current density with fast neutron fluence. AT, LT - irradiation at ambient temperature and 77 K, respectively; LTTC - irradiation at 77 K and thermal cycle to room temperature.

the results discussed in the previous section (AT is irradiation at ambient temperature), the broken lines refer to the data including a thermal cycle (LTTC), and the crosses show the few results available so far, which were obtained without warming up (LT). It will be noted immediately that the inclusion of a thermal cycle leads to a negligible deterioration of J_c over the entire range of fluences for the conductors 19 and 32 - 34. The response of the other materials is close to the behavior observed upon ambient temperature irradiation. It is interesting to note that both of these materials were annealed as the final preparation step, whereas in the other cases the last treatment consisted of cold working.

This pronounced effect of a thermal cycle provides again evidence for radiation-induced changes of the cell pinning mechanism: because of the room temperature anneal, the radiation-induced defects <u>within</u> the cell cores move towards the cell walls and

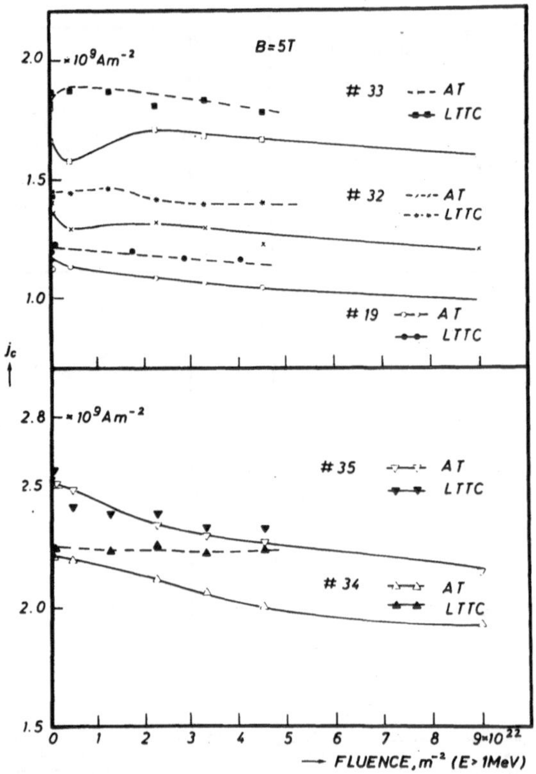

Fig. 3. Comparison of absolute J_c values obtained upon AT and LTTC irradiation.

thereby reestablish or even slightly enhance the original pinning forces.

The lack of this recovery effect may have serious consequences if the dependence of critical current densities on fluence is considered on an absolute scale (Fig. 3). A comparison of J_c values in the multifilamentary conductors 34 and 35 shows that the original gain of J_c obtained by the final annealing step is matched at a fluence of 4.5×10^{22} m^{-2} by the superior recovery effects present in sample 34.

CONCLUSIONS

The experiments of neutron irradiation on NbTi conductors of differing metallurgical starting conditions, which were made at different temperatures, establish the following aspects:

- Irradiation at ambient temperatures leads to the smallest degradation of J_c in samples with the highest Ti content. This observation is related to the pinning mechanisms and indicates significant changes of cell pinning, whereas precipitate pinning remains unaffected.

• Low temperature irradiation and a subsequent thermal cycle to room temperature cause, in most of the samples, negligible changes of J_c ($\leq 4\%$), i.e., an almost complete recovery. This aspect is, of course, of primary interest for fusion reactor applications. The lack of recovery in some of the samples has to be analyzed in more detail and may become clearer when the low temperature data without thermal cycles are available.

REFERENCES

1. S. T. Sekula, Effect of irradiation on the critical currents of alloy and compound superconductors, in: "Radiation Effects on Superconductivity," B. S. Brown, H. C. Freyhardt, and T. H. Blewitt, eds., North Holland Publishing Company, Amsterdam (1978).
2. R. K. Maix, Flußverankerung und kritische Stromdichten in Niob-Titan-Supraleitern, Thesis, Technical University Wien, Vienna, Austria, unpublished (1974).
3. F. Nardai, H. W. Weber, and R. K. Maix, Neutron irradiation of a broad spectrum of NbTi superconductors, Cryogenics 21:223 (1981).
4. F. Nardai and H. W. Weber, Computerized measuring system for critical currents in superconductors, Cryogenics 21:219 (1981).

ALPHA-TITANIUM PRECIPITATION IN NIOBIUM-TITANIUM ALLOYS

A. W. West and D. C. Larbalestier

Applied Superconductivity and Materials Science Centers
University of Wisconsin, Madison, Wisconsin

INTRODUCTION

Electron microscopy studies have been performed on both annealed and cold-worked Nb-Ti alloy samples as well as on a multifilamentary Nb-Ti superconducting composite. The morphology of the α-Ti particles that precipitate in these samples is found to be dependent on the existing Nb-Ti microstructure. Conventional transmission electron microscopy (CTEM) techniques are adequate for identifying α-Ti precipitates in annealed or lightly cold-worked Nb-Ti, but scanning transmission electron microscopy (STEM) and energy dispersive x-ray analysis (EDAX) are required for the conclusive identification of α-Ti in heavily cold-worked commercial multifilamentary composites. A preferred orientation relationship between the α-Ti and the matrix is found in annealed material but not in the cold-worked alloy. Cold-drawing alloys containing precipitates results in the elongation of the precipitates along the drawing axis.

The J_c values obtainable in Nb-Ti multifilamentary superconductors are known to be extremely dependent on the microstructures of the filaments. The two most important microstructural features are believed to be subbands and α-Ti precipitates, but the relative importance of these two factors is still very unclear.[1] Several investigations[2,3,4] have shown the importance of controlling the subband size to values between 35 and 45 nm to achieve high J_c values. However, the optimum size, density, and distribution of α-Ti precipitates in the filaments has been less easy to determine owing to the extreme difficulties in identifying the α-Ti precipitates in the heavily cold-worked filament microstructures. In this work, different electron microscopy techniques have been used to determine the distribution and morphology of α-Ti precipitates in various Nb-Ti samples.

CTEM INVESTIGATION OF Nb-Ti ALLOY SAMPLES

Under equilibrium conditions, Nb-Ti alloys in the composition range used commercially, i.e., 46.5 to 55 w/o Ti, should be two-phase at temperatures below about 500°C. The two phases are the β-phase, which is b.c.c., and the α-Ti phase, which has a c.p.h. crystal structure.

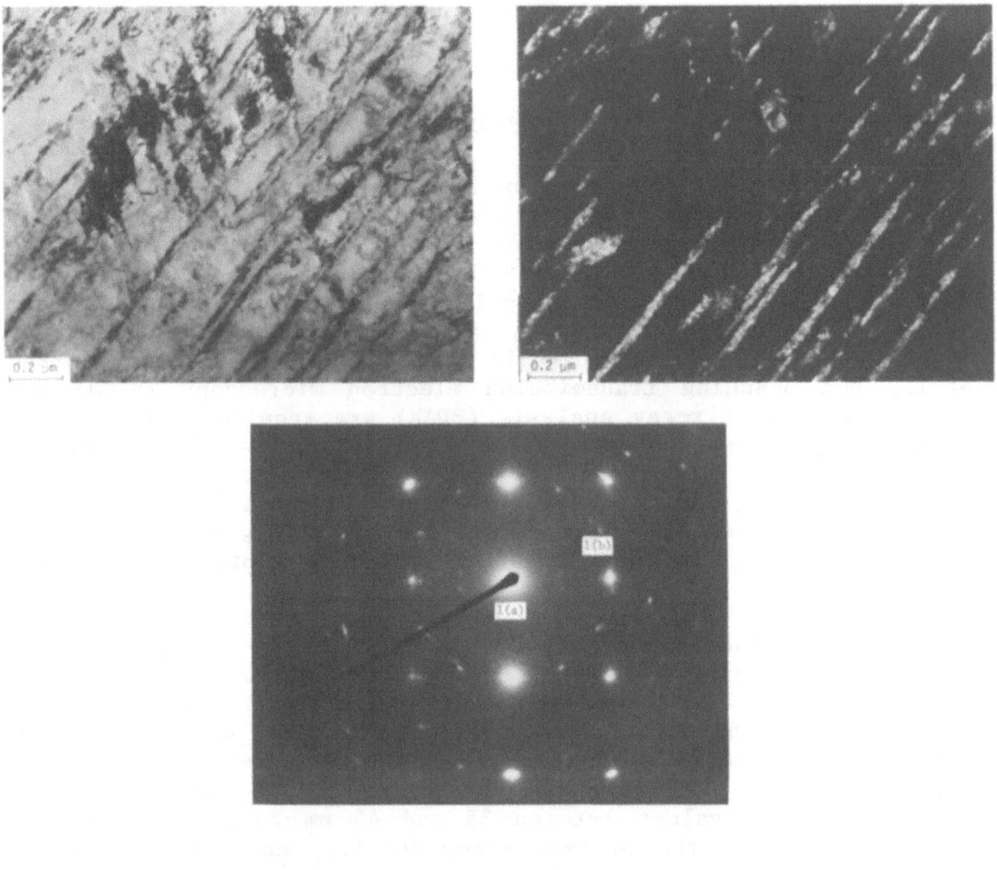

Fig. 1. CTEM micrographs of a Nb-55 w/o Ti annealed rod sample, heat-treated 100 h at 450°C. (a) Bright field. (b) Dark field. (c) Selected area diffraction pattern.

Experimental Procedure

An Nb-55 w/o Ti alloy was used for this part of the study. A length of annealed rod 3.2 mm in diameter was rolled to a 0.3-mm thick strip. Samples of this strip and the original rod were heat-treated in evacuated silica capsules under a partial pressure of argon for 100 h at 450°C. Three-mm discs were cut or punched from the samples and thin foils for TEM were prepared using a Fischione electropolishing unit. A 2% HF, 5% H_2SO_4, 90% methanol solution cooled to -40°C was used for electropolishing with an operating voltage of 35 to 45 V. The foils were examined in a JEOL 100 B electron microscope operating at 120 kV.

Results and Discussion

Figures 1(a) and (b) are micrographs taken of a foil from the rod sample. Figure 1(a) is a bright-field image, whereas Fig. 1(b) is a dark-field image obtained using diffracted illumination from the α-Ti precipitates themselves. The morphology of the α-Ti particles is seen far more clearly in Fig. 1(b) and is plate-like. The orientation of the precipitates relative to the matrix obeys the Burgers relationship, $(0001)_\alpha || \{110\}_\beta$; $\langle 1120\rangle_\alpha || \langle 111\rangle_\beta$.[5] Figure 1(c) shows a selected area diffraction pattern taken from the same foil; the spots used to form the images in Figs. 1(a) and (b) are marked.

Figure 2(a) shows a bright-field image taken from the strip sample. In this case, the α-Ti precipitates, (marked on the micrograph), have a very different morphology from those that grew in the rod sample. These precipitates have an equiaxed morphology and have no apparent orientation relationship with the matrix. These α-Ti precipitates are also less easy to identify in the cold-worked microstructure. Dark-field imaging is more difficult in these structures owing to the close positioning of the matrix and precipitate reflections in the selected area diffraction pattern, Fig. 2(b).

STEM AND EDAX INVESTIGATIONS OF MULTIFILAMENTARY Nb-Ti COMPOSITES

In CTEM, a stationary electron beam approximately 10 µm in diameter is focussed on the foil and the electrons transmitted through the sample are used to form the image on a fluorescent screen. Image contrast is determined primarily within the specimen itself and cannot be greatly altered by subsequent processing techniques. In STEM, a very narrow diameter, 20-nm beam is scanned across a raster on the foil; the transmitted electrons enter a detection system below the sample and are processed into a signal that is displayed on a CRT. Far greater possibilities exist for

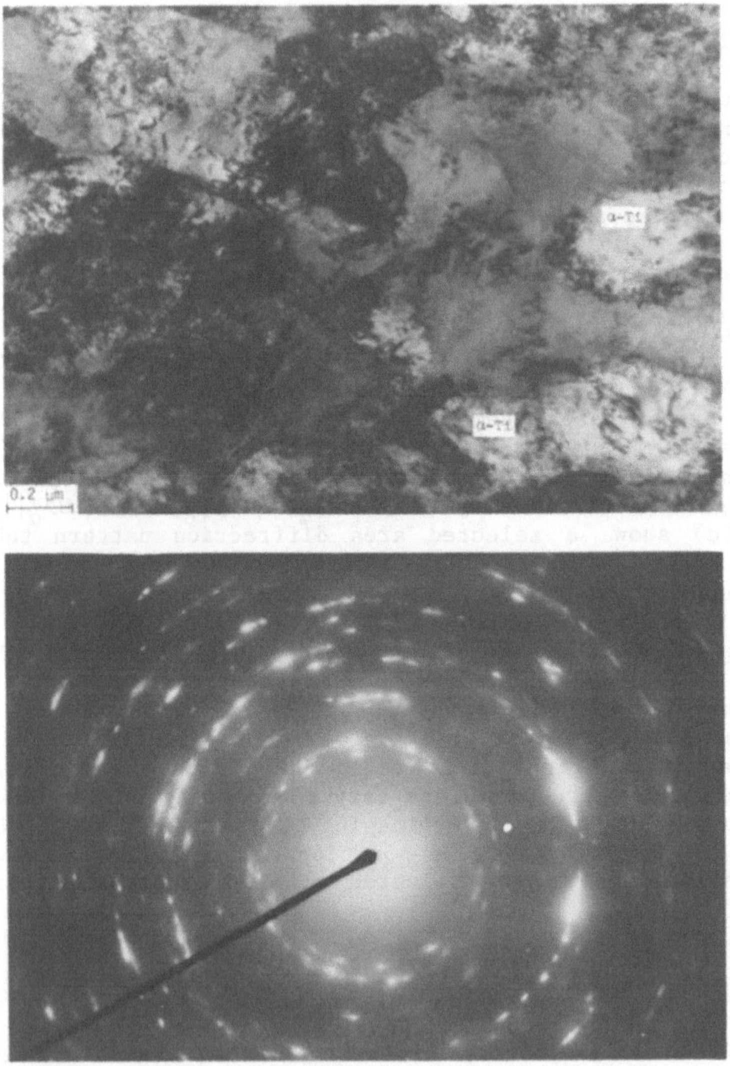

(a)

(b)

Fig. 2. CTEM micrographs of a Nb-55 w/o Ti strip sample, heat-treated 100 h at 450°C. (a) Bright-field micrograph. (b) Selected area diffraction pattern.

image processing and enhancement. It is also possible using this system to position the electron beam on a particular spot in the foil and to analyze the x-rays generated in that limited area using EDAX.

As shown above, it becomes more difficult to distinguish α-Ti precipitates in cold-worked microstructures, and this is especially true for the subband structures present in the Nb-Ti filaments of commercially produced composite wires. Several multifilamentary wire samples were examined using a STEM system together with EDAX to take advantage of the image enhancement and x-ray analysis capabilities afforded by these techniques.

Experimental Procedure

A Nb-46.5 w/o Ti Fermilab composite, which had been used in a previous study[6] into the development of the subband structure during drawing, was used for this part of the investigation. This composite had been given a two-stage heat treatment during the drawing process. The first heat-treatment of 80 h at 375°C had been carried out at 3.66-mm diameter and the composite had then been drawn to 1.5-mm diameter and heat-treated again for 40 h at 375°C before being drawn down again to a final diameter of 0.67 mm. The drawing process had been carried out with a 20% reduction in area at each pass. Thin foils were prepared from all the samples using a gravity-jet electropolishing technique.[4,6] These foils were examined using a JEOL 200CX TEMSCAN system operating at 200 kV.

Results and Discussion

A STEM micrograph of a Fermilab sample at 1.5-mm diameter after a heat treatment of 40 h at 375°C is shown in Fig. 3. Lighter, approximately equiaxed areas are clearly visible in the subband structure. X-ray analyses from these areas showed that they contained a lower concentration of Nb than the darker surrounding areas. Figure 4 is a STEM micrograph from the same composite after further drawing to a 0.85-mm diameter. In this micrograph, the lighter areas appear more elongated in the direction of drawing. Again x-ray analysis showed that these areas contained lower concentrations of Nb.

There are several contrast mechanisms that can be used to differentiate between a precipitate and the surrounding matrix. One method is based on the different electron absorption capabilities of different elements; heavier, higher atomic number elements absorb electrons to a greater extent than lighter elements. X-ray analysis confirmed that the lighter areas contained lower concentrations of the higher atomic number element, Nb. Selected area diffraction patterns showed that α-Ti precipitates were present in

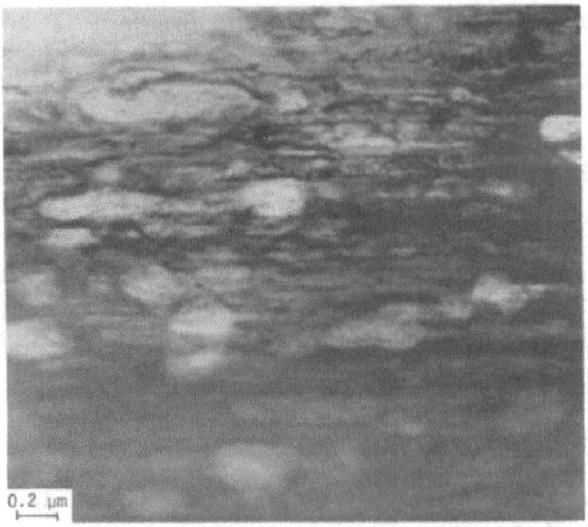

Fig. 3. STEM micrograph of a Fermilab composite at 1.5-mm diameter.

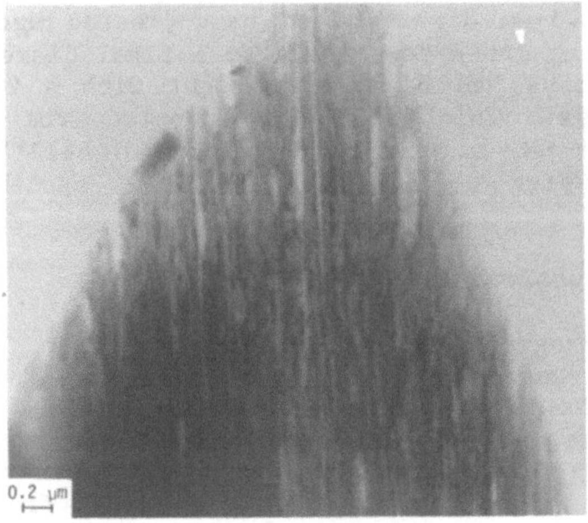

Fig. 4. STEM micrograph of a Fermilab composite at 0.85-mm diameter.

these samples, and therefore, it is believed that these lighter areas are α-Ti precipitates.

In the first Fermilab sample (Fig. 3), the α-Ti had not been deformed after precipitation and so appeared approximately equiaxed in morphology. After drawing (Fig. 4), the α-Ti precipitates were seen to have become elongated in the direction of drawing.

A more detailed discussion of the critical current density values of these Fermilab composite samples will be reported later, but some preliminary comments may be made here. The 5-T J_c value (ρ = 0.05 pΩm) of the sample shown in Fig. 4 was 1500 A mm^{-2}, a value fairly close to those obtained in common practice. It is interesting to note that the α-Ti precipitates tend to occupy whole subbands and to be several tens of nanometers wide and several hundred long. Similarly sized precipitates have been observed in wires where the whole fabrication sequence was under the control of the manufacturer.[7] Such precipitates are significantly larger than those seen in the high J_c (~2500 A mm^{-2} at 5 T) samples of Willbrandt and Schlump.[3] They are also much larger than the optimum size for flux pinning, where precipitates of 5 to 10 nm are preferred. Since larger precipitates may also tend to embrittle wires during drawing, it appears that careful control of the precipitate size and morphology is needed to obtain optimum properties. Further work into the factor affecting the precipitate size distribution is in progress.

CONCLUSIONS

The morphology and distribution of the α-Ti phase that precipitates in Nb-Ti alloys is very dependent on the microstructure of the matrix. The α-Ti that nucleates in a heavily cold-worked structure forms as equiaxed particles that have no orientation relationship with the matrix, unlike the plate-like α-Ti precipitates that nucleate in annealed alloys. In the Nb-Ti filaments in composite wires, the α-Ti grows as equiaxed precipitates, and these precipitates deform with the matrix on subsequent drawing. In these complicated microstructures, it is necessary to use STEM and EDAX in conjunction with CTEM to distinguish the α-Ti phase from the Nb-Ti matrix.

ACKNOWLEDGMENTS

We are grateful to JEOL U.S.A. for use of the 200CX TEMSCAN system and especially grateful to D. F. Harling, Applications Microscopist, JEOL for technical assistance. This work has been financially supported by the Department of Energy and the Wisconsin Electric Utilities Research Foundation.

REFERENCES

1. D. C. Larbalestier, Niobium-titanium superconducting materials, in: "Superconducting Materials," S. Foner and B. B. Schwartz, eds., Plenum Press, New York (1981), in press.
2. R. Arndt and R. Ebeling, Influence of microstructure on critical current density of niobium-titanium superconductors, Z. Metallkd. 65:364 (1974).

3. J. Willbrandt and W. Schlump, Influence of precipitation density and particle diameter on the current carrying capacity of NbTi superconductors, Z. Metallkd. 66:714 (1975).
4. A. W. West and D. C. Larbalestier, Transmission electron microscopy of commercial filamentary Nb-Ti superconducting composites, in: "Advances in Cryogenic Engineering," Vol. 26, Plenum Press, New York (1980), p. 471.
5. J. C. Williams, Precipitation in titanium-based alloys, in: "Precipitation Processes in Solids," K. C. Russell and H. I. Aaronson, eds., The Metallurgical Society of AIME, New York (1978), p. 191.
6. A. W. West and D. C. Larbalestier, Microstructure superconducting property relationships in a Fermilab Nb-46.5w/oTi filamentary superconducting composite, IEEE Trans. Magn. MAG-17:65 (1981).
7. A. W. West and D. C. Larbalestier, to be published.

NEW PERSPECTIVES ON THE PHYSICS OF HIGH-FIELD SUPERCONDUCTORS*†

M. R. Beasley

Departments of Applied Physics and Electrical Engineering
Stanford University, Stanford, California

INTRODUCTION

In making a practical high-field superconducting conductor, the manufacturer is confronted with the problem of optimizing simultaneously several properties (e.g., T_c, H_{c2}, J_c, and mechanical behavior), all of which themselves depend on a variety of material parameters. At the same time it is necessary to fabricate the conductor in the form of fine, twisted filaments imbedded in a normal metal matrix. This is no simple matter, needless to say, and in carrying out such a task it is helpful to have as clear an understanding as possible of how the physical properties of interest depend on the various material parameters. In this paper we review recent advances in the understanding of the upper critical field, H_{c2}, of real high-field superconductors in an effort to provide some improved guidance to the practical conductor community in just this regard. In particular, we address the role of disorder in determining H_{c2} of the A15 superconductors and corrections to the theory of Pauli limiting that have come to light recently. The insights obtained from these results are also used to provide an up-to-date set of ground rules for optimizing H_{c2} and to establish expected theoretical limits of H_{c2} in superconductors of practical interest.

The Physics of H_{c2}

Historically the theory of Type-II superconductors and of H_{c2} in particular was developed in the mid-1960s. It is the basis upon

*Invited paper.

†Work supported by the Department of Energy.

which practical conductors are being developed today. This theory is well known and has been described in many review articles.[1] In the theory H_{c2} is determined by the pair-breaking action of a magnetic field on the Cooper pairs that form the superconducting state. A magnetic field produces pair-breaking through its effect on the orbital motion (e.g., the existence of vortices) and the spin state of the Cooper pairs. Spin pair-breaking arises owing to the removal of the degeneracy between spin-up and spin-down members of a Cooper pair via Zeeman splitting. It is commonly referred to as Pauli paramagnetic limiting because of its close connection with the Pauli susceptibility of the material. Random spin-orbit scattering counteracts such pair-breaking and reduces Pauli paramagnetic limiting of H_{c2}.

The results of the theory can be summarized most conveniently in terms of two characteristic fields $H^*_{c2}(0)$ and $H_p(0)$, which represent the critical fields (at T = 0) due to orbital pair-breaking and Pauli limiting alone, respectively, in the absence of the other. In the dirty limit $\ell \ll \xi_o$ (but no spin-orbit scattering) the net $H_{c2}(0)$ is

$$H_{c2}(0) = \frac{H^*_{c2}(0)\ H_p(0)}{\left[2H^{*2}_{c2}(0) + H^2_p(0)\right]^{1/2}} \tag{1}$$

which shows that the smaller of the two characteristic fields tends to dominate.* The presence of Pauli limiting in practice is signaled by the condition $H_{c2}(0) < H^*_{c2}(0)$, where the orbital pair-breaking critical field, $H^*_{c2}(0)$, can always be determined experimentally from the slope of the critical field curve, $H_{c2}(T)$, near T_c, as given by Eq. (2) below. At finite temperatures and for arbitrary cleanliness and spin-orbit scattering rate, the results cannot be written in such simple analytical form, but the general idea is the same and can be found in the review articles.[1] According to the theory

$$H^*_{c2}(0) = 0.69\ dH_{c2}/dT|_{T_c} \cdot T_c \qquad \text{(dirty limit)}\ \ \ell \ll \xi_o \tag{2}$$

$$= 0.72\ dH_{c2}/dT|_{T_c} \cdot T_c \qquad \text{(clean limit)}\ \ \ell \gg \xi_o$$

*This formula is valid for second-order transitions, as is the usual case in practice. As $H^*_{c2}(0) \to 0$, the formula yields $H_{c2}(0) = H_p/\sqrt{2}$, which in this case is the supercooling field of the material. The actual transition occurs at H_p and is of first order.

where

$$dH_{c2}/dT|_{T_c} \simeq 9.6 \times 10^{24}\ \gamma^2 T_c (n^{2/3} S/S_F)^{-2} + 5.3 \times 10^4 \gamma\rho \quad (3)$$

and, as the theory is usually applied

$$H_p(0) = H_p^{BCS}(0) \equiv \frac{\Delta(0)}{\sqrt{2}\ \mu_B} \quad (4)$$

Here n is the density of electrons, S/S_F the reduced area of the Fermi surface, ρ the low temperature normal state resistivity, and γ the normal state electronic specific heat coefficient.* Note that Eqs. (2) and (4) relate the two characteristic critical fields H^*_{c2} and H_p to readily available material parameters. Except for possible strong-coupled superconductivity corrections, Eqs. (2) and (3) are found to be quite satisfactory in practice. Equation (4), however, is now believed to require substantial correction, as discussed in detail below.

Both orbital and Pauli paramagnetic pair-breaking are present in materials of practical interest. Pure Nb-Ti alloys are apparently slightly Pauli limited, Nb_3Sn is only limited by orbital effects (although the reason why is not so simple as has been believed previously), and V_3Ga is strongly Pauli limited. The routes to optimization of H_{c2} are roughly as follows: For the alloys, one increases the resistivity as much as possible in a material with a favorable T_c so as to maximize $dH_{c2}/DT|_{T_c}$ [Eq. (3)] and hence H^*_{c2}, adding a high-Z (Z = atomic number), strong spin-orbit scattering component to eliminate whatever Pauli limiting may be present. This approach has been carried out with great refinement most recently by Hawksworth and Larbalestier[2] for the Nb-Ti system. For the A15 compounds and probably also for any likely successor compounds, the situation is more complicated. The reason is that T_c and γ as well as ρ are dependent on the state of order in the compound. Hence, the dependence of H_{c2} on disorder is more complicated. Also, as already mentioned, V_3Ga is strongly Pauli limited. Consequently, with such materials one needs a detailed understanding of the effect of disorder on the compounds and a complete understanding of the Pauli limiting process in order to know how to proceed with any precision. It is in just these areas that recent advances in our fundamental understanding have occurred, and although the situation is still evolving, it seems useful to review the progress at the present time.

*The units in Eq. (3) are ohm-cm (ρ), erg cm^{-3} K^{-2} (γ), cm^{-3} (n), and K (T_c).

ROLE OF DISORDER IN THE A15 SUPERCONDUCTORS

It has been known for some time that almost any kind of disorder (e.g., quenched-in thermal disorder, deviations from stoichiometry, radiation damage, or, in the case of thin films, imperfect ordering due to deposition on a cold substrate) reduces the transition temperature of the high-T_c, A15 superconductors.[3] (Here we use the term disorder in a generic sense and not in the more specific sense of an x-ray diffraction order parameter.) A similarly long list of possible factors (e.g., broadening of peaks in the densities of states, rigid band effects, suppression of a martensitic transition, phonon softening, anisotropy, chemical effects, and the proximity effect) have been identified as possible specific mechanisms through which such disorder may affect T_c. Unfortunately, much less attention has been paid to the role of disorder on H_{c2} or on other physical properties of interest in superconductivity. In an attempt to shed some light more broadly on the situation and, specifically, to examine carefully the effect of disorder on H_{c2}, Orlando et al.[4] have carried out a systematic set of measurements on a series of well-characterized electron-beam codeposited thin films of Nb_3Sn in which disorder was introduced via various means. These included deviations from stoichiometry, tertiary additions of Aℓ and Zr, and physical disorder introduced by deposition at low substrate temperature. Thin films are nicely suited for such studies because of the ease with which they can be produced with varying material parameters using electron-beam codeposition.

The results for T_c, $dH_{c2}/dT|_{T_c}$, and $H_{c2}(0)$ are shown in Fig. 1 along with some data from other researchers.[5] In the figure, the resistivity at T_c, ρ_{T_c}, is used as a measure of the degree of disorder. As can be readily seen, when plotted in this way there is a remarkable degree of universal behavior in the data. Specifically for $\rho \lesssim 60$ μΩ-cm the dependence of T_c, $dH_{c2}/dT|_{T_c}$, and $H_{c2}(0)$ on ρ_{T_c} is independent to first order on how the disorder was introduced. Only for $\rho > 60$ μΩ-cm can one see major differences between the radiation-damaged samples of Ref. 5 and the off-stoichiometry samples. Recent measurements of T_c and $dH_{c2}/dT|_{T_c}$ on bulk Nb_3Sn crystals of varying Sn composition by Devantay et al.[6] over the entire A15 phase field are in excellent agreement with those of Fig. 1, as are the T_c measurements of Adrian et al.[7]* in which radiation damage was introduced by high-energy sulfur ions, even as regards the differences between radiation-damaged and off-stoichiometry samples at large ρ_{T_c}.

*In comparing the results of Fig. 1 with Ref. 7, we have scaled their results by an assumed room-temperature resistivity of Nb_3Sn of ≈ 90 μΩ-cm.

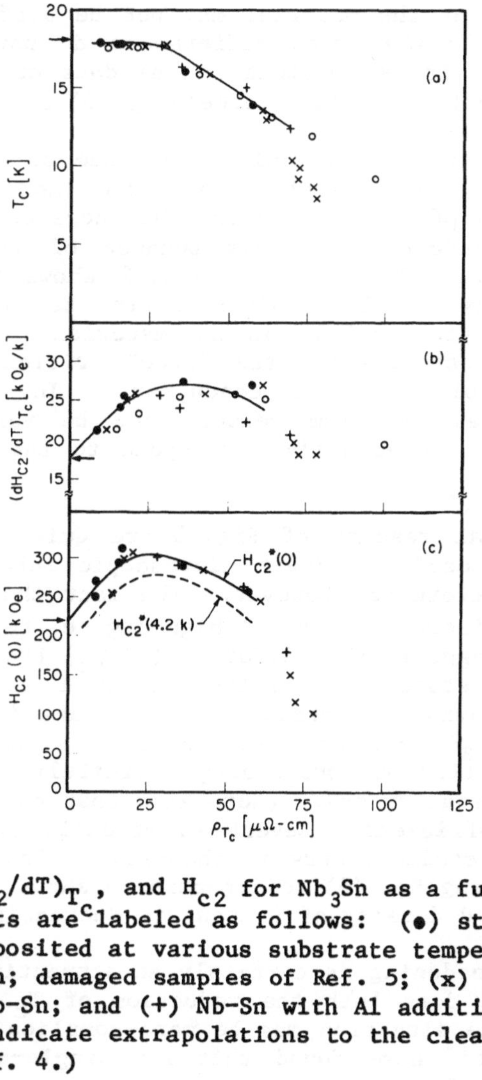

Fig. 1. T_c, $(dH_{c2}/dT)_{T_c}$, and H_{c2} for Nb_3Sn as a function of ρ_{T_c}. The points are labeled as follows: (•) stoichiometric Nb_3Sn deposited at various substrate temperatures; (o) radiation; damaged samples of Ref. 5; (x) off-stoichiometric Nb-Sn; and (+) Nb-Sn with Al additions. The arrows indicate extrapolations to the clean limit. (From Ref. 4.)

From the practical point of view, the most striking new feature of these data is the peak in $H_{c2}(0)$. Clearly H_{c2} can be increased markedly in Nb_3Sn at the expense of only a slight decrease in T_c by the introduction of a small amount of disorder. Evidently the maximum in $H_{c2}(0)$ results from the rapid rise in dH_{c2}/dT with disorder [i.e., ρ through Eq. (3)] before the disorder becomes strong enough to depress T_c. This is perhaps not so surprising, although it argues against always optimizing T_c. What is surprising is that it appears to matter very little how the amount of disorder is introduced, providing it yields a $\rho \approx 30$ μΩ-cm, which is not to suggest that the material may not be different on a microscopic scale, only that such differences do not matter to first order. Moreover, it is implicit in the data of Fig. 1 that there is a one-to-one correspondence between $H_{c2}(0)$ and T_c.

Another important practical consequence of the data in Fig. 1 is that there are necessarily very large (and possibly nonmonotonic) variations of H_{c2} through the thickness of a typical bronze-reacted Nb_3Sn conductor, if only because of inevitable compositional variations. For example, Fig. 2 shows the variations of $H_{c2}(0)$ implied by the T_c depth profiling measurements of Evetts et al.[8] on some practical conductor material. Clearly care must be exercised in establishing the "true" critical temperature and critical field of any real conductor. In particular, onset measurements (i.e., maximum values) can be very misleading because they do not necessarily correspond to the same part of the material.

The empirical results of Fig. 1 are quite satisfying, being both simple and useful. Their microscopic interpretation is not entirely straightforward, however. The fact that ρ_T is a useful measure of the disorder to such a high degree, regardless of how it is introduced, suggests that lifetime (h/τ_{tr}) broadening[9,10] of the band density of states may be the universal mechanism underlying the observed behavior. Certainly if the peaks in the band density of states of Nb_3Sn are as sharp as band structure calculations suggest,[11] then lifetime broadening definitely is playing a major role, if not a singular one. (Note also that the expected lifetime broadening is sufficiently severe to strongly attenuate any rigid band effects as stoichiometry is changed.)* Indeed analyses[5,9,12] of these data using Eq. (3) to determine γ shows that γ (and almost certainly the band density of states) is decreasing as ρ increases.

Lifetime broadening is certainly an attractive interpretation, explaining as it does both the reduction of T_c and the universal behavior, but the situation may be more complicated. For example, Foner and McNiff[13] have found that for single-crystal samples of Nb_3Sn, the H_{c2} of the cubic phase (i.e., non-martensitic-transforming) portions of the crystal have a much higher $H_{c2}(0)$ than the tetragonal (transforming) portions. Moreover Kwo et al.[14] have shown that for Nb_3Al (which is necessarily disordered even when

*There have been many dramatic predictions regarding chemical substitutions in the A15 compounds based on rigid-band extrapolations of the superconducting properties. Such predictions should probably not be taken seriously, unless lifetime broadening effects associated with the substitutions were also included, as in Ref. 10.

well made because of the metastability of the compound at low temperatures), the band density of states is not appreciably varying with disorder even though both T_c and $dH_{c2}/dT|_{T_c}$ do decrease. These results are consistent with the anomalously low heat capacity of Nb_3Al as well, suggesting there is not a high peak in the density of states in this material, as presently made in its imperfectly ordered state. Hence Nb_3Al appears to provide a clear demonstration that factors other than lifetime broadening do play a role, at least for strong disorder. We note also that the data for Nb_3Al and Nb_3Sn are very similar in this strongly disordered regime.

Further insight into the role of the martensitic transformation in Nb_3Sn is provided in Ref. 6, where it is argued that the martensitic transformation only occurs for Sn concentrations down

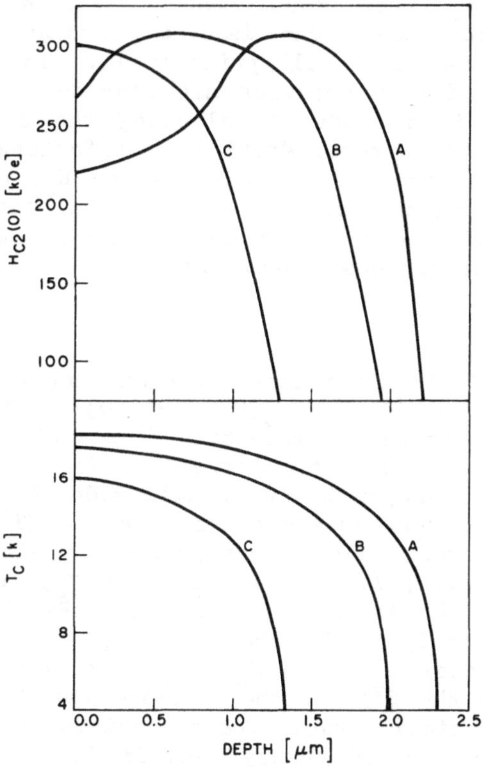

Fig. 2. Spatial variations of $H_{c2}(0)$ expected in practical bronze conductor deduced from Fig. 1 based on the T_c depth profile measurements of Ref. 8. Curves A, B, and C correspond to increasing reaction times (53, 26, and 54 h at 750°C).

to ~24.5% Sn, which corresponds to $\rho \simeq 20$ μΩ-cm in Fig. 1. Tertiary additions apparently also suppress the martensitic transformation.[3] If disorder generally suppresses the martensitic transformation, then this phenomena very likely is playing some role in the systematics of T_c and $H_{c2}(0)$ observed at low ρ in Fig. 1. In this view, disorder not only increases $dH_{c2}/dT|_{T_c}$, but also eventually favors the cubic phase, which for a given level of disorder is believed to have a slightly higher T_c (probably less than ~0.5 K). The large $H_{c2}(0)$ of the cubic phase observed by Foner and McNiff probably reflects the larger ρ's of the cubic regions (necessary to stabilize them) than an intrinsically higher $H_{c2}(0)$ (i.e., in the clean limit). Moreover, if the martensitic transformation is suppressed at a specific level of disorder (e.g., ρ) then the variations of T_c, $dH_{c2}/dT|_{T_c}$, and $H_{c2}(0)$ in Fig. 1 presumably have a slight step upward at this value of ρ. Clearly the situation could benefit from further clarification, but in any event there appears to be little doubt that a small amount of disorder, no matter how introduced, can lead to substantial increases in $H_{c2}(0)$ of Nb_3Sn at a very small price in T_c. This happy situation means that the conductor designer has considerable flexibility in how the disorder is introduced, allowing the specific choice to be determined by other considerations. Unfortunately, comparably thorough studies of the role of disorder on H_{c2} are not available for the other A15 superconductors of practical interest. Hence it is not clear to what extent the conclusions found here for Nb_3Sn will apply to other of these materials.

Pauli Limiting

From Fig. 1 it is clear that the upper critical field of Nb_3Sn is not Pauli limited [i.e., $H_{c2}(0) \simeq H^*_{c2}(0)$]. Traditionally this fact has been interpreted as resulting from a high spin-orbit scattering rate in Nb_3Sn due to the large Z of Nb. As recently pointed out by Orlando et al.,[12] however, well-made Nb_3Sn is actually a rather clean ($\xi_o/\ell \sim 1$) superconductor, not a dirty one, and consequently even assuming the maximum physically allowable spin-orbit scattering rate (i.e., $1/\tau_{so} = 1/\tau_{tr}$ where τ_{tr} is total transport scattering time), Nb_3Sn should show some Pauli limiting. The result of their analysis is shown in Fig. 3, where the maximum allowable theoretical H_{c2} with $H_p = H_p^{BCS}$ is compared with experiment. As seen in the figure, the experimental data lie well above the maximum theoretical curve. The recognition of this problem has led to a reexamination of theory, and it is now clear that large corrections to the Pauli limiting field have been neglected in previous analysis of Pauli limiting in high field superconductors. As pointed out in Ref. 12, the simple BCS approximation for H_p [Eq. (4)] must be generalized to include the electron-phonon and electron-electron interactions when analyzing the behavior of

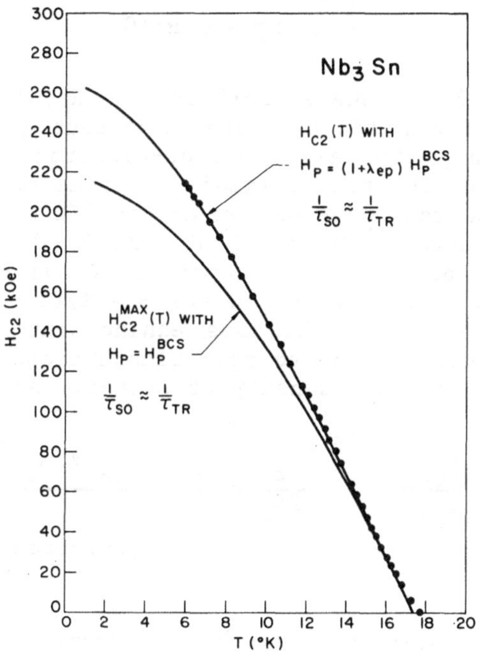

Fig. 3. Comparisons of $H_{c2}(T)$ of Nb_3Sn with various theoretical curves with and without the electron-phonon many-body corrections to H_p. (Data from Ref. 12.)

real superconductors. Note that these corrections are not strong-coupling superconductivity effects, which also exist but are small (~10-20% at most), but large corrections due to many-body renormalizations of the normal-state material parameters that determine H_p. They are never truly negligible even for weak-coupled superconductors. The existence of these corrections is implicit in the standard theory,[15] properly interpreted, and the reason they have been overlooked for so long is not entirely clear.

The origin of these corrections to H_p can be understood physically by using the original arguments of Chandrasekhar[16] and Clogsdon.[17] Here one notes that the existence of Pauli paramagnetism in the normal state but not the superconducting state (neglecting spin-orbit scattering) due to spin-up and spin-down pairing causes the free energy of the normal state to decrease in a field relative to that of the superconducting state, $F_N(H) - F_S = -(1/2)\chi_p H^2$. The Pauli limiting field is then obtained by calculating the field at which the decrease in the normal state free energy exceeds the condensation energy of the superconducting state:

$$\frac{1}{2} \chi_p H_p^2 = \frac{1}{2} H_c^2(0) \qquad (5)$$

where χ_p is the Pauli susceptibility and $H_c(0)$ is the so-called thermodynamic critical field. The important point here is that when the electron-phonon and electron-electron many-body interactions are included, the renormalized densities of states that enter $\chi_p = 2\mu_B^2 N^{\chi}(0)$ and the condensation energy $1/2\ H_c^2(0) = 1/2\ N^{\gamma}(0)\ \Delta^2$ are different. Here $N^{\chi}(0) = SN^b(0)$ and $N^{\gamma}(0) = (1 + \lambda_{ep} + \lambda_s)\ N^b(0)$ are the so-called spin susceptibility and heat capacity densities of states, respectively, where $S = 1/(1 - I)$ is the Stoner factor, $(1 + \lambda_{ep} + \lambda_s)$ is the mass enhancement factor due to the electron-phonon interaction and spin fluctuations, and $N^b(0)$ is the bare (unrenormalized) band density of states. As a result,

$$H_p(0) = \left[\frac{N^{\gamma}(0)}{N^{\chi}(0)}\right]^{1/2} \frac{\Delta}{2\mu_B} = \left[\frac{1 + \lambda_{ep} + \lambda_s}{S}\right]^{1/2} H_p^{BCS}(0)$$

(1st-order transition) (6)

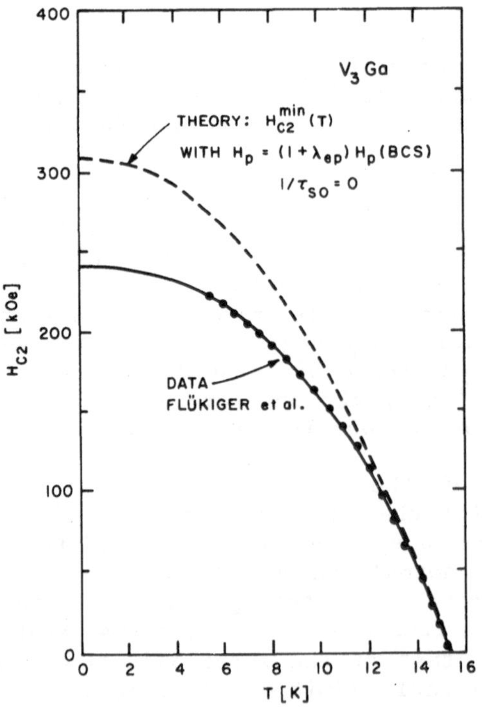

Fig. 4. Comparison of $H_{c2}(T)$ of V_3Ga with the theoretical curve assuming electron-phonon many-body corrections but not the electron-electron corrections. (Data from Ref. 19.)

This result is based on thermodynamic arguments and therefore applies only when the transition is first order. When the transition is second order, which is usually the case in practice, the full pair-breaking theory[15] must be used and yields the larger corrections:

$$H_p(0) = \frac{1 + \lambda_{ep} + \lambda_s}{S} H_p^{BCS}(0)$$

(2nd-order transition) (7)

That the inclusion of these correction factors can resolve the difficulties with Nb_3Sn is shown in Fig. 3, where the theoretical curve assumes $H_p = (1 + \lambda_{ep}) H_p^{BCS}$ (i.e., taking $S = 1$, which implies $\lambda_s = 0$) and the usually quoted value of $\lambda_{ep} \simeq 1.8$. Note, however, that a spin-orbit scattering rate $1/\tau_{so} \simeq 1/\tau_{tr}$ is still required, even though the minimum value of S was assumed. In any event, it seems clear that the absence of Pauli limiting in Nb_3Sn is due very likely in large part to its strong electron-phonon interaction and not only to spin-orbit scattering. The situation for V_3Ga is even more interesting. V_3Ga is strongly Pauli limited, and as shown in Fig. 4, the observed critical field is well below the minimum theoretical H_{c2} (i.e., assuming no spin-orbit scattering) if H_p is taken as $(1 + \lambda_{ep}) H_p^{BCS}$. Hence, in this case, a substantial Stoner factor is required to account for the data. In fact, the data can be used to place a lower limit $S \gtrsim 2$ for V_3Ga.[18]

The above analyses demonstrate the need for corrections to H_p but do not provide a means of determining the two correction factors independently. Further progress requires an analysis including other physical properties that depend on the electron-phonon and electron-electron interactions. Such an analysis has been carried out approximately by Orlando and Beasley.[18] These authors conclude that V_3Ga is not unique in having a substantial Stoner factor and that spin fluctuations associated with large Stoner factors may be playing a prominant role in the properties of the A15 superconductors generally, including a substantial deleterious effect on T_c--a remarkable result considering these are high-T_c superconductors.

The above results clearly demonstrate that the theory of Pauli limiting, as it has been traditionally applied, is seriously deficient. Qualitatively, the many-body correction factors appear to resolve the problem. A quantitative test of the theory is still lacking, however, because of lack of an independent measure of τ_{so}. Also, we note that some evidence for problems with the theory in the limit of high fields and large spin-orbit scattering, even with the many-body corrections included, has been reported by Tedrow and Meservey.[20] In a similar vein, we recall that for Nb_3Sn one still requires perhaps (but not demonstrably) an unphysically large spin-orbit scattering rate (e.g., $1/\tau_{so} \simeq 1/\tau_{tr}$) to obtain sensible agreement with theory. A possible resolution of these remaining difficulties may lie in the recent work of Schopohl and Scharnburg,[21] who have carried out an improved calculation of the effect of spin-orbit scattering on Pauli limiting. These authors have calculated the reduction in Pauli limiting as a function of the spin-orbit scattering rate for arbitrary ratios of τ_{so}/τ_{tr}, thereby eliminating the assumption $1/\tau_{tr} \gg 1/\tau_{so}$ in the usual theory. Hence, their calculations are the correct ones for cases of very strong spin-orbit scattering, such as Nb_3Sn. They find a greater reduction of Pauli limiting (i.e., a larger critical field) for a given spin-orbit scattering rate when $1/\tau_{so} \simeq 1/\tau_{tr}$ than when $1/\tau_{so} \ll 1/\tau_{tr}$. Clearly this would make the fits to Nb_3Sn, which were made using the old theory, more reasonable, but a complete quantitative study including both the many-body corrections and the proper theory of spin-orbit scattering has not yet been carried out.

Maximum Critical Fields

In light of our improved understanding of H_{c2} in real superconductors, it is of interest to reassess what the maximum H_{c2} is likely to be for materials of practical interest--in particular the

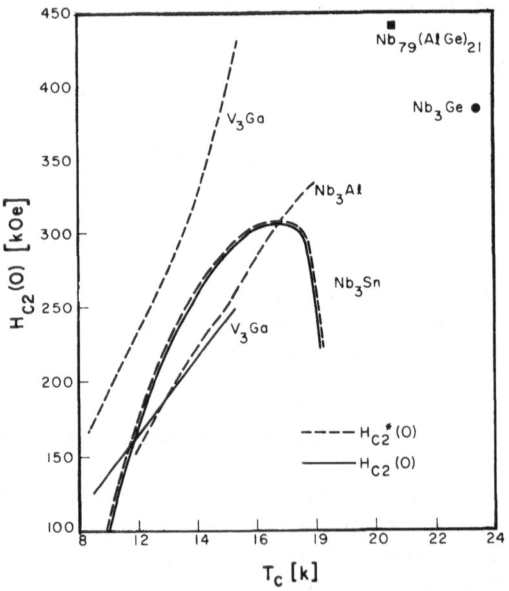

Fig. 5. $H_{c2}(0)$ versus T_c for various well-characterized A15 compound systems. Data for Nb_3Sn from Ref. 4, for V_3Ga from Ref. 22, for Nb_3Al from Ref. 14, and for $Nb_3(AlGe)$ and Nb_3Ge from Ref. 22.

high-field A15 compounds. For Nb_3Sn the situation is quite clear, as we have discussed in detail above. There is no doubt that this material can be made to have an $H_{c2}(0) \simeq 300$ kOe (~215 kOe at 4.2 K)--well above what is usually achieved in practical conductors. Whether this is because of strong spatial variations of H_{c2} (e.g., Fig. 2), or whether the Nb_3Sn is being made too well, (i.e., too ordered) is not clear. In any event, the material has more potential than has been reached in practice.

For materials other than Nb_3Sn, there has not been sufficiently thorough studies of well-characterized, homogeneous material to be so specific. One reasonable basis for comparing different materials is to compare plots of $H_{c2}(0)$ versus T_c. On the basis of the Nb_3Sn results, one might expect there to be universal relationships, at least for the better ordered materials. Such a plot is shown in Fig. 5 for various A15 superconductors of present or potential practical interest.

Beginning with Nb_3Al, we see that this superconductor does not show the peak of H_{c2} exhibited by Nb_3Sn. On the other hand, even for $T_c = 18$ K ($\rho \simeq 40$ μΩ-cm), Nb_3Al is not the ideal, well-ordered, stoichiometric compound. Also, as we have already mentioned, the origin of the decline of T_c and $H_{c2}(0)$ in Nb_3Al is apparently different from that of Nb_3Sn in the region where H_{c2} shows a maximum. Hence, there is no firm basis upon which to anticipate what the complete curve for Nb_3Al would look like. In any event, the best-made Nb_3Al is clearly better than well-ordered Nb_3Sn, but only marginally better than optimized Nb_3Sn. Any further improvements in Nb_3Al clearly require improved synthesis techniques for making the stoichiometric compound in a well-ordered form.

V_3Ga is a particularly interesting case, both because it has already been shown to have higher critical current densities at high fields than Nb_3Sn and because it is so strongly Pauli limited. The $H^*_{c2}(0)$'s of V_3Ga are very large indeed, making a tempting target for improvement. In practice, however, very dramatic improvement may be hard to achieve. Even given the evolving state of our understanding of Pauli limiting, it is clear that increased spin-orbit scattering (through tertiary additions, for example) will be required. However, because the increased disorder introduced by such additions will almost certainly reduce T_c, it is not possible to move vertically in Fig. 5. Even a modest 1.5-K drop in T_c (such as optimizes Nb_3Sn) for V_3Ga leads to an $H^*_{c2}(0)$ of only ~300 kOe. What may be of greater practical importance for V_3Ga would be to understand why it has larger critical currents near H_{c2}. Here our improved understanding of the Pauli limiting process on a microscopic scale may be of considerable relevance.

In the end, then, it seems likely that optimized Nb_3Sn and V_3Ga would have roughly the same $H_{c2}(0)$ although for different reasons. If this is so, then any applications requiring $H_{c2}(0)$'s larger than 300 kOe will probably require using the metastable A15s: Nb_3Al (in a well-ordered form), $Nb_3(AlGe)$, or Nb_3Ge. At this point one would also want to consider the Chevrel phases where even higher critical fields have been demonstrated.

ACKNOWLEDGMENTS

The author would like to acknowledge useful discussions with T. H. Geballe, R. H. Hammond, J. Kwo, S. Foner, E. J. McNiff, Jr., and R. Flükiger. Also, a special acknowledgment must go to T. P. Orlando, with whom almost all of the author's understanding of real high-field superconductors was developed.

REFERENCES

1. See, for example, A. L. Fetter and P. C. Hohenberg, in: "Superconductivity," Vol. II, R. D. Parks, ed., Marcel Dekker, New York (1969), p. 886; N. R. Werthamer, ibid, p. 321; and D. Saint-James, G. Sarma, and E. J. Thomas, "Type-II Superconductivity," Pergamon Press, New York (1969).
2. D. G. Hawksworth and D. C. Larbalestier, IEEE Trans. Magn. MAG-17:49 (1981).
3. For general reviews of the A15 superconductors see J. Muller, Rep. Prog. Phys. 43:641 (1980) and D. Dew-Hughes, Cryogenics 15:435 (1975). For a recent discussion of disorder effects and their interpretation, see the relevant articles in "Superconductivity in d- and f-Band Metals," H. Suhl and M. B. Maple, eds., Academic Press, New York (1980).
4. T. P. Orlando, J. A. Alexander, S. J. Bending, J. Kwo, S. J. Poon, R. H. Hammond, M. R. Beasley, E. J. McNiff, Jr., and S. Foner, IEEE Trans. Magn. MAG-17:368 (1981).
5. A. K. Gosh, M. Gurvitch, H. Weismann, and M. Strogin, Phys. Rev. B 18:6116 (1978).
6. H. Devantay, J. L. Jorda, M. Decroux, J. Muller, and R. Flükiger, to appear in J. Mater. Sci. (1981).
7. H. Adrian, G. Ischenko, M. Lehmann, P. Müller, H. Braun, and G. Linker, J. Less-Common Met. 62:99 (1978).
8. J. E. Evetts, J. R. Cave, R. E. Somekh, J. P. Stanton, and A. M. Campbell, IEEE Trans. Magn. MAG-17:360 (1981).
9. A. K. Ghosh and M. Strongin, in: "Superconductivity in d- and f-Band Superconductors," H. Suhl and M. B. Maple, eds., Academic Press, New York (1980), p. 305.
10. L. R. Testardi and L. F. Mattheiss, Phys. Rev. Lett. 41:1612 (1978).
11. See for example B. M. Klein, L. L. Boyer, D. A. Papaconstantapoulous, and L. F. Mattheiss, Phys. Rev. B 12:6411 (1978).

12. T. P. Orlando, E. J. McNiff, Jr., S. Foner, and M. R. Beasley, Phys. Rev. B 19:4545 (1979).
13. S. Foner and E. J. McNiff, Jr., Phys. Lett. A 58:318 (1976).
14. J. Kwo, T. P. Orlando, and M. R. Beasley, to appear in Phys. Rev. B.
15. D. Rainer, G. Bergmann, and U. Eckhart, Phys. Rev. B 8:5324 (1973).
16. B. S. Chandrasekhar, Appl. Phys. Lett. 1:7 (1962).
17. A. M. Clogston, Phys. Rev. Lett. 9:266 (1962).
18. T. P. Orlando and M. R. Beasley, Phys. Rev. Lett. 46:1598 (1981).
19. R. Flükiger, S. Foner, and E. J. McNiff, Jr., in: "Superconductivity in d- and f-Band Metals," H. Suhl and M. B. Maple, eds., Academic Press, New York (1980), p. 265.
20. P. M. Tedrow and R. Meservey, Phys. Rev. Lett. 43:384 (1979).
21. N. Schopohl and K. Scharnberg, to be published.
22. S. Foner, E. J. McNiff, Jr., B. T. Matthias, T. H. Geballe, R. H. Willens, and E. Corenzwit, Phys. Lett. 31A:349 (1970).

THE PHASE RELATIONSHIPS IN Nb_3Sn WIRES AT LOW TEMPERATURES AS DETECTED BY CRYSTALLOGRAPHICAL (NEUTRON AND X-RAY DIFFRACTION) AND BY PHYSICAL [$B_{c2}(T)$, J_c vs. ε] MEASUREMENTS

R. Flükiger, W. Schauer, W. Specking, and L. Oddi

Institut für Technische Physik, Kernforschungszentrum Karlsruhe
Karlsruhe, Federal Republic of Germany

L. Pintschovius

Institut für Angewandte Kernphysik, Kernforschungszentrum Karlsruhe
Karlsruhe, Federal Republic of Germany

W. Müllner

Institut für Kernphysik, Universität Frankfurt
Frankfurt, Federal Republic of Germany

B. Lachal

Département de Physique de Matière Condensée, Université de Genève, Geneva, Switzerland

INTRODUCTION

The effect of uniaxial tensile stress on the critical current density, J_c, of multifilamentary Nb_3Sn at 4.2 K and high magnetic fields has been studied for wires prepared by different methods, i.e., bronze route,[1,2] in situ processing,[3,4,5] powder metallurgy,[6,7] and external diffusion on a Cu matrix,[8] including wires with ternary or quaternary additions like Ga[4,5] or Hf+Ga.[9] In all cases, J_c was found to exhibit a maximum, J_{cm}, at strain values $0.2 \leq \varepsilon_m \leq 0.8$. Although it is well accepted that this behavior is connected to the compressive strain on the Nb_3Sn filaments resulting from the larger thermal contraction of the bronze, the basic mechanism leading to the pronounced maximum of J_c is not understood in detail.

From the strong field dependence of the ratio J_{cm}/J_{co}, particularly at high fields, and from the appreciable difference in B_{c2} between transforming and nontransforming Nb_3Sn,[10] it can be deduced that low temperature structural changes[11] could be at the origin of this effect. Recently, the structure of filaments and tapes under precompression has been studied by x-ray diffraction methods. Hoard et al.[12] (on tapes under various stress states), Roberge et al.,[13] and Flükiger et al.[14,15] (on multifilamentary wires prepared by different methods) reported considerable changes of the line shapes as a function of temperature. This is a strong indication for the occurrence of structural changes. However, interpretation is very difficult, due to peak broadening caused by both stress and composition gradients and to peak overlapping. To improve the unfavourable peak-to-background ratio, which is also deteriorated by the x-ray absorption in the bronze, diffraction measurements have to be made in the step scanning mode, with 400 to 1000 s per 0.02° or 0.05° step.[15] Recently, a splitting of the (400) line was observed in a bronze-processed Nb_3Sn wire, which was attributed to the presence of a stress-induced tetragonal distortion.[15]

The main problem with the x-ray analysis resides in its limited penetration depth. To study the effects of precompression in filaments lying far away from the surface, we have undertaken neutron diffraction analysis on the same wire used for the previous[15] x-ray investigation.

EXPERIMENTAL

The Nb_3Sn wire (0.2-mm diam.) used for the present investigation was prepared by the bronze route (Vacuumschmelze) and reacted at 700°C for 10 h. It contains 9072 untwisted filaments of 1.1-μm diam. Results of the recently measured $J_c(B,\varepsilon)$[15] are:

$$J_c(10^8 A/m^2) : 5.2\ (13\ T),\ 1.3\ (16\ T),\ 0.16\ (19\ T)$$

$$J_{cm}/J_{co}\ (13\ T) = 1.44;\ \varepsilon_m = 0.45\%$$

The temperature dependence of the upper critical field, $B_{c2}(T)$, was measured up to 13 T using the midpoint of the resistive transition. Specific heat was determined at the University of Geneva by an improved relaxation method.[16]

Neutron diffraction experiments were carried out on the high resolution powder diffractometer D1A[17] at the Institut Laue-Langevin in Grenoble. The diffraction patterns, measured with an incident wavelength of $\lambda = 1.909$ Å, were collected from a scattering range of $66° \leq 2\theta \leq 126°$, covering up to 7 reflections of Nb_3Sn

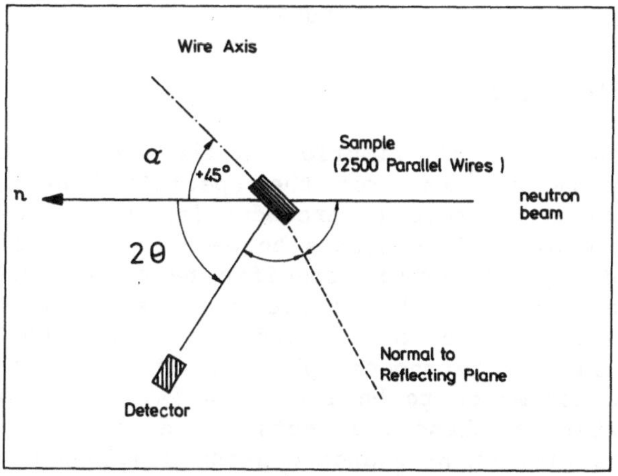

Fig. 1. Neutron diffraction experiment; wire sample orientation with respect to the incident neutron beam (angle α).

in the cubic phase. Attention was centered at the (400) line and its splitting at low temperature, which permits safe conclusions about the presence of lattice tranformations or distortions as well as approximate values of the parameters of the distorted phase. No standard has been added to the wire, so the absolute values of the peak positions may be subject to a slight correction. A refinement of the lattice parameters will be published later.[18] In the investigated range $6\ K \leq T \leq 320\ K$, the accuracy of the temperature setting was of the order of ±0.02 K.

The specimen consisted of a bundle of 2500 parallel wires of 2-cm length (Fig. 1). There are two orientations of the wire axis relative to the neutron beam, $\alpha \cong +45°$ and $\alpha \cong +135°$, that are of particular importance for the present investigation. Since the diffraction angle, 2θ, of the (400) cubic line is close to 90°, $\alpha = +45°$ is very close to the orientation for which the (400) reflecting plane is perpendicular to the wire axis. For $\alpha = +135°$, the (400) reflecting plane is parallel to the wire axis. The possibility of varying α during the neutron diffraction experiment is a considerable advantage to the x-ray case, where a new sample has to be prepared for each α angle.[15] The corresponding 2θ angle of the (400) line for CuK_α radiation is at 71.5°, i.e., quite different from 90°. However, due to the strongly anisotropic stress distribution in the multifilamentary geometry, both cases are comparable: the terms "parallel" and "perpendicular" used in the previous x-ray investigation[15] correspond roughly to $\alpha = +135°$ and $\alpha = +45°$.

RESULTS

The Concentration Gradient

The compositional distribution in the etched filaments (700°C for 19 h) can be deduced from the specific heat measurement in Fig. 2. It shows a broad superconducting transition from 15 to 18 K, corresponding to Sn contents between 23.5 and 25 at.%.[19] The linear term of the electronic specific heat was determined to be $\gamma = 8.0 \pm 0.2$ mJ/K^{2}g-at. This value and the entire dependence of C/T is very similar to that found for cubic Nb_3Sn samples.[20] Indeed, the amount of tetragonal phase in the Nb_3Sn filaments after etching can be estimated to be less than 5%. The room temperature lattice parameter of these filaments, as determined with a Debye-Scherrer camera with an Si standard using a Nelson-Riley extrapolation, was a = 5.288 ±0.001 Å. The concentration gradient can also be visualized by the fact that the $K\alpha_1$ and $K\alpha_2$ lines cannot be resolved (Fig. 3, Ref. 15).

Neutron Diffraction

It is interesting to study the effect of rotating the angle α on the shape of the diffraction lines at 300 K. In Fig. 3, we have represented the variation of the diffraction angle, 2θ, and of the half-width, Δ, for the distorted (400) line, after reaction for 10 and 19 h at 700°C. It is seen that the angle 2θ varies gradually from 92.20° for α - +135° to 92.32° for α = +45°, while Δ varies in the opposite sense from 0.62° to 0.52°.

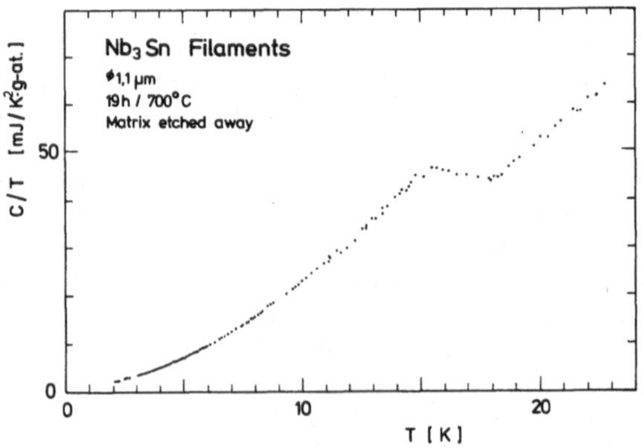

Fig. 2. Specific heat of Nb_3Sn filaments (without bronze matrix).

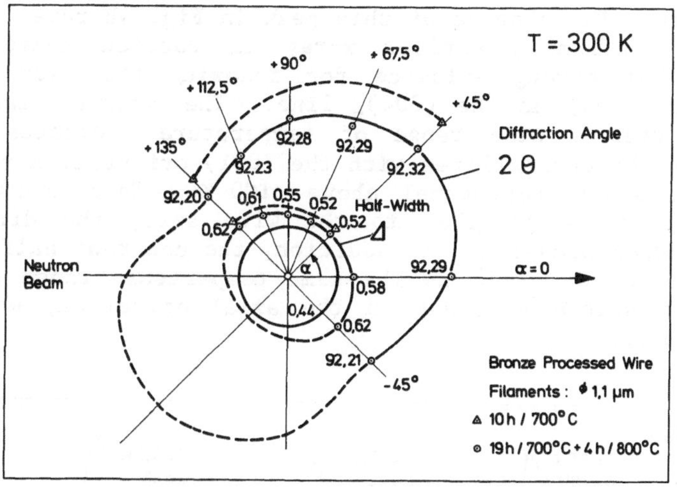

Fig. 3. Diffraction angles, 2θ, and half-widths, Δ, at 300 K as a function of the sample orientation angle, α. The Δ = 0.44° circle applies to the wire without bronze matrix.

Changing α means changing the orientation of the reflecting planes relative to the wire axis. Figure 3 thus furnishes direct evidence for a strongly anisotropic stress distribution in the Nb_3Sn filaments even at 300 K. The effect of stress is illustrated by comparing the Nb_3Sn (400) half-widths of the wire with that after etching away the matrix (Δ = 0.44° circle, Fig. 3). The compositional distribution also leads to a line broadening, but its effect is much smaller and contained in the 0.44° circle.

The stress distribution in the elastic state is correlated with a lattice distortion. The simplest way for indexing the peak positions in Fig. 3 is to assign the narrow peak for α = +45° to the tetragonal $(004)_t$ line and the corresponding one for α = +135° to tetragonal $(400, 040)_t$ line. However, the latter is considerably broader, its half-width showing a further increase with decreasing temperature, which has previously been attributed to the superposition of the $(400)_c$ and the $(400,040)_t$ lines,[15] but is rather due to an increased stress gradient in the axial direction.

In Fig. 4a and b we have represented the diffraction patterns of Nb_3Sn surrounded by the bronze matrix for $91° \leq 2\theta \leq 93.5°$ and different temperatures at orientation of α = +45° and α = +135°, respectively. The variation of the line positions and their shapes is very different for both orientations. At α = +45°, the shape and intensity of the peak are essentially unchanged with temperature, aside from an increasing asymmetry at lower angles at low temperatures. The absence of this peak in Fig. 4b reveals a strong texturing, confirming earlier x-ray diffraction measurements.[15] There is thus strong evidence for indexing the narrow line at $\alpha = +45°$ (Fig. 4a) as the $(004)_t$ line. The constant intensity of this line over a wide range of temperature indicates that the totality of the crystallites with the (001) orientation parallel to the wire axis are tetragonal above ~100 K. This means that the shorter c-axis is parallel to the wire axis, the direction of strongest precompression. In addition, the constant half-width for the orientation $\alpha = +45°$ in the same temperature range means that there is no stress gradient in the axial direction, which is an expected result.

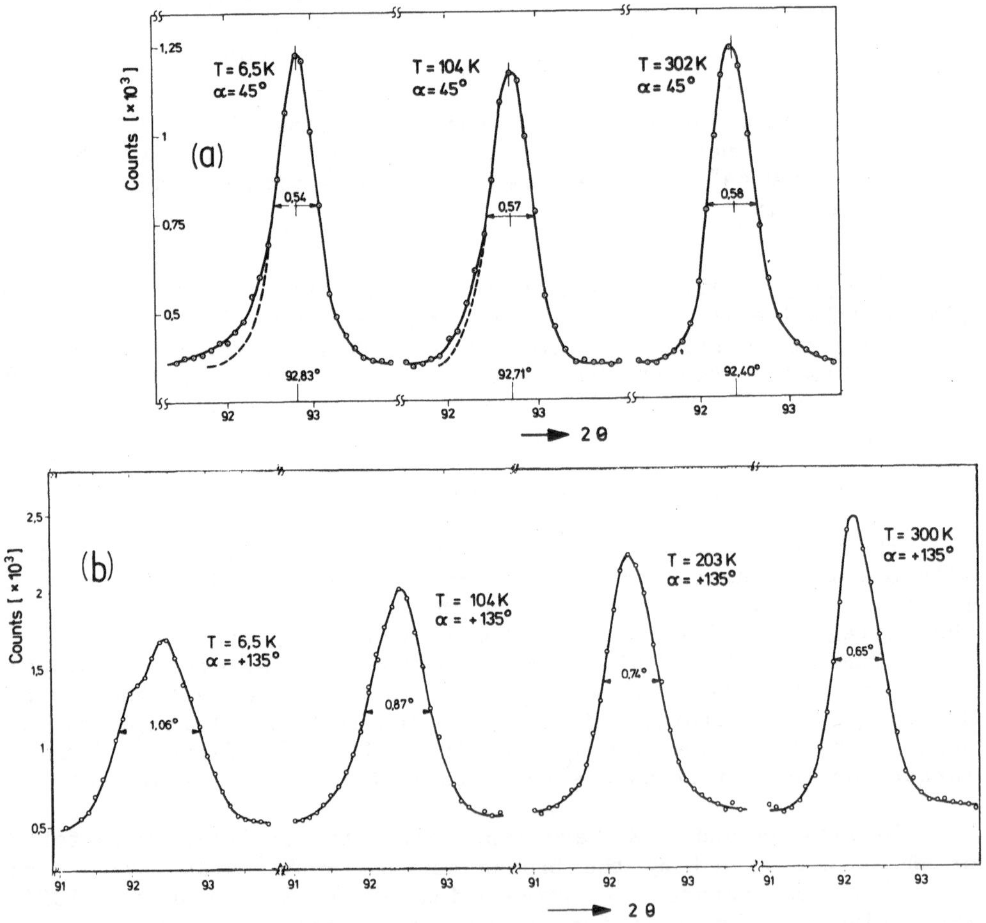

Fig. 4. Diffraction pattern for sample orientation $\alpha = +45°$ (a) and $\alpha = +135°$ (b) at temperatures between 6.5 and 300 K. The asymmetry at lower angles in (a) indicates a superposition with new, weaker line(s); for indexing see text.

It is interesting to follow the development of the line positions and half-widths as a function of temperature. There are three regions of variation for these parameters (Figs. 5, 6). In the first one, A (~100 K $\leq$ T $\leq$ 300 K), there is a stress-induced tetragonal distortion. Attributing the reflections at α = +45° and +135° to the $(004)_t$ and to the $(400, 040)_t$ lines, respectively, the tetragonality is (1-c/a) ≈ 0.0013 at 300 K and increases up to ~0.0028 at 104 K. A linear extrapolation to 6 K would give ~0.0035, while the value (1-c/a) = 0 would be reached several hundred degrees above room temperature, where the bronze is very soft. Attributing overlapping peaks at = +135°, the corresponding values of (1-c/a) would be somewhat larger, i.e., ~0.0016 at 300 K and ~0.0037 at 104 K.

Region B (~40 K $\leq$ T $\lesssim$ 100 K) is characterized by an enhanced increase of the line half-widths for the orientation α = +135°, while Δ decreases for α = +45°. In particular, an important line broadening for α = +135° was observed, from Δ = 0.85° at 100 K to 1.10° at 8 K. It is interesting that a slight broadening in region B is also observed in filaments without the matrix, the corresponding values being Δ = 0.44° and 0.47°, respectively. On one hand, this indicates that the new, unresolved kind of distortion is also stress induced. On the other hand, it appears that some stress is also present in the Nb_3Sn filaments without the matrix.[21] This distortion is connected with a change of the stress tensor[18] in

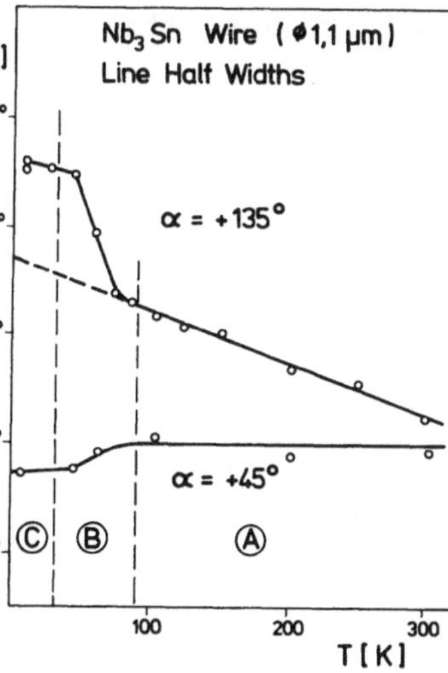

Fig. 5 Temperature dependence of the distorted (400) peak for the α = 45° and α = 135° sample orientation.

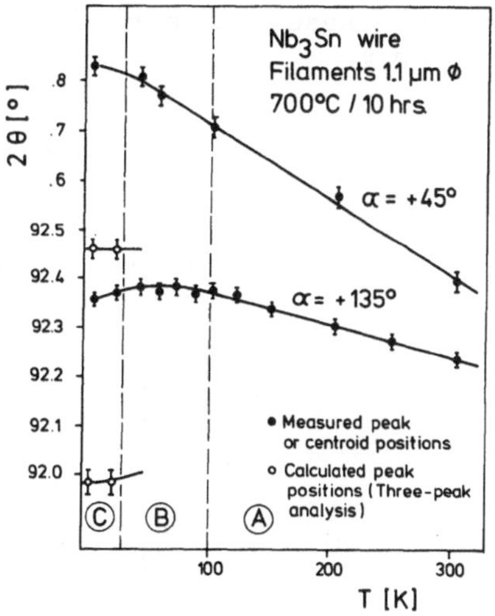

Fig. 6. Temperature dependence of the peak or centroid positions for the α = 45° and α = 135° sample orientation.

range B: the compressive stress is no longer fully uniaxial and exhibits radial components. The spectrum is now very complex, as shown by the pattern at 6.5 K in Fig. 4b resolved by three-peak analysis. The occurrence of one (or more) new peaks at lower angles is also the reason for the asymmetry of the $(004)_t$ peak in Fig. 4a.

The Upper Critical Field

As recently shown,[15] the measurement of B_{c2} close to the T_c can be used to detect the presence of tetragonal Nb_3Sn in bulk specimens as well as in multifilamentary wires. From a comparison with the specimen for the specific heat measurement (700°C/19 h; Fig. 2), it is clear that the distribution in our specimen (700°C/10 h; Fig. 7) would be even more off-stoichiometric. Therefore, the $B_{c2}(T)$ measurement shows no change in the slope dB_{c2}/dT around 2 T as observed for bulk[19] and multifilamentary Nb_3Sn[15] close to stoichiometry. With the specific heat measurement in Fig. 2 as a supplementary argument, this is interpreted as the consequence of the absence of the martensitic tetragonal phase at 43 K.[11] The latter is found at Sn contents very close to the stoichiometric composition and thus has a higher T_c value than the cubic phase, as recently shown by Devantay et al.[19c]

Since our x-ray and neutron diffraction data show the presence of stress-induced tetragonal phases only, this indicates that the latter would have a lower T_c than the martensitically transformed tetragonal phase. A more detailed investigation of these effects is under way.

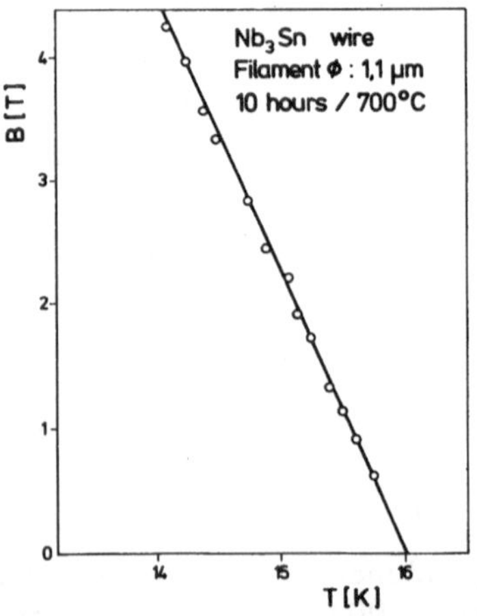

Fig. 7. Upper critical field, B_{c2}, vs temperature; the constant slope dB_{c2}/dT indicates off-stoichiometric composition showing no spontaneous tetragonal transformation.

CONCLUSION

The investigation of a particular multifilamentary Nb_3Sn wire showing no martensitic tetragonal transformation has led to a simplified study of the low temperature structure in Nb_3Sn filaments under the compressive strain exerted by the bronze matrix. At least two phases are present at 4.2 K, both obtained by distortions of the originally cubic A15 lattice. Above ~100 K, a tetragonal distortion with $(1-c/a) > 0$ is observed, corresponding to a uniaxial stress distribution for the crystallites with the (001) orientation parallel to the wire axis.

In the temperature range ~40 $\leq T \leq$ ~100 K, a new type of distortion was detected, which is still unresolved. Since the (222) line width is constant at all temperatures and there are three lines at low temperature, an orthorhombic distortion could be possible. However, more measurements have to be done to determine unambiguously the type of distortion at lower temperatures.

The ratio J_{cm}/J_{co} = 1.44 at 13 T for our wire is comparable to that of wires exhibiting a martensitic tetragonal transformation: the latter thus seems to have little influence on the strain behavior of J_c. The variation of J_c, therefore, has to be attributed to the degree of distortion of the two stress-induced phases, which obviously depends on the applied strain. From these remarks and from the x-ray diffraction data on Nb_3Sn wires with ternary additions,[13,14] it follows that the weaker strain dependence of J_c at fields $B_o > 12$ T for the latter is essentially due to their higher H_{c2} values. This result is confirmed by a recent investigation of Ekin et al.[9] on wires containing Hf+Ga additions.

ACKNOWLEDGMENTS

The authors would like to thank A. Hewat (ILL, Grenoble) for his competent help during the neutron experiments as well as M. Suenaga (Brookhaven National Laboratory) and D. Larbalestier (University of Wisconsin) for fruitful discussions. They are particularly indebted to J. Ekin (NBS, Boulder) for making available and discussing his manuscript prior to publication, to E. Springer (Vacuumschmelze) for preparing the wire, and to B. Schmidt (University of Karlsruhe) for his help in performing x-ray diffraction experiments.

REFERENCES

1. J. W. Ekin, Strain scaling law and the prediction of uniaxial and bending strian effects in multifilamentary superconductors, in: "Filamentary A15 Superconductors," M. Suenaga and A. F. Clark, eds., Plenum Press, New York (1980), p. 187 and references therein.

2. G. Rupp, The importance of being prestressed, in: "Filamentary A15 Superconductors," M. Suenaga and A. F. Clark, eds., Plenum Press, New York (1980), p. 155 and references therein.
3. J. L. Fihey, Appl. Phys. Lett. 34:241 (1979).
4. H. LeHuy, R. Roberge, J. L. Fihey, G. Rupp, and S. Foner, IEEE Trans. Magn. MAG-17:261 (1981).
5. R. Flükiger, Critical currents of Cu-$(Nb_{1-x}Ta_x)_3Sn$ in situ multifilamentary wires, in: "Filamentary A15 Superconductors," M. Suenaga and A. F. Clark, eds., Plenum Press, New York (1980), p. 299.
6. R. Flükiger, R. Akihama, S. Foner, E. J. McNiff, Jr., and B. B. Schwartz, Appl. Phys. Lett. 35:810 (1979).
7. H. C. Freyhardt, R. Bormann, and K. Mroviec, Powder metallurgically prepared A15 microcomposite superconductors, in: "Filamentary A15 Superconductors," M. Suenaga and A. F. Clark, eds., Plenum Press, New York (1980), p. 289.
8. S. E. Cogan, D. S. Holmes, J. D. Klein, and R. M. Rose, Multifilamentary Nb_3Sn by an improved external diffusion method, in: "Filamentary A15 Superconductors," M. Suenaga and A. F. Clark, eds., Plenum Press, New York (1980), p. 91.
9. J. W. Ekin, H. Sekine, and T. Tachikawa, to be published in J. Appl. Phys. 52 (Sept. 1981).
10. S. Foner and E. J. McNiff, Jr., Phys. Lett. A 58:318 (1976).
11. R. Mailfert, B. W. Battermann, and J. J. Hanak, Phys. Lett. A. 24:315 (1969).
12. R. W. Hoard, R. M. Scanlan, G. S. Smith, and C. L. Farrell, IEEE Trans. Magn. MAG-17:364 (1981).
13. R. Roberge, H. LeHuy, and S. Foner, Phys. Lett. A 82:259 (1981).
14. R. Flükiger, KfK Report Number 3140, Institut für Technische Physik, Kernforschungszentrum Karlsruhe, Karlsruhe, FRG (Oct. 1980).
15. R. Flükiger, W. Schauer, W. Specking, B. Schmidt, and E. Springer, IEEE Trans. Magn. MAG-17:2285 (1981).
16. A. Junod, J. Muller, H. Rietschel, and E. Schneider, J. Phys. Chem. Solids 39:317 (1978).
17. A. W. Hewat and A. A. Bailey, Nucl. Instrum. Methods 137:463 (1976).
18. W. Müllner, R. Flükiger, L. Pintschovius, and R. Reichert, to be published.
19. H. Devantay, J. L. Jorda, M. Decroux, J. Muller, and R. Flükiger, to be published in J. Mater. Sci.
20. B. Lachal and A. Junod, Proc. Journées de Calorimétrie et d'Analyse Thermique (June 1980), Barcelona, 9:121 (1980).
21. R. Flükiger, Phase relationships, basic metallurgy, and superconducting properties of Nb_3Sn and related compounds, in: "Advances in Cryogenic Engineering - Materials," Vol. 28, Plenum Press, New York (1982), p. 399.

LOW-TEMPERATURE STRUCTURAL TRANSFORMATIONS IN POWDERED SAMPLES OF Nb_3Sn

H. W. King

Engineering-Physics Department, Dalhousie University, Halifax, Nova Scotia, Canada

D. W. Penfold

International Union of Crystallography, Chester, England

INTRODUCTION

Recent observations of the effects of stress on the properties of superconductors[1] have focussed attention on the low-temperature martensitic phase transformations that occur in β-W compounds such as V_3Si[2] and Nb_3Sn.[3] In both of these alloy systems, the transformation is known to be sensitive to materials characteristics such as composition and the presence of strain. Thus, Mailfert, Batterman, and Hanak,[3] Vieland,[4] and King, Cocks, and Pollock[5] all failed to observe the transformation in samples of Nb_3Sn, prepared by vapour deposition or from reaction with molten Sn, when the room-temperature lattice parameter of the cubic structure was in the range 5.2950 - 5.2960 Å. After these samples were annealed in air or vacuum for long periods at 900 - 1000°C, however, significant amounts of Sn boiled off, reducing the room-temperature lattice parameter to the region of 5.290 Å. On subsequent cooling, the low-temperature phase transformation was observed.[3-5] Vieland, Wickland, and White[4,6] have also shown that the transformation in Nb_3Sn can be suppressed by the addition of H_2 or Al, and it is significant to note that both of these alloying elements increase the lattice parameter of Nb_3Sn. The addition of Sb, which lowers the lattice parameter of Nb_3Sn, is also of interest, because this causes the sign of the tetragonality of the martensite phase to be reversed.[7]

In a previous study,[8] the compositional dependence of the phase transformation was investigated, in terms of the room

*Present address: International Union of Crystallography, Chester, England.

temperature lattice parameter of the alloys, by detecting the start (A_f) and finish (A_s) temperatures of the reverse transformation (i.e., tetragonal → cubic phase). The compositional dependence of the transformation has also been studied by Devantay, Jorda, Decroux, Muller, and Flükiger,[9] who report a much narrower composition range--in fact the low-temperature transformation was only observed in a single alloy. The influence of stress due to differential thermal contractions or to the martensitic phase transformation has also been the subject of recent investigations using multifilamentary Nb_3Sn wires.[10-12] The significance of these latter results relies essentially on the value assumed for the tetragonality of the martensite structure (typically c/a = 0.994), but the compositional dependence, if any, of this parameter has not been specifically investigated. The present paper is thus concerned with determining the compositional dependence of M_s and M_f (the start and finish temperatures of the transformation during slow cooling) and of the axial ratio of the tetragonal martensite phase.

MATERIALS AND METHODS

The Nb_3Sn alloys used in this investigation are the same as those used previously in the study of the reverse transformation.[8]

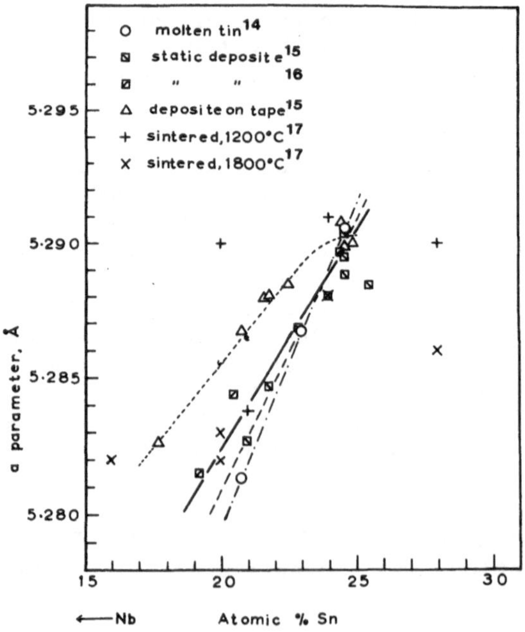

Fig. 1. Room-temperature lattice parameters of Nb_3Sn alloys prepared by different methods.

They were prepared by hot pressing mixtures of predetermined amounts of powdered Nb and Sn at 1350 - 1450°C for periods up to 3 h, followed by a 2-h annealing treatment at 900°C. All alloys were found to be single phase with the β-W structure when examined by x-rays at room temperature (25°C). Since the sintering operation invariably results in some loss of Sn, the room-temperature lattice parameters (listed in Table 1) were used to characterize the alloys, rather than their intended chemical compositions.

As indicated in Fig. 1, the reported lattice parameters for Nb_3Sn alloys fall on distinct series of plots depending on their method of manufacture. To obtain nominal compositions for the present alloys, the heavy line plot in this figure has been used, since this refers to sintered alloys, but no attempt was made to derive a nominal composition for alloy 10, which has a room-temperature lattice parameter of 5.2961 Å, because this lies outside the accepted range of the solid solution for Nb_3Sn.[13] The equation for the plot is given in Table 1.

The diffraction patterns of the alloys were recorded at low temperatures using an Oxford Instruments CF 100 continuous-flow cryostat, specially modified to incorporate the alignment device described by King and Preece.[18] This device can be brought into the path of the x-rays at any desired temperature so that any loss

Table 1. Room-Temperature Lattice Parameters and Low-Temperature Martensitic Transformations in Nb_3Sn Alloys

Alloy	a_{RT}, Å	M_s, K	M_f, K	c/a	at.% Sn*
1	5.2827	-	-	-	20.2
2	5.2858	-	-	-	22.2
3	5.2865	18	10	1.0049	22.6
4	5.2868	23	16	1.0041	22.8
5	5.2874	45	34	1.0049	23.2
6	5.2883	53	50	1.0041	23.7
7	5.2909	60	56	0.9963	25.3
8	5.2910	54	46	0.9967	25.4
9	5.2912	41	36	0.9964	25.5
10	5.2961	-	-	-	?

*Nominal compositions were derived using the following equation, which refers to the solid line plot in Fig. 1: at.% Sn = 17.5 + 621 (a_{RT} - 5.2783).

of resolution due to misalignment can be corrected at any time during the cooling experiment. The cryostat was mounted on a Siemens x-ray diffractometer, and diffraction profiles were recorded graphically using CuK_α radiation. The temperature of the sample in the cryostat was measured to ±1 K, using a resistance thermometer calibrated at 4.2 and 77 K.

RESULTS

The cubic → tetragonal phase transformation was investigated by arresting the cooling through the transformation-temperature region. The M_s temperature was detected by the first appearance of profile broadening or additional peaks; the M_f temperature was associated with no further change in the diffraction pattern. It is important to note that although the latter represents the temperature of the finish of the transformation, it does not mean that the transformation has necessarily proceeded to 100%.

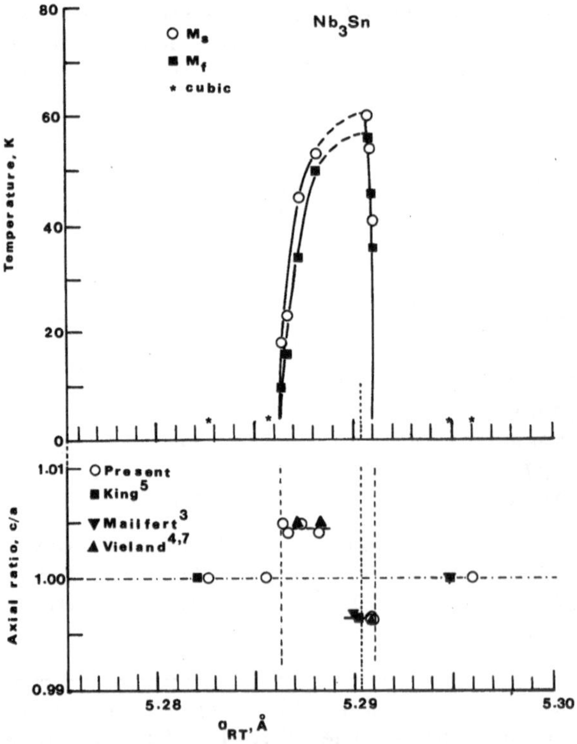

Fig. 2. Martensitic transformation temperatures, M_s and M_f, and axial ratios of tetragonal phases plotted as functions of the room-temperature lattice parameters for Nb_3Sn alloys.

The M_s and M_f temperatures observed for the various alloys are recorded in Table 1 and plotted against respective room-temperature lattice parameters in Fig. 2. No thermal hysteresis was observed to within the accuracy measurements (±1 K) and hence these M_s and M_f values can be identified with the A_f and A_s temperatures observed during the previous warm-up experiments.[8] In general, the M_s temperatures are found to lie above the respective A_f temperatures, since it is easier to observe the first trace of a lack of resolution in a profile rather than the point of disappearance of the distortion. The present results also confirm that the low-temperature martensitic transformation is only observed in alloys with room-temperature lattice parameters that lie between 5.2865 and 5.2912 Å (~ 22.5 - 25.5 at.% Sn).

The forms of the peak splitting observed at temperatures below M_s are illustrated by the diffraction profiles in Fig. 3. The low-temperature diffraction profiles for alloys 7 - 9 (Fig. 3A) were of the expected form for partial transformation to a tetragonal phase

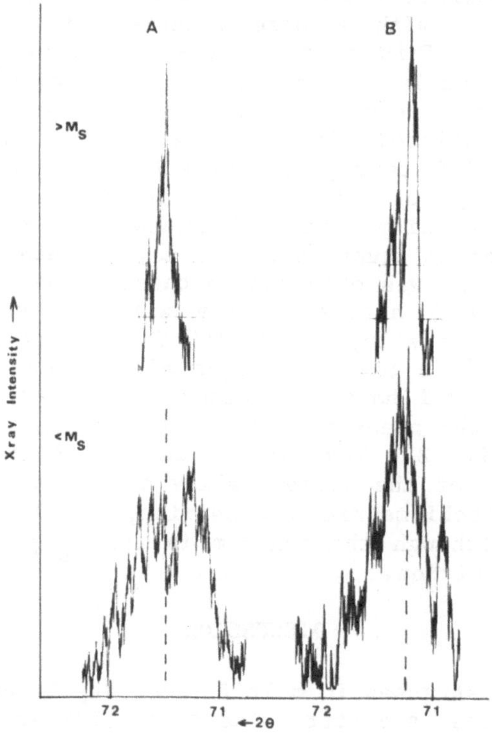

Fig. 3. X-ray diffraction profiles for {400} reflections from Nb_3Sn alloys at temperatures above and below the M_s. A = Alloy 8. B = Alloy 3.

with c/a less than unity. The profile is generally broadened compared with that obtained above M_s, and the increase in intensity on the low-angle side of the cubic peak is to be associated with the tetragonal 400,040 peak, whereas the smaller additional peak on the high-angle side refers to the 004 tetragonal peak. The axial ratio of the tetragonal phase in these alloys was determined from the splitting of the diffraction profiles using the equation:

$$c/a = 1 + \frac{\Delta\Theta}{\tan\Theta} \quad (1)$$

where $\Delta\Theta$ is the degree of splitting, Θ is the Bragg angle of the cubic profile, and the sign of $\Delta\Theta$ is taken to be negative from the direction of the peak splitting referred to above. As indicated in Table 1, the axial ratios of these alloys exhibit no distinct trend with composition, and the average value of c/a = 0.9965 agrees well with previously reported c/a values for the tetragonal phase in Nb_3Sn,[3-5] as shown by the data plotted in Fig. 2.

A different form of diffraction profile was recorded at temperatures below M_s in alloys 3 - 6, as shown in Fig. 3B. Small, but distinct, satellite peaks were observed on the low-angle side of the cubic peak, with a more intense satellite peak on the higher-angle side. This form of peak splitting is identical to that reported previously by Vieland[7] for Nb_3Sn alloys doped with Sb and, following his interpretation, the small low-angle satellite peak is to be associated with the 004 and the larger high-angle peak with the 400,040 tetragonal reflections. This means that in these alloys the sign of the tetragonality is reversed so that c/a values greater than unity were derived when the measured values of $\Delta\Theta$ were substituted in equation 1. As may be seen from Table 1, no definite trend in c/a was observed in these alloys, but the average value of c/a = 1.0045 agrees closely with the value reported by Vieland[7] for an Nb_3Sn alloy doped with Sb. It is also significant to note that Vieland's Sb-doped Nb_3Sn alloys that have c/a values greater than unity all have room-temperature lattice parameters in the same range as the present alloys 3 - 6, as may be seen from the data plotted in Fig. 2. The results in Fig. 2 also indicate that the change of sign of the tetragonality of the martensite does not occur at the stiochiometric composition, which is indicated by a_{RT} = 5.2904 Å, although the maximum in the M_s plot does occur at or near this composition.

DISCUSSION

The residual stresses in multifilamentary Nb_3Sn wires prepared by the bronze process are attributed to the larger thermal contraction of the bronze matrix with respect to that of the Nb_3Sn filaments on cooling from the reaction temperature to 4.2 K.[9,20] If the c-axis of the tetragonal phase is aligned along the filament axis,[12] the resultant stress within the material will then depend on whether this thermal strain reinforces or opposes the strain associated with the low-temperature martensitic phase transformation. The present axial ratio results indicate that for alloys with room-temperature lattice parameters between $5.2909 \leq a_{RT} \leq 5.2912$ Å, the thermal strains and the transformation strains will act in the same direction and the residual stresses will thus be increased. This is the situation normally envisaged[10-12] and would apply to alloys close to the stoichiometric composition, as shown in Fig. 3.

If the c-axis of the tetragonal phase is again assumed to be aligned with the filament axis,[12] alloys with room-temperature lattice parameters between $5.2865 \leq a_{RT} \leq 5.2993$ Å should be in a lower-stressed state, because the transformation strains will now tend to offset the thermal contraction strains. Under these conditions, the tetragonality of the martensite phase would be reduced from 1.0045 to a value closer to unity, or even reduced below unity. In theory, it should thus be possible to formulate an Nb_3Sn-based alloy that has a room-temperature lattice parameter such that multifilamentary wires would be in a stress-free condition at 4.2 K since this is conducive to a maximum critical current density.[19,20] Although this may not be feasible with the normal bronze process, which yields alloys with compositions close to stoichiometry, it may be achievable by the addition of solutes such as Sb (see Fig. 3) that are known to reduce a_{RT} and induce martensites with axial ratios greater than unity.

ACKNOWLEDGMENTS

This work was supported in part by the Natural Sciences and Engineering Research Council of Canada and in part by the United Kingdom Atomic Energy Authority. The authors are grateful to Drs. C. J. Gillham and D. C. Larbalestier for helpful discussions and to the Oxford Instrument Company for the loan of the CF 100 cryostat used in this work.

REFERENCES

1. C. C. Koch and D. S. Easton, Cryogenics 17:391 (1977).
2. B. W. Batterman and C. S. Barrett, Phys. Rev. 145:296 (1965).
3. R. Mailfert, B. W. Batterman, and J. J. Hanak, Phys. Lett. 24A:315 (1967).
4. L. J. Vieland, Phys. Lett. 34A:43 (1971).
5. H. W. King, F. H. Cocks, and J. T. A. Pollock, Phys. Lett. 26A:77 (1967).
6. L. J. Vieland, A. W. Wickland, and J. G. White, Phys. Rev. 11:3311 (1975).
7. L. J. Vieland, J. Phys. Chem. Solids 31:1449 (1970).

8. H. W. King, in: "The Mechanism of Phase Transformations in Crystalline Solids," Monograph No. 33, Institute of Metals, London (1968), p. 196.
9. H. Devantay, T. L. Jorda, M. Decroux, J. Muller, and R. Flükiger, J. Mater. Sci., in press.
10. R. Roberge, H. LeHuy, and S. Foner, Phys. Lett. 82A:259 (1981).
11. R. W. Hoard, R. M. Scanlan, G. S. Smith, and C. L. Farrell, IEEE Trans. Magn. MAG-17:364 (1981).
12. R. Flükiger, W. Schauer, W. Specking, B. Schmidt, and E. Springer, IEEE Trans. Magn. MAG-17 (1981), in press.
13. J. P. Charlesworth, I. MacPhail, and P. E. Madsen, J. Mater. Sci. 5:580 (1979).
14. L. J. Vieland, R.C.A. Rev. 25:366 (1964).
15. J. J. Hanak, K. Streter, and G. W. Cullen, R.C.A. Rev. 25:342 (1964).
16. H. Pfister, Z. Naturforsch. 20A:1059 (1965).
17. T. B. Reed, H. C. Gatos, W. J. LeFleur, and J. T. Roddy, in: "Metallurgy of Advanced Electronic Materials," G. E. Brock, ed., Interscience Publishers, New York (1963), p. 171.
18. H. W. King and C. M. Preece, Siemens Rev. (special issue on X-rays and Electron Microscopy) 35:46 (1968).
19. G. Rupp, IEEE Trans. Magn. MAG-13:1565 (1977).
20. J. W. Ekin, IEEE Trans. Magn. MAG-15:197 (1979).

SUPERCONDUCTING PROPERTIES OF $(Nb,Ti)_3Sn$ WIRES FABRICATED BY THE BRONZE PROCESS*

M. Suenaga, S. Okuda,† R. Sabatini, K. Itoh,‡ and T. S. Luhman

Division of Metallurgy and Materials Science, Brookhaven National Laboratory, Upton, New York

INTRODUCTION

Because of the interest in the use of multifilamentary superconducting Nb_3Sn composites for magnets to confine magnetically the plasma in fusion reactors, fabrication techniques as well as superconducting properties of multifilamentary Nb_3Sn wires have been extensively examined.[1,2] As pointed out earlier, the use of pure Nb_3Sn wires as currently produced appears to be limited for magnetic fields of 12 T or below. However, recent designs for future Mirror[3] and Tokamak[4] types of fusion reactors call for magnets operating at magnetic fields of 12 T or higher. Thus, it is necessary to improve the current capacity of the wires to meet these future demands.

In the past, various attempts were made to improve the critical-current density, J_c, of Nb_3Sn at high fields by alloying the core or the matrix.[1,2] Among these attempts, only the additions of Ta[5,6] to the Nb core or Hf to the core in conjunction with the addition of Ga to the matrix[7] have made significant improvements in the critical-current density at high magnetic fields. However, the cost of Ta is substantially higher than Nb, and the increased J_c with alloying Nb with Hf and the matrix with Ga appears to be only

*Work performed under the auspices of the U.S. Department of Energy under Contract No. DE-AC02-76CH00016.

†A visiting scientist from Sumitomo Electric Industry, Osaka, Tokyo, Japan.

‡Present address: National Research Institute for Metals, Ibaraki, Japan.

marginal when the current density for the overall conductor rather than Nb_3Sn itself is considered. Thus, it was felt that some other means of improving J_c at high magnetic fields should be sought. We have examined the influence of the alloying addition of Ti in the Nb core on the superconducting critical properties since (1) Ti is less expensive than Nb or Ta, (2) Ti has a lower melting temperature than Ta and Nb, thus making the alloy fabrication easier, (3) Ti-Nb alloys are also known to be easily fabricated in composite wires with Cu, and finally (4) it appears that an increased normal-state resistivity, ρ_n, and therefore the upper critical field, H_{c2}, will lead to enhanced J_c at high fields. It is expected that the addition of Ti will increase the ρ_n of the Nb_3Sn. As shown in the preliminary results,[1,8] small additions of Ti to the Nb can significantly enhance critical-current densities of Nb_3Sn at high magnetic fields (H > 12 T). In this article, a detailed account of this study is given.

SPECIMEN PREPARATION

Six monofilamentary composite wires were fabricated with a matrix and a core having nominal compositions of Cu-13 wt.% Sn and Nb-0, 0.5, 1, 2, 3, or 5 wt.% Ti, respectively. All Nb-Ti alloys were arc melted in an Ar atmosphere, homogenized for 1.5 h at 1800°C, and swaged to rods of a 6.35-mm diameter. After these alloys were annealed at 1100°C for 1 h under vacuum ($\sim 10^{-6}$ mm Hg), they were inserted into a bronze matrix of a 12.7-mm diameter. Then, the composite rods were cold-drawn to 0.635-mm wires using several intermediate annealing steps at ~500°C. These composite wires were then encapsulated in quartz tubes and heat-treated at 725°C for periods of 32, 64, and 120 h.

For determination of the thickness and the cross-sectional areas of the compound layers, optical micrographs at 150 and 1000 times magnifications were taken at the center of each 38-mm-long wire, which was tested for critical current, I_c. The areas of the compound layers were determined by multiplying the perimeter of the core (measured using an opisometer) by the thickness of the layers. The thickness was determined with a magnifier scale on the micrographs, which were taken at a magnification of 1000.

Growth of the layer thickness as a function of duration of heat treatment, t, is shown in Fig. 1. The growth for all compounds was faster than $t^{1/2}$. (If the mechanism for the growth was controlled by the bulk diffusion of either Nb or Sn, or both atoms, simultaneously through the compound, then $t^{1/2}$ growth is expected.) In the thick layers (64 h and 120 h), several cracks in the compound running radially through the layer were observed in the optical micrographs. These cracks could presumably cause faster diffusion of Sn atoms through the compound layer, short circuiting the bulk or grain-boundary diffusion and resulting in a growth of

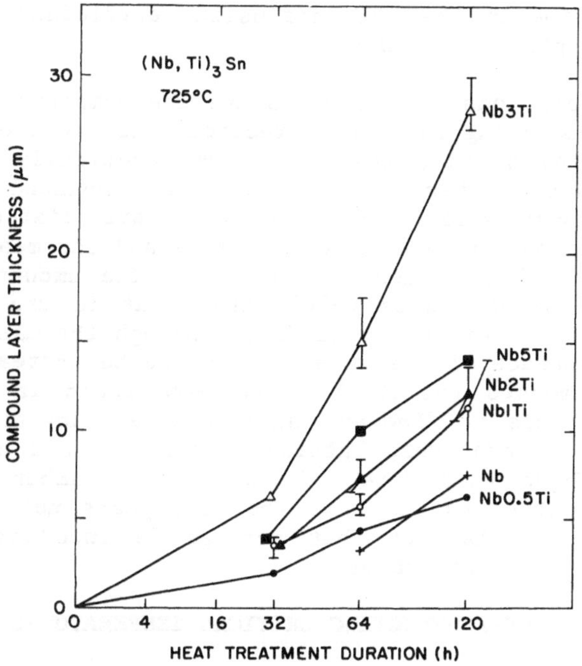

Fig. 1. The growth of $(Nb,Ti)_3Sn$ compound layers as a function of heat-treatment duration at 725°C.

the compound faster than $t^{1/2}$.[2] In addition, as shown in Fig. 1, alloying the cores with Ti also monotonically increased the growth rate of the compound with increasing alloying contents except for the Nb-5 wt.% Ti core. In this case, a compound layer between Nb_3Sn and the matrix, presumably a Sn-Ti compound, formed and slowed the growth of the A15 compound layer. Interestingly, the greater rate of the layer growth with increased Ti content was observed while the grain sizes of the compound also increased with the addition. For the wires that were heated for 64 h at 725°C, the grain size, as measured by scanning electron microscopy of the fractured surfaces of Nb_3Sn, varied from ~0.15 μm for the unalloyed Nb_3Sn to ~0.3 μm for $(Nb,Ti)_3Sn$ from the Nb-5 wt.% Ti core. This result is totally contradictory to the earlier observations on the relationship between the grain size and the growth rate and to the proposed mechanism for the growth of the compound.[2,9] The growth of the compound was previously thought to be by the grain-boundary diffusion mechanism, and the previously observed growth rates confirmed such a concept. Thus, it appears that the addition of Ti may have enhanced the bulk diffusion rate and thereby the growth in the grain size with the additions. However, because these measurements were made on cylindrical specimens, the growth rate did not follow a parabolic time dependence due to the crack formation. The assumed increase in the bulk diffusion coefficient has to be re-examined with planar specimens.

The compositions of the cores and the compounds were measured by a microprobe using the ZAF correction. To avoid erroneous x-ray counts from the matrix, the matrix was chemically etched with a H_2O-HNO_3 solution after the specimens were mounted and polished. The nominal compositions of the cores are listed in Table 1 together with their measured compositions and the measured composition of the $(Nb,Ti)_3Sn$ compound layers. The amount of Ti in the compound was somewhat less (~70%) than that in the core. It was believed that some of the Ti diffused through the matrix and formed TiO_2 at the surface of the wires during the heat-treatment process. Significant amounts (~1 at.%) of Cu were found in all cases, as presented in Table 1. However, an extensive study of superconducting critical temperatures, lattice parameters, and chemical compositions of Nb_3Sn that were alloyed with a number of transition elements indicates that most of the Cu appears not to be incorporated in the A15 lattice, but perhaps is incorporated in grain boundaries or as precipitates.[9]

SUPERCONDUCTING CRITICAL TEMPERATURES

After each heat treatment, two ~6-mm-long pieces of the wires were cut and were used for the measurements of the critical temperature, T_c. An ac technique at ~220 Hz was used to observe the superconducting transition, and the temperature was measured using a calibrated Ge thermometer. Accuracy of the measurement is ±0.1 K. Midpoint critical temperatures of these wires as heat-treated and after chemically removing the matrix are shown in Fig. 2 as a function of alloying content of Ti in the compounds. As was reported earlier,[2] the T_c of the compound generally decreased with increasing Ti content in the compound. However, as

Table 1. Compositions of the Cores and the Compound Layers

Nominal Core Compositions	Compositions of the Core (Ti)		Compositions of the Compounds (at.%)			
(wt.%)	(wt.%)	(at.%)	Nb	Ti	Sn	Cu
Nb	-	-	74.3	-	24.2	1.5
Nb-0.5Ti	0.7	1.3	74.5	0.65	23.2	1.6
Nb-1Ti	1.2	2.2	75.0	1.2	22.2	1.6
Nb-2Ti	1.7	3.3	74.8	2.0	21.3	1.9
Nb-3Ti	3.4	6.4	73.9	3.2	21.8	1.1
Nb-5Ti	4.4	8.2	74.0	4.6	20.3	1.1

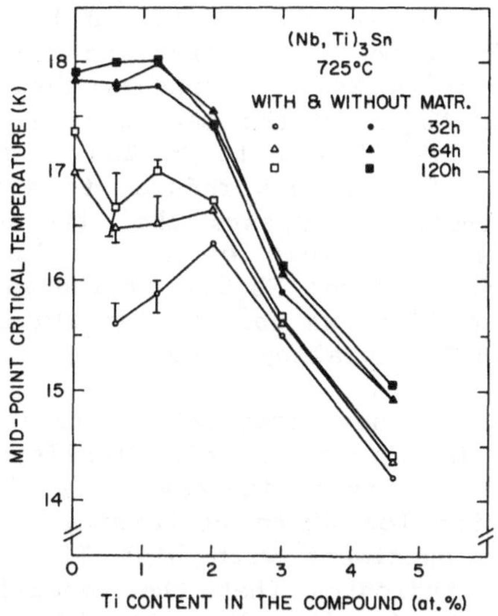

Fig. 2. The critical temperature of Ti-alloyed Nb_3Sn as a function of Ti content in the compound layer for the wires with and without the bronze matrix.

shown in Fig. 2, the details of the T_c variations with Ti content in the layers for T_c measured with or without the bronze matrix were significantly different. The T_c, which was measured as prepared, exhibited a minimum at 0.65 at.% Ti in the layer, and then increased slightly upward for 2.0 at.% Ti before it decreased drastically. On the other hand, T_c of the same specimens after the matrix was removed increased slightly (or stayed essentially the same depending on the heat-treating conditions) with the Ti content up to 1.2 at.% Ti before decreasing on further addition of Ti. This slight increase in T_c can be attributed to partial or total arrest of the martensitic phase transformation of Nb_3Sn at low temperatures (~45 K) by the alloying. Earlier, a similar observation was made that small additions of Al to Nb_3Sn resulted in a small increase in T_c as well as arrestment of the transformation.[10] Furthermore, the measurements of internal friction in similarly prepared specimens of Ti-alloyed Nb_3Sn indicated that the compound containing more than 1.2 at.% Ti retained the cubic phase down to ~7 K,[11] supporting the above possibility.

In discussing the dip in T_c with the increasing Ti content when the matrix was still intact, we note two earlier observations on the difference in T_c of these wires with and without the matrix: (1) Disordering Nb_3Sn by irradiation with high-energy neutrons resulted in the increased $\Delta T_c = [T_c(\text{without})-T_c(\text{with})]$[12] and (2) ΔT_c also decreased with the increased heat-treating duration.[13]

These results were interpreted as evidence for an enhanced sensitivity of T_c of Nb_3Sn to the interaction between nonhydrostatic strain and disorder in Nb_3Sn. It appears that the sensitivity of T_c to the strain on Nb_3Sn by the matrix is similarly enhanced by the small additions of Ti (up to ~1.2%). However, on further increase of Ti in Nb_3Sn, ΔT_c was reduced to a value below ΔT_c for the pure Nb_3Sn. Thus, it appears that the dip in T_c with the matrix still on is the result of the variation in the strain sensitivity of T_c for alloyed Nb_3Sn. Although the extent of the dip in T_c was much smaller, an observation similar to the one above was made for the $(Nb,Ta)_3Sn$ alloy system.[14]

The influence of heat-treatment duration on T_c measured without the matrix is slight for all Ti-alloyed Nb_3Sn specimens. However, T_c with the matrix on increased significantly with longer heat-treating duration for Nb_3Sn containing less than 2 at.% Ti, while T_c for those specimens with high Ti contents varied very little. Again, it indicates that the sensitivity of T_c to the strain increases with the disorder (in this case, the thermally induced disorder) when the Ti content is limited to 2 at.%.

SUPERCONDUCTING CRITICAL CURRENTS AND MAGNETIC FIELDS

Superconducting critical currents, I_c, for these wires were measured as a function of applied magnetic fields up to 19 T for most cases and preliminary measurements of I_c up to 25 T were measured for a limited number of the specimens. For these measurements of I_c, ~38-mm-long wires were used. The voltage leads were ~15 mm apart, and 1 μV was used as the criterion for the definition of the critical current. Critical-current densities were determined by dividing critical currents by the cross-sectional areas of the compounds for each wire. Examples of variations in the critical-current densities with applied magnetic fields are shown in Fig. 3 for the wires with the Nb-1 wt.% Ti core. Three heat-treating periods (32, 64, and 120 h) are shown. The critical-current densities at high fields ($H > 12$ T) for all wires tend to increase with heat-treating periods, although J_c for Nb_3Sn at low fields usually decreases with the increased heat-treating time.

Another and perhaps a more illustrative way of presenting the influence of heat-treatment time or alloying effects on the critical-current densities of these wires is to plot the values of $J_c^{1/2}H^{1/2}$ as a function of H. Using this method of comparing J_c for different wires, the influence of alloying Nb_3Sn with Ti is summarized in Fig. 4 for those wires that were heat-treated at 725°C for 64 h. As shown in the figure, the addition of Ti reduced J_c at lower fields and increased it at very high fields ($H > 13$ T). For the wires heat-treated for a shorter period (32 h), the values of J_c were significantly lower than those that are shown in Fig. 4. Perhaps this is due to the lower T_c in these wires. However, the

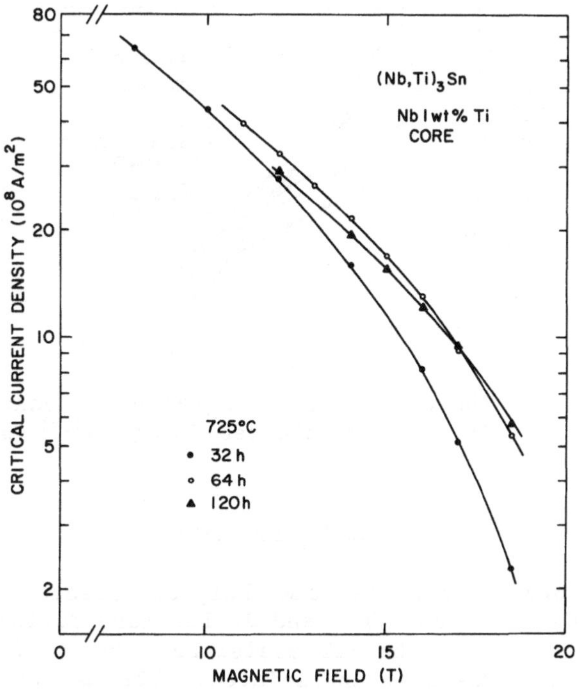

Fig. 3. The dependence of $J_c(H)$ on heat-treatment duration for $(Nb,Ti)_3Sn$ from a Nb-1 wt.% Ti core.

values of J_c for the wires heated for 120 h at 725°C were essentially the same as the wires heated for 64 h.

When the wires were heat-treated at 725°C for 64 h, it was found that J_c at 16 T increased to a maximum value of 1250 A/mm^2 at 1.2 at.% Ti in the compound and decreased on further increases in Ti in the layer. However, at 10 T, J_c decreased monotonically with increasing Ti contents. This decrease in J_c is primarily due to the increased grain size of the alloyed Nb_3Sn. The peak in J_c(16 T) as a function of the Ti content is due to two competing effects on J_c of Nb_3Sn by alloying: J_c at high fields is enhanced by an increased H_{c2} due to alloying, whereas J_c is reduced by the increased grain size with increasing Ti.

Since the plots of $J_c^{1/2}H^{1/2}$ vs H, as shown in Fig. 4, were not straight lines for the wires with Ti additions, it was difficult to extrapolate H_{c2} from these curves. However H_{c2} appears to increase with the additions up to 1 wt.% Ti in the core and to decrease upon further additions. The preliminary measurements of J_c up to 25 T for these wires indicate that H_{c2} could exceed 25 T.

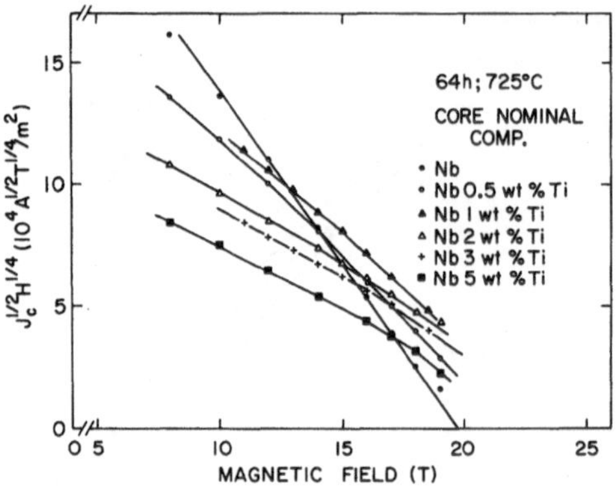

Fig. 4. The dependence of $J_c(H)$ on Ti content in the core plotted in $J_c^{1/2}H^{1/2}$ vs H.

ACKNOWLEDGMENTS

The authors appreciate the helpful discussions with Drs. D. O. Welch, C. L. Snead, Jr., and J. Bussiere during the course of this investigation. Technical assistances by F. Iseli, D. Horne, and A. Cendrowski are also appreciated. The use of the high-field magnets at the National Magnet Laboratory and very helpful assistance by its staff are greatly appreciated.

REFERENCES

1. M. Suenaga, W. B. Sampson, and T. S. Luhman, IEEE Trans. Magn. MAG-17:646 (1981).
2. M. Suenaga, to be published in: "Proceedings of NATO Advanced Study Institute on Superconducting Materials," Sintra, Portugal (1980).
3. "Mirror Fusion Quarterly Report," UCAR-10060-80-3 (July-September, 1980).
4. B. Coppi and A. Taroni, in: "Theory of Magnetically Confined Plasmas," B. Coppi et al., eds., Pergamon Press, New York (1979).
5. J. D. Livingston, IEEE Trans. Magn. MAG-14:611 (1978).
6. M. Suenaga, K. Aihara, K. Kaiho, and T. S. Luhman, Superconducting properties of $(Nb,Ta)_3Sn$ wires fabricated by the bronze process, in: "Advances in Cryogenic Engineering - Materials," Vol. 26, Plenum Press, New York (1980), p. 442.
7. H. Sekine and K. Tachikawa, Appl. Phys. Lett. 35:472 (1979).
8. A. I. Braginski, H. C. Freyhardt, R. Flükiger, H. Hillmann, D. C. Larbalestier, M. Suenaga, and M. N. Wilson, IEEE Trans. Magn. MAG-17(5):2343 (1981).
9. Unpublished work.
10. L. S. Vieland and A. W. Wicklund, Phys. Lett. 34A:43 (1971).

11. J. F. Bussière, B. Faucher, C. L. Snead, Jr., and M. Suenaga, Effects of ternary additions on Young's modulus and the martensitic transformation of Nb_3Sn, in: "Advances in Cryogenic Engineering - Materials," Vol. 28, Plenum Press, New York (1982), p. 453.
12. C. L. Snead, Jr. and M. Suenaga, Appl. Phys. Lett. 36:474 (1980).
13. K. Aihara, M. Suenaga, and T. S. Luhman, in: "Proceedings of the 8th Symposium on Engineering Problems of Fusion Research," IEEE Pub. No. 79H1441-5 NPS, IEEE, New York (1979), p. 236.
14. K. Aihara, M. Suenaga, and D. O. Welch, Bull. Am. Phys. Soc. 25:385 (1980).

EFFECTS OF THE IVa ELEMENT ADDITIONS ON COMPOSITE-PROCESSED Nb_3Sn

K. Tachikawa, T. Takeuchi, T. Asano, Y. Iijima, and H. Sekine

National Research Institute for Metals, Ibaraki, Japan

INTRODUCTION

Composite-processed multifilamentary Nb_3Sn conductors are being developed for high-field applications, such as fusion reactor magnets. However, the improvement of critical current density, J_c, of the composite-processed Nb_3Sn in high magnetic fields is required since the J_c decreases rapidly in fields above 12 T. Therefore, numerous studies have been made aiming at the improvement of J_c in high fields of the composite-processed Nb_3Sn through alloying additions to the niobium core, or to the matrix, or to both.[1-9]

It has been reported that the addition of titanium,[8,9] zirconium,[1,6] or hafnium[6,7] (all belonging to the IVa periodic group) to the niobium core enhances the Nb_3Sn growth rate, the critical temperature, T_c, the upper critical field, H_{c2}, and the J_c in high fields of the composite-processed Nb_3Sn. The addition of gallium to the matrix decreases the Nb_3Sn growth rate and the J_c in fields lower than 12 T in compensation for the increases in H_{c2} and J_c in high fields.[3,5] The simultaneous addition of hafnium to the core and gallium to the matrix significantly improves the H_{c2} and J_c in high fields.[6,7] In this paper, results recently achieved in the National Research Institute for Metals on the effects of IVa element addition to the composite-processed Nb_3Sn with Cu-Sn or Cu-Sn-Ga matrix have been summarized. Titanium has been also added to the matrix.

EXPERIMENTAL PROCEDURES

The bar-shaped ingots of pure niobium and niobium alloys containing 1-10 at.% Ti, Zr, or Hf were prepared by an arc melting. After the annealing at 1000°C for 1 h, the ingots were rolled and swaged into core rods. Cu-7at.%Sn, Cu-7at.%Sn-2at.%Ti, Cu-5at.%Sn-4at.%Ga, and Cu-3at.%Sn-9at.%Ga matrix alloys were melted in a graphite crucible under an argon atmosphere and cast into ingots. The matrix alloy ingots were machined to encase the core therein. The resulting composites were fabricated by grooved rolling and drawing with intermediate annealings at 600°C into single-core tapes ($0.25^t \times 5^w$mm, core width being 2 mm), single-core wires (0.5 mmϕ, core diam. being 0.20 mm), 19-core wires (0.5 mmϕ, core diam. being 0.05 mm) and 49-core wires (0.40 mmϕ, core diam. being 0.03 mm). Specimens cut from the tape and the wire were heat-treated under an argon atmosphere at temperatures ranging from 700°C to 800°C. The T_c was measured by a resistive method and defined as the midpoint of the transition. The H_{c2} was measured at 4.2 K and a current density of 100 A/cm^2 in pulsed magnetic fields whose rise time was 10 ms and defined as the midpoint of the transition. The critical current, I_c, at 4.2 K in fields up to 17 T was measured by a V_3Ga/Nb_3Sn superconducting magnet, and that in fields from 17 to 22 T was measured by a water-cooled copper magnet at the Francis Bitter National Magnet Laboratory of MIT. I_c was defined as the current required to induce 1 μV across a 10-mm-length of the specimen. J_c was determined by dividing I_c by the cross-sectional area of the Nb_3Sn layer. The microstructure of the Nb_3Sn layer was examined by scanning electron microscopy on the fractured surface of the specimen. Compositions of the Nb_3Sn layer were determined by an x-ray microanalyzer (XMA) using a ZAF correction.[10-12]

RESULTS AND DISCUSSION

Metallurgical Aspects of the Nb_3Sn Layer

Figure 1 shows Nb_3Sn layer thickness versus heat treatment time at 800°C for single-core composite tapes with 2 at.% IVa element addition to the niobium core and with Cu-7at.%Sn or Cu-5at.%Sn-4at.%Ga alloy matrix. The 2 at.% IVa element addition to the core increases the Nb_3Sn growth rate to about twice that of the composite with a pure niobium core. The partial substitution of gallium for tin in the matrix considerably decreases the growth rate of Nb_3Sn. However, the addition of IVa element to the niobium core can compensate for this decrease, as demonstrated in Fig. 1. The titanium addition to the matrix also increases the growth rate of Nb_3Sn as in the case of that to the niobium core; the thickness of the Nb_3Sn layer formed in the single-core Nb/Cu-7at.%Sn-2at.%Ti composite wire heat-treated at 750°C is about the same as that formed in the Nb-5at.%Ti/Cu-7at.%Sn composite wire.

The results of the XMA analysis on the Nb_3Sn layer formed in the single-core composite tapes with different core and matrix compositions are shown in Table 1. The concentration of the IVa

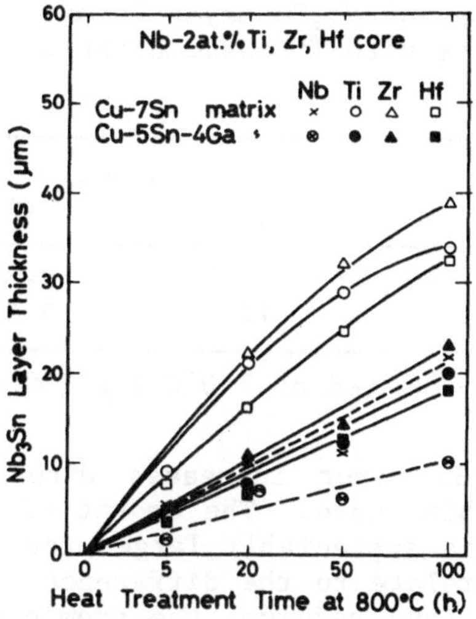

Fig. 1. Nb_3Sn layer thickness versus heat-treatment time at 800°C for single-core composite tapes with 2 at.% IVa element addition to the niobium core. The matrix is Cu-7at.%Sn or Cu-5at.%Sn-4at.%Ga alloy.

Table 1. Compositions of Nb_3Sn Layer Formed in Single-Core Composite Tapes with Different Core and Matrix.*

Composition of core/matrix (at.%)	Composition of Nb_3Sn (at.%) Nb	Ti,Zr,Hf	Sn	Cu	Ga
Nb/Cu-7Sn	77.0	-	22.5	0.5	-
Nb-2Ti/Cu-7Sn	76.0	1.4	22.0	0.6	-
Nb-2Zr/Cu-7Sn	77.1	0.3	22.0	0.6	-
Nb-2Hf/Cu-7Sn	77.3	0.6	21.5	0.6	-
Nb-5Ti/Cu-7Sn	74.0	3.9	21.6	0.5	-
Nb-5Hf/Cu-7Sn	76.8	0.8	21.5	0.9	-
Nb-10Ti/Cu-7Sn	71.5	7.6	20.1	0.8	-
Nb-10Hf/Cu-7Sn	75.7	1.8	21.4	1.1	-
Nb/Cu-5Sn-4Ga	74.8	-	23.5	0.6	1.1
Nb-2Ti/Cu-5Sn-4Ga	73.4	1.3	23.4	0.9	1.2
Nb-2Zr/Cu-5Sn-4Ga	74.7	0.2	22.9	0.8	1.4
Nb-2Hf/Cu-5Sn-4Ga	74.6	0.6	22.8	0.8	1.2
Nb-5Ti/Cu-5Sn-4Ga	71.8	3.0	23.3	0.8	1.1
Nb-5Hf/Cu-5Sn-4Ga	74.2	0.9	22.7	0.9	1.3

* The tapes were heat-treated at 800°C for 50 h.

Table 2. Compositions of Nb_3Sn Layer Formed in Single-Core Composite Wires with Titanium Addition to the Core or to the Matrix.*

Composition of core/matrix (at.%)	Composition of Nb_3Sn (at.%) Nb	Ti	Sn	Cu
Nb-4Ti/Cu-7Sn	74.0	2.3	22.9	0.8
Nb/Cu-7Sn-2Ti	72.7	3.6	22.8	0.9

*The wires were heat-treated at 750°C for 100 h.

element in the Nb_3Sn layer increases with increasing amount of addition to the niobium core. The amount of titanium incorporated in the Nb_3Sn layer is appreciably larger than that of zirconium or hafnium. This may relate to the difference of atomic radius among titanium, zirconium, and hafnium; the atomic radius of titanium is slightly smaller than that of niobium, whereas the atomic radii of zirconium and hafnium are considerably larger than that of niobium.[13] The titanium concentration dissolved in the Nb_3Sn layer increases with increasing reaction temperature. The gallium substituted for tin in the matrix is also partly incorporated into the Nb_3Sn layer. The gallium addition to the matrix slightly raises the tin concentration in the Nb_3Sn layer.

Table 2 indicates the composition of the Nb_3Sn layer formed in the single-core composite wire with titanium addition to the core or to the matrix. A larger amount of titanium is incorporated into the Nb_3Sn from the matrix than from the core. The diffusion of titanium in the Cu-Sn matrix is probably much faster than that in the niobium core. Moreover, the quantity of titanium contained in the matrix is larger than that contained in the core taking into account the bronze ratio of the wire.

The addition of more than 5 at.% zirconium to the niobium core results in the precipitation of a second phase richer in zirconium in the Nb_3Sn layer. This may relate to the small solubility of zirconium to the Nb_3Sn, indicated in Table 1. In the composite with a Cu-3at.%Sn-9at.%Ga matrix, the compound layers richer in gallium are formed on both sides of the Nb_3Sn layer.

Scanning electron microscopy on the fractured surface of the specimen reveals that the grain size of Nb_3Sn depends on both the reaction temperature and the amount of IVa element addition to the niobium core. At reaction temperatures of 700–750°C, the grain size of Nb_3Sn is not appreciably changed by the IVa element addition. On the other hand, for the composite reacted at 800°C the IVa element addition causes much difference. The pure Nb_3Sn grains show a significant coarsening, while the IVa element addition prevents the grain coarsening and produces Nb_3Sn grains appreciably finer than pure Nb_3Sn. The addition of gallium to the matrix is found to cause grain coarsening and reduce the shape anisotropy of Nb_3Sn grains.

Superconducting Properties

Figure 2 shows T_c and H_{c2} (4.2 K) of single-core wires heat-treated at 700-800°C as a function of the titanium concentration in the niobium core. The T_c and H_{c2} increase with increasing reaction temperature. The T_c curves exhibit a small maximum at 1-2 at.% titanium, which is shifted to a smaller concentration with increasing reaction temperature. The removal of the bronze matrix raises the T_c by about 0.3 K owing to the release of the prestrain in the Nb_3Sn layer.[14] The H_{c2} values of composites reacted at 700-750°C increase with increasing titanium concentration in the niobium core, whereas those of composites reacted at 800°C exhibit a maximum at 2-3 at.% Ti. The 2 at.% titanium addition results in the enhancement of H_{c2} by about 4 T. The enhancement of H_{c2} may be attributed to the increases in T_c and normal state resistivity, ρ_n,[15] by the titanium addition. The H_{c2} curve shows a maximum at a higher titanium concentration than the corresponding T_c curve, probably because the ρ_n increases with increasing titanium concentration. The suppression of martensitic phase transformation in the composite-processed Nb_3Sn by a third element addition is considered to be an alternative explanation of the H_{c2} enhancement.[16] The removal of the bronze matrix increases H_{c2} by about 3 T. This increase in H_{c2} becomes smaller when the amount of titanium addition exceeds 2 at.%.

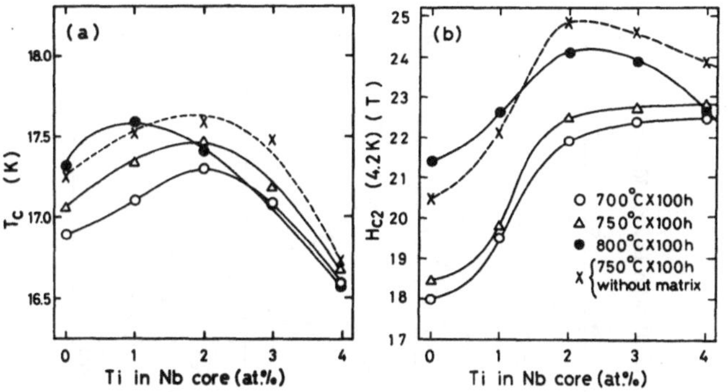

(a) (b)

Fig. 2. T_c and H_{c2} of single-core composite Nb_3Sn wires reacted at 700-800°C with titanium addition to the niobium core.
(a) T_c versus titanium concentration in the niobium core.
(b) H_{c2} versus titanium concentration in the niobium core.

Figure 3 shows the J_c (Nb_3Sn layer) at 10-16 T of single-core wires reacted at 700-800°C versus the titanium concentration in the niobium core. J_c values in high fields for composites reacted at 700-750°C increase with increasing titanium concentration, whereas those at 800°C decrease by excess titanium addition; this may relate to the variation of H_{c2} against titanium concentration shown in Fig. 2(b). Although the most appropriate amount of titanium addition to the core for enhancing the J_c depends on both the applied magnetic field and the heat-treatment condition, the addition of about 2 at.% titanium seems to be preferable. A J_c (Nb_3Sn layer) of over 1 x 10^5A/cm^2 is obtained at 15 T, which is several times larger than that of the pure Nb_3Sn. The overall critical current density of the wire is more remarkably increased than the J_c (Nb_3Sn layer), since the Nb_3Sn layer thickness is also increased by the titanium addition. An overall J_c (bronze + cores) of 2 x 10^4A/cm^2 is obtained at 15 T in the 49-core Nb-2at.%Ti/Cu-7at.%Sn wire reacted at 700°C for 192 h. These results indicate that the composite-processed Nb_3Sn with Nb-Ti alloy core may be promising for generating a magnetic field of 15 T at 4.2 K.

The titanium added to the matrix produces effects similar to those on the high-field J_c of the composite-processed Nb_3Sn when titanium is added to the core. The titanium addition to the matrix can improve high-field performances of the composite-processed Nb_3Sn without deteriorating the workability of the cores. The simultaneous addition of gallium to the matrix scarcely improves the superconducting properties of the composite-processed Nb_3Sn with Nb-Ti alloy core.

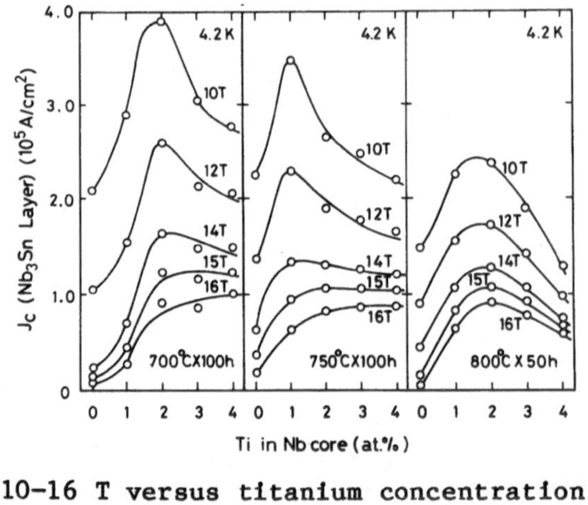

Fig. 3. J_c at 10-16 T versus titanium concentration in the niobium core for single-core composite Nb_3Sn wires reacted at 700-800°C.

For the single-core composite tape reacted at 800°C for 100 h, a zirconium addition of 2 at.% to the niobium core increases the T_c by about 0.2 K; further addition decreases it. The J_c in high fields is also improved by zirconium addition of 2 at.%; the J_c of the Nb-2at.%Zr/Cu-7at.%Sn composite at 16 T is about twice that of the pure Nb_3Sn. The simultaneous addition of gallium to the matrix is effective for enhancing the T_c and the J_c in high fields of the Nb_3Sn composite with Nb-Zr alloy core. The T_c of the composite with Nb-10at.%Zr alloy core is increased from 16.5 to 17.5 K by the gallium substitution of 9 at.% for tin in the matrix. However, the zirconium addition is not as promising as the titanium or the hafnium addition for improving high-field properties of the composite-processed Nb_3Sn.

The heat-treatment condition, T_c, and H_{c2} (4.2 K) are listed in Table 3 for the single-core composite tapes with hafnium and gallium additions. The hafnium addition to the core increases the T_c by 0.4-0.5 K and the H_{c2} by 2-4 T. The hafnium concentration at which the T_c and the H_{c2} show maxima is between 5-10 at.%, considerably larger than in the case of the titanium addition. The simultaneous addition of hafnium to the core and gallium to the matrix increases the T_c by 0.6-0.9 K and the H_{c2} by 6-7 T. The increase of H_{c2} produced by the simultaneous addition of hafnium and gallium is larger than that produced by the titanium addition.

The J_c in fields higher than 12 T is considerably improved by the hafnium addition to the core. The simultaneous addition of gallium to the matrix produces a further increase in the J_c in high fields. Figure 4 shows J_c versus magnetic field curves at 4.2 K for the 19-core composite wire. A J_c (Nb_3Sn layer) of 1.3 x

Table 3. Effects of Hafnium Addition to the Core and Gallium Addition to the Matrix on the T_c and the H_{c2} (4.2 K) of Single-Core Composite Nb_3Sn Tapes.

Composition core/matrix (at.%)	Heat treatment (°C x h)	T_c (K)	H_{c2} (4.2 K) (T)
Nb/Cu-7Sn	800 x 100	17.1	19
Nb-2Hf/Cu-7Sn	800 x 100	17.5	21
Nb-5Hf/Cu-7Sn	800 x 100	17.6	23
Nb/Cu-5Sn-4Ga	800 x 100	17.4	21
Nb-2Hf/Cu-5Sn-4Ga	800 x 100	17.7	25
Nb-5Hf/Cu-5SN-4Ga	800 x 20	17.7	25
Nb-5Hf/Cu-5Sn-4Ga	800 x 100	17.7	25.5
Nb/Cu-3Sn-9Ga	800 x 100	17.6	23
Nb-5Hf/Cu-3Sn-9Ga	800 x 20	18.0	26

$10^5 A/cm^2$ is obtained at 18 T for the Nb-5at.%Hf/Cu-3at.%Sn-9at.%Ga wire reacted at 750°C for 50 h, and that of 1.2 x $10^5 A/cm^2$ is obtained for the Nb-5at.%Hf/Cu-5at.%Sn-4at.%Ga wire reacted at 750°C for 100 h. These results indicate that the generation of magnetic fields over 15 T may be feasible by the multifilamentary Nb_3Sn wire with simultaneous addition of hafnium to the core and gallium to the matrix.

Normalized values of J_c at 16 T of single-core composite tapes are summarized in Fig. 5 as a function of IVa element concentration in the niobium core. The J_c values are normalized to that of the pure Nb_3Sn. The titanium addition to the niobium core is most effective for enhancing J_c in high fields for the composite with Cu-7at.%Sn matrix. The simultaneous addition of gallium to the matrix improves J_c in high fields for composites with Nb-Zr or Nb-Hf alloy core. The highest J_c at 16 T is obtained in the Nb-Hf/Cu-Sn-Ga composite.

CONCLUSION

IVa elements added to the niobium core increase the growth rate of Nb_3Sn 2-3 times and prevent the grain coarsening of Nb_3Sn at high reaction temperatures. Titanium is incorporated into the Nb_3Sn layer in a much larger amount than zirconium or hafnium. The T_c shows a small maximum against the titanium or zirconium concentration in the niobium core at about 2 at.%, and against hafnium concentration at 5-10 at.%. The titanium or hafnium addition increases the H_{c2} by about 4 T. The titanium addition more significantly increases J_c in high fields than the zirconium or hafnium additions; the J_c of Nb-2at.%Ti/Cu-7at.%Sn at 15 T is several times larger than that of the pure Nb_3Sn. A larger amount of titanium is incorporated into the Nb_3Sn layer from the matrix than from the core. The titanium addition to the matrix improves high-field properties of the composite-processed Nb_3Sn without deteriorating the

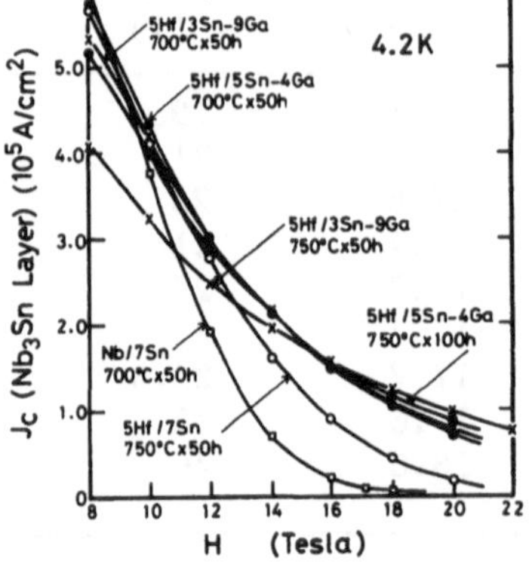

Fig. 4. J_c versus magnetic field curves at 4.2 K for 19-core composite Nb_3Sn wires with hafnium addition to the core and gallium addition to the matrix.

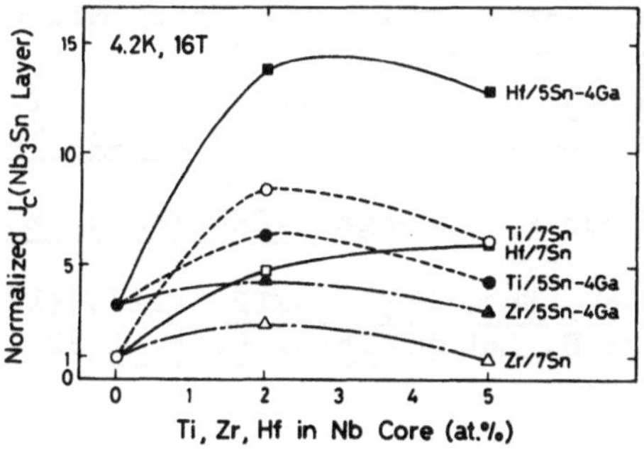

Fig. 5. Normalize J_c at 16 T of single-core composite Nb_3Sn tapes with IVa element addition to the core and gallium addition to the matrix. J_c values are normalized to that of the Nb/Cu-7at.%Sn composite. Specimens were reacted at 800°C for 100 h.

ductility of the core. The gallium addition to the matrix produces increases in T_c and J_c in high fields except for the composite with Nb-Ti alloy core. The simultaneous addition of hafnium and gallium is most effective for the enhancement of J_c in high fields. The composite-processed Nb_3Sn with titanium addition and that with hafnium and gallium addition may be promising for generating magnetic fields from 13 to 15 T and from 15 to 17 T at 4.2 K, respectively.

REFERENCES

1. M. Suenaga, T. S. Luhman, and W. B. Sampson, J. Appl. Phys. 45:4049 (1974).
2. O. Horigami, T. S. Luhman, C. S. Pande, and M. Suenaga, Appl. Phys. Lett. 27:738 (1976).
3. D. Dew-Hughes, IEEE Trans. Magn. MAG-13:651 (1977).
4. K. Togano, T. Asano, and K. Tachikawa, J. Less-Common Met. 68:15 (1979).
5. M. Suenaga, K. Aihara, K. Kaiho, and T. S. Luhman, Superconducting properties of $(Nb,Ta)_3Sn$ wires fabricated by the bronze process, in: "Advances in Cryogenic Engineering - Materials, Vol. 26," Plenum Press, New York (1980), p. 442.
6. H. Sekine, K. Tachikawa, and Y. Iwasa, Appl. Phys. Lett. 35:472 (1979).
7. H. Sekine, T. Takeuchi, and K. Tachikawa, IEEE Trans. Magn. MAG-17:383 (1981).
8. M. Suenaga, W. B. Sampson, and T. S. Luhman, IEEE Trans. Magn. MAG-17:646 (1981).
9. K. Tachikawa, T. Asano, and T. Takeuchi, to be published in Appl. Phys. Lett.
10. G. Springer, Fortschr. Miner. 45(1):103 (1967).

11. J. Philibert, "X-ray Optics and X-ray Microanalysis," Academic Press, New York (1963), p. 379.
12. S. J. B. Reed, Br. J. Appl. Phys. 16:913 (1965).
13. G. R. Johnson and D. H. Douglass, J. Low Temp. Phys. 14:565 (1974).
14. T. S. Luhman and M. Suenaga, IEEE Trans. Magn. MAG-13:800 (1977).
15. A. A. Abrikosov, Sov. Phys. JETP 5:1174 (1957).
16. R. Roberge and H. Lehuy, Phys. Lett. A 82:259 (1981).

PHASE RELATIONSHIPS, BASIC METALLURGY, AND SUPERCONDUCTING PROPERTIES OF Nb_3Sn AND RELATED COMPOUNDS*

R. Flükiger

Institut für Technische Physik, Kernforschungszentrum Karlsruhe
Karlsruhe, Federal Republic of Germany

INTRODUCTION

Since the discovery of the high T_c superconductors, V_3Si and Nb_3Sn in 1953,[1] the A15 type superconductors have been the subject of more than a thousand publications. The great interest in the A15 materials has, of course, been mainly motivated by their high T_c values, the highest known to date having been measured for Nb_3Ge: $T_c = 23$ K.[2] From the known data it follows that the maximum T_c in an A15 type compound containing a nontransition element, i.e., Nb_3Sn, Nb_3Ge, Nb_3Ga, Nb_3Al, V_3Si, or V_3Ga, can only be reached if the following conditions are fulfilled: a) stoichiometric composition, b) perfect long-range atomic order, and c) stability of the cubic A15 phase at low temperatures.

However, A15 type compounds with optimum superconducting properties tend (with one exception, V_3Ga) to be on the verge of structural instability. This empirical correlation has also been observed for materials crystallizing in other structures and seems plausible since both phenomena, superconductivity, and structural stability have a common origin--the electronic structure. There are different kinds of instabilities leading to a deviation from the conditions formulated above and thus to lower T_c values:

1. The compound does not form in the required A15 structure at equilibrium and has to be stabilized by using appropriate preparation techniques or by impurities, as $V_{\sim 3}Al$,[3] $Nb_{\sim 3}Si$,[4] or $Mo_{0.5}Re_{0.5}$.[5]

---

*Invited paper.

2. The compound crystallizes in the A15 structure, but the composition corresponding to the highest T_c is metastable (if at all, it is stable at high temperatures only) and must be stabilized by fast quenching or by impurities.

3. The compound crystallizes in the A15 phase at the stoichiometric composition, but undergoes a structural transformation at low temperatures, as Nb_3Sn[6] or V_3Si.[7]

4. All conditions a-c mentioned above can be fulfilled in Nb_3Sn multifilamentary wires, but stress-induced structural changes are introduced through the precompression resulting from the differential contraction of bronze matrix and Nb_3Sn filaments.[8,9]

These instabilities, together with the metallurgical difficulties encountered in preparing homogeneous, single-phased samples, are the main reason why such fundamental properties as the variation of T_c, ρ, and H_{c2} as a function of composition in V_3Si and Nb_3Sn are still a subject of investigation 28 years after the discovery of these materials. To determine the variation of these properties for a given A15 phase, in particular the vicinity of the stoichiometric composition, a precise knowledge of the phase diagram is needed.

In this paper, a brief review of current methods of determining phase diagrams is presented, together with the results on the most interesting A15 phases. Their superconducting properties are also presented phenomenologically. Particular attention is given to the compound Nb_3Sn because of its actual importance for the construction of high-field superconducting magnets. It is shown that the superconducting properties in the system $Nb_{1-\beta}Sn_\beta$ can only be understood on the basis of the low temperature phase diagram.

HIGH TEMPERATURE PHASE DIAGRAMS

The stoichiometric composition in most A15 phases with high T_c values exists only in the equilibrium phase fields at very high temperatures and has to be quenched at sufficiently high rates in order to be retained for a subsequent analysis at room temperature. However, quenching experiments depend on kinetics and thus give only an indirect image of the high temperature behavior: the retained state does not necessarily reflect equilibrium at the quenching temperature. The accurate determination of the temperature-dependent high temperature phase limits thus requires the use of direct observation methods at high temperatures, in addition to the room temperature analysis of quenched alloys (indirect observation), which includes optical and electron microscopy, x-ray and microprobe analysis, and other methods, one of which is the measurement of T_c.

Direct Observation Methods

A commonly used method for detecting phase transformations at high temperatures is the differential thermal analysis (DTA), which detects a signal due to the latent heat of transformation.[10] In A15 superconductors, the phase of interest always forms in the temperature range 1500 < T < 2300°C, where the contamination of the alloy by the crucible material or by impurities in the heating chamber can be a serious problem. Furthermore, the important evaporation rates of volatile components at these temperatures may render the detection of thermal arrests more difficult. In some cases, a DTA apparatus working at moderately high argon pressures (up to 20 atm.) was found to give good results.[10,11] There are, however, cases where the contamination of the analysed alloy by the crucible material induces important errors in the measurement of the solidus temperature. A technique that completely eliminates this problem is the recently developed levitation thermal analysis (LTA).[11,12] This method is based on the fact that the shape of the melting part of the sample at the solidus temperature immediately follows the electromagnetic field gradient in the rf levitation coil. The consequence of the suddenly enhanced coupling is a dramatic increase in temperature, (several hundred degrees), which is detected by means of a two-color pyrometer. To account for the unknown spectral emissivity of each alloy and for the selective absorption of all optical components (glass window, crown glass prism, filter), the pyrometer has to be calibrated for each sample,[11,12] after which the error in temperature is limited to ±10°C at 2000°C.

The occurrence of superconductivity in a given phase offers a unique possibility for determining the concentration gradient in the measured alloy. Indeed, the value of T_c, depending on composition and the width of the superconducting transition as measured by low temperature calorimetry, thus reflects the state of homogeneity. Prolonged homogenization heat treatments (up to one week) close to the solidus temperature were found to be necessary for a considerable reduction of this width to a few tenths of a degree.[11] It is obvious that such homogenized alloys lead to sharper thermal arrests, thus improving the precision. It is a somewhat unexpected result of the research in the superconductivity field that it contributed to the development of new or improved methods for the accurate determination of high temperature phase diagrams.

Phase Diagrams of High T_c Materials

The Nb-rich portion of the phase diagrams for the systems Nb-Ga,[13] Nb-Ge,[14] and Nb-Al,[15] as determined by a combination of direct and indirect methods, is represented in Fig. 1. The particular shape of the peritectically forming A15 phase is very similar for the three systems. In Nb-Ga and Nb-Al, the stoichiometric composition represents the extremum solubility limit of the A15 phase field. In Nb-Ge the highest Ge content at equilibrium is 23 at.%, i.e., the stoichiometric composition does not exist in the equilibrium phase field. The application of the same principles has led to the diagrams shown in Figs. 2 and 3, showing the V-rich portions of the V-Ga[16] and of the V-Si[17,18] systems. In contrast to the Nb-based compounds, the A15 phase in both V-based systems is formed congruently, in V_3Si from the melt and in V_3Ga from the bcc solid solution.

As shown recently,[11] the type of formation of the A15 phase in a system $A_{1-\beta}B_\beta$ is essentially governed by the B element. If successive B elements are taken, starting from VIIB up to VA elements, the formation of the A15 phase shows the sequence: peritectoid ⟶ congruent ⟶ peritectic. The limits between two

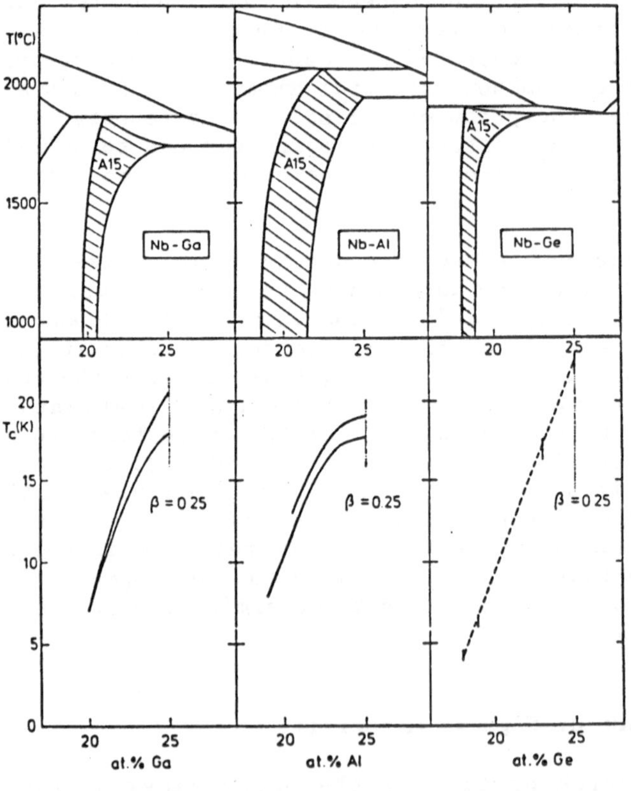

Fig. 1. A15 phase fields and superconductivity in Nb_3B compounds, where B = Ga, Al, Ge.[11-15]

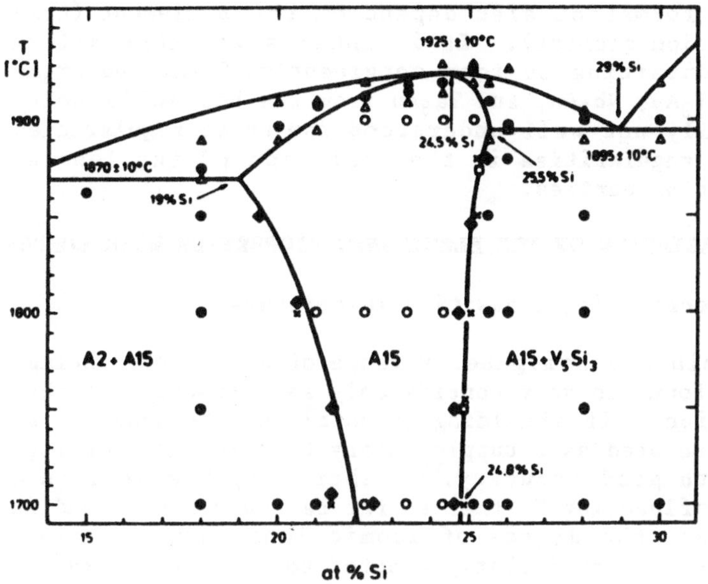

Fig. 2. The A15 phase field in the system V-Si.[18]

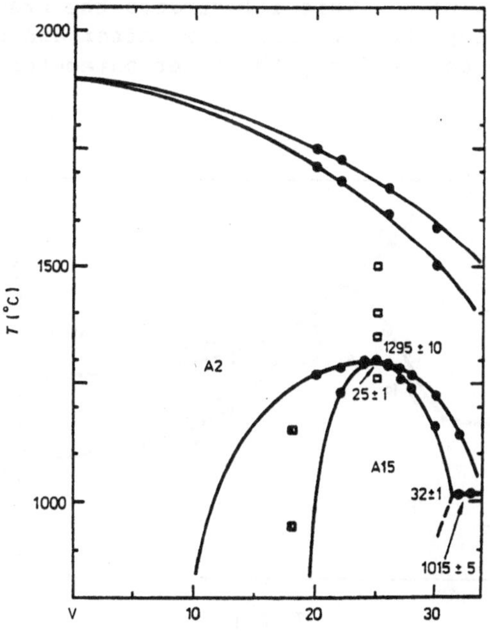

Fig. 3. The A15 phase field in the system V-Ga.[16]

types of formation also depend on the B element (transition or nontransition element). It is interesting that all known compounds containing Au form congruently, Ti_3Au and Zr_3Au from the melt and V_3Au, Nb_3Au, and Ta_3Au from the bcc solid solution. It is actually not well understood how these regularities are connected to regularities in the variation of the low temperature electronic properties.

THE VARIATION OF THE ELECTRONIC PROPERTIES WITH COMPOSITION

The Superconducting Transition Temperature

Within the homogeneity range of a superconducting phase, T_c has been found to vary considerably as a function of the chemical composition. If shielding effects[11] can be avoided, this variation can be used as a supplementary tool in determining compositions with good accuracy.[11] There is, however, an additional effect influencing T_c that so far has only been found in A15 type compounds: the degree of atomic long-range ordering.[19,20] The system V_3Ga is particularly adapted to illustrate ordering effects on T_c, the stoichiometric composition being stable at all temperatures (see Fig. 3). As shown in Fig. 4, a higher degree of ordering is correlated to a higher T_c value. The observed peak at 650 °C is only apparent: it simply indicates the temperature at which the originally arc-cast V_3Ga sample reached its equilibrium order parameter value.[21] This illustrates the ordering kinetics: even during the very short cooling time after arc melting (initial cooling rate: 50 to 100 °C/s), the order parameter still increases

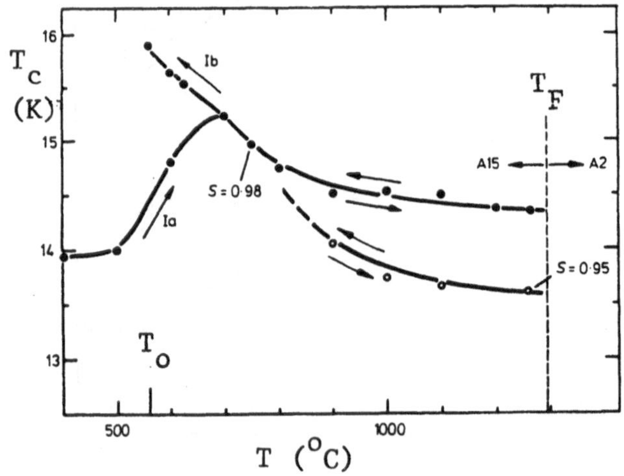

Fig. 4. Variation of T_c of V_3Ga with heat treatment. Cooling rates: • 15 °C/s, o 10^3 °C/s.[21]

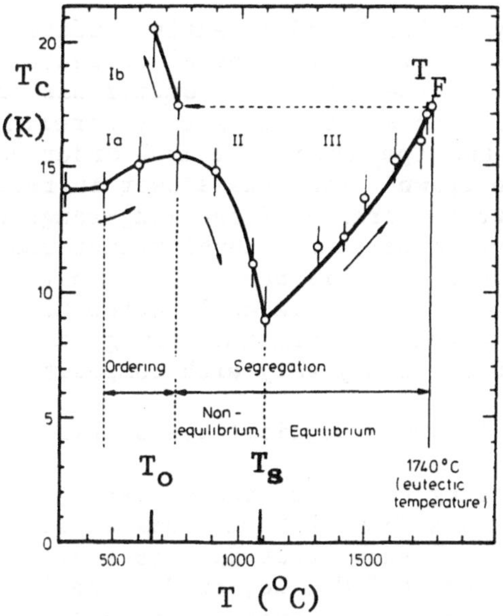

Fig. 5. Variation of T_c of the Nb-Ga phase with heat treatment.[21]

The situation is more complex if the stoichiometric composition is stable at high temperatures only. The separation of ordering and compositional changes on T_c has been studied for the system Nb-Ga.[21] Figure 5 shows the variation of T_c for an arc-cast alloy of the overall composition $Nb_{0.74}Ga_{0.26}$ that was successively annealed and quenched at increasing temperatures. In this case, the behavior of T_c as a function of the heat treatment is characterized by two activation energies--for ordering and segregation. These energies correspond to the temperatures T_o and T_s in Fig. 5. It is interesting that even an alloy at the metastable stoichiometric composition can, in principle, be ordered if $T_o < T_s$. This condition seems to be fulfilled for Nb_3Al and Nb_3Ga. However, perfect order is very rarely reached in A15 type compounds, and most order parameters are limited to values below $S_a = 0.98$,[19-20 20-21] probably as a consequence of insufficient diffusion at the temperature T_o, which is typically about 650°C. The two apparent exceptions, V_3Ga and Nb_3Sn, are discussed in the next paragraph.

The variation of T_c as a function of composition for a phase crystallizing in the A15 structure can be established if the annealing temperature and the cooling rate are the same for all compositions. In general, an essentially linear variation of T_c with composition is observed. In some cases, a "saturation" of T_c at compositions close to stoichiometry was reported, but was mostly due to shielding effects.[11] There is one exception, Nb_3Al: it was first reported by Müller[22] and recently confirmed by Moehlecke et al.[23] and Flukiger et al.[24] that for Al contents beyond 23 at.% a further decrease of the lattice parameter was observed, while T_c remained essentially constant. In a very careful investigation, which included also specific heat and order parameter measurements, it has been shown[24] that the slower increase of T_c in Nb_3Al is not connected to a decrease of the long-range order parameter. Since no low temperature structural transformation could be detected,[24] the reasons for this particular behavior in this system are still unclear. A possible explanation has been furnished by Kwo et al.,[25] who concluded from tunneling experiments that the phonon spectrum in Nb-Al may vary with composition.

Upper Critical Field and Electrical Resistivity

It is interesting to follow the variation of the upper critical field, H_{c2}, as a function of composition for different A15 compounds. Figure 6 shows that H_{c2} for Nb-based compounds is of the order of ∿6 T per at.% Ga, Ge, or Sn (the latter in the cubic phase). Like T_c or the electronic specific heat, the upper critical field of A15 compounds also depends on the order parameter, as shown in Fig. 7 for V_3Ga.[26] The corresponding variation of $\rho(T)$ for V_3Ga is represented in Fig. 8: for a higher order parameter, a smaller resistivity value was measured. The effect of ordering on ρ has been studied on Nb_3Pt, where the order parameter was changed by quench and anneal methods[26] from $S = 1$ to $S = 0.94$: $\rho_{\gtrsim T_c}$ was found to decrease from 87 to 20 μΩcm. The electronic mean free path was determined to be $\ell \simeq 100$ Å in the most ordered

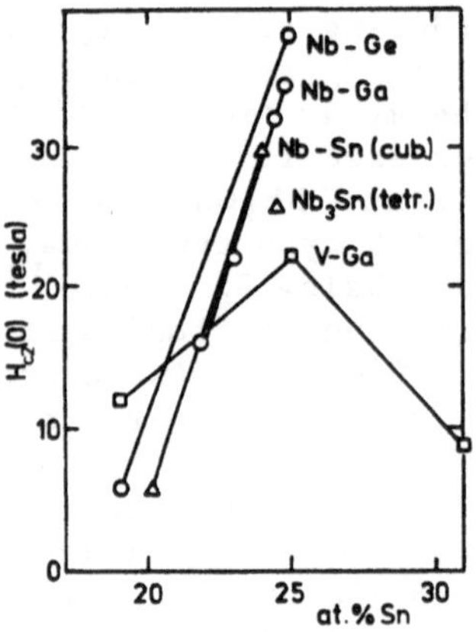

Fig. 6. Variation of the upper critical field of several A15 compounds with composition.

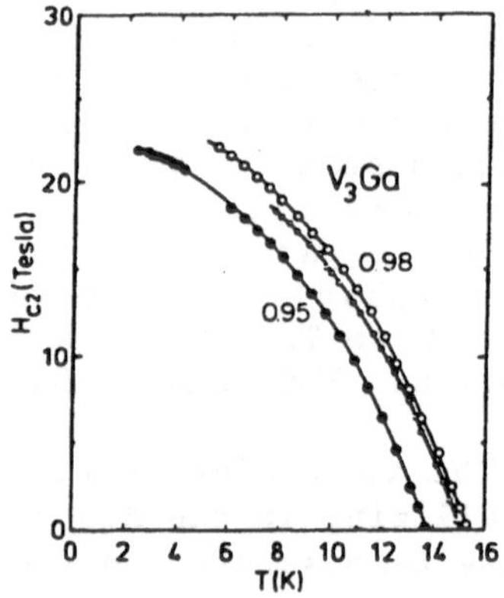

Fig. 7. $H_{c2}(T)$ for V_3Ga at two different degrees of ordering, $S = 0.98$ and $S = 0.95$.[26]

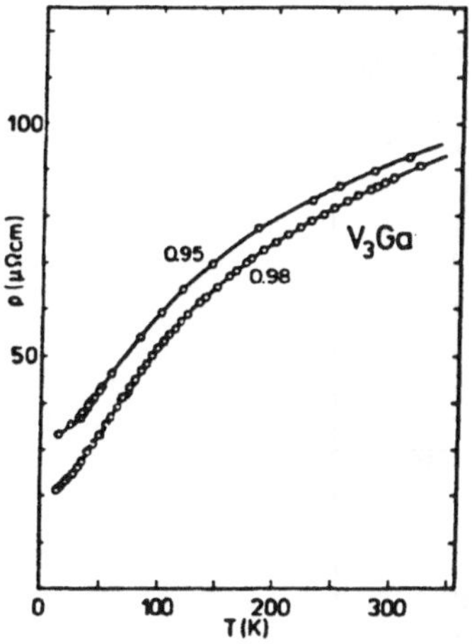

Fig. 8. ρ(T) for V_3Ga at two different degrees of ordering, S = 0.98 and S = 0.95.[26]

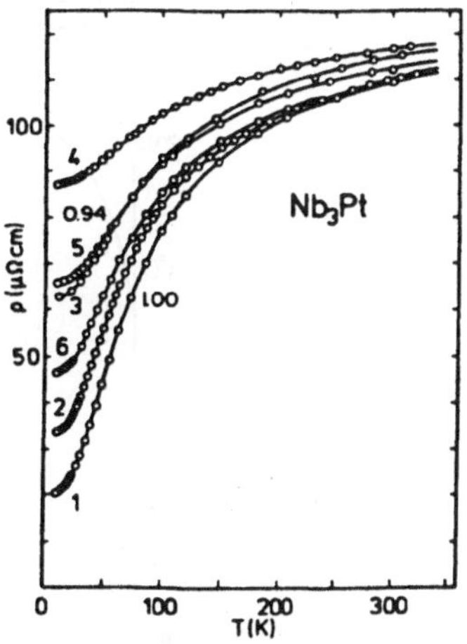

Fig. 9 ρ(T) for Nb_3Pt at different degrees of ordering.[26]

state and decreased by a factor 10 for the state at S = 0.94. The latter corresponds to 1.5% of the Nb sites occupied by Pt atoms. From the reversibility of the resistivity curves in Fig. 9, it can be deduced that the effect of the heat treatments on ρ(T) is really an order effect and is not due to microcracks, precipitations, or other effects.

THE A15 PHASE IN THE SYSTEM Nb-Sn

The Cubic and the Tetragonal Phase Fields

The high temperature phase diagram at the Nb-rich portion of the Nb-Sn system according to Charlesworth et al.[27] is represented in the upper part of Fig. 10. At T ≃ 1000°C, the A15 phase field varies from ∿19 to 25.5 at.% Sn, the stoichiometric composition being stable up to ∿1700°C. In the lower part of Fig. 10, the phase field of the low temperature tetragonal modification has been represented. The tetragonal phase forms spontaneously at T_M = 43 K and 25 at.% Sn, according to several authors.[7,33,34] This phase forms congruently from the solid A15 phase, the transformation being of the first order.[34] From a great number of observations, however, it follows that the transformation is on the verge of being second order ("weak" first order). The tetragonal phase is stable within the limits 24.5 and 25.2 at.% Sn, the uncertainty in composition being of the order of ±0.02%. This narrow phase field is confirmed by the specific heat data of Junod

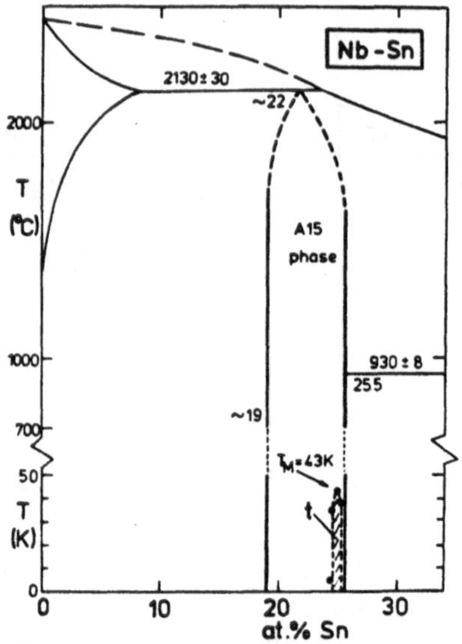

Fig. 10. The A15 and the tetragonal phase fields in the system Nb-Sn. The high temperature part is taken from Ref. 27.
o transformation temperature[30]
● not transforming[28]
The point at T_M = 43 K was taken from Refs. 33, 34, and 45.

et al.[36] on Nb-Sn samples of different compositions and with different fractions of transformed phase. The calorimetrically determined superconducting transition of the tetragonal phase (∿0.5 K) is indeed much narrower than that of the cubic phase present in the same sample (>3 K).[28,36]

Homogeneous Nb-Sn samples in the A15 cubic phase cannot be prepared by arc melting.[36] They have to be prepared either by a) levitation melting, followed by a homogenization heat treatment at 1800°C (both under high argon pressure)[28] or b) sintering of pressed fine Nb and Sn powders at 1300°C (1 week), followed by crushing, pressing, and new sintering under the same conditions.[30] Both methods, a and b restrict the distribution in composition within a range of ∿1 at.%.[28] The better sample homogeneity in the recent investigations[9,28,30] is probably the reason why the tetragonal phase field in Fig. 10 is narrower than that previously reported by King,[29] which is based on samples sintered for 3 h at 1450°C. Indeed, a larger composition gradient always leads to a certain volume fraction with Sn contents exceeding ∿24.5 at.%. This causes a line broadening at low temperature, which may be interpreted as a transformation.[28]

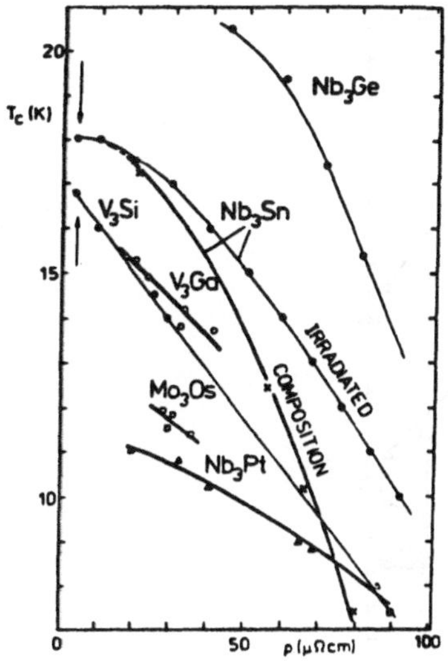

Fig. 11. Superconductivity and electrical residual resistivity for various A15 compounds.[26,28,38] The arrows show the lowest values of $\rho_{\gtrsim T_c}$ measured for the compounds V_3Si and Nb_3Sn.

Atomic Ordering and Superconductivity

It is well known that V_3Si and Nb_3Sn are the only A15 compounds showing a low temperature structural transformation.[6,7] Other properties encountered exclusively in these two compounds are: a) both exhibit in single crystals residual resistivity values $\rho_{\gtrsim T_c}$ of the order of ~ 3 μΩcm,[37] by far the lowest among all known A15 phases (see Fig. 11) and b) their measured order parameter at room temperature is $S = 1$,[31,32] i.e., both are perfectly ordered. It is interesting that T_c of V_3Si, as well as that of Nb_3Sn, is not affected by conventional quenching techniques, even at $T = 1800$[18,32] and 1700°C,[11,32] respectively. This is in contrast to the behavior in all other A15 compounds with reasonably high values of T_c and of the electronic density of states, where large variations have been reported (Figs. 4 and 5). On the other hand, T_c can be drastically changed by irradiation techniques, which were recently reviewed by Muller.[38] Recently, Pannetier et al.[39] have performed cw laser annealing on V_3Si, with quenching rates $>10^4$ °C/s, and observed a gradual decrease of T_c of several degrees from $T_o \simeq 1700$°C. This can be interpreted as the proof that the cooling rates in conventional quenching devices, $<5 \times 10^3$, were not sufficiently high for preventing complete reordering on cooling. Thus, V_3Si seems to have a very fast ordering kinetics. If this can be extended to Nb_3Sn, it follows that the chance of obtaining disordered states in multifilamentary wires, as obtained by the bronze process, is small. In this case, T_c changes after different reaction heat treatments should rather depend on composition and/or on the stress state.[11,32]

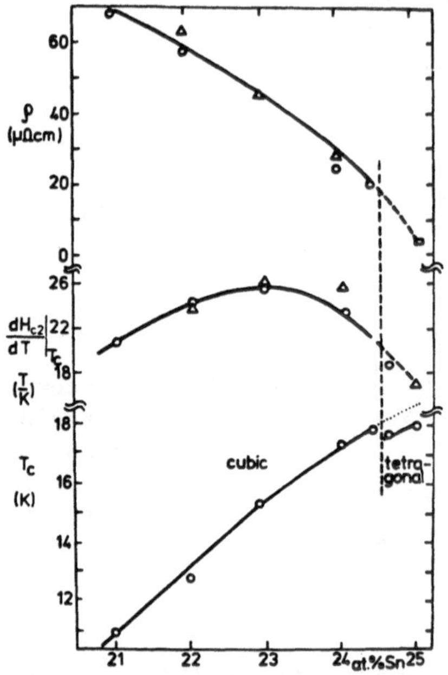

Fig. 12. T_c, $(dH_{c2}/dT)T_c$, and $\rho \gtrsim T_c$ in the system Nb-Sn as a function of the Sn content.
o data on bulk samples[28]
Δ data on coevaporated films[40]

The Electronic Properties as a Function of the Sn Content

The variation of T_c, $\rho \gtrsim T_c$, and $(dH_{c2}/dT)T_c$ as a function of the Sn content on the basis of the data of Orlando et al.[40] and Devantay et al.[28] is shown in Fig. 12. The residual electrical resistivity between 24 and 25 at.% Sn decreases by a factor of ∿5. This strong variation reflects the perfectly ordered state in Nb_3Sn: $\rho \gtrsim T_c$ is strongly affected by small changes of both composition and atomic ordering, the latter being obtained by irradiation methods (Fig. 11). The most interesting feature is the variation of T_c (Fig. 12): the reported value T_c = 18 K corresponds to the tetragonal phase; T_c = 17.8 K for the cubic phase was measured at ∿24.5 at.% Sn. If the cubic phase could be stabilized at stoichiometry (β = 0.25), values as high as ∿18.5 K would be reached.[28] The higher $\rho \gtrsim T_c$ value for the cubic phase, due to a lower Sn content, leads to higher values of the initial slope (Fig. 12) and of H_c (Fig. 6). This is particularly important for the application of Nb_3Sn in high-field magnets: for fields >12 T, where J_c is essentially governed by H_{c2}, a stabilization of the cubic phase would be preferable.

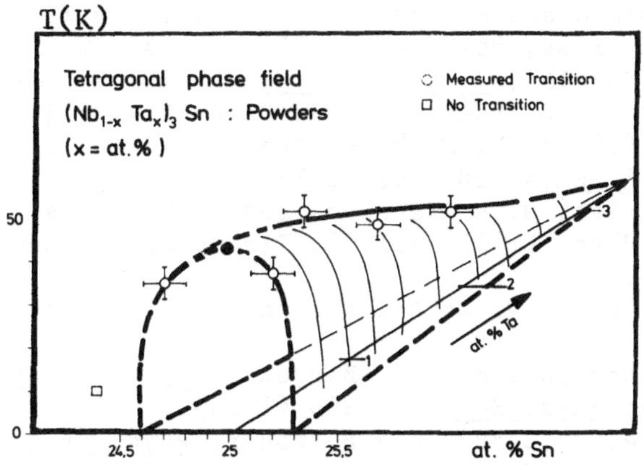

Fig. 13. The tetragonal phase field in $(Nb_{1-x}Ta_x)_3Sn$.[30]

The Suppression of the Spontaneous Tetragonal Phase in Nb_3Sn

Dissolving hydrogen[41–43] or metallic additions, such as Al[45] or Ta,[8,9,44] in Nb_3Sn leads to a suppression of the tetragonal phase, which forms spontaneously at T_M = 43 K. In both cases, the extension of the cubic phase toward higher Sn contents has led to higher T_c values, varying from 18.2 to 18.6 K.[30,42,44,45] This increase in T_c confirms the above result[28] of a higher T_c value for stoichiometric, cubic Nb_3Sn. The mechanism for $(Nb_{1-x}Ta_x)_3Sn$ is shown in Fig. 13.[30]

At the Ta-rich limit of the tetragonal phase field, shown in Fig. 13, a maximum of T_c was observed: T_c = 18.3 K,[30,44] which can be explained as a superposition of two opposite effects: a) with higher Ta contents, the limit of the cubic phase gets closer to the stoichiometric composition, which causes an increase in T_c and b) the increasing amount of Ta on the chain sites causes an increasing disorder and thus a decrease in T_c. The increase of H_{c2} as a consequence of the Ta addition is thus not only due to a shorter electronic mean free path, but mainly to the increase of T_c of the cubic phase, in the present case close to 0.5 K. The same behavior has been found for a series of other additions.[30,34]

REFERENCES

1. G. F. Hardy and J. D. Hulm, Phys. Rev. 87:884 (1953); B. T. Matthias, T. H. Geballe, S. Geller, and E. Corenzwit, Phys. Rev. 95:1435 (1954).
2. J. R. Gavaler, Appl. Phys. Lett. 23:480 (1973).
3. L. D. Hartsough and R. H. Hammond, Sol. State Commun. 9:885 (1971).

4. R. H. Hammond and S. Hazra, in: "Proc. Low Temp. Phys. LT13," Plenum Press, New York (1972), p. 465; G. R. Johnson and D. H. Douglas, J. Low Temp. Phys. 14:565 (1974) and 14:575 (1974); R. E. Somekh and J. E. Evetts, Sol. State Commun. 9:733 (1977); R. M. Waterstrat, F. Haensler, and J. Muller, J. Appl. Phys. 50:4763 (1979).
5. J. R. Gavaler, M. A. Janocko, and C. K. Jones, in: "Proc. Low Temp. Phys. LT13," Vol. 3, Plenum Press, New York (1972), p. 558.
6. R. Mailfert, B. W. Batterman, and J. J. Hanak, Phys. Lett. 22:315 (1967).
7. B. W. Batterman and C. S. Barrett, Phys. Rev. Lett. 13:390 (1964); Phys. Rev. 149:296 (1966).
8. R. Roberge, H. LeHuy, and S. Foner, Phys. Lett. 82A:259 (1981).
9. R. Flükiger, KfK Report No. 3140, Kernforschungszentrum Karlsruhe (unpublished) (Oct. 1980); R. Flükiger, W. Schauer, W. Specking, B. Schmidt, and E. Springer, IEEE Trans. Magn. MAG-17:2285 (1981). R. Flükiger, W. Schauer, W. Specking, L. Oddi, L. Pintschovius, W. Müllner, and B. Lachal, in: "Advances in Cryogenic Engineering--Materials," Vol. 28, Plenum Press, New York (1982), p.
10. W. W. Wendtlandt, in: "Thermal Methods of Analysis," 2d edition, P. J. Elving and I. M. Kotthoff, eds., John Wiley & Sons, New York (1974), and references therein.
11. R. Flükiger and J. L. Jorda, in: "Phase Diagrams in Metallurgy and Ceramics," M. Carter, ed., National Bureau of Standards Special Publication No. 196 (1978), p. 375; R. Flükiger, in: "Superconductor Materials Science," S. Foner and B. B. Schwartz, eds., Plenum Press, New York (1981), p. 511.
12. J. L. Jorda, R. Flükiger, and J. Muller, J. Mater. Sci. 13:2471 (1978).
13. J. L. Jorda, R. Flükiger, and J. Muller, J. Less-Common Met. 55:249 (1977).
14. J. L. Jorda, R. Flükiger, and J. Muller, J. Less-Common Met. 62:25 (1978).
15. J. L. Jorda, R. Flükiger, and J. Muller, J. Less-Common Met. 75:227(1980); J. Muller, Rep. Prog. Phys. 43:641 (1980).
16. R. Flükiger, J. L. Staudenmann, and P. Fischer, J. Less-Common Met. 50:253 (1976).
17. H. A. C. M. Bruning, Philips Res. Rep. 22:349 (1967).
18. J. L. Jorda and J. Muller, to be published in J. Less-Common Met.
19. E. C. Van Reuth and R. M. Waterstrat, Acta Crystallogr. B24:186 (1968).
20. R. Flükiger, Ph.D. Thesis No. 1570, University of Geneva, Geneva, Switzerland (1972).
21. R. Flükiger and J. L. Jorda, Sol. State Commun. 22:109 (1977).
22. A. Müller, Z. Naturforsch. 25A:1659 (1970).

23. S. Moehlecke, A. R. Sweedler, and D. E. Cox, Phys. Rev. B21:2712 (1980).
24. R. Flükiger, J. L. Jorda, A. Junod, and P. Fischer, to be published in Appl. Phys. Commun.
25. J. Kwo, R. H. Hammond, and T. H. Geballe, J. Appl. Phys. 51:1726 (1980).
26. R. Flükiger, S. Foner, and E. J. McNiff, Jr., in: "Superconductivity in d- and f-Band Metals," H. Suhl and M. B. Maple, eds., Academic Press, New York (1980), p. 265.
27. J. P. Charlesworth, I. Macphail, and P. E. Madsen, J. Mater. Sci. 5:580 (1970).
28. H. Devantay, J. L. Jorda, M. Decroux, J. Muller, and R. Flükiger, to be published in J. Mater. Sci.
29. H. W. King, in: "The Mechanism of Phase Transformations in Crystalline Solids," Monograph No. 33, Institute of Metals (1980), p. 196.
30. R. Flükiger and L. Oddi, to be published.
31. J. L. Staudenmann, Solid State Phys. 23:121 (1977), 26:461 (1978).
32. R. Flükiger, unpublished results.
33. C. W. Chu and L. J. Vieland, J. Low Temp. Phys. 17:25 (1974).
34. L. J. Vieland, J. Phys. Chem. Solids 31:1449 (1970).
35. L. J. Vieland, R. W. Cohen, and W. Rehwald, Phys. Rev. Lett. 26:373.
36. A. Junod, J. Muller, H. Rietschel, and E. Schneider, J. Phys. Chem. Solids 39:317 (1978).
37. A. J. Arko, D. H. Lowndes, F. A. Muller, L. W. Roeland, J. Wolfrat, A. T. Van Kessel, H. W. Myron, R. M. Muller, and G. W. Webb, Phys. Rev. Lett. 4:1590 (1980).
38. P. Muller, Ph.D. Thesis, University of Erlangen, FRG (1980).
39. B. Pannetier, T. H. Gebballe, R. H. Hammond, and J. F. Gibbone, to be published in Proc. LT16, Los Angeles (1981).
40. T. P. Orlando, J. A. Alexander, S. J. Bending, J. Kwo, S. J. Poon, R. H. Hammond, R. Beasley, E. J. McNiff, Jr., and S. Foner, IEEE Trans. Magn. MAG-17:368 (1981).
41. L. J. Vieland and A. W. Wicklund, Phys. Rev. B 11:3311 (1975).
42. P. R. Sahm, Phys. Lett. 26A:459 (1968).
43. M. Lehmann, Ph.D. Thesis, University of Erlangen, FRG (1980).
44. M. Suenaga, in: "Superconductor Materials Science," S. Foner and B. B. Schwartz, eds., Plenum Press, New York (1981).
45. L. J. Vieland and A. W. Wicklund, Physics Lett. 34A:43 (1971).

THE LAYER THICKNESS DEPENDENCE OF THE TRANSITION TEMPERATURE IN NIOBIUM-TIN

D. B. Smathers and D. C. Larbalestier

Materials Science Center, University of Wisconsin, Madison, Wisconsin

INTRODUCTION

It is commonly observed that bronze-processed Nb_3Sn composites (internal diffusion, external diffusion, or in situ) have a superconducting critical temperature (T_c) that rises as reaction time increases at any reaction temperature until a maximum value is finally reached. This has been explained by some as a reduction in the stress, which produces a degradation of the T_c of Nb_3Sn layers,[1,2,3] or, in a slightly different system, as a proximity effect of the unreacted Nb core on the Nb_3Sn.[4] We have studied two 13 wt.% Sn bronze tape composites (bronze:superconductor ~ 3:1) and found that the intrinsic T_c (i.e., without bronze) is more a function of layer thickness than reaction time. We have compared our data to that of other workers and have found a common plot of T_c versus layer thickness for diffusion-grown Nb_3Sn. Our Auger electron spectroscopy (AES) studies show that this behavior can be explained by the composition gradients existing in the diffused layers and that impurity concentrations within the layer and particularly reaction temperature also affect T_c but are found to be secondary in their effect. The common plot of T_c versus layer thickness is valid only for layers where the reaction is still proceeding, that is, for layers that have a considerable core of unreacted Nb.

RESULTS AND DISCUSSION

The thicknesses of our layers were determined from the AES measurements as we have previously described.[5,6] The T_c of each sample was measured inductively and was taken as the point of intersection of an extrapolation of the superconducting susceptibility and the normal-state susceptibility. In the present study, the T_c values are those without a surrounding layer of bronze or other material. The T_c measurements were made at a frequency of 17.5 Hz and a measuring field of 0.1 mT. The onset extrapolation procedure is used because shielding and small layer thickness effects obscure any correlation between T_c midpoints and the basic properties of the layer.

The plot of T_c onset values (measured with the bronze removed) as a function of the layer thickness is given in Figure 1. This figure also includes the data of Dickey et al.[4] whose samples were made by evaporating a thin layer of Sn onto a Nb substrate and subsequently diffusing at temperatures greater than 850°C to form Nb_3Sn. Their T_c data were determined resistively, and their thicknesses were estimated from a knowledge of the original Sn layer thickness. We have also included data for bronze-processed single-core wires,[8] for vapor-deposited (cold substrate)[9] and electron-beam co-evaporated[10] Nb_3Sn, and for in situ processed conductors.[7] The plot appears common for samples that involve the diffusion reaction and for which an excess of Nb remains, approximating an infinite sink. The deposited film samples[9,10] show that the T_c behavior is not the result of grain size or some intrinsic layer thickness effect, and the in situ samples (T_c measured with the bronze on)[7] show that T_c in diffusion-grown Nb_3Sn can be very high for extremely thin layers when the reaction is complete (no unreacted Nb remaining).

There is some evidence of a dependence of the T_c on reaction temperature,[1,8] but this is a small effect when looking at the scatter in the data of Figure 1. The more important factor contributing to the scatter of our data appears to be impurities (principally C and O) incorporated into the diffusion layer. This point will be addressed later. The shaded region in Figure 1 gives the maximum and minimum bounds over which our samples and those of other workers fall. There clearly remains a general trend irrespective of reaction temperature and impurity effects.

A logical suspect for the T_c behavior of thin Nb3Sn layers on a thick Nb core is the proximity effect. The thick-film approximation of DeGennes[11] has been previously applied,[4] but the analysis required both the coherence length (ξ_0) and the mean free path (ℓ) to be about 20 nm in order to fit the data, which is on our common plot of T_c versus layer thickness. Measurements of the coherence length for clean and dirty Nb_3Sn fall in the range 5.7 to 7.7 nm with corresponding mean free paths of 10 and 2.6 nm, respectively.[10] These values applied in the thick film approximation predict a much steeper dependence than the data of Figure 1 exhibit (see dotted line). Note that as the films go off stoichiometry, the product $\xi_0\ell$ gets smaller, implying an even smaller proximity effect from the Nb. The in situ samples[7] bear this analysis out: a 60-nm filament of Nb_3Sn surrounded by bronze can have a T_c as high as 17.9 K.

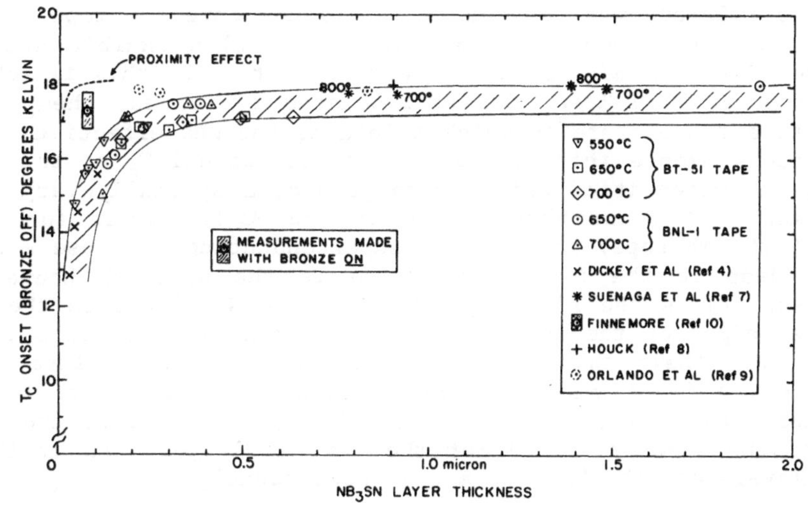

Fig. 1. The extrapolated T_c onset (bronze off) as a function of Nb_3Sn layer thickness for a variety of Nb_3Sn samples. Note the seemingly universal dependence of T_c on layer thickness for the diffusion-grown samples. The <u>in situ</u> samples of Ref. 7 are completely reacted and therefore do not fall on this curve. The dotted line shows the predicted proximity effect that the Nb will have on thin Nb_3Sn layers.

The effects of internal strain or stress on the T_c of thin layers are not believed to be significant, since the mismatch between the thermal contraction coefficients of Nb and Nb_3Sn is small.[12] The possibility of the volume expansion of Nb_3Sn over Nb or grain growth producing extra internal strains and thereby reducing T_c should not be a strong factor until the layer thickness is much greater than the grain size[3] and will, therefore, not affect thin layers.

When the proximity or stress effects are used to analyze the behavior of Nb3Sn layers, there is usually a hidden assumption made, which is that the layer has a uniform, stoichiometric composition. Our experimental data show that this is not the case, especially for thin layers. The Sn concentration profile across the Nb_3Sn layer evolves with reaction time:[5,6,13] the Sn content of the sample increases with reaction time and the gradient becomes less steep. The nearly inverse layer thickness dependence of T_c (Fig. 1) can be shown to arise from a knowledge of the Sn profiles and of some details of how our T_c measurements are made.

An inductive T_c measurement made on a sample with a substantial Nb core will usually exhibit shielding. The Nb_3Sn layer around a filament will obviously shield the core. A tape geometry where Nb_3Sn is grown on either face will also exhibit shielding because of the normal state skin effect in the Nb (this is substantial even at 17.5 Hz). Our extrapolated T_c values, then, correspond to a temperature at which shielding becomes effective. For a homogeneous sample this corresponds to the actual T_c of the Nb_3Sn. If any inhomogeneities exist which produce a spatially varying T_c, shielding will occur when enough of the Nb_3Sn layers (on either face of the Nb tape) have become superconducting to cause a significant drop in the field penetrated to the Nb_3Sn-Nb boundaries. This drop forces the field to penetrate the Nb core from the unobstructed edges of the tape. Since the width of most tapes is much larger than their thickness, this causes a drastic change in the distance the field must penetrate to reach the center of the core and therefore a larger skin effect. The result is that the magnitude of the susceptibility for the tape is larger than if the two Nb_3Sn layers and the Nb core acted independently.

The Ginzburg-Landau equations must be employed to analyze a sample in which the superconducting properties are allowed to vary spatially. To properly model our samples, a three-dimensional analysis is required. We have developed a numerical solution for the less complicated, one-dimensional case and can make a modest approximation of the real geometry by examining the behavior of a single inhomogeneous Nb_3Sn layer adjacent to a layer of Nb. The shielding in the normal core is changed by adjusting the ratio of the Nb thickness to the normal skin depth. The various effects of shielding can be qualitatively observed. Figure 2 shows an example of this type of analysis. Curve I shows the behavior of the temperature-dependent susceptibility for an inhomogeneous Nb_3Sn layer when no Nb is included. Most of the curvature arises from using a half-thickness to penetration depth ratio of about two for the sample. The extrapolated T_c onset indicates a T_c corresponding to most of the layer being superconducting. When a significant amount of Nb is included (curve II), the magnitude of the normalized susceptibility (by volume) increases. At the same time, the extrapolated T_c rises slightly. As the thickness of the Nb is increased further (curve III), the extrapolated T_c continues to rise. The upper limit of the observed T_c onset is determined by the T_c profile, the amount of shielding possible, and the sensitivity of the susceptibility experiment. The T_c of samples with equivalent shielding (equivalent geometry) corresponds to about the same thickness of Nb_3Sn being superconducting in the layers.

Given a linear gradient with a constant difference in concentration across the layer, the composition at a constant distance into the layer varies inversely with layer thickness. Taking the T_c of Nb_3Sn to be linear with composition, we have, with shielding, a mechanism to observe an inverse layer thickness dependence of the onset T_c. Our AES data show that as the reaction time increases

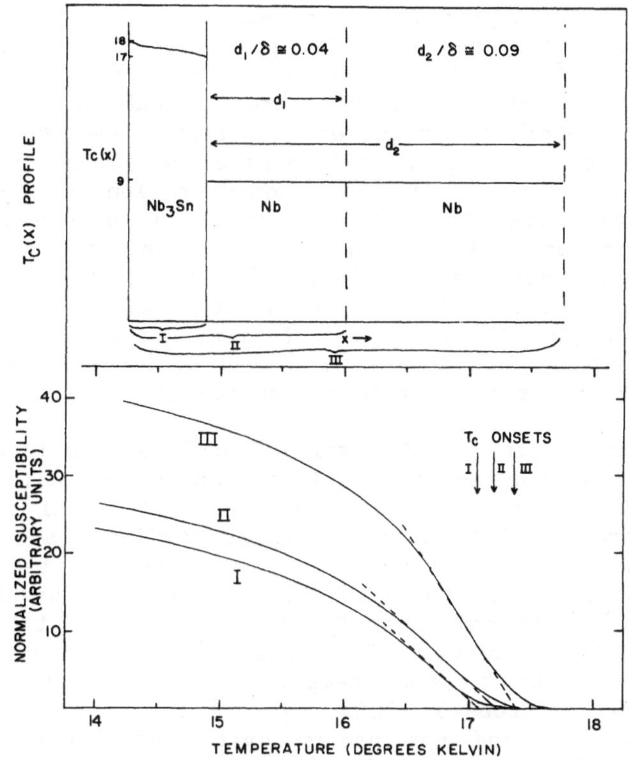

Fig. 2. The effect of shielding on the temperature-dependent susceptibility for an inhomogeneous Nb_3Sn layer. The extrapolated T_c onset and the overall magnitude rise as more Nb is added (curves II and III) because of the normal state skin effect in the Nb. δ is the skin depth. Curve I is for the case of no shielding (no Nb).

the Sn content at a given depth into the layer from the bronze-Nb_3Sn interface increases.[5,6,13] Although the observed Sn concentration profiles are not exactly linear and it is not yet clear whether the difference in concentration across the layer is constant, the evolution of the concentration profiles does suggest that it is indeed the concentration gradients and the presence of shielding that give rise to the observed T_c behavior. We believe that we have enough shielding in our samples that the T_c onset values correspond to the composition very near the bronze-Nb_3Sn interface (where the highest Sn contents are located). The current flowing in a sample in a dc resistive measurement is forced to the surface of the superconducting regions. Thus, a similar kind of argument can be made for the resistive T_c behavior of samples with composition gradients, and the onset T_c also corresponds to the high Sn regions. The T_c versus layer thickness behavior exhibited by diffusion-grown layers is a consequence of the Sn concentration gradients produced by the diffusion process. Slight changes in the phase relationships or the kinetics of the diffusion process will subtly affect the Sn concentration profile and, thus, the T_c onset value; this is where the reaction temperature and impurity effects arise. Once complete reaction is approached, by exhausting either the Sn or the Nb supply, the concentration gradients should disappear, and there should be no longer any dependence of T_c on the layer thickness. If the Sn supply is exhausted first, further annealing will remove any gradients, but the film will most likely not be stoichiometric. Thus, there are no limits imposed on the T_c of submicron Nb_3Sn filaments by diffusion, the proximity effect, or internal strains (other than that caused by the bronze) so long as the filaments are at or very near complete reaction.

Using AES we can detect levels of O, C, and Cu in our samples. These are the only detected impurities (AES cannot detect the presence of H, however). In most samples the O level is less than 1 at.%. Moore et al.[14] noted in their studies that O had little effect on the T_c of their samples. The more significant impurity is C; C in Nb_3Sn lowers the T_c.[15] In our studies, samples that had higher C levels also had lower T_c values than would fit on a smooth T_c-versus-thickness curve. This was particularly noticeable for samples reacted at the same temperature. (Note the 550°C BT-51 sample at 0.092 microns.) This suggests that the C levels are at least partly real and not vacuum system artifacts. The 550°C samples generally have lower C and O levels than the 650°C samples, and they also have higher T_c values. The C level decreases through the layer, being highest at the bronze-Nb_3Sn interface.

Incorporation of Cu in the Nb_3Sn is not surprising since the bronze-Nb couple forms a ternary diffusion system.[1,16] We observe gradients of Cu in the Nb_3Sn layers. What might be considered surprising is that the Cu content and concentration gradient change with increasing layer thickness. In addition, the levels of Cu in submicron layers are higher than those previously reported.[5,17,18] Figure 3 (also see Table 1) shows the high and low limits of the gradient, inferred from our AES measurements, as a function of layer thickness. As the layer thickness increases, the level of Cu drops and the gradient becomes smaller approaching a constant level of 0.2 to 0.5 at.% for layer thicknesses greater than one micron. The behavior is slightly temperature dependent. The 650 and 700°C results are very similar, but the 550°C samples still have 1.5 at.% Cu for samples up to 240-nm thick. Annealing the samples after etching off the bronze removes the gradient and lowers the Cu content to about 0.5 at.%.

It is tempting to correlate the Cu results with the T_c data in Figure 1, but the effect of the Cu appears to be more subtle. First, the samples containing the most Cu (550°C samples) have the

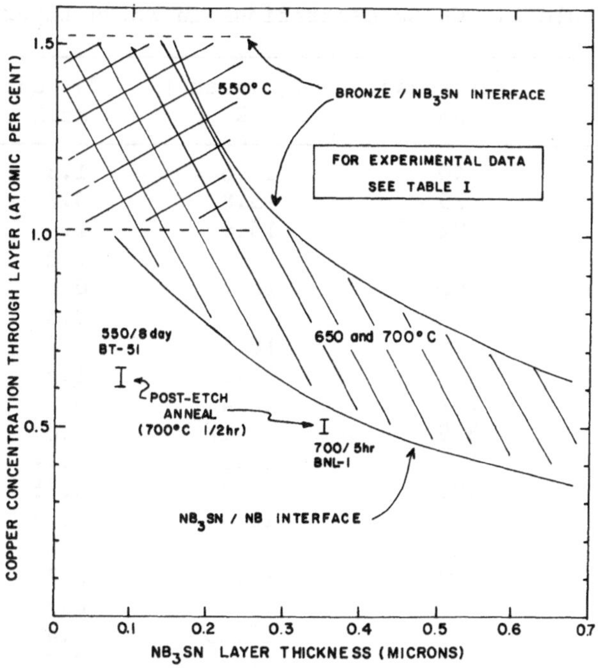

Fig. 3. Cu content as a function of layer thickness. The intersection of vertical lines with the outside edges of the shaded region gives the maximum and minimum values of the Cu gradient through the layer for that reacted layer thickness.

highest T_c values for a given thickness. Second, the data of Dickey et al.[4] fall on our curve, but there was no Cu in their system. This may be evidence to suggest that the Cu is in the grain boundaries. We also find a correlation between increasing layer growth rate and lower Cu content.[13] Thus, the effect high Cu contents have on the T_c may be to slow down layer growth thereby giving thinner layers for a given reaction time and correspondingly lower T_c values.

CONCLUSIONS

The thickness dependence of T_c commonly observed for bronze-processed Nb_3Sn is shown to be connected with the diffusion process, being primarily associated with the Sn concentration gradient across the layer rather than the reaction temperature or the impurities in the layers. In our samples, the impurity concentrations (C and Cu) appear to be related to the reaction temperature. The common plot of T_c versus layer thickness is for samples involving incomplete reaction. For layers at or near complete reaction,

Table 1. Cu Concentrations in Nb_3Sn Layers

Sample*	Thickness, nm	T_c Onset,† K	Cu Content in Layer, at.%	at x nm
550/24 BT-51	40	14.8	1.3	18
550/120 BT-51	75	15.8	0.7	81
550/192 BT-51	92	15.9	1.5	50
			1.0	73
550/192+ BT-51	92	15.7	0.7	48
700/½‡			0.6	76
550/336 BT-51	115	16.5	1.5	68
550/1348 BT-51	234	16.9	1.5	86
650/2 BT-51	162	16.4	0.9	61
650/20 BT-51	359	17.17	0.8	78
650/5 BNL-1	149	16.1	0.94	48
650/20 BNL-1	303	17.5	0.75	179
700/1 BT-51	164	16.5	0.73	118
700/2 BT-51	330	17.0	0.6	119
700/5 BT-51	493	17.1	0.7	110
			0.5	338
700/10 BT-51	633	17.15	0.4	437
700/5 BNL-1	350	17.5	0.9	50
			0.6	180
700/7 BNL-1	409	17.4	0.85	91
			0.67	188
700/5+ BNL-1	350	17.6	0.54	140
700/½‡			0.52	243

* 550/24 sample treated at 550°C for 24 h
† T_c values measured with bronze off
‡ Second anneal after bronze was etched off

the Sn composition will be nearly homogeneous, and the T_c values will not necessarily follow the present curve. The explanation of the behavior in Figure 1 also points out the need to understand how the superconducting properties are being measured in any experiment. When an inductive experiment is used to measure T_c, the ratio of the radius (or sample thickness) to the superconducting penetration depth affects the shape of the temperature-dependent susceptibility because the penetration depth is strongly temperature dependent. Thus, even though composition gradients within the layer affect the shape of the temperature-dependent susceptibility, penetration depth and shielding effects will make correlation between a T_c midpoint and chemical composition extremely difficult to elucidate. For layer thicknesses less than one micron, T_c midpoint values can be very misleading.

There is a large amount of Cu (up to 1.5 at.%) in very thin Nb_3Sn layers, the content decreasing as the layer thickness increases. Although the high levels of Cu do not appear to lower the T_c of the layers, the Cu apparently affects the growth rate. The Cu results imply that the Nb_3Sn phase field in the Cu-Sn-Nb ternary phase diagram is slightly different at 550°C than at 650°C.

ACKNOWLEDGMENTS

This work has been financially supported by the Department of Energy, Office of Fusion Energy and the University of Wisconsin Graduate School. We are grateful to Dr. Eric Gregory of Airco Superconductors (BNL-1 tape) and Dr. Jean Bussiere of the Brookhaven National Laboratory (BT-51 tape) for the supply of the unreacted bronze-Nb tapes.

REFERENCES

1. M. Suenaga, Metallurgy of continuous filamentary A15 superconductors, to be published in the proceedings of the NATO Advanced Study Institute on Superconducting Materials and Technology held in Sintra, Portugal, S. Foner and B. Schwartz, eds. (1980).
2. T. Luhman and D. O. Welch, Studies of the strain-dependent properties of A15 filamentary conductors at Brookhaven National Laboratory, in: "Filamentary A15 Superconductors," M. Suenaga and A. F. Clark, eds., Plenum Press, New York (1980).
3. K. Kwasnitza, A. V. Narlikar, H. V. Nissen, and D. Salathé, Growth and structural studies of multifilamentary Nb_3Sn formed by solid state diffusion, Cryogenics 20:715 (1980).
4. J. M. Dickey, M. Strongin, and O. F. Kammerer, Studies of thin films of Nb_3Sn on Nb, J. Appl. Phys. 42:5808 (1971).
5. D. B. Smathers and D. C. Larbalestier, Composition profiles in Nb_3Sn diffusion layers, in: "Advances in Cryogenic Engineering - Materials," Vol. 26, Plenum Press, New York (1980), p. 415.
6. D. B. Smathers and D. C. Larbalestier, An Auger electron spectroscopy study of bronze route niobium-tin diffusion layers, in: "Filamentary A15 Superconductors," M. Suenaga and A. F. Clark, eds., Plenum Press, New York (1980), p. 143.
7. D. Finnemore, Ames Laboratory, Iowa State University, Ames, Iowa, private communication.
8. M. Suenaga, T. Luhman, and W. B. Sampson, Effects of heat treatment and doping with Zr on the superconducting critical current densities of multifilamentary Nb_3Sn wires, J. Appl. Phys. 45:4049 (1974).
9. L. Houck, unpublished.

10. T. P. Orlando, E. J. McNiff, Jr., S. Foner, and M. R. Beasley, Critical fields, Pauli paramagnetic limiting, and material parameters of Nb_3Sn and V_3Si, Phys. Rev. B 19:4545 (1979).
11. P. G. DeGennes, Boundary effects in superconductors, Rev. Mod. Phys. 36:225 (1964).
12. H. W. Schadler, L. M. Osika, G. P. Salvo, and V. J. DeCarlo, The thermal expansion of Nb_3Sn (Cb_3Sn), Trans. Metall. Soc. AIME 230:1074 (1964).
13. D. B. Smathers and D. C. Larbalestier, University of Wisconsin, Madison, Wisconsin, to be published.
14. D. F. Moore, R. B. Zubeck, and J. M. Rowell, Energy gaps of the A15 superconductors Nb_3Sn, V_3Si, and Nb_3Ge measured by tunneling, Phys. Rev. B 20:2721 (1979).
15. R. E. Enstrom and J. R. Appert, Preparation, microstructure, and high-field superconducting properties of Nb_3Sn doped with group-III, -IV, -V, and -VI elements, J. Appl. Phys. 43:1915 (1972).
16. D. S. Kopecki, The physics of growth of vanadium gallide and niobium stannide superconducting phases by the bronze technique, Ph.D. thesis, University of Texas, Austin, Texas (1979).
17. J. D. Livingston, Phys. Status Solidi A 44:295 (1977).
18. P. E. Madsen and R. F. Hills, The effect of heat treatment on the superconducting properties of a multifilamentary Nb_3Sn composite, IEEE Trans. Magn. MAG-15:182 (1979).

SUPERCONDUCTING CURRENT DENSITIES IN BRONZE-PROCESSED Nb_3Sn MULTIFILAMENTARY WIRES*

S. Okuda,† M. Suenaga, and T. S. Luhman

Division of Metallurgy and Materials Science, Brookhaven National Laboratory, Upton, New York

INTRODUCTION

Since the discovery of the "bronze process" for fabrication of multifilamentary A15 superconducting wires, significant progress has been made in the production of Nb_3Sn wires on a commercial scale.[1-4] Although a number of modifications or new fabrication processes for filamentary wires have been made since the discovery, this process is the only one that is currently used to produce sufficient amounts of conductor for the construction of large-scale magnets. However, as pointed out earlier,[2] the published values of the superconducting critical-current densities, J_c (including the bronze, Nb_3Sn, and Nb), of these wires are significantly different depending on the manufacturer of the wire and on the details of heat-treatment and fabrication steps. In addition, the values of the critical-current density, J_c', of the Nb_3Sn layers in these multifilamentary wires, as reported by various groups, differ greatly, and even the best values of J_c' at 10 T are nearly a factor of 2 lower than the value measued in monofilamentary Nb_3Sn wires. Also, the utilization of these wires for the construction of magnets for future high-energy particle accelerators and magnetically confined plasmas for fusion reactors requires higher current densities, J_c, than those achieved in present wires that can be produced in large quantities. Thus, it is of interest to clarify the factors that influence J_c and J_c' of these wires. Toward

*Work performed under the auspices of the U.S. Department of Energy under Contract No. DE-AC02-76CH00016.

†Visiting Scientist from Sumitomo Electric Industry, Osaka, Tokyo, Japan.

achieving this goal we have made a study of superconducting critical-current densities in a set of commercially produced Nb_3Sn wires as a function of the bronze-to-filament ratio, R, heat-treatment conditions, and filament size. All these factore intimately connected to the critical currents of the wires. For example, the volume ratio, R, of the bronze to the filaments influences the total amount of Nb_3Sn produced in the wires and the extent of the compressive strains in the filaments.[1] Heat-treatment conditions determine the amount of Nb_3Sn formed and influence the critical-current density, J_c', of the Nb_3Sn layers through the variation of grain sizes. Finally, the filament size can also influence the total Nb_3Sn formed for a given heat treatment, and a value of $R \approx 3$ is required for total reaction of the Nb filaments. The results reported here are primarily for the critical currents measured at 9.5 T, although the measurements at higher fields up to 19 T are also reported for a limited number of wires.

EXPERIMENTAL PROCEDURE

A set of five fine multifilamentary wires with diameters of 0.41 mm and different ratios of bronze to Nb (R = 2.0 to 3.7) were purchased from the Superconductor Division of Airco. The pertinent information on these wires is tabulated in Table 1, and the cross sections of four wires, which were extensively studied, are shown in Fig. 1. In addition to the wires described in the table, lengths of wire having larger diameters (~1 mm) were received for examination of the influence of the filament sizes on J_c and J_c'.

The superconducting critical currents of these wires were measured at 4.2 K in a superconducting split-pair magnet for $8\ T \leq H \leq 9.5\ T$ and were also measured for some specimens up to 19 T in a Bitter magnet at the National Magnet Laboratory. For the measurements in the superconducting magnets, the specimens were wound on a U-shaped graphite block and heated in a vacuum furnace at temperatures from 650°C to 750°C for 48 to 384 h. The block has

Table 1. Description of the Multifilamentary Wires

Wire Types	No. of Filaments	Filament Size* (μm)	Bronze†/Nb Ratio, R	Nb (%)	Cu (%)	Ta (%)
Airco #221	1159	5.8	2.04	24	15	12
Airco #222	703	6.3	3.29	17	15	12
Airco #223	1045	4.9	3.71	15.5	15	12
Airco #224	703	7.0	2.48	21	15	12
Airco #225	361	9.7	2.48	21	15	12

*These sizes are at the wires size of 0.045 mm.
†The composition of the bronze matrix is Cu-13 wt.% Sn.

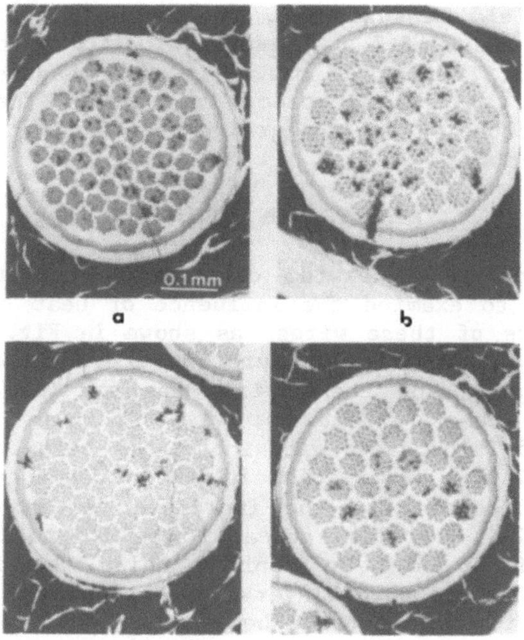

Fig. 1. Cross-sectional views of the multifilamentary Nb_3Sn wires; (a) #221 (R=2.0), (b) #222 (R=3.3), (c) #223 (R=3.7), and (d) #224 (R=2.5).

a 30-mm-radius semicircular end to form the U-shaped wires. After heat treatment, the wires were removed from the mold and placed on a specimen holder, which was made of a fiberglass composite (G-10) and which was exactly the same shape as the graphite block. Then, the holder was inserted into the split-pair magnet so that the plane of the U-shaped holder was perpendicular to the magnetic field. In this arrangement, the current in the wire was always perpendicular to the field. The magnet produces a maximum field of 10 T at the center, but it provides a uniform 9.5 T over the semicircular section of the wire. The voltage leads were attached at the ends of this section and the current connections were made at a section well outside the magnet. A criterion of 2 pΩ•cm for I_c was used. The values of J_c presented here are an average of at least two measurements. In some cases where damage to the wire was suspected, a number of wires were tested. In most cases, the difference in J_c for two sections of a wire was ~5% or less. For smaller wires (~0.2 mm), twisted triplex wires were made for heat treatment and the measurements of I_c. As shown in Fig. 1, there were longitudinal cracks in some sections of these wires, but it was not obvious whether these had any deleterious influence in the determination of I_c of these wires.

For high-field measurements, ~37-mm-long straight wires were used for measurements of I_c. The criterion for I_c that was used in this case was 1 μV over ~15 mm. For the measurements of Nb_3Sn layer thickness, optical microscopy was used. The grain size of Nb_3Sn and the morphology of the grains were studied with a scanning electron microscope.

RESULTS AND DISCUSSION

A number of heat-treating conditions for the formation of Nb_3Sn were used to examine the influence of heat treatment on the critical currents of these wires, as shown in Fig. 2. First, the influence of the heat-treating duration at 725°C on the critical-current density, J_c, of these wires was examined. As shown in Fig. 2, in all cases except for wire #223, J_c of these wires (0.41 mm) at 9.5 T increased with increasing heat-treatment time at least up to 192 h. Examination of the cross-sectional areas of these wires indicated that the increased values of J_c in these wires (#221, #222, and #224) appear to be due to increases in the total amount of Nb_3Sn. To confirm this, the critical-current densities, J_c', for the Nb_3Sn layers were calculated for these wires using the following relationship between J_c and J_c':

$$J_c = 1.37(R + 1)^{-1}(1 - x)J_c' \qquad (1)$$

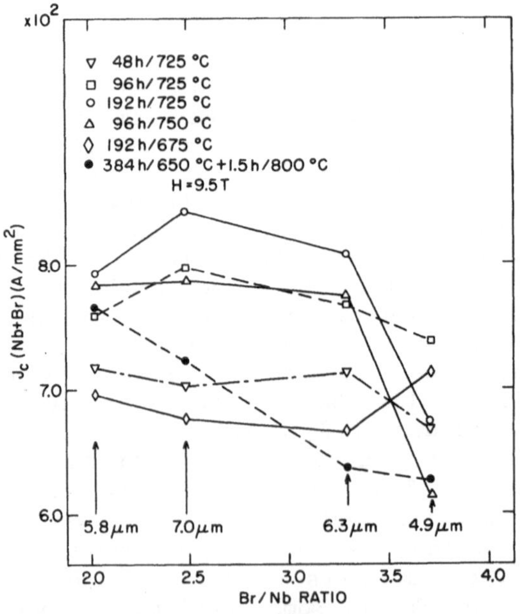

Fig. 2. The critical current densities, J_c, of the wires at 9.5 T as functions of heat treatments and the bronze/Nb ratio, R.

where 1.37 is for the volume increase when Nb is converted to Nb_3Sn, and x is the amount of unreacted Nb. The J_c' of the wires is shown in Fig. 3 as a function of the ratio, R, for the selected heat treatments. As shown in the figure, there was very little change with the heat treatments in the values of J_c' for wires #221, #222, and #224 when the heat-treatment duration was increased from 48 h (not shown) to 96 h and 192 h at 725°C. Thus, the increased J_c with the heat-treatment time discussed above is due to the increased Nb_3Sn layer thickness for these wires. However, J_c' for wire #223, which has the highest ratio and the smallest filaments (~5 µm), decreased slightly for the same heat treatment. As a result of this, J_c of this wire decreased with the heat-treatment duration, as shown in Fig. 2. However, this leaves another question: Why did the value of J_c' decrease with increased heat-treatment duration from 96 h to 192 h for #223 but not for the other wires, #221, #222, and #224? Examination of grain size and morphology of fractured Nb_3Sn filaments using a scanning electron microscope indicated very little difference in the grain sizes for two wires of #223, which were heated for 96 and 192 h at 725°C. The only difference noted was that the grains in the wire with the shorter heat treatment appeared to be more columnar than those in the other wires. It appears that, at least for these wires, the current-carrying capacity of Nb_3Sn decreases if the wire was heat-treated beyond the time required to complete the total conversion of Nb filaments, but it does not change significantly as long as the growth of the layer is progressing, as in the case of other wires.

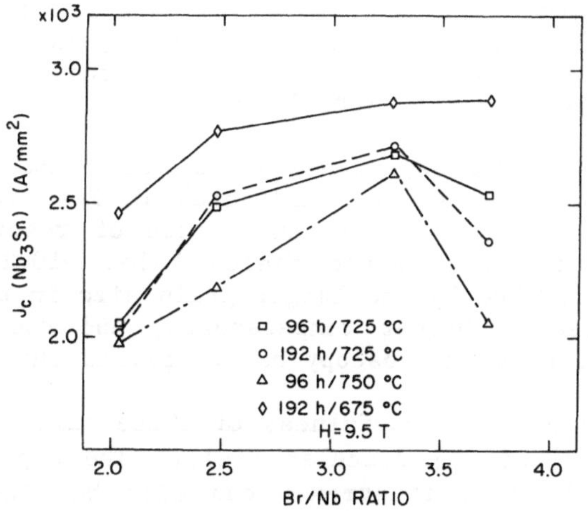

Fig. 3. The critical current densities, J_c', of the Nb_3Sn layers at 9.5 T as a function of the bronze/Nb ratio, R.

One can also note that the values of J_c' depend on the ratio of the bronze to the Nb filaments. To account for the variation we note the following: (1) The rate of Nb_3Sn layer growth increases with the ratio and this results in smaller grains and thus higher J_c'. (2) The compressive strain on Nb_3Sn filaments increases with R and this reduces J_c in as-prepared wires. Preliminary measurements on the influence of tensile strain on J_c for wire #224 (R = 2.58) indicate that very little compressive strain ($-\varepsilon < 0.2\%$) resulted on the Nb_3Sn when the wire was heated more than 48 h at 725°C.[5] Thus, the difference in J_c' between wires #221 and #224 was due to the difference in their grain size, which is the result of a smaller amount of Sn being available in wire #221 (R = 2.0). The values of J_c' measured in wire #222 (R = 3.3) were slightly higher than those of #224 (R = 2.5) for all heat treatments shown in this figure; again, this was because the grains in #222 were smaller than those in #224 and #221 owing to the faster growth of the layers in #222. Also, J_c' for wire #222 was probably reduced by the compressive strains from the matrix, therefore, keeping the difference in J_c' in these wires smaller. However, J_c for wire #223 (R = 3.7) decreased significantly from the value for #222 when they were heat-treated at 725°C and 750°C, whereas J_c' for this wire heated at 675°C for 192 h was essentially the same as that for #222. Although it is expected that there exists a substantially larger compressive strain in the Nb_3Sn for this wire than for all the others to suppress J_c' in this wire, it appears that additional sources for the lower J_c in #223 exist in addition to the prestrain. The electron micrographs of the Nb_3Sn grains in these wires showed that the smaller the ratio R, the more columnar the grains are; among wires #221, #222, and #224, it was found that the larger the ratio, the smaller the columnar width. This leads to a higher J_c for larger values of R, as discussed above. However, it was again difficult to attribute the above J_c variation of #223 to the difference in the grain size. Further studies are required to clarify the behavior of J_c' in #223.

The trend in the dependence of J_c' of these wires on heat-treatment temperatures is also indicated in Fig. 3. Generally it was found that the higher the temperature of reaction, the lower the J_c, at least for magnetic fields below ~10 T. Again, this trend can be described by the larger grain size in those wires that were heat-treated at higher temperatures, and the observation of the grains by scanning microscopy has confirmed this assumption.

What determines the usefulness of these wires is the value of J_c rather than J_c' at a particular field. The dependence of J_c on the ratio and the heat treatments can also be studied in Fig. 2. The highest J_c among these wires was achieved when wires having R = 2.5 to 3.3 were heat-treated at 725°C for 192 h. When the heat-treatment duration was shorter, the decrease in J_c was primarily due to the decreased Nb_3Sn area, and when the temperature was high, the decrease in J_c was due to the decrease in J'_c. When these wires were heated at 675°C for 192 h, J_c was ~15 to 20% lower than that in the wires that were heated at 725°C for 192 h. However, J'_c for these wires was ~10% higher than those of the latter, indicating that a longer heat-treatment time at this temperature may result in higher values of J_c for these wires.

In attempts to achieve higher values of J_c and following the suggestion by Schauer and Schelb,[6] these wires were heated at 650°C for 384 h and then given additional heat treatments at 800°C for 0.5 and 1.5 h. The J_c after one of these heat treatments (1.5 h) is also shown in Fig. 2. The J_c and J'_c of these wires without these treatments were substantially lower than the values discussed above, and the values increased by ~15% after the additional heat treatments. However, they were still lower by a significant margin than the best values achieved in wires #224 and #222 after a heat treatment of 725°C for 192 h.

To examine the influence of the filament size on J_c, three wires (#222, #223, and #224) were drawn to 0.202 mm and to 0.643 mm. In these wires the size of the filaments varied from ~3 to 11 μm. They were heat-treated at 725°C for 96 h and 192 h. Interestingly, for these heat treatments, J_c for #222 and #224 changed only slightly (less than 10%) with filament size in contrast to the earlier observation reporting an increased J_c with reduced filament size.[4] Again, the wire #223 behaved differently from the other two wires. The J_c for this wire increased as the filament size decreased from ~7 to ~5 μm; then it decreased as the size reduced to ~4 μm and to ~3 μm. The low value of J_c for the larger filaments was probably due to the presence of unreacted Nb, and for the finer filaments, due to continued heat treatment after the filaments were totally reacted. Thus, at least for these heat-treating conditions, the reduction in the filament size to ~3 μm was not effective in increasing J_c of these wires. However, it might be possible to have some other heat-treatment conditions that would increase J_c when the filament size is reduced.

The measurements of the critical currents of these wires at higher magnetic fields were made up to 19 T using a Bitter magnet. Since the criterion for I_c was not as sensitive as for the measurements at 8-9.5 T, the critical currents noted at 12 T in these measurements were higher than the values extrapolated from the 8-9.5 T measurements. However, as shown in Fig. 4, the trend in the dependence of J_c on the ratio R was essentially the same. At very high fields ($H > 15$ T), the ratio becomes an important factor in determining J_c since this influences the critical fields of the Nb_3Sn filaments.[7] Hence, the wires with a larger R tend to have a lower J_c and J'_c, as illustrated in the figure. The effect of the filament size on the critical currents was also examined for a heat treatment at 725°C for 192 h. For this treatment, very little difference was observed in J_c and J'_c for wires with filament sizes of ~5-7 μm and ~3 μm.

As pointed out earlier,[2] the critical-current densities, J_c, of commercial wires varied significantly, and the better values were nominally 1 kA/mm^2 at 10 T. The highest value of J_c at 10 T was 1.35 kA/mm^2 reported by Walker et al.[7] This particular wire had very fine filaments (<1 μm) and had a heat treatment of 650°C for 286 h with R = 3.13. Assuming all of the filaments were reacted, J'_c for this wire was 4.1 kA/mm^2, which is ~2/3 of the value found for monofilamentary wires (heat-treated at 725°C for 64 h). It is also interesting to note that a J_c of 1 kA/mm^2 and a J'_c of 4.2 kA/mm^2 were achieved in a wire with a very high ratio, R = 4.7, and heated at 750°C for 168 h.[4] The fact that two wires with such drastically different heat treatments and values of R can achieve the same value of J'_c is very intriguing.

The above values of J_c and J'_c were substantially lower than those in the wires reported here; i.e., the extrapolated values of critical current densities were J_c(10 T) = 750 A/mm^2 and J'_c(10 T) = 2.3 kA/mm^2. Why a 1 kA/mm^2 or a higher level of J_c(10 T) in these wires was not achieved is not clear. Further investigation on these wires continues in order to examine the possibility of achieving higher J_c's in these wires or to determine the reason for J_c not reaching that level.

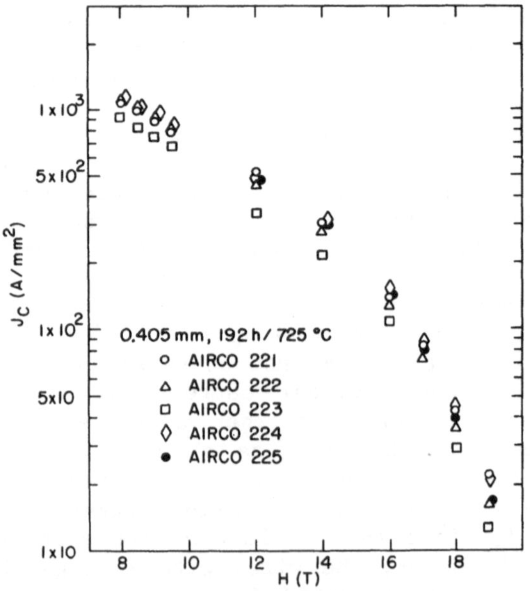

Fig. 4. The critical current densities, J_c, of the wires as a function of magnetic fields.

ACKNOWLEDGMENTS

The authors appreciate excellent technical assistance by A. Cendrowski, R. Sabatini, and O. Kammerer. The use of the high-field facility at the National Magnet Laboratory is also appreciated.

REFERENCES

1. "Filamentary A15 Superconductors," M. Suenaga and A. F. Clark, eds., Plenum Press, New York (1980).
2. M. Suenaga, W. B. Sampson, and T. S. Luhman, IEEE Trans. Magn. MAG-17:646 (1981).
3. M. Suenaga, in: "Proceedings of NATO Advanced Study Institute on Superconducting Materials," Sintra, Portugal (1981), to be published.
4. D. C. Larbalestier, IEEE Trans. Magn. MAG-17(5):1668 (1981).
5. T. S. Luhman, Brookhaven National Laboratory, Upton, New York, unpublished results.
6. W. Schauer and W. Schelb, IEEE Trans. Magn. MAG-17:374 (1981).
7. M. S. Walker, J. M. Cutro, B. A. Zeitlin, G. M. Ozeryansky, R. E. Schwall, C. E. Oberly, J. C. Ho, and J. A. Woollan, IEEE Trans. Magn. MAG-15:80 (1979).

AN INVESTIGATION ON THE ENHANCEMENT OF THE CRITICAL CURRENT DENSITIES IN BRONZE-PROCESSED Nb_3Sn

M. Hong,* I. W. Wu, J. W. Morris, Jr., W. Gilbert, W. V. Hassenzahl, and C. Taylor

Lawrence Berkeley Laboratory, University of California, Berkeley, California

INTRODUCTION

Multifilamentary Nb_3Sn A15 superconductors fabricated by the "bronze" process[1,2] or its variants[3,4] have been intensely studied and are strong candidate materials for the construction of magnets to be used for high energy physics and magnetic fusion. In the bronze process, the A15 compound is formed at the interface between a Nb filament and the Cu-Sn bronze matrix when the composite is at high temperatures. Among the superconducting properties of this system, the one of greatest engineering interest is the critical current density, J_c. It is well known that $J_c(H)$ is strongly affected by metallurgical parameters,[5,6] such as A15 grain size and Sn content in the A15 layer. There have been many attempts to improve $J_c(H)$ for the bronze process. These include the optimization of heat-treatment conditions[7,8] and the efforts to find appropriate alloying additions.[9-11] However, the fundamental relationships between processing, microstructure, and $J_c(H)$ remain poorly understood.

The work reported here addressed the problem of improving the critical current characteristic of a commercial multifilamentary Nb_3Sn strand[7] by varying its heat treatment. The work was done from the perspective that the critical current characteristic is controlled by the metallurgical state of the reacted layer, which is, in turn, fixed by the processing the wire has undergone. The research was carried out in parallel with metallographic studies,[12] which analyzed the microstructure and composition profile within

*Present address: Bell Laboratories, Murray Hill, New Jersey.

the reacted Nb_3Sn layer as a function of heat treatment. The combined results of metallographic and processing research suggest that it is possible to "engineer" the microstructure of the reacted layer to improve $J_c(H)$. The specific product of the work is a tailored double-aging treatment that introduces a favorable combination of microstructure and composition in the reacted layer and causes a substantial improvement in the critical current characteristic of the strand.

EXPERIMENTAL WORK

The multifilamentary Nb_3Sn strand used in this study was manufactured by Airco Superconductors[7] and has a diameter of 0.7 mm. It contains 2869 (19 x 151) Nb filaments in a Cu-13 wt.% Sn bronze matrix, a Ta diffusion barrier, and stabilizing Cu that comprises 64.5% of the cross-sectional area of the strand.

The superconducting A15 phase was formed by heat treatments at temperatures between 650 and 800°C for various periods of time. Samples were sealed in quartz tubes under a low-pressure Ar atmosphere with Ti rods to minimize contamination.

Critical currents, I_c, were measured using a standard four-probe technique in transverse applied magnetic fields up to 19 T. The critical current was taken to be that current at which the potential difference across voltage leads spaced 5 mm apart exceeds 1 μV, and the overall current density, J_c, was taken to be the critical current divided by the non-Cu area of the strand, 0.14 mm^2. Samples for the I_c measurement were in a U-shape with a 25-mm length in the transverse magnetic field.

In the bronze process, the Nb_3Sn A15 phase is formed by the diffusion reaction between the Nb core and the Cu-Sn bronze matrix, but the basic thermodynamics and kinetics pertaining to this reaction are inadequately characterized.[13] Owing to a lack of quantitative data about the Cu-Nb-Sn ternary equilibrium phase diagram, specifically the detailed shape of the A15 phase range and its tie lines to the Cu-Sn solid solution at various temperatures, it is not possible to predict the Sn concentration across the reacted A15 layer. Therefore, to determine the Sn composition of the A15 layer, we employed the STEM/EDAX technique to do this microchemical analysis.

The microstructure of the reacted wire was characterized through a combination of scanning electron microscopy (SEM), transmission electron microscopy (TEM), and scanning transmission electron microscopy (STEM), as described in Ref. 12. The composition measurements given below were determined from energy dispersive x-ray analysis (EDAX) in the STEM and were computed from the total integrated intensities of the characteristic x-ray peaks of Nb and

Sn using the "thin-foil" approximation.[14] The absolute accuracy of the Sn concentration is ~3 at.% in the composition range of interest. The statistical error in determining the relative Sn concentrations at two different points within an A15 layer is, however, less than 1 at.% Sn. Hence, the Sn profiles determined should be reasonably accurate in shape, but may have a small error in absolute value. The spatial resolution of the STEM analysis is limited both by beam size and foil dispersion to ~500 Å.

RESULTS AND DISCUSSION

The Influence of Aging Temperature on $J_c(H)$

The materials studied in this work were commercial superconductors. Thus, the only processing variable available for property improvement was the reaction heat treatment, which forms the Nb_3Sn layer on the Nb filament. Previous work[7] on similar strands suggests that the optimum single heat treatment lies in the range 700-800°C and requires a reaction time on the order of two days. The results reported in Ref. 7 are essentially confirmed by the data presented in Fig. 1. Good high field properties resulted from two-day reaction treatments in the range 700-800°C; the best critical current characteristic was obtained after reaction at 730°C. Other heat treatments are included in the figure for comparison.

The Relation between Aging Temperature, Microstructure, and $J_c(H)$

Although two-day heat treatments and temperatures between 700 and 800°C led to reasonable superconducting properties, subsequent metallurgical analysis[12] revealed that major differences occurred in the metallurgical state of the Nb_3Sn layer as the aging temperature was varied through this range. At lower temperatures, the reacted layer had a good structure, but a less desirable composition profile. As the aging temperature rose, the structure deteriorated, but the composition profile improved dramatically. It is apparently the balance of the two effects that causes the maximum in the critical current characteristic near 730°C.

The microstructure of the reacted layer was qualitatively similar for all heat treatments studied. It may be visualized as three concentric cylindrical shells surrounding an unreacted Nb core. The three shells are distinguished by the structure of the A15 grains they contain. They are illustrated in the schematic cross section shown in Fig. 2. The inner shell consists of large columnar grains, which emanate from the Nb core. The central shell contains fine, equiaxed grains, which probably form through the break-up of the columnar grains during film growth. The outer shell contains relatively large, equiaxed grains, which presumably represent coarsening of the finer grains in the interior. Both

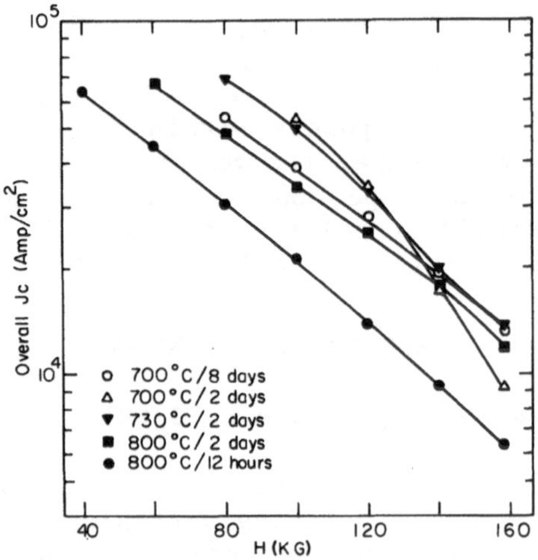

Fig. 1. The overall J_c vs. the applied transverse magnetic field for the bronze process multifilamentary Nb_3Sn superconductors with single-temperature aging.

the overall dimensions of the reacted layer and the relative dimensions of the interior shells depend on heat treatment, as shown in Table 1.

Since the critical current of the A15 phase is a strong reciprocal function of its grain size,[6] the high-field portion of the critical current characteristic should be dominated by the central shell of small, equiaxed grains. The measured thickness of this shell, 0.8 μm, was nearly insensitive to heat treatment temperature for two-day treatments in the range 700-800°C. However, the mean

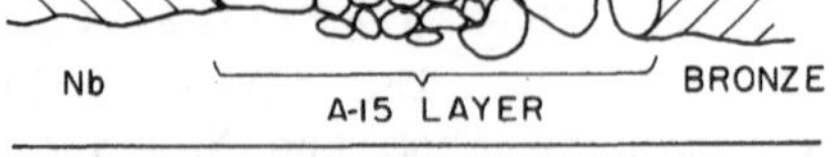

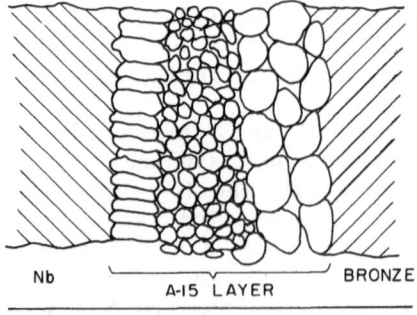

Fig. 2. Schematic of Nb_3Sn microstructure near a Nb filament in the bronze process. Columnar grains are adjacent to the Nb, fine-grained Nb_3Sn is in the center, and coarse grains are close to the bronze, which has been depleted of Sn.

Table 1. Comparison of Layer Thickness, Average Grain Size of A15 Phase, and J_c for Various Heat Treatments

Reaction Temperature Reaction Time	Reacted A 15 Layer Thickness (μm)				Average Equi-axed Grain Size	J_c(10T)	J_c(14T)
	Columnar Grains Near Nb Core	Equi-axed Grains	Large Grains Near Bronze	Total	(Å)	10^4 A/cm^2	10^4 A/cm^2
700°C/ 2 days	0.4	0.8	0.3	1.5	640	5.3	1.7
700°C/ 8 days	0.6	0.9	0.4	1.9	800	3.9	1.9
730°C/ 2 days	0.4	0.8	0.5	1.7	720	5.0	2.0
800°C/ 12 hours	0.6	0.4	0.6	1.6	1090	3.1	0.93
800°C/ 2 days	0.7	0.8	0.8	2.3	1190	3.4	1.2

Table 2. Results of Sequential Heat Treatment

Reaction Temperature Reaction Time	Reacted A 15 Layer Thickness (μm)				Average Equi-axed Grain Size	J_c(10T)	J_c(14T)
	Columnar Grains Near Nb Core	Equi-axed Grains	Large Grains Near Bronze	Total	(Å)	10^4 A/cm^2	10^4 A/cm^2
730°C/ 2 days	0.4	0.8	0.5	1.7	720	5.0	2.0
700°C/ 2 days + 730°C/ 2 days	0.4	1.0	0.5	1.9	730	6.2	2.6
700°C/ 4 days + 730°C/ 2 days	0.5	1.2	0.4	2.1	760	6.7	2.8
650°C/ 16 days + 800°C /4 hours	0.4	0.5	0.8	1.7	930	2.9	1.2

grain size within the shell increased with temperature from ~640 Å at 700°C to ~1200 Å at 800°C. The value of J_c at 10 T varied roughly with the reciprocal of the grain size, as is expected if the critical current is determined by the substructure of the equiaxed layer. At 14 T, on the other hand, the critical current is not simply related to grain size; it is largest for the intermediate grain size associated with the 730°C treatment. This behavior suggests that the composition of the central shell also affects high-field properties.

The composition profile across the reacted layer is plotted in Fig. 3 for the three 2-day heat treatments listed in Table 1. In each case there is a gradient in the Sn concentration, from a Sn-rich composition at the bronze interface to a Sn-poor concentration at the Nb interface. The composition passed through the stoichiometric value within the fine-grained shell (a phenomena which certainly contributes to the success of the bronze process). As the reaction temperature is raised, the reacted layer thickens and the difference between the two end-point compositions narrows. Both effects cause the composition profile to flatten, increasing the thickness of the near-stoichiometric shell within the reacted layer. There is, in addition, an exaggerated flattening of the Sn profile near the stoichiometric composition, which is most pronounced for the 800°C treatment and is responsible for the irregular shapes of the 730°C and 800°C profiles. This effect further increases the near-stoichiometric subvolume of the reacted layer. Since the upper critical field of Nb_3Sn decreases as deviation from chemical stoichiometry increases, a high superconducting current requires a more nearly stoichiometric compound as the field is

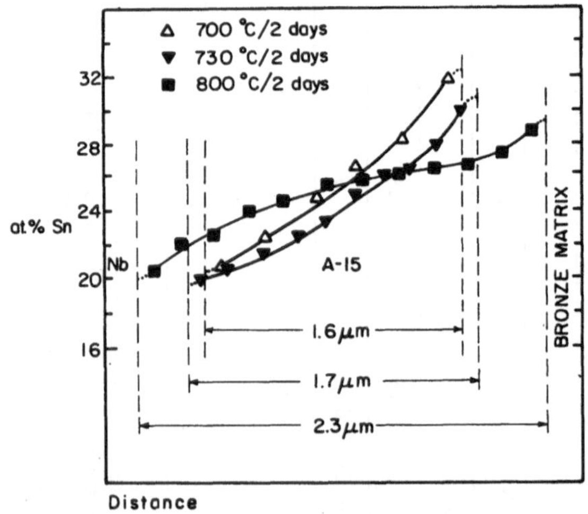

Fig. 3. Sn-concentration gradient across the reacted Nb_3Sn A15 layer for two-day, single-aging treatments of 700°, 730°, and 800°C.

raised. Because the high-field current will be carried primarily by a subshell of the reacted layer, whose thickness decreases with increasing field given a composition gradient through the reacted layer, the J_c at high fields decreases more rapidly the sharper the composition gradient in the near-stoichiometric region.

From this perspective, it is not surprising that the 700°C treatment, which establishes a sharp concentration gradient through the reacted layer, yields a strand whose critical current decreases rapidly for fields above 12 T. The 800°C treatment establishes a relatively flat composition profile at a nearly stoichiometric composition, but an unfavorable microstructure compromises its high-field properties. The 730°C aging treatment provides a favorable combination of grain size and composition profile and yields the best critical current characteristic for fields in the range 12 to 16 T.

Development of an Improved Aging Treatment

The results described in the previous section suggest that the critical current characteristic of the strand is determined by three metallurgical parameters that should be simultaneously controlled: (1) the volume of the fine-grained shell within the reacted layer, which should be as large as possible; (2) the A15 grain size within the fine-grained shell, which should be as small as possible; and (3) the composition profile across the reacted layer, which should be as flat and nearly stoichiometric as possible. Given that the composition of the reacted layer is Sn-rich at the bronze interface and Sn-poor at the Nb interface, there is always at least some near-stoichiometric material within an incompletely reacted layer. Hence, the composition profile primarily affects $J_c(H)$ at very high fields (>12 T in the present case); the lower field properties are largely determined by the grain size and the volume of the fine-grained shell.

These considerations suggest three possible approaches to improve $J_c(H)$: (1) long-time, low-temperature aging with the intent of increasing the thickness of the fine-grained shell and flattening the composition profile while retaining fine grain size; (2) short-time, high-temperature aging with the intent of achieving fine grain size by limiting the time available for grain coarsening while preserving a good composition profile; and (3) sequential aging treatments at different temperatures chosen so that one aging establishes a favorable microstructure while the second improves the composition profile. Each of these approaches was explored.

For reasons that appear to be fundamental, neither a long-time, low-temperature nor a short-time, high-temperature aging was successful in improving $J_c(H)$. Increasing the aging time to 8 days at 700°C (Fig. 1, Table I) did result in a small increase in the critical current above 14 T, but at substantial cost to the critical current at lower fields. Metallographic analysis showed that the additional heat treatment led to a coarsened grain structure with only a slight increase in the thickness of the fine-grained shell. Raising the temperature to 800°C and decreasing the aging time to 12 h caused a decrease in $J_c(H)$ at all fields. Parallel metallographic analysis (Table 1) revealed a large grain size and a thin central shell.

Given the limitations of modified single-aging treatments, a "double-aging" treatment was designed to establish a favorable microstructure and a favorable composition profile in sequential steps. The microstructure was developed by an initial heat treatment for 2-4 days at 700°C; the composition profile was then improved by an additional aging at 730°C for 2 days. The choice of the 730°C aging temperature was based on earlier results, reported above, which show that 730°C provides a reasonable composition profile without exaggerated grain coarsening. An alternate double-aging proposed in Ref. 8, of 650°C for 16 days plus 800°C for 4 h was also tried for comparison.

The critical current characteristics of two 700/730°C double-aged specimens are plotted in Fig. 4. These double-aging treatments caused a substantial increase in $J_c(H)$ over the entire range tested. The treatment of 700°C, 4 days plus 730°C, 2 days increased J_c by ~40% with respect to the best single-aging treatment. Metallographic analysis of the specimens revealed the source

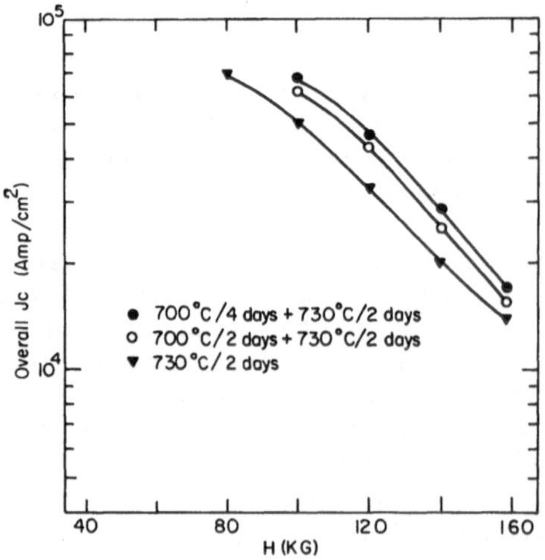

Fig. 4. The overall J_c vs. applied transverse magnetic field for bronze process multifilamentary Nb_3Sn superconductors with double-aging treatments. (Single aging of 730°C for 2 days is used for comparison.)

of these good properties. The 700/730°C double-aging treatments provided a relatively thick shell of fine A15 grains, as shown in Table 2 and simultaneously established a relatively flat Sn profile at a near-stoichiometric composition, as shown in Fig. 5.

The 650°C, 16-day plus 800°C, 4-h treatment, on the other hand, gave properties inferior to those of the single-aged specimens. Metallographic analysis showed that the reacted layer had, in this case, both a poor microstructure and a poor composition profile. Further, this reaction sequence is not practical for large magnets.

CONCLUSION

We conclude from the research reported here that the critical current characteristic of bronze-processed multifilamentary Nb_3Sn superconducting wire is largely controlled by three metallurgical parameters: (1) the volume of the fine-grained central shell within the reacted A15 layer; (2) the A15 grain size within the fine-grained shell; (3) the composition profile through the reacted layer, which passes from a Sn-rich composition at the bronze interface to a Sn-poor composition at the Nb interface. Each of these features is influenced by heat treatment. Fine grain size is promoted by low temperature reaction, whereas the composition profile is better after reaction at higher temperature. A sequential double-aging treatment at 700°C, 4 days plus 730°C, 2 days establishes a superior combination of microstructure and composition and yields critical currents (J_c) that are substantially above the best

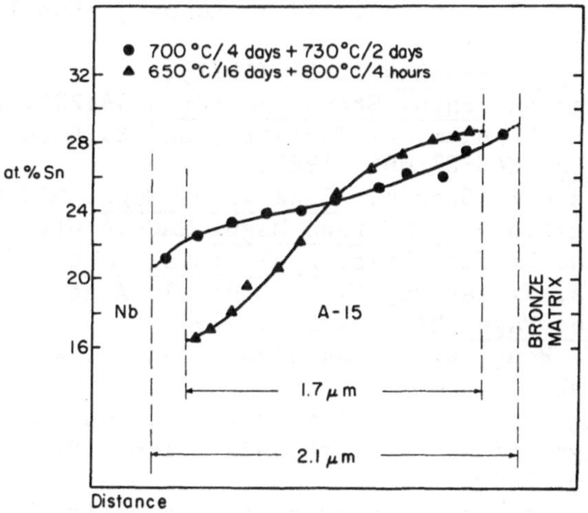

Fig. 5. Tin concentration gradient across the reacted Nb_3Sn A15 layer for superconductors with double-aging treatments.

obtained from single-aging treatments over the entire range of fields tested (10-16 T).

The results presented here suggest that still better heat treatments can be devised, as the best microstructures obtained in this work are still far from the best that can be imagined. It seems clear that additional research on metallurgical heat treatment, perhaps coupled with minor changes in manufacturing processes, will lead to important gains in the high-field superconducting properties of multifilamentary Nb_3Sn.

ACKNOWLEDGMENTS

This work was supported by the Division of High Energy Physics and by the Division of Materials Sciences, U.S. Department of Energy under Contract No. W-7405-ENG-48. The $J_c(H)$ data was obtained at the Francis Bitter National Magnet Laboratory which is supported at MIT by the National Science Foundation. The authors greatly thank J. T. Holthuis, D. R. Dietderich, L. G. Rubin, and B. Brandt for experimental assistance.

REFERENCES

1. A. R. Kaufman and J. J. Pickett, Bull. Am. Phys. Soc. 15:838 (1970).
2. K. Tachikawa, in: "Proceedings of the Third International Cryogenic Engineering Congress," Iliffe Science & Technology, Guildford, Surrey, England (1971), p. 339.
3. Y. Hashimoto, K. Yoshizaki, and M. Tanaka, in: "Proceedings of the 5th International Cryogenic Engineering Conference," IPC Science & Technology Press, Guildford, Surrey, England (1974), p. 332.
4. M. Suenaga, W. B. Sampson, and T. Luhman, IEEE Trans Magn. MAG-17:646 (1981).
5. J. S. Caslaw, Cryogenics 11:57 (1971).
6. J. D. Livingston, Phys. Status Solidi a 44:295 (1977).
7. P. A. Sanger, E. Adam, E. Ioriatti, and S. Richards, IEEE Trans Magn. MAG-27:666 (1981).
8. W. Schauer and W. Schelb, IEEE Trans Magn. MAG-17:374 (1981).
9. J. D. Livingston, IEEE Trans Magn. MAG-14:611 (1978).
10. H. Sekine and K. Tachikawa, Appl. Phys. Lett. 35:472 (1979).
11. K. Tachikawa, Y. Tanaka, Y. Yoshida, T. Asano, and Y. Iwasa, IEEE Trans Magn. MAG-15:391 (1979).
12. I. W. Wu, M. Hong, W. V. Hassenzahl, and J. W. Morris, Jr., to be published.
13. J. D. Livingston, in: "Filamentary A15 Superconductors," M. Suenaga and A. F. Clark, eds. Plenum Press, New York (1980), p. 363.
14. J. I. Goldstein, in: "Introduction to Analytical Electron Microscopy," J. J. Hren, J. I. Goldstein, and D. C. Joy, eds., Plenum Press, New York (1979), p. 83.

THE PROPERTIES OF MULTIFILAMENTARY WIRES ON Nb_3Sn BASE PREPARED BY THE HYDROEXTRUSION METHOD

A. A. Galkin, V. P. Buryak, N. I. Matrosov, A. B. Dugadko,
B. A. Shevchenko, L. A. Dereza, G. A. Korneeva, and O. N. Mironova

Physical-Technical Institute, Donetsk, USSR

V. M. Pan, V. S. Flis, Yu. I. Beletsky,
M. I. Tsypin, L. A. Malysheva, and T. D. Manchenkova

Institute of Metal Physics, Kiev, USSR

INTRODUCTION

One of the most promising methods of preparing superconductors is a solid phase diffusion method.[1-3] The method has been known long enough and has been adopted by many laboratories of the world. However, wires, including multifilamentary ones, prepared from material of this class on an industrial scale are restricted. This is mainly because bronzes, in particular the tin bronzes being used by the solid-phase diffusion method in preparing Nb_3Sn superconductors, possess a low plasticity, and therefore they are difficult to process in the traditional ways (hot pressing and drawing). Hot pressing, as the first stage of fabrication, is usually made in the temperature range where diffusion processes with formation of Nb_3Sn compounds may effectively proceed. A drawing is accompanied by frequent thermal treatments that promote additional formation of brittle intermetallic compound, making it difficult to treat the substance further.

In most cases, the difficulties of fabrication limit the tin content in the bronze to 8 to 10%, although the solubility limit of tin in copper is about 15%. This results in a decrease of tin concentration in the diffusion zone and exerts a detrimental effect on the final properties of the superconductors. Therefore, one of the problems in the preparation of multifilamentary Nb_3Sn wires is the development of such bronzes and their cold-treatment techniques.

The bronze must be able to withstand repeated cold-working applications and have high superconducting characteristics.

In this work, Cu-Sn bronzes with tin content from 8 to 14% were used. To improve their plasticity, some were alloyed with a rare-earth element. The effects of bronze compositions and alloying on their straining, kinetics of Nb_3Sn formation, and superconducting properties were investigated.

A hydrostatic extrusion method (cold strain by means of a high-pressure liquid) was used as the deformation method. An application of hydrostatic extrusion for making the superconducting multifilamentary Nb_3Sn-based cable, which is described in the literature,[2] involves the use of hydrostatic extrusion at 750°C with a 99% degree of strain in the first three of five stages during its treatment. This treatment method was used because it was estimated to be economically efficient, but the properties of the final product were not considered. However, it is quite clear that the strain accomplished at such a high temperature is sufficient to cause a reactive selective diffusion, resulting in a formation of intermetallic Nb_3Sn compound and in an essential embrittlement of composite material. It is difficult to treat embrittled composite material. Furthermore, obtaining the high operating characteristic in this material is very problematic.

Unlike the situation described above, we used cold hydrostatic extrusion without any external heating of billets, working medium, and container. The treatment method mentioned above possesses a number of advantages:

1. The possibility of obtaining multifilamentary cables, including complex combined superconductors, just as one obtains single-filament cables, without principal limits on the number of filaments and the relative position of constituent elements.

2. The elimination of the hot extrusion process at all stages of technology.

3. A decrease (by 2-5 times) in the number of stage limits corresponding to the drawing during the traditional technological process.

4. The possibility of preparing a favourable structure and stressed state of the composite material, allowing an increase in the kinetics of superconducting compound formation.

MATERIALS AND METHODS OF INVESTIGATION

Tin bronzes were used as the initial materials; they contained from 8 to 14% tin in bulk and were 300 mm in diameter in the hot extrusion state and 70 mm in the homogenized state. A bronze of Cu-9% Sn-9% Zn composition was also used. Niobium bars and tubes from a vacuum electron-beam remelting were used for preparation of bimetallic billets.

The study included an estimation of the deformability of the bronzes and bimetallic billets by the hydropressing method, metallographic investigations, electron microprobe analysis, and an investigation of superconducting characteristics.

The limit value of the strain was that strain at which billet integrity was maintained without the cracks and microcracks being observed on it. A measurement of critical current was performed by a standard four-point circuit with a sensitivity level of 1 μV/cm, and the transition temperature to the superconducting state was measured by resistive and inductive methods as accurately as ±0.05 K.

EXPERIMENTS

We made numerous experiments to determine the value of a single strain, achieving 65% for most of the bronzes under investigation. Then the billets received thermal treatments at temperatures of 450-550°C, depending on their diameter, prehistory, and bronze composition.

Along with a direct arrangement of constituent elements in bimetallic billets (niobium rods in bronze tubes), inverted and combined assemblages with the inverted and combined (niobium-bronze-niobium) arrangements of the elements were also used.

Single-filament bimetallic wires with a diameter of 1.0 mm were prepared for direct assemblage with various compositions of bronze matrix. In addition to the direct, inverted, and combined assemblages, 55-filament cables with a diameter of 0.5 mm were obtained by means of the hydrostatic extrusion method of strain with intermediate thermal treatments.

A typical view of a cross section for various cable constructions is shown in Figs. 1-3.

Microstructural investigations showed that in the wire and in its semifinished items a uniformity of bronze microstructure exists, the initial state of which is characterized by the nonuniform coarse-grain structure of fcc tin solid solution in copper. Good bonding between bronze and niobium, a conservation of the initial configuration in assembled elements, and filament integrity in the multifilamentary wires were also observed.

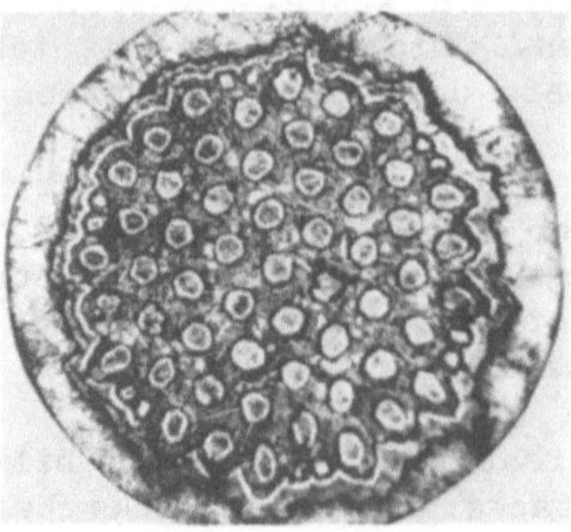

Fig. 1. Typical view of the cross section of a 55-filament wire with a 0.5-mm diameter (direct assemblage).

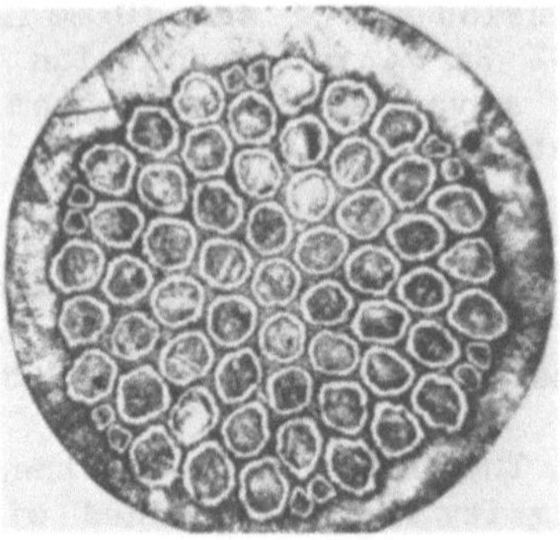

Fig. 2. Typical view of the cross section of a 55-filament wire with a 0.5-mm diameter (inverted assemblage).

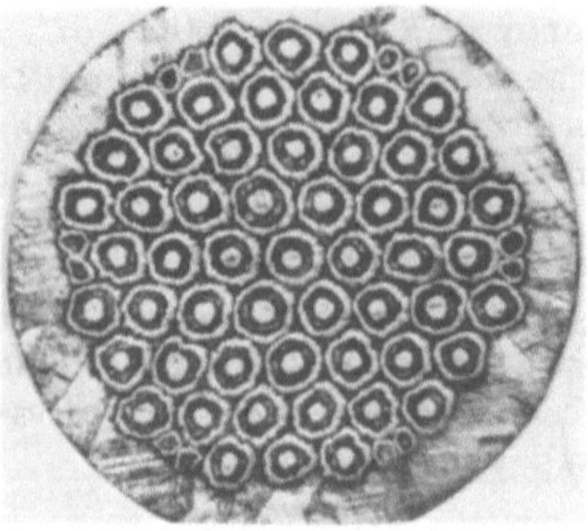

Fig. 3. Typical view of the cross section of a 55-filament wire with a 0.5-mm diameter (combined assemblage).

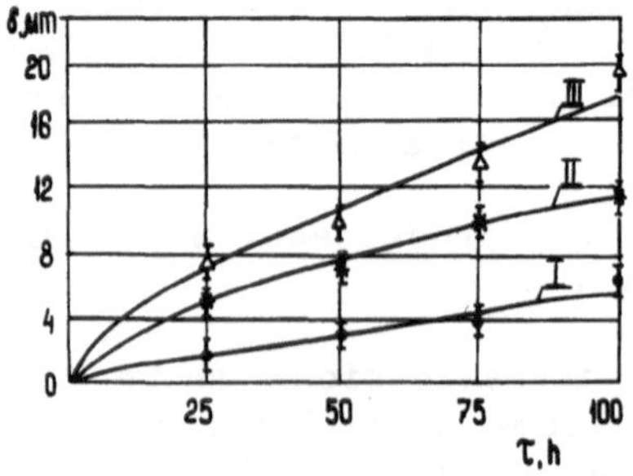

Fig. 4. Thickness of the diffusion layer as a function of type of thermal treatment for the single-filament Cu-Sn-13%-based bronze wire with a diameter of 1.0 mm; I - 700°C; II - 750°C; III - 800°C.

To investigate the kinetics of a superconducting Nb_3Sn layer formation, thermal treatment was performed in the temperature range 700–800°C for 100 h. The data for Nb_3Sn layer thickness as a function of the thermal treatment in single-filament bimetallic wire of 1.4-mm diameter (bronze Cu-13% Sn) are given in Fig. 4.

Increasing the temperature and annealing time increased the layer thickness; moreover, a saturation did not arise at maximum regimes, although the layer thickness was 20 μm and the growth velocity during the 100-h 800°C thermal treatment was considerable, 0.12 μm/h.

The data for the kinetics of Nb_3Sn compound formation in multifilamentary wires of various constructions are given in Fig. 5. The layer achieves a maximum thickness in the direct assemblage (~ 6 μm). For the combined assemblage, taking into account the formation of two layers in each construction element (niobium rod-bronze and bronze-niobium tube boundaries), the occupied area of the A15 compound is a maximum in this case, typically 10 to 13% in cross-sectional area.

As shown by electron microprobe analysis, the niobium-tin ratio in the compound layer is nearly a stoichiometric one. There is no copper in the layer. Subsequent to the thermal treatment at 750°C for 50 h and as a result of diffusion for the combined assemblage, the tin content in bronze of the initial Cu-13% Sn composition decreased to 4%, and for the inverted assemblage, it decreased to 8%. One believes there is a possibility to increase the Nb_3Sn phase layer in the second case with further thermal treatment.

The transition temperatures into the superconducting state were measured on the bimetallic and multifilamentary wire samples.

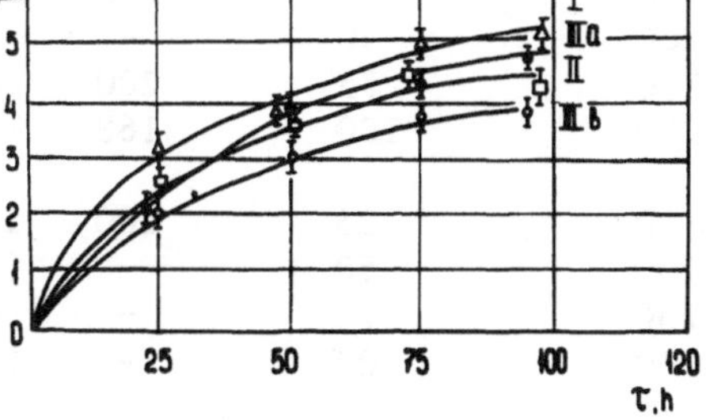

Fig. 5. The kinetics of Nb_3Sn layer growth for the multifilamentary wires of various construction at 700°C; I – direct assemblage; II – inverted assemblage; III – the combined assemblage.

In all cases the temperatures, T_c, as determined in the middle of the transition curve, were 17.04 to 17.97 K with a width of transition of 0.4 to 0.8 K. These high T_c values were close to the maximum value of T_c = 18.0 K for bulk samples and the behaviour of the transition curves testify that the compound composition formed is nearly stoichiometric.

The investigations of critical current in 1.0-mm-diameter single-filament bimetallic samples obtained by using the bronzes of various compositions have shown that the maximum critical current is a function of tin content in the bronze. Alloying of the tin bronze with rare-earth metals increases the critical current. The maximum value of the critical current for this wire is about 200 A in an induced magnetic field of 6 T, which, in terms of constructive current density, is $2.2 \cdot 10^8$ A/m^2.

The critical current data for 0.5-mm-diameter multifilamentary wires with various constructions in 2- to 6-T magnetic fields after thermal treatment at 700°C are given in Table 1. The wire samples of the combined assemblage possess the highest critical currents in induced fields above 5%. This seems to be due to obtaining the maximum filling factor on Nb_3Sn superconductor for all the types of assemblages used. Overall current density, defined as the ratio of the critical current to total section of composite wire, achieves the value of 4 to $6 \cdot 10^8$ A/m^2, but the current density of the Nb_3Sn layers is 1 to $3 \cdot 10^{10}$ A/m^2 in the 6-T magnetic field.

Table 1. The Critical Currents of Multifilamentary Samples in Magnetic Fields

Wire Construction	Annealing Temperature, °C	Annealing Time, h	Critical Current, A, in Magnetic Field				
			2 T	3 T	4 T	5 T	6 T
Direct	700	25	200	140	112	122	81
Direct	700	50	-	-	-	-	92
Direct	700	75	200	146	121	91	82
Direct	700	100	168	135	119	133	91
Inverted	700	25	200	140	106	84	64
Inverted	700	50	-	200	167	91	73
Inverted	700	75	-	-	-	-	-
Inverted	700	100	-	200	158	119	88
Combined	700	25	-	200	160	-	-
Combined	700	50	-	-	-	-	100
Combined	700	75	-	200	157	125	110
Combined	700	100	-	200	164	120	96

CONCLUSIONS

1. Single-filament, 1.0-mm-diameter Nb_3Sn superconducting wires with Cu-Sn bronze bases of different composition as well as 55-filament, 0.5-mm-diameter wires of various construction were prepared by the method of cold hydrostatic pressing.

2. The wires prepared possessed good geometry and adhesion between the constituent construction elements and preserved the filament integrity.

3. The kinetics of Nb_3Sn formation was investigated, and the compound obtained a thickness, which for the bimetallic wires was 20 μm, which is shown to be close to the stoichiometric one.

4. The transition temperature into the superconducting state was 17 to 18 K for the obtained diffusion layers, depending on the type of thermal treatment.

5. The critical currents of the wires produced were high enough. The current density in the diffusion layers was 1 to $3 \cdot 10^{10}$ A/m^2, but the overall current density was 4 to $6 \cdot 10^8$ A/m^2 in the 6-T magnetic field.

REFERENCES

1. A. R. Kaufmann and J. J. Pickett, Multifilament Nb_3Sn superconducting wire, J. Appl. Phys. 42:58 (1971).
2. H. H. Farrel, G. H. Cilmer, and M. Suenaga, Grain boundary diffusion and growth of intermetallic layers Nb_3Sn, J. Appl. Phys. 45(9):4025 (1974).
3. M. Suenaga, W. B. Sampson, and C. J. Klamut, The fabrication and properties of Nb_3Sn superconductors by the solid diffusion process, IEEE Trans. Magn. MAG-11:231 (1975).
4. E. G. Smith, R. I. Fiorentino, E. W. Collings, and F. I. Jelinek, Recent advances in hydrostatic extrusion of multifilament Nb_3Sn and Nb-Ti superconductors, IEEE Trans. Magn. MAG-15(1):91 (1979).

EFFECTS OF TERNARY ADDITIONS ON YOUNG'S MODULUS AND THE MARTENSITIC TRANSFORMATION OF Nb_3Sn*

J. F. Bussière and B. Faucher

Institut de Génie des Matériaux, CNRC, Montreal, Quebec, Canada

C. L. Snead, Jr. and M. Suenaga

Brookhaven National Laboratory, Upton, New York

INTRODUCTION

Recent measurements on bronze-processed Nb_3Sn using a vibrating reed technique have shown that Young's modulus at low temperatures decreases to 0.4 of its room temperature value and that internal friction increases dramatically below 50 K, the martensitic transformation temperature.[1] In the present paper, this technique was used to study softening and the occurrence of the martensitic transformation in bronze-processed Nb_3Sn containing ternary additions. The additions studied were Ta and Ti, which are known to cause substantial improvements of critical current density at high fields,[2-4] and Zr, which improves the growth kinetics of bronze-processed Nb_3Sn.[5] The additions were added to the Nb core and are believed to occupy Nb sites in the A15 lattice. Results show that ternary additions to Nb_3Sn can affect both its structure by suppressing the martensitic transformation and the state of stress, when included in a matrix, by causing large changes in Young's modulus at low temperatures. Both of these effects are of technological interest since the upper critical field, H_{c2}, and critical temperature depend on whether Nb_3Sn is cubic or tetragonal[6,7] and, in the case of composite conductors, on the combined strain resulting from the compression due to the matrix and externally applied stress.[8,9]

*Work supported in part by the U.S. DoE.

SAMPLE PREPARATION AND CHARACTERISTICS

Samples containing Ti and Ta additions were in the form of flattened wires ∿1.5-mm wide and ∿15-mm long. Niobium-titanium alloys having 0.5 and 3 wt.% Ti and a NbTa alloy containing 10 wt.% Ta were arc melted in an Ar atmosphere, homogenized for 1.5 h at 1800°C, and swaged to rods of 3.2-mm diameter. After annealing at 1200°C for 1 h in vacuum, the alloy rods were then inserted into a Cu-13 wt.% Sn bronze matrix of 12.7-mm diameter. The composite rods were then cold-drawn to a final core diameter of 1.5 mm and rolled to a final core thickness of 25 to 30 μm using several intermediate annealing steps at 500°C. The flattened composite wires were then heat-treated in quartz ampoules at 750°C for 50 to 100 h (see Table 1).

Samples containing pure Nb and Zr additions were of tape geometry. The samples made from pure Nb and Nb-1% Zr were prepared by Airco Industries by sputtering 25-μm layers of Cu-13 wt.% Sn bronze on each side of commercially pure Nb and Nb-1% Zr foil of 25-μm thickness. The other samples (2% and 5% Zr) were prepared by casting Cu-13 wt.% Sn bronze on each side of the NbZr core (bronze/Nb ratio = 4) and rolling to a final core size of 25 μm. The composite samples were then heat-treated in an Ar atmosphere at 750°C for 70 to 190 h (see Table 1).

The bronze matrix of all samples was etched away in dilute nitric acid; the tape samples were then cut on four sides to a final size of 1.5 x 15 mm^2 by electrical discharge machining. The flattened wire samples, being sufficiently narrow, were cut at each end only. Finally, all samples except the pure Nb and Nb-1% Zr were annealed at 750°C for 1/2 h without the bronze matrix to relieve any strain associated with plastic deformation of the unreacted core.[10] Strain associated with this effect is negligible for the Nb and Nb-1% Zr samples because the bronze/core ratio was lower (≈2) and insufficient to cause plastic yielding of the unreacted core.

The actual composition of the A15 layers was measured with energy-dispersive x-ray analysis. In all specimens, approximately 1 to 2 at.% copper was found. The Sn concentration was generally less than the stoichiometric value by 2 to 3%. The Ta additions were essentially all incorporated in the A15 layers, as reported by Livingston,[2] but substantial loss of Ti and Zr occurred as expected from the limited solubility of these additions in Nb_3Sn.[11] The approximate atomic percentages of each addition in the A15 layers are given in Table 1.

Critical temperatures, T_c, were measured either inductively or resistively on similarly prepared samples and are listed in Table 1. Accuracy is expected to be ±0.1 K.

Table 1. Characteristics of Nb_3Sn Samples

Core Composition (wt.%)	Addition in Compound (at.%)	Reaction Time @ 750°C (h)	Sample Thickness, z(μm)	Nb_3Sn Layer Thickness (μm)	E(300 K) (GPa)	E(10 K) / E(300 K)	T_c* (K)	T_m† (K)
Nb	0	189	27.6±1	7.8	130±2	0.39	17.9	49±1
Nb-10% Ta	4	64	28.1±1.2	8.4	119±14	0.61	17.8	35±5
Nb-0.5% Ti	0.6	64	38.3±2	13.2	132±14	0.50	18.0	28±2
Nb-3% Ti	2	50	28.0±0.5	12.7	144±15	0.76	16.5	NT ‡
Nb-1% Zr	0.4	100	32.8±3	7.8	140±16	0.55	17.7	48±2
Nb-2% Zr	0.3	100	31.6±0.1	11.5	150±2	0.60	17.6	48±2
Nb-5% Zr	2	72	28.0±0.8	11.5	153±9	0.74	17.3	45±3

* Midpoints
† Deduced from onset of rise in internal friction
‡ Nontransforming

INTERNAL FRICTION AND MODULUS MEASUREMENT

Internal friction and dynamic Young's modulus were measured by the vibrating-reed technique using electronics based on a phase-locked loop and frequency modulation developed by Simpson and Sosin.[12] Samples were mechanically clamped at one end and forced into flexural vibration at the other end with an electrostatic drive. For this cantilevered beam geometry, Young's modulus, E, is given in terms of the resonant frequency as

$$f_1 = 0.55966(z/12\ell^2)\,(E/\rho)^{1/2} \tag{1}$$

with overtones at $f_2 = 6.267f_1$ and $f_3 = 17.548f_1$, where z is the thickness of the reed, ℓ its length, and ρ its density.[12]

Because of the composite nature of the samples, analysis was used to deduce Young's modulus of the Nb_3Sn layers assuming the flexural rigidity and areal density of the composite to be the sum over each layer.[1] Corrections resulting from this analysis were less than 15%.

Internal friction, Q^{-1}, was determined by counting the number of cycles, N_{ij}, for decay of free vibrations between amplitudes A_i and A_j at the resonant frequency.

$$Q^{-1} = (1/\pi N_{ij})\,\ell n(A_i/A_j) \tag{2}$$

Internal friction measurements presented below are for the composite reeds and were obtained with the first harmonic, which had a frequency $f_2 \sim 1$ kHz. The amplitude of vibration was adjusted so that Q^{-1} did not change by more than 20% for an increase in amplitude of a factor of ten at liquid nitrogen temperature.

The temperature was measured with a Ge resistance thermometer between 0 and 55 K with a precision of ±0.5 K and with a Pt resistance thermometer above 55 K with a precision better than ±1 K. A lower precision was obtained in many cases near 100 K because of heating-rate-dependent temperature gradients in the cryostat.

RESULTS

The room-temperature values of Young's modulus were determined using equation (1) and the frequency of the first harmonic (f_2). For the tape samples (pure Nb and Zr additions), the sample thickness was determined by weighing, assuming a density of 8.86 g/cm^3 for Nb_3Sn. For the flattened wire samples, calculations were made by assuming uniform and elliptic cross sections and taking the average. This uncertainty in cross sections introduces an extra error of 10%. Other errors are associated with the nonuniformity along the length of the sample, especially in the case of the flattened wires.

Results, given in Table 1, indicate that the room-temperature modulus increases slightly with additions (except for the 10% Ta addition). Within the margin of error, however, most samples do not differ significantly from the polycrystalline average of 137 GPa obtained from the single crystal elastic constants of pure Nb_3Sn.[10] The present result for pure Nb_3Sn (130 GPa) is also in excellent agreement with a previously reported measurement based on static deflection of similar composite strips.[10]

The temperature dependence of Young's modulus and internal friction for pure Nb_3Sn and (Nb-10 wt.% Ta)$_3$Sn are given in Fig. 1. For pure Nb_3Sn, considerable softening of Young's modulus

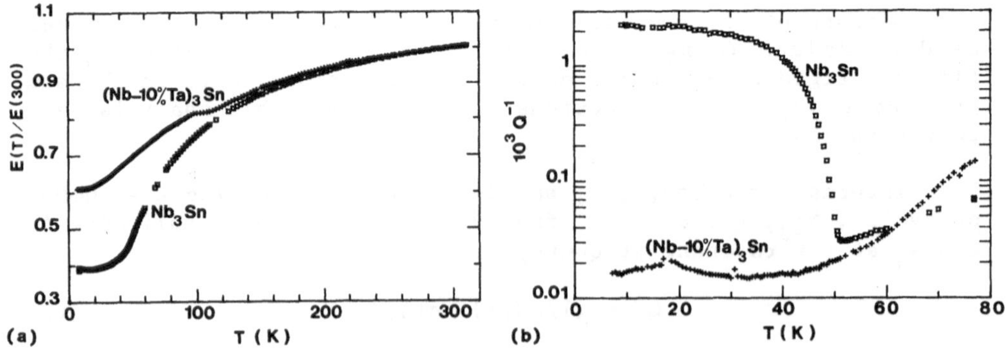

Fig. 1. Temperature dependence of Young's modulus (a) and internal friction (b) for pure Nb_3Sn (□) and Nb_3Sn processed with a Nb-10 wt.% Ta core (+). Note the logarithmic scale for Q^{-1}.

occurs, E(10K)/E(300K) = 0.39, and a large increase in internal friction is observed below 50 K, as reported earlier.[1] The increase of Q^{-1} is associated with domain wall motion in the tetragonal phase of Nb_3Sn, and its temperature dependence approximately follows the square of the tetragonal deformation, $(c/a-1)^2$. In the presence of 10 wt.% Ta softening is reduced, E(10)/E(300) = 0.6, compared with 0.39 for pure Nb_3Sn. Internal friction decreases smoothly with temperature down to ∿35 K in the temperature range shown and then increases on cooling to 18 K. The slight increase in Q^{-1} below 35 K is believed to be associated with the cubic-to-tetragonal phase transformation, with a transition temperature of ∿35 K. The level of internal friction, however, is nearly two orders of magnitude smaller than that for pure Nb_3Sn. Internal friction of a 20 wt.% Ta sample showed no sign of a transformation.

The effect of Ti and Zr additions on E and Q^{-1} are presented, respectively, in Figs. 2 and 3. Note that the presence of 3 wt.% Ti in the Nb core suppressed the martensitic transformation. The 0.5 wt.% Ti sample transformed at ∿28 K. For Zr additions of 1 to 5%, softening gradually decreased, E(10)/E(300) = 0.7 for 5% Zr, and the internal friction peak at 18 K also decreased. The transformation temperature, however, decreased by a small amount (T_m = 45 K for the 5% Zr sample). See Table 1 for T_m values. The modulus plateau observed for several of the samples near 100 K (Figs. 1-3) is an experimental artifact.

DISCUSSION

As shown above, both Young's modulus and internal friction are strongly affected by the presence of additions. Increases of up to a factor of 2 in Young's modulus were found to occur for

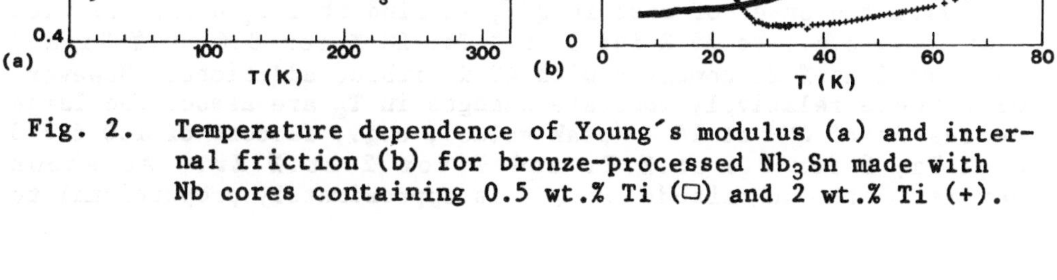

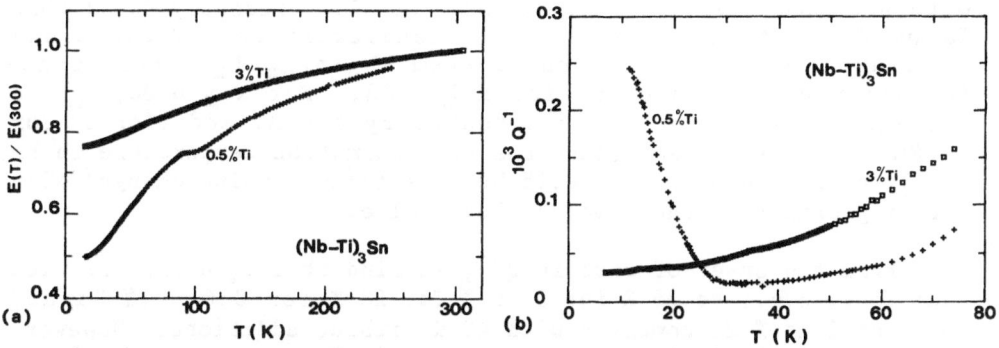

Fig. 2. Temperature dependence of Young's modulus (a) and internal friction (b) for bronze-processed Nb_3Sn made with Nb cores containing 0.5 wt.% Ti (□) and 2 wt.% Ti (+).

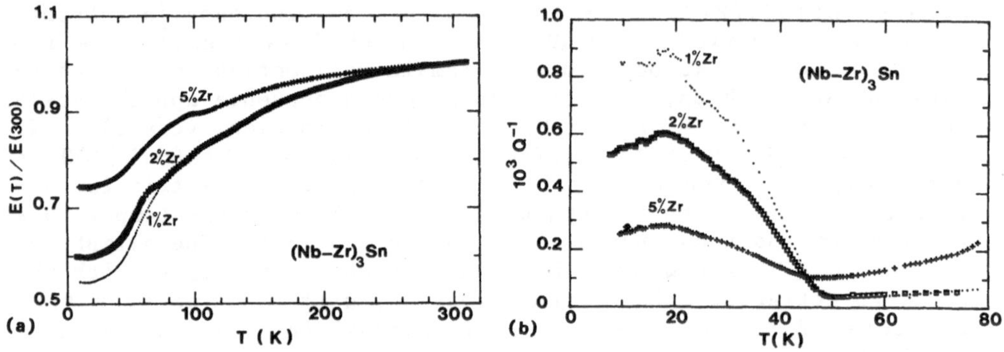

Fig. 3. Temperature dependence on Young's modulus (a) and internal friction (b) for bronze-processed Nb_3Sn made with Nb cores containing 1 wt.% Zr (•), 2 wt.% Zr (□), and 5 wt.% Zr (+).

relatively small additions (e.g., 2 at.% Ti in the compound). The main features of the internal friction results are a rapid decrease in absolute value at low temperature (up to 2 orders of magnitude) and a slower decrease in the temperature of onset of rise in internal friction, which we associate with the martensitic transformation temperature, T_m.

Although we are not aware of other work on the effect of additions on softening of A15 compounds, the above results are consistent with the well-known sensitivity of anomalous properties of A15's to defects[13-15] and the more recent observation of reduced softening in neutron-irradiated V_3Si.[16]

The currently observed reduction in T_m and/or suppression of the transformation with ternary additions to Nb_3Sn is consistent with previous observations. Roberge et al.[17] recently found that Ta and Zr additions suppressed the transformation. Vieland and Wicklund found that for the system $Nb_3Sn_{1-x}Al_x$[18] the transformation ceased to occur for $x \geq 0.08$. For $x \leq 0.06$, T_m and tetragonal splitting was unaffected by the Al additions. For $Nb_3Sn_{1-x}Sb_x$, Vieland[19] found the transformation temperature to be nearly unchanged up to $x = 0.15$ but the $(c/a-1)$ value changed sign with approximately the same absolute value.

From the onset of rise in Q^{-1}, we find that T_m decreases with additions, e.g., to 35 K for 5 at.% Ta, 28 K for 0.6 at.% Ti and 45 K for 2 at.% Zr compared with 49 K without additions. However, with these relatively moderate changes in T_m are associated large decreases of Q^{-1} at 18 K (peak value), e.g., factors of 100 for 5 at.% Ta, 8 for 0.6 at.% Ti and 7 for 2 at.% Zr. As shown earlier,[1] the magnitude of Q^{-1} is approximately proportional to

$(c/a-1)^2$; one might therefore expect the reduction in Q^{-1} to be associated with decreased values of (c/a-1). However, other factors, such as domain wall mobility and the fraction of transforming material, may also affect Q^{-1}. In the absence of x-ray measurements it is therefore impossible to conclude whether the reduced values of Q^{-1} are associated with changes in (c/a-1) or other effects.

CONCLUSIONS

Softening of Young's modulus and the occurrence of the martensitic transformation were found to be strongly affected by the presence of relatively small amounts (0-4 at.%) of Ta, Ti, or Zr with the largest effects occurring for Ti. Additions incorporating 2 at.% Ti in the Nb_3Sn causes Young's modulus to increase by a factor 2 at 10 K and the transformation to be suppressed. Since Young's modulus also increases rapidly with strain,[1] further measurements will be required to determine to what extent the stress state of Nb_3Sn incorporated in a bronze matrix is affected by ternary additions.

ACKNOWLEDGMENTS

The authors wish to thank M. Garber for helpful discussions, R. Jones for technical help, and D. Roy for preparing programs used in analysing and plotting the data.

REFERENCES

1. J. F. Bussière, B. Faucher, C. L. Snead, Jr., and D. O. Welch, Phys. Rev. B. (in press).
2. J. D. Livingston, IEEE Trans. Magn. MAG-14:611 (1978).
3. M. Suenaga, K. Aihara, K. Kaiho, and T. S. Luhman, in: "Advances in Cryogenic Engineering," Vol. 25, Plenum Press, New York (1980), p. 442.
4. M. Suenaga, S. Okuda, K. Itoh, and T. S. Luhman, Superconducting properties of $(Nb,Ti)_3Sn$ wires fabricated by the bronze process, in "Advances in Cryogenic Engineering--Materials," Vol. 28, Plenum Press, New York (1982), p. 379.
5. M. Suenaga, T. S. Luhman, and W. B. Sampson, J. Appl. Phys. 45:4049 (1974).
6. S. Foner and E. J. McNiff, Jr., Phys. Lett. A 58:318 (1976).
7. C. W. Chu and L. J. Vieland, J. Low. Temp. Phys. 17:25 (1974).
8. T. S. Luhman, Metallurgy of superconducting materials, in: "Treatise on Materials Science and Technology," Academic Press, New York (1979), p. 221.
9. G. Rupp, IEEE Trans. Magn. MAG-17:1099 (1981).
10. J. F. Bussière, D. O. Welch, and M. Suenaga, J. Appl. Phys. 51:1024 (1980).

11. Y. V. Efimov, B. P. Mikhaylov, and E. Moroz, Russ. Metall. Min. 5:168 (1979).
12. H. Simpson and A. Sosin, Rev. Sci. Instrum. 48:1392 (1977).
13. M. Weger and I. B. Goldberg, in: "Solid State Physics," Vol. 28, F. Steiz and D. Turnbull, eds., Academic Press, New York (1978), p.1.
14. L. R. Testardi, in: "Physical Acoustics," Vol. X, W. P. Mason and R. N. Thurston, eds., Academic Press, New York (1973), p. 193.
15. L. R. Testardi, in: "Physical Acoustics," Vol. XIII, W. P. Mason and R. N. Thurston, eds., Academic Press, New York (1977), p. 29.
16. A. Guha, M. Sarachik and F. W. Smith, Phys. Rev. B 18:9 (1978).
17. R. Roberge, H. LeHuy, and S. Foner, Phys. Lett.
18. L. J. Vieland and A. W. Wicklund, Phys. Lett. 34A:43 (1971).
19. J. J. Vieland, J. Phys. Chem. Solids 31:1449 (1970).

MULTIFILAMENTARY Nb-Hf/Cu-Sn-Ga COMPOSITE WIRES

K. Kamata

Metal Research Laboratory, Hitachi Cable, Ltd., Tsuchiura, Japan

K. Aihara

Hitachi Research Laboratory, Hitachi, Ltd., Hitachi, Japan

H. Sekine and K. Tachikawa

National Research Institute for Metals, Ibaraki, Japan

INTRODUCTION

Multifilamentary Nb_3Sn superconductors have been developed for large-scale applications, such as fusion reactor magnets and high-energy particle accelerators. However, the composite-processed Nb_3Sn conductors with pure niobium cores and a Cu-Sn matrix are required to have better high-field current-carrying capacities for these applications. Recently, several attempts have been made to improve the critical current density, J_c, of composite-processed Nb_3Sn in high fields by an alloying addition to the core or to the matrix. The additions of titanium, zirconium, or tantalum have been reported to increase both the Nb_3Sn layer growth rate and J_c in high magnetic fields.[1-3] Gallium additions to the Cu-Sn matrix have been reported to raise the upper critical field, H_{c2}, of the Nb_3Sn.[4] Of the various alloying additions to the core and matrix, it has been revealed through studies on single-core and 19-core composite specimens that the simultaneous addition of hafnium to the core and gallium to the matrix is most effective for increasing H_{c2} and J_c in high fields of the composite-processed Nb_3Sn.[5,6] Thus it is of great interest to examine the possibility of practical use of multifilamentary Nb-Hf/Cu-Sn-Ga composites. We report here the effect of the simultaneous addition of hafnium to the core and gallium to the matrix on high-field superconducting properties of the composite-processed multifilamentary Nb_3Sn superconductors.

In addition to large current-carrying capacities in high fields, superconductors for practical use are required to withstand large stresses arising from strong electromagnetic forces. We also report stress effects on the Nb-Hf/Cu-Sn-Ga composites.

EXPERIMENTAL PROCEDURE

The bar-shaped ingots of pure niobium, Nb-1, 2, 3 and 5 at.% Hf alloys were prepared by electron-beam melting following arc melting. After annealing at 1000°C for 1 h, the ingots were rolled and swaged to 6.4-mm-diameter core rods. Cu-7 at.% Sn, Cu-5 at.% Sn-4 at.% Ga, Cu-6.5 at.% Sn-3 at.% Ga, Cu-7 at.% Sn-2 at.% Ga matrix alloys were melted in a graphite crucible under an argon atmosphere and cast into 30-mm-diameter ingots. The ingots were homogenized at 650°C for 40 h and then machined into 11.1-mm-diameter rods. The bronze/core volume ratio was 2.5. A 6.5-mm-diameter hole was drilled in those matrix rods to encase the core rods therein. The single-core composite was fabricated into a wire of 0.8 mm in diameter with intermediate annealings at 550°C for 1 h. Three hundred thirty-one single-core wires were bundled, surrounded by a niobium diffusion barrier and encased in a copper jacket of 24 mm in outer diameter. The resulting composite was then fabricated into a wire of 0.7 mm in diameter with intermediate annealings at 550°C for 1 h. Specifications of specimens are summarized in Table 1. Specimens cut from the wire were heat-treated under an argon atmosphere at temperatures from 700°C to 750°C. The critical temperature, T_c, and the critical current, I_c, were measured by a four-probe resistive method. Critical temperature was defined as the midpoint of the transition. Values of I_c in magnetic fields up to 17 T were measured by the V_3Ga/Nb_3Sn superconducting magnet at NRIM, and those in magnetic fields from 17 to 21.5 T were measured by the water-cooled copper magnet at the Francis Bitter National Magnet Laboratory of MIT. Critical current was defined as the current where a voltage of 1 μV appeared across the 10-mm-apart terminals. For the measurement of strain dependence of I_c, 19-core specimens were prepared. Nineteen 1.4-mm-diameter

Table 1. Conductor Specifications

Hf content in core	(at.%)	0, 1, 2, 3, 5
Bronze composition	(at.%)	Cu-7Sn, C7-5Sn-4Ga, Cu-6.5Sn-3Ga, Cu-7Sn-2Ga
Overall outer diameter	(mm)	0.7
Bronze:Nb-Hf core	(vol.%)	2.5:1
Cu stabilizer	(vol.%)	39
Nb barrier	(vol.%)	8
No. of filaments		331
Filament diameter	(μm)	17

holes were drilled in a 14-mm-diameter matrix rod to encase the core rods therein. The composite was fabricated into a wire of 0.5 mm in diameter and heat-treated. The strain dependence of I_c was measured using an apparatus designed to apply tensile strain, current, and a perpendicular magnetic field simultaneously to short wire samples at 4.2 K. The x-ray microanalysis was performed to obtain the compositional profile on the cross section of the composite wires. The size and morphology of the Nb_3Sn grains were observed using scanning electron microscopy (SEM) on the tensile-fractured surface of the multifilamentary composite wire specimens.

RESULTS AND DISCUSSION

Metallographic Studies

Niobium-hafnium cores with hafnium content up to 5 at.% showed good workability. As seen in Fig. 1(a), the 331 cores in the multifilamentary Nb-5Hf/Cu-6.5Sn-3Ga composite wire keep almost the same diameter, about 17 μm. Figure 1(b) and (c) show the cross sections of the 331-core multifilamentary Nb-1Hf/Cu-6.5Sn-3Ga and Nb-5Hf/Cu-6.5Sn-3Ga composites, respectively, heat treated at 750°C for 50 h. It is apparent that the addition of hafnium to the core enhances the growth rate of the Nb_3Sn layer. The quantitative x-ray microprobe analysis (XMA) on the Nb-5Hf/Cu-6.5Sn-3Ga specimen revealed the tin concentration in the matrix decreased to about 0.5 at.% after the reaction. Figure 2 shows an XMA line scanning chart taken on the cross section of the multifilamentary Nb-5Hf/Cu-6.5Sn-3Ga composite wire heat treated at 750°C for 50 h. Although it is difficult to make quantitative analysis on the extremely small area on account of the inherent resolution of the XMA apparatus, the incorporation of hafnium and gallium in the Nb_3Sn compound layer is undoubtedly recognized. It is found that the more hafnium

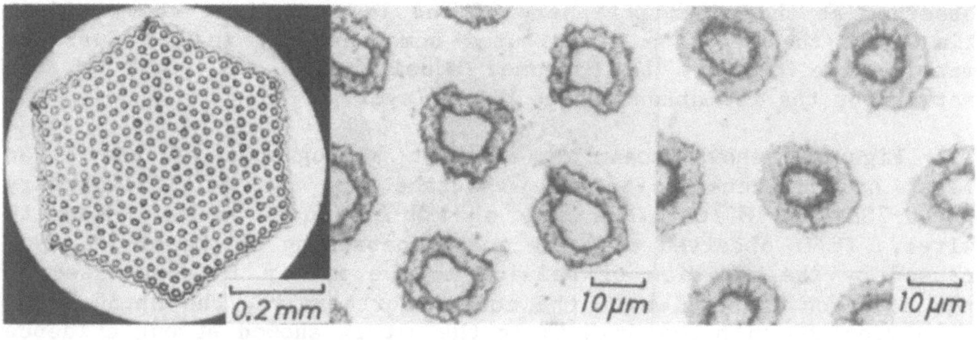

Fig. 1. Cross sections of 0.7-mm-diameter multifilamentary Nb-Hf/ Cu-Sn-Ga composite wires with 331 cores heat treated at 750°C for 50 h. (a) Nb-5Hf/Cu-6.5Sn-3Ga, (b) Nb-1Hf/Cu-6.5Sn-3Ga, (c) Nb-5Hf/Cu-6.5Sn-3Ga.

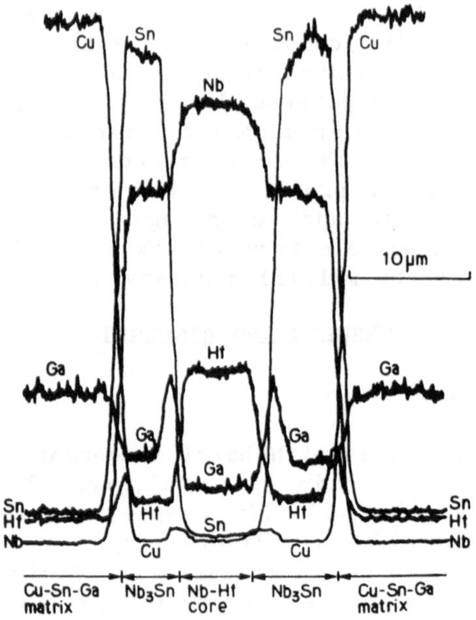

Fig. 2. An XMA line scanning chart taken on the cross section of the multifilamentary Nb-5Hf/Cu-6.5Sn-3Ga composite wire heat treated at 750°C for 50 h.

and gallium that are added to the core and matrix, the more those elements are incorporated in the Nb_3Sn layer. Sharp peaks of gallium and copper concentrations are observed at the inner periphery of the Nb_3Sn layer, and a sharp peak of hafnium concentration is observed at the outer periphery of the layer. Since the residual tin after the reaction distributes homogeneously in the Cu-Sn-Ga matrix, the tin may be consumed equally from all parts of the matrix for the formation of the Nb_3Sn layer.

Figure 3 shows scanning electron micrographs of the Nb_3Sn layer on the tensile-fractured surface for the multifilamentary Nb/Cu-7Sn, Nb-1Hf/Cu-6.5Sn-3Ga, and Nb-5Hf/Cu-6.5Sn-3Ga composite wires. It is observed that the grain coarsening of the Nb_3Sn layer caused by the addition of gallium to the matrix is suppressed by the addition of hafnium to the core. Furthermore, the Nb_3Sn layer of a specimen with 3-4 at.% Ga in the matrix showed strong evidence of transgranular fracturing on tensile loading. This may probably be due to the strengthening of grain boundary on account of the incorporation of gallium into the Nb_3Sn.

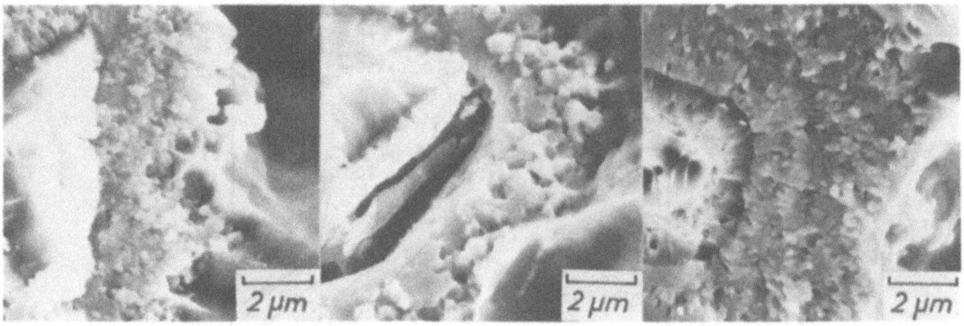

Fig. 3. Scanning electron micrographs of the Nb_3Sn layer on the tensile-fractured surface for multifilamentary composite wires heat treated at 750°C for 50 h. (a) Nb/Cu-7Sn, (b) Nb-1Hf/Cu-6.5Sn-3Ga, (c) Nb-5Hf/Cu-6.5Sn-3Ga.

Superconducting Properties

Critical temperatures, T_c, for the multifilamentary Nb-Hf/Cu-Sn-Ga composite wires are listed in Table 2. The results show that the gallium addition to the matrix increases T_c by 0.4 K, and the hafnium addition to the core produces further increase in T_c.

For the design of practical superconductors, it is convenient to show the overall critical current density, J_c (overall), which

Table 2. Heat Treatment Conditions, Critical Temperature, T_c, of the Specimens Listed in Table 1.

Specimen	T_c, K	
	Heat treated at 700°C for 150 h	Heat treated at 750°C for 50 h
Nb/7Sn	17.1	17.2
2Hf/5Sn-4Ga	17.7	17.7
5Hf/5Sn-4Ga*	17.7	17.8
Nb/6.5Sn-3Ga	17.5	17.6
1Hf/6.5Sn-3Ga	17.7	17.7
3Hf/6.5Sn-3Ga	17.7	17.8
5Hf/6.5Sn-3Ga	17.7	17.7
5Hf/7Sn-2Ga	17.6	17.7

*The abbreviation 5Hf/5Sn-4Ga means Cu-5 at.% Sn-4 at.% Ga composite.

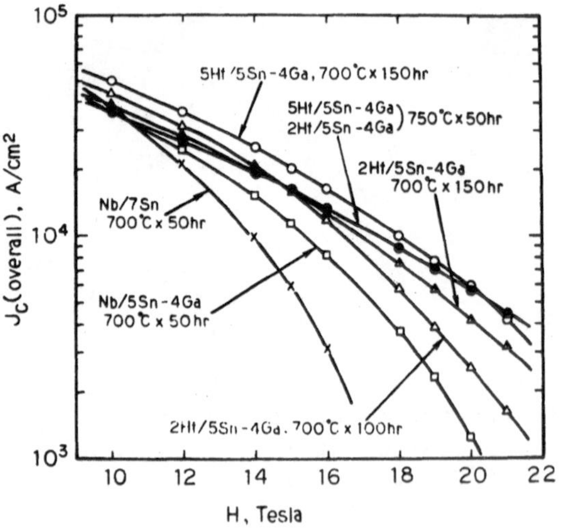

Fig. 4. J_c (overall) vs. H curves at 4.2 K for the multifilamentary Nb-5Hf/Cu-5Sn-4Ga and Nb-2Hf/Cu-5Sn-4Ga composite wires.

is obtained by dividing I_c by the cross-sectional area of the conductor except the copper for stabilization. Figure 4 shows J_c (overall) vs. magnetic field, H, curves for the multifilamentary Nb-5Hf/Cu-5Sn-4Ga and Nb-2Hf/Cu-5Sn-4Ga composite wires in fields up to 21.5 T. It is obvious that the decrease in J_c (overall) in high magnetic fields for the multifilamentary Nb-5Hf/Cu-5Sn-4Ga and Nb-2Hf/Cu-5Sn-4Ga composite wire is much less than that for the multifilamentary Nb_3Sn composite wire. The Nb-5Hf/Cu-5Sn-4Ga composite wire heat treated at 700°C for 150 h shows J_c (overall) of 1.6×10^4 A/cm^2 at 16 T and 0.6×10^4 A/cm^2 even at 20 T. The inclination of the J_c (overall) H curve of the wire heat treated at 750°C is less steep than that of the wire heat treated at 700°C.

Figure 5 shows J_c (overall) vs. H curves in magnetic fields up to 16 T for the multifilamentary Nb-1Hf/Cu-6.5Sn-3Ga, Nb-3Hf/Cu-6.5Sn-3Ga, and Nb-5Hf/Cu-6.5Sn-3Ga composite wires, heat treated at 700°C for 150 h and 750°C for 50 h. It is also apparent in this case that the more hafnium that is added to the core, the higher the J_c (overall) becomes and that the higher the heat-treatment temperature, the less steep the inclination of the J_c (overall) vs. H curve becomes.

Figure 6 shows J_c (overall) vs. H curves in magnetic fields up to 16 T for the multifilamentary Nb-5Hf/Cu-5Sn-4Ga, Nb-5Hf/Cu-6.5Sn-3Ga, and Nb-5Hf/Cu-7Sn-2Ga composite wires heat treated at 700°C for 150 h and 750°C for 50 h. It is found that the more

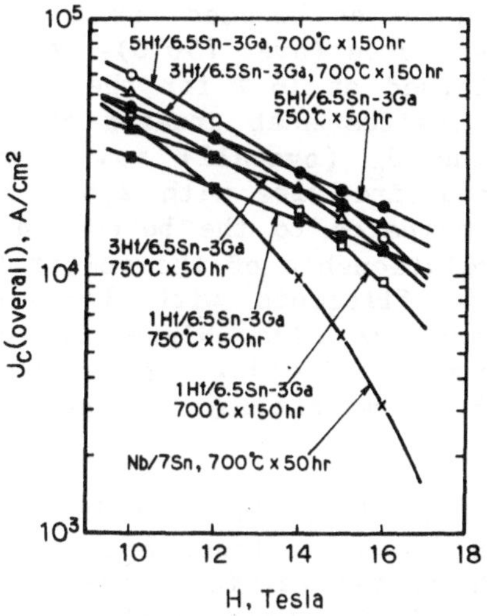

Fig. 5. The influence of hafnium concentrations in the core and heat treatment conditions on the J_c (overall) vs. H relationships for the multifilamentary Nb-Hf/Cu-6.5Sn-3Ga composite wires.

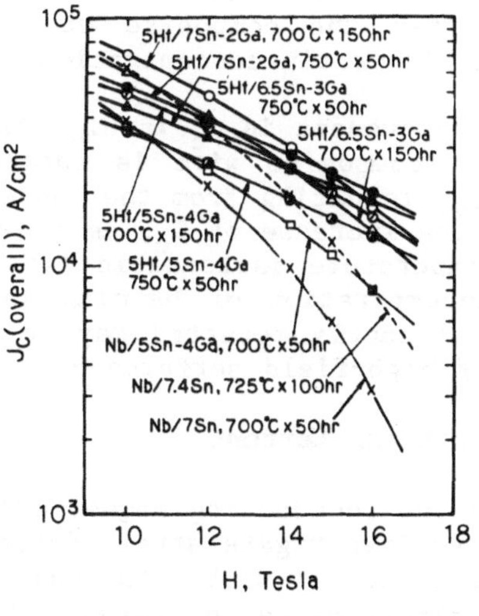

Fig. 6. The influence of gallium concentrations in the bronze matrix and heat treatment conditions on the J_c (overall) vs. H relationships for the multifilamentary Nb-5Hf/Cu-Sn-Ga composite wires.

gallium that is added to the matrix, the less steep the inclination of J_c (overall) vs. H curves becomes for the composite wires heat treated at high temperatures (over 750°C). A J_c (overall) of about 2×10^4 A/cm^2 is obtained at 16 T for the Nb-5Hf/Cu-6.5Sn-3Ga and Nb-5Hf/Cu-7Sn-2Ga composites heat treated at 750°C for 50 h. The dotted line shows the J_c (overall) vs. H curve of the multifilamentary Nb_3Sn composite wire with 2,317 filaments with 4-μm-diameter cores. As compared to the bottom curve representing the J_c (overall) vs. H relationship of the multifilamentary Nb_3Sn composite wire with 331 filaments with 17-μm-diameter cores, the dotted line shows about two times higher J_c (overall) values over the whole range of magnetic fields. The smaller core size and the higher tin content in the matrix may account for the enhancement of J_c (overall).

As for the J_c of the Nb_3Sn layer at 16 T, which is obtained by dividing I_c with the cross-sectional area of the Nb_3Sn layer, it does not depend much on the hafnium concentration in the core in the range of 2-5 at.% Hf for the multifilamentary composite wires with Cu-6.5Sn-3Ga and Cu-7Sn-2Ga matrices. However, the J_c (overall) of composite wires increases with increasing hafnium concentration in the niobium core because the Nb_3Sn layer growth rate increases with increasing hafnium concentration. This is also the case for tin concentration in the matrix. Multifilamentary Nb-5Hf/Cu-5Sn-4Ga and Nb-2Hf/Cu-5Sn-4Ga composite wires heat treated at 700°C for 150 h show the highest J_c of the Nb_3Sn layer at 16 T, but Nb-5Hf/Cu-6.5Sn-3Ga and Nb-5Hf/Cu-7Sn-2Ga composite wires show larger J_c (overall) than the composite wires with a Cu-5Sn-4Ga matrix because of the higher tin concentration in the matrix.

The remarkable improvement in J_c in high fields for multifilamentary Nb-Hf/Cu-Sn-Ga composite wire is considered to be due to the enhancement of H_{c2}, resulting from the increase of normal-state resistivity, ρ_n, and the increase of T_c, as shown in Table 1.[7] The suppression of low-temperature martensitic transformation of Nb_3Sn on account of the incorporation of hafnium and gallium into the Nb_3Sn layer that was recently reported may be another explanation for the improvement of high-field performance.[8]

Effect of Strain on Critical Current

In fabricating a superconducting magnet using Nb_3Sn conductor, it is very important to investigate strain dependence of I_c for the conductor because the conductor must withstand large strain arising from bending or the electromagnetic force. Figure 7 compares the uniaxial strain dependence of I_c of the 19-core Nb-5Hf/Cu-5Sn-4Ga composite wire heat treated at 700°C for 40 h with a typical multifilamentary Nb_3Sn composite wire.[9,10] It is a characteristic of the strain-I_c relationship of Nb-Hf/Cu-Sn-Ga composite wires that the degradation of I_c in high fields under strain is appreciably

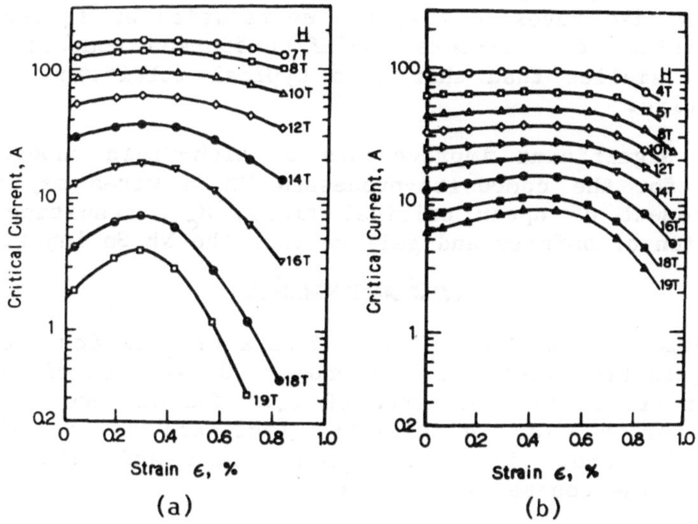

Fig. 7. Comparison of uniaxial-strain dependence of the critical current, I_c, of a Nb-5Hf/Cu-5Sn-4Ga composite wire with 19 cores (a) with a typical multifilamentary Nb_3Sn composite wire (b).

smaller, and there is a less sharp I_c peak against strain, in contrast to that for pure Nb_3Sn composite wires. This improvement in the strain-I_c relationship for Nb-Hf/Cu-Sn-Ga composite wires may be due to the enhancement in H_{c2}.

CONCLUSION

The simultaneous addition of hafnium to the core and gallium to the matrix increases critical temperature, T_c, by 0.4-0.6 K and significantly improves critical current density, J_c (overall), in high fields of the multifilamentary Nb_3Sn composite wires. A multifilamentary Nb-5Hf/Cu-5Sn-4Ga composite wires with 331 filaments with 17-μm-diameter cores heat treated at 700°C for 150 h shows J_c (overall) of 1.6 x 10^4 A/cm^2 at 16 T and 0.6 x 10^4 A/cm^2 even at 20 T. Similar multifilamentary Nb-5Hf/Cu-7Sn-2Ga and Nb-5Hf/Cu-6.5Sn-3Ga composite wires heat treated at 750°C for 50 h show a J_c (overall) of 2 x 10^4 A/cm^2 at 16 T. The addition of hafnium to the core enhances the growth rate of the Nb_3Sn layer, and the more hafnium that is added to the core, the higher the J_c (overall) becomes. The more gallium that is added to the matrix, the less steep the inclination of J_c (overall) vs. H curves for the composite wires becomes. Present results indicate that generation of magnetic fields over 16 T may be feasible with multifilamentary Nb-Hf/Cu-Sn-Ga composite wires.

It was also revealed that the degradation of I_c under strain in high fields of a 19-core Nb-5Hf/Cu-5Sn-4Ga composite wire is appreciably smaller than that of a typical multifilamentary Nb_3Sn composite wire.

These significant improvements of high-field superconducting properties for the composite-processed Nb_3Sn wires may be due to the enhancement of upper critical field, H_{c2}, resulting from the incorporation of hafnium and gallium into the Nb_3Sn layer.

ACKNOWLEDGMENTS

We would like to thank Dr. Y. Tanaka of NRIM for the measurement of I_c in high fields at MIT and Dr. J. W. Ekin of NBS for the measurement of strain dependence of I_c. The authors are also indebted to Mr. N. Tada of Hitachi, Ltd. and Mr. Y. Ishigami of Hitachi Cable, Ltd., for their encouragement and valuable suggestions during the course of this work.

REFERENCES

1. K. Tachikawa, T. Asano, and T. Takeuchi, High-field superconducting properties of the composite-processed Nb_3Sn with Nb-Ti alloy cores, Appl. Phys. Lett., to be published.
2. O. Horigami, T. Luhman, C. S. Pande, and M. Suenaga, Superconducting properties of $Nb_3(Sn_{1-x}Ga_x)$ by a solid state diffusion process, Appl. Phys. Lett. 28:738 (1976).
3. J. D. Livingston, Effect of Ta additions to bronze processed Nb_3Sn superconductors, IEEE Trans. Magn. MAG-14:611 (1978).
4. D. Dew-Hughes and M. Suenaga, Critical-current densities of bronze-processed $Nb_3(Sn_{1-x}Ga_x)$ wires up to 23.5 T, J. Appl. Phys. 49:357 (1978).
5. H. Sekine, K. Tachikawa, and Y. Iwasa, Improvements of current-carrying capacities of the composite-processed Nb_3Sn in high magnetic fields, Appl. Phys. Lett. 35:472 (1979).
6. H. Sekine, K. Togano, and K. Tachikawa, Superconducting current-carrying capacities of the composite-processed Nb-Hf/Cu-Sn-Ga in high magnetic fields, Cryogenics 21:152 (1981).
7. R. R. Hake, Upper critical field limits for bulk type II superconductors, Appl. Phys. Lett. 10:189 (1967).
8. R. Roberge, H. LeHuy, and S. Foner, Effects of added elements and strain on the martensitic transformation of Nb_3Sn, Phys. Lett. 82A:259 (1981).
9. J. W. Ekin, H. Sekine, and K. Tachikawa, Effect of strain on the critical current of Nb-Hf/Cu-Sn-Ga multifilamentary superconductors, J. Appl. Phys. 52:6252 (1981).
10. J. W. Ekin, Strain scaling law for flux pinning in practical superconductors. Part 1: Basic relationship and application to Nb_3Sn conductors, Cryogenics 20:611 (1980).

FORMATION OF MULTIFILAMENTARY V_3Ga WITH V-5 TO -7 ATOMIC PERCENT Ga ALLOYS

C. R. Spencer, E. Adam, E. Gregory, and F. T. Ormand

Airco Superconductors, Carteret, New Jersey

D. G. Howe

U.S. Naval Research Laboratory, Washington, D.C.

INTRODUCTION

Since V_3Ga has been reported to have higher critical current densities than Nb_3Sn in magnetic field intensities larger than 6 T in the temperature range 4.2-10 K,[1,2] it is seen as an alternative to Nb_3Sn in applications requiring high fields. Several possible applications of multifilamentary V_3Ga currently being considered are to wind research magnet inserts for fields above 15 T and to construct generators and motors small enough to be used on ships and airplanes. Although V_3Ga has been made in tape form, it is necessary to develop a multifilamentary V_3Ga superconductor to achieve the necessary stability for fabrication of devices that must be stable in the presence of alternating fields or temperature fluctuations.

Multifilamentary V_3Ga wire has been fabricated using the bronze process by several researchers.[3,4] In these efforts, V filaments have been successfully co-drawn in a Ga bronze matrix. It has been recently reported that additions of Ga in the V filaments can increase the V_3Ga formation rate, resulting in greatly enhanced critical current densities. The increase in critical current density is thought to occur because the V_3Ga grain size is reduced owing to the lower temperatures and shorter reaction times needed to form V_3Ga with the alloyed V-Ga filaments.[5] One added benefit is that lower reaction temperatures allow the magnet designer a wider range of insulators, which may be used in a wind and react manufacturing technique.

We describe here a program designed to characterize the material properties necessary to produce V_3Ga using alloyed V-Ga alloys and the initial efforts to scale up the volume of production of multifilamentary V_3Ga.

FABRICATION OF Ga BRONZE BILLETS

The Cu-Ga phase diagram[6] indicates that the solid solution limit of Ga in Cu is approximately 20 wt.%. Five heats of Cu-18.6 wt.% Ga have been melted in a vacuum induction melting (VIM) furnace in operation at Airco. The tapered ingots had a minimum diameter of 80 mm, were 178-mm long, and were cast from CDA-101 grade Cu and Ga of 99.999% purity. The ingots were homogenized at 710°C for 72 h to eliminate any Ga-rich phases that existed. Subsequent micrographs showed the bronze to be of homogeneous grain structure and free of second-phase regions.

Several ingots of Ga bronze were extruded, cold-drawn, and set aside for later use in 27-mm test billets. It should be noted that Cu-18.6 wt.% Ga bronze can be very readily cold-worked and lends itself very conveniently to scale-up using techniques already available in the manufacture of multifilamentary Nb_3Sn, which is made by the tonne.

CASTING OF V-Ga ALLOYS

The alloys of V and Ga initially used in this program were cast by F. A. Schmidt and associates at Ames Laboratory, DoE. Vanadium was obtained from three sources for alloying: Atomergic Chemetals Corporation (ATOM), Ames Laboratory (AMES), and the Bureau of Mines (BM). The V supplied by Ames Laboratory was purified by an alumothermic process, and the V supplied by the Bureau of Mines was purified by electrolytic refining. The history of the Atomergic material is not recorded. Table 1 lists a partial chemical analysis of these materials.

A 20-g dish of V was made for each alloy, 99.999% Ga was placed into the dish, and enough V to make V-8 at.% Ga was added. The mixture was melted using a nonconsumable arc melting technique[7] under an atmosphere of purified argon. Alloys were made using V from each of the three sources. Table 2 shows a chemical analysis of the three resulting alloys. Small quantities of each alloy were supplied to NRL for further processing.

A series of swage reductions and anneals reduced the rods to 5.13-mm o.d.; the resultant hardnesses are recorded in Table 3.

Table 1. Vanadium Chemical Analysis (ppm)

Element	ATOM	AMES Alumothermic-1969	BM Electrorefined
C	300	103	10
N	25	10	1
O	80	81	51
Si	60	200	200
Al	30	3	100
Ta	90		3.6
Fe	100	7	20
Mo	50		16
Ni	40		10
Zr	15		
P			30
W		170	
Cr		3	
S		4	

Table 2. Chemical Analysis of V-8 At.% Ga

	ATOM	AMES Alumothermic-1969	BM Electrorefined
Ga wt.% (at.%)	9.53 (7.2)	9.75 (7.3)	9.73 (7.3)
Si wt.%	0.07	0.05	0.04
O (ppm)	199	122	165
N (ppm)	1	1	1
C (ppm)	311	140	41
H (ppm)	0.4	0.5	0.5

Table 3. Hardness* Of V-7.3 At.% Ga Alloy

	EB Cast V	Alloyed†	Swaged	Annealed
ATOM	133	194		218 (9-10‡)
AMES (1969)	72	185	215	179 (6-7‡)
BM	49.4	177	194	180 (5-6‡)

* Vickers hardness number.
† Alloying done by arc melting or EB melting.
‡ ASTM grain size.

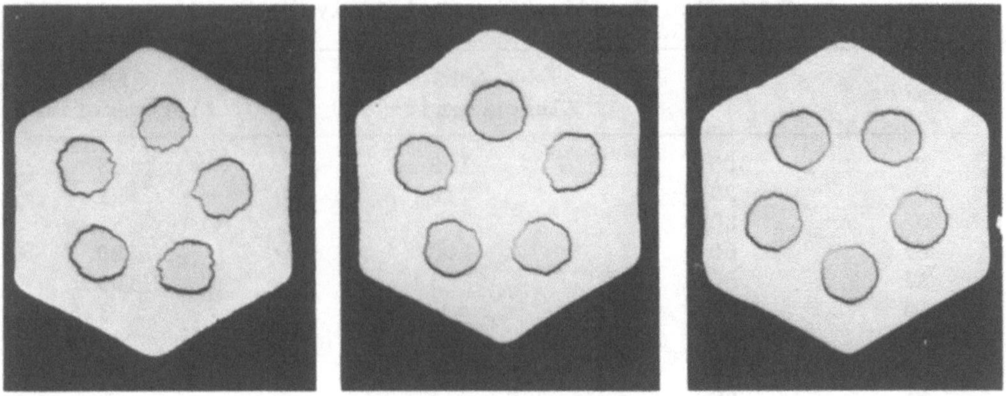

(a) (b) (c)

Fig. 1. First extrusion billets as drawn to 2.9-mm f-f hexagonal rods: (a) Atomeric; (b) Ames Lab; and (c) Bureau of Mines.

FIRST EXTRUSION BILLETS

One 27-mm diameter first extrusion billet was made from each lot of starting V-7.3 at.% Ga alloy. Each billet contained five 5.13-mm diameter rods arranged in a circular pattern in holes drilled into a Cu-18.6 wt.% Ga bronze matrix. Following evacuation and sealing in an electron beam welder, the three billets were preheated and extruded to a 10.3-mm diameter. All three billets were cold-drawn to 2.9-mm flat-to-flat hexagonal rods (Fig. 1) with intermediate anneals. Samples were taken following each 20.7% reduction in area during the cold-working process, and hardness measurements were made on the filaments. In all cases, the filaments showed a marked increase in hardness as a result of the extrusion and cold-working operations. Several heat treatments were done to anneal the rods to the greatest extent possible without significant formation of V_3Ga. The anneals were done by placing the samples in evacuated Cu tubes with Ta getters and heat-treating in an argon atmosphere furnace for 25 min at 650°C, 20 min at 750°C, and 15 min at 850°C. These anneals successfully softened the filaments. The results are listed in Table 4.

SECOND EXTRUSION BILLETS

The filaments cast from ATOM V were deemed too hard to be successfully passed through a second extrusion process, so the rods containing these filaments were set aside. Fifty-five hexes of each of the remaining two lots of V-7.3 at.% Ga alloy were packed into Cu-18.6 wt.% Ga extrusion cans of 27-mm o.d. The billets were extruded to 10.3 mm after a preheating. Cross sections (Fig. 2) of the extruded billets showed a small amount of filament diameter nonuniformity in the BM filaments and a significant amount of variation of diameter in AMES filaments.

The billets were cold-drawn to 0.8-mm diameter with intermediate anneals and without excessive breakage. The resulting cross sections (shown in Fig. 3) indicated that filament diameters in the AMES alloy became extremely nonuniform, whereas those of the BM alloy showed a much lesser degree of nonuniformity.

RESULTS OF 27-mm EXTRUSION WORK

Work-hardening rates, as reflected in the hardness of the V-7.3 at.% Ga alloys, were found to be especially dependent on the C impurity levels found in the alloys. For successful multiple extrusion of billets containing V-Ga alloys, it appears that the C levels should remain below 50 ppm. All other interstitial impurity levels must be maintained at as low levels as possible, and it would be desirable to keep oxygen under 100 ppm, since the work-hardening rate of the V-Ga alloys appears to be very sensitive to impurities. Hardness, which is impurity dependent, appears to be a useful measure of the work-hardening rate of the V-Ga alloys and is a quick and inexpensive technique to evaluate the acceptability of a heat of V-Ga alloy for extrusion. The process development done in 27-mm billets indicates that the parameters listed in Table 5 may be useful guidelines by which to judge the potential success of a heat of V-Ga alloy.

A series of wires from BM and AMES V were reacted to form V_3Ga by NRL. The results of the critical current measurements made on these specimens were reported at the 1980 Applied Superconductivity Conference.[8]

Table 4. Hardness of First Extrusion V-7.3 At.% Ga Filaments

Rod Diameter (mm)	ATOM (VHN)	AMES (VHN)	BM (VHN)
10.3	315	282	239
7.4	293	257	239
5.3	303	259	245
3.7	311	293	266
2.9 f-f hex	329	306	274
25 min @ 650°C	312	286	257
20 min @ 750°C	278	241	206
15 min @ 850°C	241	209	175

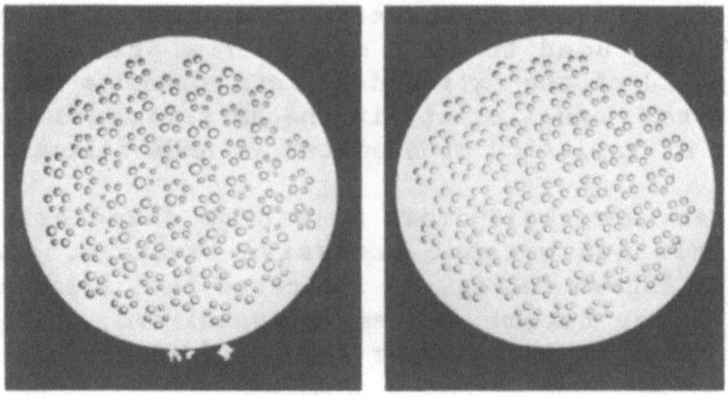

(a) (b)

Fig. 2. Second extrusion billets as extruded to 10.4-mm diameter: (a) Ames Lab, (b) Bureau of Mines.

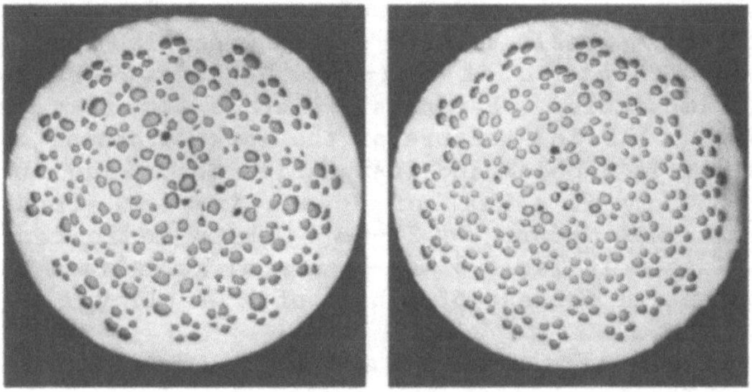

(a) (b)

Fig. 3. Second extrusion billets as drawn to 0.8-mm diameter: (a) Ames Lab, (b) Bureau of Mines.

Table 5. Guideline for Maximum Acceptable Hardness of V-Ga Alloys

Condition	Harness (VHN)
As cast	175
Swaged	220
Recrystallized	180
Extruded	240
Prior second extrusion	200

SCALE-UP OF V_3Ga PRODUCTION WITH ALLOYED FILAMENTS

Successful reduction of V-Ga filaments in a bronze matrix from billet size to wire size is dependent on the initial hardness and the work-hardening rate of the V-Ga alloy filaments. The work described in the preceeding sections indicates that the V-Ga alloy must be of very high purity. If multifilamentary V_3Ga is to be produced on a commercial scale, then a casting procedure must be available that can produce kilograms of V and V-Ga alloy of sufficient purity. Workers at AMES have developed an alumothermic process for purifying V, which produces high purity V in kilogram quantities. More importantly, this process has the potential to be scaled up to commercial quantities in industry.

The first attempt to manufacture kilogram quantities of V-Ga alloys was undertaken by AMES in the casting and swaging of 5 kg of V-5.2 at.% Ga and V-6.8 at.% Ga alloy rods. These alloys were cast in a manner similar to the V-7.3 at.% Ga alloys and reduced to rods 4.75-mm diameter x 17.5-mm long. All V was produced by AMES with the alumothermic process. The Ga used was labeled 99.999% pure. Recrystallization was done under vacuum with a Ta getter at 1090°C for 4 h. The chemical analysis is given in Table 6 for batches 1 and 2.

Two 76-mm billets (Fig. 4) were machined of Cu-18.6 wt.% Ga bronze to be used in the extrusion of the V-Ga alloys. The billet parameters are listed in Table 7.

RESULTS OF FIRST EXTRUSION SCALE-UP

Filaments in the extrusions containing V-5.2 at.% Ga and V-6.8 at.% Ga rods displayed noticeable "spikes" around the circumferences of the filaments (Fig. 5). The V-5.2 at.% Ga filament diameters were slightly nonuniform, whereas the diameters of the V-6.8 at.% Ga filaments varied by a factor of 3. The difference in flow strengths of the bronze and filaments evidently influences the degree of filament nonuniformity and the formation of "spikes," especially if the filament grain size is extremely large. An effort to test this hypothesis was made by machining a 59.6-mm diameter billet with 31 holes in a Cu-18.6 wt.% Ga bronze. Thirty-one 4.75-mm rods of V-6.8 at.% Ga were placed into the holes. Copper caps were TIG welded onto both ends of the billet, and the billet was evacuated and sealed. Instead of extrusion, this billet was swaged to 26-mm diameter and a cross section was taken. The results are shown in Fig. 6. Filament diameters are seen to be nonuniform with small "spikes" around the periphery. Apparently, the large grain size and the rapid work hardening of the rods combined to produce these phenomena, even when hot extrusion was replaced by cold-working the rods exclusively. These billets were not suitable for further processing.

Table 6. V-5 to -7 At.% Ga Alloys

	Batch 1	Batch 2	Batch 3
Ga, wt.%(at.%)	7.0 (5.2)	9.1 (6.8)	9.0 (6.8)
Y, wt.%(at.%)			0.17 (0.10)
C, ppm	96	75	90
O, ppm	80	85	68
N, ppm	5	3	5

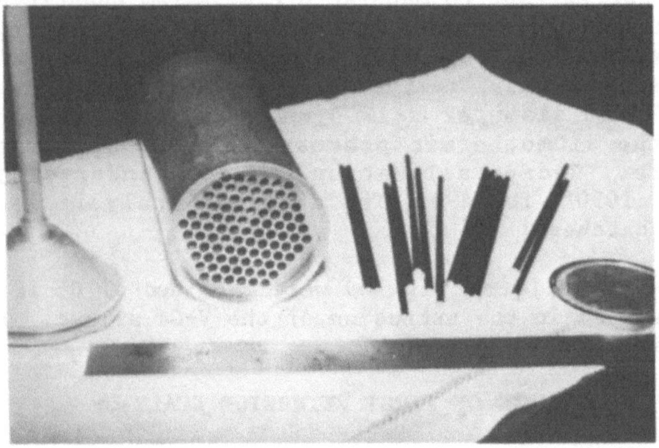

Fig. 4. First extrusion billet, 76-mm diameter of Cu-18.6 wt.% Ga bronze, prior to the insertion of V-Ga alloy rods.

Table 7. Billet Parameters for V-5.2 and -6.8 at.% Ga Alloys

Alloy	#Rods	Bz:Fil.	Diam., mm	Extrusion Ratio	Grain Size	Filament Hardness*
V-5.2 at.% Ga	91	1.92:1	76.2	10.6:1	ASTM 1	173
V-6.8 at.% Ga	55	4.1:1	79.4	10.6:1	ASTM 1	220
V-6.8 at.% Ga	31	4.1:1	59.6	Swage	ASTM 1	220
V-6.8 at.% Ga -0.1 at.% Y	55	4.1:1	79.4	6.7:1	ASTM 7-8	174

* Vickers hardness number.

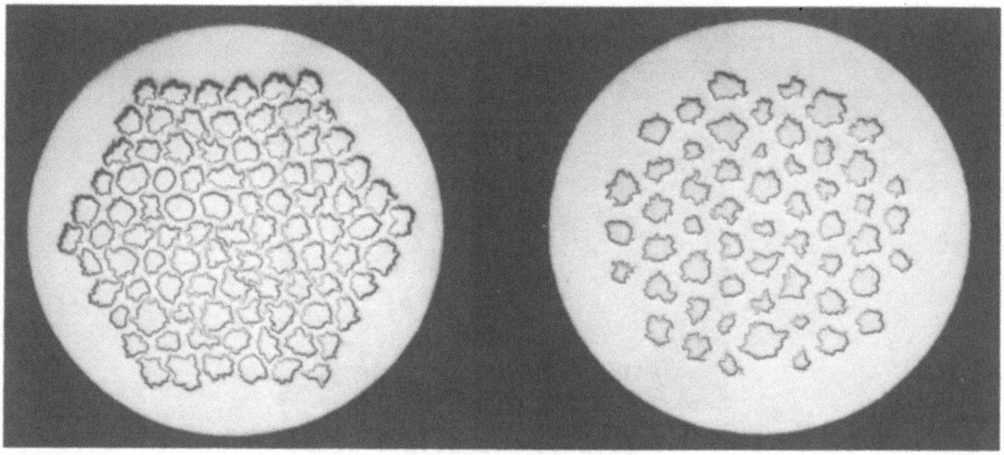

(a) (b)

Fig. 5. V-Ga alloys extruded with a nominal 10:0 extrusion ratio. Micrographs show the (a) V-5.2 at.% Ga filaments and (b) the V-6.8 at.% Ga filaments.

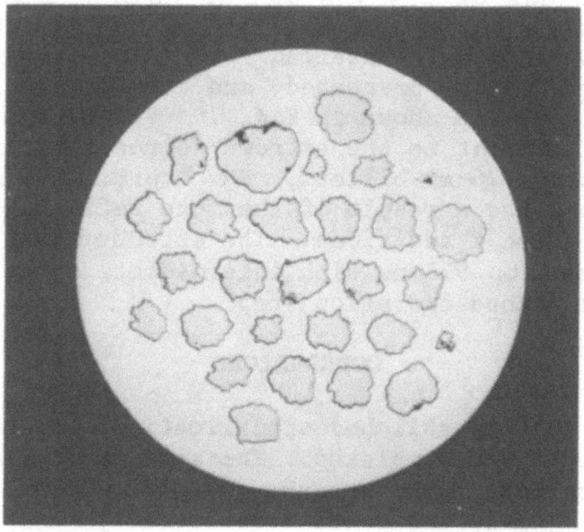

Fig. 6. Micrograph of filaments of cold-processed billet.

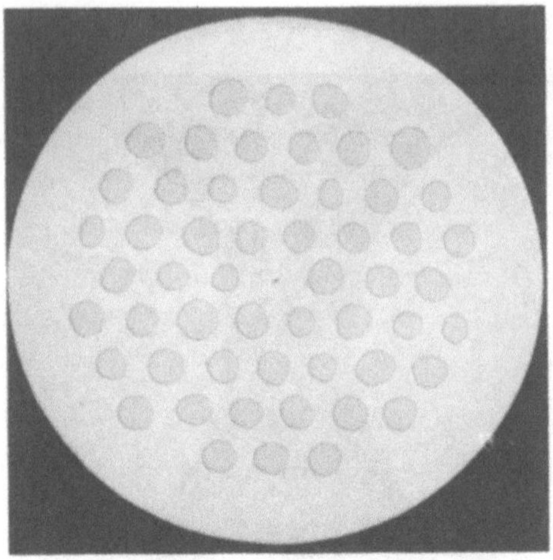

Fig. 7. V-6.8 at.% Ga-0.1 at.% Y alloy extruded with a 6.7:1 extrusion ratio.

V-Ga ALLOYS WITH Y ADDITIONS

Yttrium was added to an Ames Laboratory melt of 1 kg of V-6.8 at.% Ga alloy. Compounds of Y with C, N, and O are stable in the presence of V and Ga. The C, N, and O present in the starting V and Ga appear to be removed as Y slag during the alloy melting and casting. The resulting alloy of V-6.8 at.% Ga-0.1 at.% Y was swaged to 4.72 mm and recrystallized at 1025-1050°C for 2 h with a resulting hardness of 174 VHN and an ASTM 7-8 grain size. A chemical analysis for batch 3 is shown in Table 6. Fifty-five rods with Y additions were inserted into a 76-mm diameter Cu-18.6 wt.% Ga bronze billet, preheated, and extruded to 32-mm diameter preheat. The result is shown in Fig. 7. Filament nonuniformities were noticed, but not to the extreme degree noticed in the first attempts to extrude 76-mm billets. No "spikes" were formed and this is thought to be a result of the more refined grain structure in the rods with the Y additions. The extruded rod is currently being cold-drawn to 9.3 mm with intermediate anneals and will be evaluated for a second extrusion step.

CONCLUSIONS

A program that established specifications for the casting of V-Ga alloys has been completed. The alloy must remain very pure and special care must be made to keep C out of the V-Ga alloys. Wire with 275 V-Ga filaments was successfully made using a double extrusion process with 27-mm billets. A 76-mm billet with 55 filaments of V-Ga alloy has been successfully extruded and is being reduced for second extrusion. All the casting processes necessary for making V-Ga alloys and Ga bronze are capable of large scale-up so that quantities of V_3Ga can be produced on a scale similar to the production scale presently attained with Nb_3Sn.

ACKNOWLEDGMENTS

The authors wish to thank Mr. F. A. Schmidt and his group at Ames Laboratory, DoE for their careful preparation of the V-Ga alloys. Special thanks are in order for F. Lewicki, G. Reverri, B. Radcliffe, and A. Russo for their participation in the billet preparation and wire drawing. Part of this work was supported by the Naval Research Laboratory.

REFERENCES

1. D. G. Howe, T. L. Francavilla, and D. U. Gubser, IEEE Trans. Magn. MAG-13:815 (1977).
2. D. U. Gubser, T. L. Francavilla, D. G. Howe, R. A. Muessner, and F. T. Ormand, IEEE Trans. Magn. MAG-13:385 (1979).
3. K. Tachikawa, in: "Proceedings ICEC 3, Berlin, 1970," Iliffe Sci. & Tech., Surrey, England (1970), p. 339.
4. P. R. Critchlow, E. Gregory, and W. Marancik, J. Appl. Phys. 45:5017 (1974).
5. D. G. Howe and T. L. Francavilla, in: "Filamentary A15 Superconductors," M. Suenaga and A. F. Clark, eds., Plenum Press, New York (1980), p. 103.
6. Max Hansen, "Constitution of Binary Alloys," McGraw-Hill, New York (1958), p. 582.
7. R. A. Beall and W. J. Hurford, in: "Metallurgy of Zirconium," B. Lustman and Frank Kerze, Jr., eds., McGraw-Hill, New York (1955), p. 216.
8. D. G. Howe, D. U. Gubser, and T. L. Franscavilla, IEEE Trans. Magn. MAG-17:654 (1981).

DEVELOPMENTAL STUDIES ON POWDER-PROCESSED Nb_3Al SUPERCONDUCTING WIRE

J. M. Hong, J. T. Holthuis, I. W. Wu, M. Hong,* and J. W. Morris, Jr.

Lawrence Berkeley Laboratory, University of California, Berkeley, California

INTRODUCTION

The superconducting compounds that show greatest promise for the windings of high-field superconducting magnets are intermetallic compounds of stoichiometric composition A_3B in the A15 crystal structure. This class of compounds includes the superconducting phases Nb_3Sn, V_3Ga, Nb_3Al, Nb_3Ge, and others. They combine high critical temperature (T_c) with high upper critical field (H_{c2}) and high critical current density (J_c). These compounds are, however, brittle intermetallic phases. They are therefore difficult to form into long lengths of superconducting wires for use in high-field magnet windings.

In the case of Nb_3Sn and V_3Ga, the manufacturing difficulty has been largely overcome by utilizing the solid-state reaction between Nb or V and Sn or Ga dissolved in Cu. In this process, continuous filaments of Nb or V are inserted[1] or formed in situ within a suitable matrix. The filaments and matrix are coextruded into a fine wire. The A15 compound is then formed at the filament-matrix interface through a reaction that extracts the Sn or Ga from the matrix. This process has been successfully employed for the manufacture of high-field superconducting wires containing multifilamentary Nb_3Sn or V_3Ga for magnet windings. It is, however, inapplicable for thermodynamic reasons to the manufacture of wires containing such A15 compounds as Nb_3Al or Nb_3Ge, which have better inherent superconducting properties.

Because of the inapplicability of the "bronze" process, manufacturing research on wires based on the other promising A15 compounds has concentrated on the establishment of a fine admixture of

*Now at Bell Laboratories, Murray Hill, New Jersey.

the primary constituents and the subsequent formation of the A15 phase through some direct reaction process after the wire is drawn. At least four distinct manufacturing processes have been explored: (1) the "powder process," in which a mixture of powders of the two primary constituents is extruded into wire before reaction,[3] (2) the "jelly-roll process," in which fine foils of the two constituents are rolled into a cylindrical compact and coextruded for subsequent reaction,[4] (3) the "infiltration process," in which a porous compact of the refractory constituent is infiltrated by the liquid component and the mixture is subsequently drawn into a fine wire for high temperature reaction,[5] and (4) the "direct precipitation process," in which a supersaturated solid solution of B in A is mechanically deformed into a fine wire or a tape; the A15 compound is subsequently formed by a precipitation reaction at elevated temperature.[6] The first of these, the "powder" approach, offers the advantages of an appealingly simple and potentially economical manufacturing process, which has already been shown to yield superconducting wires having very good high-field properties (in the case of Nb_3Al).[3]

This paper reports the results of recent developmental research on powder-processed Nb_3Al superconducting wire at LBL. The development of a good superconducting wire requires the control of at least two distinct aspects of the A15 phase: its internal composition and state of order, which determine its inherent superconducting characteristics, and its microstructural distribution within the wire, including such factors as its volume fraction, grain size, and continuity, which determine the peak critical current that a phase of given internal state can transport. The research reported below concentrated on the microstructural characterization of powder-processed wire and on the modification of processing to improve the microstructure and to enhance the inherent superconducting characteristics of the A15 phase.

EXPERIMENTAL PROCEDURE

Materials and Processing

The experimental samples were made by mixing powders of Nb and Al, coextruding into wire, and reacting to form the Nb_3Al superconducting compound. The material and process characteristics for the three sample types used are presented in Table 1. The powders were hydride-dehydride Nb and Al powders that were sized by screening through appropriate sieves. The powders were machine-mixed using inert beads to prevent the Al powder from aggregating. The powder mixture was then put in a Monel 400 cladding tube, form-rolled, swaged, and drawn into a fine wire. Wire drawing proved straightforward and easy to accomplish. No intermediate annealing was required.

Table 1. Characteristics of the Wires

Chemical Composition	Initial Powder Size (μm) Nb	Al	Areal Reduction Ratio	Diameter* (mm)
5 wt.% Al	62-105	≤ 44	400	0.18
3 wt.% Al	≤ 37	≤ 10	1600	0.22
4 wt.% Al	≈ 54	≈ 20	7000	0.11

*Excluding the cladding material.

Wires were made with three different aggregate Al contents: 5, 4, or 3 wt.%. The 5 wt.% material was used only for microstructural and reaction studies. The 3 and 4 wt.% material was used both for microstructural analysis and for superconducting measurements.

The coextruded powder wires were reacted to form the A15 Nb_3Al phase at temperatures in the range 700-1100°C for various periods of time. The Monel cladding material was etched away before heat treatment. The samples were sealed in quartz tubes under an Ar atmosphere and were wrapped with Ta foil to prevent interaction with the quartz tube. As a further safeguard against contamination, Ti rods were also sealed into the quartz tube.

Superconducting Property Measurements

The principal superconducting properties of interest were the critical temperature (T_c) and the critical current density (J_c). The critical temperature was measured by an inductive method using a calibrated Ge resistance thermometer. The superconducting-to-normal transition of these wires was generally found to be sharp, with a transition width of less than 1 K. The overall critical current (I_c) was measured using a standard 4-probe technique in transverse applied magnetic fields up to 19 T. The critical current was defined to be that at which the potential difference across voltage leads spaced 5 mm apart exceeds 1 μV. The overall critical current density (J_c) was calculated by dividing the critical current by the total cross section of the wire excluding the area of the cladding.

Materials Characterization

The microstructure of the product wires was studied using a combination of x-ray analysis, scanning electron microscopy (SEM), transmission electron microscopy (TEM), and scanning transmission electron microscopy with energy dispersive x-ray spectrometer (STEM/EDS). X-ray analysis was used for an overall determination of the phases present. SEM and TEM were used to observe the microstructural state of the wire and the detailed microstructure of the reacted layer. STEM/EDS was employed to determine the chemical composition of the reacted layer. Samples for TEM and STEM analyses were made by grinding and polishing the tested wires to a thickness of 0.1 mm or less, followed by ion milling to obtain a specimen transparent to 100 kV electrons.

RESULTS AND DISCUSSION

Metallurgical Characteristics

X-ray and SEM examination of the coextruded powder wire suggests that the deformed wires contain a continuous filamentary matrix of aluminum in the interstices of the drawn Nb powder. The deformation of the Nb powder itself is reflected in the very high dislocation density in the powder, revealed by TEM studies, and by a preferred texture of the Nb, which is evident from x-ray diffraction analysis.

The reaction to form the A15 superconducting phase occurs at the interface between the Nb and Al. Since the reaction temperature is in all cases above 660°C, the melting point of pure Al, the initial reaction is between liquid and solid phases. Reference to the binary Nb-Al phase diagram[7] shows that three intermetallic compounds can be formed between Nb and Al at temperatures in the range 700-1100°C. These are the A15 phase, Nb_3Al; the σ phase, Nb_2Al; and the aluminum-rich intermetallic, $NbAl_3$. All three phases are expected to form along the liquid-solid reaction interface.

The phase distribution in the reacted samples was studied by x-ray analysis and by transmission electron microscopy. X-ray analysis of reacted 5 wt.% Al samples aged at 750-1100°C for various periods of time revealed only Nb and σ-phase reflections. This result suggests that the A15 phase is not the major constituent, but does not necessarily show an inadequate content of A15 phase because of the complexity of its diffraction pattern and the consequent low intensity of its diffraction peaks. The presence of the A15 phase is revealed by transmission electron microscopic studies.

A transmission electron micrograph of a 5 wt.% Al sample aged at 1100°C for one minute is shown in Fig. 1. Three distinct morphologies are present at the original Nb-Al interface. The residual Nb phase is characterized by a fine microstructure and a high dislocation content. The dislocations are remnants of the wire drawing process and are not recovered after one minute at 1100°C, which is less than one-half of the melting point of Nb.

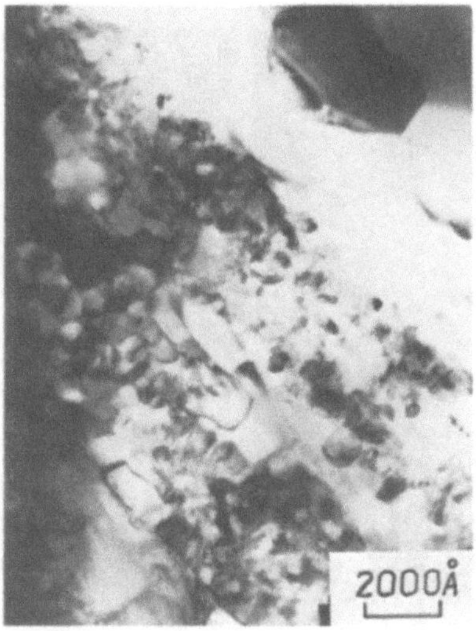

Fig. 1. TEM micrograph of 5 wt.% Al sample aged at 1100°C for 1 min. From lower left to upper right corners are regions of Nb, A15/σ, and $NbAl_3$ phases in sequence.

Adjacent to the Nb surface there is a fine-grained region, which microdiffraction shows to be a mixture of the A15 and σ phases. Beyond the fine-grained region is a coarser-grained layer of $NbAl_3$.

Because of the extremely fine grain size of the material present in the A15-σ-phase mixture, it is not possible to separate the two phases in TEM at the magnifications used in this study. If the sample is given a longer aging at 1100°C, however, the A15 and σ phases separate into distinct layers, and the $NbAl_3$ phase is consumed. The micrographs in Fig. 2 illustrate the microstructure after 1.5 h annealing at 1100°C. Figure 2a shows the A15 grains, which now have an average grain size of about 3000 Å. These are equiaxed and almost free of lattice defects. The σ grains, on the other hand, are internally defective, as shown in Fig. 2b. The dark bands within these grains are revealed to be crystallographic twins by microstructural analysis.

It appears likely that the difference in defect characteristics can be used to distinguish the A15 and σ phases in the fine-grained layer that results from the initial reaction. A reexamination of Fig. 1 shows that the grains in the fine-grained layer are a mixture of defect-free and twinned grains. It is tentatively assumed that the twinned grains represent the σ phase, and the defect-free grains represent the A15 superconducting material.

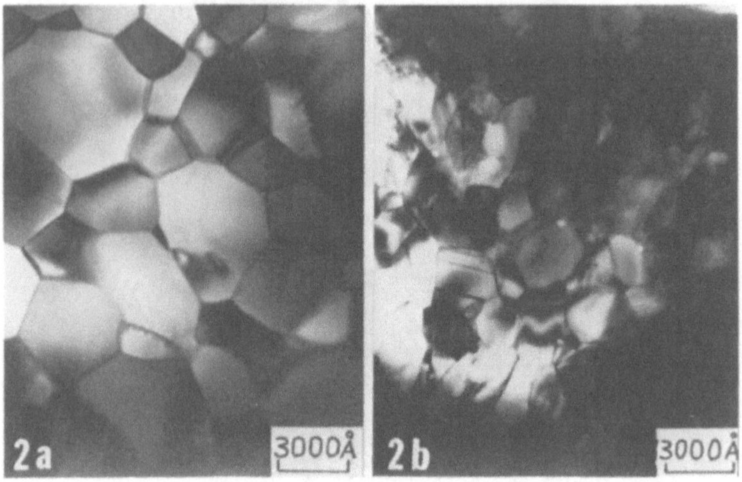

Fig. 2. Two adjacent regions of 5 wt.% Al sample aged at 1100°C for 1.5 h. (a) A15 grains which are equiaxed and nearly free of lattice defects; (b) σ grains have crystallographic twins.

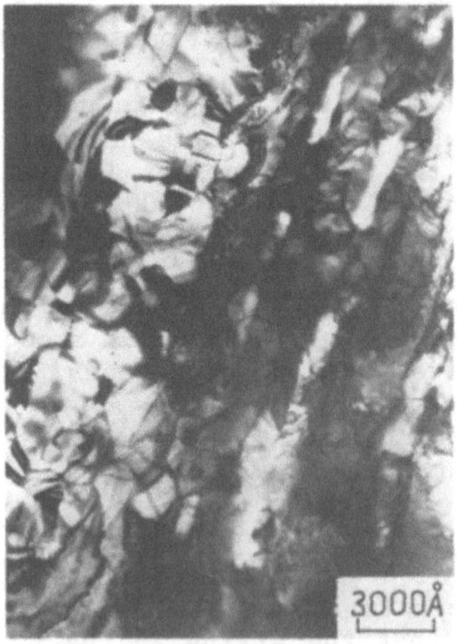

Fig. 3. A 3 wt.% Al sample aged at 800°C for 16 h. Overall J_c is high (see text), though the σ phase has a higher volume fraction than the A15 phase.

Figure 3 shows a transmission electron micrograph of a sample of 3 wt.% Al wire that was aged at 800°C for 16 h. This sample was taken from a superconducting wire, which exhibited an overall current density above 10^4 A/cm^2 in a 13-T magnetic field at 4.2 K. The microstructure adjacent to the Nb interface again contains a mixture of the A15 and σ phases. If we assume that the substructurally twinned grains constitute the σ phase, then it appears that the σ phase has a much higher volume fraction within the fine-grained layer. However, there is a sufficient volume of defect-free A15 phase to establish a reasonably interconnected, fine-grained structure. Both the fine grain size and interconnectivity of the A15 grains are reflected in the high critical current of the sample.

Measurement and Control of Inherent Superconducting Properties

The critical temperature of the powder-processed Nb_3Al was measured as a function of aging time and temperature for a 5 wt.% Al wire. The results are plotted in Fig. 4. Irrespective of aging temperature in the range 700-1100°C, the critical temperature rises slightly on initial aging and reaches a maximum at approximately 15.5 K. It then tends to decrease with longer aging time. The kinetics of this characteristic behavior depend strongly on the aging temperature. The high peak temperature exhibited by these specimens is surprising, since the A15 phase in equilibrium with σ phase at 1100°C is expected to have a maximum critical temperature of only about 14.4 K.[8,9] The reason for the high critical temperature is not yet clear, but may reflect the fine grain size of the A15 particles, features of the reaction determined by the equilibrium with the σ phase, or thermal effects occurring during the initial chemical reaction. This subject is under investigation.

It is generally accepted, however, that the critical temperature is determined by the composition and the long-range order of

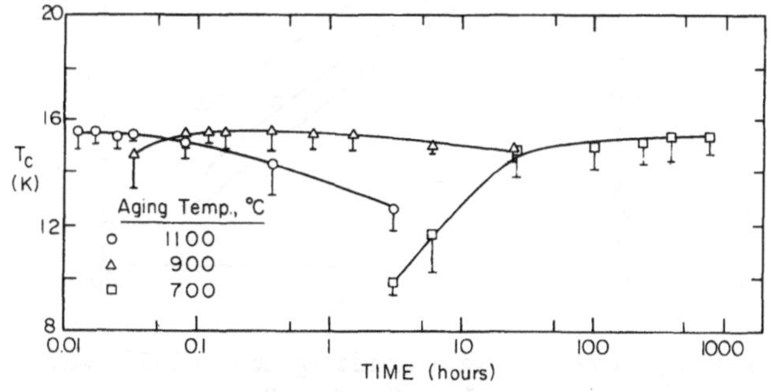

Fig. 4. Fundamental powder metallurgy approach, Nb-5 wt.% Al wires (3.0 x 10^{-2} mm^2 cross section).

the A15 phase.[10] From the equilibrium phase diagram, high temperature aging tends to promote a more nearly stoichiometric composition of the A15 phase. Aging at lower temperature, on the other hand, is expected to establish a better state of long-range order. In an attempt to improve the inherent characteristics of the A15 phase, we therefore imposed multiple-step aging treatments, beginning with a high-temperature aging to establish a good composition and following that with aging at lower temperature with the intent of improving the internal state of order. These double-aging treatments appear to be successful in substantially improving the T_c onset. For example, a treatment at 1100°C for 1 min followed by 900°C for 22 min increased the measured critical temperature from 15.5 to 16.3 K. A three-step aging treatment of 1100°C for 1 min plus 900°C for 22 min plus 750°C for 8 days raised the T_c onset 17 K. It is, of course, possible that these measured critical temperatures reflect the properties of the best of the superconducting material and are not typical of the average material within the superconducting layer. On the other hand, the superconducting-to-normal transition width remains very small, about 0.5 K, which suggests that a general improvement of the A15 material has been obtained.

Measurement and Enhancement of the Critical Current Characteristic

Samples of 3 and 4 wt.% Al were made for the purpose of measuring the critical current as a function of magnetic field. These samples were metallurgically designed to achieve an approximately

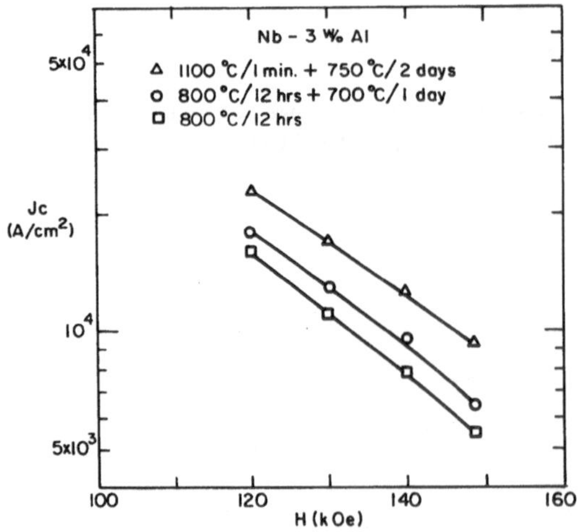

Fig. 5. Overall J_c (excluding cladding material) versus transverse applied magnetic field, H, for 3 wt.% Al sample showing the enhancement on J_c by the double-aging treatment.

50:50 ratio between the reacted product (A15 + σ phase) and the residual Nb in the cross section of the final wire. Geometric considerations suggest that for this purpose the wire should be made of initial powders having a Nb-to-Al powder diameter ratio of approximately 2.7 to 1. Samples containing 3 and 4 wt.% Al were therefore made with approximately this initial powder ratio (Table 1), and their superconducting characteristics were measured as a function of heat treatment.

The results obtained from critical current measurements to date include only a sampling of heat treatment and hence do not permit a full discussion of the influence of wire parameters on the superconducting current characteristeric. Initial results are; however, interesting and are presented in Figs. 5 and 6.

In Fig. 5 we have shown the variation of critical current with applied magnetic field for three samples of 3 wt.% Al which were given different double-aging heat treatments. The results show that a sample aged at 800°C for 12 h undergoes a significant increase in critical current at high field if it is given a subsequent aging treatment at 700°C for one day. Still better properties are obtained if the sample is given a very brief high-temperature aging, 1100°C for 1 min, followed by a more extensive low-temperature aging, 750°C for 2 days. The critical current density exceeds 10^4 A/cm^2 at 14 T.

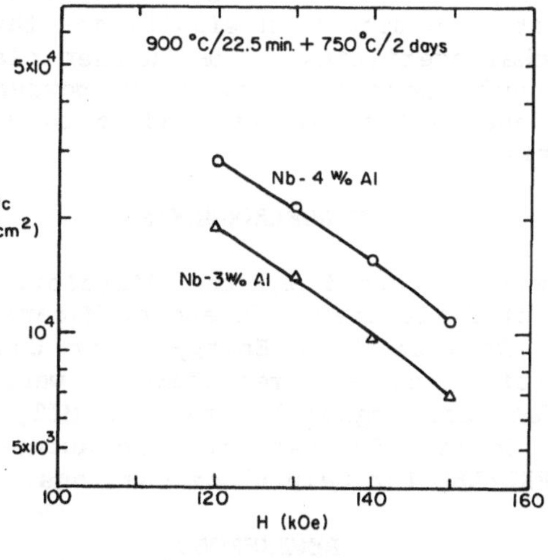

Fig. 6. Comparison of overall J_c versus transverse applied magnetic field, H, of 3 wt.% Al and 4 wt.% Al samples with the same aging treatment: 900°C for 22.5 min followed by 700°C for 2 days. The J_c of the latter exceeded 10^4 A/cm^2 at 15 T and 4.2 K.

A similar improvement in the critical current capability was obtained by increasing the Al content of the wire from 3 to 4 wt.%, as shown in Fig. 6, using similar (though not identical) wires and a standard heat treatment of 900°C for 22-1/2 min plus 750°C for 2 days. The increase of Al content from 3 to 4 wt.% resulted in amost tripling of the critical current density over the range of field from 12 to 15 T. The critical current of the 4 wt.% Al sample was above 10^4 A/cm^2 at 15 T, even though the heat treatment used with these samples was not optimal, as can be seen from the results presented in Fig. 5.

Both of these preliminary studies show that exceptionally good high-field currents can be obtained in powder-processed Nb-Al wires. Critical currents near 3 x 10^4 A/cm^2 at 12 T and above 10^4 A/cm^2 at 15 T have been measured. The increase in critical current density with alternate processing and heat treatments suggests that much higher critical current densities are possible.

CONCLUSION

The results reported here show that it is possible to obtain very promising values of the critical temperature and high-field critical current density in powder-processed Nb_3Al superconducting wires. They suggest, moreover, that much better properties are obtainable. Microstructural analysis of the reacted samples reveals a fine, complex microstructure. Its internal characteristics are not fully understood, but it responds to appropriate variations in heat treatment with significant favorable changes in superconducting characteristics. The further clarification of the microstructure should permit a choice of powder characteristics, wire processing, and heat treatment leading to very good superconducting characteristics.

ACKNOWLEDGMENTS

This work was supported by the Director, Office of Energy Research, Office of Basic Energy Sciences, Materials Sciences Division of the U.S. Department of Energy under Contract No. W-7405-ENG-48. The critical current measurements were performed at the Francis Bitter National Magnet Laboratory, MIT, which is operated for the National Science Foundation. The authors are grateful to W. Hassenzahl, AFRD/LBL for helpful discussions.

REFERENCES

1. A. R. Kaufman and J. J. Pickett, Bull. Am. Phys. Soc. 15:838 (1970).
2. J. D. Verhoeven, E. D. Gibson, C. V. Owen, J. E. Ostenson, and D. K. Finnemore, Appl. Phys. Lett. 35:270 (1979).

3. R. Akihama, R. J. Murphy, and S. Foner, Appl. Phys. Lett. 37: 1107 (1980).
4. B. Annaratone, R. Bruzzese, S. Ceresara, V. Pericoli-Ridolfini, G. Pitto, and N. Sacchetti, IEEE Trans. Magn. MAG-16:1 (1980) and the references therein.
5. M. R. Pickus, V. F. Zackay, E. R. Parker, and J. T. Holthuis, Int. J. Powder Metall. 9:3 (1973).
6. M. Hong and J. W. Morris, Jr., Appl. Phys. Lett. 37:1044 (1980).
7. C. E. Lundin and A. S. Yamamoto, Trans. Metall. Soc. AIME 236: 863 (1966).
8. J. L. Jorda, R. Flükiger, A. Junod, and J. Muller, IEEE Trans. Magn. MAG-17:557 (1981).
9. J. Kwo, R. H. Hammond, and T. H. Geballe, J. Appl. Phys. 51: 1726 (1980).
10. D. Dew-Hughes, in: "Treatise on Materials Science and Technology," Vol. 14, T. Luhman and D. Dew-Hughes, eds., Academic Press, New York (1979), pp. 156-159, and the references therein.

OPTIMIZATION OF CRITICAL CURRENTS IN Cu-Nb_3Sn MICROCOMPOSITES

J. Wecker, K. Mrowiec, R. Bormann, and H. C. Freyhardt

Institut für Metallphysik, Universität Göttingen, Göttingen, Federal Republic of Germany

INTRODUCTION

The hot powder metallurgical preparation method of Cu-Nb_3Sn microcomposite conductors is well advanced and an industrial (large-scale) application appears to be feasible. The addition of reduction components like Al or Mg to the initial Cu and Nb powder mixture enables the development of an ideal filamentary structure. These conductors exhibit critical current carrying capabilities that are comparable to those of conventionally produced A15 superconductors. Further optimization of the superconducting properties of these microcomposites can be achieved by alloying additional elements to the bronze and to the Nb.

An optimization of the critical current densities of Cu-Nb_3Sn microcomposite conductors requires the development of a uniform filamentary structure and optimum superconducting properties of the A15 phase. In the hot powder metallurgical process, both requirements can be achieved by the controlled addition of further elements to the powder mixture and to the Sn plating in the external bronze process.

FILAMENTARY STRUCTURE OF Cu-Nb_3Sn MICROCOMPOSITE CONDUCTORS

The development of a uniform filamentary structure during the (hot) powder metallurgical process is rendered possible only if an ideal co-deformation of the Nb particles to long filaments is guaranteed. This requires a purification of the Nb, which is generally hardened by interstitially dissolved oxygen. During the hot extrusion of the compacted powder billets an in situ reduction occurs if components are added that possess a larger (free) binding enthalpy to oxygen than Nb does.[1] Possible additives are Zr, Al, Hf, Mg,

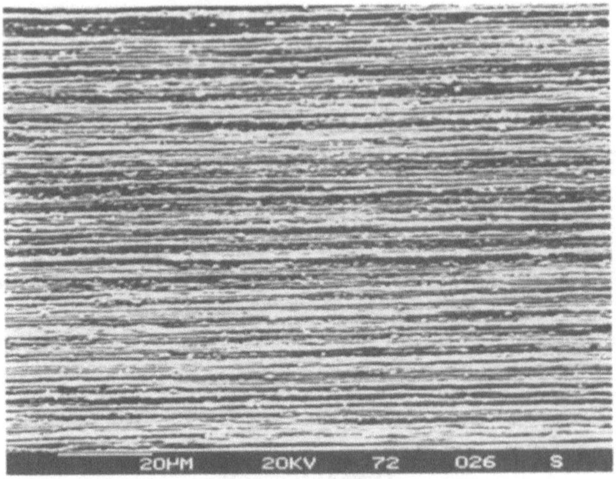

Fig. 1. Longitudinal cross section of a Cu-20 wt.% Nb-1 wt.% Al composite after a reduction, q, in cross-sectional area of about 400. The Cu matrix is slightly etched away (SEM).

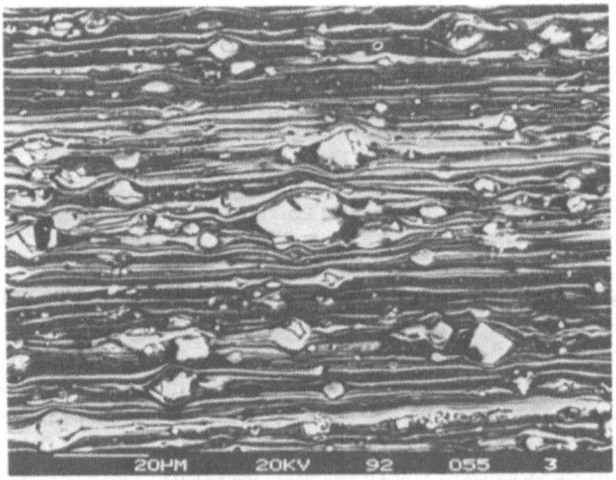

Fig. 2. Longitudinal cross section of a Cu-20 wt.% Nb-4.7 wt.% Hf composite (q = 400) (SEM).

and Ca [listed in the order of increasing (free) binding enthalpies to oxygen]. Generally less than 1 wt.% Al is sufficient to bind all the oxygen in the composite to the Al (or equivalent amounts of Mg, Zr, Hf, and Ca depending on the corresponding oxides formed).

The degree of purification achieved, i.e., the amount of oxygen retained in the niobium, can be investigated by (inductively) measuring the transition temperature, T_c, of the CuNb composite.[2] It is found that the larger the (free) binding enthalpies of the additives to the oxygen are, the better the *in situ* reduction of the Nb becomes.

However, an ideal filamentary structure and, thus, large critical current densities are observed only for those additives that are soluble in the Cu matrix (at the concentrations used in this investigation), e.g., for Al and Mg (Fig. 1). Zr, Hf, and Ca, on the other hand, form the intermetallic compounds Cu_3Zr, Cu_5Hf, and Cu_5Ca, respectively, which--because of their hardness and reduced ductility--prevent a co-deformation of the Nb powder particles (Fig. 2). They are not suited for the powder metallurgical process.

Additional annealing of the CuNb rod (after the hot extrusion) at 950-1000°C leads to a decrease of the Nb transition temperature by at least 0.1 K, indicating a (small) solubility of the additives in the Nb. Consequently they are also incorporated in the Nb_3Sn phase, which forms during the reaction treatment. This is proven by the increased (extrapolated) upper critical fields, B^*_{c2}, (~0.5 T) and transition temperatures (0.2 K) owing to the additional high-temperature annealing.

Most CuNb composites were prepared from phosphorus-free Cu powders. All conductors manufactured from these extrusion rods exhibit larger critical current densities than corresponding samples that contained small amounts (~0.2 wt.%) of phosphorus impurities. Because B^*_{c2} remains unchanged, this indicates an impeded Nb_3Sn formation due to the presence of phosphorus.[3]

SUPERCONDUCTING BEHAVIOUR

As in the case of conventionally processed multicore conductors, the superconducting properties of the Nb_3Sn phase can be improved and optimized by alloying additional elements to the bronze (In, Ga, Ge) and/or to the Nb filaments (e.g., Ta). To demonstrate this, 5 and 15 wt.% In or Ga (relative to the total weight of the composite) were added to the Sn plating of Cu-30 wt.% Nb-7(10;15) wt.% Sn microcomposite wires. The resulting changes in their superconducting behaviour were investigated[4] as a function of reaction time and temperature.

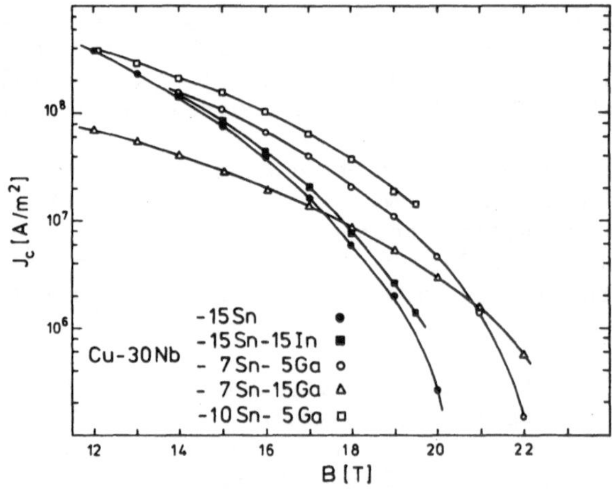

Fig. 3. Overall critical current densities of Nb_3Sn composites alloyed with different amounts of Ga or In.

Indium

Microprobe analyses of the $Nb_3(Sn,In)$ layers in monocore conductors indicate that 0.1 to 0.4 at.% In is incorporated in the A15 phase (reaction temperature, T_A, of 800°C). Nevertheless, the critical current densities of In-doped microcomposites are only slightly improved in the field regime above 15 T (Fig. 3), because the effect of the increased upper critical field (due to an enhanced) residual resistivity of the A15 phase) is almost compensated by a decreased transition temperature (T_c drops by ~1 K upon increasing the In content). The best critical current densities are obtained for Sn- and In-rich specimens (15 wt.% Sn, 15 wt.% In) heat treated at low reaction temperatures for only a short time (600°C/48 h). The compressive stresses acting on the Nb_3Sn layer due to a difference in thermal expansion coefficients of the bronze and the A15 phase are obviously not changed by the addition of In (Fig. 4); i.e., the favorable mechanical properties of the hot powder metallurgical processed composites are not deteriorated.

Gallium

Contrary to In, Ga additions cause a slight enhancement of the transition temperature ($\Delta T \approx +0.6$ K). Consequently the upper critical field, B^*_{c2}, is increased by up to 3.3 T, yielding a marked improvement of the critical current densities, J_c, in field ranges beyond 13 T relative to optimized unalloyed Cu-Nb_3Sn composite wires (Fig. 3). Enhancements of J_c by factors of 2.4 and 10 are observed at fields of 16 T and 19 T, resulting in overall J_c values of $\geq 10^8$ and 2×10^7 A/m^2, respectively. Because more Ga

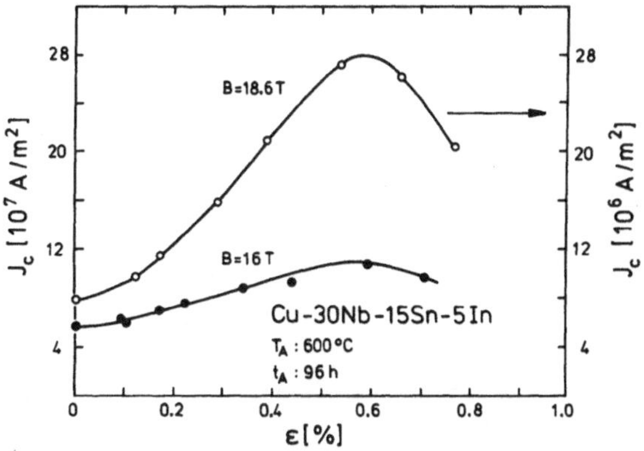

Fig. 4. Critical current density, J_c, vs applied strain, ε, at 16 T and 18.6 T for a Cu-Nb_3Sn composite with In addition.

(1.5 at.%)[5] than In can be incorporated in Nb_3Sn without deteriorating B^*_{c2}, the largest critical current densities were achieved for relatively low Sn and Ga concentrations (7 or 10 wt.% Sn, 5 wt.% Ga) and comparably high reaction temperatures. Higher Ga contents increase B^*_{c2}, however, they lead at the same time to a pronounced reduction of flux pinning in fields up to 18 T (Fig. 3).

Further improvements of the high-field critical current carrying capabilities of (hot) powder metallurgically processed Cu-Nb_3(Sn, Ga, In) microcomposite conductors are rendered possible by employing NbTa alloy powder instead of Nb powder; these investigations are currently underway.

REFERENCES

1. R. Bormann, H. C. Freyhardt, and H. Bergmann, Appl. Phys. Lett. 35:994 (1979).
2. K. Mrowiec, Diploma Thesis, University of Göttingen, FRG (1981).
3. D. Smathers, University of Wisconsin, Madison, Wisconsin, private communication (1981).
4. J. Wecker, Diploma Thesis, University of Göttingen, FRG, in preparation.
5. H. Sekine, K. Takendu, and K. Tachikawa, IEEE Trans. Magn. MAG-17:383 (1981).

ON THE OPTIMIZATION OF *IN SITU* Nb_3Sn-Cu WIRE

J. J. Sue,* J. D. Verhoeven, E. D. Gibson,
J. E. Ostenson, and D. K. Finnemore

Departments of Materials Science and Engineering and of Physics
Ames Laboratory—USDoE, Iowa State University, Ames, Iowa

INTRODUCTION

In this paper we report on two aspects of our efforts to optimize large-scale in situ Nb_3Sn-Cu superconductor wire. First, results will be presented that show that the as-drawn Nb filament size is critical in optimizing J_c values and depends upon the temperature of the Sn diffusion anneal. Second, laboratory-size experiments show that utilizing C containers for ingot preparation results in a small C contamination and significantly lower J_c values. To scale-up the in situ process, we have previously developed arc-casting techniques capable of producing 30-kg ingots utilizing graphite-lined Cu molds. Results are presented here on efforts to reduce C contamination in these large castings by utilizing ceramic coatings on the graphite liner.

OPTIMIZATION OF AREA REDUCTION RATIO

In a previous study[2] we have shown that for a given as-cast Nb dendrite size in the Cu-Nb ingot an optimum reduction ratio, R_{opt}, exists for maximizing J_c values. These results are presented by the open symbols in Fig. 1. (Note: All J_c values in this paper are based on the entire wire area.) At values greater than R_{opt}, the as-drawn Nb filaments become sufficiently small so that coarsening occurs during the Sn diffusion anneal and causes J_c to drop owing to loss of continuity in the Nb_3Sn filaments, whereas, at R less than R_{opt} the values of J_c probably rise owing

*Present address: Union Carbide Corporation, Linde Division, Indianapolis, Indiana.

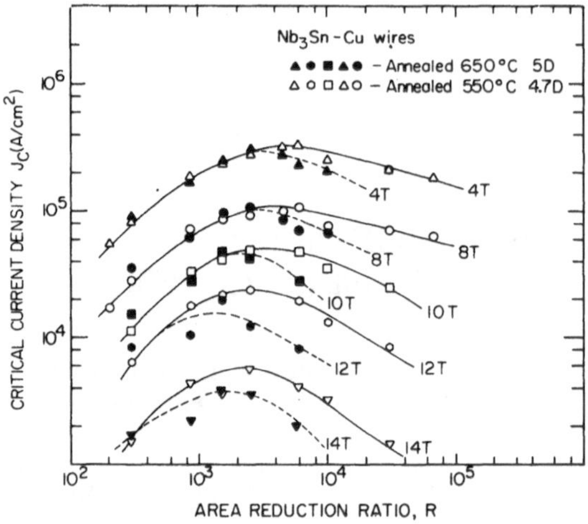

Fig. 1. Critical current density versus area reduction ratio at magnetic fields from 4 to 14 T.

to either a decreasing Nb_3Sn grain size, and/or increasing filament length with decreasing filament spacing. In previous work it was shown that the as-drawn Nb filament thickness followed the relation $t(\text{Å}) = (1.71 \times 10^4)\ R^{-0.421}$, so that the optimum as-drawn Nb filament thickness varied from 470 Å at 4 T to 630 Å at fields greater than 12 T.

There is some indication in bronze wire[3] that utilizing higher Sn diffusion temperatures increases H_{c2} and, hence, J_c at high fields. Therefore, we have repeated the previous work utilizing a Sn diffusion temperature of 650°C rather than 550°C. In both studies the alloys, Cu-20 wt.% Nb, were prepared by induction melting 80-g samples in Y_2O_3 crucibles of 2.5-cm diameter. Identical cooling rates were employed in all experiments and the as-cast Nb dendrites were found to be uniformly distributed in the casting with a mean diameter of 8.2 ±1.5 μm. The diameters were measured in a calibrated SEM by deep-etching away the Cu to reveal the true diameter of individual dendrite branches. All wires were reduced to a diameter of 0.15 mm, and the various area reductions, R, were achieved by machining ingots to different sizes prior to reduction to wire. The machined ingots were swaged to around 1.3-mm diameter and then drawn to 0.15 mm. In all cases, Sn was electroplated on the wires, which were then drawn through a die to produce a coating of 9.3 vol.%, which is the stoichiometric amount of Sn for the 20 wt.% Nb content.

The J_c data for these samples annealed at 650°C are presented by the filled symbols in Fig. 1. The values of R_{opt} at a given field are consistently reduced, thus indicating that coarsening is more severe at 650°C. This conclusion has been confirmed by metallographic evidence.

The data of Fig. 1 reveal that the increased annealing temperature of 650°C does not increase the high-field J_c properties, as had been expected. This result is further illustrated by Fig. 2, which presents $\mu_0 H^*_{c2}$ values as a function of R, as determined from plots of the Kramer function. It is seen that values of $\mu_0 H^*_{c2}$ for these in situ wires are around 16.5 T and drop off significantly as coarsening becomes significant. Hence, the results of this study indicate that optimum high-field J_c values are achieved with a 550°C Sn diffusion anneal and require an as-drawn Nb filament thickness of around 630 Å, which corresponds to the observed R_{opt} = 2500 at H $\geq$ 12 T for cast material with 8.2-μm dendrites.

EFFECT OF CARBON IMPURITIES

Previous work[4] has shown that carbon impurities in the melt can produce a significant change in Nb dendrite size. The above results indicate that this effect alone could produce a change in J_c properties. Therefore, experiments were done to produce carbon-free and carbon-contaminated alloys of Cu-20 wt.% Nb of identical dendrite size. Eighty-gram ingots of 2.5-cm diameter were prepared in Y_2O_3 and graphite crucibles utilizing identical heating and cooling procedures. As shown in Table 1, the average Nb dendrite size was essentially identical in the two ingots. The

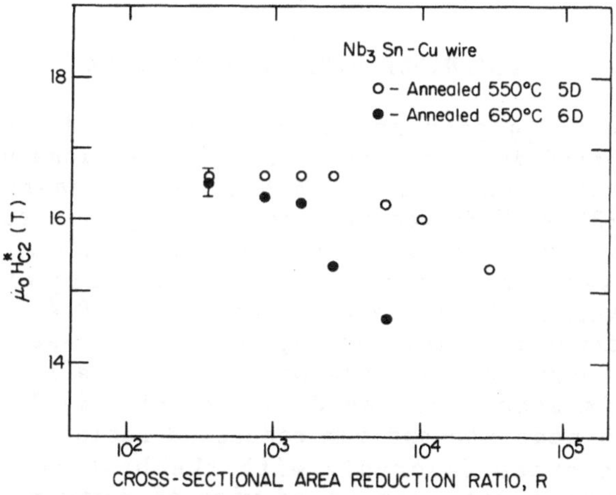

Fig. 2. Values of H^*_{c2} versus area reduction ratio for the wires annealed at 550°C for 4.7 days and 650°C for 5 days.

Table 1. Measured C Compositions and Nb Dendrite Diameters

Crucible Material	Average Dendrite Diameter (Å)	Average Carbon Composition (ppma)
yttria	8.2 ±1.5 μm	85 ± 11
graphite	8.3 ±1.6 μm	2140 ±670

morphology of the dendrites was altered a bit, with the C producing shorter primary dendrite stalks, but the effect was much less than previously found[4] at higher C levels. The carbon composition was determined by combustion analysis of 1-g samples cut from 3 different positions in the ingots. The compositions listed in Table 1 represent the average value of the three samples with composition given as parts per million atomic, ppma.

Wires were prepared using an area reduction ratio of 4600. Critical current measurements were made up to 8 T, and the results show J_c for the higher C samples decreased by a factor of 2 to 5, e.g., 0.23 x 10^5 versus 0.97 x 10^5 A/cm^2 at 8 T. Examination of several deep-etched transverse sections of both wires revealed essentially identical Nb_3Sn filamentary microstructures. Hence, it is concluded that preparation of in situ wire in graphite crucibles produces a C impurity level of around 2000 ppma, which significantly lowers J_c, and this reduction of J_c is not a result of a change in dendrite diameter, nor probably, a change in dendrite morphology. The cause of the reduced J_c is not understood and is currently being investigated further.

COMPARISON WITH PREVIOUS WORK

To evaluate in situ and powder-processed wire, Suenaga et al.[5] have compared data from several investigations by normalizing to a Cu-Nb composition of 30 wt.% Nb. For comparison, an 80-g casting of Cu-30 wt.% Nb was prepared in an yttria crucible and it yielded an average Nb dendrite size of 8.4 ±1.3 μm. Wires for J_c measurements were made using a reduction ratio of 4550, which yielded an average Nb filament thickness of 590 Å, and a slightly excess stoichiometric volume of Sn, 14.7 vol.%, was employed. The data for these wires are superimposed on Suenaga's summary plot in Fig. 3. It is seen that the J_c values of this 30% in situ wire prepared with near the optimum reduction ratio found for 20 wt.% alloys compare quite favorably with the best results found by Suenaga for the powder-processed wire of Bormann and the ribbon material of Bevk. The present results offer an explanation for the low values of most of the data of Fig. 3. Finnemore and Flükiger both used a chill casting technique, which produced Nb

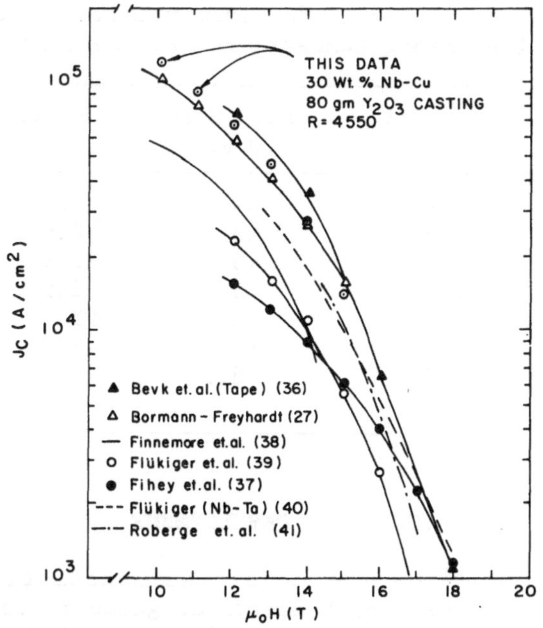

Fig. 3. A comparison of J_c for in situ and powder-processed wire. All data normalized to Cu-30 wt.% Nb. (Reference numbers are from Ref. 5).

filaments too small to avoid coarsening. For example, in the work of Finnemore, the as-drawn Nb filament thicknesses were only around 70 Å, compared with the optimum size found here of around 600 Å. The experiments of both Fihey et al. and Roberge et al. employed graphite containers for the molten Cu-Nb alloys, which would account for the lowered J_c values.

Suenaga et al.[5] have also presented J_c data for several commercially produced multifilamentary bronze processed wire and these results at $\mu_0H = 12$ T are reproduced in Table 2. The lower values of J_c in the table are for typical wires and the higher values are the best from a number of specimens. Figure 4 shows the spread of these commercial wire results in a plot of the J_c values for our 30 wt.% Nb wire. We have included measurements of a second 30 wt.% wire evaluated out to only 8 T, which indicates the scatter band expected in our laboratory-size experiments. It is seen that the 30 wt.% insitu wire compares favorably with the commercial wire. The best laboratory test results that we have found on bronze-processed wire are those of Schauer and Schelb[3] using a two-step heat treatment. Their data are plotted in Fig. 4 for an assumed Br/Nb_3Sn area ratio of 2.5, and it is seen that our best data lie at the lower limit of their 13-T results.

Table 2. Summary of J_c (12 T) for Commercially Prepared MF-Nb_3Sn Wire[5] (Reference numbers are from Ref. 5)

Source (Ref.)	No. Fil.	Fil. Size (μm)	Br/Nb	J_c (12 T) (A/cm^2)
AERE (28)	5143	∿4	4.26	5.9×10^4
Airco (29)	8965	∿3	2.9	5.6×10^4
Airco (29)	2869	∿3	2.9	3.3×10^4
Airco (29)	2869	∿3	2.9	3.5×10^4
Vac. Sch. (30)	∿10^4	∿2	∿3	5.5×10^4
I.G.C. (5)	217	∿7.5	3.5	4.3×10^4
I.G.C. (5)	78	8.2	3.4	4.9×10^4
Our 30% wire			∿2.3	6.8×10^4

SCALE-UP DEVELOPMENT

In previous work[1] we have developed consumable arc melting techniques using a Nb-plate/Cu electrode geometry to prepare 15-kg (7.5-cm-diam.) and 30-kg (10-cm-diam.) ingots of Cu-16.8 wt.% Nb. The technique employed a double melting into graphite-lined Cu molds, and the results indicated adequate chemical and microstructural homogeneity, but a pickup of 900-1800 ppma C, which was thought to be detrimental to the measured J_c properties. The above results using graphite crucibles confirm that C at the

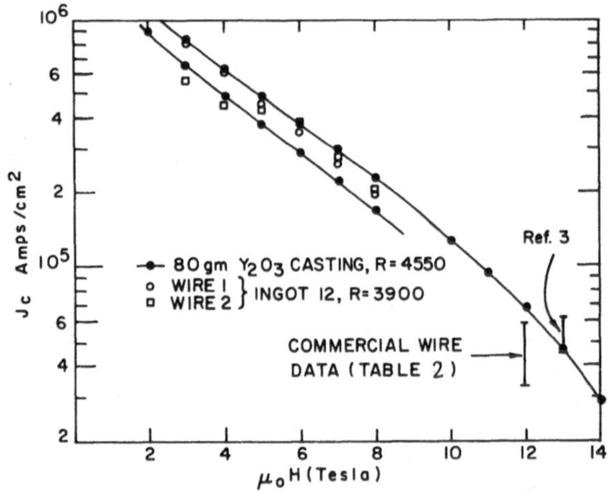

Fig. 4. Critical current vs. magnetic field for 2 wires prepared with Cu-30 wt.% Nb with R = 4550 and 550°C, 5d diffusion anneal. Data points from arc cast ingot 12 also included.

2000 ppma level is detrimental to J_c properties, and efforts have been made to reduce the C pickup of the arc-casting technique by coating the graphite liners with ZrO_2. Plate Nb/Cu electrodes, similar to that shown in Ref. 1, were designed to produce two 22.5 wt.% Nb ingots (#10 and 11) and one 30 wt.% Nb ingot (#12). Fifteen-kg ingots were produced by a single melt into 7.5-cm diameter molds utilizing currents of 2200 A at 28 V. The graphite liner of ingot 10 was flame sprayed with ∿0.2 mm of calcia-stabilized zirconia, and the liners of ingots 11 and 12 were plasma sprayed with ∿0.2 mm of yttria-stabilized zirconia. The 6-mm-thick graphite liners were split longitudinally to allow uniform coating along their length. An uncoated graphite pad was utilized at the bottom surface of the Cu mold so that some contamination of the ingots with C was expected.

When the split graphite molds were removed from the ingots, it was found that the ZrO_2 coating pulled away from the graphite and thus prevented the reuse of the sprayed graphite liners. The mechanical integrity of the plasma-sprayed coating subsequent to casting was superior to that of the flame-sprayed coating.

After removal of the ingots, they were sectioned and polished longitudinally along their entire length. A fairly uniform Nb dendrite array was found with complete wall-to-wall melting and the absence of any unmelted Nb or globular Nb particles. As previously found,[1] small but distinct transverse striations revealing the shape of the solid/liquid interface were present as a result of a microsegregation of the Cu/Nb dendrite mixture. The Nb dendrite size was measured at the ingot surface and center near the bottom, middle, and top, and the average values were found to be 7.5 ±1 μm in ingot 11 (22.5 wt.% Nb) and 7.4 ±1 μm in ingot 12 (30 wt.% Nb). These values are slightly smaller than the dendrite diameters of the laboratory-size castings used to obtain the data of Fig. 1, and R values to produce equivalent as-drawn Nb filament thicknesses were taken to be smaller for the large castings by the ratio of the square of the as-cast Nb dendrite diameters.

Complete chemical analyses were carried out on ingots 11 and 12, and the results are presented in Table 3. Analyses are presented for all elements having an impurity level in either ingot above 1 ppma. Elements O, N, and H were measured by vacuum fusion, C by combustion, Nb by wet chemical, and the remaining elements by mass spectrographic techniques. The starting materials were metallurgical grade 99.98% Nb from Wah Chang and C101 Cu. We have reported similar results[6] for 80-g castings in Y_2O_3, ThO_2, and C crucibles, and a comparison reveals that we were picking up small amounts of Zr and Y from the ceramic coating and perhaps, also, a limited amount of S and Si. The Ta comes from the original Nb, and the O is a contamination produced in casting, which is comparable to that obtained in the 80-g Y_2O_3 crucible

Table 3. Chemical Analysis* of Two 15-kg Ingots

Element	Ingot 11	Ingot 12	Element	Ingot 11	Ingot 12
S	10	10	Mo	4	2
P	<60	<100	Zr	340	340
Si	100	90	Ag	4	5
Cl	3	2	Sb	3	3
K	3	5	Hf	4	3
Ca	2	6	Ta	90	100
Fe	14	14	W	3	2
Ni	7	6	Pb	1	2
Se	0.4	1	Y	150	150
C (mid)	260	260	O	550	810
C (top)	210	260	N	24	33
Nb (mid)	22.4 wt.%	30.5 wt.%	H	6 ppmw	6 ppmw
Nb (top)	24.1 wt.%	31.3 wt.%			

* All analyses are parts per million atomic, ppma, except where noted.

castings. The C level, 260 ppma, is higher than the 60 to 80 ppma found in the 80-g castings, but is significantly reduced from the 900-1800 ppma level found with uncoated graphite liners.[1] Carbon analysis at the bottom of the ingot was significantly increased (680 ppma) over the values reported in Table 3 at the middle and the top of the ingots. This results from the use of the uncoated graphite pad at the bottom of the Cu mold. The higher C level at the bottom had no measurable effect on the Nb dendrite size. The Nb composition at the midpoint of the ingot was very close to the unmelted electrode compositions, but it was slightly higher at the top of the ingot. This small positive segregation is larger than previously found[1] and may be associated with small changes in radial segregation versus height, which were not averaged out by our sampling technique, but which would average out upon wire drawing.

Wires were prepared for J_c measurements by machining small cylinders from the top region of the ingots and reducing these to 0.15-mm wire and coating and reacting with Sn using the same procedures as described above. The solid lines in Fig. 5 are for 2 wires from a 80-g Y_2O_3 crucible-cast ingot of Cu-20 wt.% Nb that was reduced by R = 5900, near the optimum value at 8 T as shown in Fig. 1. Results for wires from the two 15-kg ingots, produced with an equivalent reduction, R = 5100, are given by the symbols and are essentially the same at the high fields. Because the 15-kg castings were 22.5 wt.% Nb compared with 20 wt.% Nb for the 80-g castings, one would have expected their J_c values to be a bit

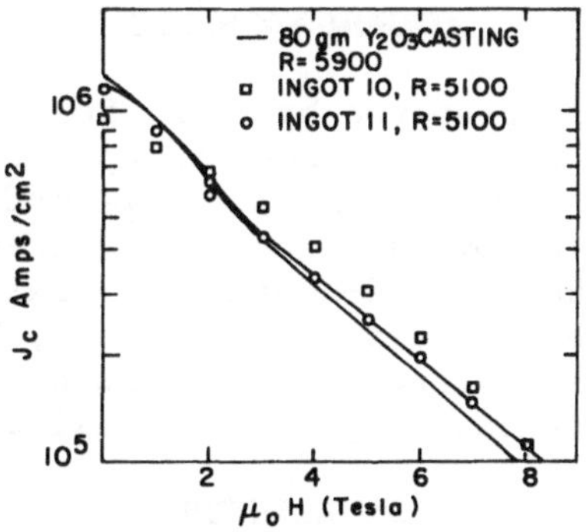

Fig. 5. J_c vs. μ_0H on Cu-20 wt.% Nb 80-g Y_2O_3 crucible castings and Cu-22.5 wt.% Nb 15-kg consumable arc-melted ingots.

higher than found here. Results on the 30 wt.% Nb arc castings are compared with the 80-g Y_2O_3 crucible castings in Fig. 4. Data are presented for 2 wires from ingot 12, and it is seen that the results are equivalent within the scatter of the data at the highest fields that have currently been measured. Hence, these results show that: (a) 15-kg ingots produced by consumable arc melting into ZrO_2-coated graphite liners will produce <u>in situ</u> wire with J_c values comparable to laboratory-size castings and to commercial bronze-processed wire, and (b) the plate Nb/Cu electrode design appears to produce adequate homogeneity in the Cu-Nb castings with only one melt required.

CONCLUSIONS

1. The J_c values of <u>in situ</u> wire may be improved by employing an optimum area reduction ratio, R_{opt}, in reducing the Cu-Nb ingot to wire size. This work shows that increasing the Sn diffusion temperature from 550 to 650°C does not increase J_c or H_{c2}, but only shifts R_{opt} to lower values.

2. The use of C containers in the casting step of <u>in situ</u> wire causes sufficient C contamination to produce a significant decrease in J_c properties.

3. The 12-T J_c values of C-free <u>in situ</u> wire produced with optimum reduction ratios are as good or better than the best reported values on powder-prepared <u>in situ</u> and bronze-processed wire.

4. It has been demonstrated that the maximum J_c values found for <u>in situ</u> wire prepared from small laboratory-size castings may also be achieved in consumable arc castings scaled up to 15 kg through the use of ZrO_2-coated graphite liners.

ACKNOWLEDGMENTS

Arc castings were prepared by F. A. Schmidt, A. Johnson and J. Wheelock. The Ames Laboratory is operated for USDoE by Iowa State University, and this work was done under contract number W-7405-Eng-82 supported by the Director of Energy Research, Office of Basic Energy Sciences, WPAS-KC-02-01. Part of this work was performed while three of the authors (J.J.S., D.K.F., J.E.O.) were guest scientists at the Francis Bitter National Magnet Laboratory, which is supported at M.I.T. by the National Science Foundation.

REFERENCES

1. J. D. Verhoeven, F. A. Schmidt, E. D. Gibson, J. J. Sue, J. E. Ostenson, and D. K. Finnemore, IEEE Trans. Magn. MAG-17:251 (1981).
2. J. J. Sue, J. D. Verhoeven, E. D. Gibson, J. E. Ostenson, and D. K. Finnemore, Acta Metall. (in press) (1981).
3. W. Schauer and W. Schelb, IEEE Trans. Magn. MAG-17:374 (1981).
4. J. D. Verhoeven, D. K. Finnemore, E. D. Gibson, J. E. Ostenson, and L. F. Goodrich, Appl. Phys. Lett. 33:101 (1978).
5. M. Suenaga, W. B. Sampson, and T. S. Luhman, IEEE Trans. Magn. MAG-17:646 (1981).
6. J. D. Verhoeven, E. D. Gibson, F. A. Schmidt, and D. K. Finnemore, J. Mater. Sci. 15:1449 (1980).

TEMPERATURE DEPENDENCE OF J_c FOR *IN SITU* SUPERCONDUCTORS

J. E. Ostenson, D. K. Finnemore, J. J. Sue,* E. D. Gibson, and J. D. Verhoeven

Ames Laboratory, USDoE, Iowa State University, Ames, Iowa

INTRODUCTION

Many performance characteristics of in situ prepared superconducting materials look very favorable for use in large scale magnets in the 8 to 14 T range. The critical current density,[1] J_c, the strain tolerance,[2] and the high ultimate tensile strength[3] all look encouraging, and recently there have been substantial advances in the casting[4] and drawing[5] processes needed for large scale production. The purpose of the experiments reported here is to determine the temperature dependence of J_c in order to be able to predict the response of magnets when they are subjected to temperature excursions well above 4.2 K.

One distinguishing characteristic of in situ materials is that the Nb_3Sn filaments are discontinuous and they are much smaller in size and spacing than bronze process wire. The thickness of the Cu barriers separating adjacent Nb_3Sn filaments is typically a few hundred to a few thousand angstroms, so the proximity effect may play a substantial role. From studies of flux pinning in these composites at 4.2 K it has been determined that core pinning at grain boundary surfaces is the dominant factor controlling J_c.[5] It appears that both the shear of the flux line lattice and the coupling across the Cu barriers between filaments is strong enough that grain boundary pinning is the weak link. As the temperature increases, however, the decay length for superconductivity in the normal metal, K_N^{-1}, decreases exponentially at $T^{-1/2}$. This means

*Present address: Union Carbide Coating Service Department, Indianapolis, Indiana.

that proximity coupling may become a more important factor at large T. To test whether the breakdown of the proximity effect degrades J_c at high temperature, we have chosen samples in which the proximity effect may be most important, that is, those that have rather short filaments due to coarsening.

EXPERIMENT

The temperature dependence of J_c was measured for two different small magnets that had been wound for ac loss measurements.[7] A 20-m length of core process wire[1] with 0.025-cm diameter was wound on a 1.27-cm diameter mandrel to form a coil having a volume of about 1 cm^3. This coil was then placed in a heat leak chamber, and the entire assembly was put in the bore of a Bitter solenoid. The axis of the coil was parallel to the applied field so that the axis of the individual wire strands were perpendicular to the field. The criterion for J_c was 2 μV across the coil or 1 nV/cm of wire. The samples were prepared from a dendritic Cu-Nb alloy that contained 20 wt.% Nb. The reduction area ratio was approximately 30,000 for the extrusion and drawing process. After reaction this gives a fairly heavily coarsened wire, so the filaments are relatively short. If there is to be a serious degradation of J_c due to the breakdown of the proximity effect, it should occur in these samples.

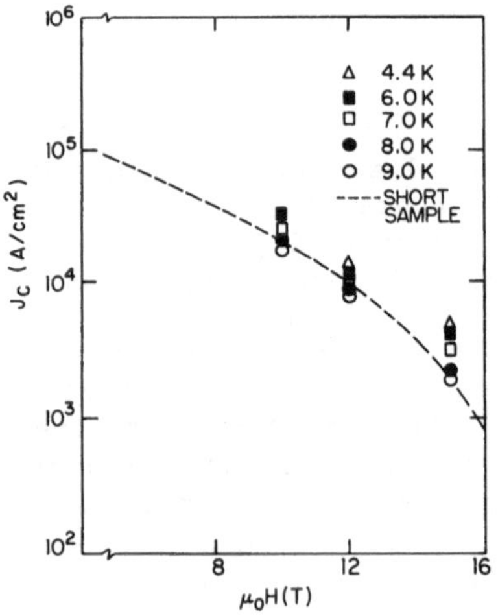

Fig. 1. J_c vs H curves for a small coil. The dashed line curve marks the normal range of J_c vs H curves at 4.2 K for short samples having the same coarsening as the coil.

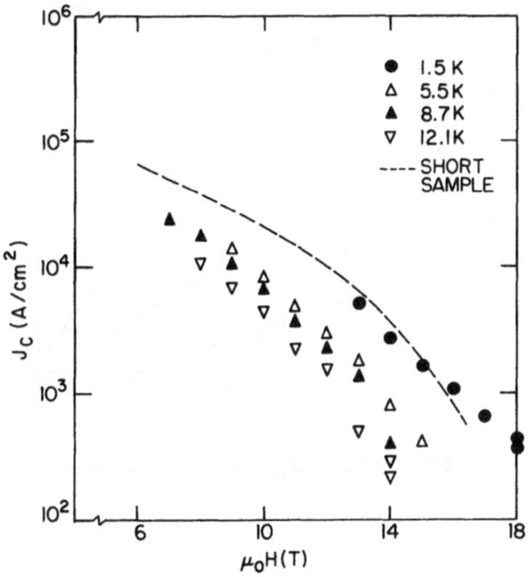

Fig. 2. J_c vs H curves for a small coil.

RESULTS AND DISCUSSION

The temperature dependence of the J_c vs H curve is illustrated for two different coils shown in Fig. 1 and Fig. 2. The overall magnitude of J_c is comparable to other short samples having a similar degree of coarsening. A typical short sample at 4.2 K is illustrated by the dashed line. For both samples, J_c values are systematically suppressed with increasing temperature and the pinning force follows $H^{1/2}(1-h)^2$ behavior. In broad terms, a doubling of the temperature reduces J_c by about a factor of two. In this field and temperature range, the overall shape of J_c vs H curve remains the same, so there is a good chance that flux pinning on grain boundaries remains the dominant factor controlling J_c. There is no evidence of breakdown in the Cu barriers.

To show the temperature dependence more explicitly, the J_c data of Fig. 1 have been plotted as a function of temperature at constant field in Fig. 3. Here it is seen that the data are nearly linear in T at 10, 12, and 15 T, just as was found for bronze process wire.[8] The overall behavior of the temperature dependence of J_c for in situ wire is similar to that of bronze process wire.

CONCLUSIONS

The temperature dependence of the critical current of in situ Nb_3Sn-Cu superconducting wire is nearly the same as bronze process

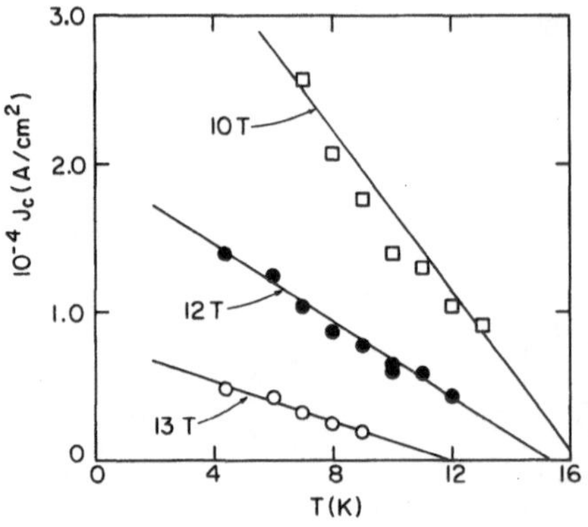

Fig. 3. J_c vs T curves.

wire. There is no evidence of new factors, such as the temperature dependence of the proximity coupling, entering the problem. At any given magnetic field, J_c drops by about 10% per degree kelvin as the temperature rises.

REFERENCES

1. J. D. Verhoeven, E. D. Gibson, C. V. Owen, J. E. Ostenson, and D. K. Finnemore, Appl. Phys. Lett. 35:270 (1979).
2. J. W. Ekin, IEEE Trans. Magn. 15:197 (1979).
3. J. W. Ekin, National Bureau of Standards, Boulder, Colorado, private communication.
4. J. D. Verhoeven, F. A. Schmidt, E. D. Gibson, J. E. Ostenson, and D. K. Finnemore, Appl. Phys. Lett. 35:555 (1979).
5. C. R. Spencer, E. Adam, E. Gregory, W. Marancik, and C. Z. Rosen, IEEE Trans. Magn. 17:257 (1981).
6. J. J. Sue, D. K. Finnemore, J. D. Verhoeven, E. D. Gibson, and D. K. Finnemore, Flux pinning for in situ prepared superconducting composites, in: "Advances in Cryogenic Engineering - Materials," Vol. 28, Plenum Press, New York (1982), p. 323.
7. J. E. Ostenson, D. K. Finnemore, J. D. Verhoeven, and E. D. Gibson, Appl. Phys. Lett. 37:662 (1980).
8. C. R. Spencer, P. A. Sanger, and M. Young, IEEE Trans. Magn. 15:76 (1979).

HIGH-FIELD CRITICAL CURRENT AND MECHANICAL PROPERTIES OF *IN SITU* PROCESSED V_3Ga SUPERCONDUCTORS

H. Kumakura, K. Togano, and K. Tachikawa

National Research Institute for Metals, Ibaraki, Japan

INTRODUCTION

Recently, the in situ technique for fabricating superconducting multifilamentary composites has become of technical interest as a possible alternative to the conventional bronze process. Since the first successful demonstration by Tsuei,[1] numerous studies on in situ processed superconductors have been made. However, the efforts have mainly concentrated on the Nb_3Sn composites,[2-5] and only a few studies on the V_3Ga composites have been reported in the literature.[6-9] V_3Ga has a superior current-carrying capacity at high magnetic fields to Nb_3Sn and is useful where high magnetic fields are crucially important.[10,11] We have been carrying out systematic experiments on the superconducting and mechanical properties of in situ processed Cu-V_3Ga composite superconductors.[12,13] In this paper, we report recent results on the developments of in situ Cu-V_3Ga composites. The effects of various processing parameters, such as vanadium and gallium concentrations, heat-treatment conditions, area reduction by wire drawing, and the morphology change of the vanadium filaments on the superconducting and mechanical properties have been studied to optimize the process. Scaling up of the in situ process has also been studied.

EXPERIMENTS

Cu-(20-65) at.% V binary alloy button ingots of about 20 g in weight were prepared by arc-melting under an argon atmosphere. The purities of the starting materials were 99.99% and 99.8%, respectively. The ingots were then cold-rolled and cold-drawn to the final wires with diameters of 0.5, 0.3, and 0.2 mm. The cross-sectional reduction ratios, R, are approximately 500, 1500, and 3000, respectively. All these alloys were ductile enough to be cold-worked without intermediate annealing. The wires were then coated with 10-20 at.% Ga and heat-treated at 450°C to diffuse the gallium inward. The reaction heat treatments were carried out at temperatures ranging 500-700°C for 2-400 h.

For several compositions, the fabrication of long, in situ processed Cu-V_3Ga superconductors has been also attempted. Preparation of Cu-V ingots of a few kilograms has been studied by several methods. First, chill casting was carried out using an induction furnace and an alumina crucible. However, sufficiently homogeneous and fine Cu-V structure was not obtained, probably owing to the gravity segregation that occurred during the solidification process after pouring the melt into a chill mold. Successful results were obtained by the continuous casting method using an arc-melting furnace. Both the consumable electrode and nonconsumable electrode methods were examined in this study. In both, the materials were continuously cast into a rod of 40 mm in diameter using a water-cooled copper mold of circular section. The details of the nonconsumable electrode method were described in a previous paper.[12] In the consumable electrode method, the electrode was a composite of copper and vanadium, and two different designs were used in this experiment. The electrode shown in Fig. 1(a) was fabricated by bundling the single-core copper-vanadium composite rods in a copper sleeve. The electrode shown in Fig. 1(b) can be prepared more readily by making a laminated composite of copper and vanadium plates. The consumable electrode arc melting was carried out at 1/2 argon atmosphere and at current densities of 300-500 A/cm^2. Cast ingots were rolled and drawn into long wires with diameters of 0.3 and 0.2 mm and continuously coated with gallium by a dipping process. The wires were then heat-treated at several temperatures.

The superconducting transition temperature, T_c, was measured resistively using a standard four-probe technique. The T_c was defined as the temperature at which the resistivity reached one-half of the normal resistivity. The overall critical current density, J_c, normalized to the total cross-sectional area of the wires was measured in magnetic fields of 10-17 T and 17-21.5 T using a 17.5-T Nb_3Sn-V_3Ga hybrid superconducting magnet at NRIM and a Bitter type hybrid magnet at Francis Bitter National Magnet Laboratory of MIT, respectively. The J_c was taken when the voltage across a 10-mm length of the sample reached 1 μV.

The effect of strain on J_c of the in situ processed Cu-V_3Ga superconductors was studied in the following way: The tensile load was applied using an Instron Testing Machine, and J_c measurement was carried out at 4.2 K in the magnetic fields applied perpendicularly to the sample. Strain was detected by a strain gauge attached directly to the sample; the accuracy of the measurement

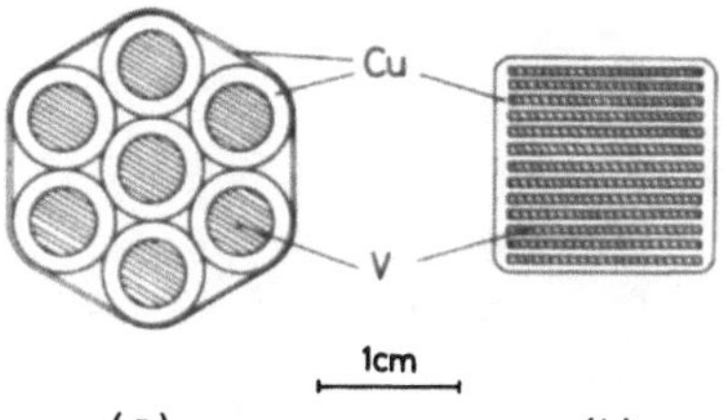

Fig. 1. Cross-sectional view of the Cu-V consumable electrodes.

was ±0.03%. The J_c under bending strain was measured at 4.2 K by mounting the sample on U-shaped brass plates with different curvatures in magnetic fields applied parallel to the bend axis. Tensile stress-strain curves were also measured at room temperature for several samples and compared with that of the commercial V_3Ga composite wire.

RESULTS AND DISCUSSION

Microstructure

The microstructure of the small button ingots prepared by conventional arc melting shows a quite uniform distribution of fine vanadium dendrites, as described in the previous paper.[13] Large ingots with fairly uniform distribution of vanadium dendrites were also prepared by the continuous casting method using both the nonconsumable electrode and consumable electrode methods. However, in the case of Fig. 1(a), a very small amount of undissolved vanadium particles was observed in the ingot, probably due to the large vanadium cores in the electrode. Figure 2(a) gives the outer view

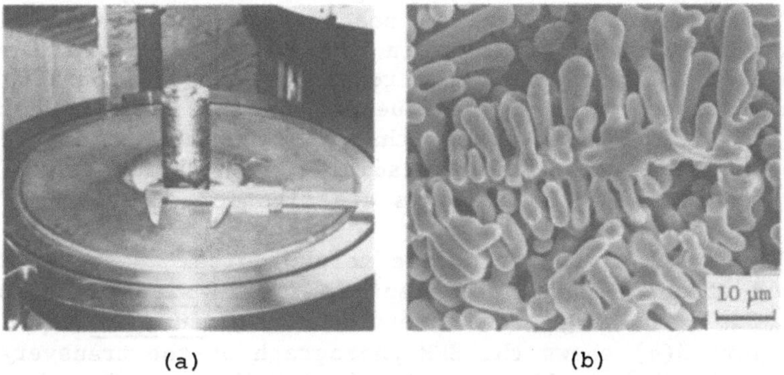

Fig. 2. (a) Cu-35 at.% V ingot prepared from the consumable electrode. (b) SEM micrograph of vanadium dendrites in the ingot.

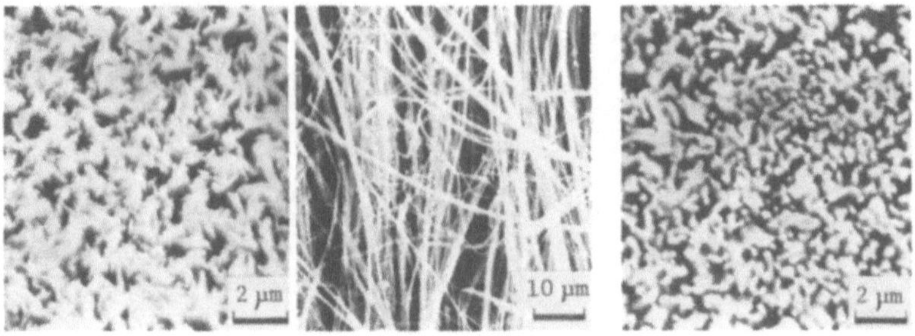

Fig. 3. SEM micrographs of the Cu-35 at.% V wire. (a) transverse and (b) longitudinal cross section; (c) transverse cross section after annealing at 700°C for 5 h.

of the ingot of Cu-35 at.% V alloy prepared from the consumable electrode, and Fig. 2(b) shows the scanning electron micrograph of the vanadium dendrite particles observed after etching away the copper matrix. Figures 3(a) and (b) show the transverse and longitudinal microstructure of the same alloy after the large area reduction by the cold working (R ≈ 3000). The vanadium dendritic particles were deformed into ribbon-like filaments (1000 ≈ 3000 Å in thickness and 1-2 µm in width), which have been formed by the plane-strain deformation of the vanadium filaments with <110> fiber texture.

The gallium diffusion in the composites was investigated using XMA. Although the diffusion distance of the gallium increased with annealing time at 450°C, the gallium did not reach the center even after annealing for 400 h. During the following reaction heat treatment at 500-600°C, gallium penetrated further inward. However, the gallium distribution in the radial direction was still not completely uniform after the reaction at 550°C for 100 h. The diffusion distance was also influenced by the vanadium concentration in the wire. It is likely that the vanadium filaments act as a diffusion barrier: Higher vanadium concentration makes it more difficult to achieve a homogeneous distribution of the gallium.

Verhoeven et al. reported the morphological change of the niobium fiber after the heat treatment in the <u>in situ</u> Cu-Nb composite wire.[14] Similar change was observed in the <u>in situ</u> Cu-V composites. Figure 3(c) shows the SEM photograph of the transverse cross section of the 0.3-mm diameter Cu-35 at.% V wire after heating at 700°C for 5 h. Ribbon-like filaments were converted into the more cylindrical shape. This was probably caused by the surface tension of the vanadium filaments. Study on the effect of the morphological change on the gallium diffusion is now in progress.

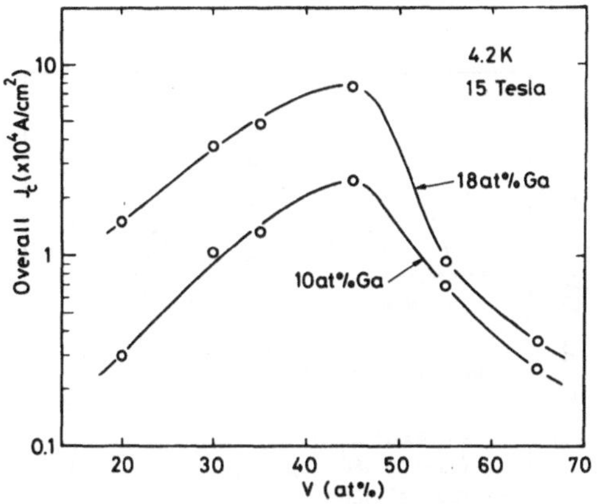

Fig. 4. Overall J_c at 4.2 K and 15 T of the (Cu-V)+Ga wire as a function of the vanadium concentration for the 10 and 18 at.% Ga coated wires.

Superconducting Properties

The optimizations of various factors, such as vanadium concentration, amount of the gallium coated on the wire, and heat-treatment conditions, were carried out on short wire samples prepared from small ingots. Figure 4 shows J_c at 15 T and 4.2 K as a function of vanadium concentration for the 10 and 18 at.% Ga coated wires (0.3-mm diameter). The samples were annealed at 450°C for 100 h followed by a reaction heat treatment at 500°C for 100 h. The J_c reached a maximum at around 45 at.% V where the volume fraction of the vanadium reached half of the wire, and then rapidly decreased, probably owing to the deterioration of the gallium penetration. The maximum overall J_c values are 3.5×10^4 A/cm^2 and 1.1×10^5 A/cm^2 for 10 and 18 at.% Ga samples, respectively.

The gallium content required to convert all the vanadium into V_3Ga is 9 at.% for the Cu-30 at.% V alloy. In Fig. 4, J_c increases beyond this value, maybe owing to insufficient gallium diffusion. It is speculated that the excess gallium contributes to maintain a steep concentration gradient in the wire along which the gallium diffuses inward. However, such an excess of gallium is not favorable from the standpoint of the mechanical tolerance, as mentioned later. For a given heat-treatment condition and alloy composition, J_c increases when the wire diameter decreases from 0.5 mm to 0.2 mm. This increase in J_c can be explained by the increased volume ratio of the reacted vanadium filaments to the total filaments in the wire.

For compositions of 20, 35, and 45 at.% V, samples were prepared using the continuous arc-casting and continuous gallium

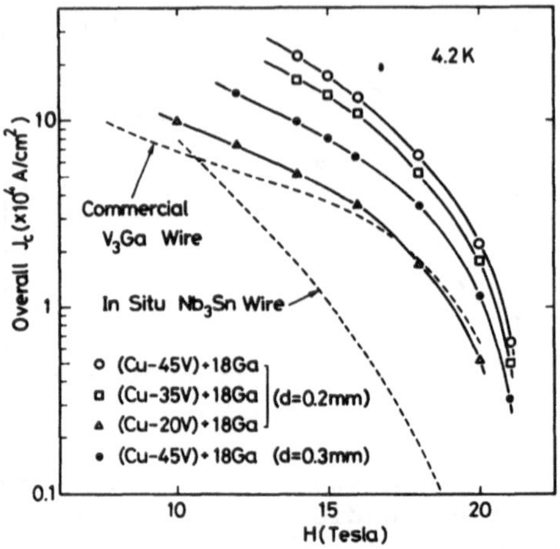

Fig. 5. Typical J_c-H curves for the _in situ_ Cu-V_3Ga wires prepared using the continuous arc-melting and continuous gallium coating, as compared with those of the commercial V_3Ga wire[15] and the _in situ_ Cu-Nb_3Sn wire.[3]

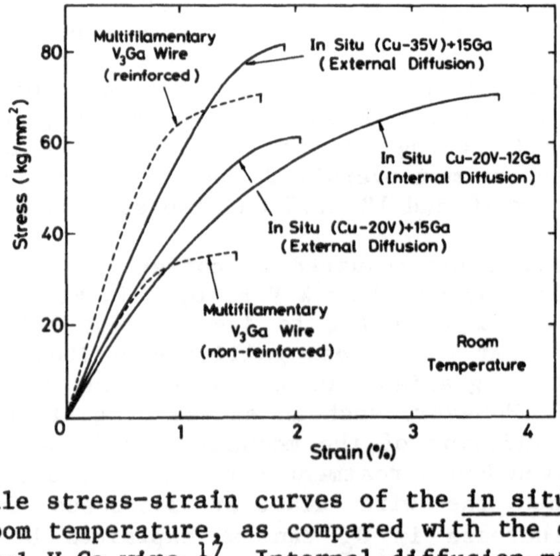

Fig. 6. Tensile stress-strain curves of the _in situ_ Cu-V_3Ga wire at room temperature, as compared with the data of the commercial V_3Ga wire.[17] Internal diffusion means the _in situ_ process which involves direct casting of ternary Cu-V-Ga alloy and subsequent cold working and heat treatment.[12]

coating. Figure 5 shows the typical results of J_c versus applied magnetic field up to 21.5 T. The heat treatments were carried out at 500°C for 100 h. Overall J_c of the commercial multifilamentary V_3Ga wire (without the stabilizing and reinforcing materials) fabricated by the conventional bronze process[15] and that of the in situ processed $Cu-Nb_3Sn$ composite wire[3] are also shown for comparison. The in situ processed $Cu-V_3Ga$ wires show overall J_c-H properties superior to those of other superconductors. The J_c of the (Cu-45 at.% V)+18 at.% Ga samples is about five times as large as that of the commercial V_3Ga wire at all magnetic fields. The highest critical temperature, T_c, and upper critical field, H_{c2} (4.2 K), are 15.2 K (midpoint) and ~22 T, respectively. Those values are slightly better than the best values of 15.1 K and 20.8 T for the bronze-processed $Cu-V_3Ga$ superconductors.[16]

The effect of the heat-treatment condition on J_c was studied in the temperature range 500-700°C. For a given reaction temperature, J_c at 15 T increases with the increase of reaction time and reaches a maximum. At higher reaction temperature, the maximum occurs in a shorter time. However, the maximum J_c value is significantly lowered at the higher reaction temperature. At 500°C, the maximum occurs at about 100 h. The reaction temperature of 500°C, which gives the best J_c value, is appreciably lower than that of the bronze-processed $Cu-V_3Ga$ composites.

Stress Effects

The in situ processed $Cu-V_3Ga$ composites have large mechanical strength. Figure 6 shows the typical tensile stress-strain curves of the in situ processed V_3Ga composites measured at room temperature. In Fig. 6, the data of the commercial bronze-processed multifilamentary V_3Ga wires are also shown for comparison.[17] The ultimate tensile strength of the in situ composites ranges from 60 to 80 kg/cm^2, which is significantly higher than that of the non-reinforced multifilamentary V_3Ga wire (55 cores). The strength of the in situ processed sample tends to increase with the volume fraction of the vanadium filaments, and the tensile strength of the (Cu-35 at.% V)+15 at.% Ga wire exceeds that of the tungsten-reinforced multifilamentary V_3Ga wire. Such a higher strength of the in situ composites is probably attributed to the smaller spacing of the filaments, which act as barriers for the motion of dislocation in the matrix.

Our previous experiments on J_c degradation under tensile load indicated that the in situ processed $Cu-V_3Ga$ composites do not show a J_c maximum in the J_c-strain curve and exhibit a complete recovery of J_c when the load is removed.[13] The J_c degradation of the Cu-20 at.% V-12 at.% Ga tape fabricated by the internal diffusion process began at a strain of about 1.2%.

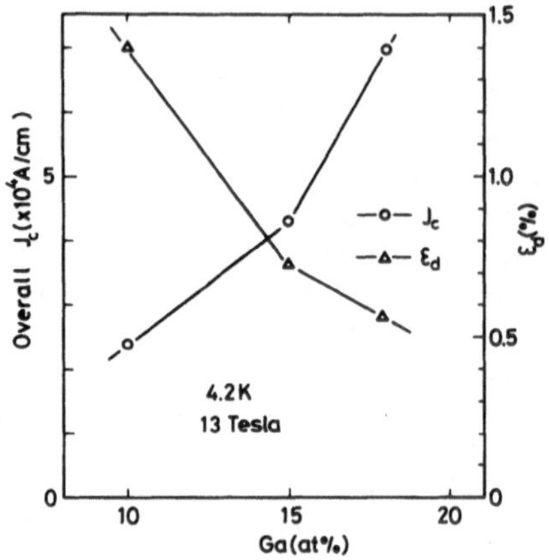

Fig. 7. Overall J_c and ε_d as a function of the gallium content in the (Cu-35 at.% V)+Ga composite wire (0.3-mm diameter).[12]

In the *in situ* composites fabricated by the external diffusion method, J_c degradation under the bending strain is strongly dependent on the vanadium and gallium concentrations. Figure 7 shows the bending strain, ε_d, and J_c at 13 T as a function of the gallium content in the (Cu-35 at.% V) composites (0.3-mm diameter) fabricated by the external diffusion method. The ε_d is the surface strain at the beginning of the J_c degradation and is calculated by the formula $\varepsilon_d = d/D_d$; d is the diameter of the samples and D_d is the bending diameter where J_c degradation begins. The results indicate that ε_d is a strong function of the gallium concentration. A large amount of gallium increases J_c but, on the contrary, reduces the mechanical tolerance. The (Cu-35 at.% V)+10 at.% Ga wire, whose J_c is comparable to that of the bronze-processed single-core V_3Ga wire, shows no J_c degradation up to 1.4% bending strain, whereas the J_c degradation of the bronze-processed V_3Ga wire begins at 0.6%.[13] The ε_d is also dependent on the vanadium content in the composites. The ε_d of the (Cu-20 at.% V)+Ga composites is somewhat better than that of the (Cu-35 at.% V)+Ga composites.[12] These ε_d behaviors can be mostly explained by the incomplete gallium diffusion, namely a larger amount of gallium or vanadium produces a gallium-rich phase near the surface and reduces ε_d. The above results suggest that a proper choice of gallium and vanadium concentrations is necessary for the practical application of the *in situ* processed Cu-V_3Ga composite.

CONCLUSIONS

Superconducting properties of the *in situ* processed Cu-V_3Ga composites wire are strongly affected by the alloy compositions, heat-treatment conditions, reduction ratio, etc. The overall critical current density, J_c, increases with increasing vanadium concentration up to 45 at.% V and then decreases rapidly owing to the decrease of gallium penetration velocity. High gallium concentration and large cross-sectional area reduction also increase J_c. The maximum J_c value obtained was over 10^5 A/cm^2 at 17 T for the (Cu-45 at.% V)+18 at.% Ga wire of 0.2-mm diameter, which is about five times as large as that of the commercial multifilamentary V_3Ga wire. The highest critical temperature, T_c, and upper critical field, H_{c2} (4.2 K), were 15.2 K and ~22 T, respectively. These values are slightly better than the best values of the bronze-processed V_3Ga. Copper-vanadium ingots of a few kilograms were prepared by the continuous arc-casting method, and sufficient uniform distribution of fine vanadium dendrites was achieved throughout the ingot. The superconducting properties of wires prepared from the large ingot were almost identical to that of the in situ V_3Ga prepared from the small button ingot.

The mechanical properties of the in situ processed Cu-V_3Ga composites also depend on the alloy composition. The in situ processed (Cu-35 at.% V)+10 at.% Ga wire showed no J_c degradation up to 1.4% bending strain, which is significantly higher than that of the bronze-processed wire whose overall J_c is nearly equivalent. However, the bending tolerance of the in situ processed wire deteriorates as the gallium concentration increases. This is considered to be mostly caused by the incomplete gallium diffusion.

ACKNOWLEDGMENTS

The authors would like to thank Mr. T. Takeuchi and Dr. Y. Tanaka of the National Research Institute for Metals for the fabrication of the long-length wires and for the J_c measurements at 17-21.5 T at MIT, respectively.

REFERENCES

1. C. C. Tsuei, Science 180:57 (1973).
2. C. C. Tsuei, M. Suenaga, and W. B. Sampson, Appl. Phys. Lett. 25:318 (1974).
3. J.-L. Fihey, M. Neff, R. Roberge, M. C. Flemings, S. Foner, and B. B. Schwartz, Appl. Phys. Lett. 35:715 (1979).
4. D. K. Finnemore, J. E. Ostenson, J. D. Verhoeven, and E. D. Gibson, J. Appl. Phys. 51:1714 (1980).
5. J. P. Harbison, and J. Bevk, J. Appl. Phys. 48:5180 (1980).
6. W. Y. K. Chen, and C. C. Tsuei, J. Appl. Phys. 47:715 (1976).
7. J. Bevk, F. Habbal, and J. P. Harbison, Appl. Phys. Lett. 35:93 (1979).
8. J.-L. Fihey, R. Roberge, S. Foner, E. J. McNiff, Jr., and B. B. Schwartz, in: "Advances in Cryogenic Engineering - Materials," Vol. 26, Plenum Press, New York (1980), p. 350.

9. J. Bevk and F. Habbal, Appl. Phys. Lett. 36:336 (1980).
10. K. Tachikawa, and Y. Iwasa, Appl. Phys. Lett. 16:230 (1970).
11. K. Tachikawa, Y. Tanaka, K. Inoue, K. Itoh, and T. Asano, Cryogenic Engineering (in Japanese) 11:252 (1976).
12. K. Togano, H. Kumakura, and K. Tachikawa, IEEE Trans. Magn. MAG-17:985 (1981).
13. H. Kumakura, K. Togano, and K. Tachikawa, J. Less-Common Met. 79:181 (1981).
14. J. D. Verhoeven, J. J. Sue, D. K. Finnemore, E. D. Gibson, and J. E. Ostenson, J. Mater. Sci. 15:1907 (1980).
15. Y. Furuto, T. Suzuki, K. Tachikawa, and Y. Iwasa, Appl. Phys. Lett. 24:34 (1974).
16. Y. Tanaka, and K. Tachikawa, J. Jap. Inst. Met. (in Japanese) 40:509 (1976).
17. Y. Tanaka, M. Ikeda, S. Meguro, and Y. Furuto, Teion Kogaku (in Japanese) 13:165 (1978).

DEVELOPMENT OF CRYOSTABILIZED Nb_3Sn-Cu SUPERCONDUCTING WIRE USING THE *IN SITU* PROCESS

E. D. Gibson, J. E. Ostenson, J. J. Sue,* J. D. Verhoeven, and D. K. Finnemore

Departments of Materials Science and Engineering and of Physics
Ames Laboratory—USDoE, Iowa State University, Ames, Iowa

INTRODUCTION

This paper presents two contributions to the development of the in situ process for production of high-field Nb_3Sn-Cu superconducting wire. First, in the utilization of multidie drawing equipment for reduction of Cu-Nb alloy castings to wire size, even though the alloy is quite ductile, it is desirable to utilize some intermediate anneals to insure zero breakage.[1] It is shown here that these anneals can be controlled to improve the J_c properties of the wire. Second, a technique is presented for the production of a cryostabilized in situ wire that is compatible with large-scale production methods. The technique utilizes a combination of rolling and extrusion to produce wire with a stabilizing Cu core.

HEAT TREATMENT

The in situ process begins with production of a Cu-Nb casting in which the Nb is present as randomly arrayed dendrites. Because of the nature of the solidification process, a size distribution of dendrites exists, and this distribution will be carried over to the aligned Nb filaments produced by the wire drawing step. During the Sn diffusion step it is necessary to heat the alloy, and experiments have shown[2] that for annealing temperatures of only 550°C, Nb filaments of thicknesses less than 400-500 Å will coarsen sufficiently to reduce J_c. Consequently, the possiblity exists that one

*Present address: Union Carbide Corporation, Linde Division, Indianapolis, Indiana.

may not be utilizing the Nb originally present as the smaller sized dendrites because of preferential coarsening of the smaller Nb filaments during the diffusion anneal. In general, coarsening rates are inversely proportional to the square of the particle diameter,[3] so that one expects the smaller particles to coarsen more rapidly.

It has occurred to us that it may be possible to utilize the coarsening in a controlled manner to preferentially eliminate smaller sized Nb filaments. Such controlled coarsening would consist of heat-treating the Cu-Nb alloy one or more times during the drawing sequence, sufficiently to cause substantial coarsening of the deformed Nb dendrites. This may reduce the size distribution of the Nb filaments by reducing the relative number of smaller sized filaments, which in turn should result in a greater utilization of Nb for carrying supercurrents in the final reacted wire. In designing such a process, one must be careful to allow sufficient mechanical reduction following the coarsening to produce adequate elongation of the coarsened Nb filaments. The amount of mechanical reduction prior to coarsening is also important because the Nb filaments must be sufficiently reduced in size to produce a coarsening effect. The optimum design of the controlled coarsening process is very complex because the final microstructure is controlled by the amount of reduction both prior and subsequent to the coarsening treatment as well as by the coarsening temperature. We report here on initial exploratory experiments.

Alloys were prepared from 15-kg arc castings of 22.5 wt.% Nb-Cu (ingot #11 of Ref. 2). Cylinders of 25-mm diameter were swaged and drawn to 0.79-mm diameter, heat-treated at 750°C for 1 day, drawn to 0.15-mm diameter, coated with 14.7 vol.% Sn, and reacted for 5 days at 550°C. Critical current measurements (based on total wire area) are compared with those of wire prepared from the same alloy without the 750°C coarsening step in Fig. 1. The uncoarsened wire was prepared using an area reduction ratio, R, of 5010 (10.6-mm cylinders reduced to 0.15-mm-diameter wire), which has been determined in previous work[2] to be near the optimum reduction. As shown in Fig. 1, a small but significant improvement in J_c was produced by the 750°C heat treatment. In addition, the figure illustrates that the data compare favorably to data of wire made from 20 wt.% Nb-Cu alloys prepared from 80-g ingots cast in Y_2O_3 crucibles and reduced to wire with a reduction ratio close to the optimum value for high field performance,[2] R = 2500. Metallographic examination revealed that the mean Nb filament size increased from 1020 to 2800 Å during the 750°C heat treatment, thus indicating a significant coarsening effect. The mean Nb thickness in the final 0.15-mm wire was only 350 Å, which is smaller than the optimum value of 600 Å found in previous work on non-heat-treated wire for high-field critical currents, $H \geq 12$ T. The favorable high-field J_c values found for the heat-treated wire may indicate

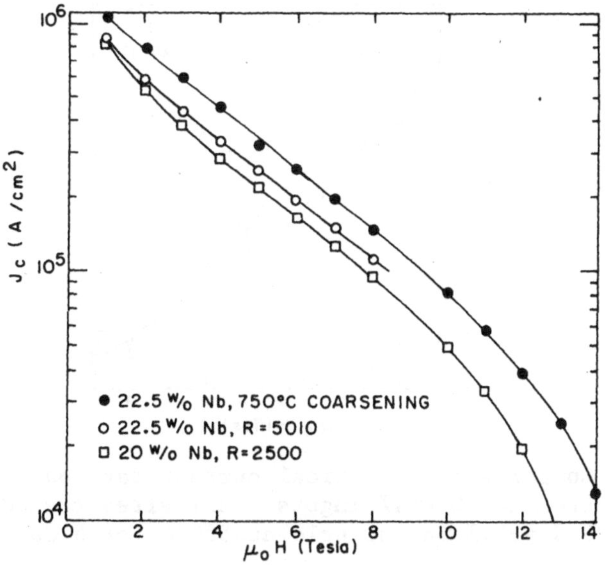

Fig. 1. Comparison of critical current data on wires made with and without heat treatment. All wires reacted at 550°C for 5 days after Sn plating.

that they are less susceptible to degradation in J_c as a result of the coarsening inherently produced by the 550°C Sn diffusion anneal as R becomes large.

Although the above picture seems to indicate that the 750°C heat treatment has improved J_c by a coarsening effect, there is another possible explanation of these results. It may be that the 750°C heat treatment forms impurity precipitate particles that act to pin Nb_3Sn grain boundaries and thereby increase J_c by reducing the Nb_3Sn grain size. Data from two sets of experiments lend qualitative support to this view.

In the first set of experiments, 9.5-mm cylinders were machined from adjacent locations in a 15-kg Cu-30 wt.% Nb ingot (ingot #12 of Ref. 2), and one of them was heated to 1000°C for 3 h, cooled to 445°C at 12.5°C/h and then to room temperature at 35°C/h. Previous work has shown that the as-cast Nb dendrites do not coarsen at 1000°C so that, following heat treatment, the primary difference in these two cylinders is that the heat-treated cylinder might contain impurity precipitates as a result of the very slow cool from 1000°C. Since the alloy contained Zr, Y, C, and O impurities at the 200-800 ppma range,[2] formation of Zr and Y carbides and oxides is possible. After drawing to wire, plating, and reacting with Sn using identical procedures, the J_c (8 T) values of two heat-treated wires were 1.91 and 1.98 x 10^5 A/cm^2 and

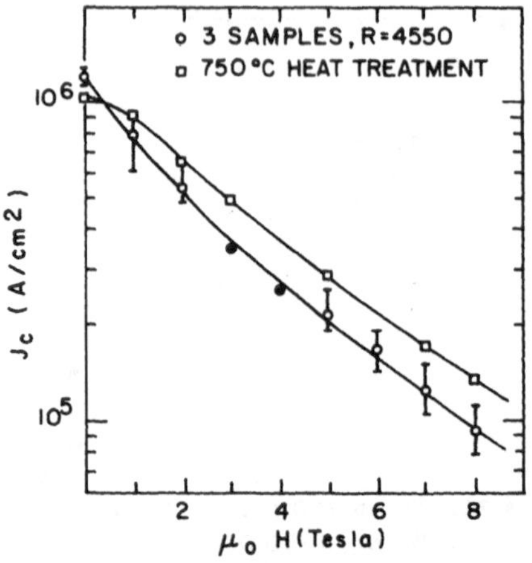

Fig. 2. Comparison of critical current data on wires made from Cu-20 wt.% ingots. All wires coated with 9.3 vol.% Sn and held at 550°C for 5 days.

1.35 and 1.41 x 10^5 A/cm^2 for two non-heat-treated wires. The higher J_c values for the heat-treated alloy supports the idea that heat treatment increases J_c by a precipitation effect rather than a coarsening effect. In the second set of experiments, wire was prepared from 80-g Cu-20 wt.% Nb ingots cast in Y_2O_3 crucibles. A 750°C heat treatment was designed to enhance precipitation but not coarsening. It is known that alternate drawing and heat treatment enhances precipitation from Cu alloys.[4] Therefore, wire was prepared by heat treating four times at 750°C, after area reduction ratios of R = 25.6, 2.5, 2.6, and 26.7. The total reduction of 4500 was the same as for the non-heat-treated wires of Fig. 2 for which the vertical bars indicate the spread of the data points on 3 wires. The initial reduction was only sufficient to produce Nb filaments of 4400 Å, which is well above the mean coarsened size of 2800 Å obtained by the 750°C anneal. The data for the heat-treated wire shows small but significant enhancement of J_c, as seen in Fig. 2, and because one expects enhancement of precipitation and reduction of coarsening in these wires, the results offer qualitative support to the view that the improvement on heat treatment may be due, at least in part, to precipitation effects.

CRYOSTABILIZED WIRE

To produce an in situ processed wire with sufficient Cu for cryostability, the following scheme has been investigated. A cast Cu-Nb billet is rolled to sheet and wrapped around a Cu cylinder. This assembly is then canned in Cu, hot extruded at 750°C, partially drawn, control coarsened at 750°C for 1 day, drawn to final wire size, plated with Sn, and reacted at 550°C to form Nb_3Sn. This scheme has a number of potential advantages:

1. Production of large-diameter in situ wire is limited by problems associated with the Sn diffusion step. As the diameter increases, the diffusion times become quite long, but more importantly, the thickness of the tin layer increases, and at sufficient thicknesses the tin "balls up" when it melts during the diffusion step. Both of these problems are reduced in this cryostabilized wire because all of the Nb is contained in a thin layer near the wire surface. To illustrate this point, the stoichiometric thickness of tin was calculated for the cryostabilized configuration having a Cu/Cu-Nb ratio of 5:1 for the cases of 5- and 10-cm extrusion cans with a 0.79-mm wall thickness. The diameter of a solid Cu-Nb wire that requires the same tin thickness was then calculated and the equivalent diameters are shown in Fig. 3, where in both cases the amount of Sn was taken as 14 vol.% of the Cu-Nb alloy, roughly stoichiometry for a Cu-30 wt.% alloy. Figure 3 illustrates that relatively large cryostabilized wires are equivalent to quite small solid wires. For example, the commonly fabricated 0.25-mm (10-mil), solid, in situ wire is equivalent to a 1.56-mm cryostabilized wire, whereas the 2.88-mm strand diameter of the IGC design for the 12-T ETF-TF coil[5] would require the same tin plate thickness as a 0.47-mm (18.5-mil) solid wire.

2. A much larger flexibility is available in reduction in area ratio, R, because of the rolling step, and larger values of R are beneficial since they promote coarsening and/or precipitation

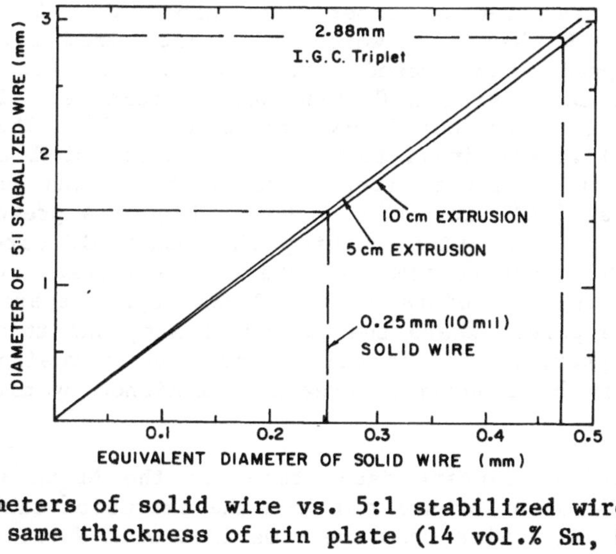

Fig. 3. Diameters of solid wire vs. 5:1 stabilized wire requiring the same thickness of tin plate (14 vol.% Sn, Cu-30%Nb).

Table 1. Area Reduction Ratios, R, for Various Schemes to Produce 2.88-mm Cryostabilized Wire with a Cu/Cu-Nb Ratio of 5:1

Reduction Step	10-cm Extrusion Can		5-cm Extrusion Can	
	1 wrap	8 wraps	1 wrap	8 wraps
Rolling-R_1	5.83	46.5	11.8	94.8
Extrusion-R_2	9.43	9.43	14.2	14.2
Drawing-R_3	132.3	132.3	22	22
R (Total)	7,300	5,8000	3,700	29,600

in the controlled coarsening step. One may increase R of the rolling step by simply making more wraps of the Cu-Nb sheet on the Cu core. This result is illustrated in Table 1, where it is seen that the total R may be increased from 7,300 to 58,000 on a 2.88-mm cryostabilized wire by simply increasing the number of wraps from 1 to 8.

3. It may be possible to eliminate the use of a Ta diffusion barrier between the Cu-Nb alloy and the Cu core by employing a small excess of the Nb-Cu alloy. Because the Nb is so finely divided, it may absorb the Sn sufficiently to inhibit diffusion of a significant amount of the Sn into the Cu core.

We present here results from some initial experiments to evaluate the critical current characteristics of the cryostabilized wire. The first set of experiments employed a 22.5 wt.% Nb-Cu alloy (ingot 11, Ref. 2), which was rolled from 25.4 to 1.09 mm and then wrapped twice around a Cu core and extruded in a 50.8-mm Cu extrusion can having a 0.79-mm wall thickness. This geometry should produce a Cu core/Cu-Nb ratio of 5:1. The billet was extruded to 13.6 mm, drawn to 1.27 mm, heated for 1 day at 750°C, drawn to 0.5-mm diameter wire, coated with various amounts of Sn, and reacted at 550°C for 5 to 6 days. Figure 4 presents a micrograph of the 0.5-mm wire showing the inner Cu core, the Cu/Nb layer, and the outer Cu rim from the extrusion can. One sees discontinuities at two points in the Cu-Nb layer separated by 180°. These are the points where the two wraps met, and they result from an initial upsetting of the billet into the extrusion chamber and could probably be removed or greatly diminished by minimizing this effect.

The critical current data, based on the Nb_3Sn + bronze area (excluding the outer bronze rim from the extrusion can) is compared with data on solid wire from the same 22.5 wt.% Nb alloy in Fig. 5.

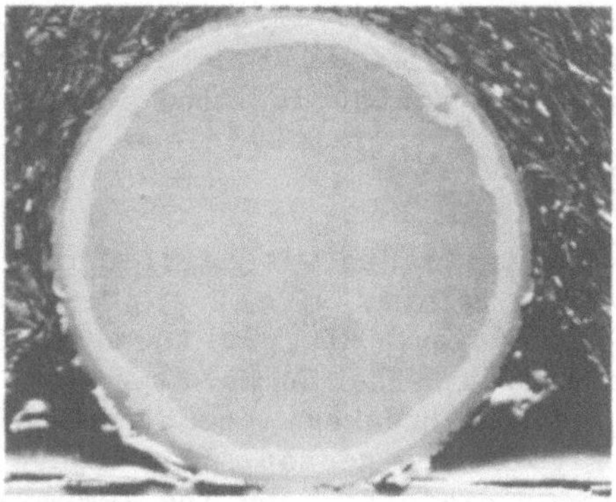

Fig. 4. Transverse section of cryostabilized wire having a 0.5-mm overall diameter (160X). (Reduced 28%.)

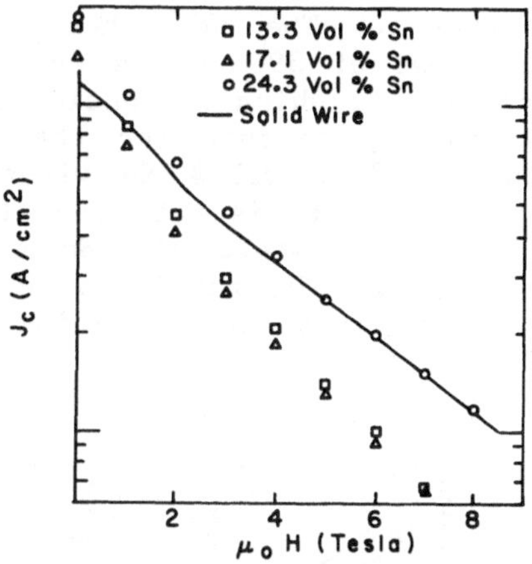

Fig. 5. Comparison of critical current data on cryostabilized vs. solid wires, both made from Cu-22.5 wt.% 15-kg arc castings.

It may be seen that the data for the highest vol.% Sn agree favorably with the solid wire results. The vol.% Sn is calculated over the volume of only the Sn plus the Cu-Nb layer and is 10.3% for a stoichiometric amount of Sn in a 22.5 wt.% Cu-Nb alloy. The much higher vol.% Sn required here to give the best J_c data must result because much of the Sn remains in the Cu rim. This was qualitatively confirmed by evaluating the Sn content of the rim using a microprobe, where it was found to contain 6.8 wt.% Sn in wires with 24.3 vol.% Sn. The effectiveness of the Cu-Nb layer in blocking diffusion of Sn into the core has not been evaluated. The 1-day,

750°C heat treatment at the 1.27-mm diameter increased the mean Nb filament size from 360 to 2500 Å, thus indicating that a significant coarsening had been achieved. The Nb filament size in the final 0.5-mm wire was 710 Å, fairly close to the optimum desired value[2] of around 600 Å.

A second set of experiments was carried out on a Cu-30 wt.% Nb alloy (ingot #12, Ref. 2) and wire was prepared in the same way with the following exception: It was found that drawing this 30% alloy produced breakage in the Cu-Nb layer at wire diameters of less than 1.73 mm. (This breakage occurred at 0.99 mm in the Cu-22.5 wt.% alloys.) For this reason, the 1-day heat treatment was done on wire of 1.73-mm diameter, and the temperature was increased to 800°C to compensate for the increased reduction in the final drawing step to 0.5-mm diameter wire. In Fig. 6 the critical current data for this wire is compared with results on solid wire prepared from 80-g Cu-30 wt.% Nb ingots cast in Y_2O_3 crucibles, and it is seen that the cryostabilized wire falls just below the lower band of these data at 8 T. The 800°C heat treatment produced a mean Nb filament thickness of 3900 Å and the Nb thickness in the final 0.5-mm wire was 1100 Å, which is considerably above the optimum value of 600 Å and may account for the slightly reduced J_c values. The reason the 32.8 vol.% Sn wires produced lower J_c values than the 26.7 vol.% wires is not clear to us. An identical set of experiments was done where the Sn diffusion anneal was extended from 6 to 11 days. The results were nearly identical to those shown in Fig. 6, so that the reduction of J_c with high Sn appears to be a real effect.

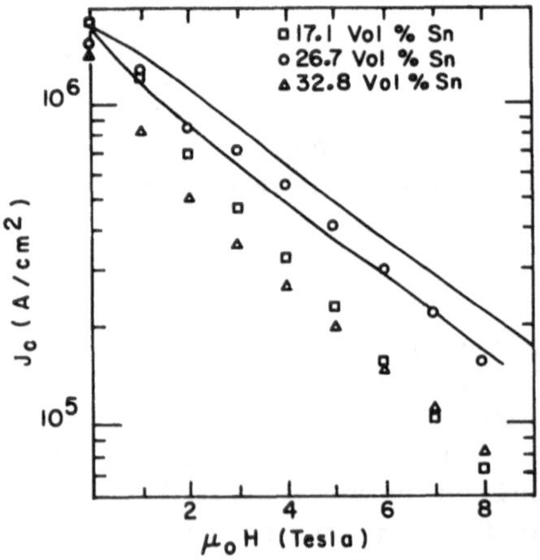

Fig. 6. Comparison of critical current data on cryostabilized wire made from Cu-30 wt.% Nb 15-kg arc castings to solid wire made from Cu-30 wt.% Nb 80-g Y_2O_3 crucible castings (solid lines).

The results of these initial studies indicate that we will be able to achieve J_c values on cryostabilized in situ wire comparable to that of solid wire out to fields of 8 T. However, additional work needs to be done to optimize the processing variables and to evaluate these wires at higher magnetic fields.

CONCLUSIONS

1. Small improvements in the J_c values of in situ Cu-Nb_3Sn wire are obtained by heat treatment of the Cu-Nb alloy during the mechanical reduction step.

2. It is thought that this improvement results from one or a combination of two reasons:

   a. Coarsening during the heat treatment reduces the relative number of small Nb filaments that would have coarsened during the Sn diffusion step and thereby not have contributed effectively to J_c.

   b. Impurity elements, such as Y, form precipitates during the heat treatment that are effective in pinning Nb_3Sn grain boundaries and thereby reducing the Nb_3Sn grain size.

3. The beneficial effect of heat treatment after some wire drawing is attractive for large-scale production of wire using multidie drawing machines because intermediate anneals are required to produce zero breakage and these same intermediate anneals can be used to simultaneously enhance J_c values.

4. A method has been developed to prepare a cryostabilized wire from in situ material utilizing a rolling-extrusion combination. The technique has three potential attractions:

   a. One may go to much larger wire diameters than with solid wire before encountering problems with the Sn diffusion step.

   b. An improved flexibility in controlling Nb filament size is obtained due to the rolling step.

   c. It may be possible to avoid a Ta diffusion barrier by simply employing an excess of Nb-Cu alloys in the wire.

5. Preliminary results on 0.5-mm diameter wires indicate results equivalent to solid test wires at fields out to 8 T.

ACKNOWLEDGMENTS

The arc castings were prepared for us by F. A. Schmidt, A. Johnson, and J. Wheelock. Extrusions were done by C. Owens and L. Reed. The Ames Laboratory is operated for U.S. DoE by Iowa State University and this work was done under contract No. W-7405-Eng-82 supported by the Director of Energy Research, Office of

Basic Energy Sciences, WPAS-KC-02-01. Part of this work was performed while three of the authors (J.J.S., D.K.F., J.E.O.) were guest scientists at the Francis Bitter National Magnet Laboratory, which is supported at M.I.T. by the National Science Foundation.

REFERENCES

1. C. Spencer, Airco Superconductors, Carteret, New Jersey (1979).
2. J. J. Sue, J. D. Verhoeven, E. D. Gibson, J. E. Ostenson, and D. K. Finnemore, On the optimization of in situ Nb_3Sn-Cu wire, in: "Advances in Cryogenic Engineering--Materials," Vol. 28, Plenum Press, New York (1982), p. 501.
3. C. Wagner, Z. Elektrochem. 65:581 (1961).
4. W. Hodge, R. I. Jaffee, J. G. Dunleavy, and H. R. Ogden, Trans. Metall. Soc. AIME 180:32 (1948).
5. J. P. Heinrich et al., IEEE Trans. Magn. MAG-17:634 (1981).

TESTING RESULTS OF MULTIFILAMENTARY Nb_3Sn COMPOSITES MADE BY A MODIFIED JELLYROLL METHOD*†

S. S. Shen

Oak Ridge National Laboratory, Oak Ridge, Tennessee

W. K. McDonald

Teledyne Wah Chang, Albany, Oregon

INTRODUCTION

A new method to produce multifilamentary (MF) Nb_3Sn superconductors called "modified jellyroll" has been recently developed by Teledyne Wah Chang, Albany. The novel technique shows real promise to simplify the manufacture of MF superconducting wires. Unlike the conventional, internal bronze process, which employs niobium in either rod or tube form, modified jellyroll (MJR) uses an expanded niobium foil in its billet assembly. Simply by rolling thin expanded niobium foil along with other metal sheets, as desired (e.g., copper, bronze, tantalum), into a jellyroll form, the resultant billet can then be extruded and processed as conventional wires. More importantly, the method does not require a high area reduction to achieve fine filaments; therefore, significant cost and time savings are expected in this fabrication process. However, because of the "multiply connected" filamentary structure, there is an immediate concern about whether the final drawn wires would behave as a solid wire or as independent filaments.

---

*Patent for this product is pending. It is made by a patented process of Teledyne Wah Chang, Albany.

†Research sponsored by the Office of Fusion Energy, U.S. Department of Energy, under contract W-7405-eng-26 with the Union Carbide Corporation.

This paper presents results of systematic evaluation tests performed on such composites. The results are analyzed and discussed in terms of critical current density, diffusion time constant, ac losses (magnetization), and the effective filamentary diameter.

SAMPLES AND EXPERIMENTAL TECHNIQUES

We have performed magnetization measurements on a series of specifically prepared sample wires (0.5 mm in diameter). All composites are stabilized with copper shells (20-30%) that are protected by either a tantalum or a niobium diffusion barrier. Figure 1 shows a cross section of samples with (a) niobium and phosphor bronze as the diffusion barrier and (b) niobium as the diffusion barrier. As shown in Table 1, test samples vary with parameters such as reaction condition, twisting, and diffusion barriers.

Samples were heat-treated on a metal mandrel with the same diameter (38 mm) as that of the sample holder used in the measurements. This arrangement avoided any possible winding stress on the composites, which could have affected the superconducting properties of the Nb_3Sn structures. Samples of more than 6 m in length were co-wound with an insulating spacer into a single layer, open

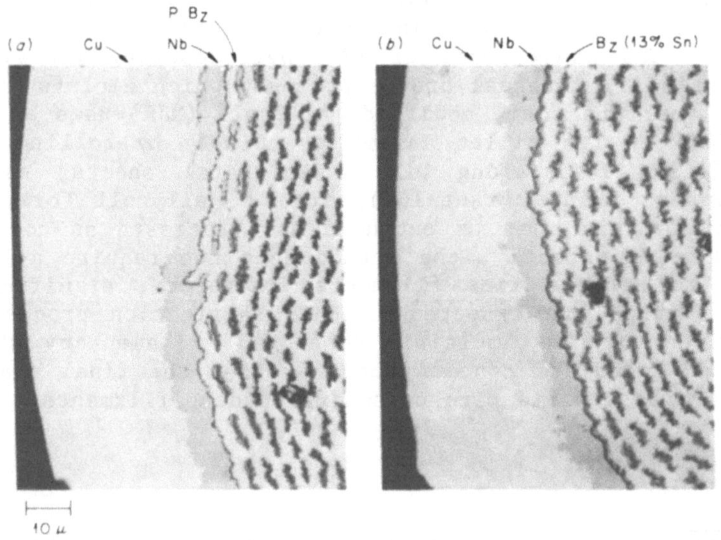

Fig. 1. Cross section of typical MJR composite. (a) Phosphor bronze is used with niobium as the diffusion barrier, (b) niobium as the diffusion barrier.

Table 1. Parameters of MJR Samples

Notation	Nb %	Twist parameter $X_o = \frac{2\pi r_o}{\ell_p}$	Diffusion barrier: Ta	Diffusion barrier: Nb P-bronze	Diffusion barrier: Nb Bronze	Reaction condition: Temp. °C	Reaction condition: Time h
M7A	12	0.063	•				0
M7B	12	0.063	•			735	25
M10A	17	0		•		735	25
M10B	17	0.31		•		735	25
M10C	17	0.063		•			0
M10D	17	0.063		•		735	25
M11A	16	0			•	735	25
M11B	16	0.31			•	735	25
M11C	16	0.063			•		0
M11D	16	0.063			•	735	25

circuited coil to be placed in the bore of a low-loss superconducting pulsed magnet. A magnetization measurement system and its calibration technique have been described previously in detail.[1]

Results are presented in both static and transient M-H curves from which critical current density and filamentary properties are deduced.

DATA ANALYSIS

It has been shown[2] that the measured magnetization of the superconducting composite can be fully correlated to the detailed structure of a multifilamentary conductor in terms of superconductor distribution and matrix properties. In this study, data analysis will be carried out primarily to determine the steady-state, effective diameter, d_{eff}, and coupling time constant, τ_o, of MJR conductors.

For the external field $\mu_o H_e \gg \mu_o H_p$, i.e., a "fully penetrated region" where a simple critical-state model applies, the magnetization of a circular composite can be expressed as[2]

$$-\mu_o M = \left(\frac{2}{3\pi}\right)\mu_o\left(\lambda J_c\right)d_{eff}, \tag{1}$$

where λJ_c represents an overall transport critical current density and d_{eff} takes a value between d_f (filamentary diameter) and d_o (composite diameter) depending on the coupling conditions. Since the quantities $\mu_o M$ and λJ_c can be determined experimentally, d_{eff} is thus defined for any practical, external field conditions.

RESULTS AND DISCUSSION

Effect of Twisting

The main purpose of studying untwisted composites is to determine their effective current density directly from magnetization measurements. For example, Fig. 2 illustrates M-H curves of MJR sample M10A and a commercial bronze-process wire[3] whose current density has been measured by short-sample measurements. Since the untwisted composite exhibits strong shielding as a solid wire, its magnetization is determined only by its average current density, as shown in Eq. (1). The fact that they displayed similar magnetization thus leads one to conclude that both wires possess equal overall current densities. Assuming that the twisting process would not cause any ill effect on the electrical properties of the composite, the M-H curve (trace 1 in Fig. 3) can thus serve as a reference of magnetization that corresponds to a solid conductor with the same current density.

As the same sample is twisted with ℓ_p = 5 mm, its magnetization (trace 2 in Fig. 3) decreases drastically in a manner similar to what was observed in a conventional wire. Although for an untwisted sample, flux distribution is not uniform inside the wire in such a low field, we still can roughly estimate the effective diameter, d_{eff}, of the twisted sample. This, for example, can be done by measuring the ratio $(\mu_o M_{\ell p\,=\,\infty})/\mu_o M_{\ell p}$ at $\mu_o H_e$ = 1 T. From digitally recorded data, the ratio is found to be 25. Although the diameter of the filamentary region, d_o, is 450 μm, one finds that

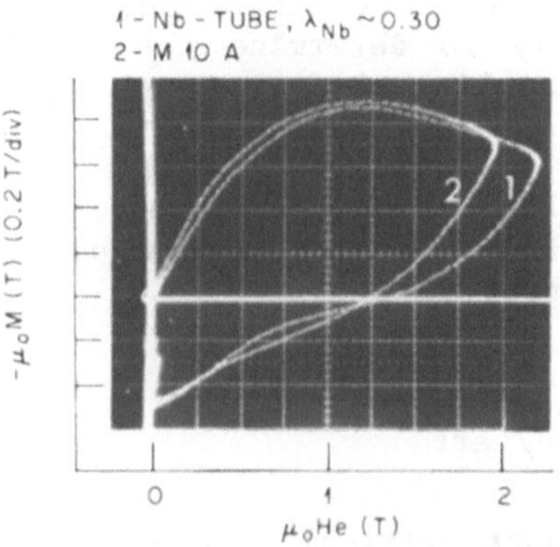

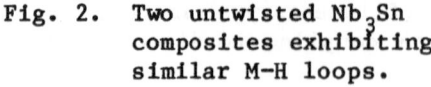

Fig. 2. Two untwisted Nb_3Sn composites exhibiting similar M-H loops.

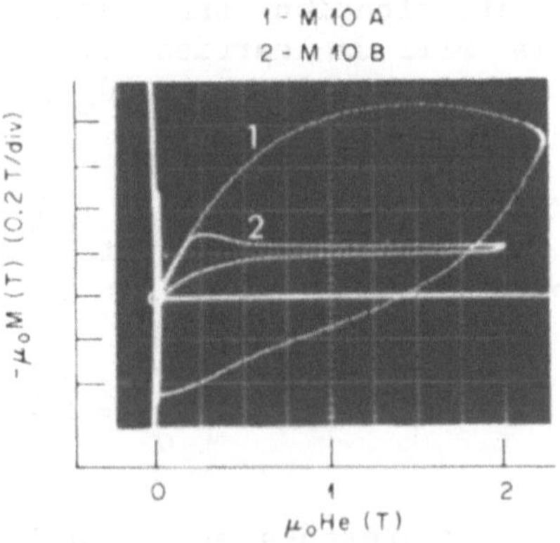

Fig. 3. Effect of twisting in MJR composites.

the effective diameter, d_{eff}, is 18 μm, as compared with a typical filament dimension 2 μm x 10 μm (Fig. 1). The difference can be accounted for by the shielding effect in untwisted wire and the effect of a niobium barrier, which will be discussed in a later section. Also, in a later section, it will be shown that d_{eff} can be calculated directly from magnetization measurements of twisted conductor using Eq. (1).

The findings that twisting effectively decouples filaments is indeed promising. Apparently, the nodes that connect elongated filaments have been separated far enough (>> 10 m) compared with the twist pitch (~5 mm); this ensures that shielding (coupling) currents can be limited within the twist pitch.

Diffusion Barrier

To protect the copper stabilizer from tin poisoning during heat treatments, an inert diffusion barrier is required in Nb_3Sn composites. In this study, we have investigated three types of barriers: (a) tantalum, (b) niobium, and (c) niobium with a phosphor bronze. The last approach is illustrated in Fig. 1(a), where the reaction of niobium and bronze has been suppressed by a local concentration of phosphor.

In Fig. 4 the M-H curve of M7B with a tantalum barrier is shown as trace 1, exhibiting the magnetization as expected for a Nb_3Sn filamentary conductor. However, if tantalum is replaced by niobium, a large Nb_3Sn shell will result, which in turn would give rise to additional magnetization. This effect is clearly shown in Fig. 4, where trace 2 represents the M-H curve of sample M10B with

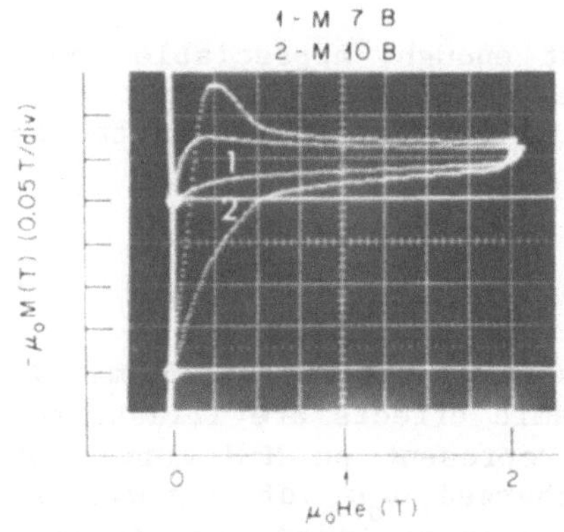

Fig. 4. Magnetic effect of niobium diffusion barrier.

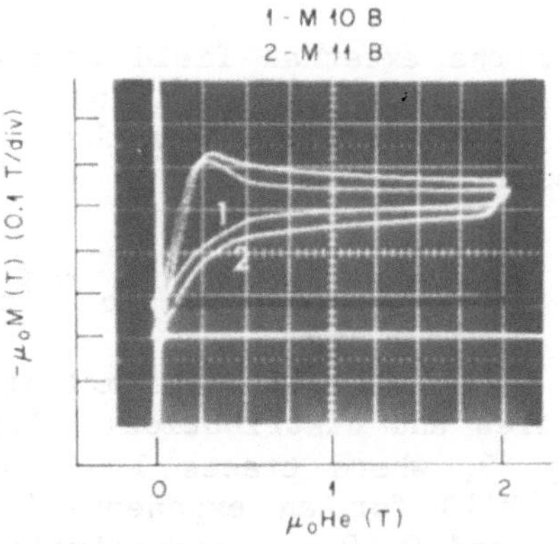

Fig. 5. Effect of phosphor poisoning on the formation of Nb 3Sn shell.

Table 2. Steady-State Effective Filamentary Diameter.*

Sample Identification	Barrier	d_{eff} (μm)	Shielding Field $2\mu_o M_s$ (10^{-4} T)
M7B	Ta	7	124
M10B	Nb+P	14	223
M11B	Nb	36	580

*A typical Nb_3Sn filament (approximately rectangular in shape) measures about 2 μm x 10 μm.

niobium and phosphor bronze as a barrier. Although the formation of an Nb_3Sn shell has been retarded, the resulting thin (~0.2 μm) Nb_3Sn shell still dominates the overall M-H curve with its own magnetic features, i.e., a long tail at the low field and an additional fully penetrated magnetization at the high field. Without the phosphor poisoning, a thicker (~1 μm) Nb_3Sn shell results. Its magnetic effect is illustrated in Fig. 5, where trace 2 shows appreciable increases in the M-H loop area. Nevertheless, one should note that the unwanted Nb_3Sn shell has not introduced any magnetic instability in such composites.

Steady-State Effective Diameter of MJR Composites

Effective diameters have been deduced for MJR wires on the basis of a steady-state magnetic field of $\mu_o H_e$ = 3 T. Assuming an average current density, $\lambda J_c \sim 3$ GA/m^2, one can use Eq. (1) to calculate d_{eff}. Some of the results are listed in Table 2.

Coupling Time Constant, τ_o

If the external field changes fast enough, appreciable additional shielding current can be induced with a certain time constant. It has been shown that this time constant, τ_o, can be expressed as[2]

$$\tau_o \sim \mu_o(\ell_p^2/\rho_e), \tag{2}$$

where ρ_e is the effective matrix resistivity determined by matrix properties and distribution. Such dynamic effects are illustrated in Fig. 6, where traces 1, 2, and 3 represent an M-H curve for sample M11D for an exponentially discharged $\mu_o H_e$ of τ = 0.33 s, 0.25 s, and 0.20 s, respectively. The area enclosed, in addition to that of the hysteretic loop, can be attributed to a transient shielding current flow between filaments and the copper stabilizer.

Table 3. Coupling Time Constants of MJR Conductors

Sample Identification	Twist Pitch, ℓ_p (cm)	Coupling Time Constant, τ_o (s)
M7B	2.5	10^{-3}
M10B	0.5	10^{-3}
M10D	2.5	11×10^{-3}
M11B	0.5	10^{-3}
M11D	2.5	40×10^{-3}

Time constants for MJR wires are analyzed and presented in Table 3. The results indicate that tightly twisted MJR wires, with a time constant no larger than 1 ms, are suitable for any practical, fast field-change applications.

Effect of Heat Treatment

The effect of heat treatment was also studied by comparing M-H curves before and after the heat treatment (735°C, 25 h). Typical results of M7A and M7B are shown in Fig. 7, where M7A shows appreciable magnetization for fields up to 2 T, not expected for non-reacted niobium filaments that should have an upper critical field, $\mu_o H_{c2}$, much lower. For samples built with a niobium barrier, the results are even more surprising, as displayed in Fig. 8, where the

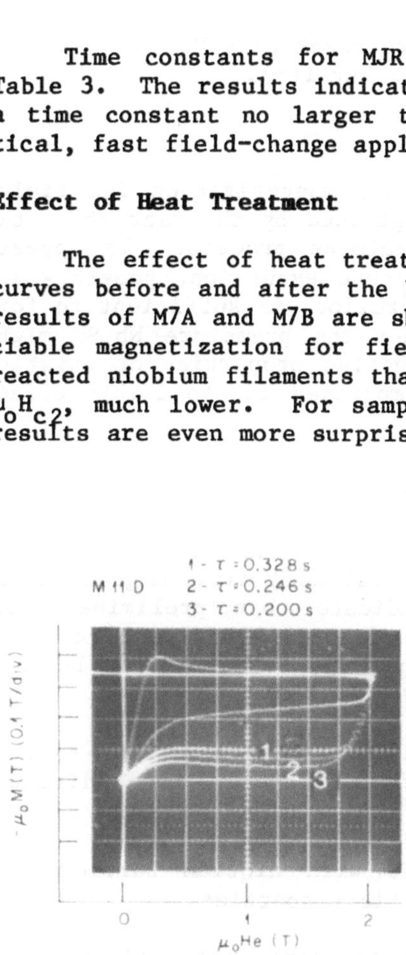

Fig. 6. Dynamic magnetization due to exponentially decayed external fields.

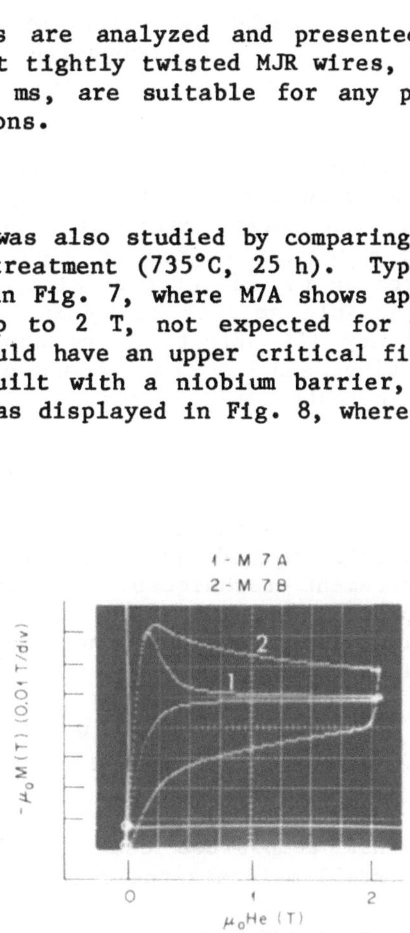

Fig. 7. Effect of heat treatment on a MJR composite (tantalum barrier).

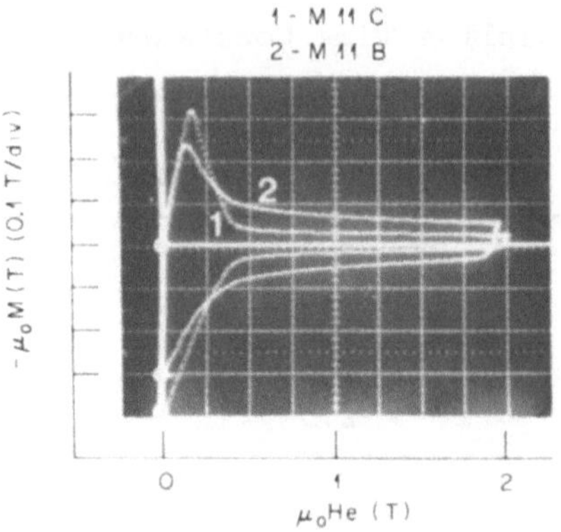

Fig. 8. Effect of heat treatment on a MJR composite (niobium barrier).

nonreacted sample M11C exhibits definite magnetization for fields higher than 1 T. This can only be explained by the fact that some Nb_3Sn was formed before the heat treatment process. It appears that a certain degree of high temperature excursion, possibly incurred during billet preheating or extrusion, is sufficient to form appreciable amounts of Nb_3Sn. Whether the premature Nb_3Sn might have any ill effect on the superconducting properties of existing MJR wires and how the process can be modified to avoid this effect remain to be investigated.

CONCLUSIONS

Multifilamentary Nb_3Sn superconductors made by a "modified jellyroll" method were systematically evaluated using magnetization measurement techniques. Results indicate that preliminary MJR samples possess excellent magnetic properties equivalent to those of conventional Nb_3Sn conductors. The major findings can be summarized as follows:

1. MJR composites made with a tantalum diffusion barrier exhibit magnetic properties equivalent to those of individual filaments.

2. MJR composites can also be made with niobium barriers and still maintain their stable magnetic properties.

3. The process shows promise of further improvement and optimization to produce a practical MF Nb_3Sn conductor in a most economical way.

ACKNOWLEDGMENTS

The authors wish to thank M. S. Lubell and W. A. Fietz for their interest and encouragement, J. P. Rudd for his technical assistance, and the Research and Development personnel at Teledyne Wah Chang, Albany, for preparing the sample conductors.

NOTATION

B	magnetic induction field
d_{eff}	effective diameter
d_f	filament diameter
d_o	composite diameter
H_{c_2}	upper critical field
H_e	applied field
H_p	penetration field
ℓ_p	twist pitch
M_p	magnetization
M_s	shielding magnetization
X_o	twist parameter

Greek Symbols

λJ_c	overall critical current density
ρ_e	effective matrix resistivity
τ	applied field change time constant
τ_o	magnetic diffusion time constant or coupling time constant
μ_o	permeability of free space

REFERENCES

1. R. E. Schwall, S. S. Shen, J. W. Lue, J. R. Miller, and H. T. Yeh, Superconductors for tokamak poloidal field coils, in: "Advances in Cryogenic Engineering," Vol. 24, Plenum Press, New York (1978), p. 427.
2. S. S. Shen, Magnetic properties of multifilamentary Nb_3Sn composites, in: "Filamentary A15 Superconductors," M. Suenaga and A. F. Clark, eds., Plenum Press, New York (1980), p. 309.
3. S. Murase, Y. Koike, and H. Shiraki, J. Appl. Phys. 49:6020 (1978).

HYDROSTATIC EXTRUSION OF *IN SITU* PROCESS SUPERCONDUCTIVE WIRE FOR USE IN NMR MAGNETS*

D. W. Hazelton and R. E. Schwall

Intermagnetics General Corporation, Guilderland, New York

B. Avitzur

Lehigh University, Bethlehem, Pennsylvania

J. D. Verhoeven

Ames Laboratories, Iowa State University, Ames, Iowa

INTRODUCTION

This research was undertaken to develop a technique for processing wire to be used for high field persistent mode NMR magnets. Early superconducting NMR spectrometers utilized magnets wound of NbTi superconducting wire yielding a magnet performance in the 8.5-9 T (360-400 MHz) range. Operation above this range requires Nb_3Sn superconductors, and several magnets with persistent mode multifilamentary (MF) Nb_3Sn joints in the 450-500 MHz range have been produced. Operation above 12 T is barred by the fundamental limitation that the critical current density of MF Nb_3Sn lies below a useful threshold value for fields above 12 T. The fabrication of an economical magnet with sufficient homogeneity for NMR operating at fields of 14 T and above thus requires current densities higher than those exhibited by conventional MF Nb_3Sn.

In the past few years, research work at Iowa State University,[1,2] Harvard,[3,4] and MIT[5,6] has shown a ductile MF Nb_3Sn superconducting wire prepared by the in situ process exhibits critical current density comparable to or better than wire prepared by the

*This material is based upon work supported by the National Science Foundation under award number DAR 8009738.

more conventional internal bronze or external bronze techniques. The investigators referenced have produced such wire from an *in situ* Cu-Nb ingot containing dendritic second phase Nb. When extruded and drawn to a considerable reduction in area, these Nb dendrites transform into elongated ribbons. After a diffusion and reaction heat treatment in the presence of an Sn source, these Nb ribbons are transformed into elongated whiskers with the compound Nb_3Sn on the surface.

Most wire produced by the investigators above was prepared by coating the surface of a wire produced from a Cu-Nb ingot with Sn and then diffusing the Sn into the wire. This resulted in a wire containing Nb_3Sn filaments in a bronze matrix. The maximum wire diameter that may be produced by this technique is limited by the diffusion step. Wire much larger than 0.38 mm in diameter requires special plating and diffusion techniques to maintain good wire stability. This limitation can be avoided by introducing the Sn internally into the Cu-Nb billet. The presence of Sn in a number of holes distributed through the cross section of wire leads to a reduction in diffusion distance and, hence, an increase in the diameter of the wire that can be produced.

The major difficulty with this internal diffusion process involves using a cold extrusion step. It is essential that the Sn cores do not melt during this step or the pressure behind the die will be transmitted to the material outside the die. The extruded rod will then burst open causing the local molten Sn to blast out of the extruded rod. In view of the low melting point of Sn (231.9°C), it is clear that conventional extrusion processes are of little use for these composites. The process of hydrostatic extrusion (Chapter 7 of Ref. 7) offers an attractive alternative. Forced through the die by means of liquid pressure, the billet is surrounded by fluid that essentially eliminates the frictional contact and resultant heating.

In this investigation, a total of 33 billets were hydrostatically extruded, a safe extrusion range was characterized by extrusion ratio, and die semi-cone angle was mapped. Various heat treatments were utilized to optimize the Sn diffusion and critical current.

PREPARATION OF Cu-Nb BILLETS

The Cu-Nb material used for this project produced by Ames Laboratories, Iowa State University contained ~17% Nb.* Material preparation has been described fully in Reference 8. To summarize,

*Production funded by the U.S. Department of Energy and supported by the Director of Energy Research, Office of Basic Energy Science.

Cu and Nb were arc melted to form a 4-in diameter billet. This billet was hot extruded to 1.3 in and then cold drawn to 0.74 in at which point material was supplied to IGC for this project.

The material was annealed in Ar at 450°C for 4 h, and then it was cold-drawn to 0.375 in. The diameter of 0.375 in was chosen because it would allow the evaluation of a large number of extrusion parameters with the available in situ material and because the results can be directly extrapolated to larger billets. A number of billet designs were tested at the beginning of this program and that shown in Figure 1 was found to yield satisfactory results. The design does leave some exposed Nb in the machined nose of the billet, but this does not appear to cause any problems during the subsequent extrusion and drawing operations.

PREPARATION AND INSERTION OF Sn-Cu CORE

Three alloys were initially chosen to be used in the processing of the billets; Sn-0, -5, and -10 wt.% Cu. Sn-Cu alloys were chosen since the addition of Cu increases the liquidus temperature of the alloy allowing greater flexibility in extrusion parameters. However, the ductility of the alloy decreases with increasing Cu content.

The alloys were prepared by heating under vacuum to temperatures above the alloy liquidus for times sufficient to form a homogeneous melt. The billet cores were then cast using pressurized Ar to force molten SnCu into evacuated quartz tubes.

Conventional air or furnace cooling resulted in a structural orientation radiating from the core axis perpendicular to the later drawing direction. The cores were, therefore, directionally

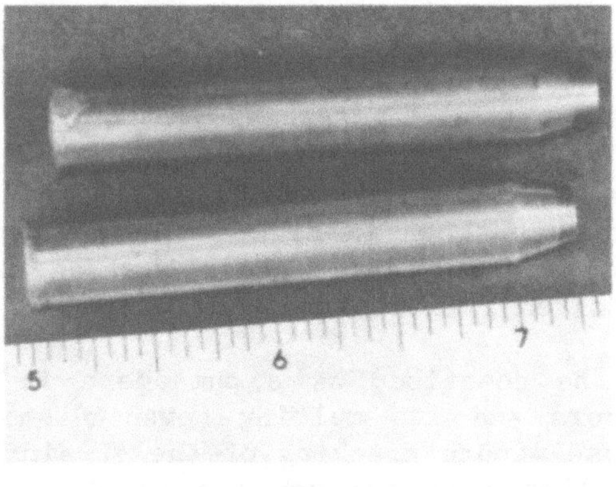

Fig. 1. Typical billets prior to extrusion.

solidified to orientate the alloy structure preferentially along the core axis. The cores were then cold drawn to 0.144 in.

The core material was then cut to size and inserted into 0.144-in diameter holes predrilled in the billets. Enough Sn was provided for complete reaction of the Nb to Nb_3Sn and Cu to a Cu 6 wt.% Sn bronze. The backs of the billets were then sealed using low melting point solder.

BILLET EXTRUSIONS

Although three alloys were chosen for evaluation, only Sn-5 wt.% Cu was used in the extrusion experiments. The Sn-10 wt.% Cu proved too brittle to draw and since the Sn-5 wt.% Cu displayed sufficient ductility, there was no incentive to use the lower melting point Sn.

After completion of the filling and sealing operations, the billets were hydrostatically extruded at Lehigh University. Prior analysis showed that two out of seven possible independent process parameters would have the most effect on the extrusion properties. These two, reduction in area and die semi-cone angle, were varied to define a processing range yielding a uniform flow pattern. The remaining five process parameters (core size ratio, % Cu in Sn, length of die bearing, extrusion speed, and receiver pressure) were held constant. A description of the extrusion unit and its operation is provided in Chapter 7 of Reference 7.

A total of 22 parametric combinations were utilized in this investigation. A plot of press force vs. ram displacement was automatically recorded during each billet extrusion. As expected, the steady-state extrusion pressure increased with increasing die angle at constant reduction and also increased with increasing reduction at constant die angle.

Extrusion of a composite material consisting of a soft core and a hard sleeve may develop a unique flow pattern. The problems encountered with some of these unique flow patterns are described fully in Section 7.5.4.3.5 of Reference 7.

Of the 22 combinations processed, 16 exhibited a sound flow pattern similar to that shown in Figure 2. The remaining six combinations exhibited a flow pattern described as sweating, and two of these also exhibited a flow pattern described as swelling.

Sweating can be described as a phenomena in which the inner pressure of the core and its melting (even a small layer on the interface) may cause minute cracking of the strain-hardened sleeve. Core material may then pass through the cracks and emerge on the outer surface of the sleeve.

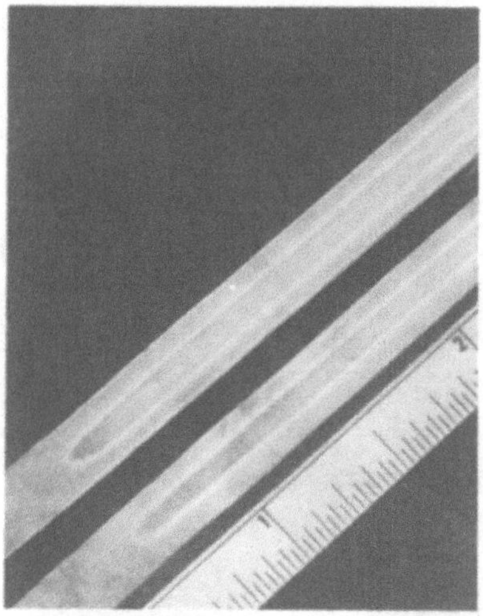

Fig. 2. Longitudinal section of extruded billet exhibiting sound flow from billet nose back.

Swelling can be described as a phenomena in which a soft core subject to a nominally uniform radial pressure is squeezed outwardly towards the exit. In these cases, the outside diameter of the extrudate remained uniform while the soft core swelled. The swelling starts immediately upon exiting the die where the inner core pressure is still high and proceeds to develop as the extrudate and the swelling move forward. As the extrudate moves forward, the inner core pressure drops to values insufficient for further swelling. This results in a periodic initiation, growth, and demise of swelling. Figures 3 and 4 show examples of both sweating and swelling.

Sweating was detected by examination of the billets upon removal from the extrusion chamber. To detect swelling, it was necessary to section each billet longitudinally in order to examine the internal billet structure. Figure 5 plots the extrusion results as a function of die angle and reduction of area. From these results, a safe zone has also been mapped on Figure 5.

DRAWING

For drawing purposes, a billet that had been extruded at 7° semi-cone die angle and 74% reduction in area was chosen. This billet was used since billets extruded under similar conditions exhibited a sound flow pattern. The billet was cold drawn without intermediate anneals from the extruded diameter of 0.192 in to the final diameter of 0.0204-in. During drawing there were no signs of wire instability. Figure 6 shows a cross section of the wire at the final 0.0204-in diameter. In this micrograph, the three distinct layers of the drawn material are clearly visible: the Cu shell, the Cu-Nb in situ material, and the Sn-Cu core.

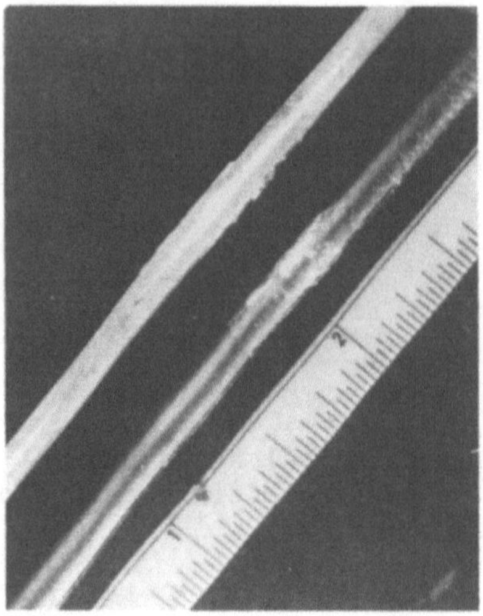

Fig. 3. Example of sweating and swelling.

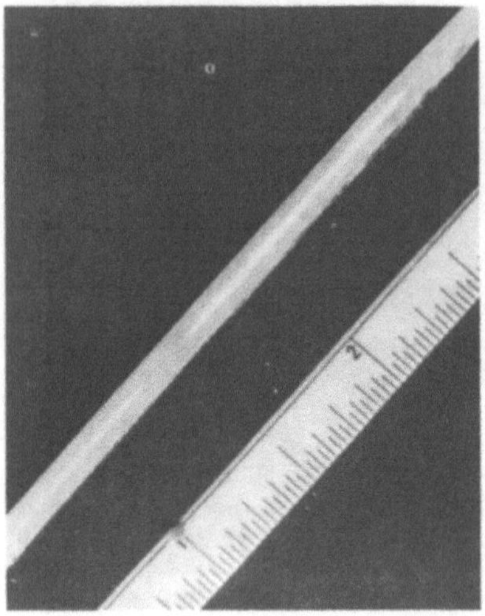

Fig. 4. Example of swelling.

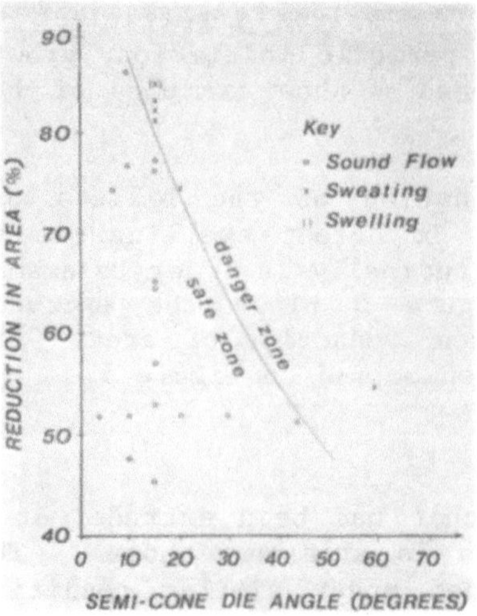

Fig. 5. Extrusion results as a function of semi-cone die angle and reduction in area.

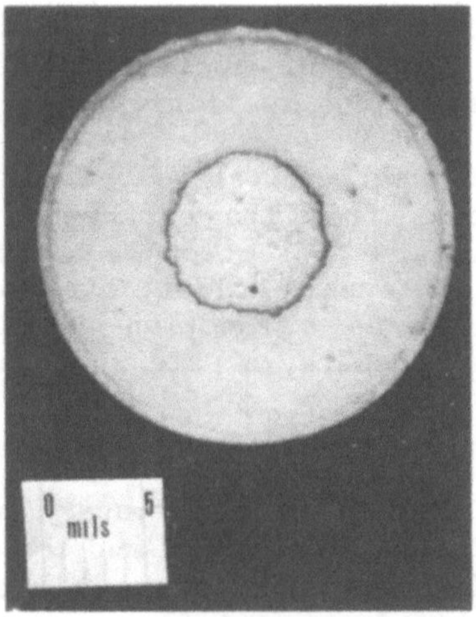

Fig. 6. Cross section of 0.0204-in-diameter cold-drawn wire.

HEAT TREATMENTS

A variety of heat treatment schedules were used to form the final 0.0204-in diameter superconducting wire. In the first step, heat treatments at temperatures from 200°C to 400°C were employed to diffuse the Sn from the core into the surrounding Cu-Nb without reaction with the Nb. A variety of time-temperature profiles were employed with the objective of obtaining reasonably rapid diffusion of the Sn while minimizing the local gradient of the Sn concentration. Control of this gradient is necessary since rapid diffusion of Sn through a large concentration gradient will result in the formation of large Kirkendahl voids.

Following diffusion, a higher temperature heat treatment was used to form the Nb_3Sn compound. Temperatures from 500°C to 580°C and times from 72 h to 312 h were used. Temperatures greater than 580°C were not used due to critical current density degradation resulting from in situ Nb dendrite coarsening. A cross section of a reacted billet is shown in Figure 7.

CRITICAL CURRENT

Critical currents were measured in fields of 4-12 T in the bore of an IGC Nb_3Sn solenoid. Straight samples ~5-cm long were used with voltage taps located approximately 2 cm apart at the sample center. The values of critical current density (J_c) plotted in Figure 8 represent the critical current density calculated using the area of the final reacted wire (including the core and Cu shell) with naturally occurring prestrain and at a voltage sensitivity of 0.5 μV/cm. This corresponds to an equivalent resistivity of 1 to 10 pΩ•cm, depending upon the current.

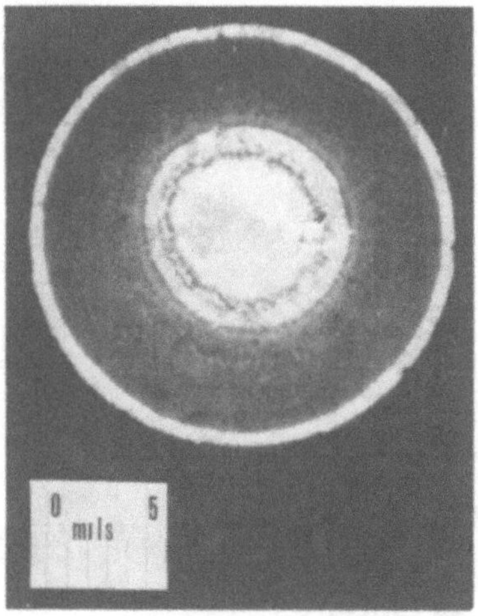

Fig. 7. Reacted billet.

As can be seen in Figure 8, the critical current density increases with diffusion and reaction times, indicating that the wire is not completely transformed to Nb_3Sn. Nevertheless, the best results obtained to date are comparable to those obtained by the Ames Group[9] on 0.010-in diameter wire of the same material, which is included on Figure 8 as a reference. Current generation in situ material produced by Ames achieves higher current density through control of the Nb dendrite size.[10] The techniques here are expected to yield similar results.

CONCLUSIONS

This investigation has provided the following results:

- A method of billet fabrication that is compatible with hydrostatic extrusion has been demonstrated.
- Through analysis of 22 different extrusions, a range of extrusion parameters has been defined within which these composites can be hydrostatically extruded, yielding a sound product suitable for further processing.
- The extruded composites have been successfully drawn to 0.0204-in diameter wire.
- The wires have been heat-treated successfully, yielding an in situ superconductor with critical current density comparable to that of smaller wires prepared from the same Cu-Nb ingot.

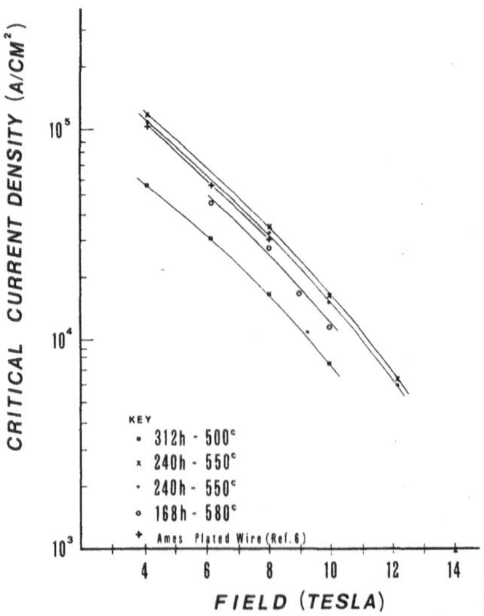

Fig. 8. Critical current density vs. field for reaction heat treatments as noted.

REFERENCES

1. J. D. Verhoeven, F. A. Schmidt, E. D. Gibson, J. E. Ostenson, and D. K. Finnemore, Appl. Phys. Lett. 35(7):555 (1979).
2. D. K. Finnemore, J. D. Verhoeven, E. D. Gibson, and J. E. Ostenson, IEEE Trans. Magn. MAG-15:693 (1979).
3. J. P. Harbison and J. Bevk, J. Appl. Phys. 48:5180 (1977).
4. J. Bevk and J. P. Harbison, J. Met. Sci. 14:1457 (1979).
5. R. Roberge, S. Foner, E. J. McNiff, Jr., and B. B. Schwartz, Appl. Phys. Lett. 31:853 (1977).
6. R. Roberge, S. Foner, E. J. McNiff, Jr., B. B. Schwartz, and J. L. Fihey, Appl. Phys. Lett. 34:111 (1979).
7. B. Avitzur, "Metal Forming Processes," John Wiley & Sons, New York (1982).
8. J. D. Verhoeven, E. D. Gibson, F. A. Schmidt, and D. K. Finnemore, J. Mater. Sci. 15:1449 (1980).
9. J. D. Verhoeven, F. A. Schmidt, E. D. Gibson, J. J. Sue, J. E. Ostenson, and D. K. Finnemore, IEEE Trans. Magn. MAG-17:251 (1981).
10. J. J. Sue, J. D. Verhoeven, E. D. Gibson, J. E. Ostenson, and D. K. Finnemore, On the optimization of in situ Nb_3Sn-Cu wire, in: "Advances in Cryogenic Engineering--Materials, Vol. 28," Plenum Press, New York (1982), p. 501.

FABRICATION OF STABILIZED *IN SITU* Nb_3Sn WIRE BY CONSUMABLE ARC MELTING AND ITS MAGNETIC BEHAVIOR

K. Yasohama, H. Ohkubo, T. Ogasawara, and K. Yasukōchi

Atomic Energy Research Institute
College of Science and Technology, Nihon University, Tokyo, Japan

INTRODUCTION

Recent advances have shown that the in situ technique is one of the most promising fabrication processes for Nb_3Sn and V_3Ga wires from the standpoint of both superconducting and mechanical properties. In this technique, two-phase alloys (Cu-Nb, Cu-V) as starting materials have been produced by means of several casting methods. For various reasons all of these casting methods cannot be used in the industrial fabrication process. Consumable arc casting, first applied by Verhoeven et al.,[1] is well developed as an industrial process and can be used to produce a large amount of starting two-phase alloys. This technique offers the prospect of improved superconducting wires for practical applications. In the practical use of a new superconducting wire, however, a full investigation of ac loss and stability is needed. In this study we have performed an experimental work on the fabrication of a stabilized in situ Nb_3Sn wire by the consumable arc casting technique with external tin diffusion and measured the magnetic properties of twisted and untwisted wires in transverse and parallel external magnetic fields.

EXPERIMENTAL PROCEDURES

Sample Preparation

The in situ Nb_3Sn wires were fabricated from Cu-35 wt.% Nb alloys cast by consumable electrode arc melting. The consumable electrode composed of a niobium rod inserted in a copper tube of 1.3-cm o.d. and 25-cm long was melted into a graphite crucible of

3.0-cm o.d. by 2.0-cm i.d. and 15-cm overall height. To avoid contamination of the alloy, the graphite crucible was degassed at a high temperature in a vacuum. The crucible was set on a water-cooled copper plate.

Arc melting was carried out in a high purity argon gas atmosphere with a pressure of 400 torr, and a constant arc voltage of 21 to 22 V was maintained by lowering the electrode at an arc current of 580 A. Under this condition the arc was stable without any side arcing between the electrode and the inside wall of the crucible, and the metal was cast at a constant rate of 190 g/min. The cast Cu-Nb alloy was 2 cm in diameter and 10 cm in length.

In the small-scale consumable arc melting applied in this study, it is rather difficult to maintain the molten pool at the top of the ingot unless the heat flow is appropriately limited. In our case, the crucible was cooled only at its lower part, and the graphite bottom of the crucible (0.5-cm thick) reduced the longitudinal heat flow to prevent excessive cooling.

Dendritic niobium was revealed in the alloys after the copper matrix was removed by deep etching, and the average diameter of dendrite arms was found to be 6 μm. This microstructure was confirmed to be homogeneous throughout the alloy by optical microscope examination.

The surface of the ingot was machined off to a diameter of 1.9 cm and about 1 cm at both ends was cut off. A hole of 0.73 cm in diameter was drilled to insert a pure copper rod of 0.5 cm in diameter as a stabilizer. We used a tantalum tube of 0.7-cm o.d. by 0.52-cm i.d. as a diffusion barrier. This Cu-Nb alloy with copper stabilizer was then swaged down to a diameter of 0.49 cm. During this process when the rod diameter became 0.76 cm, it was jacketed with a copper tube of 0.81-cm i.d. by 0.91-cm o.d. Then it was drawn to a final diameter of 0.269 mm with three intermediate anneals at a temperature of 500°C for a period of 4 h. The reduction in cross-sectional area of the in situ shell region was 6200.

A part of the wire was twisted at a pitch of l_p = 2.3 mm and the other was used as untwisted wire. These wires were then electroplated with tin by a small-scale electroplating line built by us. On this line the actual electroplating was done in a commercially available stannous sulphate solution in which anodes of a pure tin plate were symmetrically located. A cross section of the electroplated wire is shown in Fig. 1(a). The plated tin content was 13 wt.% of the copper except the stabilizer, which was close to the stoichiometric value to transform all the niobium filaments into Nb_3Sn.

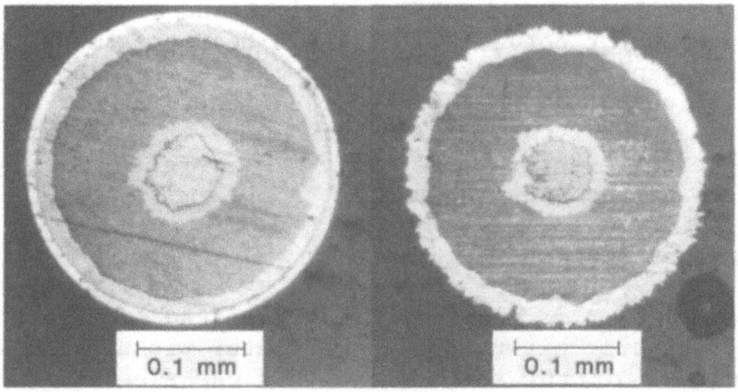

(a) (b)

Fig. 1. Cross-sectional views of the copper-stabilized in situ Nb_3Sn wire: (a) as electroplated with Sn and (b) after reaction treatment. Outside diameter was 0.29 mm. Inside and outside diameter of the in situ region were 0.09 and 0.242 mm, respectively. The copper stabilizer had a diameter of 0.06 mm.

Prior to the reaction treatment to form Nb_3Sn, the tin was diffused in and homogenized at a temperature of 200°C for a period of 1 h and then at 400°C for 3 h. The reaction treatment was carried out for period of 3 to 8 days at temperatures of 550 and 600°C in a vacuum. Figure 1(b) shows a cross section of the untwisted wire reacted at 550°C for 8 days. The outermost diameter of the wire after the reaction treatment was 0.29 mm.

For the magnetization measurements in a transverse magnetic field, an open-ended sample coil of three layers and 1.9 cm in length was made by winding the wire on a quartz tube of 2.0-cm o.d. prior to the reaction treatment; the coil was not rewound onto a different bobbin after the reaction in order to prevent any possible winding stress on the wire. The wire was co-wound with an insulating spacer of a thin glass-fiber yarn, and for an insulator between the layers, a few turns of a glass tape were wound. Total length of the wire of the sample coil was 5.0 m.

For the parallel-field magnetization measurements, the wire was cut into 2-cm-long sections and each piece was inserted into a thin quartz tube to avoid the contact among the pieces during reaction treatment. Both ends were polished after the reaction treatment to eliminate contact between Nb_3Sn filaments that might be caused by cutting the wire. A total number of 80 pieces were mounted in parallel spacing on a bakelite bobbin with an adhesive glass tape.

Both samples were heat-treated at one time to maintain the same heat-treatment condition. In this heat treatment, the diffusion and homogenization treatment was carried out at a temperature of 400°C for period of 6 h and the reaction treatment at 550°C for 106 h. Several of the pieces were used as samples to examine the possible difference of the critical current density between the twisted and the untwisted wire.

Method of Magnetization Measurements

The magnetization was measured by a standard electric method with two concentric pickup coils. An outer pickup coil was wound on the outside of the sample, and this was sleeved on a bakelite mandrel on whose surface a groove was cut and an inner pickup coil was wound. The magnetization, M, was measured by integrating the difference between the voltages induced in the two pickup coils when the external magnetic field, H, was cycled between 0 and 5 T at sweep rates ranging from 1.1 to 11 T/min.

RESULTS AND DISCUSSION

Critical Current

The critical current, I_c, of the short samples was measured at 4.2 K by a standard four-probe technique in a transverse magnetic field, H. The I_c was defined by the criterion of the sample voltage of 0.1 μV/cm. The critical current density, J_c, refers to the total cross-sectional area (including the copper stabilizer, the tantalum diffusion barrier, and the outermost

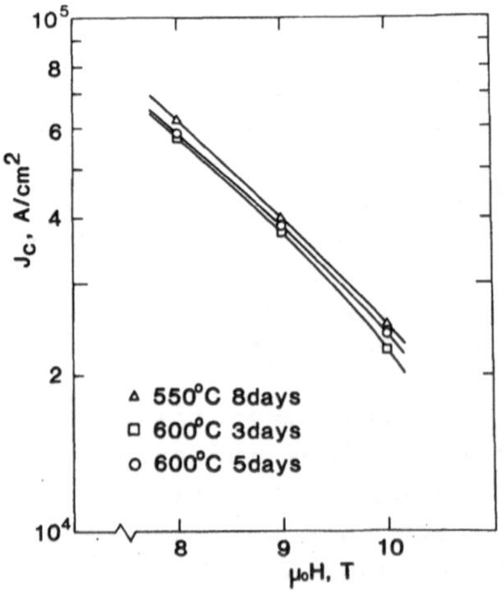

Fig. 2. Overall critical current density, J_c, versus applied transverse magnetic field, H, at 4.2 K for Cu-35 wt.% Nb-13 wt.% Sn wires.

bronze layer). In Fig. 2 the critical current density, J_c, is plotted as a function of the field, H, from 8 to 10 T. An overall critical current density of 2.5×10^4 A/cm^2 was obtained at 10 T. At a fixed reaction temperature of 550°C, the critical currents reached nearly their maximum values in 6 or 7 days; at the higher reaction temperature, 600°C, the critical currents saturated within 4 days; and a longer reaction time, up to 8 days, at least had no ill effect. These maximum values are nearly the same, as illustrated in Fig. 2.

The J_c value of our sample wire was lower than the data obtained in other studies. We guess that the main cause might be the carbon contamination[3] by the graphite crucible. Since a, surface/volume ratio becomes high for a small ingot like the one in this study, the relative content of the carbon pickup would be higher than for large ingots. Verhoeven et al.[3] have proposed that by lining the inner wall of the crucible with a ceramic material, the carbon contamination can be eliminated. We carried out a preliminary experiment: the inner wall of a graphite crucible was plasma sprayed with a 0.2-mm coating of zirconia. But the zirconia was entirely reacted with a Cu-Nb alloy by single casting, and furthermore, the reacted surface of the alloy was very hard to machine off.

Magnetization

In Fig. 3, M-H curves of the twisted (dashed line) and the untwisted (solid line) samples in a transverse field are given for comparison. In this figure the M-H curves of the virgin run and the second run, in which the samples had already been cycled to high field at least once, are presented. From the magnetization

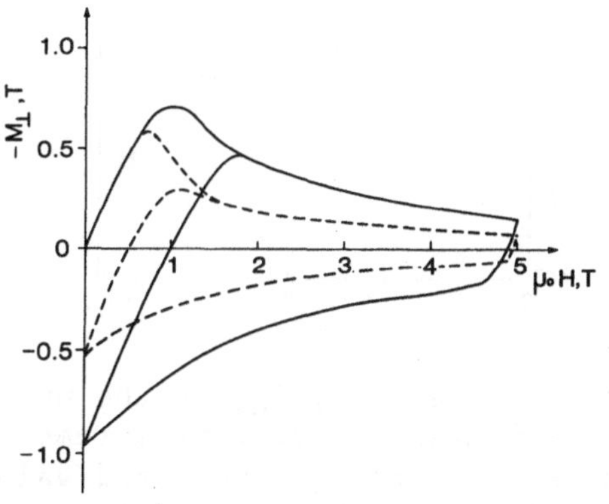

Fig. 3. M-H curves for a twisted (dashed line) and an untwisted (solid line) samples in transverse magnetic field. The twist pitch, ℓ_p, = 2.3 mm.

curve of the untwisted wire, it was found that the sample was almost perfectly diamagnetic up to the field of about 1 T and the magnetization was hysteretic. Moreover, no time-dependent magnetization was observed when the sweep rate of the field was varied from 1.1 to 11 T/min. The measured hysteretic magnetization is large compared with that of the conventional multifilamentary wires, and it is inferred that the electromagnetic coupling among the Nb_3Sn filaments is strong. The above results are similar to those of by Shen and Verhoeven[4] and indicate that the *in situ* wire behaves like a solid wire. When we consider the *in situ* shell region of the wire to be a solid hollow superconductor, the transverse magnetization above the penetration field can be shown as

$$-M_{\perp} = \frac{4}{3\pi}\cdot\mu_o\cdot J_c\left(\frac{r_o}{r_2}\right)^2\left[1+\frac{r_1}{r_2}+\left(\frac{r_1}{r_2}\right)^2\right]\left[1+\frac{r_1}{r_2}\right]^{-1} \tag{1}$$

In Eq. (1), r_o is the outermost radius of the wire, and r_1 and r_2 are the inner and outer radius of the *in situ* region. Substituting $\mu_o = 4\pi \times 10^{-7}$, $r_o = 0.145 \times 10^{-3}$m, $r_1 = 0.045 \times 10^{-3}$m, $r_2 = 0.121 \times 10^{-3}$m, and the measured $M_{\perp}$, J_c can be calculated and is illustrated in Fig. 4. The dashed line was extrapolated using the Kramer[5] relation from the measured J_c at high fields. As shown in

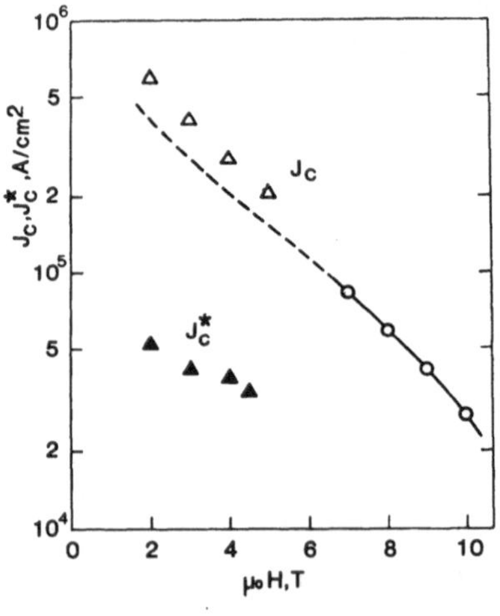

Fig. 4. Critical current density, J_c, of the untwisted wire calculated from magnetization measurements (open triangles; cf. Fig. 3) and the measured values at high field (open circles). The dashed line represents a curve extrapolated from the measured J_c by the Kramer relation. J_c* (closed triangles) are also shown for comparison.

Fig. 4, a good agreement was found between the calculated and the extrapolated J_c.

In the twisted wire, the magnetization became small in comparison with that of the untwisted wire, as shown in Fig. 3. It was confirmed from the critical current measurement that the twisting had almost no effect on the J_c-H properties of the wire. Therefore, the decrease of magnetization was caused by twisting, but the effect was not enough to decouple the Nb_3Sn filaments completely.

As shown in Fig. 5, the magnetization of the untwisted wire in the parallel field was much smaller than that obtained in the transverse field. The hysteretic magnetization curve was again not time dependent when the sweep rate of the field was varied. This supports the assumption that coupling of a persistent nature exists for the circular current in the wire. When a solid hollow conductor is considered, the parallel magnetization can be expressed as

$$-M_{\parallel} = \frac{1}{3} \mu_o \cdot J_c{}^* \cdot r_o \left[1 + \frac{r_1}{r_2} + \left(\frac{r_1}{r_2} \right)^2 \right] \tag{2}$$

where $J_c{}^*$ is the circular shielding supercurrent density. The notations of r_o, r_1, and r_2 in Eq. (2) are the same as those in Eq. (1). The $J_c{}^*$ can be calculated by applying the measured $M_{\parallel}$ into Eq. (2) and is shown in Fig. 4. The $J_c{}^*$ is smaller by one order of magnitude than that obtained in the transverse field. The magnetization curve of the twisted wire is illustrated in Fig. 6. The magnitude of the magnetization was almost the same as that of the untwisted wire, and the twisting did not effectively decouple the Nb_3Sn filaments. It should be noted in Fig. 6 that the magnetization is unusual; the second-run curve does not

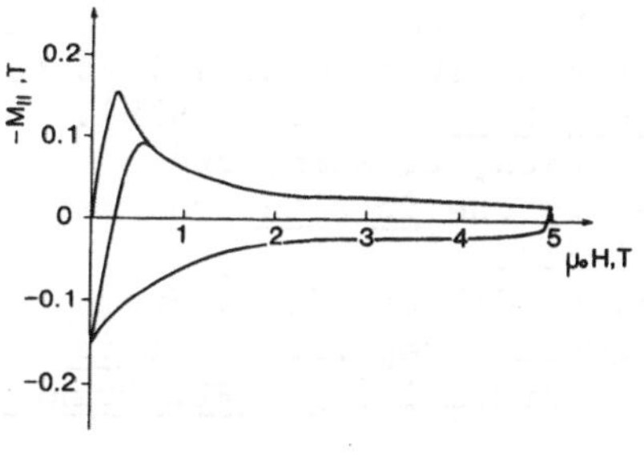

Fig. 5. M-H curves of the untwisted wire in a parallel magnetic field.

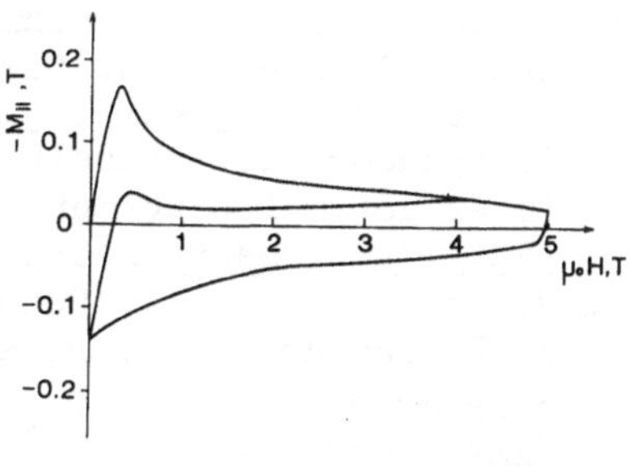

Fig. 6. M-H curves of the twisted wire in a parallel magnetic field. The twist pitch, ℓ_p = 2.3 mm.

trace the virgin run until the field is raised to 4 T. This unusual behaviour of the magnetization cannot be explained at present, but it may be ascribed to the flows of the helical shielding current in the twisted wire.

CONCLUSION

Experimental work was carried out on the fabrication of the stabilized Nb_3Sn wire by consumable electrode arc casting. The small ingots of Cu-Nb alloy were produced by controlling the heat flow of the graphite crucible. However, this technique seems to include a problem of the carbon contamination. Further investigation is needed to solve the problem by using an improved method of lining the graphite crucible or by using other suitable materials.

From the results of the magnetization measurements, Nb_3Sn filaments in *in situ* composites were found to be strongly coupled electromagnetically. Twisting significantly reduced magnetization loss in a transverse field. While in a parallel field, the twisting also influenced the magnitude of the magnetization, but unusual magnetic behavior was observed.

ACKNOWLEDGMENTS

The authors wish to thank E. Suzuki of Showa Electric Wire & Cable Co., Ltd. and T. Noguchi of Vacuum Metallurgical Co., Ltd. for their important contributions to this work. The authors also wish to thank Dr. Y. Kubota of our laboratory for his assistance in the magnetization measurements and its analysis. The present study was supported by the Grant in Aid for Scientific Research of the Education Ministry of Japan.

REFERENCES

1. J. D. Verhoeven, F. A. Schmidt, E. D. Gibson, J. E. Ostenson, and D. K. Finnemore, Casting of dendritic Cu-Nb alloys for superconducting wire, *Appl. Phys. Lett*. 35:555 (1979).
2. C. R. Spencer, W. Marancik, C. Z. Rosen, E. Adam, and E. Gregory, Production of Nb_3Sn by the *in situ* process, *IEEE Trans. Magn*. MAG-17:257 (1981).
3. J. D. Verhoeven, F. A. Schmidt, E. D. Gibson, J. J. Sue, J. E. Ostenson, and D. K. Finnemore, Preparation of *in situ* Nb_3Sn-Cu wire by consumable arc melting, *IEEE Trans. Magn*. MAG-17:251 (1981).
4. S. S. Shen and J. D. Verhoeven, Magnetization of *in situ* multifilamentary superconducting Nb_3Sn-Cu composites, *IEEE Trans. Magn*. MAG-17:248 (1981).
5. E. J. Kramer, Scaling laws for flux pinning in hard superconductors, *J. Appl. Phys*. 44:1360 (1973).

CONTINUOUS HIGH TEMPERATURE GRADIENT SOLIDIFICATION OF *IN SITU* Cu-Nb ALLOYS FOR LARGE-SCALE DEVELOPMENT*

H. LeHuy,† J. L. Fihey, and R. Roberge

IREQ, Institut de Recherche d'Hydro-Québec, Varennes, Québec, Canada

S. Foner

Francis Bitter National Magnet Laboratory‡ and Plasma Fusion Center
Massachusetts Institute of Technology, Cambridge, Massachusetts

INTRODUCTION

The in situ technique for preparing multiply connected Nb_3Sn (V_3Ga) superconducting wires consists of melting and casting a Cu-Nb (V) alloy, mechanically deforming the casting into wires, and finally reacting the Nb (V) filaments into Nb_3Sn (V_3Ga) by diffusing Sn (Ga). For its relative simplicity, but also for the improved mechanical properties of the resulting composites, the in situ process is a promising alternative to the bronze process. Several groups are evaluating various scale-up approaches[1-6] (see also other papers in this volume). We had developed an arrangement for high temperature gradient solidification of small diameter rods.[4-5] The choice of a relatively small diameter had the advantages of:

- low power and inexpensive equipment
- good dispersion of the Nb along the rod
- control of the form and the size of the Nb dendrites.

The process has been upgraded to permit the solidification of larger diameter (9.5-mm) rods in virtually unlimited lengths. In

* Supported in part by the Department of Energy.
† IGM, Conseil National de Recherches, Canada.
‡ Supported by the National Science Foundation.

this paper we focus on the Nb size effects in composites, particularly in the range of small dendrite size: 1 to 3 μm. This work is in progress and this paper should be considered to be an interim report.

EXPERIMENTAL DETAILS

Material Preparation and Microstructure

Most experimental details have been published in the last ICMC proceedings.[4] A Cu-Nb rod was continuously melted in a graphite tube, then resolidified in high temperature gradients. At steady state the molten zone was long enough to provide homogeneity in the liquid phase. Upon cooling, aligned dendrites grew in the direction of the heat flow, i.e., parallel to the length of the rod.

In the modified equipment the Cu-Nb rod was 9.5 mm in diameter in a 12.7-mm graphite tube. To obtain long lengths, a number of rods had to be joined end to end. An appropriate geometry for the graphite tube ends, as illustrated in Fig. 1a, was used. Therefore, the Cu-Nb was continuously fed. A macrograph of a Cu-Nb rod junction with microhardness indentations through the interface is shown in Fig. 1b. The Nb dendrites form and the size uniformity of the microhardness indentations are sharp indicators of the cleanliness of the joint.

Characteristics of materials studied are shown in Table 1. Three different growth rates, 0.015 cm/s, 0.07 cm/s, and 0.15 cm/s, were used to investigate the finer range of the dendrite size. Figure 2 shows the microstructure of the Cu-Nb rod solidified at 0.07 cm/s. By maintaining unidirectional heat flow, dendrites of the first order are forced to grow as an aligned array in the heat

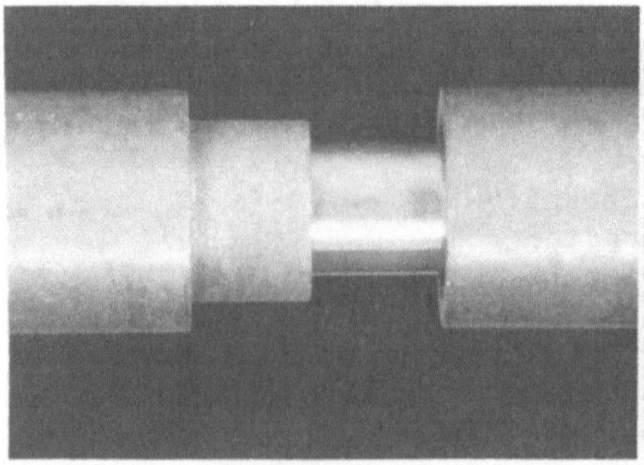

Fig. 1a. Two graphite tubes to be joined over a Cu-Nb rod.

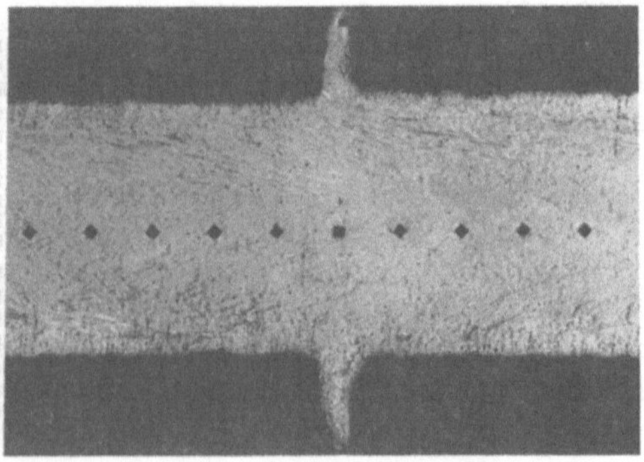

Fig. 1b. Macrograph of the Cu-Nb rod joint after solidification.

Table 1. Characteristics of Directionally Solidified In Situ Materials*

Identification	Solidification Rate	Dendrite Branching	Dendrite Size
DS 130581	0.015 cm/s	3rd order	20–40 μm
DS 290481	0.07 cm/s	2nd order	5–10 μm
DS 270581	0.15 cm/s	1st order	1–3 μm

*Nominal composition: Cu–30 wt.% Nb; dimension: ϕ 9.5–mm rod.

flow direction (Fig. 2a). The dendrites of the second order are branched perpendicularly to the first-order dendrites (Fig. 2b), i.e., perpendicular to the heat flow direction.

Decreasing the growth rate increases the dendrite size and multiplies the dendrite branching. The microstructure of a Cu-Nb rod solidified at 0.015 cm/s shows coarser dendrites and third-order branches (Fig. 3). On the other hand, increasing the growth rate reduces side branching and produces fiber-like dendrites, as shown in Fig. 4. Rough estimates of the dendrite size for each growth rate are 20–40 μm, 5–10 μm, and 1–3 μm, respectively.

Specimens of each rod were drawn without intermediate annealing to the final wire size of 0.25 mm, with an area reduction of about 1400. During the processes of swaging and drawing, the Nb

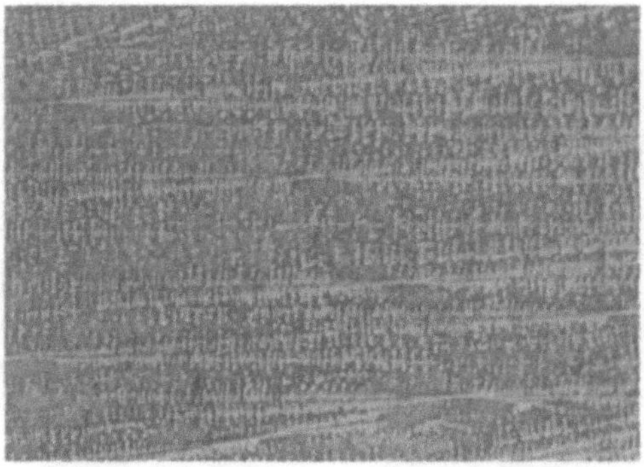

Fig. 2a. Aligned dendritic microstructure of a Cu-Nb alloy solidified at 0.07 cm/s.

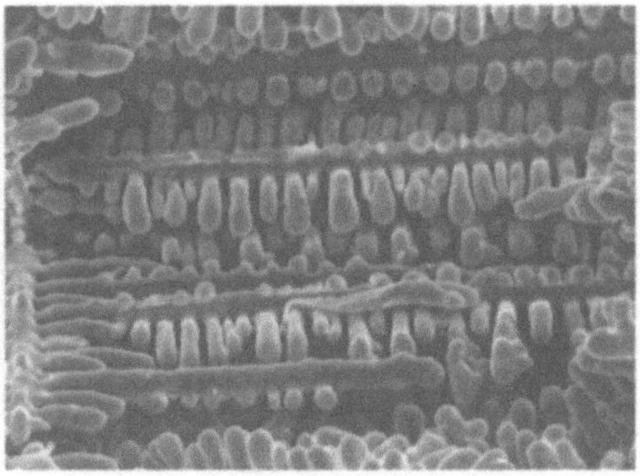

Fig. 2b. Enlarged microstructure showing second-order dendrites.

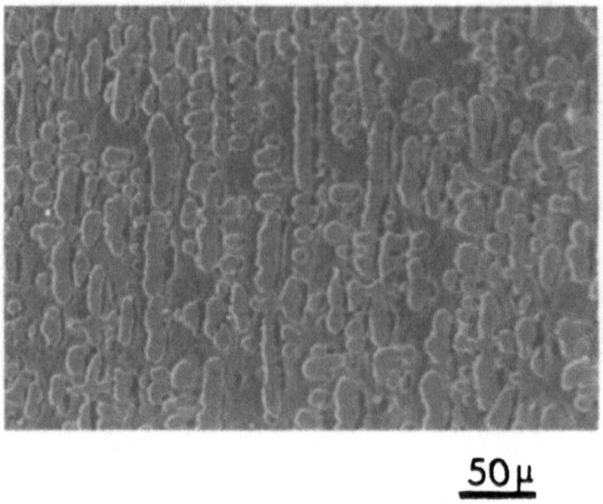

Fig. 3. Micrograph of a Cu-Nb alloy solidified at 0.015 cm/s; third-order dendrites visible.

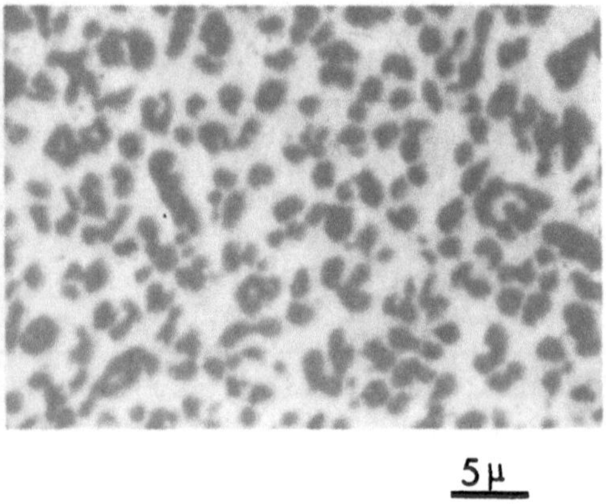

Fig. 4. First-order dendrites in a Cu-Nb alloy solidified at 0.15 cm/s.

dendrite branches were reshaped and aligned in the direction of the mechanical deformation. Figures 5a and 5b show TEM micrographs of the extracted Nb filaments. As previously explained[7] and observed,[8-9] the dendrites become ribbon shaped during severe mechanical deformation. Figure 5b shows a twisted ribbon at high magnification revealing the high ratio of width-to-thickness dimensions.

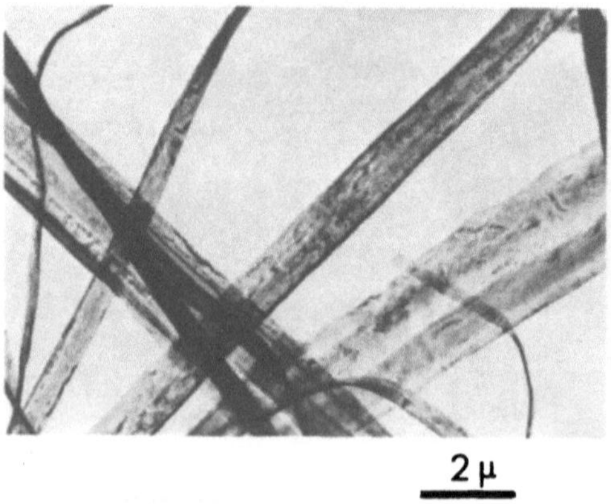

Fig. 5a. TEM of extracted Nb filaments.

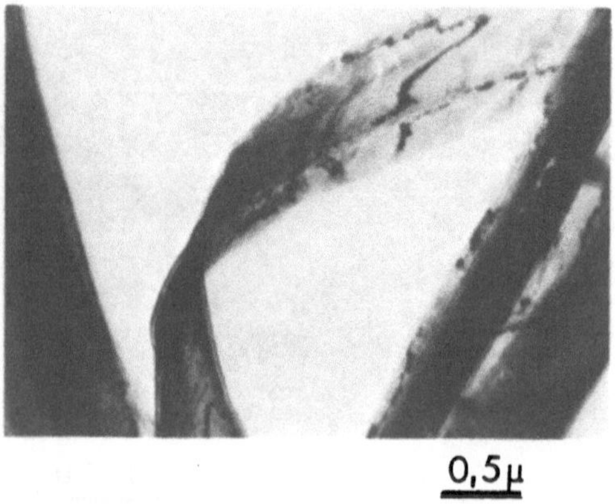

Fig. 5b. TEM of twisted Nb filaments showing high ratio of width to thickness.

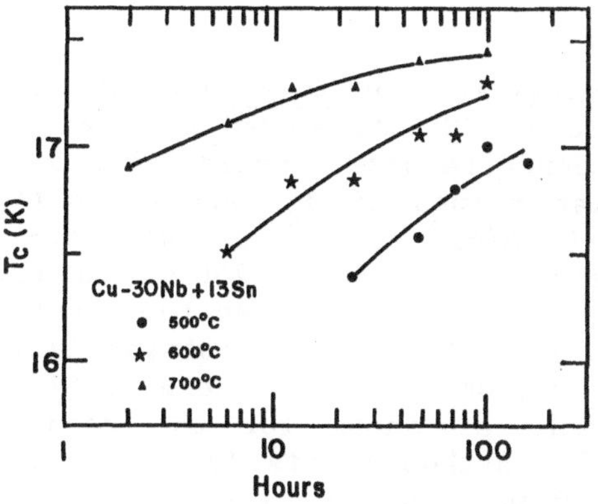

Fig. 6. Critical temperature as a function of reaction time at different temperatures.

The wires were then Sn coated, heated to produce diffusion of the Sn, and reacted with different treatment schedules (time and temperature).

Superconducting Properties

The superconducting transition temperature, T_c, was measured resistively using the standard four-probe technique. The T_c was defined as the value at the midpoint of the resistive transition.

The critical current was measured in a transverse magnetic field up to 18.5 T and was defined as the current at which the voltage exceeded 1 μV/cm. Overall critical current density, J_c, was normalized to the total cross-sectional area of the wires.

RESULTS AND DISCUSSION

Critical Temperature

The measured superconducting critical temperatures, T_c, are presented in Fig. 6 as a function of the treatment schedules. These specimens were electroplated with 13 wt.% Sn calculated with respect to the Cu only. Three reaction temperatures, 500°, 600°, and 700° were investigated. The T_c increases with increasing heat treatment temperature and seems to approach a maximum after 48 h at 700°C. In a previous paper,[10] we reported the dependence of T_c on Sn content; for materials reacted at 700°C for 1 day, T_c increased with increasing Sn level and approached a maximum at 13-15 wt.% Sn. Systematic studies[11-12] of the dependence of T_c on heat-treatment schedules have been made on multifilamentary bronze process composites, and results similar to ours were reported. The composition of the transformed Nb_3Sn tends to be more stoichiometric for a high temperature reaction with increasing time. Generally, between 500 and 700°C the optimal T_c is obtained at the highest temperature.

Critical Currents

Examples of critical current densities for in situ Cu-30 wt.% Nb-16 wt.% Sn following 3 days at 650°C are shown in Fig. 7. Directionally solidified rods with dendrite sizes ranging from 5 to 10 μm and 20 to 40 μm yield similar values of J_c versus applied field, B_0. An overall J_c of about 8 kA/cm^2 at 14 T was also typically observed for a heat treatment for 1 day at 750°C and for 1 week at 550°C. In Fig. 7 the narrow band encompasses the results of measurements on these dendrite size ranges. A substantially higher J_c, 20 to 30 kA/cm^2 at 14 T, was obtained for the directionally solidified 1 to 3 μm material, as shown for the results defining the narrow band in Fig. 7. Although this is one of the highest reported values of J_c for 30 wt.% Nb, it is not yet clear whether further improvement can be achieved by careful microstructural control.

Further results (measurement of the mechanical properties and evaluation of small coils) will be reported later.

CONCLUSIONS

Using a high temperature gradient solidification (HTGS) technique the production of long lengths was demonstrated. This technique offers several advantages, in particular the control of the microstructure, form, and size of the Nb dendrites.

Critical current densities of 10 kA/cm^2 overall at ~16.5 T have been measured for the fine dendrite materials.

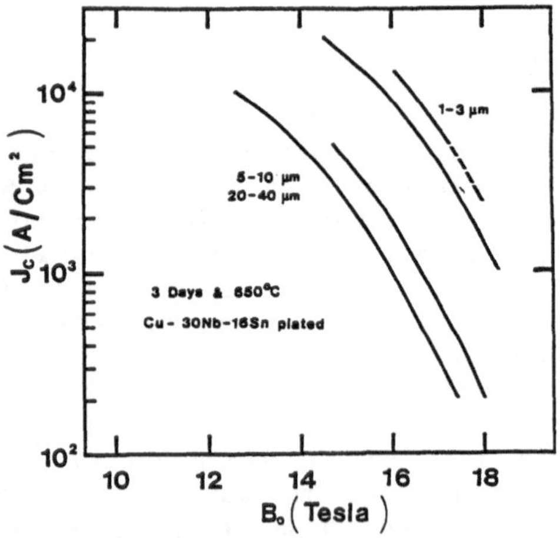

Fig. 7. Overall critical current densities of in situ Cu-30 wt.% Nb-16 wt.% Sn (3 days at 650°C) as a function of field.

ACKNOWLEDGMENTS

We wish to acknowledge the valuable assistance of R. Dubuc with the directional solidification technique and of C. L. H. Thieme, H. Zhang, S. Pourrahimi, J. Otubo, and E. J. McNiff, Jr. with the superconducting properties measurements.

REFERENCES

1. J. D. Verhoeven, F. A. Schmidt, E. D. Gibson, J. E. Ostenson, and D. K. Finnemore, Appl. Phys. Lett. 35:555 (1979).
2. J. D. Verhoeven, E. D. Gibson, F. A. Schmidt, and D. K. Finnemore, J. Mater. Sci. 15:1449 (1980).
3. K. Togano, H. Kumakura, and K. Tachikawa, IEEE Trans. Magn. MAG-17:985 (1981).
4. J. L. Fihey, M. Neff, R. Roberge, M. C. Flemings, S. Foner, and B. B. Schwartz, High-temperature-gradient casting of in situ multifilamentary superconductors, in: "Advances in Cryogenic Engineering," Vol. 26, Plenum Press, New York (1980), p. 343.
5. J. L. Fihey, M. Neff, R. Roberge, M. C. Flemings, S. Foner, and B. B. Schwartz, Appl. Phys. Lett. 35:715 (1979).
6. C. R. Spencer, E. Adam, E. Gregory, W. Marancik, and C. Z. Rosen, IEEE Trans. Magn. MAG-17:257 (1981).
7. W. F. Hosford, Jr., Trans. AIME 230:12 (1964).
8. J. P. Harbison and J. Bevk, J. Appl. Phys. 48:5180 (1977).
9. J. D. Verhoeven, J. J. Sue, D. K. Finnemore, E. D. Gibson, and J. E. Ostenson, J. Mater. Sci. 15:1907 (1980).
10. H. LeHuy, R. Roberge, J. L. Fihey, G. Rupp, and S. Foner, IEEE Trans. Magn. MAG-17:261 (1981).
11. M. Suenaga and W. B. Sampson, Appl. Phys. Lett. 20:443 (1972).
12. I. L. McDougall, LT-15, J. Phys. (Paris), Colloque C 6 39:402 (1978).

EFFECT OF TWIST PITCH ON SHORT-SAMPLE V-I CHARACTERISTICS OF MULTIFILAMENTARY SUPERCONDUCTORS*

L. F. Goodrich, J. W. Ekin, and F. R. Fickett

Electromagnetic Technology Division, National Bureau of Standards, Boulder, Colorado

INTRODUCTION

Precise determination of the critical current of practical superconductors requires measurement of the voltage-current (V-I) characteristic of the conductor at various magnetic fields. The measurement usually requires the detection of quite small voltages since very sensitive critical current criteria are necessary for the design of practical devices. Furthermore, most laboratories have only relatively small-bore solenoidal magnets, leading to the common use of very short sample lengths for routine critical current measurements. This situation may lead to some difficulties, as we show here.

Data taken on short samples of commercial multifilamentary superconductors have uncovered anomalous V-I characteristics. A voltage was detected at currents well below the sharp upturn in the V-I characteristic near I_c. It was apparently due to current transfer, but larger in magnitude than would be expected from previous current-transfer analyses.[1] Further data indicated that the voltage was strongly dependent on the voltage tap location. In fact, the voltage measured below I_c in the current direction between some taps was negative. In all cases, as I_c was approached, the V-I characteristic returned to "normal." Extensive experiments have shown that there are two extreme anomalous shapes of the V-I curves. These are illustrated in Fig. 1. Depending on the test geometry, the magnitude of many of these anomalous voltages can be on the order of commonly used critical-current criteria and may significantly affect the determination of I_c.

*Partially funded by the Department of Energy.

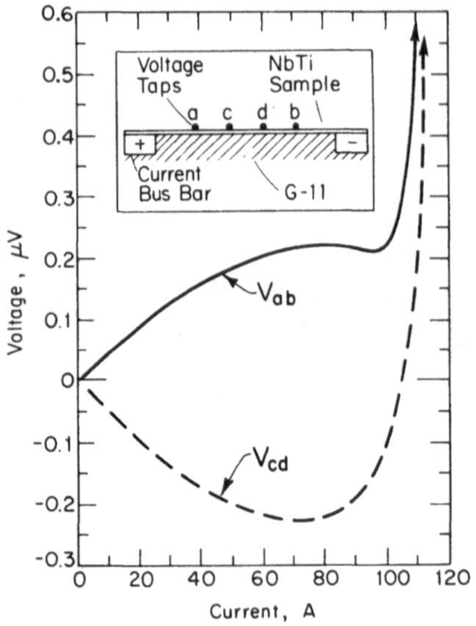

Fig. 1. Experimental data at 8 T showing anomalous V-I characteristic of short NbTi sample ($\ell_{ab} \simeq 1.5$ cm, 2.5 filament twists between current contact centers). The horizontal dimensions of the inset are approximately to scale.

In this paper, the experimental investigation of the anomalous behavior and a phenomenological model developed to account for the observations are presented. Several techniques are discussed that minimize the effect and, thus, allow precise critical current determination in short samples.

APPARATUS AND SAMPLE PREPARATION

The apparatus used in this experiment was typical for a short, straight sample critical current measurement. It consisted of the following: a Dewar with a 9-T superconducting solenoid (3.8-cm bore), a cryostat with 600-A vapor-cooled leads, a series-regulated 600-A battery current supply, an analog nanovoltmeter, and an X-Y recorder. The magnetic field measurements were made to a precision of 0.1% with a calibration accuracy of 0.2%. The voltage and current measurements had an accuracy of 2% and 0.4% and a precision of 1% and 0.2%, respectively. Typical noise voltages were ±5 nV. Thermal voltages were checked at zero current and usually did not vary by more than ±10 nV. The sample holder was made of NEMA G-11 epoxy-fiberglass. Superconducting bus bar current contacts were set flush with one surface of the G-11 (see the inset in Fig. 1) such that the Lorentz force on the sample could be supported by the G-11, either directly or by a thin layer of varnish. Small slots were routed into the G-11 for voltage taps on the underside of the wire where needed.

The measurements were made principally on two samples: a twisted multifilamentary NbTi (twist pitch 1.27 cm, Cu:NbTi of 1.8:1, RRR of the copper ~70, 0.53 x 0.68 mm, 180 filaments) and an untwisted multifilamentary Nb_3Sn (0.70-mm diameter, 2869 filaments). The Nb_3Sn wire had an outer copper jacket separated by a tantalum diffusion barrier from the core of bronze, niobium, and Nb_3Sn.

Sample preparation was typical for short sample testing except for the two following techniques. A technique was developed to spot-solder a pair of voltage taps directly across the wire from each other to allow measurement of transverse voltages. The alignment of these taps was checked by measuring the room temperature resistivity and the voltage polarity of the pair and thus deducing the approximate misalignment. The worst case misalignment was ~0.2 mm, but more usually $\lesssim$ 0.1 mm. The other special technique was selective etching of the copper jacket from the Nb_3Sn wire. This was accomplished using an enamel insulating paint as a mask and a nitric acid etch. Small copper islands were left on the sample for ease in soldering voltage taps and current contacts.

EXPERIMENT

To investigate the anomalous voltage seen on the NbTi critical current sample as described in the Introduction, tests were made with two pairs of voltage taps spaced 0.5 and 1.5 cm apart. The V-I characteristics are shown in Fig. 1. Here the voltage definition $V_{ab} = V_a - V_b$ was used. These curves were reproducible and reversible to within 1%.

It was observed that both V_{ab} and V_{cd} changed sign when the current was reversed, but not when the field direction was reversed. There were slight differences (~10%) in the magnitudes of these voltages, especially close to I_c, as the direction of the current and the field were changed. These are attributed to the Hall effect and are discussed below, but they do not significantly affect the unusual shape of the V-I characteristics.

During the development of the phenomenological model, experiments were made using several unique sample configurations of both the twisted NbTi and the untwisted Nb_3Sn. These data and the voltage tap and current lead arrangements are presented in the appropriate places in our discussion of the model.

PHENOMENOLOGICAL MODEL AND SUPPORTING DATA

The unusually shaped V-I characteristics may be understood in terms of the interaction between current transfer and the twist pitch of the superconductor. Filaments nearest the current contacts carry current near their critical current density and exhibit a flux-flow resistivity. Conversely, filaments on the opposite side of the superconductor from the current contacts carry very little current because the resistive matrix separates them from the point of current injection. Therefore, it is possible for a voltage tap to be sampling either a resistive or nonresistive group of filaments and, thus, the voltage between taps may be quite different, depending on the relationship of tap spacing to twist pitch. Also, significant transverse voltages should be observed across the wire.

Results of transverse voltage measurements made on the NbTi sample are shown in Fig. 2. Notice that these curves are similar in shape and size to those in Fig. 1 except near I_c. As I_c is approached, the transverse voltage tends to go to zero as the current distribution among the filaments becomes more uniform. The distance between current injection points (approximately center to center) for the data shown in Fig. 2 was 2.6 times the twist pitch. Thus, the group of filaments nearest to the current bus bar on one end of the sample are not the same as the group nearest to the bus bar at the other end. Current must therefore transfer between the two groups of filaments by flowing through the resistive matrix material of the wire. This generates the large transverse voltages. The unusual V-I characteristics shown in Fig. 1 are simply the sum of the usual flux-flow V-I characteristic and

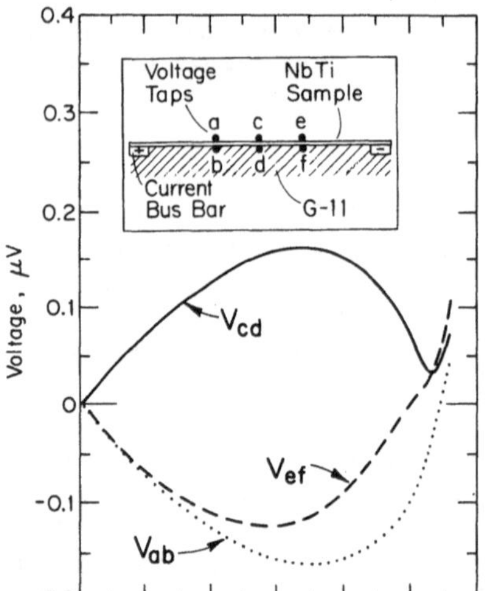

Fig. 2. Experimental data at 8 T on transverse voltages of short NbTi sample ($\ell_{ae} \simeq 1.25$ cm, 2.6 filament twists between current contact centers).

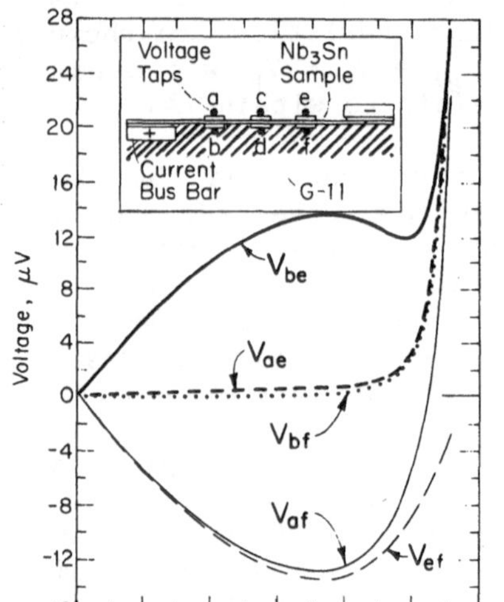

Fig. 3. Experimental data at 9 T on a selectively etched Nb_3Sn conductor ($\ell_{ae} \simeq 1.25$ cm, no filament twists). Note the bus bar locations.

the transverse voltages caused by current transfer through the resistive matrix from one group of filaments to another.

Understanding of the effect is simplified considerably if the conductor used to obtain the V-I characteristics in Figs. 1 and 2 is untwisted. In this arrangement one current bus bar was placed on top, the other on the bottom to provide a half-integral number of twists between the current injection points. Data were obtained on the untwisted Nb_3Sn conductor in this geometry and are shown in Fig. 3. These data can be explained by the model shown in Fig. 4. Note that in multifilamentary conductors the equipotential lines at low currents are much more closely aligned with the conductor axis than in a conductor with an isotropic resistivity. The voltage between taps e and f starts from zero at $I \ll I_c$ (Fig. 4A), rises in magnitude as I increases (Fig. 4B), and decreases toward zero as all the filaments become resistive near I_c (Fig. 4C). Similarly, the voltage between taps a and f starts from zero at $I \ll I_c$ (Fig. 4A), rises in magnitude to a negative peak at an intermediate value of I (Fig. 4B), and then becomes positive as the entire conductor becomes resistive at $I \cong I_c$ (Fig. 4C). The voltage between taps b and e rises from a low value at $I \ll I_c$ to an intermediate high at an intermediate value of I, back to a low value as the

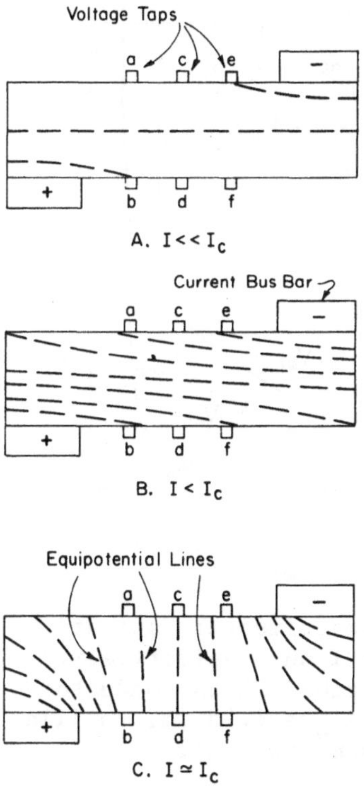

Fig. 4. Model of equipotential lines in a superconductor with an odd half-integral number of filament twists between the current contacts: (A) current much less than I_c; (B) intermediate currents less than I_c; and (C) current near I_c.

resistivity becomes more isotropic at $I \approx I_c$, and finally to a high value as I exceeds I_c. Note that the curves in Fig. 3 have about the same shape as those in Fig. 1.

In Fig. 3 it is easy to separate the anomalous V-I characteristics (V_{be}, V_{af}) into an essentially intrinsic characteristic (V_{bf}, V_{ae}) and an anomalous current-transfer characteristic (transverse voltages V_{ab}, V_{ef}). These separations, $V_{be} = V_{bf} - V_{ef}$ and $V_{af} = V_{ae} + V_{ef}$, were demonstrated experimentally with agreement of about 1% ±10 nV. So it is possible to have V-I characteristics with these shapes (and everything in between) depending on voltage tap location. Remember that this discussion and the data shown in Figs. 1, 2, and 3 correspond to a sample with a half-integral number of twist lengths between the points of current injection.

When there is an integral number of twist lengths between the points of current injection, the current transfer pattern is altered. Data obtained on the NbTi conductor with a current contact spacing of about two twist lengths are shown in Fig. 5. In this geometry, the V-I characteristic of the resistive filament,

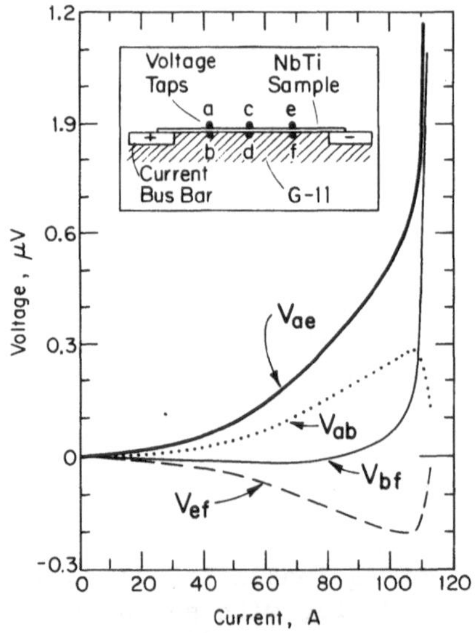

Fig. 5. Experimental data at 8 T on a NbTi conductor ($\ell_{ae} \approx 1.25$ cm, 2 filament twists between current contact centers).

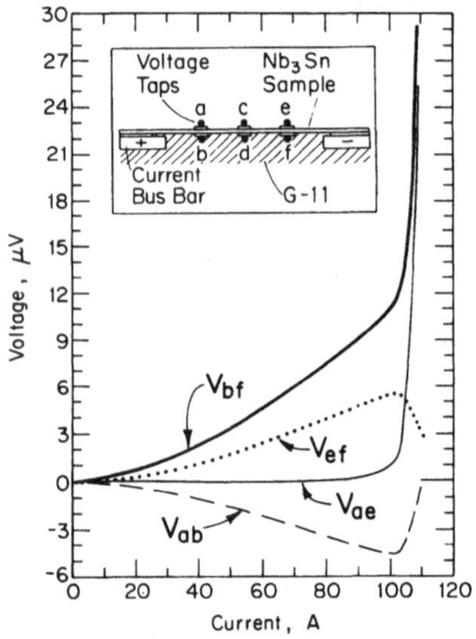

Fig. 6. Experimental data at 9 T on a selectively etched Nb_3Sn conductor ($\ell_{ae} \approx 1.25$ cm, no filament twists).

V_{ae}, is the sum of the V-I characteristics of nonresistive filaments, V_{bf} (essentially intrinsic characteristic), and the transverse voltages, V_{ab}, V_{ef} (anomalous current-transfer characteristic). The equation $V_{ae} = V_{bf} + V_{ab} - V_{ef}$ was demonstrated experimentally with an agreement of about 1.5% ±15 nV. Data obtained on the Nb_3Sn conductor in this geometry are shown in Fig. 6. Note that these curves are very similar in shape to the curves in Fig. 5 and can also be separated into an intrinsic V-I characteristic and an anomalous current-transfer characteristic, $V_{bf} = V_{ae} + V_{ef} - V_{ab}$, with experimental agreement of about 1% ±10 nV. The model for this configuration is shown in Fig. 7. The current is injected and extracted from the same group of filaments. The voltage between taps b and f is indicative of this group of filaments and rises much more rapidly than the voltage between taps a and e. Current does not transfer through the matrix to the far group of filaments sampled by taps a and e until I approaches I_c. In fact, the voltage taps on the top of the conductor remain at about the same potential until I reaches I_c.

TECHNIQUES TO MINIMIZE THE ANOMALOUS CURRENT TRANSFER VOLTAGE

Note from the above discussion that these unusually shaped V-I characteristics result from nonsymmetric current injection. If the

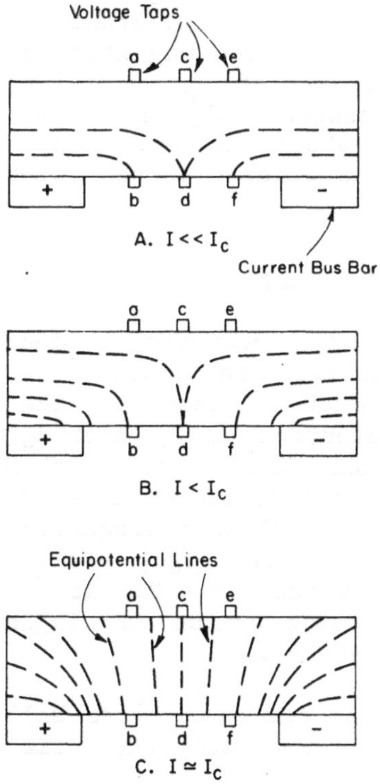

Fig. 7. Model of equipotential lines in a superconductor with an integral number of filament twists between current contacts: (A) current much less than I_c; (B) intermediate current less than I_c; and (C) current near I_c.

current is uniformly introduced around the circumference of the superconductor, the current transfer voltage should be independent of twist pitch considerations and reduced in magnitude.

There are at least two possible methods of making the current injection symmetric in short sample testing. One method is to make the solder joint at the current contact longer, at least one twist length, so that the current is introduced into all of the outer filaments. In the second method, current injection is made more uniform by having a symmetric current contact. Both of these techniques have been shown experimentally to reduce the anomalous voltages greatly, thus permitting a more accurate determination of the critical current.

The long current contact method was tested on the twisted NbTi wire. First the voltages were measured with the current contacts covering more than one twist length, then the wire was cut to shorten the current contact, and the same voltage taps were measured again. The current-transfer voltages were about a factor of 10 lower for the long current contacts and of a magnitude consistent with symmetric current-transfer analysis.[1]

Measurements on the Nb_3Sn wire with the copper jacket intact illustrate the second method. The copper jacket gives a more uniform injection of the current into the superconducting filament region because of the relatively high resistivity of the bronze in that region. The results are shown in Fig. 8. The magnitude of the current transfer voltage is reduced over that of V_{bf} in Fig. 6, again consistent with symmetric current-transfer analysis. Also, the voltage drop along the top filaments is about the same as that along the bottom filaments, which indicates a removal of the twist pitch dependence.

A COMMENT ON GALVANOMAGNETIC EFFECTS

Observations of strange voltages in a current-carrying conductor at low temperatures in a high magnetic field are often explained by arguments involving one or more of the many classical galvanomagnetic effects. It is our contention that, although several of these effects are present in our data, none of them cause the unusual shape of the V-I curves. Most of the effects are seen only in metals where the product of the cyclotron frequency, ω, and the electron collision time, τ, is quite large, usually ~100. Even in the copper stabilizer on our wires, $\omega\tau \lesssim 2.4$ at 4 K and 10 T. Furthermore, large effects are most common to single crystals and our metals are highly polycrystalline.

Two galvanomagnetic effects do appear in our data: the Hall effect and transverse magnetoresistance. Of these, only the former

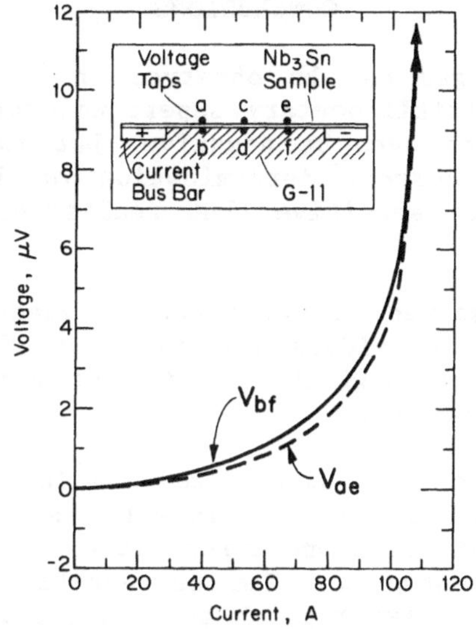

Fig. 8. Experimental data at 9 T on Nb_3Sn conductor with copper jacket giving uniform current injection ($\ell_{ae} \simeq 1.25$ cm, no filament twists).

can be seen directly in I-V curves made at fixed field. The Hall voltages appear most strongly on the probe pairs transverse to the current. They reverse with both field and current and are developed almost entirely in the normal metal components of the wire. In the wires measured here, at currents well below I_c, we noted Hall voltages of ~10 nV in the NbTi at 8 T (on top of a ~200 nV current transfer voltage) and ~100 nV in the Nb_3Sn at 9 T (on top of a ~15 μV transfer voltage). These values appear to be consistent with the known Hall coefficients for the matrix materials[2] and the other parameters, but even an order of magnitude calculation requires a much more detailed model than we have space for. Suffice it to say that the Hall effect, although observable, represents only a small contribution to the measured transverse voltages.

Magnetoresistance is an even effect, it does not depend on current and does not reverse with field, thus it does not affect the shape of the V-I curves. It shows up most strongly in measurements on voltage taps along the sample as an increase in the resistance of the relatively pure copper stabilizer as the field is increased. For our wires this effect causes the zero field resistivity of the stabilizer at 4 K to increase by about a factor of 3 in going to 9 T.

CONCLUSIONS

Anomalous voltages may be observed in short sample critical current tests on multifilamentary superconductors. These voltages may in some instances even be negative, but in any case they can interfere with the correct determination of I_c (even for NbTi) especially when using sensitive (but realistic) electric field or resistivity criteria.

All of the observed behavior can be adequately explained by use of a model that considers the combined effects of the twist pitch of the filaments and the details of current injection and transfer within the sample.

The anomalous voltages can be reduced in short sample testing by insuring that the current is injected symmetrically into the conductor by long (at least one twist pitch) current contacts or by symmetric current contacts. The current-transfer voltages then will be independent of twist pitch and reduced in magnitude, making the usual methods of treating current-transfer voltages applicable.[1]

ACKNOWLEDGMENTS

The authors have benefited greatly from numerous discussions with A. F. Clark. E. Pittman has been most helpful in the care and feeding of the instrumentation. Ms. V. Grove prepared the manuscript--many times. Our thanks to them all.

REFERENCES

1. J. W. Ekin, J. Appl. Phys. 49:3406 (1978).
2. C. M. Hurd, "The Hall Effect in Metals and Alloys," Plenum Press, New York (1972).

END EFFECTS ON THE LOSS FOR SHORT SUPERCONDUCTORS*

W. J. Carr, Jr.

Westinghouse Electric Corporation, Research and Development Center, Pittsburgh, Pennsylvania

INTRODUCTION

Although superconducting wire is generally used in long lengths, loss measurements are sometimes made on short sections and therefore an understanding of end effects is important. Even more important is the information that can be obtained about the superconductor from such measurements. End effects result from the necessity for induced circulating current to cross the conductor near the ends, and therefore they yield information on the anisotropy of the superconducting material. For an untwisted filamentary material it was recently shown[1] that domains of current density form near the ends when the middle of the conductor is fully penetrated by a magnetic field. At the end of the conductor the axial electric field must vanish since no current flows out of the end, but this component of field does not change gradually from the center to the end. It goes to zero at a domain boundary near the end, and this boundary separates regions in which the axial electric field is essentially zero on one side of the boundary and nonzero on the other side. In the calculation cited, the current domains near the end of a cylinder were described, and their effect on the magnetic moment was computed. A strip made up of untwisted filaments is considered here, and the effect on the loss is examined. In addition, a precise "critical length" for the conductor is calculated. Although the calculations apply for a continuous filamentary material, a similar domain structure would be expected for other types of superconductor.

*Supported in part by the Air Force Office of Scientific Research Contract F49620-78-C-0031.

CALCULATION OF THE CURRENT DOMAINS

Consider a strip of multifilamentary material in which the filaments run parallel with the length of the strip along the z axis. The applied magnetic field is transverse to the length and in the plane of the strip along the x axis. The rate of change of field, $\dot{H}_A$, is constant, and the field is assumed to have fully penetrated the strip. A Bean model is used for the current density, and near the middle of the strip, current flows down the length of the strip in the lower half of the thickness and back in the other direction in the upper half.

The electric field from Maxwell's equations is

$$\underset{\sim}{E} = -\dot{H}_A \, y \, \hat{a}_z - \nabla\phi \tag{1}$$

where ϕ is some potential, $\hat{a}_z$ is a unit vector along the length, and y measures distance through the thickness, as in Fig. 1. The current density is

$$\underset{\sim}{j} = j_z \, \hat{a}_z + \sigma_\perp \underset{\sim}{E}_\perp \tag{2}$$

where $\sigma_\perp$ is the transverse conductivity, and for vanishing divergence

$$\frac{\partial j_z}{\partial z} = -\sigma_\perp \operatorname{div} \underset{\sim}{E}_\perp . \tag{3}$$

For $E_z > 0$, $j_z = \lambda j_c$ and for $E_z = 0$, j_z is between $\pm \lambda j_c$, where λ is the volume fraction of superconducting material. The problem is to find a solution to these equations, and it turns out that the solution is given by the formation of domains as shown in Fig. 1. In the regions labeled A

$$\underset{\sim}{E} = -\dot{H}_A \, y \, \hat{a}_z \tag{4}$$

$$\underset{\sim}{j} = -\lambda j_c (\operatorname{sgn} \dot{H}_A \, y) \, \hat{a}_z \tag{5}$$

where sgn x indicates the sign of x. In the regions B

$$\underset{\sim}{E} = -\dot{H}_A \left[y - \frac{dz}{z_o} \left(1 - \frac{z}{2z_o}\right) (\operatorname{sgn} y) \right] \hat{a}_z \tag{6}$$

$$\underset{\sim}{j} = -\lambda j_c (\operatorname{sgn} \dot{H}_A \, y) \, \hat{a}_z \tag{7}$$

while in C

$$\underset{\sim}{E} = \dot{H}_A \left[z - z_o \left(1 - \sqrt{1 - 2\,|y|/d} \right) \right] \hat{a}_y \tag{8}$$

$$\underset{\sim}{j} = \frac{-\lambda j_c (1 - z/z_o)\,(\mathrm{sgn}\,\dot{H}_A\, y)\,\hat{a}_z}{\sqrt{1 - 2\,|y|/d}} + \sigma_\perp \underset{\sim}{E} \tag{9}$$

The length of the end effect, z_o, is given by

$$z_o = \sqrt{\frac{\lambda j_c d}{\sigma_\perp |\dot{H}_A|}} \tag{10}$$

and the equation for the boundary between B and C is

$$|y| = \frac{dz}{z_o}\left(1 - \frac{z}{2z_o}\right). \tag{11}$$

It is readily shown that these solutions satisfy (1) and (3) and that $\underset{\sim}{E}$ and $\underset{\sim}{j}$ are continuous at the boundaries between A, B, and C.

CALCULATION OF THE POWER

In terms of macroscopic vectors, the loss in a multifilamentary conductor over a closed cycle is

$$\oint P\,dt = \oint dt \int \underset{\sim}{E} \cdot \underset{\sim}{j}\,dV + \int dV \oint \underset{\sim}{H} \cdot d\underset{\sim}{M} \tag{12}$$

where P is the power, t is time, V is volume, H is the magnetic field, and $\underset{\sim}{M}$ is the magnetization due to shielding currents flowing in the filaments. In the current-saturated regions, the filaments have no magnetization, and therefore, in Fig. 1 M is nonvanishing only in the region C. The power loss $\underset{\sim}{E} \cdot \underset{\sim}{j}$ that exists in A and B occurs in the filaments, while in C it occurs in the matrix. For simplicity, only the instantaneous loss for the conditions of small filament size (small M) and large $\dot{H}_A$ will be considered here. In this case

$$\frac{P}{V} \approx \frac{1}{V}\int \underset{\sim}{E} \cdot \underset{\sim}{j}\,dV. \tag{13}$$

In region A the power per unit total volume is

$$\frac{P_A}{V} = \frac{\lambda j_c d\,|\dot{H}_A|}{4}\left(1 - \frac{2z_o}{\ell}\right) \tag{14}$$

since $\ell - 2z_o$ is the length of this zone, where ℓ is the length of the conductor. In regions B, counting both ends,

$$\frac{P_B}{V} = \frac{\lambda j_c\, d\, |\dot{H}_A|}{10} \frac{z_o}{\ell} \tag{15}$$

while for C

$$\frac{P_C}{V} = \frac{4\sigma_\perp}{15} \dot{H}_A{}^2 \frac{z_o{}^3}{\ell}$$

$$= \frac{4}{15} \lambda j_c\, d\, |\dot{H}_A| \frac{z_o}{\ell}. \tag{16}$$

The ratio of the total power per unit volume to that for an infinite strip is

$$\frac{P}{P_\infty} = 1 - \frac{8}{15} \frac{z_o}{\ell}, \tag{17}$$

which applies for $\ell \geq 2z_o$. The power per unit volume, P_∞/V, is $\lambda j_c\, d\, |\dot{H}_A|/4$, which, except for the factor λ, is the same as the power dissipated in a solid superconductor of infinite length. Measurements of the power can be used to obtain z_o.

CRITICAL LENGTH AND $\dot{H}_A$

The field required for full penetration for a strip of infinite length is

$$H_p = 2\pi\, \lambda j_c\, d, \tag{18}$$

and in terms of this field

$$z_o{}^2 = \frac{H_p}{2\pi\, \sigma_\perp\, |\dot{H}_A|} \tag{19}$$

where the units are in e.m.u. For a given $|\dot{H}_A|$ it is reasonable to define a critical length by

$$\ell_c = 2z_o. \tag{20}$$

For lengths shorter than ℓ_c, the middle of the strip is not fully penetrated, and the calculations here do not apply. It follows from (19) that

$$\ell_c = \sqrt{\frac{2H_p}{\pi\,\sigma_\perp\,|\dot{H}_A|}}. \tag{21}$$

For lengths longer than ℓ_c, where the picture of Figure 1 applies, it is observed, in agreement with Wilson et al.,[2] that normal current flows across the matrix only within a length $\ell_c/2$ at each end. This observation formed the original argument for twisting a multifilamentary conductor, since in a qualitative sense a twisted conductor behaves like a finite length, in a transverse field. However, the domain picture, here, gives a precise ℓ_c, and it is, of course, related to H_p for the conductor rather than the filaments.[3]

Conversely, for a fixed ℓ, a critical $\dot{H}_A$ for the current to fully penetrate the conductor at the middle under steady-state conditions may be defined by

$$\dot{H}_c = \frac{2H_p}{\pi\,\sigma_\perp\,\ell^2}. \tag{22}$$

In a twisted multifilamentary wire, a relaxation time τ exists that is given by $\sigma_\perp L^2/2\pi$, where L is the twist length. In analogy, one can refer to $\pi\,\sigma_\perp\,\ell^2/2$ as a relaxation time and write (22) as

$$\dot{H}_c = \frac{H_p}{\tau}. \tag{23}$$

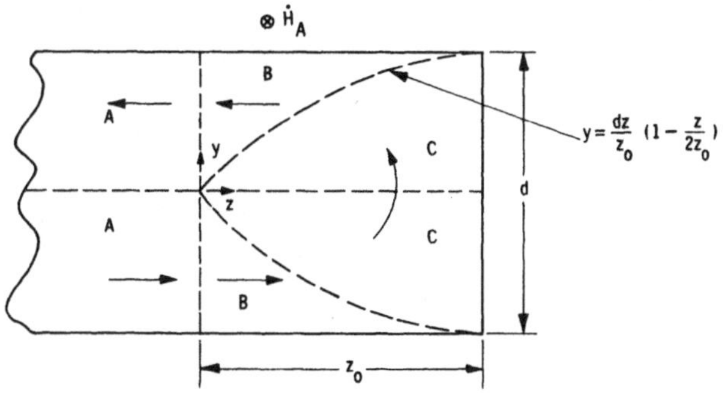

Fig. 1. Pattern of current flow near the end of a fully penetrated multifilamentary strip. $\dot{H}_A$ is positive and in the x direction. The filaments run along the z axis.

REFERENCES

1. W. J. Carr, Jr., to appear with Proceedings of Intermag. Conference in IEEE Trans. Magn.
2. M. N. Wilson, C. R. Walters, J. D. Lewin, P. F. Smith, and A. H. Spurway, J. Phys. D: Appl. Phys. 3:1517 (1970).
3. J. H. Murphy, M. S. Walker, and W. J. Carr, Jr., IEEE Trans. Magn. MAG-10:868 (1974).

TWO-DIMENSIONAL ANALYSIS OF AC LOSS IN SUPERCONDUCTORS CARRYING TRANSPORT CURRENT*

J. V. Minervini †

Cryogenic Engineering Laboratory
Massachusetts Institute of Technology, Cambridge, Massachusetts

INTRODUCTION

Previous treatments of hysteresis loss in cylindrical type II superconductors have been based primarily on slab models owing to the difficulty in solving the two-dimensional field problem for the flux and current distribution in the conductor cross section. Solutions have recently been found from numerical techniques for the two-dimensional transverse magnetic field penetration,[1,2,3] although these solutions do not include the effect of the transport current. To date, this effect has only been included by approximations and empirical interpolations from the slab model.[4] Since most cases of practical interest include filaments carrying transport current, greater accuracy in the loss calculation can only come from a two-dimensional solution that takes account of the transport current.

In this paper two extensions of the loss model are presented. First, the results are given from a mathematical computation of the two-dimensional current distribution in a superconducting filament exposed to a time-varying transverse magnetic field. The mathematical technique is based upon a method previously used by Beth[5] and Morgan[6] for calculating two-dimensional fields due to arbitrarily shaped regions of constant current density. The solution utilizes functions of complex variables and the Cauchy integral formula.

*Work supported by U.S. DoE under Contract No. EX-76-A-01-2295 Task Order No. 11.

†Present address: National Bureau of Standards, Boulder, Colorado.

Kato et al.[7] have used this technique to compute approximate current distributions and the hysteresis loss. In the present treatment, the contours that define the limits of flux penetration are given as a function of the external applied field. The loss model is then extended by inclusion of the transport current in the two-dimensional solution of the current distribution. It is shown that for a filament carrying a dc transport current in a small ac field superimposed on a large dc bias field, the transport current fills the inner core of the filament while surrounded by time-varying shielding currents. The flux penetration contours for the zero transport case are directly applicable.[8] Once the field and current distributions are specified, the hysteresis loss can be computed for all operating conditions.

MATHEMATICAL TECHNIQUE

The problem is to determine the magnetic field inside and outside an arbitrarily shaped region of constant current density, J. The magnetic fields are solutions of the magneto-quasi-static Maxwell equations,

$$\nabla \times H = J, \quad \nabla \cdot H = 0. \tag{1}$$

First it is necessary to define a field point in the complex plane, $z = x + iy$, and the complex field $H(x,y)$ where

$$H(x,y) = H_y(x,y) + iH_x(x,y). \tag{2}$$

The field, H, is a complex function of position in which the real part of the function defines the y component of magnetic field and the imaginary part defines the x component of magnetic field. A suitable function is given by

$$f(z) = H - \frac{1}{2}Jz^* = u + iv \tag{3}$$

where $z^* = x - iy$ is the complex conjugate of z, $u = H_y - (1/2)Jx$, and $v = H_x + (1/2)Jy$.[5] This function can be evaluated by application of the Cauchy Integral Formula, which, when applied to (3), gives

$$f(z) = \frac{iJ}{4\pi} \oint_c \frac{\zeta^*}{\zeta - z}\, d\zeta = \begin{cases} H_{in} - \frac{1}{2}Jz^* & \text{for } z \text{ inside } C \\ H_{out} & \text{for } z \text{ outside } C \end{cases} \tag{4}$$

where the variable ζ denotes a point along the contour C enclosing the current-carrying region. Evaluation of (4) yields the magnetic field components everywhere.

PARTIAL PENETRATION FLUX PROFILES

This mathematical formulation has been used to compute the region of flux penetration into a cylindrical, superconducting, nonideal, type II superconducting filament. The model is based upon the following assumptions: (a) the constant background field is uniform and much greater than $H_p(0)$, the field required to fully penetrate the filament at zero transport current, (b) the filament carries no transport current, and (c) the filament is in the critical state and the Bean model holds.[9]

Consider the filament to be in a transverse field, H_e. The induced current distribution creates a uniform magnetic field in an interior region that is equal in magnitude and antiparallel to the external field such that the interior region is completely shielded. The net magnetic field distribution is just the superposition of the external field and the field generated by the induced current distribution. However, the function ζ, which defines the exact contour that delineates the inner region of zero net flux and current, is not known beforehand. This function was determined from the boundary condition that requires the field to be zero everywhere within and on the contour. An iterative technique was used to determine these functions for a range of values of applied field. The curve was assumed to have the general form of a polynomial,

$$\zeta = x + i(a_0 + a_2x^2 + a_4x^4 + a_6x^6 + a_8x^8). \qquad (5)$$

Table 1. Coefficients of Polynomial Functions that Define Flux Penetration Curves

$\Delta H_e/H_p(0)$	a_ϕ	a_2	a_4	a_6	a_8	X_m
0.1	0.8695	-0.432	-0.050	-0.383	-0.0045	1.0
0.2	0.760	-0.492	-0.209	-0.0559	-0.0031	1.0
0.3	0.640	-0.467	-0.180	-0.0100	0	0.9899
0.4	0.529	-0.450	-0.095	-0.038	0	0.9627
0.5	0.423	-0.427	-0.050	-0.026	0	0.9295
0.6	0.328	-0.420	-0.001	0	0	0.8828
0.7	0.236	-0.370	-0.001	0	0	0.7993
0.8	0.150	-0.315	-0.010	0	0	0.6850
0.9	0.0705	-0.265	-0.010	0	0	0.5132
1.0	0	0	0	0	0	0

The coefficients were determined iteratively by computing the residual fields at several points on and within the assumed boundary. The coefficients that resulted in a minimum of the sum of squares of the residual fields at these points were chosen to represent the flux front. The results of this analysis are shown in Table 1, which contains the coefficients of the polynomial functions shown plotted in Fig. 1. These curves are very similar to the curves computed numerically by Ashkin,[1] Pang et al.,[2] and Zenkevitch et al.[3]

CURRENT DISTRIBUTION WITHOUT TRANSPORT CURRENT

The flux penetration curves allow the current distribution to be determined for a changing external field. For the initial condition, assume the dc component of the external field, H_0, is much greater than $H_p(0)$. The initial distribution is as shown in Fig. 2a. Now consider what occurs as the external field is cycled by $H_e = H_0 \pm \Delta H_e$ where $\Delta H_e < H_p << H_0$. Shielding currents are induced to shield the interior region of the filament from the field change. On a macroscopic scale these appear as regions of positive and negative critical current density, J_c. The induced field exactly cancels ΔH_e everywhere inside the region circumscribed by the currents, as shown in Fig. 2b and 2c. The actual time-dependent current and flux patterns are just a series of quasi-static steps from one critical state configuration to the next. During the increasing portion of the cycle, the current distribution appears as in Fig. 2d, finally returning to the initial distribution (Fig. 2e) at the end of the cycle. Full

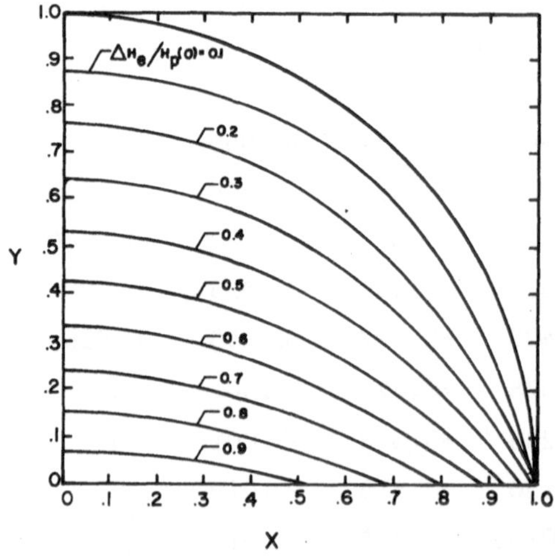

Fig. 1. Limits of flux penetration into a superconducting filament for different values of external field.

penetration occurs if ΔH_e exceeds $H_p(0)$. At this point the filament can no longer shield itself and the current distribution remains constant as the flux continues to penetrate uniformly.

THE EFFECT OF TRANSPORT CURRENT

The transport current destroys the symmetry of the problem owing to the addition of the self-field of the filament. The net current must always satisfy the condition,

$$i = (A_+ - A_-)/(A_+ + A_-) \tag{6}$$

where $A_\pm$ = area of $\pm J_c$, $i = (I_t/I_c)$ = ratio of transport to critical current, and d is the diameter of the filament. In addition, the field required to fully penetrate the filament must become a function of the transport current such that $H_p(i) \leq H_p(0) = {}_c d/\pi$, where $H_p(i)$ is a function yet to be determined.

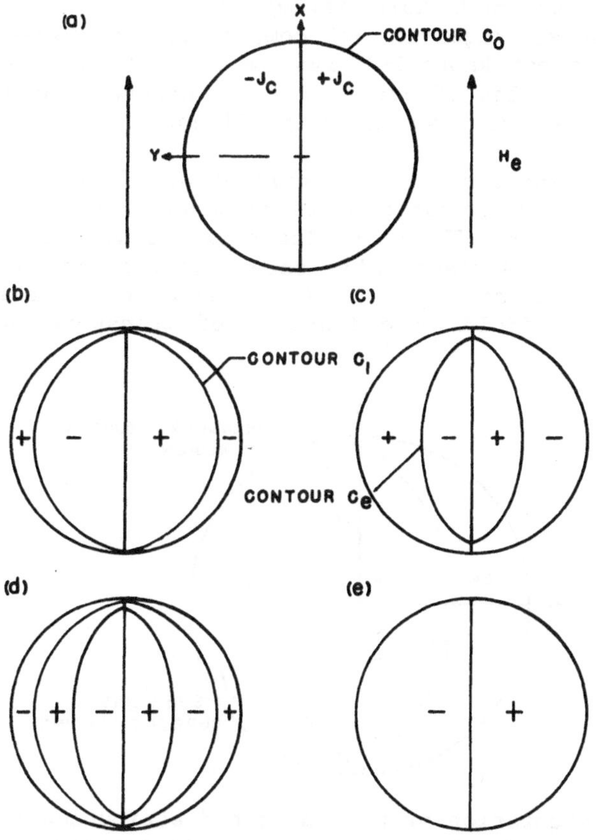

Fig. 2. Current distribution as H_e is cycled by $\pm\Delta H_e$.

Assume the filament carries a current i (0 < i < 1) in a uniform external field $H_0 \gg H_p(i)$. The problem is to compute the current distribution as the external field is cycled between $H_0 \pm \Delta H_e$ while the current remains constant for two conditions: (1) $\Delta H_e \leq H_p(i)$ and (2) $\Delta H_e > H_p(i)$. The exact initial current distribution is not known. In fact, there is not any one true initial state since the current pattern is determined by the most recent history of current and field change. The crucial element in this process is that only the most recent history of change is important.

Thus, after several cycles of external field change with constant transport current, the current distribution of the initial state will be wiped out or, more accurately, rearranged. The new current distribution at a given time in the cycle will appear as in Fig. 3 with the transport current redistributed into a central core of the filament and surrounded by positive and negative shielding currents near the outer surface. This pattern occurs because flux either enters or leaves the filament only from the surface. The condition of (6) must hold throughout the process. The exact number of cycles required to completely rearrange the current distribution is not known but must be a function of i and ΔH_e. If there are many cycles of external field change, the details of the transition stage will have an insignificant effect on the loss.

The time dependence of the current and flux distribution will be a series of quasi-static steps from one critical state configuration to the next. The shielding currents will alternate sign in a region near the surface while screening the transport current in the core as long as $\Delta H_e \leq H_p(i)$. Thus, the definition of the full penetration field as a function of transport current, $H_p(i)$,

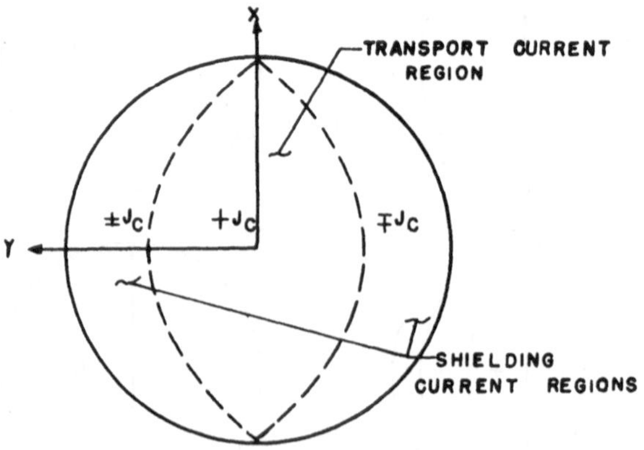

Fig. 3. Representation of the current distribution in a filament carrying dc transport current after several cycles of external field.

follows in that it is just the magnitude of the external field change, ΔH_e, which, if exceeded, will no longer change the current distribution in the filament. The curves that define the boundary between shielding currents and transport current are also known. They are the same curves that define the field penetration profiles for the case when $i = 0$. This follows because the current distributions defined by those profiles totally shield the region within them. A curve of the function $H_p(i)$ normalized to $H_p(0)$ is shown in Fig. 4. The value of $H_p(i)$ goes to zero at $i = I$ because the entire filament is filled with the critical transport current.

TWO-DIMENSIONAL AC LOSS CALCULATION

The loss is computed for the cylindrical filament as the external field is cycled while the transport current is held constant.[8] The current distributions of Fig. 1 are used in the calculation. The problem has been divided into four parts to include partial and full penetration with and without transport current.

No Transport Current, Partial Penetration

The hysteresis loss per cycle per unit volume is computed for this case by integration of the magnetization, $M(H_e)$, over a complete cycle. The magnetization is computed from the current distributions of Fig. 1, but Pang et al.[2] have shown that M can be expressed with reasonable accuracy in the simple form,

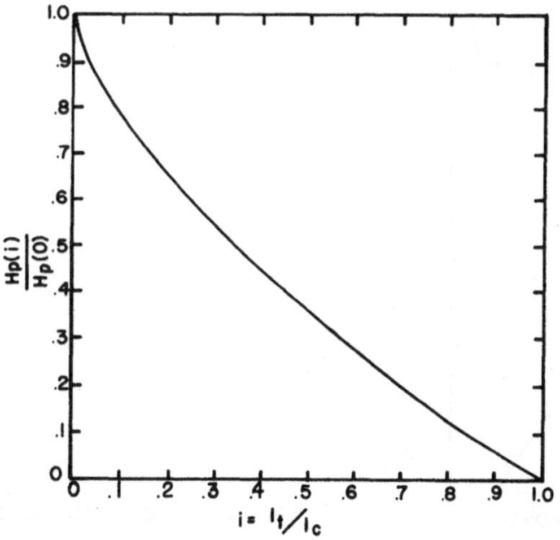

Fig. 4. Full penetration field as a function of the reduced transport current.

$$\frac{M}{H_p(0)} = \frac{2}{3}\left[\frac{\Delta H_e}{H_p(0)}\right]^3 - 2\left[\frac{\Delta H_e}{H_p(0)}\right]^2 + 2\,\frac{\Delta H_e}{H_p(0)} \qquad (7)$$

for $\Delta H_e \leq H_p(0)$. Integration of (7) over a cycle from $H_0 + \Delta H_e$ to $H_0 - \Delta H_e$ results in

$$\frac{W_{p0}}{W_0} = 2\left[\frac{\Delta H_e}{H_p(0)}\right]^3 - \left[\frac{\Delta H_e}{H_p(0)}\right]^4 \qquad (8)$$

where $W_0/V = 4/3\ \mu_0 H_p(\phi)$ is the loss per unit volume for $\Delta H_e = H_p(\phi)$. Equation (8) was also derived by Zenkevitch et al.[3]

No Transport Current, Full Penetration

In this case, the current distribution cannot change once the full penetration field is exceeded. Thus, the field penetrates uniformly. The total loss over a full cycle is the sum of the partial penetration loss plus the loss when ΔH_e exceeds $H_p(0)$ and is given by

$$W_{fp0} = W_{p0}(H_p(0)) + \int_{cycle} \int_{volume} J_c \cdot E \; dV \, dt \qquad (9)$$

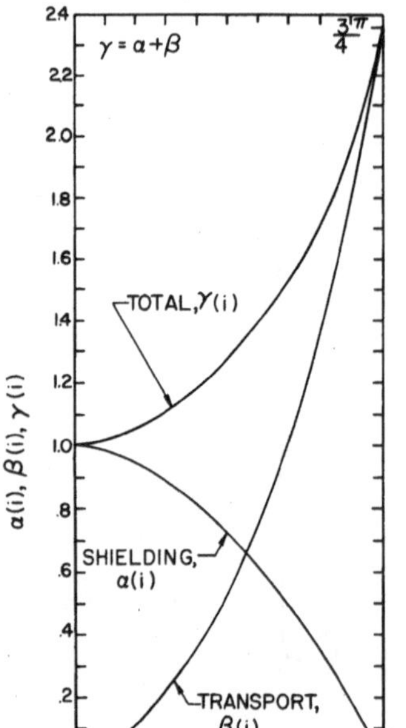

Fig. 5. Nondimensional loss functions α, β, and γ versus reduced transport current.

where the electric field is found from the Maxwell equation,

$$\nabla \times E = - \mu 0 d(\Delta H_e)/dt. \tag{10}$$

Solving for E in (10) and substituting it into (9) gives for the loss per cycle,

$$\frac{W_{fp0}}{W0} = \frac{\Delta H_e}{H_p(0)} - 1 \qquad \Delta H_e > H_p(0). \tag{11}$$

Partial Penetration with Transport Current

The loss calculation for this case is quite simple for the current distribution defined in Fig. 5 and the full penetration field as given in Fig. 6. Since the transport current is completely shielded from the external field during all portions of the cycle, the transport current contributes nothing to the loss. The loss is identical to the loss for partial penetration without transport current as long as the condition $\Delta H_e \leq H_p(i)$ holds, i.e.,

$$\frac{W_{pt}(\Delta H_e)}{W_0} = 2\left[\frac{\Delta H_e}{H_p(0)}\right]^3 - \left[\frac{\Delta H_e}{H_p(0)}\right]^4 \tag{12}$$

Full Penetration with Transport Current

Once ΔH_e exceeds $H_p(i)$, the flux penetrates the filament uninhibited, and the energy dissipation for this portion of the cycle is given by

$$\frac{W_{fpt}}{V} = \frac{W_{pt}(H_p(i))}{V} + \frac{1}{V}\int_{\text{cycle}}\int_{\text{volume}} J_c \cdot E \, dv \, dt \tag{13}$$

The electric field is determined from (10) but the computation is somewhat more complicated than the zero current case because the position of zero electric field no longer lies on a line of symmetry along the x-axis. It now coincides with the boundary separating regions of positive and negative current (Fig. 3). This boundary is a function of the transport current and its shape is given by the appropriate curve in Fig. 1 for $\Delta H_e = H_p(i)$. Integration of (13) over a complete cycle to obtain the full penetration loss per unit volume for $\Delta H_e > H_p(i)$ results in

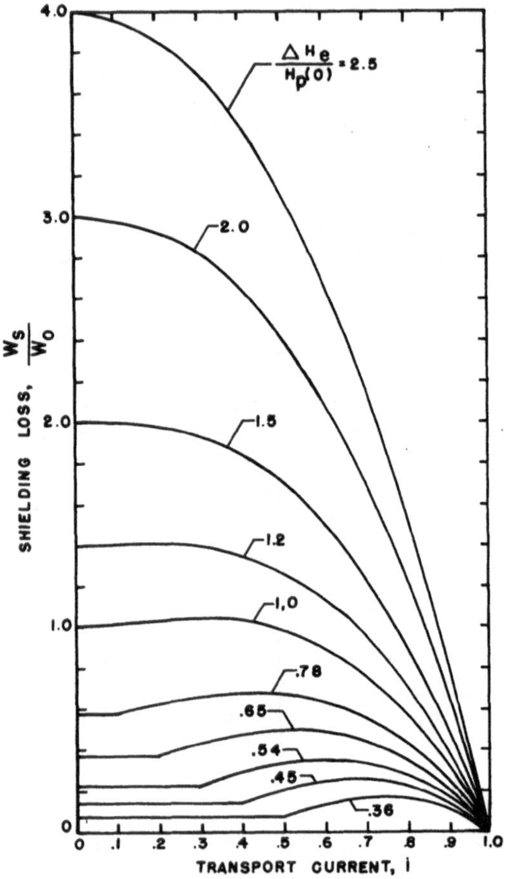

Fig. 6. Shielding loss per cycle versus transport current for different values of external field change.

$$\frac{W_{fpt}(\Delta H_e)}{W_0} = \left\{2\left[\frac{H_p(i)}{H_p(0)}\right]^3 - \left[\frac{H_p(i)}{H_p(0)}\right]^4\right\} + 2\left\{\frac{\Delta H_e}{H_p(0)} - \frac{H_p(i)}{H_p(0)}\right\}\gamma(i) \quad (14)$$

The function $\gamma(i)$ is shown in Fig. 5. It is very similar to a function $g(i)$ computed by Carr et al.,[4] but it is slightly different, since they assumed the line of zero electric field was parallel to the x-axis but displaced owing to the transport current. A similar term is found by Ogasawara et al.,[10] but they give it as $(1 + i^2)$, which is compatible with the one-dimensional slab model.

More interesting results can be obtained if it is noted that a portion of the loss occurs in the transport current region and a portion in the shielding region during the part of the cycle when ΔH_e exceeds $H_p(i)$. Thus, $\gamma(i)$ can be written as $\gamma(i) = \alpha(i) + \beta(i)$, where $\alpha(i)$ represents the fraction of the loss that occurs in the shielding current region and $\beta(i)$ represents the fraction of the loss in the transport current region. The functions α and β are shown along with γ as a function of current in Fig. 5. Since the first term in braces on the right hand side of (14) is also a shielding loss, the net effect can be separated out and summarized as follows:

$$\frac{W_{ft}}{W_0} = \frac{W_s}{W_0} + \frac{W_t}{W_0} \tag{15a}$$

where, for $\Delta H_e > H_p(i)$

$$\frac{W_s}{W_0} = \left\{ 2\left[\frac{H_p(i)}{H_p(0)}\right]^3 - \left[\frac{H_p(i)}{H_p(0)}\right]^4 \right\} + 2\left\{\frac{\Delta H_e}{H_p(0)} - \frac{H_p(i)}{H_p(0)}\right\} \alpha(i) \tag{15b}$$

$$\frac{W_t}{W_0} = 2\left\{\frac{\Delta H_e}{H_p(0)} - \frac{H_p(i)}{H_p(0)}\right\} \beta(i) \tag{15c}$$

and for $\Delta H_e \leq H_p(i)$

$$\frac{W_s}{W_0} = 2\left[\frac{\Delta H_e}{H_p(0)}\right]^3 - \left[\frac{\Delta H_e}{H_p(0)}\right]^4 \tag{15d}$$

$$\frac{W_t}{W_0} = 0 \tag{15e}$$

The shielding loss is shown in Fig. 6, the transport loss in Fig. 7, and the total loss in Fig. 8.

CONCLUSIONS

The two-dimensional current distributions have been computed for a circular filament of type II superconductor carrying a transport current in a changing transverse magnetic field. The mathematical technique of using complex variables with an iterative solution is adaptable to filaments of any regular geometric shape. These current distributions have been used to compute the full penetration field as a function of the transport current and to compute the cyclic hysteresis loss for the full range of field penetration. The portion of the loss that occurs in the shielding

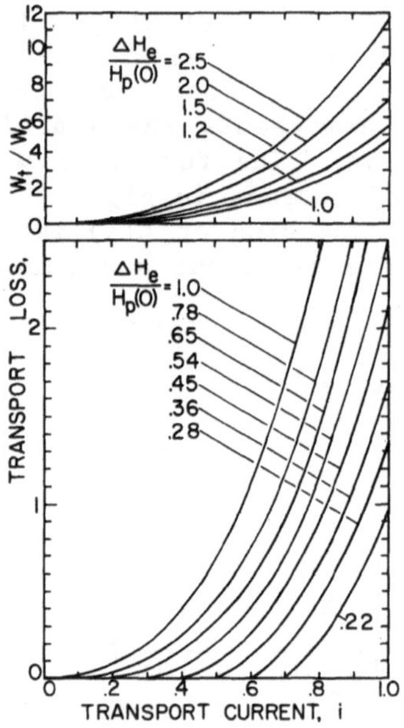

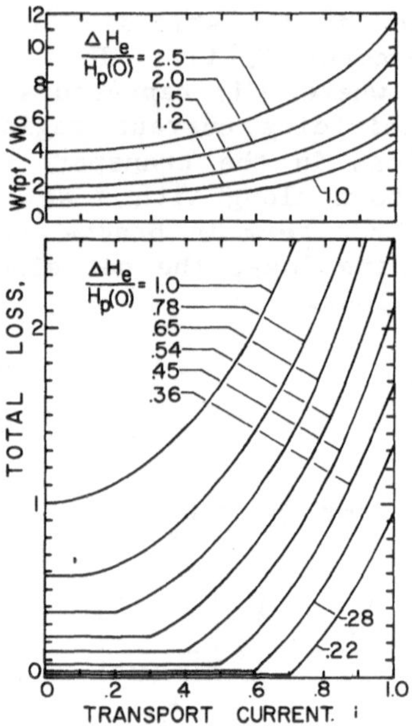

Fig. 7. Transport loss per cycle versus transport current for different values of of external field change.

Fig. 8. Total loss per cycle versus transport current for different values of external field change.

region has been clearly distinguished from the loss in the transport current region.

REFERENCES

1. M. Ashkin, J. Appl. Phys. 50(11):7060 (1979).
2. C. Y. Pang, P. G. McLaren, and A. M. Campbell, in: "Proceedings of the 8th International Cryogenic Engineering Conference," IPC Science & Technology Press, Guildford, Surrey, England (1980).
3. V. B. Zenkevitch, A. S. Romanyuk, and V. V. Zheltov, Cryogenics 20(12):703 (1980).
4. W. J. Carr, Jr., J. H. Murphy, and G. R. Wagner, in: "Advances in Cryogenic Engineering," Vol. 24, Plenum Press, New York (1978), p. 415.
5. R. A. Beth, J. Appl. Phys. 38(12):4689 (1967).
6. G. H. Morgan, IEEE Trans. Nucl. Sci. 16:768 (1969).
7. Y. Kato, M. Hanawaka, and K. Yamafuji, Jap. J. Appl. Phys. 15(4):695 (1976).

8. J. V. Minervini, Analysis of loss mechanisms in superconducting windings for rotating electric generators, Ph.D. Thesis, Department of Mechanical Engineering, M.I.T., Cambridge, Massachusetts (1981).
9. C. P. Bean, Rev. Mod. Phys. 36:31 (1964).
10. T. Ogasawara, Y. Takahashi, K. Kanbara, Y. Kubota, Y. Yasohama, and K. Yasukochi, Cryogenics 20:216 (1980).

THE SPECIFIC HEAT OF NbTi FROM 0 TO 7 T BETWEEN 4.2 AND 20 K*

S. A. Elrod, J. R. Miller, and L. Dresner

Oak Ridge National Laboratory, Oak Ridge, Tennessee

INTRODUCTION

Reasonably reliable zero-field specific heat data for commercial NbTi have been reported.[1,2] Also reported are specific heat measurements at magnetic field.[3] However, existing data are not well known in the technical community, and there appear to be significant uncertainties associated with the measurements at field, as discussed in this paper. We report the results of specific heat measurements made at field on an annealed Nb-44.6% Ti rod. A simple analytic formula suitable for engineering calculations has been derived by thermodynamic arguments for the specific heat, c_p, of Type II superconductors as a function of temperature, T, and applied magnetic field, H. The formula relates the specific heat in the superconducting state to the Sommerfeld constant, γ, the critical temperature, T_c, the normal-state phonon contribution, βT^3, and the upper critical magnetic field, $H_{c2}(0)$:

$$c_p = [\beta + (3\gamma/T_c^2)]T^3 + [\gamma H/H_{c2}(0)]T \tag{1}$$

This formula is used to interpret the data of this paper as well as those previously reported.

THEORY

If one makes the simplifying assumption that hard Type II superconductors exhibit thermodynamic reversibility, one can apply

*Research sponsored by the Office of Fusion Energy, U.S. Department of Energy under contract W-7405-eng-26 with the Union Carbide Corporation.

known thermodynamic relationships[4] to a cylindrical Type II superconductor in a paraxial magnetic field to get:

$$G_s(T,H) = G_n(T) + \frac{\mu_o}{\rho}\int_H^{H_{c2}(T)} M\, dH \tag{2}$$

(A list of symbols is given at the end of the paper.) Measured magnetization curves of Type II superconductors look like those sketched in Fig. 1. The upper and lower critical fields as a function of temperature can be very nearly represented by:

$$H_{ci}(T) = H_{ci}(0)[1 - (T/T_c)^2] \ , \quad i = 1,2 \tag{3}$$

According to Eq. (3), the ratio $H_{c1}(T)/H_{c2}(T)$ is independent of temperature.

As the temperature changes from T to T', both the base and height of the magnetization curve are transformed by the factor $[1 - (T'/T_c)^2]/[1 - (T/T_c)^2]$. As a simplifying approximation, we apply this factor to both ordinate and abscissa of the entire magnetization curve; i.e., we assume that the magnetization curves corresponding to different temperatures are geometrically similar. If so, then

$$\int_0^{H_{c2}} M\, dH = -a\left(1 - \frac{T^2}{T_c^2}\right)^2 \ ; \qquad a = \left|\int_0^{H_{c2}(0)} M(T=0)\, dH\right| \tag{4}$$

Using Eqs. (2) and (4), and the thermodynamic relationship $c_{p,H} = -T(\partial^2 G/\partial T^2)_{p,H}$, one can show that

$$c_s = c_n - (4\mu_o a/\rho)(T/T_c^2)[1 - (3T^2/T_c^2)] \ , \qquad H = 0 \tag{5}$$

The constant a can be determined from the empirical observation that the specific heat at zero field in the superconducting phase vanishes faster than linearly with vanishing temperature. Since this is so, the term $-(4\mu_o a/\rho)(T/T_c^2)$ in Eq. (5) must cancel the linear term in $c_n = \gamma T + \beta T^3$, so that $a = \rho\gamma T_c^2/4\mu_o$. From this, it follows that

$$c_s = [\beta + (3\gamma/T_c^2)]T^3 \ ; \qquad (c_s - c_n)_{T=T_c} = 2\gamma T_c \tag{6}$$

when H = 0. These formulas, derived by Görter and Casimir using the two-fluid model,[5] will result from any model in which the magnetization curves at different temperatures are similar to one another and scale according to Eq. (3) in M and H.

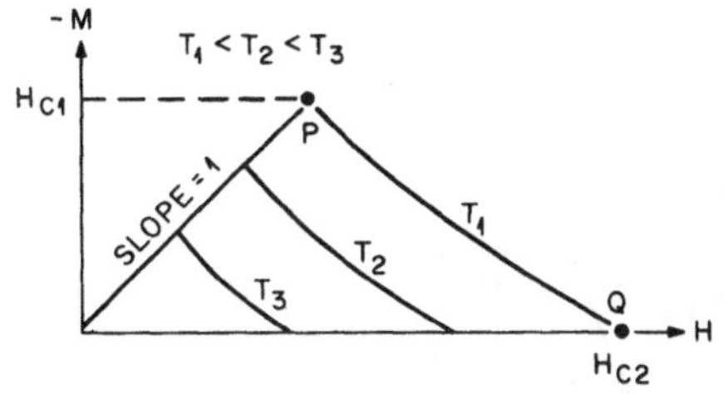

Fig. 1. Typical form of magnetization curves for a Type II superconductor.

To study the field dependence of c_p, one must go further, approximating the curve PQ in Fig. 1 by a straight line. The slope of segment PQ then equals

$$\left(\frac{dM}{dH}\right)_{H_{c2}} = \frac{H_{c1}}{H_{c2} - H_{c1}} = \frac{H_{c1}(0)}{H_{c2}(0) - H_{c1}(0)} \approx \frac{H_{c1}(0)}{H_{c2}(0)} \tag{7}$$

the second equality following from Eq. (3). Furthermore,

$$\frac{1}{2} H_{c1}(0) H_{c2}(0) = a \tag{8}$$

Then $(dM/dH)_{H_{c2}} = \rho\gamma T_c^2/2\mu_o H_{c2}^2(0)$, and it follows from Eq. (2) that

$$c_s = [\beta + (3\gamma/T_c^2)]T^3 + [\gamma T H/H_{c2}(0)] \tag{9}$$

Thus, the presence of a magnetic field is expected to increase c_s by roughly $[\gamma H/H_{c2}(0)]T$ over its zero-field value. This result, derived in the microscopic theory of superconductivity, has been cited in Refs. 4 and 5.

EXPERIMENTAL PROCEDURE

The low temperature section of the cryostat is shown diagrammatically in Fig. 2. The sample, a 10-mm-diameter, 75-mm-long, annealed NbTi rod weighing 35 g, was hung by nylon threads inside an evacuated, temperature-controlled copper can. An evacuated outer can isolated the copper can from the bath. Table 1 summarizes the characteristics of our sample and of samples used in Refs. 1 and 2. The value of T_c for our sample is taken from recent data of Larbalestier[6] for alloys of this composition.

A 42-gage Manganin wire, tightly wrapped around the sample, served as the sample heater. A thin layer (≈0.15 g) of GE 7031 varnish bonded the heater to the sample.

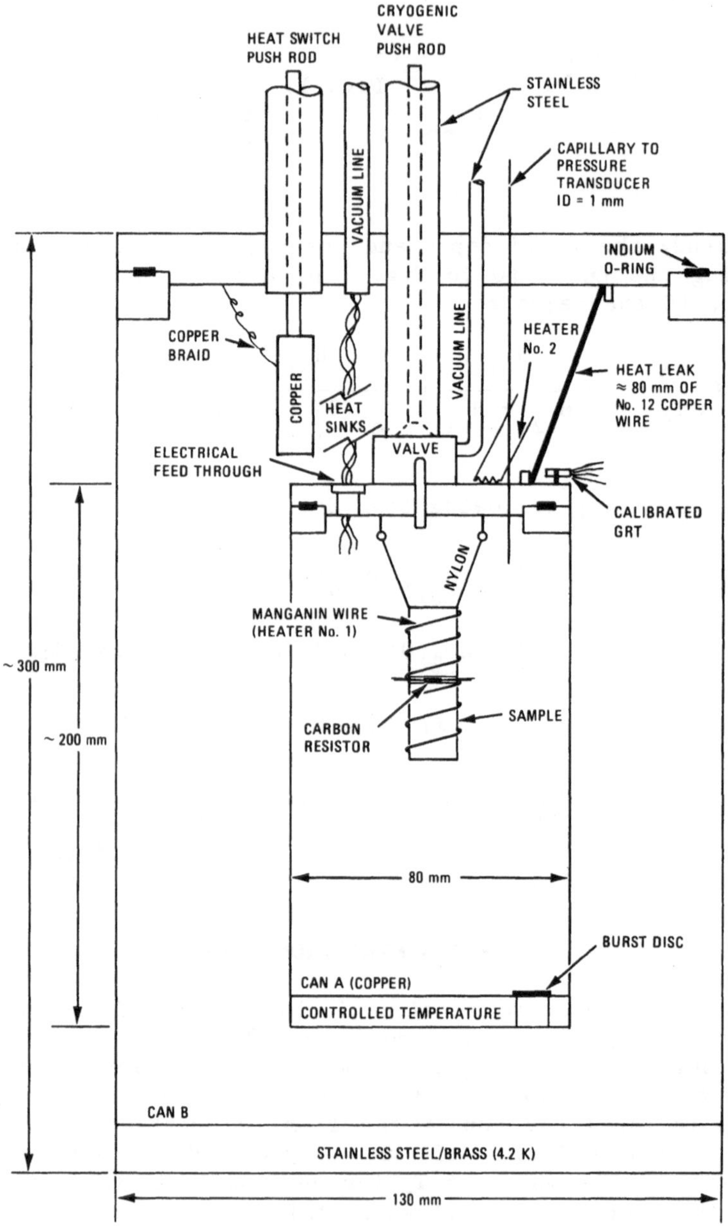

Fig. 2. Schematic of the cryostat used for the calorimetric measurements of the specific heat of the NbTi sample at field.

Table 1. Sample Specifications

Measurement	Composition	T_c
Corsan[1]	Nb-44% Ti	9.2*
Zbasnik[2]	Nb-48.8% Ti	9.0*
Elrod, Miller, and Dresner, this paper	Nb-44.6% Ti†	9.1

*Midpoint of transition (determined graphically from experimental data).
†Provided by Kawicki Berylco Industries; 270 ppm Fe, 400 ppm Ta.

Pulses of 1- to 100-ms duration (E ≈ 0.1-1.0 mJ) to the heater caused temperature excursions of 1-10 mK. The record of a particular pulse is shown in Fig. 3. The temperature excursion was measured by a carbon resistor embedded in the sample (Allen-Bradley, 1/8 W, 430 Ω at 300 K, 11 kΩ at 4.2 K). An ac bridge technique (15 Hz) was used to monitor the change in resistance resulting from the temperature excursions.

Heater 2 (see Fig. 2) controlled the temperature of the inner can to within $\lesssim$5 mK/min. A calibrated germanium resistance thermometer (GRT) (Lake Shore Cryotronics GR-200A-2500) gave the temperature of the inner can to within ±5 mK. Electrical leads running to the sample and to the GRT were singly heat sunk to the outer can and doubly heat sunk to the inner can.

After a series of data points was taken at field, it was necessary to transfer the GRT factory calibration to the carbon resistor. The GRT's highly field-dependent resistance complicated the situation. In order to transfer the zero-field GRT calibration to the carbon resistor at field, we filled the inner can with 2.66×10^3 Pa (20 torr) of helium gas at 4.2 K and used its gas pressure as an intermediate temperature scale. A pressure transducer (sensitivity = 3×10^{-5} V output/mm Hg) communicated with the inner can via a capillary (see Fig. 2).

The following polynomials are least-squares fitted to the calibrations, along with the maximum estimated errors associated with each procedure.

$$\text{GRT vs T (factory calibration):} \quad \log T = \sum_{n=1}^{5} A_n (\log R)^n \, , \qquad \text{error } \Delta T_{max} = 2 \text{ mK} \tag{10}$$

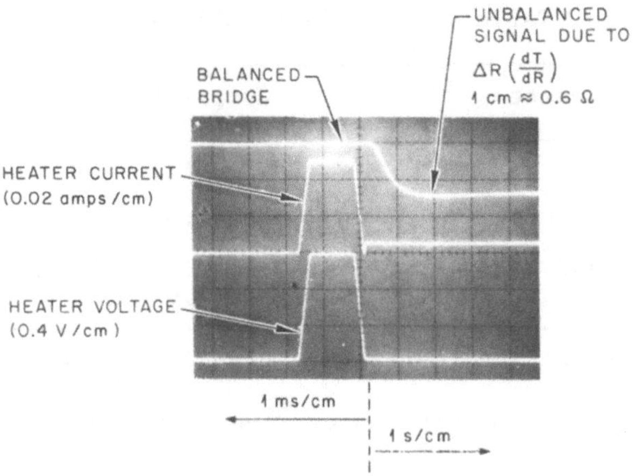

Fig. 3. An example of the transient record of the response to a particular heat pulse. This record provides one datum point.

$$\text{Carbon resistor vs T:} \quad \frac{1}{T} = \frac{A}{\log R} + B + C \log R,$$

$$\text{error } \Delta T_{max} = 10 \text{ mK} \qquad (11)$$

To obtain ΔT corresponding to a change in the resistance of the carbon resistor, Eq. (11) is explicitly differentiated to give dT/dR. The error involved in this procedure is estimated both graphically and by comparing dT/dR for Eq. (11) with that for $1/T = \sum_{n=-2}^{2} A_n(\log R)^n$. The error appears to be no greater than 2%.

ANALYSIS

Shown in Fig. 4 are the experimental results of this paper. Addenda corrections, involving $\lesssim$8% of the total specific heat, were made based on specific heat values reported in the literature.[7] The error bars shown in Fig. 4 represent the maximum random errors due to resolution limits in nulling the bridge and in reading the transient recorder. Error bars for $T < T_c$ are smaller than 2%.

The zero-field data, both for the superconducting state and for the normal state, are fit by straight lines. The temperature range of the data is at least an order of magnitude below the Debye temperature for this alloy. Thus the normal-state specific heat, although possibly not showing a pure T^3 dependence at the highest temperatures, should be nearly enough so that serious errors are

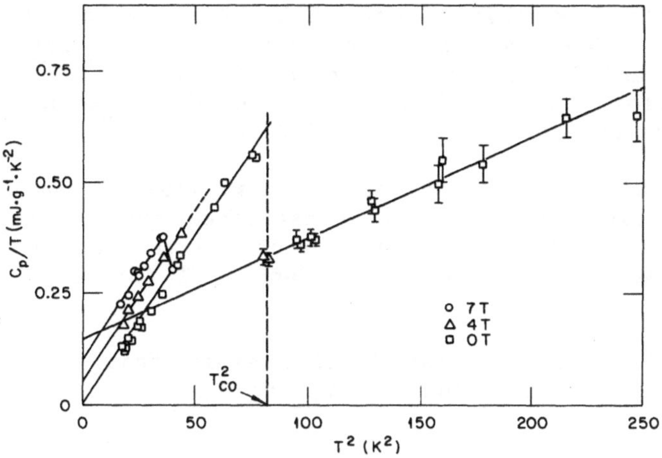

Fig. 4. Experimental results for c_p/T vs T^2 for Nb-44.6% Ti at fields from 0 to 7 T.

not introduced by this fitting procedure. Note that the normal-state line must have a slope of β and must intercept the vertical axis at γ. Similarly, the superconducting-state line must have a slope of $\beta + (3\gamma/T_c^2)$ and at $T = T_c$ should show a step of 2γ over the normal-state line. That these lines can be fitted to give a consistent set of these parameters, as shown in Table 2, suggests that the fitting procedure is reasonable and that the assumption in the derivation of self-similarity of magnetization curves was a fairly good one.

In Fig. 5, c_p/T versus T^2 is plotted for what we believe to be the best zero-field data from the literature (Refs. 1, 2, and 8)

Table 2. Experimental Values of γ and β

	γ (mJ·g^{-1}·K^{-2})	β (mJ·g^{-1}·K^{-4})	$\beta + (3\gamma/T_c^2)$ (mJ·g^{-1}·K^{-4})
Elrod, Miller, and Dresner, this paper	0.145	2.3×10^{-3}	7.5×10^{-3}
Corsan[1] and Zbasnik[2]	0.175	2.6×10^{-3}	9.0×10^{-3}
Schmidt[8]*			8.3×10^{-3}

*Composition of sample not given.

and for data from this paper. Table 2 summarizes the numerical results of this best-fit procedure. Also for comparison, Hawksworth and Larbalestier give $\gamma = 0.15\ mJ\cdot g^{-1}\cdot K^{-2}$ for Nb-64.2 at.% Ti alloy (Nb-48 wt.% Ti), obtained from measurements of the normal state resistivity, and dH_{c2}/dT at $T = T_c$.[9]

By requiring the c_p/T versus T^2 at field be parallel to the zero-field line in Fig. 4, we constructed the 4-T and 7-T lines shown in the figure. A comparison of the theoretical and experimental values for the field-dependent c_p/T intercept is given in Table 3.

The discrepancies most likely result from the assumed linearity of the magnetization from H_{c1} to H_{c2}.

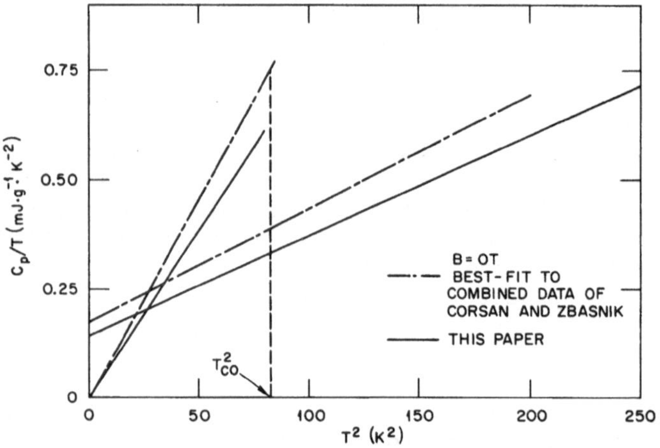

Fig. 5. Comparison of results from this paper with others from the literature for c_p/T vs T^2 at zero field.

Table 3. Comparison of Theory and Experiment

$\mu_o H$	Value of c_p/T Intercept ($mJ\cdot g^{-1}\cdot K^{-2}$) Theory $[\gamma H/H_{c2}(0)]$*	Experiment
4 T	0.041	0.053
7 T	0.073	0.105

*$\mu_o H_{c2}(0)$ = 14 T (paramagnetically limited value; see Ref. 6, p. 58).

We do not include the data of Turck and Duchateau (Ref. 3) in Table 1, since they appear to involve significant uncertainties. The authors do not give the composition of NbTi; more seriously, their field-dependent curves do not show differences proportional to temperature.

ACKNOWLEDGMENTS

We would like to thank Paul Rudd and Ron Dixon for their help in preparing the cryostat and instrumentation.

SYMBOLS

c_p	Constant pressure specific heat ($mJ \cdot g^{-1} \cdot K^{-1}$)
c_s, c_n	Superconducting and normal state specific heats ($mJ \cdot g^{-1} \cdot K^{-2}$)
G_s, G_n	Superconducting and normal state Gibbs free energy per unit mass ($mJ \cdot g^{-1}$)
H	Magnetic field strength ($A \cdot m^{-1}$)
H_{c1}, H_{c2}	Lower and upper critical field strengths ($A \cdot m^{-1}$)
M	Magnetization ($A \cdot m^{-1}$)
T	Temperature (K)
T_c	Critical temperature (K)
β	Proportionality constant in Debye T^3 law of phonon contribution to specific heat ($mJ \cdot g^{-1} \cdot K^{-4}$)
γ	Sommerfeld constant ($mJ \cdot g^{-1} \cdot K^{-2}$)
μ_o	Permittivity of free space ($= 4\pi \times 10^{-7}\ H \cdot m^{-1}$)
ρ	Density ($kg \cdot m^{-3}$)

REFERENCES

1. J. M. Corsan, private communication (July 1981).
2. J. Zbasnik, cited as private communication in: Y. Iwasa, D. Weggel, D. B. Montgomery, R. Weggel, and J. R. Hale, Prediction of transient stability limits for composite superconductors subject to flux jumping, J. Appl. Phys. 40:2008 (1969).
3. J. J. Duchateau and B. Turck, J. Appl. Phys. 46:4989 (1975).
4. R. D. Parks, "Superconductivity," Marcel Dekker, Inc., New York (1969), pp. 19-20, 891.
5. G. D. Cody, Phenomena and theory of superconductivity, in: "Superconducting Magnet Systems," H. Brechna, ed., Springer-Verlag, Berlin (1973).

6. D. C. Larbalestier, Niobium-titanium superconducting materials, in: "Superconducting Materials," S. Foner and B. B. Schwartz, eds., Plenum Press, New York (1981), p. 44.
7. R. B. Stephens, The thermal properties of sample addenda used in T < 1 K specific heat measurements, Cryogenics 15:481 (1975).
8. C. Schmidt, "The Induction of a Propagating Zone (Quench) in a Superconductor by Local Energy Release," Internal Report SUPRA/78-26 EG, Centre d'Etudes Nucléaires de Saclay (April 1978), p. 19.
9. D. G. Hawksworth and D. C. Larbalestier, Enhanced values of H_{c2} in Nb-Ti ternary and quaternary alloys, in: "Advances in Cryogenic Engineering--Materials," Vol. 26, Plenum Press, New York (1980), p. 479.

FIELD EMISSION AND SECONDARY ELECTRON EMISSION FROM Nb_3Sn SURFACES

G. Arnolds-Mayer

Wuppertal University, Wuppertal, Federal Republic of Germany

N. Hilleret

CERN, Geneva, Switzerland

INTRODUCTION

High-energy electron storage rings require powerful acceleration systems in order to cope with the large energy loss by synchrotron radiation. Because of their lower surface resistance, superconducting accelerating cavities offer a substantial gain in terms of the total power consumption, which makes their use in future large electron storage rings very interesting. At present, cavities made from niobium are under study for this purpose.[1-4]

Following the BCS theory, higher T_c materials result in lower surface resistances at a given temperature, and vice versa, the same surface resistance at higher temperatures.

Among the high T_c alloys, Nb_3Sn was generally chosen and studied for superconducting cavity application.[5-8] At 500 MHz, the proposed frequency for the superconducting LEP project,[1] the first measurements were made on a Nb_3Sn-coated reentrant cavity, machined from bulk niobium. The surface resistances of this cavity measured before Nb_3Sn deposition (i.e., for a Nb surface[9]) and after deposition (i.e., for a Nb_3Sn surface) are shown in Fig. 1. Measurements of the niobium cavity with a Nb_3Sn-deposited surface showed that the R_s^{BCS} at 4.2 K of $1 \cdot 10^{-7}$ Ω was already reached at 7 K. The reduction of $R_s(T)$ at temperatures higher than 4.2 K will allow safe operation of large-scale cavities at 4.2 K instead of 1.8 K, resulting in a considerable reduction of refrigeration costs.

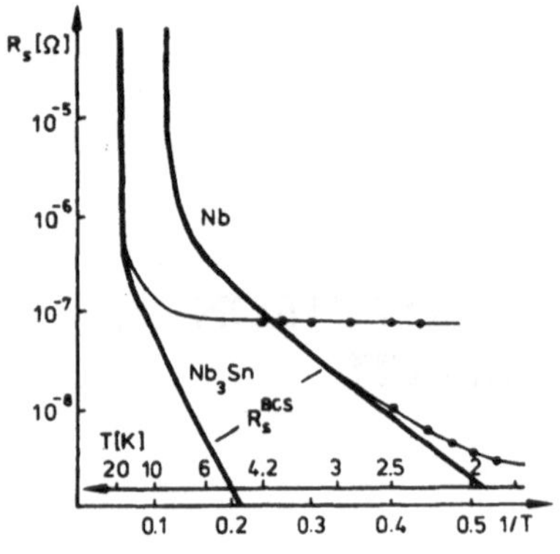

Fig. 1. Surface resistance of a 500 MHz rentrant cavity with Nb and Nb_3S surfaces.

For accelerating cavities, a very important factor is the accelerating field. As all measurements of large-scale niobium structures show field limitation by electron loading, i.e., field emission and/or secondary electron emission, Nb_3Sn can only be a competitive material if its properties as an electron emitter are at least comparable to those of niobium. To study this, the secondary electron yields and field emission characteristics of several Nb_3Sn samples after different surface treatments were measured under dc conditions, and the results were compared with those obtained from pure niobium.

SAMPLE PREPARATION

The Nb_3Sn samples were prepared by the diffusion of vaporized tin into niobium sheet material.[10] Prior to this, the samples were chemically polished, rinsed in water and methanol, and then mounted inside an 8-GHz cavity. The vapour diffusion proper was carried out in a specially constructed oven described elsewhere,[8] at a temperature of 1130 ±20°C. Under these conditions, the Nb_3Sn layer thickness grows with the reaction time, t, proportionally to $t^{0.36}$. The Nb_3Sn layers of the samples studied varied from 0.5 to 2 μm in thickness.

Some of the samples prepared above were oxidized anodically in 8% HNO_3 solution using a niobium cathode. Different oxide layer thicknesses were obtained by increasing the anodizing voltage up to 100 V.

The samples could be cleaned by <u>in situ</u> argon glow discharge (AGD).[11] During AGD cleaning, the measurement vessel was filled with argon gas at a pressure of about 1 Pa and continuously evacuated. A negative bias of 1 kV was applied to the sample. Non-anodized samples were baked to 150 or 300°C for 24 h together with the experimental setup.

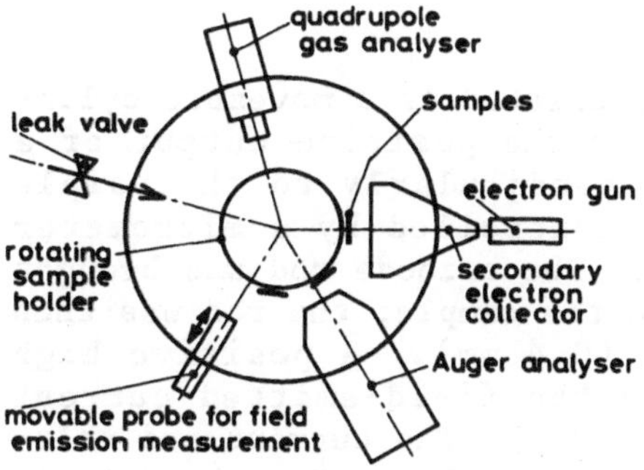

Fig. 2. The experimental system.

EXPERIMENTAL SYSTEM AND MEASUREMENT PROCEDURE

The measurements were performed in a bakeable UHV system (Fig. 2), incorporating an Auger electron spectrometer, a secondary electron collector, field emission measuring equipment, a residual gas analyzer (RGA), and a gas injection system. Several samples could be examined simultaneously. The setup was linked to a dedicated minicomputer, which performed the secondary electron yield (δ) measurements and processed the data collected during the field emission measurements.

Without baking, the pressure levels off at 10^{-6} Pa. After bakeout, the ultimate pressure reached by the system was typically 10^{-8} Pa (N_2 equivalent) and the gas composition monitored by the RGA was: 88% H_2, 8% CH_4, 4% CO. After an AGD cleaning treatment, the total pressure inside the system stabilized at a higher value ($4 \cdot 10^{-8}$ Pa) with the following gas composition: 94% H_2; 5% CH_4; 1% CO_2, CO, Ar.

Secondary Electron Yield Measurement

The secondary electron yield, δ, was measured for primary electron energies (E_p) ranging between 50 eV and 1.8 KeV at normal incidence. As the secondary electron emission (SEE) is very surface dependent, it is especially important to limit the electron dose required for the measurements to avoid surface changes induced by electron bombardment.[12] For this reason, low electron currents were used (typically $3 \cdot 10^{-9}$ A; beam size, 1 mm^2) in 30-ms pulses. The value of δ was calculated from the expression:

$$\delta = \frac{i_c}{i_c + i_s} \quad (1)$$

where i_c is the current measured on the secondary electron collector and i_s, the current measured on the sample.

Field Emission Measurement

The field emission equipment consisted of a moveable cylindrical stainless steel rod connected to the positive output of a high-voltage supply and moving perpendicularly to the sample plane. The translation movement was controlled by a micrometer screw with an accuracy of ±0.01 mm. The cathode rod was brought up to the point where it just touched the sample; the rod was then withdrawn to the desired distance (0.4 mm). A positive high voltage was applied to the rod, and the field-emitted current measured between some 10^{-12} A up to 10^{-6} A. The current and voltage readings were fed into the computer, which plotted the data in the Fowler Nordheim plot (FNP) with log I/E^2 versus $1/E$, where I is the measured field emitted current and E is the applied electric field.

Determination of the Oxide Layer Thickness

The thickness of the oxide layer was determined by bending samples, which were anodized with voltages from 80 V to 130 V, and measuring the height of the oxide layer from photographs taken with a scanning electron microscope. In the voltage range given above, the oxide appears to grow linearly with the voltage as 1.8 nm/V ±10%. After AGD the thickness of the remaining oxide layer was estimated via the characteristic colours of the oxide films, giving a sputter rate of 2 nm/min for the Nb_3Sn oxide under the given conditions.

EXPERIMENTAL RESULTS AND DISCUSSION

Superconducting Parameters

The Nb_3Sn samples were prepared in the same manner, and partly together with, the 8 GHz cavities. Therefore, the superconducting parameters determined by radio frequency measurements

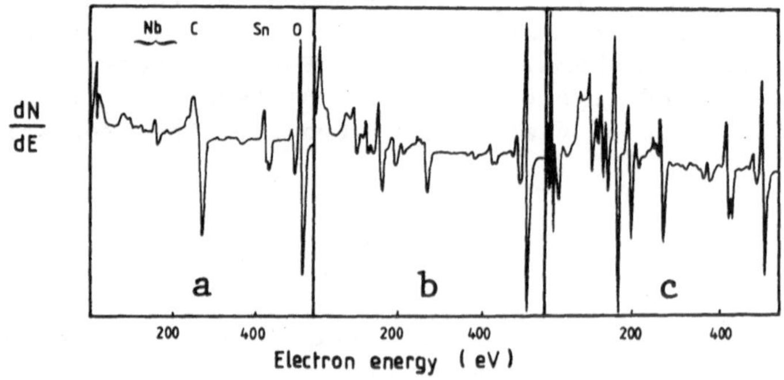

Fig. 3. Auger spectra.

of those should also be valid for the samples. The critical temperature measured was 17.9 ±0.2 K. From the change of the resonance frequency with temperature in the range $T_c > T > T_c/2$, the penetration depth was determined to be 115 ±10 nm. The BCS slope of the surface resistance gave an energy gap of 4.3 ±0.2 kT_c, a value in good agreement with tunnelling measurements,[13] confirming that Nb_3Sn is a strong coupling superconductor.

Auger Surface Analysis

The Auger spectra of anodised Nb_3Sn samples as received (a), after 1 min AGD (b), and after the complete removal of the oxide layer (c) are given in Fig. 3. After AGD, the graphite-like carbon peak is turned into a carbide one. The ratio of the niobium to the tin peaks indicates an understoichiometric content of tin on the surface, owing to the preferential sputtering for tin.[14]

Secondary Electron Emission

It has been shown[15,16] that in niobium cavities, one side multipacting may be the dominant limiting mechanism. This type of resonant electron loading can only take place if δ is larger than unity. Hence the energy (E_{p1}) of the primary electrons at which δ goes higher than unity is of special importance. With increasing E_p, δ passes a maximum (δ_{max}), decreases, and falls below unity at an energy E_{p2}.

For different oxide layers as received (a) and after an additional 1 min AGD (b), $\delta(E_p)$ is given in Fig. 4. By comparing the curves for the clean anodized and nonanodized surfaces, it can be seen that the increasing as well as the decreasing slopes are steeper for the oxide-covered sample. This results in a decrease of E_{p1} from 130 eV to 100 eV and a decrease of E_{p2} from 1500 to 920 eV when Nb_3Sn is anodized. A similar shift has been observed on Nb and Nb_2O_5[12] and is likely related to the presence of the anodic oxide.

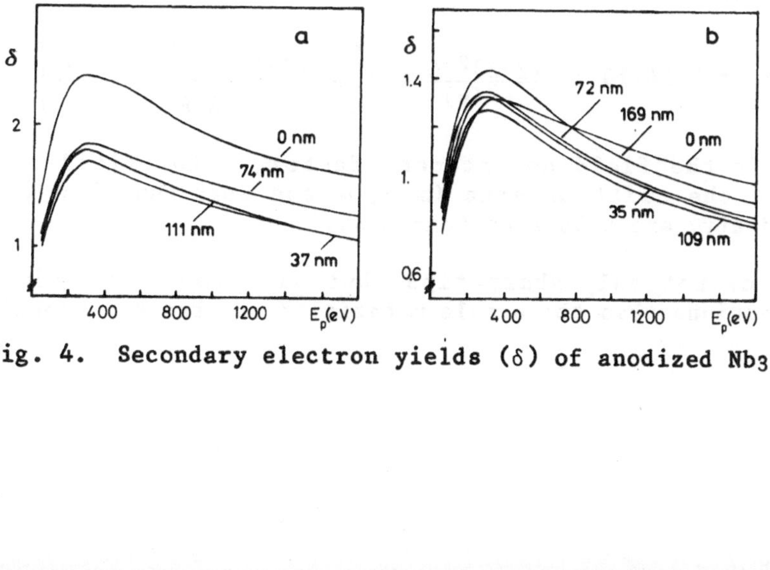

Fig. 4. Secondary electron yields (δ) of anodized Nb_3Sn.

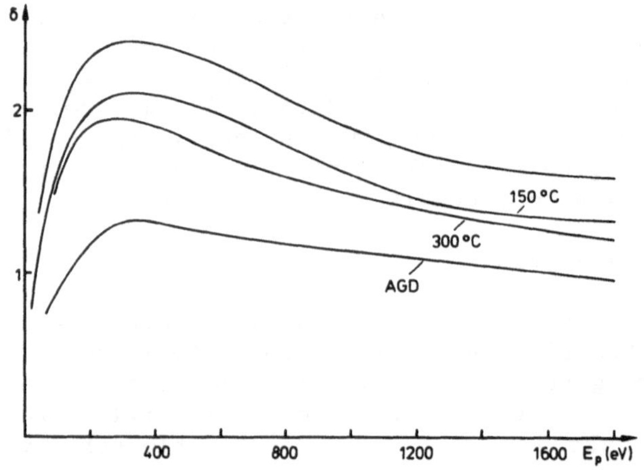

Fig. 5. Secondary electron yields of Nb_3Sn.

Figure 5 shows $\delta(E_p)$ in the as received, baked, and AGD-cleaned state. Baking lowers δ; further improvement is obtained after an AGD, which leads to δ comparable to sputter-cleaned Nb_3Sn.[15] The evolution of $\delta(E_p)$ with the cleaning procedure can be understood by attributing the enhanced SEE of non-AGD-cleaned surfaces to the presence of surface layers. A bakeout removes the relatively loosely bound absorbed layers (e.g., water vapor) and/or modifies the natural oxide. It brings δ to a value comparable to that of the anodic oxides. The more tightly bound surface species are efficiently removed by AGD cleaning, which lowers δ to the values characteristic of the clean unoxidized metal.

Field Emission

Even if multipacting can be overcome by special designs of the cavities,[17,18] the achievable accelerating fields are limited by electron loading (i.e., field emission). Following the Fowler Nordheim theory,[19] the field-emitted current, I, is given by:

$$I(A) = 1.54\cdot 10^{-6} \; \frac{(\beta\cdot E)^2\cdot A}{\phi . t^2(y)}\cdot \exp\left[\frac{-6.83\cdot 10^9\cdot\phi^{1.5}\cdot v(y)}{\beta\cdot E}\right]$$

where β is the field enhancement factor, E the field strength (V/m), A the emitting area (m^2), ϕ the work function (eV), and v(y) and t(y) are tabulated functions.[19]

Since not only sharp tips, but also impurities with lower work functions than the sample material[20] or absorbed atoms[21] are potential field emitters, β cannot be only considered as a geometrical factor, but more generally as a fitting parameter. For calculations, ϕ was kept constant (4 eV). Since the surface treatments may modify ϕ mostly between 3 and 5 eV, the influence of this variation has been investigated: for the same set of data, β changes proportionally to $\phi^{1.5}$ and A is nearly constant.

During field emission, a forming of the surface takes place that appears as discontinuities in the FNP. This can be explained by the activation or inactivation of different electron emitters. As an example, the FNP of a first measurement cycle is given (Fig. 6a). The dots indicate I(F) when rising, and the crosses when lowering the fields. The last curve was obtained several times. Beside this type of forming, sparking was observed, resulting in a current increase for the entire voltage range. Figure 6b gives the FNP of a first measurement cycle before (left curve) and after sparking (right curve). Again the curve, measured when lowering the field, was reproducible.

In more than one hundred measurements made on nonanodized and anodized samples, with oxide layers between 0 and 185 nm, no clear dependence was found between the oxide thickness and breakdown voltages or the voltage at which activation or inactivation occurred. The same is true for the currents and the currents normalized to the emitting areas. For these voltages and currents, however, a statistical analysis showed distinct Gaussian distributions with the maximum between 30 and 40 MV/m and $5 \cdot 10^{-9}$ and $5 \cdot 10^{-8}$ A, respectively.

Stable and reproducible field emission currents from samples with variable oxide layers, measured after conditioning but without sparking, are given in Fig. 7 (a and b show the FNP of a 0.5-μm and a 2-μm sample before and aa and bb, those after AGD).

The values calculated for β and A together with the measured fields at the critical current of 10^{-8} A are listed in Table 1.

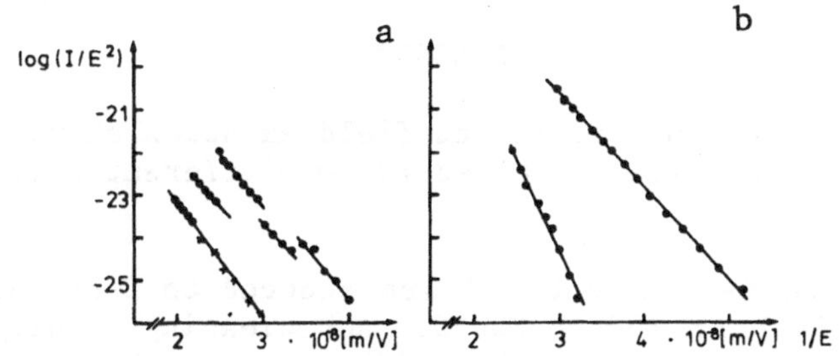

Fig. 6. Forming (a) and sparking (b) during field emission.

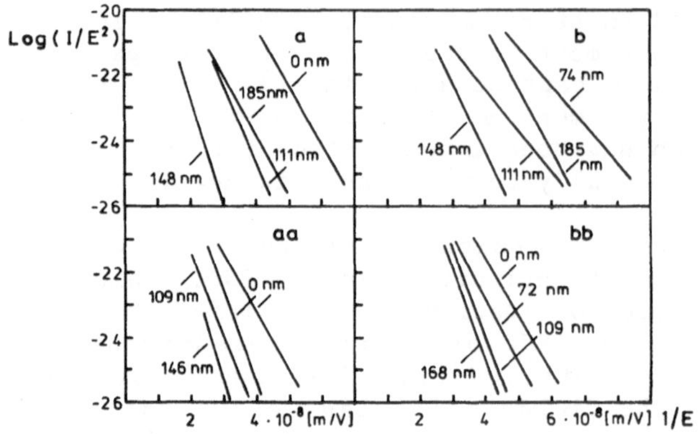

Fig. 7. Fowler-Nordheim plots of Nb_3Sn covered with various oxide layers.

After AGD both samples show the same tendency: as the layer thickness increases, β and/or A decreases and the field needed to emit 10^{-8} A increases. As in the case of the secondary electron yields, these results are quite similar to those obtained on niobium.[22] The thicker Nb_3Sn layer emits larger electron currents, possibly owing to a different surface structure.

For nonanodized, as-received samples, the βs were calculated to be 140 ±20, and currents of 10^{-8} A were measured at fields of 20 ±3 MV/m. The 150°C heat treatment resulted in β of 60 ±10 and an increase of the voltage at the same current to 27 ±3 MV/m. After the 300°C baking, the β was roughly the same, but smaller emitting areas were calculated, leading to voltages of 30 to 35 MV/m at 10^{-8} A. An additional AGD resulted in βs and fields at 10^{-8} A stabilizing to about the same values as in the initial state.

This measurement confirms the statement made previously that bakeout leads to an oxide surface comparable to the one obtained by anodic oxidation.

CONCLUSION

Secondary electron yields and field emission characteristics of Nb_3Sn have been investigated after different surface treatments.

Highest values of δ (2.4) are reduced to 2 by anodizing or heating for 24 h at 150 to 300°C. AGD cleaning results in δ_{max} of about 1.3, independent of the oxide layer thickness. However, oxides lead to a rising of δ over unity at lower primary electron

Table 1. Field Emission Characteristics of Nb_3Sn

0.5-μm Sample		After AGD		2-μm Sample		After AGD	
Oxide (nm)	β	Em. Area (m^2)	E(10^{-8}A) (MV/m)	Oxide (nm)	β	Em. Area (m^2)	E(10^{-8}s) (MV/m)
0	134	4.10^{-17}	26.5	0	136	1.10^{-15}	21.6
0	86	6.10^{-15}	31.6	72	127	2.10^{-16}	25.0
109	92	7.10^{-17}	37.0	109	92	3.10^{-14}	27.6
146	67	4.10^{-15}	41.3	168	90	9.10^{-15}	29.6

energies. This might entail lower breakdown fields in oxide-covered cavities affected by multipacting.

Anodic oxide layers result in lower field-emitted currents. After AGD cleaning, the currents at a fixed field are in inverse proportion to the oxide thickness and are decreased by roughly 2 orders of magnitude. The same reduction is found for nonanodized samples after bakeout, whereas an additional AGD on these samples increases the emitted currents. In all cases, sparking leads to a large enhancement of the field-emitted currents. A similar degradation has often been observed in superconducting cavities.

Comparing our results to those obtained on niobium, similar behaviour for Nb_3Sn and Nb cavities with respect to electron loading is to be expected.

ACKNOWLEDGMENTS

The authors wish to express their gratitude to Dr. J. D. Adam, J-P. Bacher, G. Dominichini (CERN), U. Klein and M. Peiniger (Wuppertal University) for helpful support in the experimental work. Many valuable discussions with Drs. R. S. Calder, H. Lengeler, A. G. Mathewson (CERN), and Prof. H. Piel (Wuppertal University) are gratefully acknowledged. One of us (G.A-M.) wishes to thank CERN for its hospitality and support.

REFERENCES

1. P. Bernard, G. Cavallari, E. Chiaveri, E. Haebel, H. Lengeler, J. M. Maugain, E. Picasso, V. Picciarelli, J. Tückmantel, W. Weingarten, and H. Piel, "First results on a superconducting 4-cell cavity at 500 MHz," CERN Internal Report CERN/EF/RF 81-2, CERN, Geneva, Switzerland (1981).
2. W. Bauer, A. Brandelik, A. Citron, F. Graf, J. Halbritter, W. Herz, S. Noguchi, R. Lehm, W. Lehmann, and L. Szecsi,

Measurements on a superconducting accelerating cavity for DORIS, IEEE Trans. Nucl. Sci. NS-28:3272 (1981).

3. H. Padamsee, J. Kirchgessner, J. Mioduszewski, R. Sundelin, and M. Tigner, Superconducting cavities for a large e^+e^- colliding beam accelerator, IEEE Trans. Nucl. Sci. NS-28:3240 (1981).
4. F. Furuya, S. Hiramatsu, T. Nakazato, T. Kato, P. Kneisel, Y. Kojima, and T. Takagi, First results on a 500 MHz superconducting test cavity for TRISTAN, IEEE Trans. Nucl. Sci. NS-28:3225 (1981).
5. B. Hillenbrand and H. Martens, Superconducting Nb_3Sn cavities with high quality factors and high critical flux densities, J. Appl. Phys. 47:4151 (1976).
6. P. Kneisel, H. Kupfer, O. Stoltz, and J. Halbritter, Properties of superconducting Nb Sn layers used in RF cavities, in: "Advances in Cryogenic Engineering," Vol. 24, Plenum Press, New York (1978), p. 442.
7. J. B. Stimmel, Microwave superconductivity of Nb_3Sn, Thesis, Cornell University, Ithaca, New York (1978).
8. G. Arnolds and D. Proch, Measurement on a Nb Sn structure for linear accelerator application, IEEE Trans. Magn. MAG-13:500 (1977).
9. U. Klein, Thesis (in prep.), Wuppertal University, Wuppertal, FRG (1981).
10. E. Saur and J. Wurm, Preparation and superconducting properties of niobium wire samples with Nb_3Sn coating, Naturwissenschaften49:127 (1962).
11. A. G. Mathewson, "The Absorbed Gas on Nb Surfaces at Ambient Temperature, CERN Internal Report ISR-VA/AGM (unpublished), CERN, Geneva, Switzerland (1981).
12. M. Lavarec, P. Bocquet, and A. Septier, Lowering of the secondary electron emission coefficient of real surfaces due to the primary electron bombardment, in: "Proc. 8th Int. Symposium on Discharges and Electr. Insul. Vac.," Albuquerque, (1978).
13. D. F. Moore and M. R. Beasley, Tunneling observations of delayed flux entry and surface barriers in Nb_3Sn, Appl. Phys. Lett. 30:494 (1977).
14. D. B. Smathers and D. C. Larbalestier, An Auger spectroscopy study of bronze route niobium-tin diffusion layers, in: "Filamentary A15 superconductors," M. Suenaga and A. F. Clark, eds., Plenum Press, New York (1980), p. 143.
15. H. Padamsee and A. Joshi, Secondary electron emission measurements on materials used for superconducting microwave cavities, J. Appl. Phys. 50:1112 (1979).
16. C. M. Lyneis, H. A. Schwettman, and J. P. Turneaure, Elimination of electron multipacting in superconducting structures for electron accelerators, Appl. Phys. Lett. 31:541 (1977).

17. U. Klein and D. Proch, Multipacting in superconducting RF Structures, in: "Proceedings of the Conference on Future Possibilities for Electron Accelerators," Charlottesville (January 1979).
18. H. Padamsee, D. Proch, P. Kneisel, and J. Mioduszewski, Field strength limitation in superconducting cavities - multipacting and thermal breakdown, IEEE Trans. Magn. MAG-17:947 (1980).
19. R. H. Good Jr. and E. W. Müller, Field emission, in: "Handbuch der Physik," Bd. XXI, Springer Verlag, Berlin (1956).
20. I. Brodie, Studies of field emission and electrical breakdown between extended nickel surfaces in vacuum, J. Appl. Phys. 35:2324 (1964).
21. H. A. Schwettman, J. P. Turneaure, and R. F. Waites, Evidence for surface-state-enhanced field emission in RF superconducting cavities, J. Appl. Phys. 45:914 (1974).
22. C. Sayag, Nguyen Tuong Viet, H. Bergeret, and A. Septier, Field emission from oxidized niobium electrodes at 295 and 4.2 K, J. Phys. E10:176 (1977).

INVESTIGATION ON THE PHYSICAL MECHANISM UNDERLYING THE FUNCTION OF A FAST GOLD-TIN METAL FILM SECOND-SOUND DETECTOR

H. Borner, T. Schmeling, and D. W. Schmidt

Max Planck Institut für Strömungsforschung, Göttingen, Federal Republic of Germany

INTRODUCTION

To detect second-sound shock waves in He II, suitable thermometers are needed that respond to small temperature variations, about 1 mK, at frequencies of more than 1 MHz. High sensitivity combined with short response time is obtained by using the discontinuity of the resistance of an evaporated superconducting metal film at the transition from superconductance to normal conductance. Since there are no superconducting materials available with a transition temperature in the interesting region between about 1.2 K and T_λ = 2.17 K, Laguna[1] tested evaporated films composed of different materials, and he was most successful with layers consisting of gold successively evaporated on tin. The transition temperature of such layers was decreased to about 2 K. To adjust the transition temperature in each experiment exactly to the temperature of the He II bath, Laguna applied an external magnetic field. Later on, Cummings, Schmidt, and Wagner[2] made second-sound experiments with gold-tin films that required no external magnetic field. In the work presented here as an abridged version of a more detailed paper,[3] the electrical properties and the structure of such zero-field sensors were investigated to optimize their sensitivity at a given temperature.

EXPERIMENTS

The Voltage-Current Characteristics

The electrical properties of the films are characterized by the voltage vs. current dependence. To study the influence of various gold-to-tin ratios on the superconducting transition, the variation of the voltage-current characteristics with temperature was investigated with the help of several compositions of the film. It was found that considerable changes of the critical current density at constant temperature and crucial distortion of the curves were caused by even small variations of the thickness of the evaporated tin layer as recorded by a conventional film thickness monitor. As seen in Fig. 1, the film with a nominal 1000 Å tin layer shows an abrupt increase in voltage, indicating the breakdown of superconductance below 1.9 K. This increase is smoothed more and more to a gentle transition with decreased slope at thinner tin layers (see Fig. 2). At a ratio of 200 Å gold to 600 Å tin, finally, the voltage-current characteristics almost coincide with the residual resistance line, R_n. As can be expected, the slope of

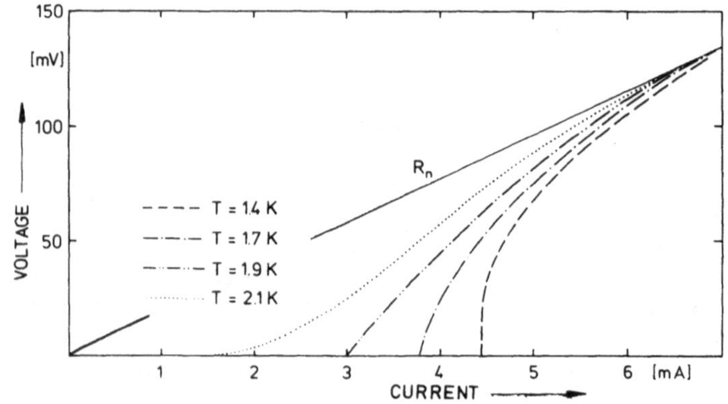

Fig. 1. Voltage vs. current dependence at different temperatures measured on a 1000 Å tin/200 Å gold layer; R_n = residual resistance.

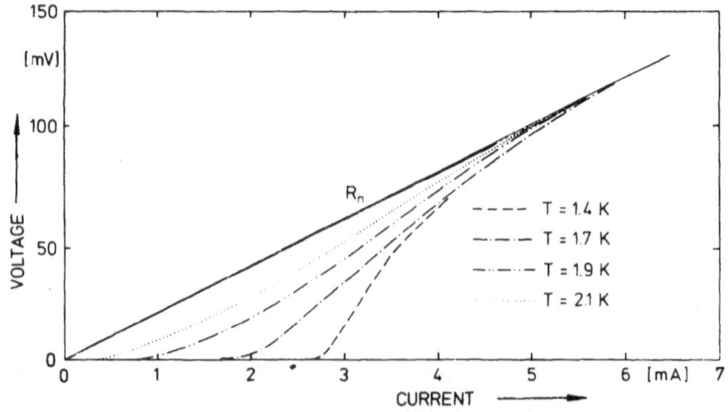

Fig. 2. Voltage-current characteristics of a 900 Å tin/200 Å gold layer.

R_n, which is approached by all curves asymptotically at a sufficiently high current density, increases with decreasing thickness of the total layer.

On the other hand, variation of the thickness of the gold layer at a fixed thickness of the tin layer affects the shape of the characteristics in just the opposite sense. Doubling the gold film thickness from 200 Å to 400 Å on a 1000 Å tin layer suppresses the superconductivity completely. This mutual influencing of both layers can be explained by proximity effects discussed below.

Figure 3 presents the voltage-current characteristics of a 1000 Å tin/200 Å gold layer that has been vacuum annealed after evaporation for 15 min at 52°C. Comparison with Fig. 1 shows that, after annealing, there is no abrupt voltage drop across the sensor even at the lowest temperatures, but a steady transition from the superconducting to the normal conducting state.

To understand the shape of the curves and the smoothing of the transition to superconductance, the structure of the gold-tin films was also investigated.

Investigation of the Structure of the Layer by Electron Microscopy

To obtain more detailed information about the evaporated films, electron microscopy photographs of the surfaces of the layers were taken. The surface structures turned out to be strewn with small crystallites. The form and size of these crystallites depend, among other things, on the substrate temperature and on the nuclei present on the surface of the substrate before evaporating.

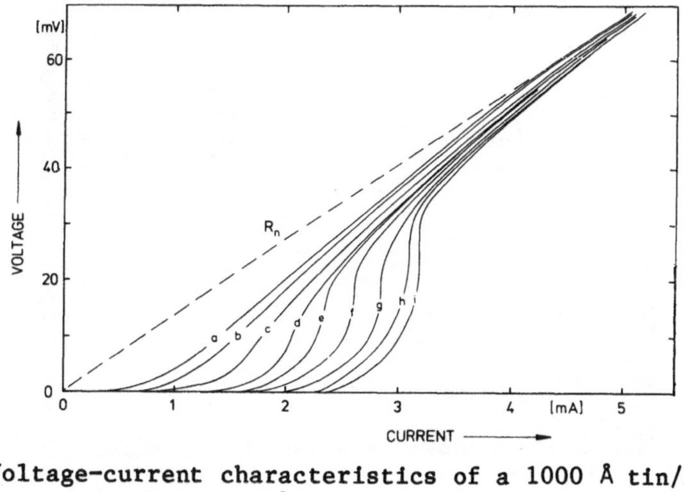

Fig. 3. Voltage-current characteristics of a 1000 Å tin/200 Å gold layer annealed at 52°C; a = 2.1 K, b = 2.0 K,...i = 1.3 K.

If a film is produced by quenching condensation of evaporated metal onto a cold substrate, it shows a higher degree of disorder; in this case the crystallites obtained are smaller and more regular. Figure 4 shows, as an example, the surface structure of a layer consisting of 1000 Å tin evaporated onto 200 Å gold at T = 20°C. From such pictures of the surface structures, it can be estimated that the thickness of the "effective" homogeneous layer of metal is only one-tenth to one-fifth of the value that is recorded by the film thickness monitor.

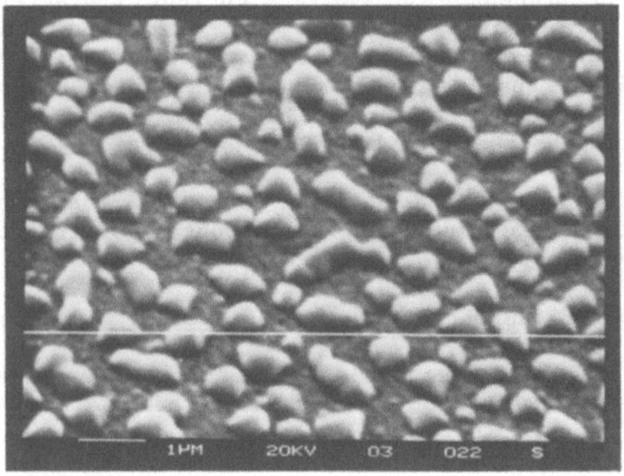

Fig. 4. Surface structure of a 1000 Å tin/200 Å gold layer.

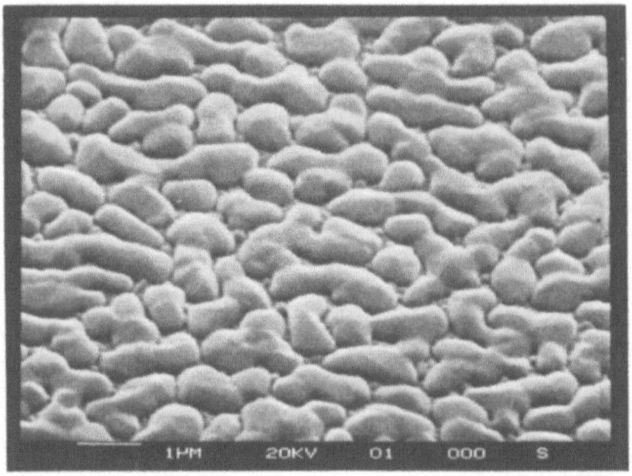

Fig. 5. The same as in Fig. 4, but after annealing.

If the substrate is heated after evaporating, the crystalline appearance of the surface will vanish. The crystallites are melted forming larger and more coherent regions. A photo of such a layer is shown in Fig. 5. It is known that the degree of disorder has a crucial effect on the superconducting behaviour of the films.[4] This effect on the shape of the voltage-current characteristics (compare Fig. 3 with Fig. 1) is discussed below.

Dependence of the Resistance of the Layer on the Thickness of Tin

To become more acquainted with the composition of the gold-tin layers, tin was evaporated onto gold films of different thickness, and the dependence of the resistance of the resulting layer on the thickness of the evaporated tin was determined (see Fig. 6). The resistance was always measured just after the evaporation, and its variation with time was then observed by further measurements. Depending on the ratios of gold to tin, the resistance increases or decreases asymptotically to a final value. In Fig. 6, one of the two curves consists of measured points taken just after each step of evaporation, the other one of points measured five minutes later. With the aid of its phase diagram, the different alloys of the system can be attributed to the points of crossover between these curves, the up and down variations of the curves resulting from the different specific resistances of these alloys. This means that the gold-tin layers consist of a mixture of fine crystalline regions of pure tin and the alloys of the gold-tin system, such as AuSn x Au, AuSn, and $AuSn_2$. Thus, the structure of the films is very heterogeneous.

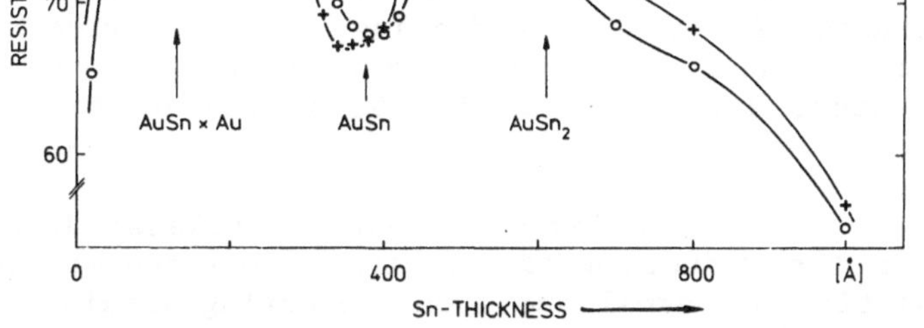

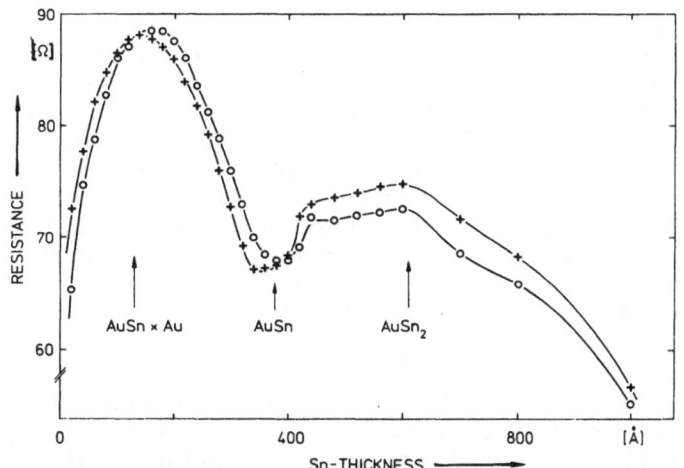

Fig. 6. Resistance vs. thickness dependence of tin layer evaporated onto a 200 Å gold layer; 0, just after evaporation, +, 5 min later.

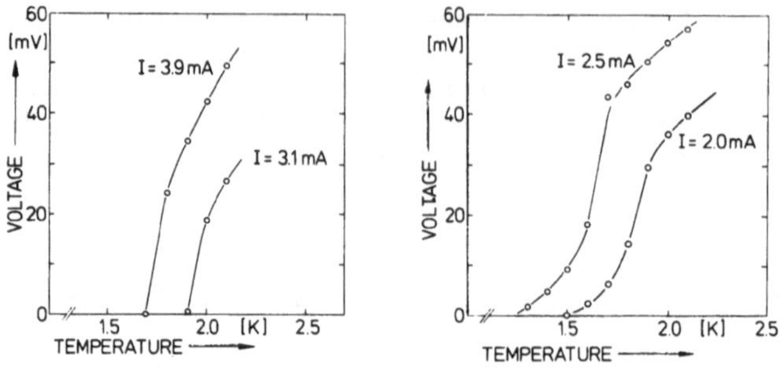

Fig. 7. The voltage drop dependence on temperature across a probe consisting of 1000 Å tin/200 Å gold: left side without annealing, corresponding to Fig. 1; right side after annealing at 52°C, corresponding to Fig. 3.

DISCUSSION

At a fixed measuring current, the sensitivity of a gold-tin layer is given by the maximum slope of its voltage vs. temperature dependence. The operating range corresponds to the steep linear portion of this curve. The V(T) graph is obtained from the respective voltage-current characteristic, considered at the fixed measuring current. The left plot in Fig. 7 shows the V(T) graph of a 1000 Å tin/200 Å gold film at two different measuring currents. By annealing such a film, its operating range is, at smaller measuring currents, shifted towards lower temperatures, as can be seen on the right side of Fig. 7.

The superconducting behaviour of the layers and, in particular, the lowering of the transition temperature to the temperature range of He II can be explained in terms of proximity effects between the normal and superconducting components of the films mentioned above. The mutual influence of these components depends on the contact surfaces and on the thickness ratio of these components. In general, the transition temperature of the film decreases with increasing thickness of the normal conducting layer at fixed thickness of the superconducting layer and vice versa. The result of combining two superconductors analogously shows a transition temperature lying between the T_c values of the pure superconductors.

Based on a phenomenological theory, Douglass[5] deduced an expression for the reduction of the transition temperature of thin films consisting of normal and superconducting metals. As is well known, the free energy, F, of a superconductor is smaller at temperatures below the transition point than in the normal conducting state. The difference of these energies, the so-called stabilization energy, is $\Delta F = -H^2_{cs}(T)/8\pi$; it becomes zero at T_c. The thermodynamic critical field is H_c; the index, s, designates the superconductor. By analogy, Douglass regarded a normal conductor (index n) as a metal with a free energy that is higher in the superconducting than in the normal conducting state. To describe both the normal conductor and the superconductor in the same terms of free energy and in the same way as above, a "critical field," H_{cn}, is attributed to the normal conductor by $\Delta F = -H^2_{cn}/8\pi$. Since ΔF is now positive, H_{cn} becomes imaginary. This is equivalent to regarding a normal conductor as a metal with a negative transition temperature.

Starting from this consideration and assuming the thickness of the layers to be smaller than the penetration depth and the coherence length, Douglass calculated the transition temperature of the composed layer as follows:

$$T_c = T_{cSn}(1 - \alpha\, d_{Au}/d_{Sn})^{1/2}/(1 + \alpha\, d_{Au}/d_{Sn})^{1/2}$$

with $\alpha = [-H^2_{cAu}/N_{Sn}(0)]/[H^2_{cSn}(0)/N_{Au}]$ and N = electron concentration. The electron concentrations of tin and gold as well as the critical field for tin are known. Once the transition temperature of a gold-tin film is known, one is able to calculate the coefficient α and the critical field, H_{cAu}. In this theory no assumption is made about the structure of the layers. The transition temperature depends only on the thickness ratio, d_{Au}/d_{Sn}.

To determine the transition temperature of the evaporated layers investigated, the critical currents, I_c, of four different gold-tin films are plotted vs. temperature in Fig. 8. The transition temperatures of the different layers are obtained by extrapolation of I_c to 0. The values of the transition temperatures obtained in this way, as well as the resulting values of α and H_{cAu}, are listed in Table 1. The phenomenological theory gives an adequate description of the lowering of the transition temperature, as can be seen from the good constancy of the values of α and H_{cAu}. The structure of the layers determines only the slope of the $I_c(T)$ curves. Annealing of the films results in a reduced slope (see Fig. 8). The transition temperature, however, remains unaffected, in agreement with the assumption of the theory. The smaller critical currents of the annealed film needed to adjust the superconducting to normal conducting transition of the film to a given temperature of the He II bath can be explained by a diminished degree of the layer's disorder.[4] The imaginary field, H_{cAu}, is a measure of the influence of the gold on the superconductor. The higher the strength of such an imaginary field, the thinner the layer can be in order to obtain the same lowering of the transition temperature. In comparison, the strength of the field of silver,

Table 1. Determination of the Imaginary Field of Gold with the Help of the Extrapolated Transition Temperatures of Different Composed Films

d_{Au}[Å]/d_{Sn}[Å]	d_{Au}/d_{Sn}	T_c[K]	α	H_{cAu}[10^{-4} T]
200/850	0.24	2.05	2.27	729 i
200/900	0.22	2.15	2.25	726 i
200/1000	0.20	2.29	2.25	726 i

obtained by Simmons and Douglass,[6] is $246 \cdot 10^{-4}$ i T, whereas the corresponding value of cobalt, which is expected to have a strong influence because of its ferromagnetism, amounts to $1400 \cdot 10^{-4}$ i T. Hence, the influence of gold is rather strong, especially if the absence of magnetic properties is taken into account. At a film thickness ratio of $d_{Au}/d_{Sn} = 1/\alpha$, the superconductivity of the bimetal film vanishes. Thus 220 Å of gold is sufficient to turn a 500-Å-thick tin layer into a normal conductor at all temperatures. In contrast, 310 Å of silver (α = 160)[6] was necessary to obtain the same effect.

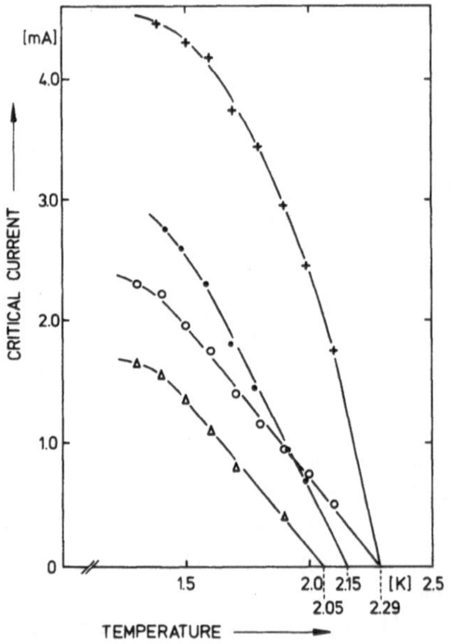

Fig. 8. Critical current of various gold-tin layers plotted versus temperature: – Δ – Δ – 850 Å Sn/200 Å Au; – • – • – • 900 Å Sn/200 Å Au; – + – + – + 1000 Å Sn/200 Å Au; – o – o – o 1000 Å Sn/200 Å Au, after annealing at 52°C.

CONCLUSION

On the basis of the strength of the critical field of gold, $H_{cAu} = 727 \cdot 10^{-4}$ i T, determined in the work presented, it is now possible to predict the conditions for a desired lowering of the transition temperature of a gold-tin film, to be used as a detector for second-sound shock waves, to a given temperature range. The fine adjustment of the transition point to the temperature of the He II bath is done by suitable adjustment of the measuring current. For measurements at low temperatures, for example at about 1.3 K, the measuring current can be reduced by annealing the layer after evaporation.

REFERENCES

1. G. A. Laguna, Second sound attenuation in a liquid helium counterflow jet, Ph.D. Thesis, California Institute of Technology, Pasadena, California (1975).
2. J. C. Cummings, D. W. Schmidt, and W. J. Wagner, Experiments on second-sound shock waves in superfluid helium, Phys. Fluids 21:713 (1978).
3. T. Schmeling, Untersuchungen über die Funktionsweise von supraleitenden Gold-Zinn-Filmen zur Messung von second sound-Stoßwellen in He II-Strömungen, Report No. 8, MPI Strömungsforschung, Göttingen (1981).
4. W. Buckel and R. Hilsch, Supraleitung und Widerstand von Zinn mit Gitterstörungen, Z. Phys. 132:420 (1952).
5. D. H. Douglass, Phenomenological theory of superimposed films of normal- and superconducting metals, Phys. Rev. Lett. 9:155 (1962).
6. W. A. Simmons and D. H. Douglass, Superconducting transition temperature of superimposed films of tin and silver, Phys. Rev. Lett. 9:153 (1962).

EFFECT OF THE DIFFUSION BARRIER ON THE MAGNETIC PROPERTIES OF PRACTICAL Nb_3Sn COMPOSITES*

S. S. Shen

Oak Ridge National Laboratory, Oak Ridge, Tennessee

INTRODUCTION

In conventional, bronze-process wire a tantalum barrier has been used to protect stabilizing copper from tin diffusion. Although the practice has been quite successful, considerable research efforts are still being devoted to search for other materials for the barrier; the one that receives the most attention is niobium. The use of niobium would simplify the fabrication process and result in appreciable cost savings in large-scale applications. (This approach was first attempted by AERE Harwell and Rutherford Laboratory, England.[1]) However, it has been well-recognized that the one major shortcoming of the process that may limit its application is that an unwanted large Nb_3Sn shell could be formed that may give rise to additional losses and result in instability.

This paper presents results of systematic experimental studies to quantify the above effects on samples that are identical except for different barriers. The composites were heat-treated under the same conditions, and the measurements were performed under no-stress conditions. These results are presented in static M-H curves, illustrating the effects of the unwanted Nb_3Sn shell.

SAMPLES AND EXPERIMENTAL TECHNIQUES

The primary objective of this study was to investigate and quantify the magnetic effect of a niobium diffusion barrier. The

*Research sponsored by the Office of Fusion Energy, U.S. Department of Energy, under contract W-7405-eng-26 with the Union Carbide Corporation.

effect of heat treatment on a conventional, bronze-process composite employing a niobium diffusion barrier is demonstrated in Fig. 1. A series of samples was specifically prepared, and the major parameters are listed in Table 1. Samples M7, M10, and M11 composites (0.5 mm in diameter), made by a "modified jellyroll"[2] method developed by Teledyne Wah Chang, Albany, consist of irregular rectangular filaments that measure about 2 μm x 10 μm. Westinghouse Large Coil Program (LCP) samples,[3] conventional bronze-process composites (0.73 mm in diameter) developed for the Westinghouse LCP coil[4] by Airco, contain 2869 filaments with a 3.7-μm diameter.

Techniques and calibration of the magnetization measurement have been described previously.[1] In this study, only static (hysteretic) magnetizations were measured, i.e., the rate of external field change was kept low enough to avoid any rate-dependent effect.

RESULTS AND DISCUSSIONS

Magnetic Effect of a Nb_3Sn Shell

The effect of a Nb_3Sn shell that resulted from tin diffusion into a niobium barrier can be seen in Fig. 2, where M-H curves are displayed for Westinghouse LCP wires with a tantalum barrier (trace 1) and a niobium barrier (trace 2). The differences between the loops can be attributed mostly to the presence of an unwanted Nb_3Sn shell. It is also observed that the M-H loop of a Nb_3Sn shell has two distinguished features: (a) a long tail at a low-field region representing a large amount of trapped flux associated with the superconducting shell, and (b) definite additional magnetization in a high-field region, primarily determined by the thickness of the shell and its critical current density. Most

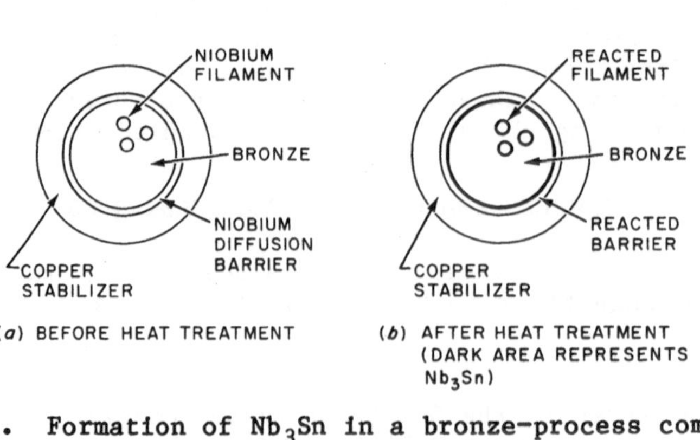

Fig. 1. Formation of Nb_3Sn in a bronze-process composite with a niobium diffusion barrier.

Table 1. Sample Parameters

Notation	Nb %	Twist parameter $X_o = \frac{2\pi r_o}{\ell_p}$	Diffusion barrier			Reaction condition	
			Ta	Nb			
				P-bronze	Bronze	Temp. °C	Time h
M7A	12	0.063	•				0
M7B	12	0.063	•			735	25
M10A	17	0		•		735	25
M10B	17	0.31		•		735	25
M10C	17	0.063		•			0
M10D	17	0.063		•		735	25
M11A	16	0			•	735	25
M11B	16	0.31			•	735	25
M11C	16	0.063			•		0
M11D	16	0.063			•	735	25
W/LCP A	10	0.2*	•			720	30
W/LCP B	10	0.2		•		720	30
W/LCP C	10	0.2			•	720	30

importantly, no instability was observed in these samples. Similar results were also observed in modified jellyroll (MJR) composites.[2]

It has been shown that the shielding magnetization of a superconducting shell in a "fully penetrated" region, can be expressed as[5]

$$-M_\delta = \frac{4}{3\pi} J_c r_2 \left(1 - \frac{r_1}{r_2}\right)^3 \tag{1}$$

where J_c is the critical current density of a Nb_3Sn shell and r_1 and r_2 represent the inner and outer radius. For a thin shell geometry, i.e., $\delta = r_2 - r_1$ and $\delta/r_2 \ll 1$, Eq. (1) becomes

$$-M_\delta \sim \frac{4}{3\pi} J_c \cdot 3\delta \tag{2}$$

Equation (2) will be used in this study to quantify the effect of a Nb_3Sn shell.

Effect of Phosphor Poisoning

To suppress the formation of Nb_3Sn inside a niobium barrier, a local concentration of phosphor has been introduced to a barrier-bronze interface[1] simply by sheathing commercial phosphor bronze inside the niobium barrier. This paper reports such efforts in both MJR composites and Westinghouse LCP wires.

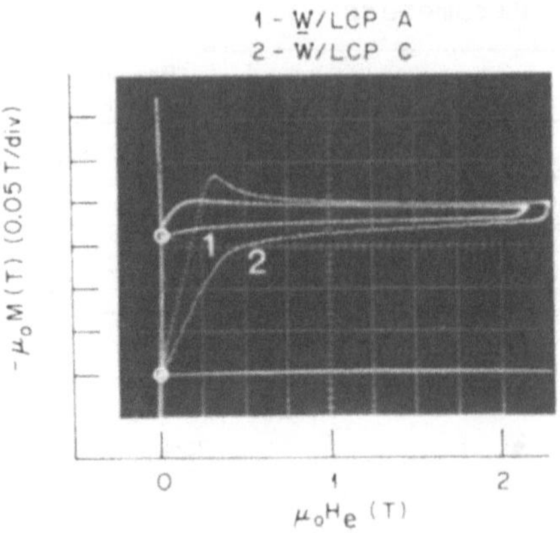

Fig. 2. M-H curves of Westinghouse LCP composites with (1) a tantalum barrier and (2) a niobium barrier.

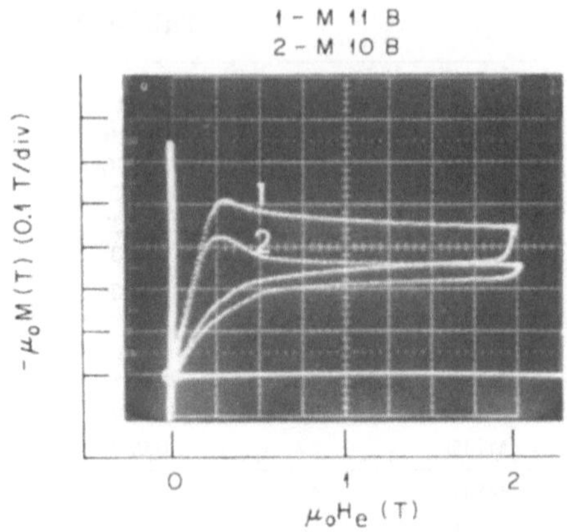

Fig. 3. Effect of phosphor poisoning in a MJR composite.

Figure 3 illustrates a quite satisfactory result of such attempts in MJR composites, where a phosphor-poisoned sample (trace 2) displays smaller amounts of trapped flux in the low-field region and decreased magnetization at the high-field region. This leads one to believe that the formation of the Nb_3Sn shell has been retarded by phosphor. These results can also be presented in the plot of hysteresis losses per unit volume versus maximum external field (see Fig. 4). Also shown, for reference, is sample M7B with a tantalum barrier.

As for the Westinghouse LCP composite, the effect is much less pronounced (see Fig. 5). It is still not known what metallurgical parameters are critically responsible for this result.

Summary of Results

As observed earlier, the niobium diffusion barrier does introduce additional magnetization that can be categorized into two aspects: "shielding region" and "penetrated region." For the shielding region, it is observed that a large amount of magnetic flux (~0.2 T) can be trapped in the superconducting shell, but such a thin shell can easily be penetrated by an external field ($\mu_o H_e > 0.3$ T) and left only with a limited fully penetrated magnetization. Therefore, it is important to note that for most high-field applications, the effect of the trapped flux would be negligible. As for a penetrated region, the Nb_3Sn shell contributes a definite magnetization determined by its thickness and current density. Applying Eq. (1) at $\mu_o H_e = 3$ T, with $J_c(Nb_3Sn) = 1.7 \times 10^{10}$ A/m^2, $\delta \sim 1$ μm for the niobium-barrier case and $\delta \sim 0.2$ μm for

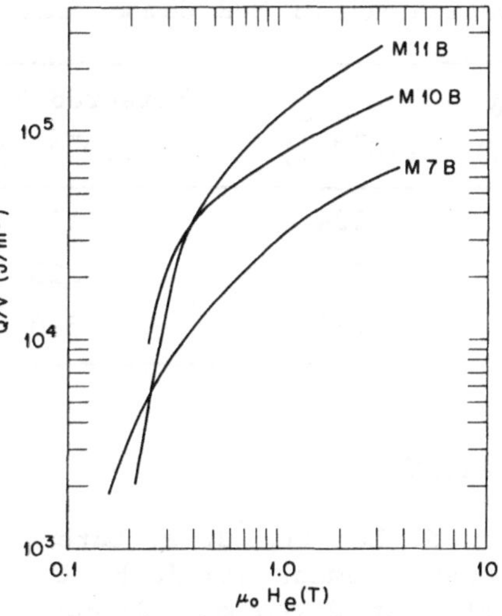

Fig. 4. Hysteresis losses (M-H loop area) for MJR composites with different diffusion barriers.

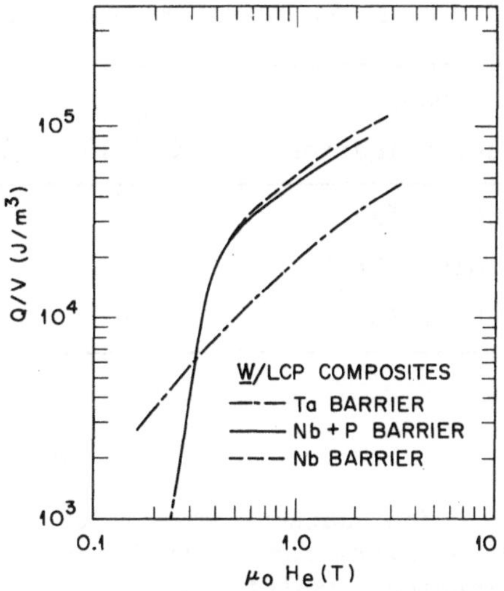

Fig. 5. Hysteresis losses (M-H loop area) for Westinghouse LCP composites with different diffusion barriers.

the poisoned case, shielding magnetization field 2 $\mu_o M_s$ is presented for MJR composites in Table 2. It is assumed that

$$\mu_o M_s = \mu_o M_f + \mu_o M_\delta \tag{3}$$

where M_f and M_δ represent magnetization due to filaments and diffusion barrier.

The last column of measured magnetization is in reasonably good agreement with theoretical results. Based on these findings, one may infer that a niobium plus phosphor bronze barrier can be a viable, cost-effective substitute for tantalum without detriment to the overall composite properties.

CONCLUSIONS

Experimental results of magnetization measurements on Nb_3Sn composites that employ niobium as the diffusion barrier are presented. In direct comparison with conventional wires in which tantalum is used, the results indicate that a niobium barrier introduces additional, but limited, magnetization that would be compensated by the cost saving and ease of wire fabrication for most applications.

Table 2. Data Analysis of Magnetization of MJR Composites

Sample Identification	$2\mu_o M_f$ $(10^{-4}$ T)	$2\mu_o M_\delta$ $(10^{-4}$ T)	$2\mu_o M_s$ $(10^{-4}$ T)	Measured $2\mu_o M_s$ $(10^{-4}$ T)
M7B (Ta)	124	0	124	124
M10B (Nb+P)	124	110	234	223
M11B (Nb)	124	550	674	580

ACKNOWLEDGMENTS

The author wishes to thank M. S. Lubell, J. N. Luton, and W. A. Fietz for their interest and encouragement and J. P. Rudd for technical assistance. The cooperation of Airco and Teledyne Wah Chang, Albany for providing sample conductors is greatly appreciated. The author would also like to express his gratitude to P. A. Sanger (Airco), J. A. Lee (Harwell), G. R. Wagner (Westinghouse), and W. J. Carr, Jr. (Westinghouse) for their helpful discussions.

REFERENCES

1. J. A. Lee and C. A. Scott, Work in the U.K. on filamentary A15 conductor development, in: "Filamentary A15 Superconductors," M. Suenaga and A. F. Clark, eds., Plenum Press, New York (1980), p. 35.
2. S. S. Shen and W. K. McDonald, Testing results of multifilamentary Nb_3Sn composites made by a modified jellyroll method, in: "Advances in Cryogenic Engineering - Materials" Vol. 28, Plenum Press, New York (1982), p. 535.
3. P. A. Sanger, E. Adam, E. Gregory, W. Marancik, E. Mayer, G. Rothschild, and M. Young, IEEE Trans. Magn. 15:789 (1979).
4. P. N. Haubenreich, IEEE Trans. Magn. 17:31 (1981).
5. W. J. Carr, Jr., Westinghouse R&D Center, Pittsburgh, Pennsylvania, private communication.
6. S. S. Shen, Magnetic properties of multifilamentary Nb_3Sn composites, in: "Filamentary A15 Superconductors," Plenum Press, New York (1980), p. 309.

STRAIN TOLERANCES IN Nb_3Sn CONDUCTORS*†

T. Luhman

Department of Energy and Environment, Brookhaven National Laboratory, Upton, New York

Knowledge of strain related effects in Nb_3Sn conductors has improved in recent years.[1,2] Early critical current - strain measurements showed variations in strain tolerances among conductors and even for individual ones after different reaction heat treatments. It was soon recognized that relative thermal contractions of the constituent materials in conductors combined to place a residual compressive strain on the Nb_3Sn compound. Other work demonstrated that whenever Nb_3Sn was strained in compression or tension, the critical current, I_c, decreased. Strain-induced critical current changes are now commonly interpreted as consequences of changes in the internal strain state of a conductor and variations from one conductor to another are attributed to differences in individual residual strain states.

Continued investigation elucidated other strain-related aspects of Nb_3Sn conductors. Universality of the intrinsic strain dependence of I_c led to the development of strain scaling rules to predict $I - \varepsilon$ behavior. Furthermore, it was shown that the upper critical field, H_{c2}, and the superconducting transition temperature, T_c, exhibit a parabolic dependence on strain and that this dependence was consistent with the microscopic theories of Type II superconductors. As yet however, there is no fundamental understanding of how the critical current, which is sensitive to metallurgical structure, varies with strain.

There has also been work to improve the effective strain tolerances of Nb_3Sn conductors. Some of the approaches used

---

* Invited paper.

† This work was performed under the auspices of the U.S. Department of Energy under Contract No. DE-AC02-76CH00016.

include metallurgical innovations to increase residual compressive strains, alloying of the Nb_3Sn compound to decrease its strain sensitivity, and external cladding on the conductor to enhance compression. Another approach involves development of improved magnet construction techniques. This paper reviews the literature pertinent to the above topics.

The effect of uniaxial strain on the critical currents of a variety of conductors is presented in Fig. 1. The data are generally characterized by a maximum in I_c. However, the critical currents of individual conductors respond differently to applied strain. The strain associated with the maximum value of I_c and the relative increase of I_c over the conductor's unstrained value differ from one conductor type to another. Similar kinds of I_c - ε behavior are also seen as a function of heat treatment on a single conductor. There are large variations in I_c with applied strain, but in all cases the changes are reversible with strain up to a certain limit. Typically this strain limit, which denotes

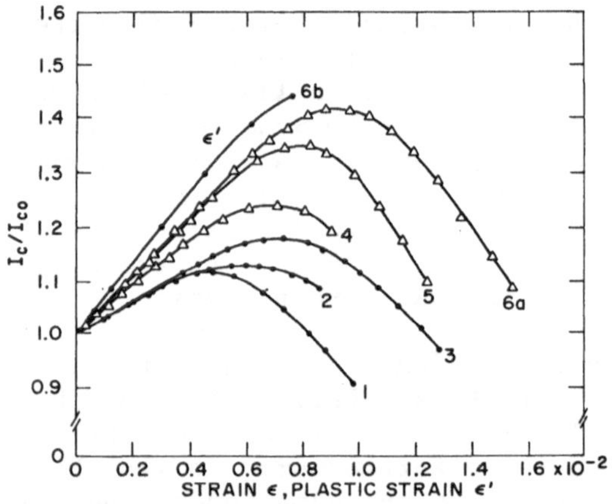

Fig. 1. Critical current, I_c, relative to the zero applied stress value, I_{co}, as a function of strain, ε, of differently heat-treated multifilamentary Nb_3Sn conductors. The ε' represents plastic strain remaining after load removal (after G. Rupp[3,4]).

Number in Fig. 1	Filament	Heat Treatment
1	1615	64 h/700 °C
2	1615	16 h/700 °C
3	1615	4 h/700 °C
4	61	64 h/750 °C
5	61	4 h/750 °C
6	61	24 h/750 °C

the onset of irreversible I_c behavior, is between 0.7 and 1.4%. The kinds of reversible changes observed when the applied load is removed can be seen in Fig. 2. Metallographic examination of conductors strained beyond their limits of reversibility has confirmed the association of the onset of irreversible behavior with cracking of the Nb_3Sn compound.[5]

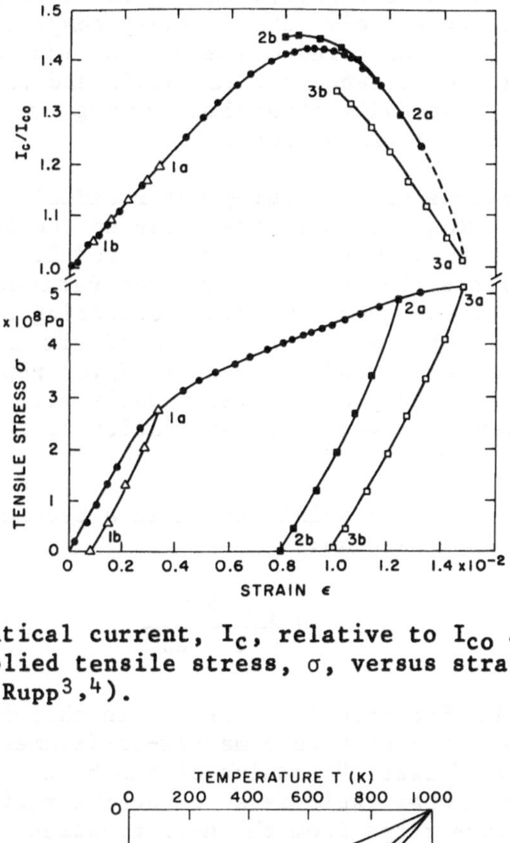

Fig. 2. Critical current, I_c, relative to I_{co} and externally applied tensile stress, σ, versus strain, ε (after G. Rupp[3,4]).

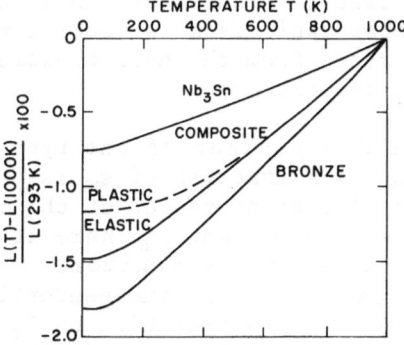

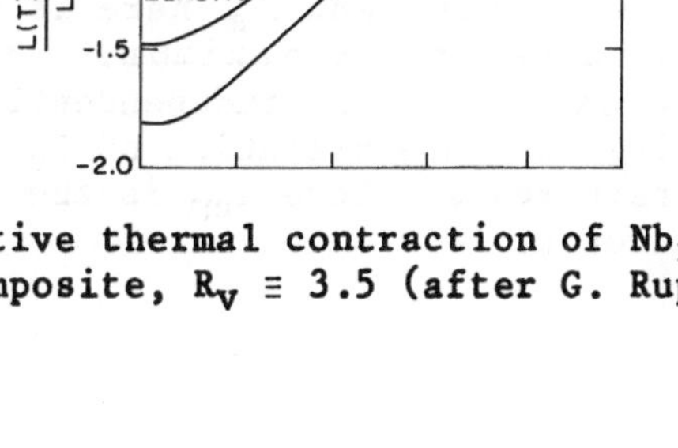

Fig. 3. Relative thermal contraction of Nb_3Sn, bronze, and a composite, $R_v \equiv 3.5$ (after G. Rupp,[1] p. 155).

Variations in the strain state of the Nb_3Sn filaments are responsible for the I_c behavior noted in Figs. 1 and 2. In as-reacted conductors, the filaments are compressively strained by the relatively large thermal contraction of the bronze matrix. A schematic representation of these relative thermal contractions is presented in Fig. 3. Two features of the composite conductor affect the net compressive strain on Nb_3Sn: the bronze-to-Nb ratio, R_v, and the mechanical strength of the bronze. An increase in the bronze-to-Nb ratio enhances the residual compressive strain on the compound. Conversely, the residual compressive strain can be decreased by virtue of a reduction in the effective relative thermal contraction difference between Nb_3Sn and the bronze matrix if the bronze is plastically stretched during cooldown from the reaction heat-treatment temperature.

The above discussion concerning the residual strain state in bronze-processed Nb_3Sn conductors leads to the following interpretation of the I_c changes in Figs. 1 and 2. Applied tensile strain counterbalance the thermally induced residual compressive strains in such a way that the superconducting critical current passes through an extremum as a function of the applied strain. Straining conductors beyond the peak in $I_c(\varepsilon)$ represents applied tensile strain on the compound and eventually will result in compound cracking and irreversible degradation of the critical current.

The initial residual axial strain in a bronze-processed wire conductor can be expressed as:

$$\varepsilon_m \simeq -\left|\frac{\Delta \ell}{\ell}\right| \frac{R_v C}{1 + R_v C} \tag{1}$$

where $\Delta\ell/\ell$ is the fractional difference in thermal contractions, R_v is the volume ratio of bronze matrix-to-filament, and C is the ratio of effective "elastic" modulus of the bronze to that of the filament. This latter ratio is an effective ratio of the moduli over the temperature range from the heat-treatment temperature to the cryogenic testing temperature.

Equation (1) is independent of the type of conductor used. The quotient I_{cm}/I_{co} as a function of ε_m is presented in Fig. 4. I_{cm} and I_{co} are critical current values at the maximum and initial strain states, respectively, and ε_m here is the experimentally determined strain value at the maximum. It is evident that I_{cm}/I_{co} indeed increases with ε_m independently of the conductor type, thus supporting the interpretation of $I_c - \varepsilon$ behavior based on the internal strain states where I_{cm} is the intrinsic "strain-free" I_c of the conductor.

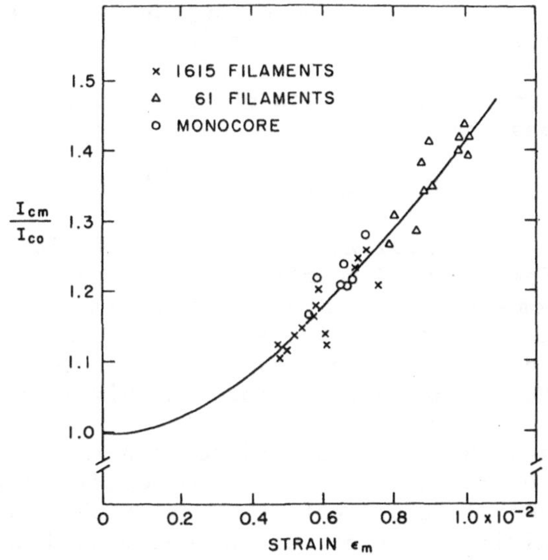

Fig. 4. Maximum value, I_{cm}, of the critical current relative to the initial value, I_{co}, as a function of applied strain of the maximum, ε_m, in the I_c-ε curve of differently heat-treated conductors. Symbols correspond to samples in Fig. 1; crosses: 1-3, circles: 4-6, triangles: monocore sample (after G. Rupp[3,4]).

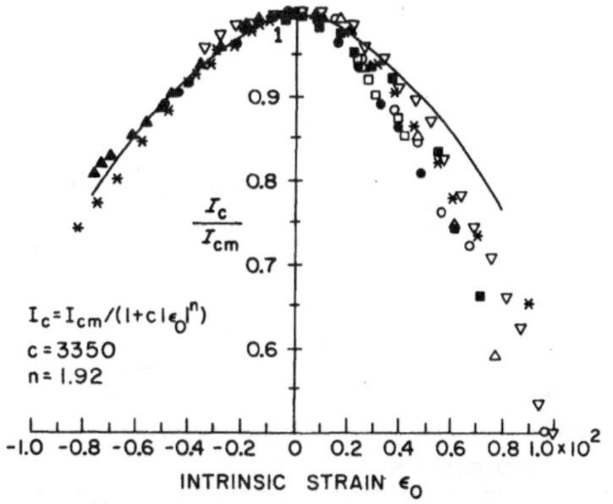

Fig. 5. Critical current, I_c, normalized by its maximum value, I_{cm}, vs. intrinsic strain, ε_o. See Fig. 1 for sample details (after G. Rupp[3,4]).

To delineate the compressive and tensile strain ranges, it is useful to introduce the term intrinsic strain, ε_o, defined as $\varepsilon_o = \varepsilon - \varepsilon_m$. For $\varepsilon_o < 0$, the Nb_3Sn compound is under compression; when $\varepsilon_o > 0$, it is under tensile strain. The universality of the critical current - strain relationship is seen in Fig. 5 where I_c/I_{cm} is plotted as a function of ε_o. The following empirical equation describes the data:

$$I_c = I_{cm}/(1 + C|\varepsilon_o|^n) \tag{2}$$

For $\varepsilon_o < 0$, n = 1.92 and C = 3350; this is the compressive strain range. The measured points can be fitted with Eq. (2) for $\varepsilon_o > 0$ using somewhat different parameters. The intrinsic or reversible strain dependence of the critical current is difficult to formulate theoretically since many metallurgical factors affect flux pinning. Much remains to be done in this area.

Further research is also needed to answer questions regarding microchemical aspects of the bronze-processed conductors. For example, when the volume percent of Nb_3Sn reaches 16-20%, compound cracking will initiate fracture of the whole conductor, as illustrated in Fig. 6. It is of interest to know what mechanisms are responsible for compound fracture. Another question that arises is: What is the extent of the reversible tensile range? For monofilamentary conductors, the extent of the tensile range is fixed at 0.6%; an illustration is presented in Fig. 7. Although the extent of the tensile range is fixed, the effective applied strain required to produce compound cracking is still a function of R_v through an increase in the initial compressive strain.

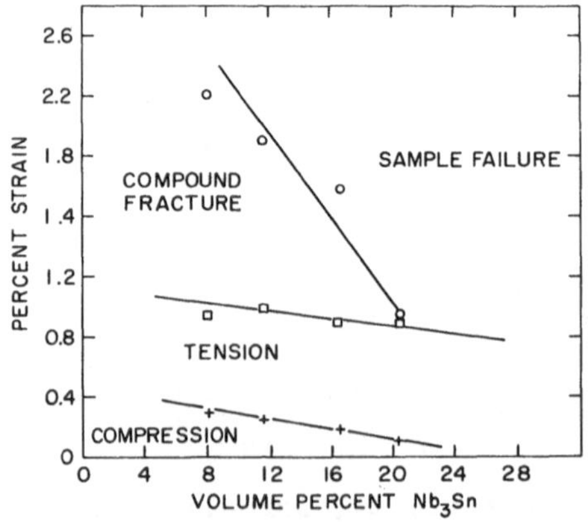

Fig. 6. Dependence of critical strain parameters on Nb_3Sn volume fraction; crosses: ε_m, squares: irreversible strain, circles: sample failure; 12-mil, 703-filament conductor, 5-μm filaments, R_v = 2.48.

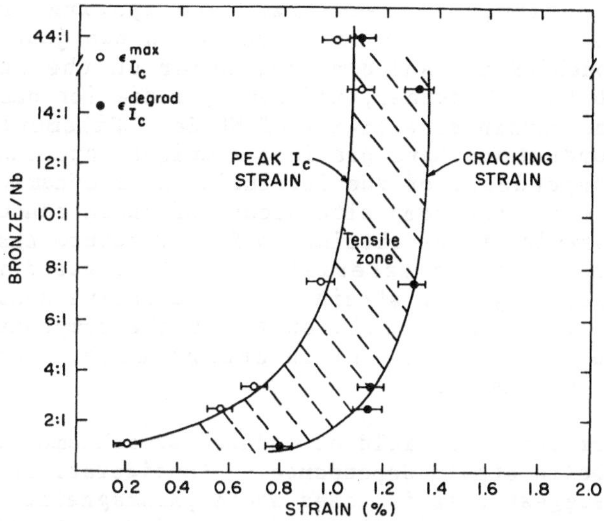

Fig. 7. Strains associated with peak I_{cm} and cracking values (irreversible behavior) as a function of the bronze-to-Nb ratio for Nb_3Sn monofilamentary conductors heat treated 15 h at 725°C, 4.0 T (after T. Luhman et al.,[1] p. 171).

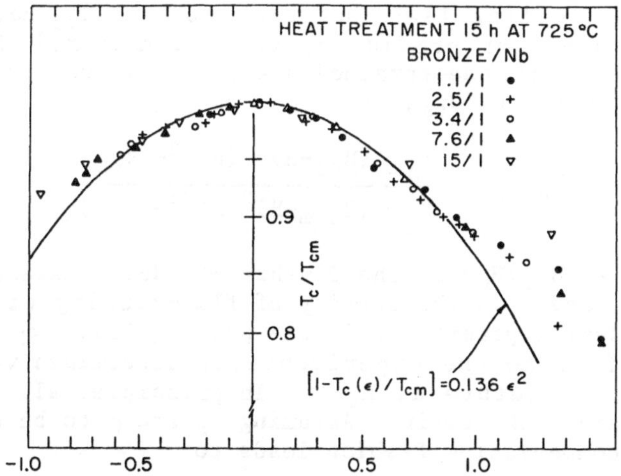

Fig. 8. Normalized superconducting transition temperature data, T_c/T_{cm}, where T_{cm} is the value of T_c at its maximum, as a function of intrinsic strain in monofilamentary conductors (after T. Luhman et al.,[1] p. 171).

Questions, such as how the extent of these strain regions depends on conductor design, filament size, and spacing, remain to be studied. Another area in which research has just begun is the effect of stoichiometry and compound order on the strain sensitivity of Nb_3Sn.[6] Disorder, induced by radiation damage, appears to enhance the strain sensitivity of Nb_3Sn. Filamentary bronze-processed conductors have provided a unique opportunity to study the strain dependence of the intrinsic thermodynamic properties, H_{c2} and T_c, since the composite nature of these conductors allows much larger strains to be obtained before fracture than is possible with free-standing material. As illustrated in Fig. 8, T_c varies quadratically with strain. A quantitative analysis, which included the triaxial strain state of the composite, concluded that this functional form is in accord with predictions of phenomenological theories.[7]

The upper critical field of ordered stoichiometric Nb_3Sn also shows a parabolic strain dependence.[7] Considerations based on the GLAG theory suggest that in cases where paramagnetic limitation is not a factor, the functional dependence of H_{c2} should be stronger than the dependence of T_c on strain. A still larger dependence would be expected if the Ginsburg-Landau parameter, κ, varies with strain. Such behavior, exhibited by Nb_3Sn and shown in Fig. 9, suggests agreement with the theory. Theoretical interpretation of the strain dependence of the intrinsic thermodynamic properties is, therefore, consistent with experimental findings.

Existing scaling laws that relate J_c and H_{c2} have been used to develop rules for predicting $I_c - \varepsilon$ behavior.[2] The critical current density in the strained state, J_c, is related to that in the strain-free state, J_{cm}, by:

$$J_c/J_{cm} = \frac{\kappa_{1m}^2}{\kappa_1^2}\,\frac{(H_{c2}-H)^2\,(H^{1/2}-\Phi\rho_m)^2}{(H_{c2m}-H)^2(H^{1/2}-\Phi\rho_o)^2} \qquad (3)$$

where κ_1 ($\equiv \sqrt{2}\,H_{c2}/H_c$) is the Ginsburg-Landau constant, Φ_o is the flux quantum, and ρ is the density of flux-pinning sites. There are three material parameters in the scaling law: κ_1, H_{c2}, and ρ. Here H_{c2} refers to the experimentally determined value usually seen in the literature as H_{c2}^*. In principle, all three parameters can vary with strain. Assuming H_c and ρ to be dependent on strain in a compensating fashion leads to

$$J_c/J_{cm} \quad \left(\frac{H_{c2}}{H_{c2m}}\right)^{1/2}\frac{(1-H/H_{c2})^2}{(1-H/H_{c2m})^2} \qquad (4)$$

If ε_m is known, then the value of Eq. (4) is its ability to predict $J_c(\varepsilon)$ without having to measure $H_{c2}(\varepsilon)^2$. One major

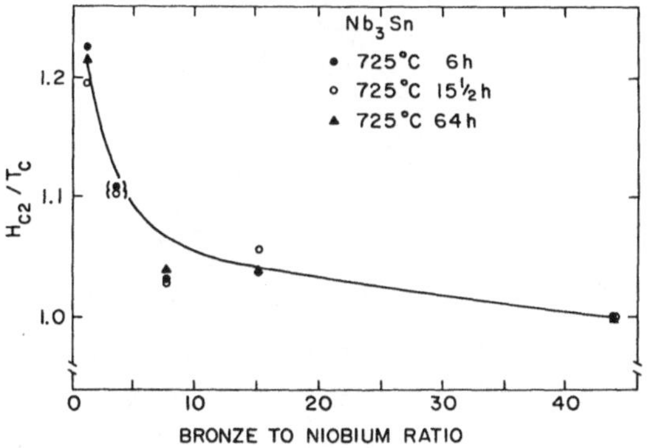

Fig. 9. Variation of the ratio of the upper critical field at 4.2 K to the critical temperature for bronze-processed monofilamentary Nb_3Sn conductors with the bronze-to-Nb ratio, and thus the degree of strain (after M. Suenaga et al.[8]).

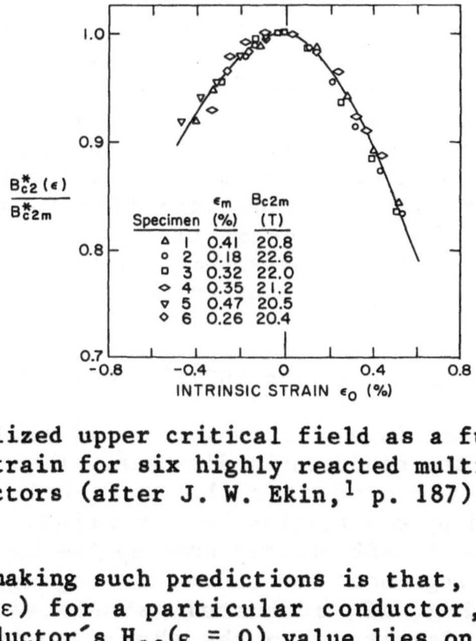

Fig. 10. Normalized upper critical field as a function of intrinsic strain for six highly reacted multifilamentary Nb_3Sn conductors (after J. W. Ekin,[1] p. 187).

restriction in making such predictions is that, without actually measuring $H_{c2}(\varepsilon)$ for a particular conductor, it is not known whether the conductor's $H_{c2}(\varepsilon \equiv 0)$ value lies on an assumed curve, such as the one in Fig. 10. This means use of Eq. (4) must be restricted to fully optimized conductors where the following assumptions are made: $H_{c2}(\varepsilon \equiv 0) \simeq 21$ T, conductors have solid rather than tubular filaments, and Nb_3Sn is unalloyed. Equation (4), evaluated for a conductor meeting these criteria, is illustrated in Fig. 11, and the agreement is remarkably good.

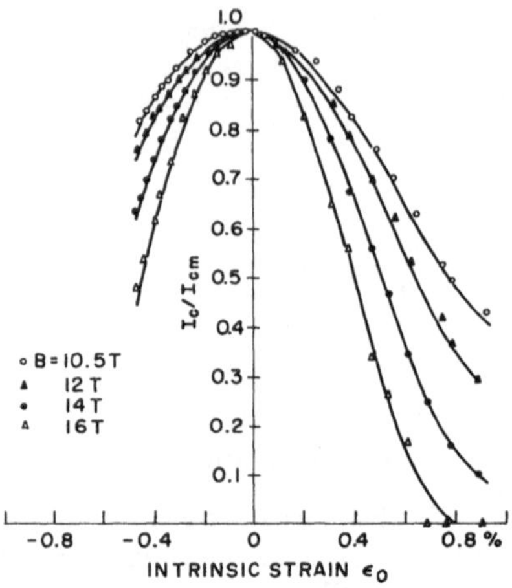

Fig. 11. Critical current, I_c, normalized to the maximum value, I_{cm}, as a function of intrinsic strain, ε_0, for different flux densities (after G. Rupp[10]).

Since the use of these conductors always involves some bending, it is helpful to be able to predict bend-strain tolerances. The approach is straightforward and involves averaging the strain distribution across a bent conductor, such as in the schematic illustration presented in Fig. 12. Twisted multifilamentary conductors that have effective current sharing require a different averaging scheme than do conductors that do not current share. Effective current sharing occurs when the conductor's current transfer length, X_{min}, is less than its twist pitch, ℓ. Current transfer lengths in multifilamentary Nb_3Sn conductors can be estimated from[2]

$$X_{min} = (4.1 \times 10^{-2})(\rho_m/\rho^*)^{1/2}D \tag{5}$$

Here ρ_m is the matrix resistivity, ρ^* is the detection sensitivity, and D is the conductor's diameter. The current transfers from regions of the conductor that have had their current capacities decreased by compressive bend strains. In other words, the transfer process effectively uses up the increased capacities of the tensioned regions. Therefore, averaging from point A to point B, as in Fig. 12, adequately simulates the conductor's overall current density when the above conditions prevail. Without effective current transfer, the current in individual filaments is limited to the minimum value of current density set by the compressive region. Then I_c is represented by an averaging scheme from point A to point O. A comparison of bend-strain current behavior for long- and short-twist-pitch regimes, $\ell > X_{min}$ or $\ell < X_{min}$, is presented in Fig. 13. The short-twist-pitch conductor shows the greatest degradation with bend strain.

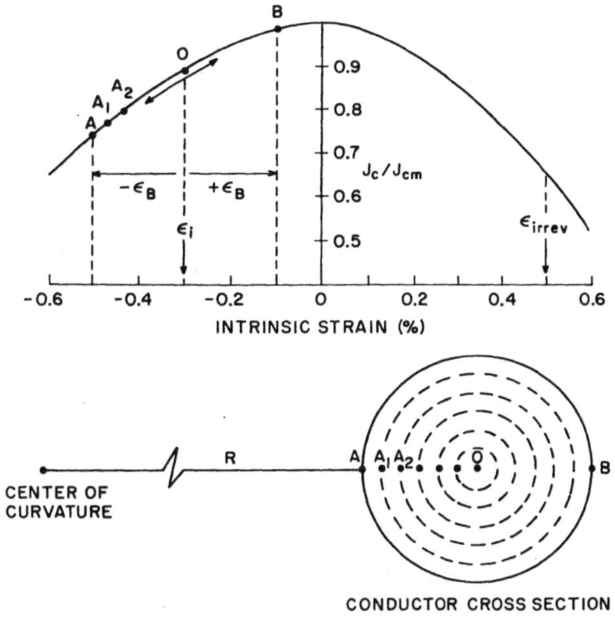

Fig. 12. Schematic diagram of the effect of bending strain on the critical current of multifilamentary Nb_3Sn superconductors (after J. W. Ekin,[1] p. 187).

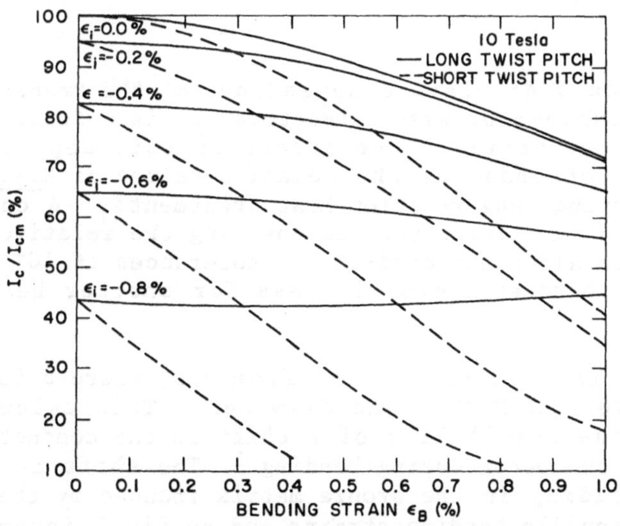

Fig. 13. Comparison of bending-strain degradation for long-twist-pitch ($\ell > X_{min}$) and short-twist-pitch ($\ell > X_{min}$) multifilamentary Nb_3Sn superconductors (after J. W. Ekin,[1] p. 187).

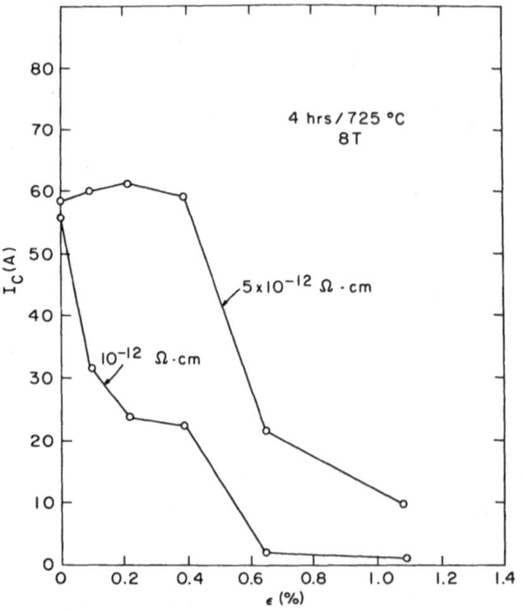

Fig. 14. The critical current vs. bend-strain relationship for a conductor whose ℓ/X_{min} values are 0.55 and 1.2 at overall resistivities of 10^{-12} and 5×10^{-12} Ω·cm, respectively. Heat treatment: 4 h at 725°C, Sn content in bronze: 12 wt.%, $\rho_m = 2 \times 10^{-6}$ Ω·cm (after T. Luhman et al.[11]).

Equation 5 displays a dependence of the transfer length on ρ_m. For bronze-processed conductors, ρ_m is a function of the Sn content of the matrix. One effect of this dependence is the possibility of changing the relative values of X_{min} and ℓ by Sn depletion during long reaction heat treatments. A comparison of Figs. 14 and 15 illustrates how changing the relative values of ℓ and X_{min} can alter the bend-strain tolerances at 10^{-12} Ω·cm. The tolerance to bend straining is less for shorter heat-treatment times.

Bend strain is calculated from d/D, where d is the conductor's diameter and D the bend diameter. This calculation must allow for the possibility of a shift in the conductor's neutral axis, which can occur during bending.[1] The shift is a result of plastic yielding in the bronze matrix induced by the combination of applied tensile bending strains and residual internal tensile strains in the bronze. The total strain resulting from a sum of the applied and residual strains can produce plastic flow in the tensile region prior to the plastic flow in the compression side.

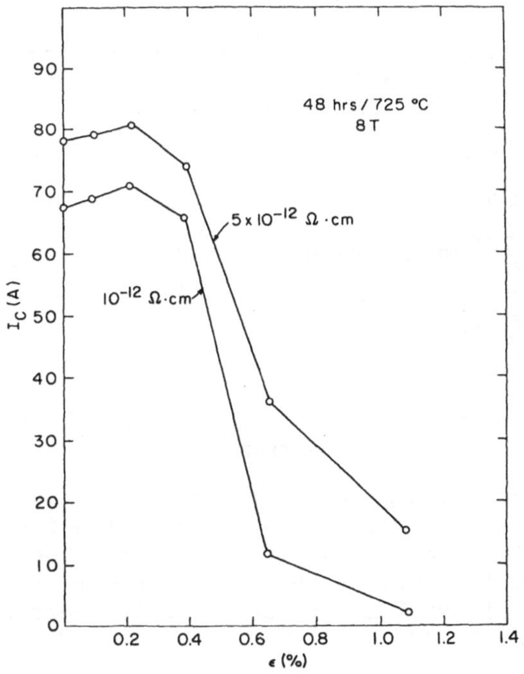

Fig. 15. The critical current vs. bend-strain relationship for a conductor whose ℓ/X_{min} values are 1.9 and 4 at overall resistivities of 10^{-12} and 5×10^{-12} Ω·cm, respectively. Heat treatment: 48 h at 725°C, Sn content in bronze: <1 wt.%, $\rho_m = 1.6 \times 10^{-7}$ Ω·cm (after T. Luhman et al.[11]).

Nonuniform plastic flow creates an imbalance of the forces across the conductor with a resultant shift of the neutral axis towards the center of bend curvature. Inclusion of the neutral axis shift enables the calculation to provide for features such as the peak observed in I_c-bend-strain data; see for instance Fig. 14.

It is evident from the preceding discussions that the composite nature of Nb_3Sn conductors, and their residual strains in particular, play an important role in determining their tensile and bend-strain tolerances. Residual strains also appear to affect these conductors' responses to fatigue deformation.[2,11] Plastic deformation and work hardening during fatiguing have been found to affect the residual strain distribution, with a resultant change in I_c. Plastic deformation removes initial residual strains, and the work hardening eventually leads to a more elastic-like deformation. The I_c changes, therefore, can generally be anticipated from known tensile behavior.

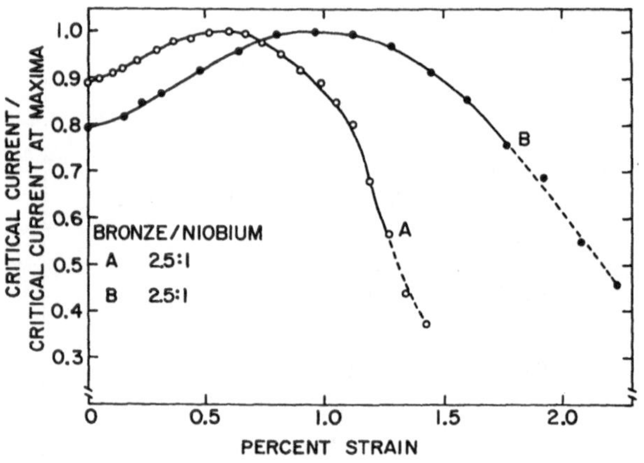

Fig. 16. Superconducting critical current, normalized to the maximum critical current value, as a function of applied tensile strain at 4.2 K and 4.0 T. Sample A represents a Cu + 11.2 wt.% Sn bronze; sample B, a Cu + 11.2 wt.% Sn + 0.18 wt.% Be bronze. Samples were heat-treated for 15 h at 725°C.

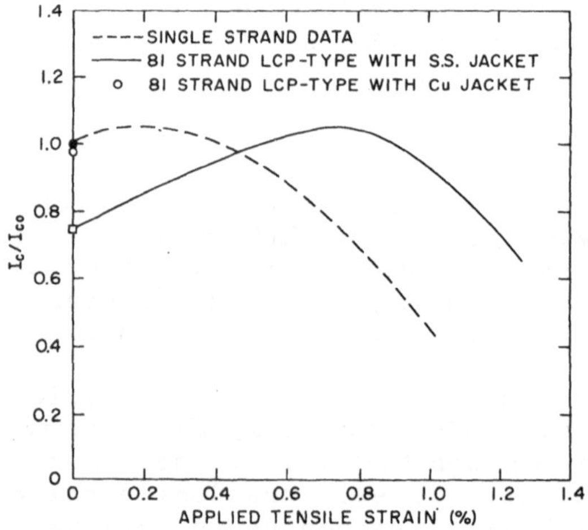

Fig. 17. The critical current, normalized to the value for zero applied strain on a single strand, plotted as a function of applied tensile strain for a single strand and for 81-strand LCP-type conductors (after R. M. Scanlan et al.,[1] p. 221).

Specific efforts have been made to enhance the strain tolerances of Nb_3Sn conductors. One metallurgical innovation takes the approach of minimizing any accommodating plastic flow in the bronze during cooldown from the reaction heat treatment temperature by alloying the bronze with Be. This approach is useful where the bronze-to-Nb ratio must be kept small, i.e., $R_v \leq 10$. For small bronze-to-Nb ratios, Be alloying can provide maximum compressive strains of the order of ∿1%. Figure 16 illustrates the effect of a 0.18 wt.% Be addition to a Cu + 11.2 wt.% Sn bronze matrix on conductor strain tolerances. Moreover, alloying of the Nb_3Sn compound itself can lead to a reduced dependence of J_c on strain.[12] In this case, Ga additions appear to be useful.

Another method has been to jacket a conductor with stainless steel, thereby increasing the effective R_v. An example of strain tolerances achieved using this method is shown in Fig. 17.

The 'wind and react' approach to magnet construction using Nb_3Sn has been receiving attention.[13-15] The usual procedure is to vacuum impregnate the coil windings with epoxy following the reaction heat treatment. It has now been established that there are several insulating materials, such as S-glass and certain ceramics, capable of withstanding typical Nb_3Sn reaction heat-treatment conditions. This approach appears to have had success in those magnets tested to date. A magnetic field of 7.5 T was achieved in a 0.9-m-long, 10-cm-bore beam-line quadrupole magnet. Overall current density in the winding was 300 A/mm^2.[12] It is notable that very minimal training occurred in this magnet.

The previous discussion has highlighted the current status of strain tolerances in Nb_3Sn conductors and identified several areas needing further research. It is the opinion of the author that sufficient information is available to support the contention that Nb_3Sn conductors will increasingly be used in a wide variety of high magnetic field applications.

ACKNOWLEDGMENTS

The work reviewed here was conducted by many scientists, including M. Suenaga, D. O. Welch, J. W. Ekin, G. Rupp, and R. M. Scanlan.

REFERENCES

1. For reviews see "Filamentary A15 Superconductors," M. Suenaga and A. F. Clark, eds., Plenum Press, New York (1980).

2. J. W. Ekin, Mechanical properties and strain effects in superconductors, in: "Superconductor Materials Science," S. Foner and B. B. Schwartz, eds., Plenum Press (1981), p. 455.
3. G. Rupp, IEEE Trans. Magn. MAG-13:1565 (1977).
4. G. Rupp, J. Appl. Phys. 48:3858 (1977).
5. T. Luhman, M. Suenaga, D. O. Welch, and K. Kaiho, IEEE Trans. Magn. MAG-15:699 (1977).
6. C. L. Snead, Jr. and M. Suenaga, Appl. Phys. Lett. 37:659 (1980).
7. D. O. Welch, in: "Advances in Cryogenic Engineering--Materials," Vol. 26, Plenum Press, New York (1980), p. 48.
8. M. Suenaga, T. Onishi, D. O. Welch, and T. S. Luhman, Bull. Am. Phys. Soc. 23:229 (1978).
9. E. J. Kramer, J. Appl. Phys. 44:1360 (1973).
10. G. Rupp, IEEE Trans. Magn. MAG-15:189 (1979).
11. T. Luhman, D. O. Welch, and M. Suenaga, IEEE Trans. Magn. MAG-17:662 (1981).
12. J. W. Ekin, H. Sekine, and K. Tachikawa, J. Appl. Phys. 52:6252 (1981).
13. A. Asner, C. Becquet, D. Hagedorn, C. Nigueletto, and W. Thomi, IEEE Trans. Magn. MAG-17:416 (1981).
14. K. Ishibashi, M. Koizumi, K. Hosoyama, M. Kobayashi, and T. Horigami, IEEE Trans. Magn. MAG-17:468 (1981).
15. K. J. Best and B. Rothe, IEEE Trans. Magn. MAG-17:478 (1981).

ON THE INTERACTION OF STRESS WITH THE MARTENSITIC PHASE TRANSITION IN A15 COMPOUNDS*

D. O. Welch

Division of Metallurgy and Materials Science, Brookhaven National Laboratory, Upton, New York

INTRODUCTION

Recently there has been a resurgence of interest in the effect of the martensitic phase transition that occurs in many A15 compounds[1] on superconductivity[2] and on elastic and anelastic behavior.[3] Since in many practical applications, A15 compounds are subject to considerable stress and strain, it is of interest to examine the interaction of stress with the martensitic transition; this paper is an examination of the effects of stress predicted by a simple Landau model, which successfully describes many features of the transition, and the related temperature dependence of the elastic modulus $(d_{11} - c_{12})/2$.[1] Earlier, Pietrass[4] discussed some of the effects of stress on the phase transition in the context of a Landau model, and the present paper is an extension and development of this theoretical approach. We focus on the effect of stress on the temperature ranges of stability and metastability of various types of martensitic domain and briefly discuss the nonlinearity of the stress-strain relation in a polycrystalline A15.

SYMMETRIZED STRESSES AND STRAINS

The lattice distortions associated with various types of martensitic domain and the effect of different types of stress are most efficiently characterized in terms of symmetrized linear combinations of strain components, e_{xx}, e_{yy}, etc. (relative to cubic crystal axes x, y, z):

*Work performed under the auspices of the U.S. Department of Energy under Contract No. DE-AC02-76CH00016.

$$e_1 = (2e_{zz} - e_{xx} - e_{yy})/\sqrt{6} \qquad (1)$$

$$e_2 = (e_{xx} - e_{yy})/\sqrt{2} \qquad (2)$$

$$e_v = (e_{xx} + e_{yy} + e_{zz})/\sqrt{3} \qquad (3)$$

Symmetrized stress components are similarly defined. In a linearly elastic cubic phase, e_1 couples only to σ_1, e_2 only to $\sigma 2$, and e_v only to σ_v. However, in a martensitically transforming A15, nonzero values of e_1 and e_2 arise below the transition temperature, T_m, owing to the tetragonal lattice strain of the transition. The relative values of e_1 and e_2 in a transformed domain depend on whether the tetragonal axis is parallel to the x, y, or z axis: in a stress-free crystal $(e_1,e_2) = (e,0)$, in a z-domain, $(-e/2, \sqrt{3}e/2)$ in an x-domain, and $(-e/2, -\sqrt{3}e/2)$ in a y-domain, where e varies with temperature. The values of the symmetrized stress components in a crystal depend on the orientation of the crystal axes relative to the principal stress axes; for example, in a cylindrical filamentary composite the σ_1 and σ_2 values in a given crystallite are $(\sigma_L - \sigma_R)(3n_z^2 - 1)/\sqrt{6}$ and $(\sigma_L - \sigma_R)(n_x^2 - n_y^2)/\sqrt{2}$ respectively, where σ_L and σ_R are the longitudinal and radial stress components and n_x, etc., are the direction cosines between the crystal axes and the longitudinal axis of the composite. Thus, for general orientations, σ_1 and σ_2 both have nonzero values, although certain experiments, such as x-ray measurements, may select crystallites whose orientation yields only nonzero values of σ_1; for example, all crystallites with the crystal z-axis parallel to the filament axis have $\sigma_2 = 0$.

A SIMPLE MODEL

The simplest model that illustrates the essential features of the interaction of stress with the transition is a simple Landau model[1] in which the free-energy density relative to the cubic phase is:

$$F = \frac{1}{2} Ae_1^2 + \frac{1}{4}Ce_1^4 - \sigma_1 e_1 \qquad (4)$$

where $A = (c_{11} - c_{12})$ is a function of temperature and C is sensibly independent of temperature. Near the transition temperature, A varies essentially linearly with temperature, and Weger and Goldberg[1] showed that the elastic properties of a Nb_3Sn single crystal can be described accurately over a wide range of temperature with a temperature dependence of the form:

$$A(T) = A'[T - T^*]\{(1 + a_\infty^2)/[(T/T^*)^2 + a_\infty^2]\}^{1/2} \qquad (5)$$

where A' and a_∞ are constants and a_∞ is chosen to give the correct ratio of the modulus $(c_{11} - c_{12})$ at low and high temperatures

($a_\infty \simeq 1$ for Nb_3Sn). The equilibrium strain at fixed T and σ_1 is obtained from $\partial F/\partial e_1 = 0$. Typical results are shown in Fig. 1; the essential features are easily obtained from the properties of quadratic and cubic equations.

The results shown in Fig. 1 illustrate the main properties of the simple model. With no applied stress, there is a second-order phase transition at T* (a model for which the transition is first order is discussed below) from a cubic phase to one of two degenerate tetragonal "ferroelastic" phases with tetragonality of opposite sign. For illustrative purposes we can regard the phase with positive e_1 as a z-domain and the phase with negative e_1 as a composite of an x- and a y-domain. The application of a σ_1 stress "parallel" to the tetragonality of the z-domain (i.e., with the same sign as e_1 for the z-domain) has several consequences: (1) the sharp second-order transition is replaced by a continuous, diffuse transition between "paraelastic" and "ferroelastic states; (2) the degeneracy in energy of the z- and xy-domains is removed with the z-domain being thermodynamically stable while the xy-domain becomes

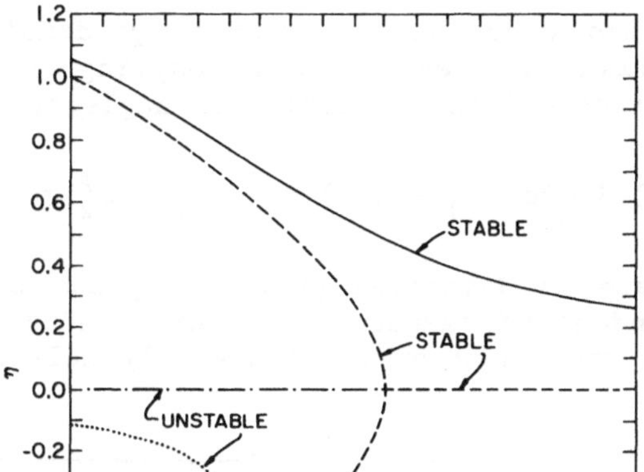

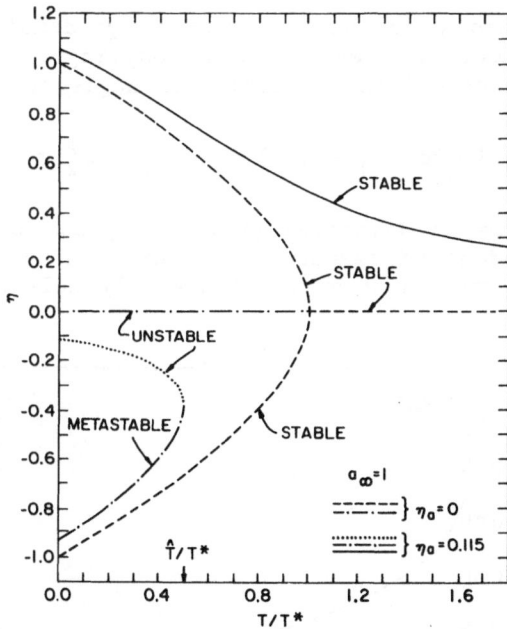

Fig. 1. Effect of applied stress on values of strain for which $\partial F/\partial e_1 = 0$ (F given by equation 4); $\eta \equiv e_1(T,\sigma_1)/\ |e_1(0,0)|$. Applied stress is expressed as an equivalent strain at high temperature: $\eta_a \equiv \sigma_1/[A(T = \infty)|e_1(0,0)|$. The temperature dependence of the modulus $A = (c_{11} - c_{12})$ is that of equation 5 with $a_\infty = 1$.

metastable; (3) the temperature of the continuous transition, as measured by the location of the inflection point $\partial^2 e_1/\partial T^2 = 0$, is independent of stress, remaining pinned at T*; and (4) the temperature, $\hat{T}$, above which the metastable domain becomes absolutely unstable is depressed, steeply at first, by increasing σ_1, $\hat{T}$ being given by the condition $A(\hat{T}) = -[27/4)C\sigma_1^2]]^{1/3}$. Taking the temperature dependence of A given in equation 5 yields the results shown in Fig. 2.

A MORE REALISTIC MODEL

The most serious deficiencies of the simple model above are: (1) it exhibits a second- rather than a first-order transition accompanied by a small strain discontinuity, as is actually the case for V_3Si and Nb_3Sn,[1] and (2) it does not properly distinguish between x-, y-, and z-domains. The first deficiency is removed by including a small term cubic in strain in the free-energy density (equation 4); the strain for which $\partial F/\partial e = 0$ is then of the form shown in Fig. 3 rather than that in Fig. 1. The second deficiency is remedied by including the symmetrized strain component e_2 as well as e_1 (see equations 1 and 2).

Pietrass[4] has described a free-energy density function with the correct symmetry; when a small coupling between hydrostatic strain and the martensitic transition is neglected, this becomes:

$$F = \frac{1}{2}A(e_2^1 + e_2^2) + \frac{1}{3}Be_1(e_1^2 - 3e_2^2) + \frac{1}{4}C(e_1^2 + e_2^2)^2 - e_1\sigma_1 - e_2\sigma_2 \quad (6)$$

where A and C are as described previously, and B is assumed to be sensibly independent of temperature. For fixed T, σ_1, and σ_2 equilibrium strains are found by simultaneously solving $\partial F/\partial e_1 = 0$ and

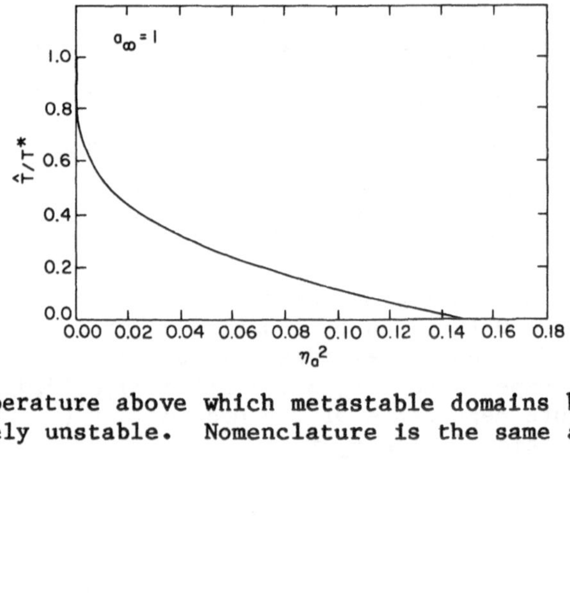

Fig. 2. Temperature above which metastable domains become absolutely unstable. Nomenclature is the same as in Fig. 1.

$\partial F/\partial e_2 = 0$; in general, this must be done numerically. However, if we restrict consideration to only nonzero σ_1 stresses, the essential features can be derived analytically, and we can thus obtain a qualitative understanding of the interaction of the stress with the transition. (The basic effect of nonzero σ_2 is to lift the energetic degeneracy of x- and y-domains.) However, our previous discussion shows that $\sigma_2 = 0$ only for crystallites of special orienta ion in, for example, a filamentary conductor.

When a stress σ_1 with the same sign as e_1 in a transformed z-domain is applied, the effect at low stresses is: (1) to raise the temperature of the transition between the distorted cubic (paraelastic) phase and a ferroelastic z-domain and simultaneously to reduce the discontinuity at the transition, as shown in Fig. 3 and (2) to make the x- and y-domains metastable, with a simultaneous reduction in the temperature of the onset of instability (the point f of Fig. 3). The amount of these changes at small stress is $\Delta T_m \simeq +2\sigma_1/A'e_m$ and ΔT_f(x- or y-domain) $\simeq -2\sigma_1/3A'e_m$ where $e_m \equiv -2B/3C$ is the spontaneous change in e_1 at the transition and A' is as in equation 5. As the stress increases, the points p and f in Fig. 3 move closer together; at the point of coalescence, the first-order character of the transition has been destroyed, and thereafter the strain-vs-T curve resembles that of Fig. 1. The critical applied stress for this event is $\sigma_{crit} = Ce_m^3/8$. This condition may be expressed more understandably as a critical value of the equivalent normalized applied strain (defined in the caption of Fig. 1): $\eta_a^{crit} \simeq e_m(B^2/A'C)/[18T^*|e_1(0,0)|]$. For typical values[1,4]

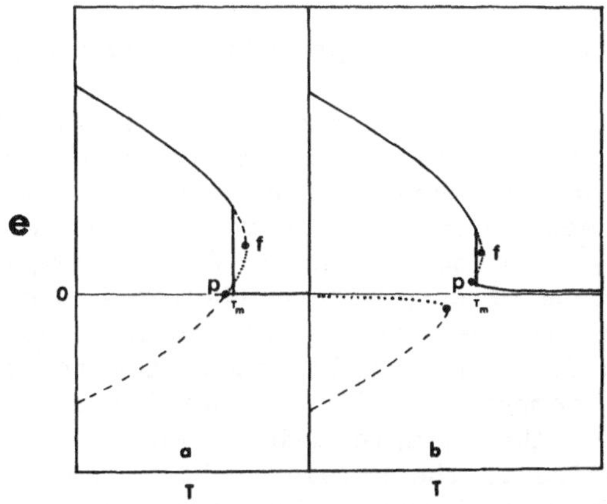

Fig. 3. Values of strain, e, at which $\partial F/\partial e = 0$ for a Landau model, including terms cubic in strain. In (a) $\sigma_1 = 0$ and in (b) $\sigma_1 \neq 0$. Stable phase ——; metastable phase ---; absolutely unstable ... . Points p and f mark, respectively, the onset of instability in the "paraelastic" or cubic phase and the "ferroelastic" phase.

of the various parameters of Nb_3Sn, $\eta_a^{crit} \simeq 0.003$: an applied strain of only ~0.3% of the spontaneous strain at $\sigma_1 = 0$ and $T = 0$ serves to destroy the first-order character of the transition. The concomitant temperature rise at the critical stress is only about 0.5–0.8 K. For larger stresses, the temperature of the diffuse transition, as measured by the inflection point in the e_1-vs-T curve, rises slowly with stress, given by $A(T_{i.p.}) = 3e_m(C^2\sigma_1)^{1/3}/2$ and equation 5.

When a stress σ_1 of sign opposite to that of e_m is applied, at low stresses a first-order transition from the tetragonal paraelastic phase to a ferroelastic x- or y-domain occurs; the symmetry of the ferroelastic domain distorted by the stress is orthorhombic. Initially, the transition temperature rises at a rate given by $\Delta T_m \simeq -\sigma_1/A'e_m$. As before, a critical value of the stress, $\sigma_{crit} = -81Cd_m^3/128$, the first-order character of the transition is destroyed and the transition is continuous thereafter. (This critical stress is equivalent to a strain at high temperatures of only ~ −1.5% of the spontaneous strain at $\sigma_1 = 0$ and $T = 0$.) Unlike the previous case ($\sigma_1 e_m > 0$), a change of symmetry from tetragonal to orthorhombic occurs at the temperature of the continuous transition when $\sigma_1 e_m > 0$. The temperature of the continuous transition, given by $A(T_{ct}) = 2B(\sigma_1/3B)^{1/2} - C(\sigma_1/3B)$ and equation 5, passes through a maximum value, which is higher than the unstressed transition temperature by $\sim(7/9)(B^2/A'C)\{1 + [(B^2/A'C)/T^*(1 + a_\infty^2)]\}$ or only by about 3–5 K for V_3Si and Nb_3Sn. Finally, for a stress with $\sigma_1 e_m < 0$, z-domains are metastable and the stress reduces the temperature of the onset of instability by $\Delta T_f \simeq [-e_m\sigma_1{}^{1/3}C^{2/3} - (27C\sigma_1{}^2/4)^{1/3}]/A'$ for stress larger than σ_{crit}.

NONLINEAR STRESS-STRAIN RELATIONS

The physical phenomena that give rise to the martensitic phase transition also give rise to important nonlinear terms in the stress-strain relation, above and beyond the normal anharmonic and plastic flow effects occurring in most materials. In principle, such nonlinearity must be accounted for in computing the thermoelastic stress state of composite conductors. Landau models of the phase transition include strong nonlinear stress-strain effects. This can be seen in Fig. 4, which shows the normalized strain, η, as a function of temperature for various levels of applied stress, as calculated with the "simple model" embodied in equation 4. If the system were linear, the strain at high temperature, denoted as η_∞ in the figure, would equal the effective applied strain, η_a; however it clearly does not, and the importance of nonlinear effects is greatest near the transition temperature.

A more realistic description of the stress-strain relation can be derived from equation 6, with σ_1 obtained from $\partial F/\partial e_1$ and σ_2

from $\partial F/\partial e_2$. Thus, for example, the stress, σ_1, arising from a pure tetragonal strain, e_1, applied to a single crystal is:

$$\sigma_1 = A(T)e_1 + [(-3e_m/2) + e_1]Ce_1^2 \tag{7}$$

where $e_m = -2B/3C$ is the spontaneous strain at the temperature of the phase transition; the second term is the nonlinear contribution. In actual conductors, the A15 compound is present as a polycrystalline aggregate with the individual crystallites being oriented at random with respect to the principal axes of stress and strain (the specimen axes). It is straightforward, on the assumption that the strain throughout the polycrystal is constant, to calculate the resulting stress state in a crystallite of arbitrary orientation, and to average over orientation to obtain the average stress; this is equivalent to the Voigt polycrystal average of elastic moduli.[5] It is much more difficult to obtain the average strain accompanying an assumed constant stress state, the equivalent of the Reuss polycrystal average of elastic constants,[5] since

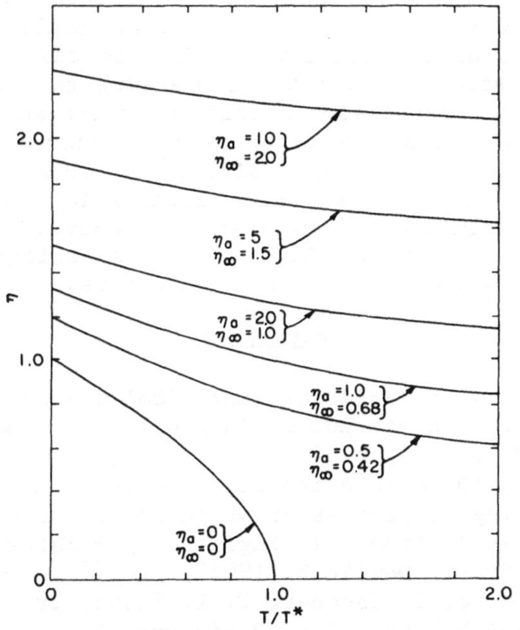

Fig. 4. Normalized strain in the thermodynamically stable state for the model described by equation 4. The effective applied strain, η_a, is defined in the caption of Fig. 1. The limit of η as $T \to \infty$ is η_∞ and would equal η_a if the system were linearly elastic.

this involves solving coupled cubic equations. We conclude by presenting the "Voigt-averaged" stress in a random polycrystalline aggregate (each crystallite of which obeys equation 6 as well as the usual Hooke's law for shear strains of the "c_{44} type" subjected to a pure tetragonal strain e_I (see equation 1), referred to specimen axes, such as is present due to differential thermal contraction in a cylindrical composite conductor:

$$(\sigma_1)_{avg} = [2A(T) + 6c_{44}]e_I/5 + [(-12e_m/35) + (3e_I/10)]Ce_I^2 \qquad (8)$$

In principle, such an equation should replace Hooke's law of linear elasticity in estimating the thermoelastic stress state in a composite conductor.

CONCLUSIONS

Analysis of Landau models[1,4] of the type that have successfully described a number of features of the martensitic phase transition and elastic softening in A15 compounds shows that the principal effect of applied stress in such models is to destroy the sharp transition, which is replaced by a smooth continuous transition, for stresses in excess of a rather small critical stress (which, expressed as an effective strain, is only a few percent of the spontaneous strain at T = 0 accompanying the transition). The temperature of the transition is shifted by stress by a few degrees at most. Stress lifts the degeneracy in energy of domains with their tetragonal axis along different <100> directions and depresses the temperature at which metastable domains become unstable. Strong nonlinear stress-strain behavior is a feature of the Landau models and should be included in estimates of the thermoelastic stress state of composite conductors.

REFERENCES

1. M. Weger and I. B. Goldberg, in: "Solid State Physics," Vol. 28, F. Seitz and D. Turnbull, eds., Academic Press, New York (1973), p. 1.
2. R. Flükiger, Phase relationships, basic metallurgy, and superconducting properties of Nb_3Sn and related compounds, in: "Advances in Cryogenic Engineering - Materials," Vol. 28, Plenum Press, New York (1982), p. 399.
3. J. F. Bussie re, B. Faucher, C. L. Snead, Jr., and D. O. Welch, Phys. Rev. B (in press).
4. B. Pietrass, Phys. Status Solidi b 68:553 (1975).
5. H. B. Huntington, in: "Solid State Physics," Vol. 7, F. Seitz and D. Turnbull, eds., Academic Press, New York (1958), p. 213.

THE EFFECT OF HIGH COMPRESSIVE STRESS ON THE CRITICAL CURRENT IN MULTISTRAND Nb_3Sn CONDUCTORS

W. Hassenzahl, M. Hong, D. R. Dietderich, C. Taylor, and W. Gilbert

Lawrence Berkeley Laboratory, University of California, Berkeley, California

R. Scanlan

Lawrence Livermore National Laboratory, Livermore, California

INTRODUCTION

The materials in superconducting accelerator magnets are subjected to high local stresses even though the coils are relatively small. A cross section of the 4.2-T Fermilab doubler dipole is shown in Fig. 1. The maximum compressive stress due to Lorentz forces is 9300 psi (64 MPa). Because the coils are precompressed during fabrication, the maximum stress in the windings is increased to about 12,000 psi (83 MPa) or higher. The NbTi conductors used today withstand this force with no apparent degradation.

Accelerators in the future will likely operate at fields of about 10 T. The stress in 10-T coils having designs similar to Fig. 1 could approach 40,000 psi (276 MPa). However, by redistributing the conductors[1] the maximum stress, including precompression, may be held to slightly more than 20,000 psi (138 MPa).

Three types of conductors have been proposed for 10-T accelerator magnets: niobium titanium (NbTi) and niobium titanium tantalum (NbTiTa) at 1.8 K and niobium tin (Nb_3Sn) at 4.2 K.[2] All three materials have acceptable current densities at 10 T in a stress-free state, but the performance of the brittle compound Nb_3Sn is known to be degraded by high tensile stress.[3,4]

The effects of tensile stress on multifilamentary Nb_3Sn wire have been thoroughly investigated.[3-7] This superconductor, which is formed by a reaction between niobium and tin at temperatures

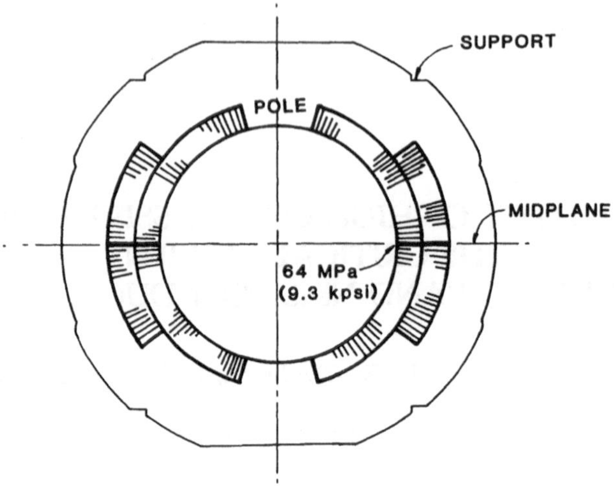

Fig. 1. Cross section of the 76.22-mm (3-in) bore Fermilab doubler magnet. The maximum load due to Lorentz forces is 64 MPa (9.3 kpsi) at the midplane.

between 600 and 800°C, has a low coefficient of thermal contraction relative to the rest of the conductor, i.e., the depleted bronze matrix and copper. As the conductor cools to room temperature, and eventually to 4 K for operation, the contraction of the matrix causes the Nb_3Sn to be under an axial compressive load. The magnitude of the compressive load depends on the area ratio of Nb_3Sn to bronze and copper[8] and significantly affects the performance of multifilamentary Nb_3Sn conductors subjected to tensile stress. This compression causes a reduction in J_c, but small tensile loads applied to the composite structure decrease this compression and improve the critical current.

The conductors in accelerator dipoles are subjected to some tensile stress, but the major stress in these conductors is a circumferential compressive load. The effects of transverse compressive loads on Nb_3Sn conductors have not been studied previously because conductors in most large, high-field magnets are subjected primarily to tensile loads, whereas the compressive load is either inherently small or is limited through the design and choice of structural materials.

During the final stage of construction of an accelerator dipole, the conductors are subjected to a large compressive load at ambient temperature. Ideally most of this load remains on the conductor as it is cooled to liquid helium temperatures. Then, as the coil is charged the local pressure will change. In the pole or high-field region the load decreases; at the midplane or low-field region (the fields at these two areas differ by only a few percent) the force increases. At maximum field the load varies from nearly zero to almost twice the initial precompression.

This paper describes a study to determine the irreversible effects of transverse compressive loads applied at room temperature and then released on a 23-strand Nb_3Sn Rutherford cable and its constituent strands.

EFFECTS OF COMPRESSIVE LOADS ON Nb_3Sn WIRE

The 0.027-in (0.69-mm) diameter wire used in these tests is similar to the Westinghouse LCP conductor.[9] Airco supplied the wire to LBL and also fabricated a 23-strand Rutherford cable that had nearly the same final dimensions as the Fermilab doubler conductor. The wire has a copper jacket occupying 64% of the cross section. The remaining 36% is niobium, bronze, and a tantalum barrier. Studies at Airco[9] showed that reacting at 730°C for 48 h gave nearly optimum performance at 10 T. (Possible improvement in critical current based on a double heat treatment is described in a separate report at this conference.)[10]

A series of U-shaped samples of this conductor were reacted at 730°C for 48 h at LBL and then subjected to different compressive forces. The samples were 45-mm long and had a 25-mm-long central straight section. The 10-mm-long side sections were soldered to the current leads. During critical current measurements, voltage taps were soldered 5 mm apart near the center of the straight section. Each sample was subjected to one compression cycle at room temperature, so that any observed effects were irreversible.

The forces were converted to an equivalent pressure on a complete cable and correspond to pressures as great as 40,000 psi (276 MPa). The method used to determine the relationship between pressure on the cable and the force on the sample was to calculate the force per unit length of conductor in the cable as a function of pressure, neglecting any stress enhancement due to conductor contact at the cable midplane. For example, at 10,000 psi (69 MPa) the force on 25.4 mm of cable was 3100 lbf (14,000 N). Since the cable was 11.5 conductors wide, the force per millimeter of conductor was 10.6 lbf (47.2 N). The force applied to the sample for 10,000 psi (69 MPa) was 480 lbf (2,100 N).

After compression at room temperature, the samples were taken to the Francis Bitter National Magnet Laboratory at MIT where the critical current was measured in fields up to 15 T. The results of these measurements are given in Fig. 2 and Table 1. The variations in the critical current in several unstressed samples give the range of data in the figure and the table and show that the measurement precision is ~3 A at 15 T and ~7 A at 7 T. The critical current values for the compressed samples are the average of two or three samples. The critical current values listed are for a voltage of 0.1 μV/mm, independent of sample current or field.

The critical currents were reduced for stress as low as 69 MPa, especially at the lower fields and higher currents. At 10 T and higher, however, the degradation was only about 25% at 138 MPa. Samples subjected to pressures of 276 MPa showed resistance at zero current, indicating severe damage. Nevertheless, these conductors showed some superconducting properties and carried 40 to 50% of the unstressed sample current at 0.1 μV/mm.

EFFECTS OF COMPRESSIVE LOADS ON Nb_3Sn CABLE

Some measurements have been made on the effects of compressive loads on the reacted cable to 276 MPa. The preliminary quantitative

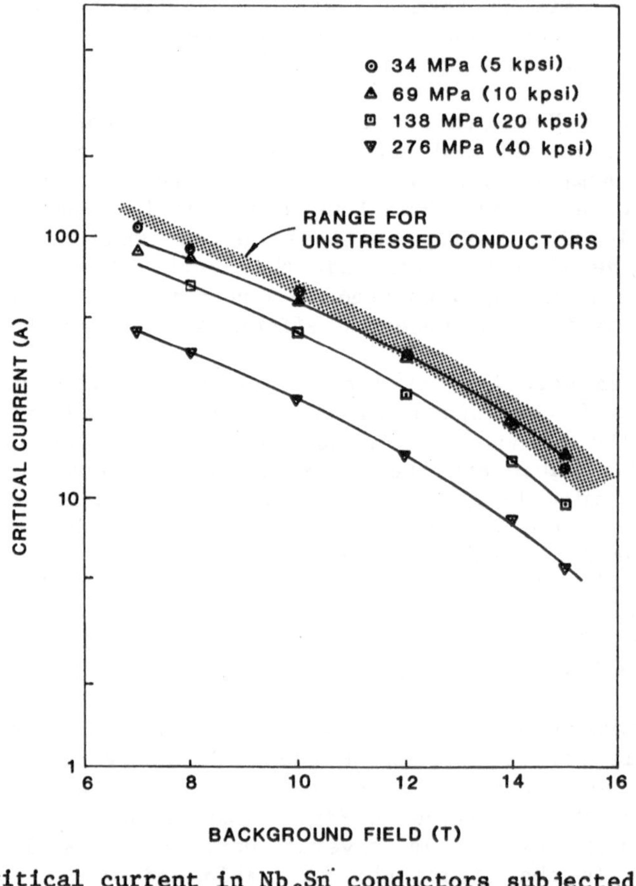

Fig. 2. Critical current in Nb_3Sn conductors subjected to compressive loads up to 276 MPa (40 kpsi). Little degradation is observed for pressures up to 138 MPa (20 kpsi).

Table 1. Critical Current in Amperes at 0.1 μV/mm Sensitivity in Nb_3Sn Strands Compressed after Reacting

Field, T	Compressive Loads, psi (MPa)				
	0	5,000 (35)	10,000 (69)	20,000 (138)	40,000 (276)
15	11-18	13	15	9.6	5.4
14	19-25	19.3	19.6	14	8.3
12	34-42	34.5	35.0	25	14.8
10	60-68	62	58.5	44	24.2
8	93-104	90	82.8	66	36.5
7	112-129	108	87		44.1

Table 2. Comparison of Observed Critical Current in Nb_3Sn Cable and Predicted Current Based on Single, Uncabled Strand Data

Field, T	Expected Current, A	Measured Current	
		No Compression, A	15,000-psi (104-MPa) Compression, A
12	780-970	760-800	850
10	1380-1565	1100-1350	1300
8	2140-2390	1600-2000	1900

results available from these measurements are given in Table 2 and show no degradation in the sample compressed with 104 MPa. (The higher current at 12 T in the compressed sample is not considered meaningful at this time.) A sample subjected to 207 MPa for 10 cycles was severely damaged and too resistive to measure, as was a single-cycle 276-MPa sample.

A COMPARISON BETWEEN CABLE AND STRAND PERFORMANCE

Most multistrand conductors exhibit some reduction in performance from that expected on the basis of individual strand characteristics. Reasons for this effect include physical damage to the strand, reduced cooling, and increased eddy currents. For our short sample tests on Nb_3Sn cables, we believe any reduction in performance would have to be due to physical effects in the strands.

Table 2 shows the expected short sample characteristics of individual strands. The cable I_c was 15 to 25% lower than expected. To determine the cause of this decrease in critical current, two types of samples were made from individual strands of conductor taken from the cable. One consisted almost entirely of the straight section of conductor from the cable; the other included the highly deformed, kinked section that occurs at the edge of the cable. These samples were the same size as and reacted in the same way as the uncabled wires. Their short sample critical currents are given in Table 3. The kinked samples showed some degradation, but the results varied by a factor of two. This variation may be a result of sample preparation or it may be a real variation in the properties of the materials. Both the straight and kinked samples showed consistently poorer current sharing characteristics than the uncabled samples. Again this may be caused by sample preparation, but it appears to be a characteristic of the cabling process.

The reduced I_c of the kinked section approximately accounts for the reduced I_c of the cable. To determine the cause of this effect, we mounted a sample of conductor with a kinked section in plastic and sliced sections perpendicular to the axis of the conductor in the region of and adjacent to the kink. After polishing,

Table 3. Comparison of the Critical Current in Amperes at 0.1 μV/mm Sensitivity of Superconducting Strands from a Nb_3Sn Cable with Identical Uncabled Conductors

Field T	Uncabled	Cable Straight Section	Cable Kinked Section
15	11-18	13-18	9-16
14	19-25	19-25	13-21
12	34-42	35-43	19-38
10	60-68	56-67	32-52
8	93-104	89*-105	44-83
7	112-129	97*-120	44-91

*Conductor transitions with no current sharing.

these samples were observed with an optical microscope. There was considerable distortion of the copper matrix, but the bronze and superconductors were relatively unaffected. The bronze/superconductor area was reduced by 1 to 7% over a length of about 5 mm. The section reduced 7% in area was less than 1-mm long.

Although this deformation may affect the performance slightly, it should not lead to a 10-20% reduction in I_c. To further pursue the possibility of some damage occurring during the fabrication process, we took several kinked sections of unreacted cabled strands and mounted them on copper bars with the kinked region free of the bar and unstressed. The outer copper shell was etched away with dilute nitric acid, the tantalum barrier was removed with a mixture of dilute nitric and hydrofluoric acids, and finally, the bronze was removed with dilute nitric acid. The exposed filaments of unreacted niobium were observed with a scanning electron microscope (SEM). Some groups of broken niobium filaments were observed.

The exact cause of these breaks is not known at present, but several possibilities are being considered. First, if there are irregularities in the filaments due to the drawing process and the reaction between niobium and tin is significant during the final annealing heat treatment, then the production of Nb_3Sn might weaken the filaments and cause them to break when the conductor is severely deformed. A second possibility is that the bronze fails under the severe deformation at the edge of the cable during the cabling process and causes the niobium to fail.

Further study will be required to determine which, if either of these effects, is the cause of degradation. A first step in this study will be to look at conductors with more moderate deformations in the final sizing operation.

CONCLUSIONS

This study indicates that multifilamentary Nb_3Sn bronze-process conductors exhibit little irreversible damage for compressive loads up to 15,000 psi (104 MPa), indicating there is no significant damage to the brittle Nb_3Sn compound. Higher stresses are expected in 10-T accelerator dipoles, so further study will be required, but this preliminary result is promising.

The Nb_3Sn cable did not perform as well as expected on the basis of individual strand tests. It appears that a major part of the degradation was associated with the highly deformed area near the edge of the cable though the cabled strands did not exhibit a good current sharing region. Reducing the degree of compaction may reduce the size of this effect.

REFERENCES

1. C. E. Taylor and R. B. Meuser, IEEE Trans. Nucl. Sci. NS-28(3): 3200 (1981).
2. W. V. Hassenzahl, IEEE Trans. Nucl. Sci. NS-28(3):3277 (1981).
3. A. F. Clark, Cryogenics 16:632 (1976).
4. J. W. Ekin, Appl. Phys. Lett. 29:21b (1976).
5. D. S. Easton and R. E. Schwall, Appl. Phys. Lett. 29:319 (1976).
6. J. W. Ekin, in: "Advances in Cryogenic Engineering," Vol. 24, Plenum Press, New York (1978), p. 306.
7. D. W. Deis, D. N. Cornish, A. R. Rosdahl, and D. G. Hirzel, in: "Sixth International Conference on Magnet Technology (MT-6)," ALFA, Bratislava, Czechoslovakia (1977), p. 1028.
8. G. Rupp, in: "Filamentary A15 Superconductors," M. Suenaga and A. F. Clark, eds., Plenum Press, New York (1980), p. 155.
9. P. A. Sanger, E. Adam, E. Ioriatti, and S. Richards, IEEE Trans. Magn. MAG-17(1):666 (1981).
10. M. Hong, I. W. Wu, J. W. Morris, Jr., W. Gilbert, W. V. Hassenzahl, and C. Taylor, An investigation on the enhancement of the critical current densities in bronze-process Nb_3Sn, in: "Advances in Cryogenic Engineering--Materials," Vol. 28, Plenum Press, New York (1982), p. 435.

EXPERIMENTAL TEST OF STRESS SCALING LAW IN SUPERCONDUCTING V and Nb*

M. Fukumoto, M. Kawamura, and T. Okada

The Institute of Scientific and Industrial Research, Osaka University, Suita, Osaka, Japan

INTRODUCTION

The degradation in superconducting characteristics induced by large axial tensile stress is one of the key problems in the application of superconductors to large scale magnets, such as fusion reactors. Consequently, numerous studies of the stress effects have been made for Nb_3Sn, V_3Ga, NbTi, and other promising materials.[1-5] However, the mechanism of the effects is not well understood yet.

The critical current density, J_c (or pinning force density, F_p), of a superconductor is, in general, determined by the several parameters shown in Fig. 1. The application of stress gives rise to the shifts in these quantities, and the change in J_c (or F_p) is caused by them. Therefore, the first step in the elucidation of the mechanism of the stress effect on F_p may be to determine which of these parameters plays the essential role in the effect.

These parameters can be divided into two groups. One is the group of the quantities related to flux pinning structure, such as the density of pins, ρ_p. Another is that of the variables related to the electronic state, such as the coupling constant, g.

In this study, we first investigated the stress (strain) scaling law[4] among F_p vs. h (= H/H_{c2}) curves at various tensile stresses using vanadium and niobium as test samples, and discussed which of the two groups is essential in the stress effect on F_p from the result. In the second place, we attempted to find the

*This work is supported by Grant in Aid for Fusion Research Number 56055028 Ministry of Education in Japan.

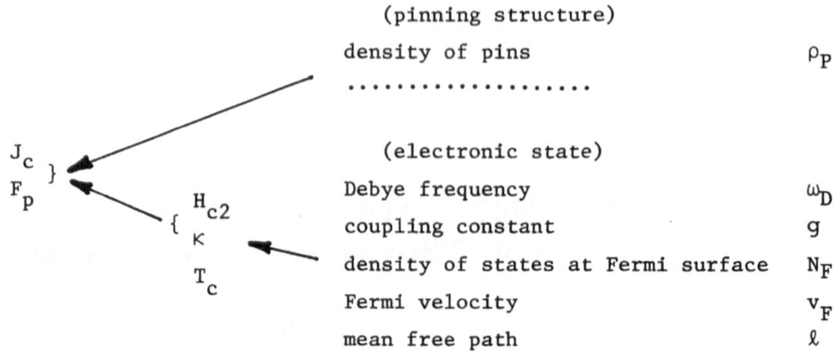

Fig. 1. Schematic relation among various quantities in superconducting materials.

essential parameter in the second group by measuring T_c, H_{c2}, and residual resistivity, ρ_r, as a function of tensile stress using a vanadium sample.

SCALING LAW

When the electronic state in a superconducting material is modified with the pinning centers unchanged, e.g., by varying the temperature of a material, scaling is normally observed among F_p-h curves.[6-8] On the other hand, if the pinning structure is changed, e.g., by varying annealing time or temperature, the peak in $F_p(h)$ shifts to lower (or higher) h and increases (or decreases) its height leaving $F_p(h)$ on the high h side of the peak relatively unchanged.[8,9] If this holds for strain-induced changes in the pinning structure as well, then it would be possible to distinguish which group of the two described in the previous section is essential in the stress effect on F_p through the comparison among F_p-h relationships at various stresses. Samples of vanadium and niobium tapes were prepared, which are monoatomic metals with the simplest composition and belong to type II superconductors. A vanadium sample was obtained by cold rolling from a 2-cm-diameter ingot to 0.045-mm thickness, cutting out the silhouette form of a dumbbell

Table 1. Impurities in the V and Nb Samples (%)

V	Na	Al	Si	Fe	Ni	Ge	Nb	Ta
>99.5	<0.01	<0.01	<0.01	0.16	<0.01	0.016	<0.01	<0.01

Nb	H	C	N	O	Si	Ti	Fe	Zr	Ta
>99.8	0.005	0.007	0.0052	0.0105	0.002	0.005	0.015	0.01	0.05

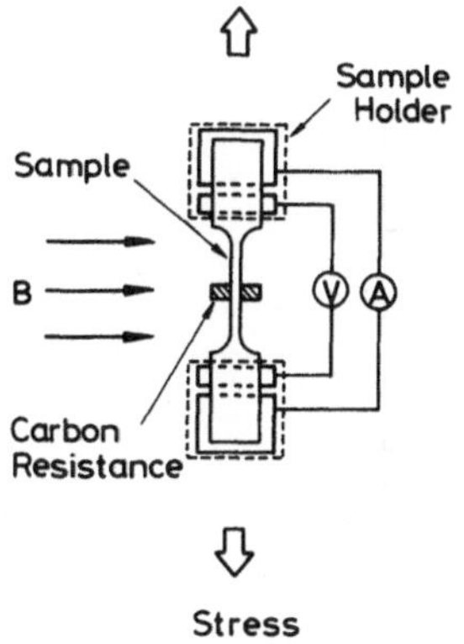

Fig. 2. Schematic diagram of the experimental setup for the measurements of $J_c(H)$ and H_{c2} under the application of tensile stress.

with a 0.5-mm width at the narrowest section, and annealing at 700°C for 2 h in a vacuum of 10^{-6} torr. A niobium sample was prepared by cold rolling from 2-mm-thick plate to 0.07-mm thickness, cutting it into the same shape as the vanadium sample, and annealing at 800°C for 2 h in a vacuum of 10^{-6} torr. The purities of samples are shown in Table 1.

The critical current, I_c, and the critical field, H_{c2}, were measured as a function of tensile stress using the standard four-probe technique, as shown in Fig. 2. The H_{c2} measurement was performed under 10-mA sample current. In both I_c and H_{c2} measurements, the normal transition was defined by a potential drop of 1 μV/cm. The magnetic field was applied perpendicular to the current and parallel with the surface of a sample. The flux pinning density vs. reduced field, h, curves for the vanadium sample are presented in Fig. 3. The F_p increases with the increase in stress, σ. Each curve has its peak somewhat below h = 0.5. The group of curves seems nearly to scale with strain, which is consistent with the results provided by Ekin for alloys and compounds.[10,11] It is, therefore, suggested that the stress-induced change in the electronic state contributes to the stress effect on F_p more essentially than that in the pinning structure. However we can also recognize slight deviation from strict scaling in Fig. 3. At the first stage ($\sigma/\sigma_f \lesssim 0.5$, σ_f = fracture stress) of the increase in stress, the lower h side of a curve increases a little more than the higher side. On the other hand, the higher side begins to rise up a little more rapidly than the lower side at the second stage ($\sigma/\sigma_f \gtrsim 0.5$). The position of the maximum peak shifts slightly to the left at the first stage and to the right at the second. This deviation from exact scaling behavior is small, but is systematically observed for all ten samples we tested. This deviation may suggest a small variation in the pinning structure. From Figs. 3 (and 6), we find that these data obey the stress (strain) scaling law[4] with n = 1.6:

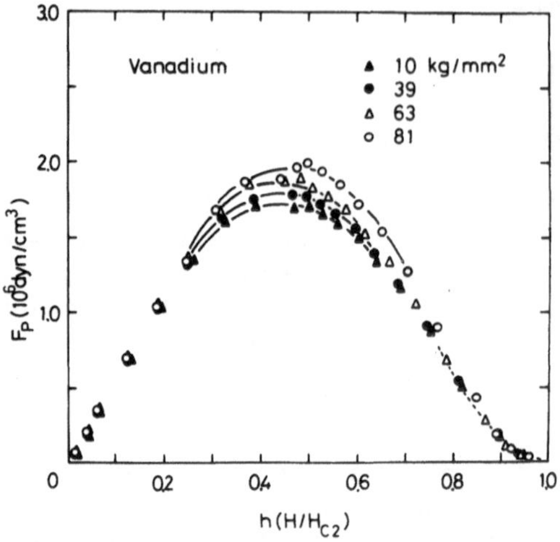

Fig. 3. Flux pinning force density, F_p, as a function of reduced field, h, and tensile stress for the vanadium sample.

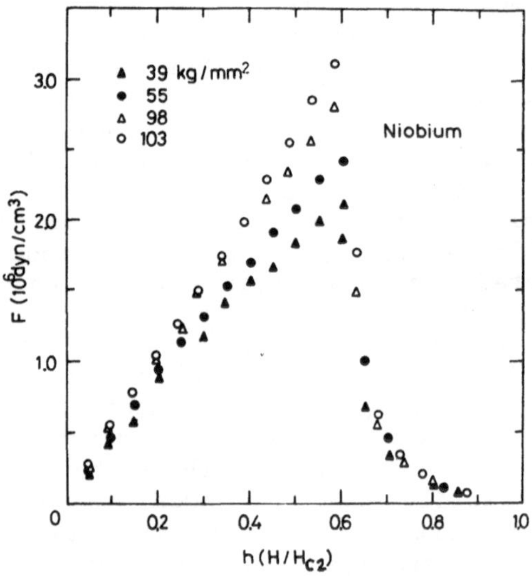

Fig. 4. Flux pinning force density, F_p, as a function of reduced field, h, and tensile stress for the niobium sample.

$$F_p(\sigma) \simeq H_{c2}(\sigma)^{1.6} f(h) \tag{1}$$

A dashed curve in Fig. 3 represents $F_p(h)$ calculated from Eq. (1) using for f(h) the form obtained in Kramer's FLL shearing model.[8]

$$F_p(h) \propto h^{1/2}(1 - h)^2 \tag{2}$$

The proportional coefficient is obtained as a fitting parameter. The h dependence of F_p is found to agree with this model from the comparison.

We present the results for a niobium sample in Fig. 4. The stress enhancement of F_p is observed and nearly obeys the stress scaling law for this sample as well as for the vanadium sample, but f(h) does not have the form predicted by the Kramer FLL shearing model. It is hard to discuss the small deviation in this case because of the drastic change in F_p on the higher side of the peak.

PARAMETERS RELATED TO ELECTRONIC STATE

As we discussed in the previous section, it is considered that the stress effect on F_p is principally brought about by the change in the electronic state. In the following, we attempt to determine which of the parameters related to the electronic state plays the principal role in the stress effect on F_p. We first examine the Debye frequency, ω_D. From the definition ω_D is given as:

$$\omega_D = v_s(6\pi^2 N/V)^{1/3} \tag{3}$$

where v_s is the sound velocity, N is the number of atoms in V, and V is the volume of the specimen.[12] The sound velocity is given by:

$$v_s = (CV/MN)^{1/2} \tag{4}$$

where C is the elastic modulus and M is the mass of an atom.[12] Hence, ω_D is written as:

$$\omega_D = (4\pi^2)^{1/3}(C/M)^{1/2}(V/N)^{1/6} \tag{5}$$

The strain ε dependence of the volume is expressed as:

$$V = V_0(1 + \varepsilon)(1 - \nu\varepsilon)^2 \simeq V_0(1 + 1/3\cdot\varepsilon) \tag{6}$$

where V_0 is the volume under $\varepsilon = 0$ and ν is the Poisson ratio ($\simeq 1/3$). The Debye frequency ω_D is, therefore:

$$\omega_D \simeq (6\pi^2)^{1/3}(C/M)^{1/2}(V_0/N)^{1/6}(1 + 1/18\cdot\varepsilon) \qquad (7)$$

The stress-strain relation of our vanadium sample was measured and shows a linear curve up to the fracture point. This means that C is independent of strain. Thus, the strain dependence of ω_D is given only by the factor $(1 + 1/18\cdot\varepsilon)$ and ω_D is changed only by less than 0.03% during the strain increase from 0 to the fracture strain ($\simeq 0.5\%$). Therefore, the strain or stress dependence of ω_D can be neglected, and we can exclude ω_D as a parameter that is essential in the stress effect on F_p.

The dependence of the rest of the four variables g, N_F, v_F, and ℓ on stress are considered next. There are four unknown quantities independent of each other. Therefore, we need four independent measurable variables. Since ω_D is already a known variable, three other independent measurable quantities are necessary.

It seems that the variables, which reflect the electronic state, are expected to lead to the above four quantities, are capable of being measured under the application of stress, and are restricted to T_c, H_{c2}, and the residual resistivity, ρ_r. Our procedure to obtain the four variables from these three measurables is based on G-L theory and BCS theory. The measurements of T_c and ρ_r were carried out for the vanadium sample using the four-probe technique described in a previous section. The resistivity-temperature curves obtained as a function of stress are shown in Fig. 5.

The normal resistivity was observed to increase by 2% during the stress increase from zero to 100 kg/mm^2, which corresponds to about 80% of the ultimate tensile strength. We define T_c as the temperature at which the resistivity is half of the normal one. In Fig. 6, T_c, and H_{c2} are shown as a function of stress. The T_c and H_{c2} vs. σ curves are nearly parabolic, as theoretically predicted by Testardi.[13]

In Table 2 we present the results of a calculation based on the procedure described above. The shift in the mean free path, ℓ, is only about 2% upon increasing σ from zero to 100 kg/mm^2 and is relatively small. On the other hand, the changes in v_F, N_F, and g are 4%, 6%, and 7%, respectively, and are much larger than ℓ. The contribution of the changes in these variables to the change in F_p is dependent on the relationship of the variables to F_p. From G-L theory and BCS theory we obtain the approximate relation of H_{c2} and the G-L parameter, κ, to T_c, v_F, and ℓ as follows:

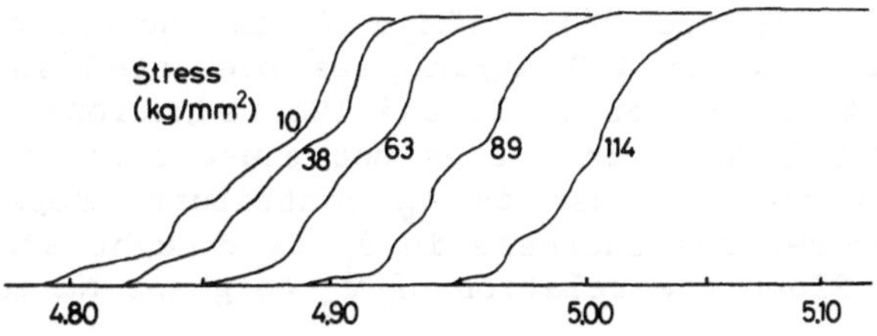

Fig. 5. Plot of resistivity vs. temperature at various tensile stresses for the vanadium sample.

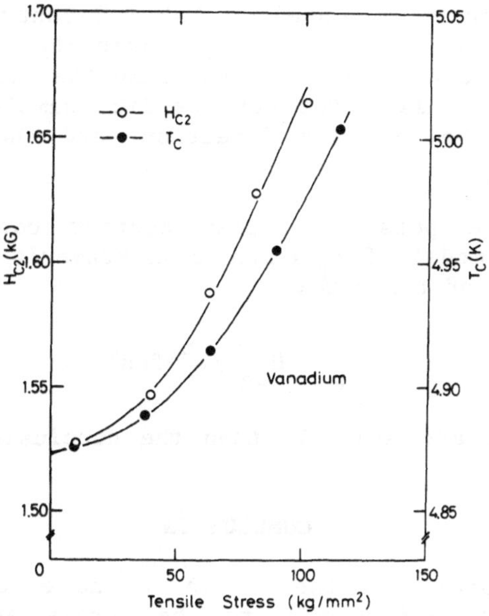

Fig. 6. T_c and H_{c2} vs. tensile stress for the vanadium sample.

$$H_{c2} \propto (T_c - T)/v_F \ell \tag{8}$$

$$\kappa \propto \ell^{-1} \tag{9}$$

If we employ Kramer's model:

$$F_p \propto H_{c2}{}^{5/2}/\kappa^2 \cdot f(h) \tag{10}$$

as the expression of F_p, F_p is related to T_c, v_F, and ℓ by

$$F_p \propto (T_c - T)^{5/2}/v_F{}^{5/2}\ell^{1/2} \cdot f(h) \tag{11}$$

from Eqs. (6), (7), and (8). While F_p is increased about 20% by increasing σ from 0 to 100 kg/mm^2, as presented in Fig. 3, a 2% rise in ℓ and 4% in v_F bring 1% and 10% reductions in F_p, respectively. The small shift in ℓ does not cause a significant alteration in F_p and the increase in v_F contributes negatively to the change. Therefore, the increase in F_p is brought about by the increase in T_c. Since the relation of T_c to g and N_F is expressed by BCS theory as:

$$kT_c = 0.567\Delta_0 = 1.14\ \hbar\omega_D e^{-1/gN_F} \tag{12}$$

the 7% increase in g presented in Table 2 contributes to the increase in T_c and 6% reduction in N_F, conversely, prevents that. It is, therefore, concluded that the variable that is essential in the stress effect on F_p is g for our vanadium sample, both N_F and v_F influence F_p in the opposite direction, and there is little influence of ℓ on F_p.

There is the possiblity that another conclusion would be obtained if some model of F_p other than Kramer's formula is valid. In the expression of the form:

$$F_p \propto H_{c_2}{}^m/\kappa^n \cdot f(h) \tag{13}$$

however, if $m > 0$ and $n/m \simeq 1$, then the conclusion is the same as that above.

CONCLUSION

1. It is suggested that the change in the electronic state contributes more effectively to the stress effect on the flux pinning force density, F_p, than the change in the pinning structure. There also seems to be a small alteration in the pinning structure.

2. In the variables related to the electronic state, the calculations also suggest that the coupling constant g contributes most essentially to the stress-induced increase in F_p. The density of states at the Fermi surface, N_F, and the Fermi velocity, v_F, bring the decrease in F_p. The contribution of ℓ is small.

ACKNOWLEDGMENTS

The authors are indebted to Dr. M. Nagata of Sumitomo Electric Industries Ltd. for preparing the vanadium tape and to Dr. T. Horiuchi of Kobe Steel Co. Ltd. for preparing the niobium plate.

Table 2. Stress Dependence of Various Quantities Calculated for the Vanadium Sample from the Measurements of H_{c2}, T_c, and ρ_r Using G-L Theory and BCS Theory

Variable X	Stress (kg/mm^2) 0	Stress (kg/mm^2) 100	$\frac{X(100)-X(0)}{X(0)}$ (%)
H_{c2} (kG)	1.52	1.67	10
T_c (K)	4.87	4.98	2
ρ_r (μΩcm)	3.54	3.61	2
ξ (μm)	0.0464	0.0443	-5
λ (μm)	0.153	0.143	-7
κ (1)	3.30	3.23	-2
H_c (kG)	0.327	0.366	12
H_0 (kG)	1.31	1.30	-1
Δ_0 (10^{-15}erg)	3.66	3.74	2
N_F (10^{33}erg^{-1}cm^{-3})	10.2	9.62	-6
g (10^{-35}erg cm^3)	2.42	2.58	7
ω_D (10^{14}s^{-1})	1.00	1.00	0
ξ_0 (μm)	0.127	0.129	2
v_F (10^8cm/s)	1.38	1.44	4
ℓ (10^{-2}μm)	0.996	1.02	2

u: coherence length, λ: penetration length, κ: G-L parameter, H_c: thermodynamic critical field, H_0: H_c at T = 0 K, $2\Delta_0$: gap energy, N_F: density of states at Fermi surface, g: coupling constant, ω_D: Debye frequency, ξ_0: Pippard length, v_F: Fermi velocity, ℓ: mean free path.

REFERENCES

1. C. C. Koch and D. S. Easton, A review of mechanical behaviour and stress effects in hard superconductors, Cryogenics 17:391 (1977).
2. T. Okada and R. Terada, Stress effects on the critical current in superconducting V_3Ga tape, Cryog. Eng. (Jpn.) 13:157 (1978).
3. G. Rupp, Stress induced normal - superconducting transition in multifilamentary Nb_3Sn conductors, IEEE Trans. Magn. MAG-15:189 (1979).
4. J. W. Ekin, Strain dependence of the critical current and critical field in multifilamentary Nb_3Sn composites, IEEE Trans. Magn. MAG-15:197 (1979). J. W. Ekin, Strain scaling law for flux pinning in practical superconductors. Part 1: Basic relationship and application to Nb_3Sn conductors, Cryogenics 20:611 (1980).

5. S. Foner, R. Roberge, E. J. McNiff, Jr., B. B. Schwartz, and J. L. Fihey, Mechanical properties of in situ multifilamentary Nb_3Sn superconducting wires, Appl. Phys. Lett. 34:241 (1979).
6. A. W. Fietz and W. W. Webb, Hysteresis in superconducting alloys--Temperature and field dependence of dislocation pinning in niobium alloys, Phys. Rev. 178:657 (1969).
7. A. M. Campbell and J. E. Evetts, Flux vortices transport currents in type II superconductors, Adv. Phys. 21:199 (1972).
8. E. J. Kramer, Scaling laws for flux pinning in hard superconductors, J. Appl. Phys. 44:1360 (1973).
9. A. J. Marker, R. W. Reed, F. G. Brickwedde, R. L. Schuyer, and W. R. Bitler, Fluxoid pinning by vanadium carbide precipitates in superconducting vanadium, J. Low Temp. Phys. 31:175 (1978).
10. J. W. Ekin, Strain scaling law for flux pinning in NbTi, Nb_3Sn, Nb-Hf/Cu-Sn-Ga, V_3Ga, and Nb_3Ge, IEEE Trans. Magn. MAG-17:658 (1981).
11. J. W. Ekin, H. Sekine, and T. Tachikawa, Effect of strain on the critical current of Nb-Hf/Cu-Sn-Ga multifilamentary superconductors, J. Appl. Phys. 52:6252 (1981).
12. C. Kittel, "Introduction to Solid State Physics," 4th ed., John Wiley & Sons, New York (1971).
13. L. R. Testardi, Unusual strain dependence of T_c and related effects for high-temperature (A-15-structure) superconductors: sound velocity at the superconducting phase transition, Phys. Rev. B 3:95 (1971).

SUPERCONDUCTING AND MECHANICAL PROPERTIES OF COLD HYDROSTATICALLY EXTRUDED MONOFILAMENTARY Nb_3Sn WIRES

V. Thadani

Department of Metallurgy and Materials Engineering, Lehigh University, Bethlehem, Pennsylvania

T. S. Luhman

Metallurgy and Materials Science Division, Brookhaven National Laboratory, Upton, New York

B. Avitzur and Y. T. Chou

Department of Metallurgy and Materials Engineering, Lehigh University, Bethlehem, Pennsylvania

INTRODUCTION

In the conventional bronze technique of manufacturing multifilamentary Nb_3Sn superconducting wires (Fig. 1), a composite billet having niobium cores in a bronze sleeve is hot-extruded and drawn down to a rod. The number of filaments is increased by assembling these rods into a starting billet, which is again hot-extruded and drawn. After the cycles of assembly and reduction are completed, the rod is drawn down to a fine multifilamentary wire. Intermediate anneals must be given frequently to soften the bronze matrix. Finally, a diffusion anneal produces Nb_3Sn at the niobium-bronze interface.

This manufacturing technique has limitations. The wire drawing process is very lengthly because only small reductions in area per pass are possible and frequent intermediate anneals are necessary.

Since the above processing involves work at high temperatures, i.e., hot extrusions at 750°C, and numerous intermediate anneals at

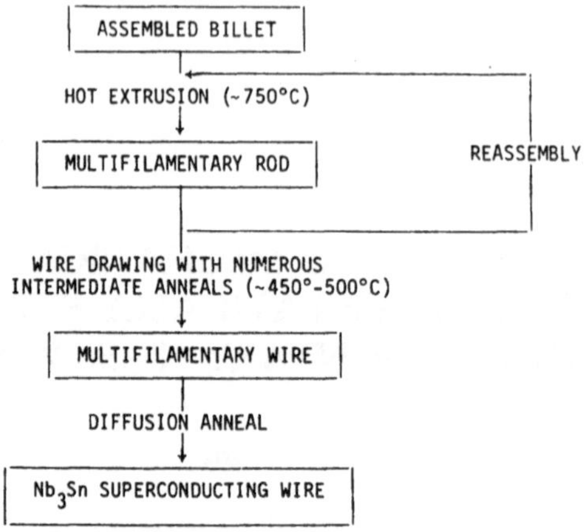

Fig. 1. Conventional bronze process of manufacturing multifilamentary Nb_3Sn superconducting wires.

450-500°C, there is some premature diffusion that leads to the formation of Nb_3Sn.

Cold hydrostatic extrusion may overcome these limitations.[1,2] The present study had two objectives: The first was to assess the merits of cold hydrostatic extrusion as an alternative metal-forming process and to compare it with conventional wire drawing. The second was to evaluate the superconducting parameter J_c of the hydrostatically extruded wires as a function of strain and magnetic field. This was supplemented by an examination of the interface, the Nb_3Sn growth rate, and the microhardness of the constituents.

FABRICATION OF WIRES

Nb_3Sn monofilamentary wires were prepared both by cold hydrostatic extrusion and wire-drawing techniques.[3]

Procedure

The outer sleeve of the billet was made of a Cu + 13 wt.% Sn bronze. The core was niobium of 99.85% purity. Care was taken to thoroughly clean the surfaces before assembly.

The Model 6 fluid extrusion unit (maximum safe pressure, 1380 MPa) and a draw bench at the Institute for Metal Forming, Lehigh University were used for fabricating the wires.[3]

Results and Discussion

A cold hydrostatic extrusion technique was used to prepare Nb_3Sn monofilamentary wires having bronze-to-niobium ratios of 2.25, 4.0, 9.0, and 17.75. The pass schedule is presented in Table 1. If a 40% reduction in area was not obtainable, the wire was given an intermediate anneal for 20 min at 450°C.

Another series of conductors was prepared by wire drawing (Table 2). These wires had the same materials, dimensions, bronze-to-niobium ratios, and diffusion treatment as the hydrostatically extruded wires. If a 15-20% reduction in area was not obtainable, the wire was given an intermediate anneal for 20 min at 450°C.

Important highlights from Tables 1 and 2 are compared in Table 3. In both hydrostatic extrusion and wire drawing, the initial billet had a diameter of 9.53 mm, and the final wire had a diameter of 0.6071 mm--a 99.6% reduction in area. Whereas the wire drawing process required 25 passes averaging 19% reduction in area per pass, the cold hydrostatic extrusion process required only 10 passes averaging 40% reduction in area per pass. Furthermore, the wire drawing technique needed 13 intermediate anneals compared with only 4 when cold hydrostatic extrusion was used.

This direct comparison shows that the cold hydrostatic extrusion technique can serve as a good alternative to the wire drawing technique for the manufacture of composite superconducting wire because cold extrusion allows for a shorter processing procedure, i.e., fewer passes and fewer intermediate anneals.

Table 1. Data on the Cold Hydrostatic Extrusion of Bronze-Clad Niobium Billets

Extrusion No.	Diameter before extrusion (mm)	Diameter after extrusion (mm)	Reduction in area (%)	Extrusion pressure (MPa)			
				Billet A	Billet B	Billet C	Billet D
1	9.53	5.79	63	800	972	1007	972
2	5.79	4.70	34	634	676	876	876
3	4.70	3.76	36	710	758	965	965
4	3.76	3.00	36	779	958	972	1048
	*Intermediate Anneal						
5	3.00	2.25	44	669	724	731	731
6	2.25	1.73	40	745	779	876	883
	*Intermediate Anneal						
7	1.73	1.33	41	717	724	772	745
8	1.33	1.03	40	786	910	958	1048
	*Intermediate Anneal						
9	1.03	0.795	41	683	738	772	821
	*Intermediate Anneal						
10	0.795	0.607	42	827	841	1055	1048
Total No. of Extrusion	Initial diameter	Final diameter	Total reduction in area	Average Pressure			
10	9.53	0.607	99.6	738	807	896	910

Table 2. Data on Wire Drawing of Bronze-Clad Niobium Billets

Draw No.	Diameter before the draw (mm)	Diameter after the draw (mm)	Redn. in area (%)
1	9.53	8.33	23
2	8.33	7.34	22
3	7.34	6.53	21
4	6.53	5.92	18
*Intermediate Anneal			
5	5.92	5.16	24
6	5.16	4.42	27
7	4.42	3.96	20
*Intermediate Anneal			
8	3.96	3.51	22
9	3.51	3.18	18
*Intermediate Anneal			
10	3.18	2.87	18
11	2.87	2.54	22
*Intermediate Anneal			
12	2.54	2.28	19
13	2.28	2.05	19
*Intermediate Anneal			
14	2.05	1.85	19
15	1.85	1.66	19
*Intermediate Anneal			
16	1.66	1.49	19
17	1.49	1.34	19
*Intermediate Anneal			
18	1.34	1.21	19
19	1.21	1.09	19
*Intermediate Anneal			
20	1.09	0.978	19
*Intermediate Anneal			
21	0.978	0.881	19
*Intermediate Anneal			
22	0.881	0.793	19
*Intermediate Anneal			
23	0.793	0.711	19
*Intermediate Anneal			
24	0.711	0.645	18
*Intermediate Anneal			
25	0.645	0.607	11
Total no. of draws	Initial diameter	Final diameter	Total redn.
25	9.53	0.607	99.6

SUPERCONDUCTING PROPERTIES

Even though cold hydrostatic extrusion has been shown to reduce the number of processing steps, it is also important to evaluate the superconducting properties of the final product.

The strain tolerance of hydrostatically extruded material was studied. It has been established that variations of J_c can occur with straining.[4] These variations are caused by internal strains on the Nb_3Sn compound. Since the coefficient of expansion of niobium and Nb_3Sn (~7 x 10^{-6} K^{-1}) is lower than that of bronze (~16 x 10^{-6} K^{-1}), the Nb_3Sn is placed under compression. Application of a tensile strain relieves this compressive prestrain till

Table 3. Comparison of Pass Schedule

	Wire Drawing	Hydrostatic Extrusion
1. Total reduction in cross-sectional area	99.6%	99.6%
2. No. of passes	25	10
3. Average reduction in cross-sectional area per pass	18-22%	40-45%
4. No. of intermediate anneals	13	4
5. Average reduction in cross-sectional area between intermediate anneals	32%	67%

ε_{Jc}^{max}, at which point there is no net strain. Beyond this value, a net tensile strain accounts for the decrease in J_c.

Figure 2 compares the strain tolerance of a hydrostatically extruded sample with that of a drawn sample having a bronze-to-niobium ratio of 4.0. The J_c measurements were made at 4.2 K and 8 T. The curve shows an increase in J_c till ε_{Jc}^{max}, followed by a decrease.[4] This curve is representative of all the examined samples having different bronze-to-niobium ratios.

Figure 3 illustrates the effect of the bronze-to-niobium ratio on ε_{Jc}^{max}. The ε_{Jc}^{max} increases until it reaches a ratio of 10:1, after which it drops slightly. Raising the bronze-to-niobium ratio increases the compressive prestrain, which, in turn, is reflected by a higher value of ε_{Jc}^{max}. As Fig. 3 shows, wires prepared by both techniques have similar values and trends of ε_{Jc}^{max}.

For the examined bronze-to-niobium ratios, the J_c values under strain for the hydrostatically extruded wires compared favorably with the drawn material. All the J_c values under strain fell in the 10^5-10^6 A/cm^2 range. These values compare favorably with those published in the literature.[5]

For cold hydrostatically extruded wires, the effect of a magnetic field on J_c at 4.2 K is presented in Table 4. This, too, compares well with the published work.

SUPPLEMENTARY OBSERVATIONS

Interface examination before diffusion showed that at low magnifications (~100x) both the cold hydrostatically extruded and wire-drawn samples had a concentric niobium core with a uniform

Table 4. Effect of a Magnetic Field on J_c at 4.2 K.

Applied Magnetic Field (T)	J_c at 4.2 K under the Applied Magnetic Field (A/cm^2)
6	1.15×10^6
6	6.4×10^5

Hydrostatically extruded sample; bronze-to-niobium ratio = 9; diffusion treatment: 24 h at 725°C.

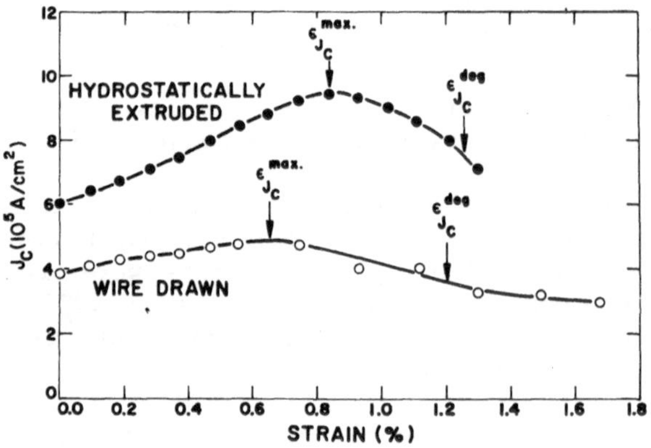

Fig. 2. Strain tolerance of monofilamentary Nb_3Sn wires with bronze-to-niobium ratio of 4.0.

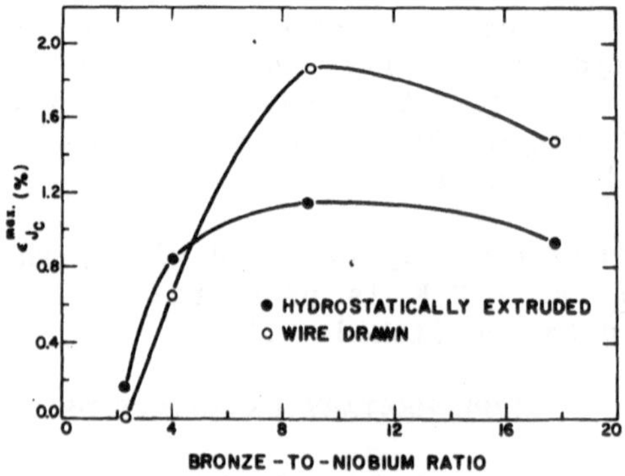

Fig. 3. Effect of bronze-to-niobium ratio on $\epsilon_{J_c}^{max}$.

diameter. At higher magnifications (~1000x), the interface was very irregular--a sign of good mechanical bonding of the constituents. Examination after diffusion showed the bronze-Nb_3Sn interface to be more irregular than the niobium-Nb_3Sn interface. Furthermore, the Nb_3Sn layer had considerable variations in thickness.

A study was carried out to determine if the bronze-to-niobium ratio or the manufacturing technique affected the growth rate of Nb_3Sn. No characteristic trend was observed. All the values fell within a narrow range. The ranges of thickness of Nb_3Sn in both hydrostatically extruded and wire-drawn samples are indicated by error bars in Fig. 4.

Microhardness measurements were taken before the diffusion anneal on the bronze sleeve and niobium core of both the extruded and drawn wires. The measurements were taken across transverse sections of the specimens. Neither the metal forming technique nor the bronze-to-niobium ratio was found to affect the microhardness.

Microhardness of the components was also measured after 3, 10, 24, 48, and 100 h of diffusion at 725°C for each preparation technique. Within 3 h, the Vickers Diamond Pyramid Hardness (16-g load) of the bronze sleeve dropped from 206 kg/mm^2 to 103 kg/mm^2, while the niobium's hardness dropped from 149 kg/mm^2 to 102 kg/mm^2. Beyond 3 h there was no change in microhardness. It is speculated that the behavior after 3 h was due to the fact that the grain size of the bronze became larger than the indentation size.

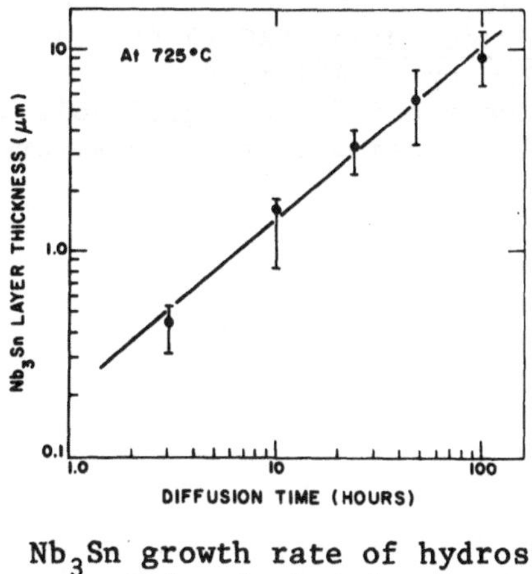

Fig. 4. Nb_3Sn growth rate of hydrostatically extruded and wire-drawn samples.

CONCLUSION

Cold hydrostatic extrusion allowed for a shorter manufacturing process (i.e., fewer passes and fewer intermediate anneals) than the wire drawing technique.

The effect of strain and magnetic field on J_c for the cold hydrostatically extruded wires was comparable with the wire drawn material. Interface characteristics, Nb_3Sn growth rate, and microhardness values of the hydrostatically extruded and wire drawn samples were also found to be similar.

ACKNOWLEDGMENT

This work was sponsored by the Division of Material Sciences, U.S. Department of Energy under contract DE-AC02-79ER10367.

REFERENCES

1. J. Breme and C. H. Massat, The effect of hydrostatic extrusion on the superconducting characteristics of tin-bronze/niobium polymer conductors, Metallwiss. Tech. 33 (1979).
2. "Experimental Evaluation of Hydrostatic Extrusion for the Fabrication of Multifilamentary Superconducting Wire," Report submitted by Battelle to the Division of Magnetic Fusion Energy, Energy Research and Development Administration (1976).
3. V. Thadani, B. Avitzur, Y. T. Chou, and T. Luhman, A comparison of hydrostatic extrusion versus wire drawing as a method of producing Nb_3Sn superconducting wire, in: "9th NAMRC Proceedings," SME, Dearborn (1981).
4. T. Luhman, M. Suenaga, D. O. Welch, and K. Kacho, Degradation mechanism of Nb_3Sn composite wires under tensile strain at 4.2 K, IEEE Trans. Magn. MAG-15 (1979).
5. T. Luhman, Metallurgy of A15 conductors, in: "Treatise on Materials Science and Technology," Vol. 14, T. Luhman and D. Hughes, eds., Academic Press, New York (1979).

RECENT DEVELOPMENTS ON METHODS FOR SUPERCONDUCTOR JOINING*

R. D. Blaugher

Westinghouse Electric Corporation, Research and Development Center
Pittsburgh, Pennsylvania

INTRODUCTION

The subject of superconductor joints and their influence on measured superconducting properties was essentially coincident with the discovery of high-current, high-field (type II) superconductors. It was immediately apparent that a low resistance, almost lossless, connection must be made to the superconductor. Early attempts to solve this problem produced a fairly straightforward approach, which even now provides an excellent method for superconductor joining. A lap joint was implemented between two long lengths of conductor that were soldered with pure indium and then clamped between two copper plates. The recent Westinghouse 5 MVA superconducting generator field winding, the Air Force four-pole superconducting generator, and the 300 kJ pulsed energy coil all utilized soldered and clamped joints.

Every superconducting magnet, almost without exception, incorporates some form of an electrical joint in its construction. A normal conductor-to-superconductor transition is generally required that completes the external current connection for energizing the magnet. In addition, manufacturing constraints either in winding or in conductor manufacture may necessitate additional joints between lengths of superconductor.

All "superconductor joints" must satisfy specific mechanical, electrical, and thermal constraints that depend on the individual magnet design. The joint resistance should be almost as lossless

*Supported in part by the Union Carbide Corporation, Contract No. 22X31747C; Electric Power Research Institute, Contract No. RP 1473-1; and U.S. Air Force, Contract No. F33615-79-C-2026.

as the superconductor itself. Otherwise at design current a high heat load presented by the joint may produce excessive cooling requirements or more seriously initiate a conductor normalization and subsequent quench of the magnet. Superconductor joints, in addition, must be as mechanically strong as the actual superconductor since a mechanical failure at the joint would open the winding with possible catastrophic results.

The design and fabrication methods used to implement superconductor joints have shown a number of varied approaches. These approaches, for the most part, depend intimately on the desired resistance level and the exact conductor configuration. A laboratory solenoid for nuclear magnetic resonance (NMR) requires an extremely low resistance joint since the magnet is normally operated in the persistent mode. Typical joint resistances for NMR magnets must be less than 10^{-12} Ω for optimum operation. It is thus not surprising that some of the reported joint resistances for NMR magnet applications are among the lowest observed. Leupold and Iwasa[1] produced a joint in a Nb/Ti multifilamentary monolithic conductor that was less than 10^{-14} Ω. This joint was produced by etching away the copper matrix to expose the filaments. The individual filaments were then crimped together in separate copper sleeves. Other variations on this approach have actually welded the individual filaments by microelectron beam welding. The copper stabilizer was then restored through the use of copper foil.[2]

Superconductor joints for large-scale applications, such as generators and fusion magnets, are usually more difficult due to higher current, higher field, and a more complicted conductor configuration. The Lawrence Livermore mirror-fusion test facility (MFTF) conductor for the Yin-Yang coil is an excellent example. A 6.5 mm x 6.5 mm monolith of multifilamentary Nb-Ti is wrapped within a perforated copper stabilizer, which is then soldered by heat treatment to the superconductor core.[3]

Joint development programs at Lawrence Livermore evaluated methods for joining this large multifilamentary superconductor core. Both explosive welding[4] and cold welding[5] demonstrated completely acceptable mechanical and electrical properties. The explosive approach, however, presented serious fabrication problems, which prompted the selection of cold-welding as the preferred method. Recent tests[5] on cold-welded MFTF Nb-Ti superconductors have shown joint resistances at 4.2 K of ~10^{-8} Ω at ~6-kA operating current and 7.5-T field.

The preceeding examples of superconductor joining demonstrate a wide range in magnet applications from small laboratory magnets to large-scale fusion devices. These different joining approaches also present examples of the two basic methods used for fabricating a superconductor joint, i.e., mechanical or welded. Mechanical joining covers all clamped techniques, which may be soldered in addition to the clamping. As mentioned earlier, the clamped and soldered joining technique has been widely used with nominal joint resistance easily obtained of better than 10^{-8} Ω. A soldered and clamped joint can be readily utilized where there are no space constraints. A large access area is required for assembly, and the resulting joint generally occupies a volume much larger than the adjoining conductor. The DEALS fusion magnet design presents a recent large-scale concept directed at a clamped or pressure joint with a large surface area to minimize the resistance.[6]

The welded joint basically covers all of the other joining approaches that have been investigated: welding of each filament, butt resistance welding, cold welding, ultrasonic and explosive welding. Some joint designs considered for the MFTF-B solenoid coils incorporate both of the basic methods, cold welding of the stabilizer followed by a soldered lap joint of the superconductor.

The superconductor joint design for most large-scale applications has traditionally followed the conductor selection, structural design, and other details of the magnet construction. As a result, the joining problems are often more difficult and usually represent a critical area with respect to magnet stability and performance.[7] The following sections review two recent joint development programs that demonstrate areas of critical joining.

LARGE COIL PROGRAM Nb_3Sn JOINT DEVELOPMENT

The Westinghouse Large Coil Program (LCP) magnet design requires a specific joint resistance and helium flow rate to insure superconducting stability of the conductor. The stability analysis, based on a minimum propagating zone (MPZ) model specifies a joint resistance of less than 1.7×10^{-9} Ω at 2 T and 18-kA operating current. Detailed experimental investigation of the joint and header under actual forced-flow conditions verified this analysis. Acceptable stability with even higher joint resistance of $\sim 10^{-8}$ Ω could be obtained by an increased helium flow rate.[8]

The Westinghouse (LCP) conductor is fabricated by cabling 486 strands of multifilamentary Nb_3Sn wire and then compacting and jacketing this cable with a welded steel enclosure. The steel outer jacket is completely sealed over its entire length with respect to helium leakage. A header welded to the end of each conductor provides space for the electrical joint and attachment of helium flow tubes.

This force-cooled, steel-enclosed Nb_3Sn cable thus presented a formidable joining problem. A joint development program to connect the lengths of full-size conductor and individual strands during wire manufacture was pursued by Westinghouse and Airco to evaluate various joining procedures. The development program concentrated on three main approaches: butt-resistance welding, lap joining, and cold welding.

Cold welding, which works quite well for Nb-Ti multifilamentary conductors is not readily adaptable to Nb_3Sn. The unreacted bronze core containing the niobium filaments becomes extremely brittle during cold welding. In addition, the upset during cold welding almost completely destroys the diffusion barrier and pushes the high-resistivity bronze to the outside of the wire. Annealing, which would be necessary to provide sufficient ductility for the cabling process, would thus contaminate the copper stabilizer. Cold welding, after reaction, would severely degrade the highly fragile A15 Nb_3Sn.

Lap joining of the multistranded cable would require uncabling and placement in a support fixture to minimize bending stresses following reaction. The strands from mating conductors would be overlapped and soldered in a grooved copper cylinder. A copper sleeve would be pressed over the entire structure for mechanical integrity. The resulting lap joint would require a long time for assembly and occupy a fairly large volume. This method was evaluated by constructing a simple 5-cm lap joint as detailed above. Nine strands of 0.7-mm wire were reacted in a special fixture at 700°C for 30 h and subsequently soldered with indium between a copper cylinder and sleeve. The observed resistance at 4.2 K in zero field was $R = 2.6 \times 10^{-8}\ \Omega$. Additional joint tests were performed on a termination prepared by swaging a copper tube around the unreacted wire. The ends of two terminations were then prepared with different bias angles and reacted. The two surfaces were then joined by an extremely thin interface of indium solder. The low-temperature resistances were measured in a special fixture fabricated from G10 laminate, which contracted during cooldown, providing slight pressure on the joint. A 45° and 30° bias showed $R_{45°} = 8.6 \times 10^{-8}\ \Omega$ and $R_{30°} = 8.9 \times 10^{-8}\ \Omega$ at zero field and 4.2 K for ~500 A current). These soldered bias tests were directed at a method for reducing the current transfer length through the highly resistive tantalum barrier and bronze matrix.

Previous work by Wilson[9] and more recent work by Ekin[10] established minimum length requirements for lap joints to reduce current transfer resistance. Using the relationship developed by Ekin[10]

$$\frac{X_{min}}{D} = 0.14 \left(\frac{\rho_t}{\rho_{min}}\right)^{1/2}$$

where $\rho_{min} = 10^{-11}\ \Omega\cdot\text{cm}$ (the resistivity detection criterion), $\rho_t = 10^{-6}\ \Omega\cdot\text{cm}$ (the transverse or matrix resistivity for bronze), X_{min} =

current transfer length, D = the conductor effective diameter or 0.762 mm for LCP strands,

$$X_{min} = D \times 44.3 = 3.38 \text{ cm}.$$

A Nb_3Sn lap joint would thus require approximately a 4-cm overlap to obtain a low-resistance joint. The lap joint tested above easily met this criterion. The soldered bias joints showed a factor of three higher resistance than the lap joint. This result indicates that a reduced current transfer length was, in fact, obtained. Its full potential, however, was not realized owing to the difficulty in obtaining good contact between the soldered surfaces. The interface resistance was thus high and contributed added resistance.

Two of the reacted six-strand terminations were prepared with perpendicular surfaces that were then butt-resistance welded. Airco had explored this technique with satisfactory results.[11] This joint showed at zero field and ~400 A current a resistance of $R = \sim 4.5 \times 10^{-9}\ \Omega$. This latter test was extremely encouraging and indicated that the larger full-size 486-strand conductor could be similarly terminated and butt-welded with acceptable properties.

As a result, additional subsize tests were performed coupled with tensile tests, which demonstrated acceptable electrical resistance and mechanical strength. Full-size terminations were then prepared and tested at full-design conditions, forced-flow helium, 2 T, and ~14 kA current. The observed resistances at 4.2 K for the full-size butt-welded joint ranged from 1.5×10^{-9} to $1 \times 10^{-8}\ \Omega$ at the above field and current.[8] There were no obvious visual differences in the welded joints that might explain the observed range in resistance. Full penetration x-rays of the joint region also showed no obvious difference. A possible explanation was developed, however, from analysis of the individual strand results explained below.

A parallel effort to joining the full-size cable was directed at the individual strands. The Nb_3Sn (0.69-mm) multifilamentary wire, fabricated by Airco, had 2869 filaments (~3.7-μm diam.) with a 63 vol.% copper stabilizing shell.[12] The strand-joining effort concentrated entirely on butt-resistance welding of the unreacted Nb-Sn wire. Sufficient strength, ductility (before reaction), and acceptable electrical properties (after reaction) were observed. Random strand breakage during drawing or cabling could thus be solved by this welding method. The Westinghouse Phase II report for the Large Coil Program contains additional information on the strand and full-size joint development effort.[13]

Table 1. Average Mechanical Strength for Welded and Unwelded Nb_3Sn Wire in kg/mm^2

Diameter (mm)	Before Reaction Unwelded	Before Reaction Welded	After Reaction Unwelded	After Reaction Welded
0.91	54	40	--	--
0.97	69	32	23	26

Additional tests have been performed on butt-resistance welds of Nb_3Sn wire. This wire was supplied by Intermagnetics General Corporation (IGC) for the Air Force Materials Technology Program. The wire we received will be eventually tested under simulated generator cooling conditions in a rotating test facility. Butt-resistance welds of 0.91-mm diameter wire were performed and reacted. The mechanical strength of the welded and unwelded wire was tested before and after reaction, as summarized in Table 1. Welding degraded the strength by as much as 50% of the strength of the unreacted wire, which nevertheless should be strong enough for most cabling operations. The reaction produced no marked difference in strength between the welded and unwelded wire.

Critical current tests using 1 μV/cm as the electric field criterion on welded and unwelded wire are compared in Figure 1. This figure also presents short sample data on the Airco LCP Nb_3Sn strands.[12] The IGC welded strands showed ~27% of the current-carrying ability of the unwelded strands.

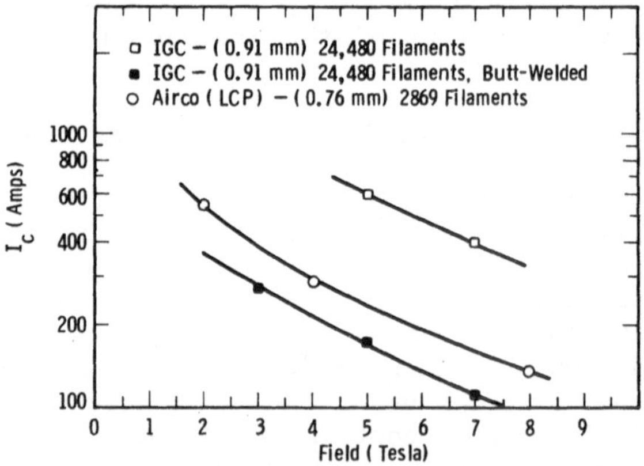

Fig. 1. Short sample critical current for welded and unwelded Nb_3Sn wire.

This data was considerably better than the data of the Airco welded strands. Typical strand welds of the Airco wire could only carry 7 to 10% of the critical current of the unwelded wire.[11,13] Electron microscopy of the IGC wire was performed to evaluate the weld morphology. A SEM cross section (shown in Figure 2) indicates a number of interesting features. The heat-affected zone extended ~1.0 mm with marked overlap of the tantalum stabilizer. No evidence for melting of the niobium or tantalum components was observed except at the fusion interface. The filaments in the heat-affected zone were observed to buckle or ripple owing to the upset pressure. Subsequent etching with dilute HNO_3 confirmed these features. At the fusion interface shown in Figure 3, a large number of filaments were actually joined end to end over almost the entire cross section. The possibility also exists that some filaments were interspersed from the joining sections. Subsequent reaction actually formed a continuous network of Nb_3Sn across the weld interface. The etched SEM studies also showed no evidence of tantalum or niobium melting except where the filaments were fused.

The observed critical current properties for the welded IGC wire are thus attributed to the high number of filaments actually joined during the welding process. The reduced values for the Airco wire simply follow as a direct result from the decreased number of filaments. If we normalize the two wires for the different cross-sectional areas, percents of copper stabilizer, and filament numbers and diameters, the IGC wire has roughly three times more filaments for an equivalent cross section. Hence the probability of welding filaments for the IGC wire is three times

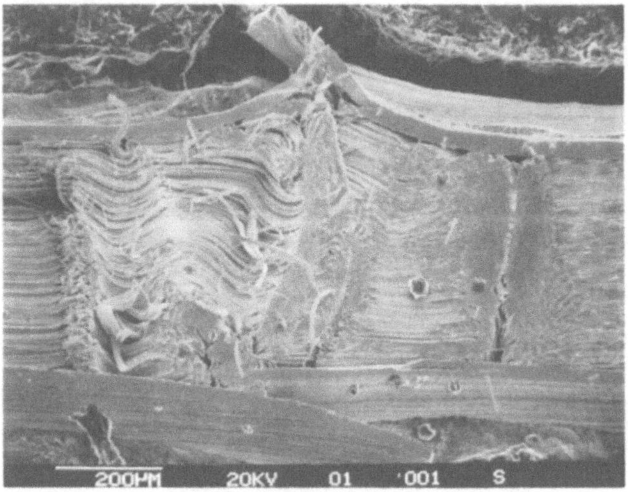

Fig. 2. Cross section of butt-welded Nb_3Sn wire (SEM).

greater, which is almost exactly the difference in critical current between the two welded wires.

These results thus imply that improved critical currents can be obtained by a higher concentration of filaments and good alignment of strands during welding. This conclusion applies directly to the larger 9-strand, 162-strand, and 486-strand butt welds performed for LCP. As we move away from ideal strand-to-strand alignment, the joint properties degrade.[13]

At 2 T, the LCP 162-strand subsize cable should carry 90 kA. The observed critical current for the 162-strand butt weld was 400 A or less than 0.5%. The full-size 486 LCP strand cable should carry 270 kA, yet at 2 T the termination showed a resistance of 10^{-8} to 10^{-9} Ω.

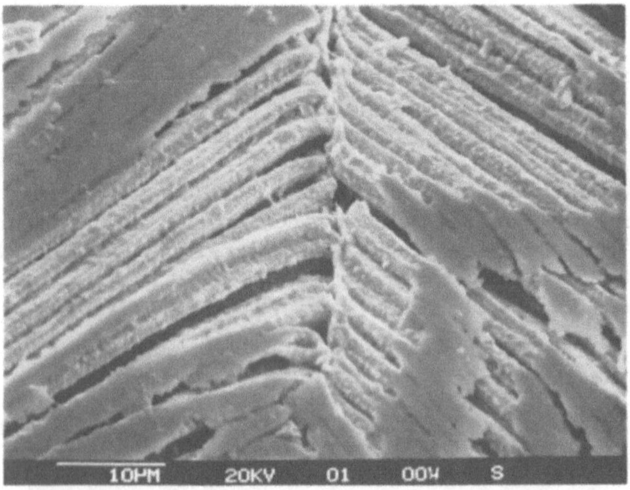

Fig. 3. Filament joining at butt-welded interface (SEM).

Table 2. 300 MVA Superconductor Specification

Dimension	0.203 cm (0.080 in) x 0.384 cm (0.151 in)
Copper to SC Ratio	2.5:1
Critical Current	3500 A at 4.2 K and 5 T
Resistivity Ratio ρ(300 K)/ρ(20 K)	150
Twist Pitch	4 cm
Filament Size	50 µm

As discussed earlier, no obvious differences for the full-size welded terminations could be detected. The possibility exists that better strand-to-strand alignment was simply obtained for the lower observed resistance. It is interesting to note that the IGC weld data extrapolated to 2 T (in Figure 1) would suggest a fully superconducting joint could be obtained for LCP at 2 T and 18 kA. This result could only occur for perfect strand alignment and comparable filament count.

SUPERCONDUCTING GENERATOR Nb-Ti JOINT DEVELOPMENT

The superconductor for the Westinghouse/EPRI 300 MVA generator is a copper-stabilized Nb-Ti multifilamentary rectangular superconductor, as specified in Table 2. The design operating current and maximum field are 1800 A and 5 T, respectively.

During winding of the field coil, the transitions from adjacent stacks and adjacent slots are facilitated by the ability to splice the conductor. These splices on the ~2.0-mm thickness are located in the end turns at the winding outer diameter. This end-turn location presents an extremely confined area for joining the conductors. Approximately 0.60 cm (0.25 in) is available on each side of two overlapped conductors. Since the splice is located at the winding o.d., unlimited space is available above the joint region in the radial direction. In addition, owing to the winding tension and angled slot configuration, there is no ability to lift the conductor away from the winding surface for improved access. A joint development program was implemented that considered three joining approaches: ultrasonic, butt-resistance welding, and lap/mechanical joining.

Preliminary results on the electrical and mechanical properties of butt-welded joints were encouraging. The limited access for accomplishing the weld, however, essentially ruled out this method.

The lap/mechanical technique was also explored and could be used if absolutely necessary. Such a joint, however, would take up most of the available space and definitely presented an assembly problem.

The ultrasonic welding method, on the other hand, showed considerable potential. This technique offered fast, easy welding with the confined area representing the only major development problem. The ultrasonic joining process was successfully used in the construction of the superconducting MHD (U-25) magnet built by Argonne National Laboratory.[14]

The ultrasonic technique works extremely well with a lap-type configuration. The two conductors are placed between a clamping fixture, pressure is applied, and a cold diffusion bond is effected through a combination of ultrasonic vibration and the applied pressure. The welding time is of the order of 1.0 s, and temperatures at the joint rarely exceed 50% of the melting point of the copper matrix. The physical strength is excellent, and when weld parameters are established the joint strength normally exceeds the ultimate strength of the adjoining conductor. This technique, like the lap/mechanical technique, does not affect the cold-worked state of the Nb-Ti superconductor. The process is limited only by the thickness of the joint and accessibility. The present commercial ultrasonic welders have a maximum thickness limitation for copper of ~2.0 mm. The 300 MVA conductor thickness of 2.03 mm is, thus, just within the limits of present welders. A bias or scarf configuration, however, would allow these thickness limits to be extended. Prototype welders are also under development with higher power and thickness capability.

The magnetic field levels at the joint locations in the end-turn region were calculated by different computer programs. These programs indicated that the field intensity for the joints would not exceed 4.5 T. Thus, the joint resistance, on the final configuration, will be measured at field levels up to 4.5 T and operating current of 1800 A.

Using the current transfer relation developed by Ekin[10] with $\rho_t = 10^{-8}\ \Omega\cdot\text{cm}$ as the transverse resistivity

$$X_{min} = 1.4\ \text{cm} \quad \text{for } \rho_{min} = 10^{-11}\ \Omega\cdot\text{cm}$$

and

$$X_{min} = 4.4\ \text{cm} \quad \text{for } \rho_{min} = 10^{-12}\ \Omega\cdot\text{cm}$$

Because a large number of joints were projected, a 4-cm overlap length was selected.

Trial welds at this length were first attempted by two different welder manufacturers. Both units produced satisfactory welds with respect to electrical and mechanical performance.

A Branson unit, however, was eventually selected and purchased for use on the 300 MVA program, since its basic design offered more potential for the confined area application. The Model 3301 Branson unit operates at 20 kHz and generates ~3000 W output power. A special confined area welding tip was constructed out of a high-strength, high-toughness, lightweight alloy of Ti-6Al-4V. This alloy was selected after a number of other candidates were eliminated for reasons of weight or poor fatigue characteristics.

Weld trials were then conducted on an actual superconductor with specifications nearly identical to those of the 300 MVA conductor. These trials established optimum welding parameters with respect to power, weld time, and applied pressure. A series of welds were then mechanically tested at room temperature. All of the welds exceeded the physical strength of the parent conductor. Subsequent mechanical tests at 4.2 K showed an approximate 40% improvement in mechanical strength over the room temperature value. Failure was again observed in the parent conductor, which approached brittle failure at 4.2 K (ductile failure was at room temperature). The point of failure generally occurred adjacent to the overlapped region. Electrical tests at 4.2 K at 5.0 T and ~1000 A showed typical resistances of $\sim 1.0 \times 10^{-9}\ \Omega$. Additional tests at higher currents (~2000 A) and 5.0 T are planned in a special fixture representative of the end turn geometry.

As discussed in the previously published paper by Hofstrom et al.,[14] ultrasonic welded joints offer what may be the only joint configuration for quantitative NDT testing. The present ultrasonically welded joints were NDT tested with a 20-MHz focus ultrasonic transducer using immersion and oil contact method. An accurate evaluation of the bond integrity was possible with this procedure. Areas that indicated satisfactory and poor bond integrity were subsequently sectioned and studied metallographically. This latter test completely verified the NDT ultrasonic tests.

In summary, an ultrasonic welding method has been developed for joining the 300 MVA superconductor. This technique offers high mechanical strength with joints actually stronger than the parent material, acceptable electrical properties at low temperature, and the ability to perform quantitative NDT testing.

CONCLUSIONS

These two recent superconductor joint development programs both represent an intensive effort to obtain an acceptable joint. Both programs were highly dependent on a successful joint demonstration. It is not always possible, but it is recommended that more attention be directed to joining problems during the initial magnet design analysis. Trade-offs in conductor design may be possible that would facilitate the joint fabrication with possibly improved electrical properties. The large fusion magnets of the future will require low overall thermal losses. An optimum superconductor joint design would certainly reduce some of the loss.

ACKNOWLEDGMENTS

I would like to acknowledge the experimental assistance of J. Buttyan and P. Vecchio and mechanical tests of W. A. Logsdon.

REFERENCES

1. M. J. Leupold and Y. Iwasa, Superconducting joint between multifilamentary wires, Cryogenics 16:215 (1976).
2. G. Luderer, P. Dullenkopf, and F. Laukien, Cryogenics 14:518 (1974).
3. E. Adam, E. Gregory, and W. Marancik, in: "Proceedings of the 7th Symposium on Engineering Problems of Fusion Research," 77CH1267-4-NPS, IEEE, New York (Oct. 1977), p. 1329.
4. D. Cornish, J. P. Zbasnik, and H. E. Pattee, in: "Proceedings of the 6th Symposium of Engineering Problems of Fusion Research," 75CH1097-5-NPS, IEEE, New York (Nov. 1975), p. 106.
5. D. H. Cornish, D. W. Deis, and J. P. Zbasnik, in: "Proceedings of the 7th Symposium on Engineering Problems of Fusion Research," 77CH1267-4-NPS, IEEE, New York (Oct. 1977), p. 1267.
6. S. Y. Hsieh, G. Danby, J. R. Powell, P. Bezler, D. Gardner, C. Laverick, M. Finkleman, T. Brown, J. Bundy, T. Bolders, I. Zatz, R. Verzera, and R. Herberman, in: "Advances in Cryogenic Engineering," Vol. 25, Plenum Press, New York (1980), p. 207.
7. H. Fujino, A. Ishihara, K. Veda, Y. Shindo, and S. Nose, "Proceedings of ICEC Eight," IPC Science & Technology Press, Guildford, Surrey, England (1980), p. 600.
8. R. D. Blaugher, M. A. Janocko, P. W. Eckels, A. Patterson, J. Buttyan, and E. J. Shestak, IEEE Trans. Magn. MAG-17:467 (1981).
9. M. Wilson, Internal Report SMR/1, Rutherford Laboratory, Chilton, Didcot, England.
10. J. W. Ekin, J. Appl. Phys. 49:3406 (1978).
11. P. A. Sanger and E. Gregory, 8th Symposium on Fusion Engineering (1979), abstract only.
12. C. R. Spencer, P. A. Sanger, and M. Young, IEEE Trans. Magn. MAG-15:76 (1979).
13. "Superconducting Magnet Coils for the Large Coil Program, Phase 2 Final Report," Westinghouse Electric Corporation, Research and Development Center, Pittsburgh, Pennsylvania (Mar. 1980).
14. J. W. Hofstrom, D. H. Killpatrick, R. C. Nieman, J. R. Purcell, and H. R. Tresch, IEEE Trans. Magn. MAG-13:94 (1977).

ELECTRICAL BOUNDARY RESISTANCE IN AN ALUMINUM-STABILIZED SUPERCONDUCTOR

D. Yu,* Y. M. Eyssa, and P. Zolliker †

Applied Superconductivity Center, University of Wisconsin, Madison, Wisconsin

INTRODUCTION

There has been considerable effort to develop aluminum-stabilized superconductors.[1-4] One possible way to form a metallurgical bond between an aluminum stabilizer and a NbTi in Cu superconductor is soft soldering.[2,4,5] The technique of manufacturing such composite conductors in lengths long enough for practical applications, however, is still in the developmental stage.

The electrical boundary resistance between the stabilizer and the superconductor is a conductor property that reflects bond quality and influences stability.[6] Measurements of this electrical contact resistance in Cu-Cu, Al-Al, and Cu-solder-Cu joints have been made[7-10] but are not in complete agreement.

This paper reports on an experimental method of determining the electrical boundary resistance by measurements of current transfer length. Moreover, three theoretical current transfer models are presented and compared.

EXPERIMENTAL PROCEDURE

A NbTi/Cu multifilamentary superconducting wire (diameter, 0.086 cm; ratio of copper to NbTi, 1.8; number of filaments, 2046;

---

*Visiting scholar from the General Research Institute for Non-ferrous Metals, the Ministry of Metallurgical Industry, Beijing, China.

†International Association for the Exchange of Students for Technical Experience (IAESTE) visitor from the Swiss Federal Institute of Technology (ETH), Zurich, Switzerland.

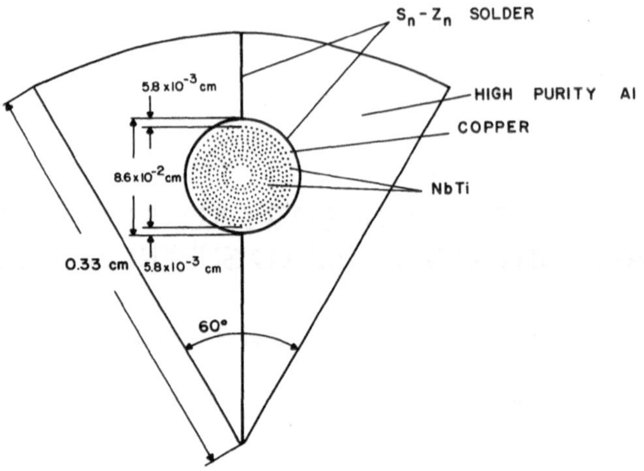

Fig. 1. Sketch of the conductor cross section.

average diameter of filaments, 11 μm) was joined to two pieces of 260-RRR (RRR ≡ $\rho_{298K}/\rho_{4.2K}$) aluminum by an 89Sn-11Zn solder. The cross section of the conductor, as shown in Fig. 1, has the shape of a 60° sector with a 0.33-cm radius.

Samples were cut from a 5-m-long piece of conductor. Sample lengths ranged from 6 to 8 cm except for one 20-cm sample. Aluminum was stripped off one end of the samples, the two halves of aluminum were peeled open at the other end, a portion of superconducting wire was removed, and the two aluminum wedges were clamped together. Current leads and voltage taps were soldered to the sample with an In-Sn alloy solder.

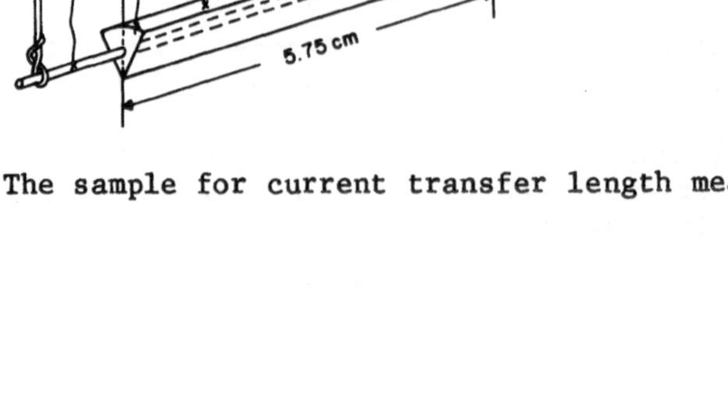

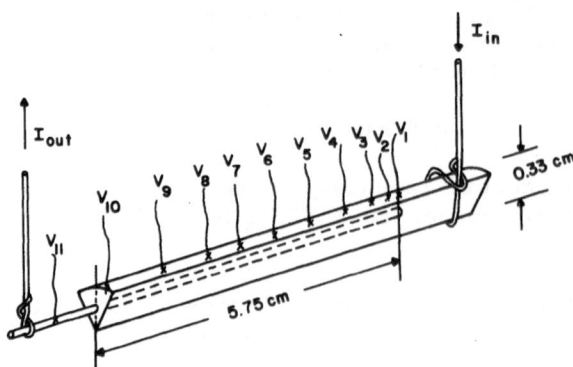

Fig. 2. The sample for current transfer length measurement.

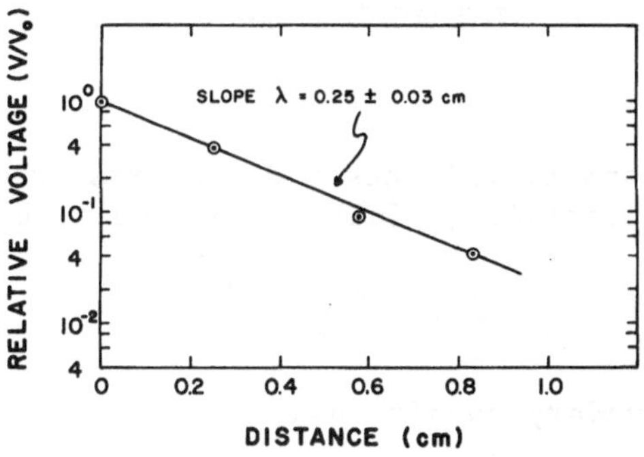

Fig. 3. Electrical potential profile along the conductor.

It was difficult to measure the distances between voltage taps accurately by a mechanical means, since these distances were quite short (0.3–0.4 cm) and about the size of the solder spot. To improve the accuracy of measuring these distances, a dc current was passed through the sample. The electrical potentials of the voltage taps with a common reference at the superconducting wire (tap #11 in Fig. 2) were measured at room temperature or in a liquid nitrogen bath. The distances between taps were determined by assuming the voltage drops between taps were directly proportional to the distances between them.

The current transfer length was measured at 4.2 K by passing a constant dc current through the sample, as shown in Fig. 2. Current reversal was used to eliminate the thermal emf effect. Figure 3 is a typical semilogarithmic plot of the electric potential profile along the conductor. The profile is exponential; its slope gives a characteristic length, λ, the current transfer length.[6,7]

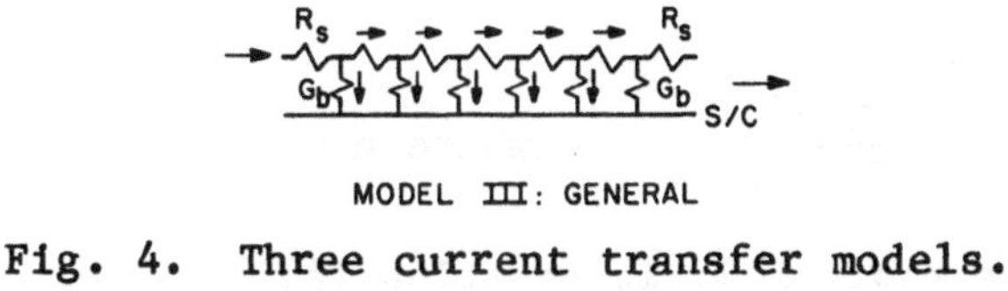

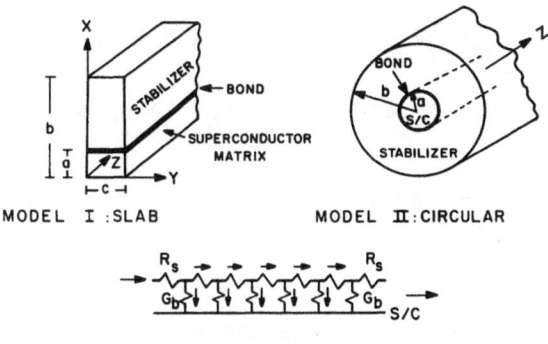

Fig. 4. Three current transfer models.

THEORETICAL ANALYSIS

Slab Model

Assuming a bond of thickness, d, and resistivity, ρ_b (Fig. 4), the electrical potential, ϕ, follows Laplace's equation

$$\nabla^2\phi = \frac{\partial^2\phi}{\partial x^2} + \frac{\partial^2\phi}{\partial z^2} = 0 \qquad (1)$$

subject to the boundary conditions,

$$z \to \infty \ , \quad \phi = 0 \qquad (2)$$

$$x = a \ , \quad \phi = \frac{\rho_b d}{\rho_s}\frac{\partial\phi}{\partial x} = q\frac{\partial\phi}{\partial x} \qquad (3)$$

$$x = b \ , \quad \frac{\partial\phi}{\partial x} = 0 \qquad (4)$$

$$z = 0 \ , \quad \frac{\partial\phi}{\partial z} = -\rho_s J_0 = -\frac{\rho_s I_0}{(b-a)c} \qquad (5)$$

where ρ_s is the resistivity of the stabilizer material, $q = \rho_b d/\rho_s$, and I_0 is the current in the stabilizer at $z = 0$.

The following transcendental equation can be derived

$$\cot\frac{b-a}{\lambda_n} = \frac{q}{\lambda_n} = \frac{\rho_b d}{\rho_s \lambda_n} \qquad (6)$$

In the case of a perfect bond ($\rho_b d = 0$),

$$\lambda_n = \frac{b-a}{\pi(n-1/2)} \qquad \text{where } n = 1,2,\ldots \qquad (7)$$

If the bond is not perfect and $\rho_b d \gg \rho_s(b-a)$, then

$$\lambda_1^2 = \frac{\rho_b d}{\rho_s}(b-a) \qquad (8)$$

and

$$\lambda_n = \frac{b-a}{\pi(n-1)} \qquad \text{where } n = 2, 3, \ldots \qquad (9)$$

Figure 5 is the plot of λ_1 in equation (6).

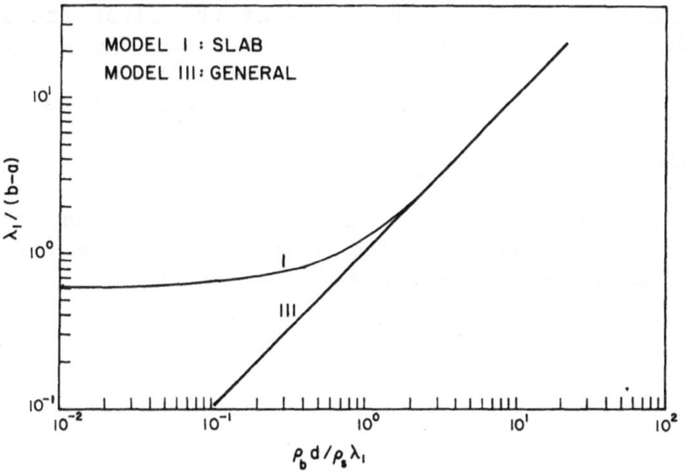

Fig. 5. Comparison between the slab and the general models.

Circular Model

In cylindrical coordinates, Laplace's equation assumes the following form.

$$\nabla^2\phi = \frac{\partial^2\phi}{\partial r^2} + \frac{1}{r}\frac{\partial\phi}{\partial r} + \frac{\partial^2\phi}{\partial z^2} = 0 \tag{10}$$

The boundary conditions can be expressed as follows:

$$z \to \infty \quad , \quad \phi = 0 \tag{11}$$

$$r = a \quad , \quad \phi = \frac{\rho_b d}{\rho_s}\frac{\partial\phi}{\partial r} = q\,\frac{\partial\phi}{\partial r} \tag{12}$$

$$r = b \quad , \quad \frac{\partial\phi}{\partial r} = 0 \tag{13}$$

$$Z = 0 \quad , \quad \frac{\partial\phi}{\partial z} = -\rho_s J_0 = -\frac{\rho_s I_0}{\pi(b^2-a^2)} \tag{14}$$

Solving equation (10) subject to (11) and (14), we have

$$Y_1\left(\frac{b}{\lambda_n}\right)J_0\left(\frac{a}{\lambda_n}\right)-J_1\left(\frac{b}{\lambda_n}\right)Y_0\left(\frac{a}{\lambda_n}\right)=\frac{q}{\lambda_n}\left[J_1\left(\frac{b}{\lambda_n}\right)Y_1\left(\frac{a}{\lambda_n}\right)-Y_1\left(\frac{b}{\lambda_n}\right)J_1\left(\frac{a}{\lambda_n}\right)\right] \tag{15}$$

where J_0 and Y_0 are Bessel functions of the first and second kind of order 0, respectively, whereas J_1 and Y_1 are Bessel functions of the first and second kind of order 1, respectively. In Fig. 6, the λ_1 in (15) is plotted as a function of ρ_b, ρ_s, d, a, and b.

General Model

This idealized model does not take into account the detailed geometry of the conductor. The profiles of electrical potential and current in the stabilizer are found to be:

$$I = I_0 \exp\left(-\frac{z}{\lambda}\right) \tag{16}$$

$$V = V_0 \exp\left(-\frac{z}{\lambda}\right) \tag{17}$$

$$\lambda^2 = 1/R_s G_b = A\rho_b d/\ell\rho_s \tag{18}$$

where R_s and G_b are the stabilizer resistance per unit length and the electrical boundary conductance per unit length of bond, respectively. The length, ℓ, is the bonding circumference length between the stabilizer and the superconducting matrix, and A is the cross-sectional area of the stabilizer.

RESULTS AND DISCUSSION

The experimental current transfer length measurement of four samples and the related calculations in accordance with three theoretical models are listed in Table 1. The calculated electrical

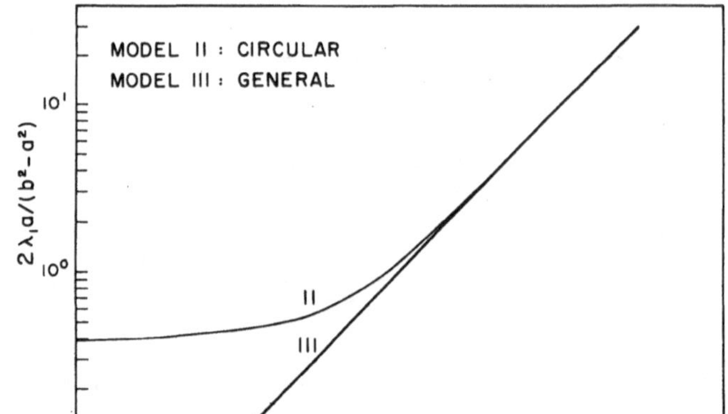

Fig. 6. Comparison between the circular and the general models.

Table 1. Current Transfer Length and Boundary Resistance

Sample No.	λ_1, cm (measured)	$\rho_b d$, $n\Omega{\cdot}cm^2$ Model I*	Model II† (calculated)	Model III
1	0.37	6.9	7.3	7.6
2	0.25	2.8	3.2	3.5
3	0.28	3.7	4.1	4.3
4	0.31	4.7	5.1	5.3

*b−a = 0.19 cm
†a = 4.3×10^{-2} cm, b = 1.35×10^{-1} cm

boundary resistances were found to be in the range 3-8 $n\Omega{\cdot}cm^2$. This result is almost one order of magnitude lower than the bonding resistances in samples prepared by crimping copper tube to copper-coated NbZr wire.[7] Although most previously reported contact resistances[8,9] are higher than our present values, the reported 4.5 $n\Omega{\cdot}cm^2$ of the eutectic PbSn solder joint[10] is in the range of our present results.

The measured boundary resistance is the sum of the resistances of the bulk solder, the copper skin in the superconducting wire, and the interfaces (Al-solder, solder-Cu, and Cu-NbTi). According to our measurements, the resistivity of 89Sn-11Zn solder at 4.2 K, ρ_{SnZn}, was 2.3×10^{-7} $\Omega{\cdot}cm$; the solder layer thickness d_{SnZn}, was 1.8×10^{-4} cm; the resistivity of copper at 4.2 K, ρ_{Cu}, was 2.5×10^{-8} $\Omega{\cdot}cm$, and the copper skin thickness, d_{Cu}, was 5.8×10^{-3} cm. The contribution of the two bulk terms to the total boundary resistance, $\rho_{SnZn}d_{SnZn} + \rho_{Cu}d_{Cu}$, was approximately 1.9×10^{-10} $\Omega{\cdot}cm^2$. Considering the average transverse resistivity data obtained from the self-field distribution measurement,[11] we estimate that the boundary resistance caused by the interface barrier layer between the copper matrix and NbTi filaments in the superconducting wire was about 1×10^{-10} $\Omega{\cdot}cm^2$. It is also reasonable to assume that there was no current transfer from the outer filaments to the inner ones in the superconducting wire[12] owing to the small current (1-5 A) used in our measurement. As expected, the two interfaces formed between the Sn-Zn solder and the two normal metals (Al and Cu) were the major sources of the measured boundary resistances.

According to equation (6) and (15) or Figs. 5 and 6, all three models gave nearly the same resistances when $\lambda_1/(b-a)$ in a slab conductor or $2a\lambda_1/(b^2-a^2)$ in a circular conductor reached 2.6, where the differences among the boundary resistances calculated with the three models were within 5%. Generally, if $\lambda_1 \ell/A$ in a conductor with slab, circular, or any similar configuration approaches 3 or $\rho_b d\ell/\rho_s A$ is close to 9, the general model can be used to determine the boundary resistance within 5%.

In our case, $\lambda_1\ell/A$ was in the range 1.3 to 2.0, and the geometry is important. By comparing the boundary resistances listed in Table 1, it is found that the slab model gave the lowest resistance values, and the general model yielded the highest. Differences between the general and the circular models were 4 to 8%, whereas those between the slab and circular models were 5 to 13%. Assuming that the circular model is the best representation of our real case among the three models, we consider the general model to be a good approximation for our conductor.

Table 2 lists the calculations in the ideal ($\rho_b d = 0$) and practical perfect bond cases. There is about 0.1-cm current transfer length even if only the bulk solder, the copper skin, and the Cu-NbTi interface resistances exist (practically perfect bond) in the bonding region of the conductor.

CONCLUSIONS

1. A boundary resistance of 3-8 $n\Omega \cdot cm^2$ in an aluminum-stabilized conductor has been determined by measurement of the dc current transfer length.

2. More than 90% of the measured boundary resistance comes from the solder bonding interfaces.

3. The simple current transfer model can be reliably applied to the boundary resistance calculation if $\lambda_1\ell/A \geq 3$. Otherwise, more detailed geometry must be considered.

ACKNOWLEDGMENTS

The authors are grateful to Dr. K. T. Hartwig for his valuable discussions. The research work is supported by the U.S. Department of Energy and the Wisconsin Electric Utility Research Foundation.

Table 2. Current Transfer Lengths for Perfect Bonding

Model	λ_1	
	$\rho_b d = 0$	$\rho_b d = 2.9 \times 10^{-10}\ \Omega \cdot cm^2$
I	0.12 cm	0.14 cm
II	0.07	0.10
III	0	0.07

REFERENCES

1. R. W. Boom et al., in: "The Wisconsin Superconductive Energy Storage Project," 1976 Annual Report, EES No. 47, University of Wisconsin, Madison, Wisconsin (1977).
2. M. Morpurgo and G. Pozzo, Cryogenics 17(2):87 (1977).
3. H. Nomura, M. Obata, and S. Shimanota, Cryogenics 11(5):396 (1971).
4. P. Genevey and J. Le Bars, in: "Proceedings of the 6th International Conference on Magnet Technology," MT-6, ALFA Publishing Co., Bratislava, Czechoslovakia (1977), p. 1033.
5. K. T. Hartwig, P. Zolliker, D. Yu., S. W. Van Sciver, and A. Khalil, Fabrication and performance of an aluminum-stabilized composite superconductor, in: "Advances in Cryogenic Engineering--Materials," Vol. 28, Plenum Press, New York (1982), p. 805.
6. V. A. Altov, V. B. Zenkevich, M. G. Kremlev, and V. V. Sychev, in: "Stabilization of Superconducting Magnetic Systems," Plenum Press, New York (1977), p. 163.
7. E. J. Lucas, Z. J. J. Stekley, C. Laverick, and G. Pewitt, in: "International Advances in Cryogenic Engineering" (Proceedings of the 1964 Cryogenic Engineering Conference, Sections M-U), Plenum Press, New York (1965), p. 113.
8. J. L. Zar, in: "Advances in Cryogenic Engineering," Vol. 13, Plenum Press, New York (1968), p. 95.
9. K. T. Hartwig and S. W. Van Sciver, in: "Proceedings of the 8th Symposium on Engineering Problems of Fusion Research," IEEE Pub. No. 79CH1441-5NPS, Vol. 4, IEEE, New York (1979), p. 1794.
10. L. F. Goodrich and J. W. Ekin, IEEE Trans. Magn. MAG-17:69 (1981).
11. B. Turck, M. Wake, and M. Kobayashi, Cryogenics 17(4):217 (1977).
12. J. W. Ekin, J. Appl. Phys. 40(6):3406 (1978).

STRUCTURAL ASPECTS OF CABLE CONDUCTORS IN HIGH-FIELD-STRENGTH, HIGH-CURRENT-DENSITY SUPERCONDUCTING DIPOLES

Z. J. J. Stekly

Magnetic Corporation of America, Waltham, Massachusetts

INTRODUCTION

Because of their versatility and modularity, cabled conductors have been selected for many superconducting coil applications. The fact that the conductor is made up of a smaller size component wire has many advantages. Conductors can be made in much larger lengths than would otherwise be possible. Eddy currents within the conductor can be reduced through the use of insulation, solder, or other types of surface coatings or treatments on the individual wires.

From a manufacturing point of view, the component wires can be made in much longer lengths and individually tested prior to incorporation into the final conductor. The assumption has generally been made that the performance of the overall cable should be equal to the sum of that of the performance of the component wires. In most applications, this may or may not be true, depending on how much the individual wires are deformed either during manufacture of the cable or as a result of magnetic loads during operation.

For high-energy physics applications, the cabled conductors are compacted prior to winding for reasons of mechanical integrity and dimensional stability. This compaction occurs by flattening the initially round wires in the regions where they are in contact either with other wires or the external forming tool. This paper presents a review of the mechanical behavior of wires in a cable. First, the expected average compressive stresses within the windings of dipole windings with simple external supports are estimated as a function of magnetic field and winding current density. These stresses are then related to the expected elastic as well as inelastic behavior of the cable.

ESTIMATED COMPRESSIVE STRESSES (ROUND WIRES)

It can be shown[1] that for a dipole winding around a circular bore that has a radial winding build which is small compared with the average radius, the magnetic body force is purely tangential. This tangential body force is reacted by a radial pressure at the winding external radial structure interface and a tangential compressive stress within the windings.

The body force as well as the stresses are shown in Fig. 1. For dipole windings with radial builds that are large compared with the bore or that have external iron or large amounts of harmonic content, the expressions in Fig. 1 need to be modified. However, even in these cases the expressions can be regarded as relatively simple, reasonably accurate, approximations.

In a continuous winding with purely external radial support (as shown in Fig. 1), the radial and compressive winding stresses build up to a maximum value at a plane perpendicular to the magnetic field. The maximum radial stress is B^2/μ_o and the maximum tangential stress is $jBr_{av}/2$. Both of these are plotted in Fig. 2 as a function of magnetic field for a range of current densities and for internal diameters of 3.81 cm and 6.5 cm, which correspond to the Fermilab Doubler dipoles and the Brookhaven Isabelle magnets. Note that the stresses in Fig. 2 assume a simple winding

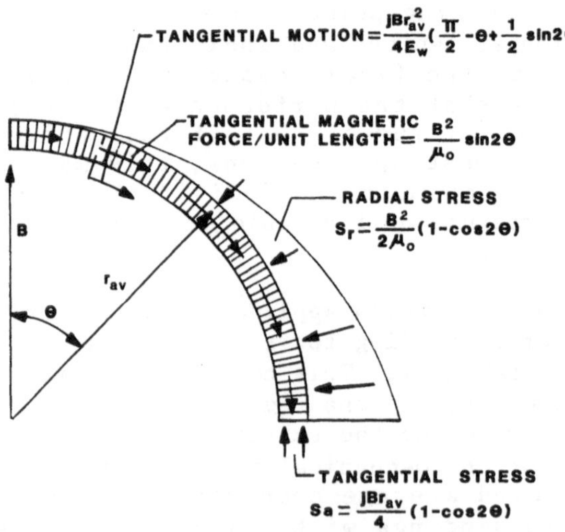

Fig. 1. Simplified forces and stresses in one quadrant of a dipole winding.

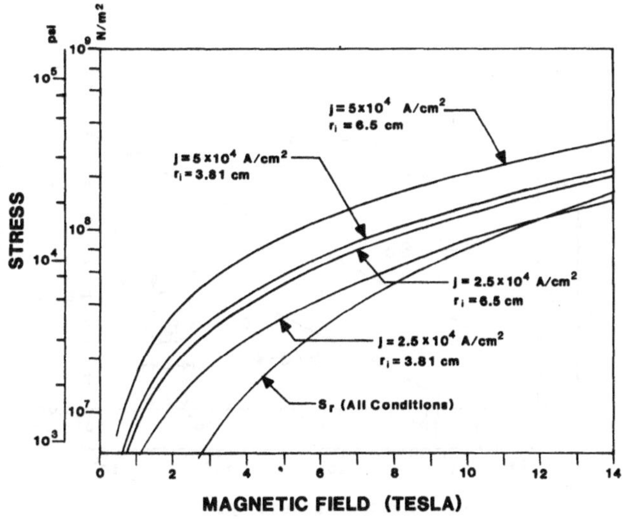

Fig. 2. Maximum radial and tangential stresses as a function of magnetic field. The two sizes shown correspond to Doubler and Isabelle dipoles.

with an external support located radially next to the windings. Other types of support may yield lower winding stresses. As such, the stresses in Fig. 2 are estimates of the maximum values that the windings could be subjected to. The tangential deflection of the winding as a result of the compression is shown in Fig. 1. The maximum value occurs at the $\Theta = 0$ position and is equal to $\pi j B r_{av}^2 / 8E_w$. It is interesting to note that the deflection varies as the square of the average winding radius. From the conductor design point of view, the deflection of the winding is inversely proportional to the effective Young's modulus of the winding, E_w. The lower the E_w, the higher the winding deflection.

ELASTIC BEHAVIOR (ROUND WIRES)

To gain some idea of the behavior of a cable under compression, it is informative to examine the elastic behavior of two parallel round wires under compression. Reference 2 gives the equations for the size of the contact region, the maximum compressive stress in the contact area as well as the center-to-center deflection of the two elastic cylinders under compression. Assuming that the cylinders remain elastic and using a Young's modulus for copper of 10.34×10^{11} N/m^2 (15×10^6 psi), the following were computed and plotted in Fig. 3 as a function of average applied stress, S_a: the maximum compressive stress in the contact, S_c, the normalized width of the contact region, and the ratio of the effective Young's modulus divided by that of copper. This latter

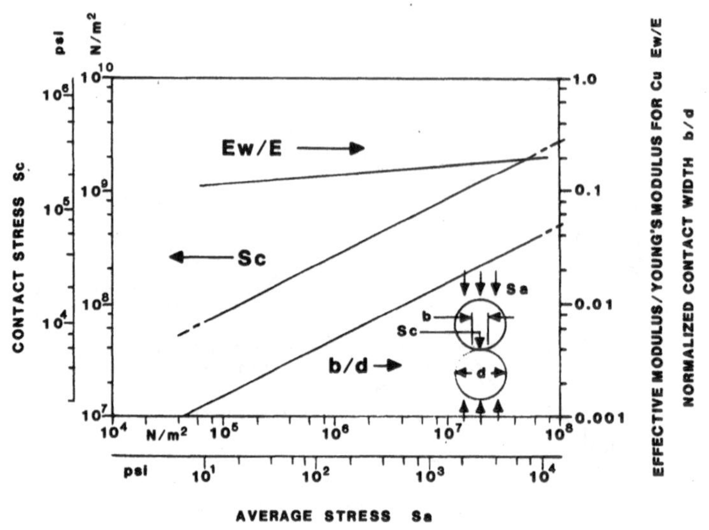

Fig. 3. Contact stress normalized as a function of average applied stress, contact width, and effective Young's modulus.

quantity was arrived at by taking the ratio of the deflection of the two cylinders and dividing by the deflection of a solid piece of copper occupying 100% of the volume under the same average applied stress.

Several important points need to be made. Using a value of 6.895×10^7 N/m^2 (10,000 psi) as the yield for OFHC copper, it takes only an average stress of 9.5 psi for the maximum compressive stress to reach this yield value in the contact area. Even for hardened alloy copper, the yield of 2.07×10^8 N/m^2 (30,000 psi) is reached for an average stress of 5.93×10^5 N/m^2 (86 psi). At these values of applied stress the effective Young's modulus of the winding ranges from 6% to 13% of that of solid copper.

The conclusion to be reached from this brief look at the elastic behavior is that the wires must yield in the region of contact. If the yielding of round wires does not take place during manufacture, then yielding will occur on the application of the magnetic loads, a situation that is very undesirable from a stability point of view. The ideal case is for all the yielding to take place during manufacture and none during the application of magnetic loads. In this case, the behavior of the compacted conductor would be purely elastic. The following section presents a simplified model of a compacted strand, which relates the degree of compaction to the bearing area, ratio of internal to external average stress, effective Young's modulus as well as internal cooling surface.

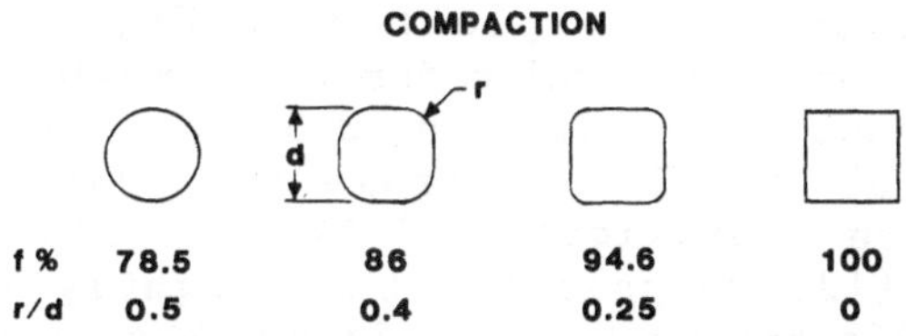

Fig. 4. Wire shape as a function of compaction.

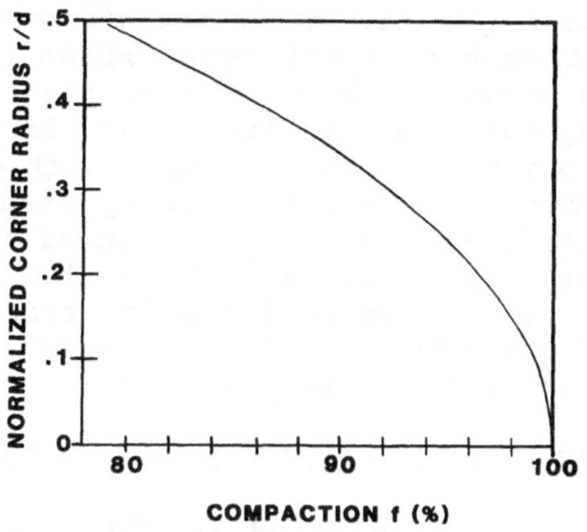

Fig. 5. Corner radius as a function of wire compaction.

COMPACTED CONDUCTOR

Various wire shapes are shown in Fig. 4. The wires have sides, d, with corner radii, r. Adjacent wires are assumed to be in contact over the flat faces.

The compacting faction, f, defined as the ratio of solid volume to total volume, is given by:

$$f = \left\{ d^2 - [(2r)^2 - \pi r^2] \right\} / d^2 = 1 - (r/d)^2 (4 - \pi) \qquad (1)$$

This can be solved for the ratio r/d:

$$r/d = [(1 - f)/(4 - \pi)]^{1/2} \qquad (2)$$

This relationship is shown plotted in Fig. 5. Note that even for a compaction of 90%, the corner radius is 0.341 d and the flat dimension is 0.317 d. The ratio of internal stress, S_c, to average stress, S_a, is simply the inverse ratio of:

$$S_c/S_a = [d/(d-2r)] = 1/\left\{1 - 2\,[(1-f)/(4-\pi)]^{1/2}\right\} \qquad (3)$$

This relationship is plotted in Fig. 6. At about 90% compaction (the approximate value for the Fermilab Doubler cable), the maximum contact stress is about 3.5 times the average stress. Figure 2 shows that at 5 T the average stresses in the Fermilab Doubler coils are approximately 5.52×10^7 N/m^2 (8,000 psi) and 8.27×10^7 N/m^2 (12,000 psi) for the slightly larger Isabelle coils. Using this factor of 3.5, the maximum contact stresses in each of these coils should be 1.93×10^8 N/m^2 (28,000 psi) and 2.89×10^8 N/m^2 (42,000 psi), respectively, both of which are well above the yield stress for copper.

The next property of interest is the effective modulus of the cable itself. Although an exact computation of the effective Young's modulus is beyond the scope of this paper, an estimate can be made by assuming the stress is uniform in planes perpendicular to the load and replacing the round corners with 45° beveled edges while still maintaining the same bearing area. This latter approximation is based on examination of published stress distributions under surface loads.[3] With this assumption, the local elongation can be computed and integrated over the whole conductor to yield the ratio of effective Young's modulus to that of copper as a function of the variable r/d:

$$E_{eff}/E = 1/\left\{1 - 2r/d + \ln[1/(1 - 2r/d)]\right\} \tag{4}$$

This relationship is plotted in Fig. 7 as a function of compaction, f.

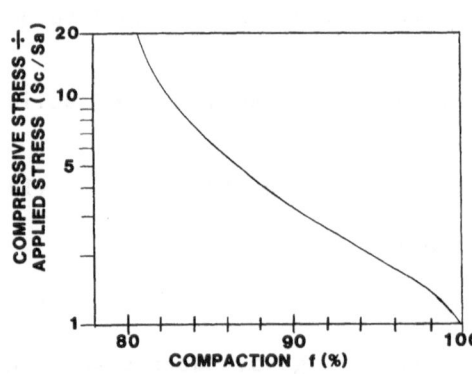

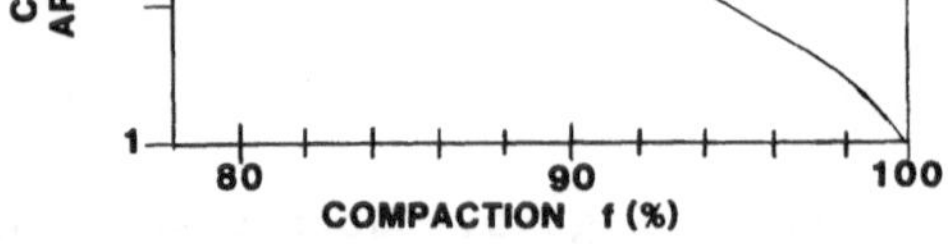

Fig. 6. Compressive stress concentration as a function of wire compaction.

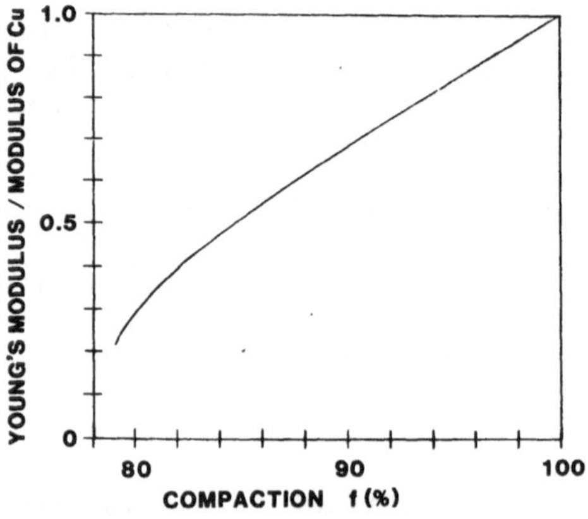

Fig. 7. Effective Young's modulus as a function of wire compaction.

We can conclude that the most severe problem in cables is the high ratio of contact stress to average stress. If conductor stress is to be limited to 6.895×10^7 N/m^2 (10,000 psi), then at a compaction of 90% the average stress is limited to 1.97×10^7 N/m^2 (2,860 psi). If the applied average magnetic stresses are higher than this, the wire will yield and become more compact under load until the required combination of compaction and work hardening is reached. It is, therefore, important that either the cable itself or the manufactured winding be subjected to the full operating loads prior to being energized. This will avoid at least one of the major potential causes of extensive training in reaching operating field.

CONCLUSIONS

Cabled conductors have excellent characteristics: ability to be manufactured in long lengths, ability to limit or eliminate wire-to-wire eddy currents, and the large internal surface area and internal helium heat capacity. However, because they have reduced bearing area, high internal stresses may be present. It is important either to keep the average operating stresses low or, as a minimum, to precompact the cable by applying the operating stresses.

REFERENCES

1. Z. J. J. Stekly et al., "Magnetohydrodynamic (MHD) Magnet Modeling," AFAPL-TR-79-2045 (June 1979).
2. R. J. Roark, "Formulas for Stress and Strain," McGraw-Hill, New York (1965).
3. S. Timeoshenko and J. N. Goodier, "Theory of Elasticity," McGraw-Hill, New York (1951).

TRAINING STUDIES OF EPOXY-IMPREGNATED SUPERCONDUCTOR WINDINGS

J. W. Ekin and E. S. Pittman

Electromagnetic Technology Division, National Bureau of Standards, Boulder, Colorado

M. J. Superczynski and D. J. Waltman

David W. Taylor Naval Ship Research and Development Center
Annapolis, Maryland

INTRODUCTION

When a superconducting magnet is first energized, thermal runaway (or "quenching") often occurs at current levels well below design operating current. This is usually accompanied by the dissipation of large amounts of energy. On reenergizing the magnet such quenching usually occurs at successively higher currents. Through a "training" process of repeatedly quenching a magnet, the maximum current can usually be increased, although the procedure is usually very costly both in terms of time and liquid helium.

To understand the material and construction factors that affect training, a series of tests have been performed on epoxy-impregnated superconductor windings. The difficulty with a systematic study of training is that a large number of magnets have to be constructed in order to individually vary the parameters that might affect training. Such experiments with full-scale magnets would be very expensive. To circumvent this problem, an apparatus was constructed to apply large hoop stresses to relatively small potted coils of superconducting wire.[1] The coil structure was reduced to the equivalent of a small section of a large magnet -- an 18-cm-diameter ring consisting of about one hundred turns of superconducting wire, co-wound with fiberglass and impregnated with various materials. A schematic diagram of the Superconductive Composite Rings (SCCR) is shown in Fig. 1.

Several series of such rings have been constructed in which different components of the winding structure have been systematically varied. This has permitted a comparative study of many factors that affect training at considerable savings over full-scale magnet experiments.

Several preliminary results obtained using this apparatus have been reported earlier.[1] In particular, it was shown that there is a significant increase in the frequency of quench events as the operating current approaches the critical current, as expected from stability theory.[2] Also, it was shown that including fiberglass in epoxy-impregnated structures significantly reduces the amount of training required to reach a given design level; the fiberglass fills the epoxy and appears to serve as a crack arrestor.[3] Finally, it was shown that mechanically prestressing a superconducting winding at low temperatures prior to energizing the winding effectively removes all training (i.e., all quenching is eliminated).

This paper reports the results of a series of tests on additional factors and their influence on the training of NbTi superconducting magnets. The objective here is to systematically study the effects of various construction techniques; consequently, a single epoxy system and the same superconductor were used throughout the tests. In particular, results are reported on the influence of the epoxy-superconductor bond, the effect of adding a milled fiberglass filler to the epoxy impregnant, the effect of

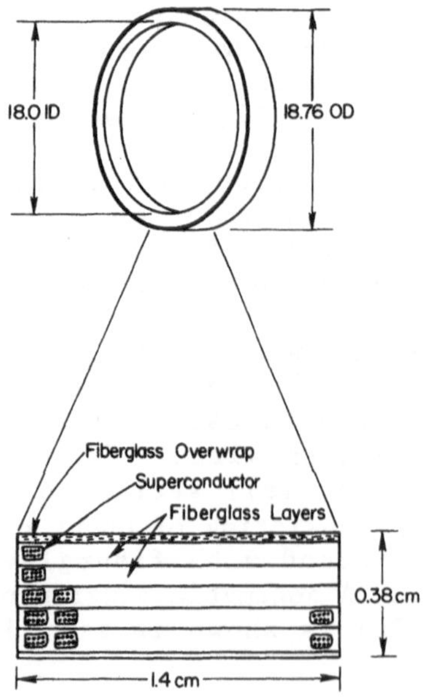

Fig. 1. Schematic diagram of a typical Superconducting Composite Ring (SCCR).

prestressing the superconducting wire at room temperature prior to coil fabrication, the influence of cycling a winding trained at 4.2 K to room temperature and back, and the effect of impregnating superconductor windings with wax instead of epoxy.

CONSTRUCTION DETAILS OF THE SUPERCONDUCTIVE COMPOSITE RINGS

The Superconductive Composite Rings (SCCR) are shown in Fig. 1. For this study, they were all wound from NbTi-Cu superconductor manufactured from the same billet. The wire was rectangular in cross section, 0.685 mm x 0.521 mm, with Formvar* insulation approximately 0.020-mm thick. There were 180 filaments having a twist pitch of 1 cm and a copper-to-superconductor ratio of 1.8 to 1.

The rings were constructed in sets of five to provide five rings of identical construction with exactly the same potting material and curing procedure. Such an approach provided a statistical base for each type of construction or impregnation system. Initially, two layers of 0.089-mm thick, type E, fiberglass cloth with an A1100 finish were wound over the mandrel on a 45° bias angle with the warp. The SCCRs were built up from five layers of wire (pretensioned at 20 N to a strain level of 0.15%) separated by dry fiberglass cloth with a 0.46-mm-thick fiberglass overwrap. The entire assembly was put into an open container, placed in a vacuum chamber, preheated to 358 K, and evacuated to approximately 1 Pa.

The impregnant used for the preliminary data reported here was a mixture of Ciba 6004 resin and an anhydride hardener formed by mixing Lindride 12 and Lindride 16 in equal parts. The proportion of resin to hardener was 100 parts to 85 parts by weight. The epoxy components were first heated to 358 K and then degassed to approximately 5 Pa. They were then mixed at 358 K for approximately 15 min and then again degassed to 5 Pa. This mixture was then piped under vacuum into the chamber, filling the container and covering the ring assembly. After the ring was covered, a pressure of 690 kPa (100 psi) was applied to the chamber containing the rings to force the impregnant into the windings. This condition, 690 kPa and 358 K, was maintained for 12 h until the epoxy had cured.

*Certain commercial materials are identified in this paper to specify the experimental study adequately. Such identification does not imply recommendation or endorsement by the National Bureau of Standards, nor does it imply that the material identi- ied is necessarily the best available for the purpose.

After the impregnation was completed, the temperature and pressure were reduced to 293 K and atmospheric pressure and the rings were removed.

EXPERIMENTAL PROCEDURE

The SCCRs were tested in a perpendicular background magnetic field. For each sample ring, the critical current, I_c, was first determined from a measurement of the V-I characteristics of the sample in this background field. The I_c was taken as the point where the resistive electric field along the superconductor reached 6 μV/m.

The current in the SCCR was then set at a given percentage of its critical value (I/I_c = 90% for these experiments) before external stress was applied. The introduction of current into the ring produced a small amount of hoop strain (about 0.1% at I_c) because of the Lorentz force developed in the 5-T combined field of the ring and background magnet. After raising the ring current to the preset level, further hoop strain was introduced into the SCCR by a split-cone loading apparatus[1] (keeping the current constant). The hoop strain was increased at a rate of 2×10^{-4} s^{-1} until eventually thermal runaway was initiated (i.e., a "quench" occurred). The quench electronically triggered a shutdown of the current to the ring. The servohydraulic loading system was also immediately reversed, ramping the applied load to zero. After thermally settling, the ring was reenergized with current to the preset level and again mechanically loaded until a second quench occurred. This was repeated for many load-quench cycles.

The open symbols in Fig. 2 show the result for a typical SCCR. The number of the load-quench cycle is plotted along the vertical axis; the hoop strain at which the quench occurred is plotted along the horizontal axis. Generally, each subsequent quench occurred at a higher hoop strain than the previous quench, i.e., the SCCRs exhibited training behavior very similar to actual magnet operation. In fact, the training shown by the data in Fig. 2 indicates a remarkably monotonic, smooth dependence of training on hoop strain. Training started at a threshold value of about 0.2%. The coil then rapidly trained to a higher hoop strain on subsequent load-quench cycles, but the increase in strain between cycles diminished as higher strain was attained. Eventually, the ring reached the point where very little gain in strain was obtained by further quenching, and the training curve approached infinite slope. The training curve for different rings fabricated in the same way was reproducible to within about ±0.05% strain.

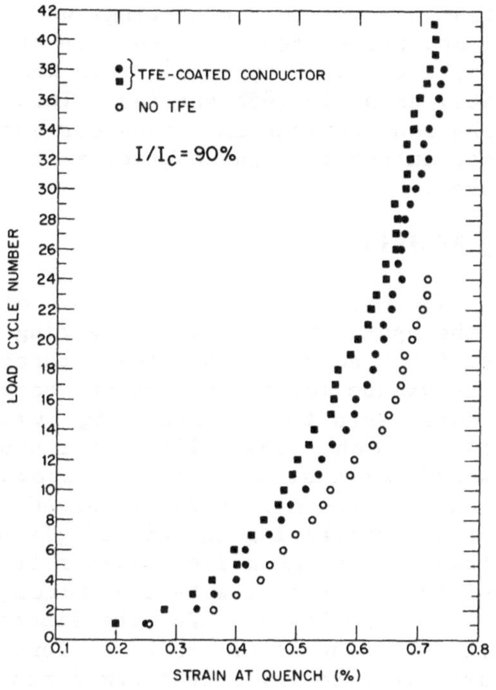

Fig. 2. Strain dependence of training showing no effect of coating the superconductor with an epoxy release agent.

RESULTS

Using the above procedure, training curves were generated on several series of superconducting coils, each set designed to systematically test the effect on training of a single construction variable.

Effect of Epoxy-Conductor Bond

A series of tests were performed to study the effect of the bond between the conductor surface and the epoxy impregnant. In the past, magnets that have had the conductor wrapped with polyimide film seemed to show less training than windings fully impregnated with epoxy without the polyimide wrapping. It has been hypothesized that the polyimide film kept the epoxy from bonding to the conductor, removing this interface as a crack initiation site and isolating the conductor from the epoxy.

To test at least part of this hypothesis, three SCCRs were tested, two wound with superconductor that had been sprayed with TFE mold release, and one wound with untreated superconductor. The rings were fabricated and vacuum impregnated with epoxy at the same time to produce an identical epoxy mix and cure cycle.

Considerable care was taken to test the entire set of rings under identical experimental conditions and procedures. The results are shown in Fig. 2. The training curves for both types of rings were nearly the same within experimental error (±0.05% strain). These data suggest that the nature of the bond between the conductor and epoxy does not have a large effect on the training characteristic of these impregnated ring structures.

Effect of Milled Fiberglass Epoxy Additive

A series of rings was tested to investigate the effect of milled-fiberglass additions to the epoxy impregnant. Five rings were tested, two with milled glass fiber added to the epoxy, three reference rings without. All were vacuum impregnated at the same time, with the milled fiber introduced into two of the rings via preimpregnation of the fiberglass cloth. The milled fiber was expected to fill epoxy-rich areas and serve as a crack arrestor. Somewhat surprisingly, the rings without the milled fiber showed a better training performance than those containing the milled glass (see Fig. 3). It is speculated that adding the milled glass fiber decreased the strain capability of the epoxy at low temperatures, causing cracks to occur at lower strain levels. But the fibers arrested the cracks at a small size, causing other cracks to form. If the crack size was still sufficient to initiate thermal runaway, the greater number of cracks would produce a greater training problem.

Effect of Conductor Prestraining at Room Temperature Prior to Coil Winding

Another possible source of training is microplastic deformation of the superconductor composite. If it affects a sufficiently large area of the superconducting composite, this rapid yielding can deposit enough energy internally to initiate thermal runaway. If such a mechanism is playing a role, then prestraining the conductor prior to winding might lead to a reduction in training.[4]

To see if this has any effect, four rings were tested, two wound from NbTi wire that had been prestrained, two without. In the first two, the superconductor was prestrained to about 1% by winding the conductor onto an auxiliary spool under 111 N of tension. The prestrained conductor was then wound into sample coils under the standard pretension of 20 N used to fabricate all the other coils.

The results are shown in Fig. 4. The training curves for both types of rings showed close agreement with each other, differing by no more than ±0.04% strain. It is concluded that, in these windings at least, prestraining NbTi conductor by 1% prior

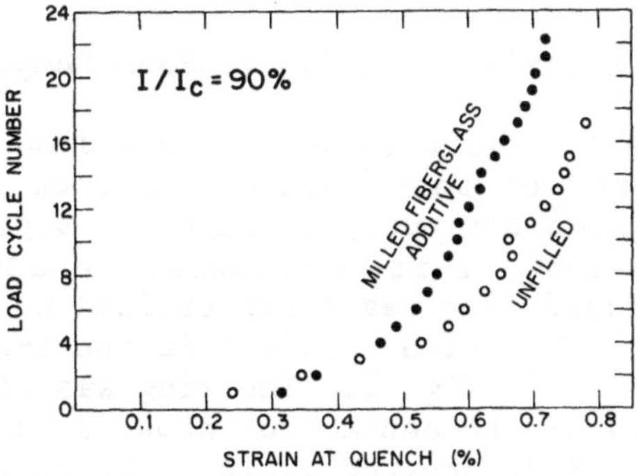

Fig. 3. Effect of adding milled fiberglass to the epoxy impregnant. Averaged values are shown for each type of coil.

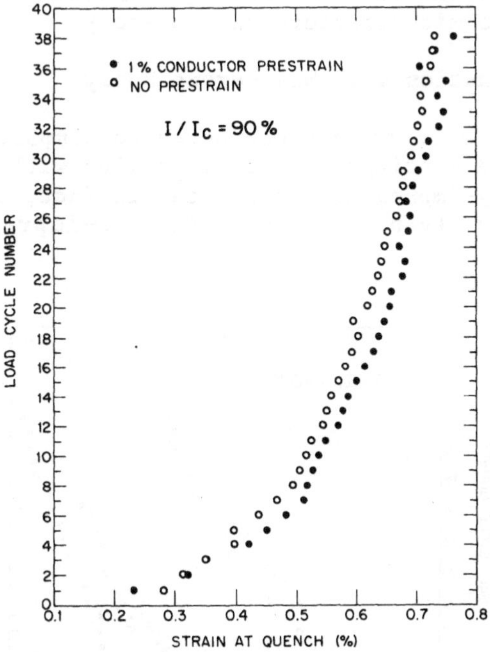

Fig. 4. Strain dependence of training, showing no significant effect of prestraining the conductor by 1% at room temperature prior to fabrication. averaged values are shown for each type of coil.

to coil fabrication produces no significant improvement in training performance.

Effect of Cycling a Trained Winding to Room Temperature

The question has been asked whether a superconducting coil loses its "training" on being warmed to room temperature. To answer this question for epoxy-impregnated superconducting windings, a systematic test of room temperature anneal effects was conducted. A standard ring was first trained to an elastic strain of about 0.7% at 4.2 K, which resulted in the training curve shown by the open symbols in Fig. 5. The ring was then warmed to room temperature and allowed to anneal for about a month. The ring was then recooled to 4 K and tested again. The second training curve is shown by the solid symbols in Fig. 6. Note that training continued in the second test, but the second training curve is shifted to much higher strain than the first. The data indicate that all but ∿0.05% of the original strain training is "remembered." The 0.05% that was lost was recovered in relatively few training steps. These results would indicate that training in epoxy-impregnated windings is both permanent and irreversible, probably owing to microfracture of the epoxy.

Effect of Impregnation with Wax versus Epoxy

Another experiment was conducted to compare the training of epoxy-impregnated windings with coils filled with wax. If training in the epoxy-impregnated coils is, in fact, due to microfracture of the epoxy, then the results for wax-impregnated windings

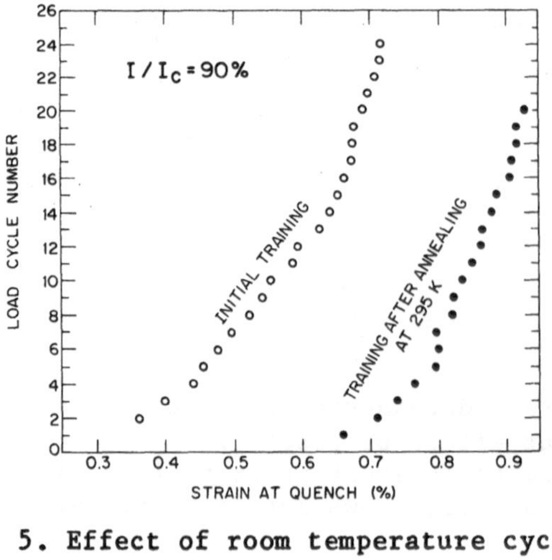

Fig. 5. Effect of room temperature cycling.

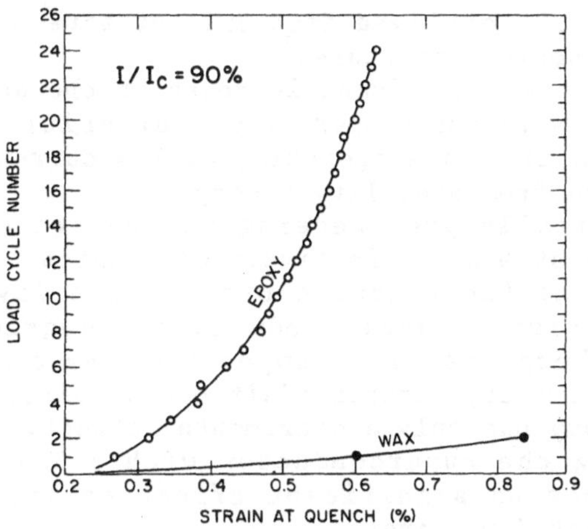

Fig. 6. Dependence of training on impregnation material.

should be very different. Wax has a much greater strain capability than epoxy at 4 K. If the yield strength is low enough, then the energy released during mechanical deformation of the wax will be insufficient to initiate thermal runaway and very little or no quenching should occur. An SCCR was constructed in the same manner as above (same number of superconductor turns and fiberglass layers), but it was impregnated with beeswax rather than epoxy. It was then tested in exactly the same way as the epoxy-filled SCCRs. The results are shown in Fig. 6. Note the almost total disappearance of training in the wax-filled ring compared with those impregnated with epoxy. This is not to suggest that wax is a superior impregnant to epoxy, since it has long-term durability problems in preventing conductor movement. But this result does point strongly to the epoxy impregnant as the source of training in these coils.

SUMMARY

These data, taken together with previous results,[1] lead to the following combined set of conclusions regarding training in epoxy-impregnated superconducting coils:

1. The simple fact that these coils exhibit training behavior monotonically dependent on hoop strain (at fixed current) suggests that training is primarily a mechanical phenomenon with strain a controlling parameter.
2. The data are explained well by a stress-relief model in which the training process is assumed to arise from a

series of localized fractures at stress concentrators in the composite structure.[1]

3. There is a significant increase in the degree of training in coil structures operating near critical current compared with those operating at low currents,[1] as would be expected from stability theory.
4. Mechanically prestressing the entire epoxy-impregnated winding at 4.2 K effectively eliminates training.[1]
5. Including fiberglass cloth between layers of superconductor significantly reduces the degree of training.[1] The fiberglass cloth appears to serve as a crack arrestor. However, further additions of milled fiberglass to the epoxy had only a detrimental effect.
6. Coating the superconductor with a Teflon mold release agent has no significant effect on training in epoxy-impregnated windings.
7. Prestressing the superconductor prior to coil winding produces no significant effect.
8. When a trained epoxy-impregnated winding is cycled to room temperature, the training is "remembered", i.e., minimal retraining is required.
9. Impregnation with wax instead of epoxy essentially eliminates training.

Although there are other causes of training, especially in unpotted magnet structures, these data consistently indicate that, at least within the winding itself, the major source of training is permanent, irreversible microfracture of the epoxy.

REFERENCES

1. J. W. Ekin, R. E. Schramm, and M. R. Superczynski, Training of epoxy-impregnated superconductor windings, in: "Advances in Cryogenic Engineering--Materials," Vol. 26, Plenum Press, New York (1980), p. 667.
2. See for example M. N. Wilson and Y. Iwasa, Cryogenics 18:17 (1978).
3. See also M. A. Green, D. E. Coyle, P. B. Miller, and W. F. Wengel, Vacuum impregnation with epoxy of large superconducting magnet structures, in: Nonmetallic Materials and Composites at Low Temperatures, A. F. Clark, R. P. Reed, and G. Hartwig, eds., Plenum Press, New York (1979), p. 409.
4. C. Schmidt and B. Turck, Cryogenics 17:695 (1977).

STABILITY MEASUREMENTS OF ALUMINUM-STABILIZED Nb-Ti AND BRONZE MATRIX Nb_3Sn POTTED SUPERCONDUCTING MAGNETS

D. J. Waltman, F. E. McDonald, and M. J. Superczynski

David Taylor Naval Ship Research and Development Center, Annapolis, Maryland

INTRODUCTION

Stable and reliable superconducting magnets for use as the field windings of electric motors and generators is an important consideration for high machinery availability. Because of their compact size, ruggedness of construction, and large energy density, epoxy-impregnated superconducting magnets are being considered for the field windings of superconductive electric machinery.

A major concern with the use of fully potted and trained super-onducting magnets is their stability when operating in an adverse environment. Heat energy can be imparted to the superconductor of the magnet from external mechanical disturbances, such as shock and vibration. In addition, energy stored in the magnet composite due to stress concentrations developed during magnet construction and cool down to 4.2 K can be released in the coil winding when the coil is energized. If the heat energy released from the various possible sources is sufficiently large, the local temperature rise will drive the superconductor normal, causing the magnet to quench.

Previous work[1,2] performed at David Taylor Naval Ship Research and Development Center (DTNSRDC) has provided stability measurement data for fully potted, superconducting test coils wound with copper-stabilized, multifilamentary NbTi superconducting wire. As a continuation and extension of this work, additional stability experiments have been performed using fully potted test coils wound with aluminum-stabilized, multifilamentary NbTi superconducting wire and test coils wound with bronze matrix (Cu-Sn) multifilamentary Nb_3Sn superconducting wire. The NbTi/Al and Nb_3Sn/Cu-Sn wires used in the test coils are experimental conductors fabricated

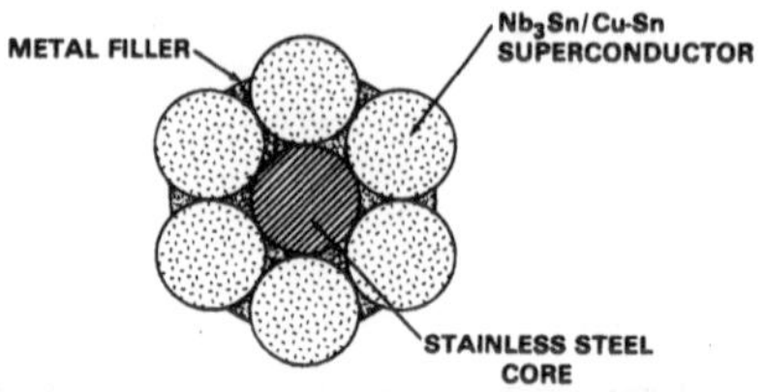

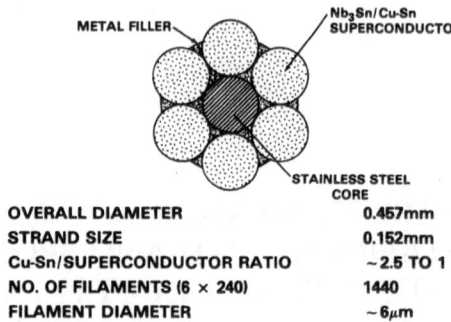

OVERALL DIAMETER	0.457mm
STRAND SIZE	0.152mm
Cu-Sn/SUPERCONDUCTOR RATIO	~2.5 TO 1
NO. OF FILAMENTS (6 × 240)	1440
FILAMENT DIAMETER	~6μm

Fig. 1. Cross-sectional view of Nb_3Sn/Cu-Sn superconducting wire.

to meet specifications developed by DTNSRDC researchers. The Navy interest in these conductors is based on the expected improvement in magnet stability that each conductor should provide compared with copper-stabilized NbTi. The A15 compound Nb_3Sn/Cu-Sn superconductor possesses enhanced values of critical temperature, current density, and magnetic field relative to NbTi. The aluminum matrix material of the NbTi/Al wire has improved thermal conductivity, heat capacity, and electrical conductivity at 4.2 K compared with standard copper matrix material. Therefore, the purpose of these experiments was to measure the stability of potted test magnets wound with these conductors and compare the results obtained with the previously measured stability characteristics of NbTi/Cu potted superconducting test coils.

BRONZE Nb_3Sn SUPERCONDUCTOR FABRICATION

A cross-sectional view of the bronze matrix, multifilamentary Nb_3Sn superconducting wire is shown in Fig. 1. The wire is of cabled construction composed of 6 strands of conductor each having 240 filaments of Nb_3Sn superconductor with a twist pitch of 2 per 2.54 cm in a bronze matrix. The cabled conductors are reinforced with a center strand of stainless steel wire and are bonded together with a metal solder filler. The twist pitch of the cabled superconducting strands is 14 per 2.54 cm. The wire is electrically insulated with a spiral wrap of fiberglass cloth.

The fabrication of the Nb_3Sn/Cu-Sn strands used conventional processing technology for Nb_3Sn superconductors. The general steps of this process are shown in Fig. 2.

ALUMINUM MATRIX NbTi SUPERCONDUCTOR FABRICATION

A cross-sectional view of the aluminum-stabilized, multifilamentary superconducting wire is shown in Fig. 3. The method used to manufacture this wire[3] involved the processes shown in Fig. 4.

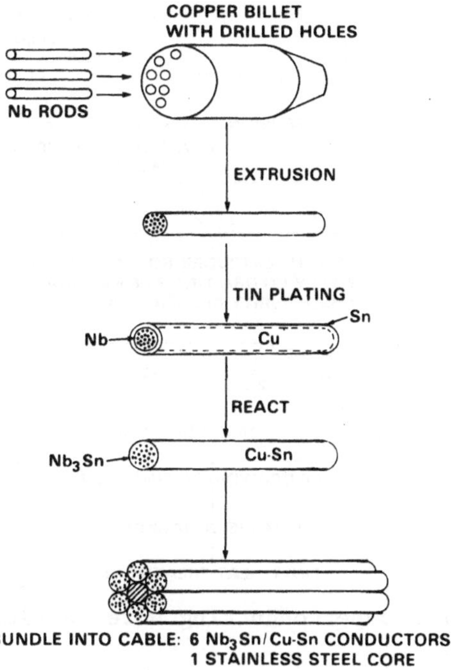

Fig. 2. Nb_3Sn/Cu-Sn superconducting wire fabrication process.

First, a composite of multifilamentary NbTi superconductor embedded in an aluminum alloy was prepared using conventional techniques. A total of 121 holes were drilled in a 76-mm-diameter, 24-mm-long, 1100 Al billet. Rods of Nb-53Ti, 4.8 mm in diameter, were inserted into the holes of the billet. The billet was then extruded to an extrusion ratio of 10. The extruded composite was then inserted in a high purity (> 99.995%) aluminum tube (24.4-mm o.d. and 19.8-mm i.d.) that had been fabricated by extrusion from bar stock. The NbTi-1100 Al core and high purity aluminum tube were then assembled in a copper tube of 31.8-mm o.d. and 25.4-mm i.d. The entire

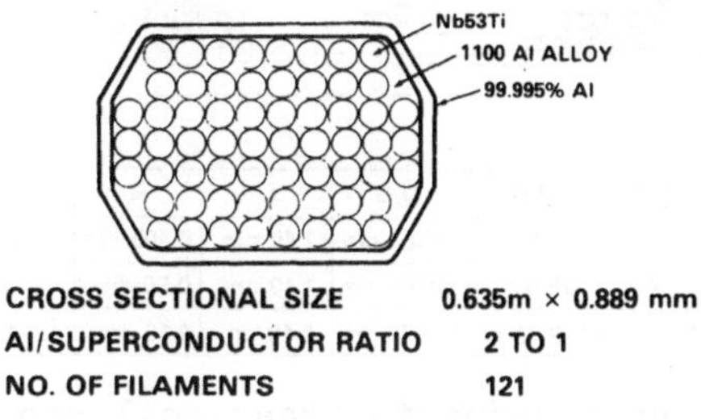

Fig. 3. Cross-sectional view of NbTi/Al superconducting wire.

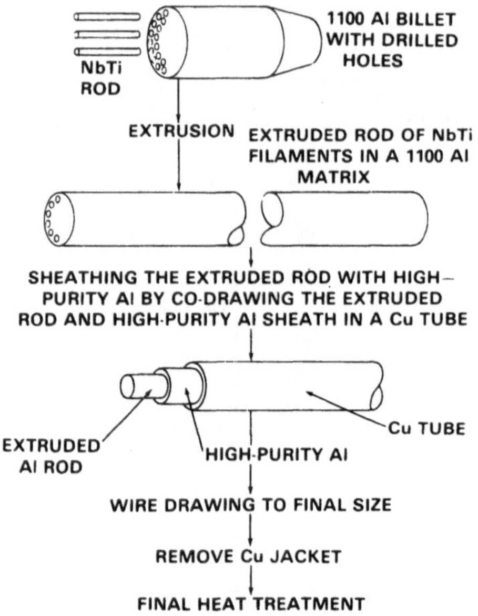

Fig. 4. NbTi/Al superconducting wire fabrication process.

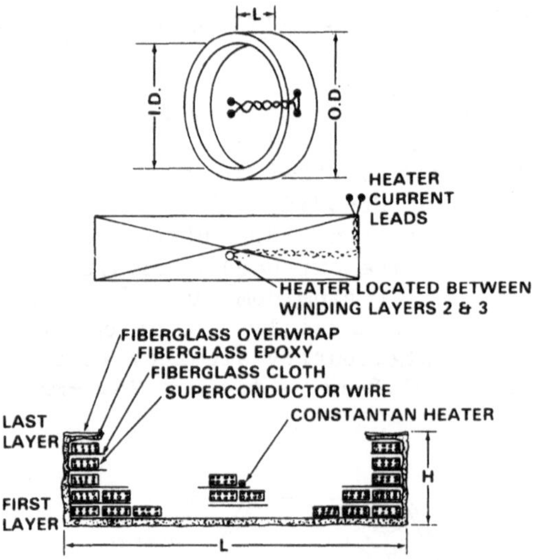

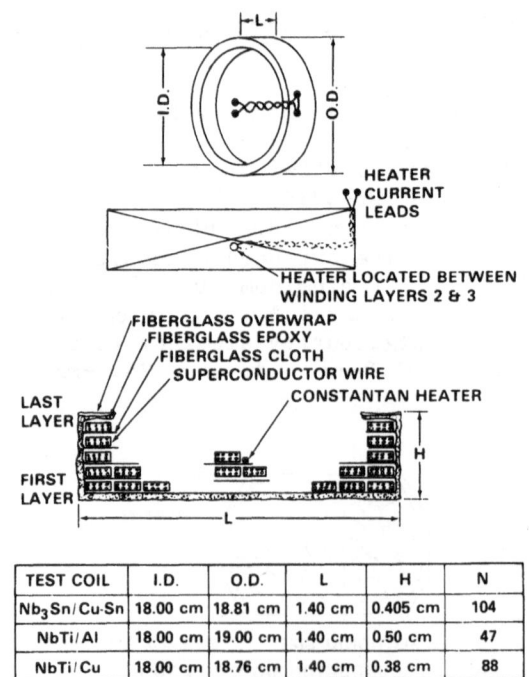

TEST COIL	I.D.	O.D.	L	H	N
Nb_3Sn/Cu-Sn	18.00 cm	18.81 cm	1.40 cm	0.405 cm	104
NbTi/Al	18.00 cm	19.00 cm	1.40 cm	0.50 cm	47
NbTi/Cu	18.00 cm	18.76 cm	1.40 cm	0.38 cm	88

Fig. 5. Superconducting test coil construction and heater location.

composite was then drawn down to its final size of 0.91-mm o.d. The outer copper sheath was stripped from the wire using nitric acid, and the wire was heat-treated at 375°C for several hours and then squared to a nominal rectangular cross section of 0.635 mm x 0.889 mm. The wire, as received from the manufacturer, was not uniform in cross-sectional size and varied in size along the entire length of the conductor. In addition, the wire was not electrically insulated by the manufacturer. Therefore, prior to winding the test coil for the stability measurements, it was insulated by hand coating the wire with varnish insulator, which resulted in a wire of highly nonuniform overall cross section along its length.

TEST MAGNET CONSTRUCTION

Two test magnets, one wound with the Nb_3Sn/Cu-Sn wire and the other wound with the NbTi/Al wire, were constructed. Both test coils are epoxy impregnated, solenoidal coils, as shown in Fig. 5. The Nb_3Sn/Cu-Sn magnet is a five-layer coil having a total of 104 electrical turns. The input current lead and first turn of the coil as well as the output current lead and last turn of the coil are made up of one wire of Nb_3Sn/Cu-Sn and NbTi/Cu soldered together. This was done to ensure the mechanical and electrical integrity of the current leads of the coil, which are subjected to mechanical abuse during the handling of the coil and installing it in the experimental setup. Similarly, the current leads and first and last turn of the NbTi/Al coil were made with NbTi/Cu wire soldered together with the NbTi/Al wire of the coil. The NbTi/Al coil is a four-layer winding having a total of 47 electrical turns. During the winding of both the Nb_3Sn/Cu-Sn and NbTi/Al coils, a multitap electrical heater fabricated from constantan wire (0.339 Ω/cm) was embedded in each coil. For both coils, the heater was located between winding layers 2 and 3 and centered along the length of each of the coils. The multitap heater configuration allowed for the potential use of various lengths of heaters of up to 2.54 cm in length in 0.635-cm increments.

Both coils were constructed using the same methods of winding, fiberglass cloth reinforcement, and vacuum impregnation that were used to construct the NbTi/Cu coils used for previous stability measurements.[2] As shown in Fig. 5, the physical size differences in these coils (NbTi/Cu coil is included for comparison) is a result of the differences in overall cross-sectional size of superconducting wire used for each coil.

EXPERIMENTAL METHOD

To measure the stability of each of these test coils at magnetic flux densities of 3 to 5 T, each coil was placed in the inner bore of a background magnet, as shown in Fig. 6. The superconducting NbTi background magnet had an inside diameter of 19.4 cm, an outside diameter of 24.0 cm, and a length of 9.4 cm. When installed for measurement, each test magnet was concentrically positioned at the midlength of the inner bore of the background magnet.

For the experiments to measure the energy to quench, the assembly of the test coil and background coil was placed in a liquid helium Dewar and cooled from room temperature to 4.2 K. After the background and test magnets reached their superconducting state, they were energized to predetermined operating currents using separate power supplies, as shown in Fig. 7.

The objective of the experiment was to measure the minimum energy required to quench the Nb_3Sn/Cu-Sn and the NbTi/Al magnets for various levels of fixed field strength and for various values of test-coil operating current. Therefore for the experiments, the background was energized to provide a constant background field level, and the coil under testing was energized to an operating current that was a desired percentage of its critical current at the background level.

To quench the test coil, a pulse generator and power amplifier system, as illustrated in Fig. 7, were used to deliver a single pulse of known energy to the resistor heater embedded in the test coil. A dummy resistor approximately equivalent to the heater resistance was used to allow measurement of the amplitude and width of the electrical pulse with a calibrated oscilloscope prior to applying the pulse to the heater. The actual test procedure

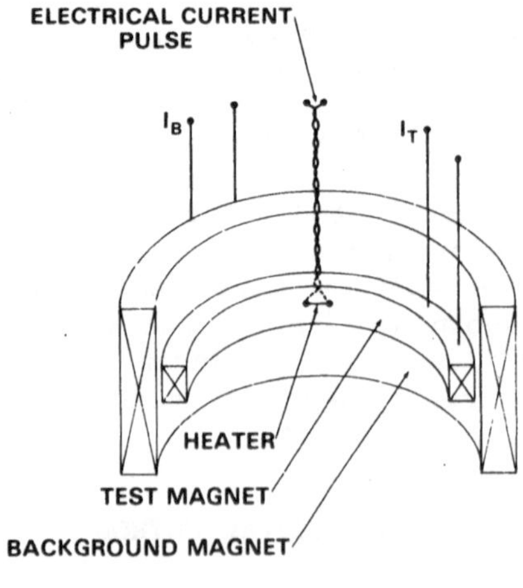

Fig. 6. Magnet assembly for energy to quench measurements.

involved maintaining the pulse width at a fixed value of 100 to 300 μs and adjusting the pulse amplitude to obtain the minimum energy required to quench the coil. Measurements previously reported by Superczynski[1] showed that energy to quench remained constant over a pulse width of 10^2 to 10^4 μs. For each run of the experiment, the amplitude of the constant-width pulse was initially set to a value estimated to be less than that needed to quench the magnet. The pulse amplitude was then increased in small increments until the test magnet quenched.

DISCUSSION OF RESULTS

The results of the measurements of the minimum energy required to quench the Nb_3Sn/Cu-Sn and the NbTi/Al test coils are shown in Fig. 8. In the figure, energy to quench the test coil having a 0.635-cm heater, is plotted as a function of the ratio of the test coil operating current to its critical current (I/I_c) for the magnetic field strengths shown. For purposes of comparison, a plot of previous measurements[2] of the energy to quench a NbTi/Cu test coil having a 0.635-cm heater is also shown in Fig. 8. The results show that the energy to quench as a function of I/I_c for both the Nb_3Sn/Cu-Sn and NbTi/Al coils is greater than that to quench the NbTi/Cu coil. In the I/I_c region of 0.3 to 0.9, the energy to quench the Nb_3Sn coil is approximately a factor of 3 ($I/I_c = 0.3$) to a factor of 10 ($I/I_c = 0.9$) greater than the energy to quench the NbTi/Cu

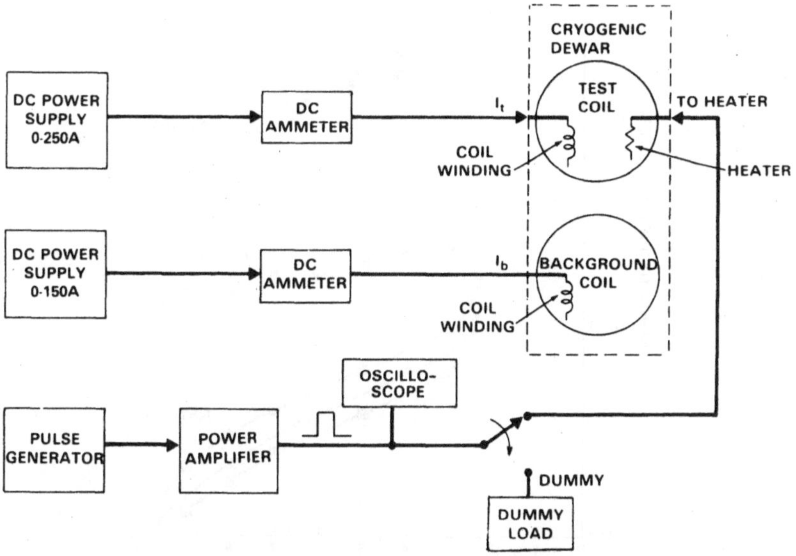

Fig. 7. Experiment instrumentation and power supply connection diagram.

coil. For the NbTi/Al coil, the energy to quench is approximately 30 times that of the NbTi/Cu coil over the I/I_c range of 0.3 to 0.9. But, as can be seen, the measured results for both the Nb_3Sn/Cu-Sn and NbTi/Al coils do not show a sharp decrease in the magnitude of the energy to quench at the higher value of I/I_c (0.9 and greater), as would be expected. Since there is insufficient data to indicate that the long-length current and magnetic field characteristics of the Nb_3Sn/Cu-Sn or NbTi/Al superconducting wires agree with the short-sample characteristics, the I/I_c values used in Fig. 8 may not be correct. Because of possible nonuniformity in the cross-sectional area and filament size of the Nb_3Sn/Cu-Sn wire, the known construction nonuniformity in the cross-sectional area and filament size of the Nb_3Sn/Cu-Sn wire and the known construction nonuniformity of the NbTi/Al wire, the critical current of each wire at the heater location could be greater than that measured for the entire coil winding. If this is the case, the actual I/I_c values would be less than those shown in Fig. 8, and the energy to quench plots for both coils would shift to the left. This would, therefore, indicate that measurements were not actually obtained at the higher I/I_c values, where the energy to quench is expected to rapidly decrease. Another observation from

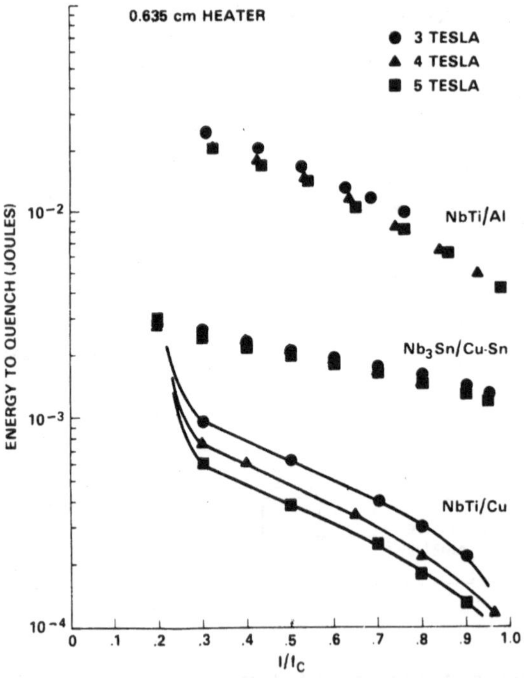

Fig. 8. Energy to quench vs. I/I_c for Nb_3Sn and NbTi/Cu test coils.

the data shown in Fig. 8 is that the measured values of energy to quench for any particular value of I/I_c did not vary greatly with magnetic field strength. This differs from the results obtained for the NbTi/Cu test coil, which indicates that high-field magnets are less stable than low-field magnets. The reason for this difference is not known.

CONCLUSIONS

The measurements of the energy to quench a potted test coil wound with Nb_3Sn/Cu-Sn superconducting wire and a potted test coil wound with NbTi/Al wire show that both these magnets are more stable than coils wound with NbTi/Cu wire. The Nb_3Sn/Cu-Sn test coil required 3 to 10 times more energy to quench over an I/I_c range of 0.3 to 0.9 than the NbTi/Cu test coil. The NbTi/Al test coil required 30 times more energy to quench than a NbTi/Cu test coil over an I/I_c range of 0.3 to 0.9. Possible nonuniformities in the construction of the Nb_3Sn/Cu-Sn wire and the known nonuniformities in the construction of the NbTi/Al wire could modify the results presented owing to the possibility that the critical current value of the wire at the heater location of each coil can be higher than the measured value of each coil. The results presented are also based on measurements made with only one coil of each type of wire. A larger data base is needed, with measurements for several coils wound with Nb_3Sn/Cu-Sn and NbTi/Al wire, to verify the results presented.

REFERENCES

1. M. J. Superczynski, Heat pulses required to quench a potted superconducting magnet, IEEE Trans. Magn. MAG-15(1) (1979).
2. D. J. Waltman, M. J. Superczynski, and F. E. McDonald, Energy pulses required to quench potted superconducting magnets at constant field, IEEE Trans. Magn. MAG-17(1) (1981).
3. M. Young, E. Gregory, E. Adam, and W. Marancik, Fabrication and properties of an aluminum-stabilized NbTi multifilament superconductor, in: "Advances in Cryogenic Engineering," Vol. 24, Plenum Press, New York (1978), p. 383.

TRANSIENT PERFORMANCE OF A PREREACTED, FULLY IMPREGNATED MULTIFILAMENTARY Nb_3Sn COIL*

B. B. Gamble, E. T. Laskaris, and T. A. Keim

Corporate Research and Development, General Electric Company, Schenectady, New York

INTRODUCTION

As part of a program to develop a lightweight high performance superconducting generator for transient duty, a development program has been carried out to prove the capability of multifilamentary Nb_3Sn conductor in coils suitable for generators. This paper describes the outcome of the coil development program.

A superconductor selection study has been performed in which generator performance goals were weighed against the conductor and coil development work necessary to achieve these goals. A prereacted, cabled, multifilamentary Nb_3Sn conductor has been selected, with a copper sulfide strand insulation applied prior to reaction and a glass-braid cable insulation applied after reaction.

A fully impregnated coil has been constructed using this conductor. The coil is wound dry with glass cloth reinforcement and then vacuum-pressure impregnated with epoxy resin, following techniques that have produced training-free coils with NbTi superconductor.

The coil achieved short-sample performance on the first and subsequent excitations. The quench current was measured for ramp rates up to 20.7 T/s, at which rate the quench current was 74% of the value obtained at slow ramps. A pulsed current wave, to 88% of the short sample value of current along the loadline with a 5-s period, was successfully applied for 11 min.

*Supported by the U.S. Air Force Aero Propulsion Laboratory, Wright Patterson Air Force Base, Ohio, under U.S. Air Force contract F33615-76-C-2167.

SUPERCONDUCTOR SELECTION FOR THE ADVANCED SUPERCONDUCTING GENERATOR

Since 1976, GE has been engaged in an Advanced Superconducting Generator Program to design, construct, and test a lightweight high-power-density superconducting generator for airborne applications. This generator is intended for intermittent duty;[1] the design goal is a start-up from the cold, unexcited condition to full power in 1 s. Such rapid start-up capability requires a substantial transient heating tolerance. Because the source of the field winding heating is in this case the self-field of the winding, shielding is less effective than in other instances. Other program goals, especially 0.1 lb/kW power density, do not permit compromises in winding flux density or current density.

Early in the course of the program, the decision was made to develop coils for the generator using commercially available multifilamentary Nb_3Sn. A subsequent program redirection has resulted in the selection of niobium-titanium coils for the generator, but a small Nb_3Sn coil development program has been retained. This paper presents the outcome of the coil development work.

The superconductor for coil development was selected on the basis of its projected performance in the advanced superconducting generator. Table 1 summarizes the conductor selection criteria that were arrived at on the basis of a generator performance study. The authors' favorable experience with epoxy-impregnated, racetrack-shaped niobium-titanium coils has led to the selection of this coil construction for the Advanced Superconducting Generator and also for the Nb_3Sn coil development. Conductor selection has been heavily influenced by a desire to minimize the modifications required in the impregnated coil manufacturing process,

Table 1. Requirements for Advanced Superconducting Generator

Conductor Requirements
Process: Prereacted
Insulation: Compatible with Impregnation Process
Bend Radius: 1.5 in
Continuous Length: Long enough to make 300-in^3 modules
Peak Field: 6.8 T
Losses not strongly dependent in field direction

Coil Performance Requirements	Typical Conductor Requirements to Meet Coil Performance Requirements
Superconducting following 1-s ramp to design field	300 A minimum current at at 6.8 T and 8.5 K
Peak temperature following quench: 115 K	30% copper module cross section
Peak voltage following quench: 3.5 kV	60 mJ/cm^3 maximum for 1-s ramp to 6.8 T
15 000 A/cm^2 on module cross section	

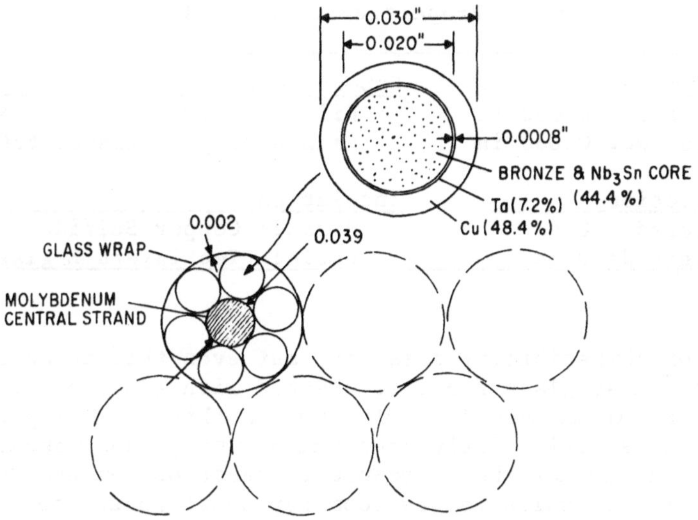

Fig. 1. Conductor Configuration.

which has produced the favorable results to date. The introduction of a reaction step into the coil manufacturing process is deemed to be a major departure from present practices, so a prereacted conductor was selected.

The conductor insulation must be compatible with the impregnation process. Polyvinyl formal insulation would have presented the fewest problems, but this insulation cannot withstand the temperature of the Nb_3Sn reaction. A conductor can be cabled only prior to a reaction, so polyvinyl formal insulation is not obtainable on a cabled conductor. The bend radius and continuous length requirements, on the other hand, are more readily met by a cabled conductor.

Figure 1 shows the conductor that was selected for use in the generator. The cabled configuration has enough advantages to merit the risk of selecting an unproven insulation system. One of the fundamental purposes of a coil development exercise is, thus, the evaluation and qualification of an insulation system.

CONDUCTOR SELECTION FOR COIL DEVELOPMENT

The principal concern with a novel insulation system is the quality of the bond between resin and conductor. The more complicated and greatly increased surface of the cable may not wet as readily as the simple rectangular conductors previously used. The designation of glass insulation superficially seems to resolve the

Table 2. Properties of Airco Conductor

Strand diameter		Superconductor	
as ordered:	0.030 in	Filament diameter:	3×10^{-6}m
as received:	0.036 in	Filaments per strand:	8305
Copper fraction		**Insulation**	
as ordered:	0.4	Strand:	Copper Sulfide
as received:	0.46	Cable:	Two layers E-glass braid

insulation compatibility question, but even this selection raises questions. Most glass fiber is treated with a sizing that serves as a lubricant in manufacture and handling. The presence and nature of this sizing influences the wetting and bonding of the resin to the glass, but exposure to reaction temperature carbonizes the sizing, which leaves room for doubt about the quality of the subsequent epoxy bond. The alternative of using unsized glass creates problems in the application of the glass, but provides no assurance of bonding as good as that achieved with sized glass.

As circumstances developed, an opportunity arose to investigate an alternative insulation system that does not require glass to be exposed to reaction temperature. Before the Nb_3Sn conductor that had been ordered for the generator was received, Airco offered to develop a quantity of wire generally in conformance with our needs, using a strand insulation based on developments for the Department of Energy's Large Coil Program. The proposal was technically attractive and fit well with our coil development plans, so we sought and were granted Air Force approval to acquire and test the Airco conductor.

Table 2 describes the Airco conductor. The basic strand insulation is a thin layer of copper sulfide, which is applied by a process developed by Westinghouse for the Large Coil Program. The sulfide, which is applied before cabling, is not destroyed by the temperature of the Nb_3Sn-forming reaction. The completed cable is further insulated by double-layer glass braid, applied after reaction. This insulation system neatly avoids problems with the strand overwrap, leaving only the one major potential difficulty at the conductor-to-resin bond.

The specified exterior dimensions of the wire were very similar to the conductor of Fig. 1. The wire received was substantially larger in cross section than that ordered, but the short-sample capability was about what might be expected from the conductor as received, and, as shown by results to be presented here, the 1.5-in bend radius was met; so for purposes of coil development, the wire size was not of great consequence. Certainly conductor size can be held substantially more nearly to the nominal value in production orders.

All along the wire there were sharp protuberances projecting through the glass overwrap; low-power magnification revealed these protuberances to be copper. The origin of these protuberances is uncertain. Their appearance suggested to some observers that copper particles have been pressed into the wire surface by mechanical action. Other observers contended that the defects were formed chemically during the application of the sulfide film. Low-power magnification also showed that the sulfide coating was not uniform; in particular, there tended to be bare copper along the lines of contact between strands.

COIL WINDING

Figure 2 is a sketch of the wound coil, showing the relevant dimensions. The coil internal diameter and height model the coils for the Advanced Superconducting Generator: the radial build was limited by the length of unbroken wire.

Due to the many sharp protuberances on the wire surface, a 0.015-in-thick glass braid was wound between turns, to prevent turn-to-turn shorts. A thick layer of glass insulation was placed between layers for the same reason. The addition of glass did not reduce the packing factor as much as might be expected. If the conductor had been a round wire with three times the strand diameter, a maximum of 561 turns could have been achieved by close packing with no clearance between wires. If the spacing between round wires is assumed to be the full thickness of the glass in both directions, 491 turns are possible. The coil contains 535 turns. The material introduced to improve insulation evidently occupies, at least in part, space inside the smallest circle that

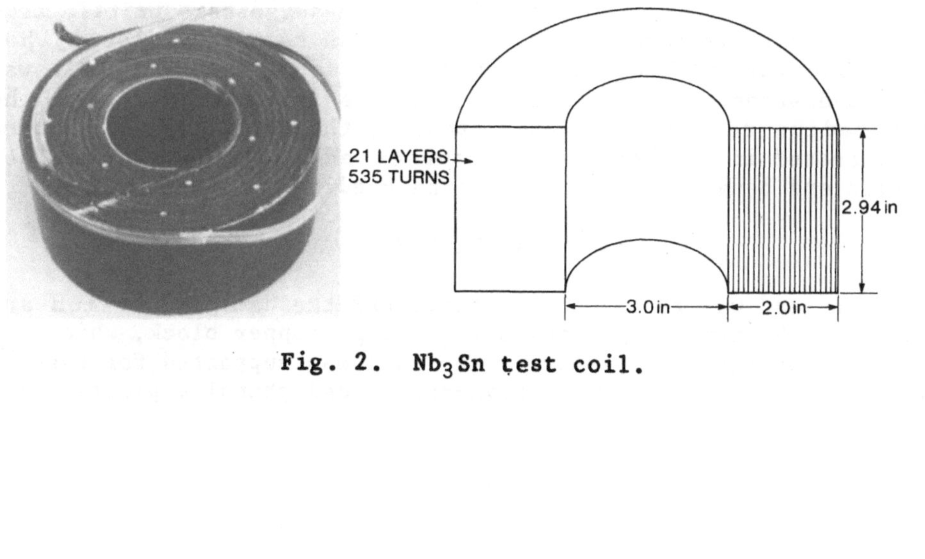

Fig. 2. Nb_3Sn test coil.

circumscribes the cable. It is not known whether it would have been possible to pack the cable tighter than indicated by the circumscribing circle if no extra glass had been used.

Impregnation was done by Intermagnetics General Corporation to our usual specifications. The cured resin is transparent, showing a clear view of the conductor, as evidenced by Fig. 2. Such transparency is an indication of complete wetting and bonding at the epoxy/fiberglass interface and also at the conductor surface.

As the coil was being removed from the impregnation forms, the broken lead (evident in Fig. 2) was first discovered. The lead was formed by filling the cable for some length with indium solder. Rectangular copper conductors were attached to the lead on two sides, using the same solder. These copper conductors provide a parallel path for current and also serve to reinforce the superconductor. The composite lead thus formed was brought from the inner to the outer radius of the coil in a channel in the winding form. Beyond the outer radius of the coil, an edge bend was made in the composite lead to carry the lead circumferentially around the winding form. This edge bend was made with due care to maintain a safe 1-1/2-in bend radius in the superconductor. After the edge bend was formed, it was observed that the solder between the copper and the superconductor had failed and that the superconductor had bent out of the plane of the main bend. Only one side of the conductor was visible; the other side was obscured by the winding form. The radius of the out-of-plane bend was large, so the coil was sent for impregnation in this condition.

After impregnation, the coil form was removed. Some excess epoxy was removed from the lead region, and gentle pressure was exerted on the superconductor to move it back into the plane of the lead. In this process, still more epoxy came loose from the back (concave) side of the out-of-plane bend, and it was observed that the superconducting strands were broken. The fractures appear clean, with little necking, which suggests a brittle tensile failure. It cannot be determined with certainty exactly how the cable was broken, but it seems possible that the cable was first overstressed when the edge bend was formed. Although the bend radius restriction was respected, the cable could have been overstressed if the bending tool subjected the wire to combined tension and bending, rather than pure bending.

COIL TESTING

The lead was repaired by cutting off the damaged portion and soldering the remaining portion to a large copper block, which was supported by the test fixture. The coil was supported for test by clamping it between two cotton-reinforced phenolic plates. The assembly of coil end plates was suspended near the bottom of a wide-mouth test Dewar by rods attached to the top plate. All tests were performed with the coil and the lead blocks completely immersed in liquid helium.

Initial coil excitation was performed with a bank of three paralleled 500-A, 10-V, adjustable thyristor power supplies. These supplies are provided with large capacitive output filters that produce a ripple-free output. Slow quench data was obtained by the following procedure: A relatively rapid ramp to 1200 A in 2 min or less was followed by a slower ramp of 5 A/s or less to quench. Quenches were observed at 1354 A, 1340 A, and 1348 A. These values were obtained by noting the highest reading of a digital multimeter, so some of the scatter may have been due to variability of the timing between the last prequench sample and the quench. It is clear that the coil showed no evidence of training.

The coil was then ramped at a rate corresponding to the maximum output voltage of the thyristor supplies. The supplies do not have the capability to accept a step voltage command, so the supplies were turned on and off by a contactor in the ac power leads. The power supply controls were all set to maximum. The duration of the on-time was set by a time delay relay operating the supply contactor. The voltage waveform achieved by this method is not a pure square wave. The initial rise time is about 0.4 s, and the wave top is modulated by a nonsinusoidal oscillation of about 2 V peak to peak. The current ramps are nevertheless reasonably straight. A series of ramps was applied, each longer than the previous one. A current of 1200 A was achieved in 1.68 s without quench; a ramp to 1290 A resulted in a quench.

The coil was then connected to a 100-V, 5000-A, 12-pulse, thyristor power supply. The output of this supply is unfiltered, so the output voltage contains 720 Hz ripple. The ripple amplitude varies with output voltage, but at zero volts dc, the waveform is approximately a sawtooth with 50 V peak to peak. With this coil, the resulting peak-to-peak current ripple is about 1/2 A, and peak-to-peak flux density ripple is about 26 gauss. At 720 Hz, this implies a root-mean-square rate of change of flux density of about 2 T/s.

The power supply was used as a power amplifier to follow a command generated by a signal generator. The signal input was a single pulse consisting of a step from a negative to a positive value, a dwell at the positive value, and a step back to a negative value. The negative value was chosen so the coil would slowly reset to zero current over about 10 s. The positive value was adjusted to determine the ramp rate, and the dwell time was adjusted to set the peak current. The power supply response to

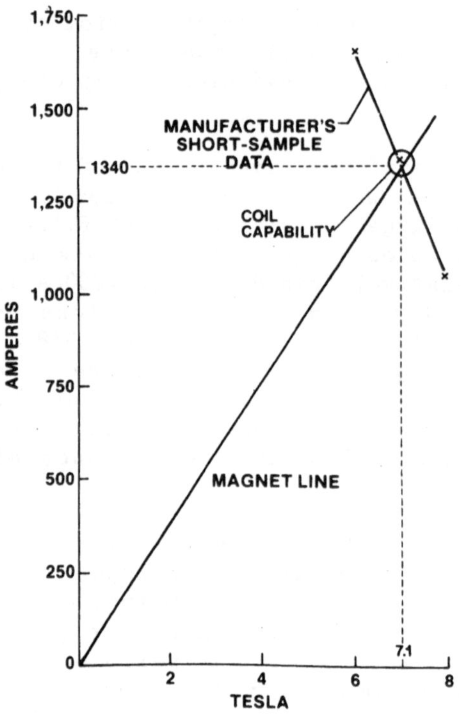

Fig. 3. Predicted vs observed quench current.

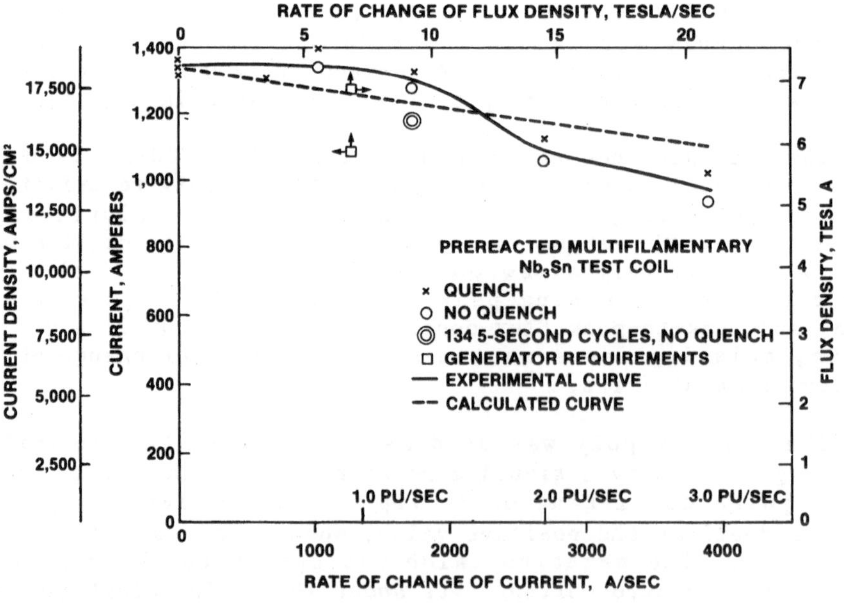

Fig. 4. Transient capability of Nb_3Sn coil.

the step input consists of a substantial initial overshoot, followed by a 0.1-s decay to the command value. For each of several voltage levels, the pulse duration, and thus the peak current, was increased by steps until a quench occurred. It proved possible to reach 1290 A in 1.2 s, 1280 A in 0.78 s, 1060 A in 0.39 s, and 940 A in 0.24 s, all without inducing a quench.

By switching the waveform generator from single cycle to continuous mode, it was possible to apply a periodic series of ramps, each followed by a slower reset to zero current. Ramps to 1150 A in 0.69 s, followed by a 6-s reset, were applied with a 10-s period. Then ramps to 1170 A in 0.7 s, followed by a 3-s reset, were applied with a 5-s period. Both series of ramps were applied for 11 min; in neither case did a quench occur.

Figures 3 and 4 summarize the test results. Figure 3 shows the current and the peak magnetic field achieved in the slow-quench test. Also shown is a prediction based on data obtained from a short sample test by the conductor supplier. The short sample test was performed on one strand of the cabled conductor at no conductor strain.

Figure 4 shows current achieved as a function of ramp rate. The actual current capability at any ramp rate may lie anywhere between the open circle and the corresponding X. The point corresponding to the repetitive ramps is shown. Also indicated are two points representative of the design point of the advanced superconducting generator. One corresponds to the design value of the flux density ramp rate and current density, and the other, to design values of flux density ramp rate and flux density.

CALCULATED TRANSIENT PERFORMANCE

Previous work has described a method of transient thermal analysis that accurately predicts the quench current of an impregnated superconducting coil as a function of ramp rate. The analysis models the coil as a single orthotropic composite material. A large amount of information is required to characterize the coil, including: the magnetic field distribution within the coil, conductor losses as a function of rate of charge of field conductor, quench current as a function of field and temperature, specific heat as a function of temperature, and thermal conductivity in various directions. When all this information is accurately available, the predictions made by this method can be very good.[2]

Except for the magnetic field distribution, and possibly the quench current vs field and temperature, none of the properties of this cabled Nb_3Sn coil are as well characterized as they are for monolithic NbTi. The thermal conduction path across a module with a round cable conductor is complicated, and a detailed calculation of composite thermal conductivity would be tedious, even if the component properties were known with complete accuracy. In the absence of values measured on a composite, we have elected to use estimates derived from a very simplified model. These estimates are $k_r = 0.0031$ W/cm K in the radial direction and $K_a = 0.0040$ W/cm K in the axial direction. No measurements of the composite specific heat have been made, so a value was estimated by averaging values for the constituents obtained from a variety of sources. The resulting function of temperature, θ, is

$$\rho\, c_p = 4 \times 10^{-5}\ \mathrm{J/cm^3\ K}(\theta/1\mathrm{K})^{2.83} \qquad 4\ \mathrm{K} \leq \theta \leq 8\ \mathrm{K}$$

The conductor loss estimate based on strand volume, as a function of rate of change of flux density, B, is

$$q = (2.8 \times 10^{-3}\ \mathrm{JT^{-1}cm^{3}})\dot{B} + (2.7 \times 10^{-4}\ \mathrm{Js^{2}T^{-2}cm^{-3}})\dot{B}$$

The functional dependence of this estimate was selected so that the first term represents hysteresis losses and the second, eddy current effects. The numerical values of the coefficients were obtained by fitting an equation of this form to another manufacturer's quoted values for a similar conductor and scaling the result to account for different conductor sizes. The authors recognize that this procedure does not reflect the state-of-the-art in superconductor loss estimation, but in view of the uncertainties in the other property estimates, the present method is regarded as satisfactory for this problem. The superconductor characteristics are generated from data produced in conjunction with the manufacturer's short-sample data.[3]

Figure 5 shows the result of a typical calculation for a constant-voltage ramp. The local critical current was determined for points along the coil midplane at various instants of time. Both the fields and the temperatures in this plane are higher than in any other, so this is where a quench will initiate. The local critical current was determined by taking into account the local temperature and the local flux-density vs current characteristic. The range of r in Fig. 5 is only about 0.25 in.

As time progressed, points away from the surface of the coil rose to higher temperatures than the point on the inner surface. After a few tenths of a second, heating caused the point of lowest critical current to move away from the surface of the coil, even though the flux density was lower. The rate of change of the critical current diminished with time as the specific heat increased, but the coil current continued to rise at a constant rate. When the coil current and the lowest local critical current coincide, a quench is predicted. Quench currents calculated by this method are shown in Fig. 4. The results are seen to predict the observed quench current generally within 10%. Considering the

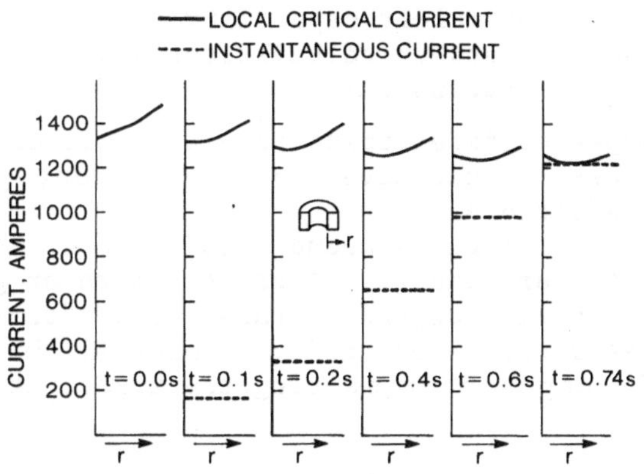

Fig. 5. Current margin vs radius and time.

substantial number of assumptions necessary to make the computa tion, this accuracy is acceptable.

The predicted transient capability decreased at a constant rate with increasing ramp rate, but the observed behavior exhibited an inflected curvature. The inflected curve is representative of results obtained with other coils and has been successfully predicted in other cases by similar analysis. The reasons for the difference in the shape of the predicted curve in this case are not completely clear, but this effect, too, probably arises from inaccuracies in the estimated properties.

CONCLUSIONS

The close agreement between short-sample performance and the experimental results permits several important conclusions to be drawn:

o The absence of training supports the observation that conductor support is adequate in this coil. In view of the radically different construction, this result is especially gratifying.

o The fact that performance of the cable was predictable from that of a single strand strongly suggests that the conductor is of uniform quality and also that nearly perfect sharing of current among the strands has been achieved.

o The elastic strain of bending around a 3-in diameter (0.78% strain based on the nominal conductor size) has had no effect on the coil capability. This observation was not completely expected, since other published results[4] have indicated a detectable strain effect at this level.

o The absence of any reduction of performance of the coil relative to short sample also implies that the winding techniques employed have not overstressed the wire.

The excellent transient capability summarized in Fig. 4 and the general agreement between calculations and the measured values support several more conclusions:

o The copper sulfide strand insulation is adequate. Otherwise, large ac losses would have caused transient performance to lie far below predictions.

o The additional glass wound into the coil was effective in preventing turn-to-turn shorts. A shorted turn or group of turns would have substantially degraded transient performance.

o A generator with a multifilamentary Nb_3Sn field winding should be fully capable of meeting the transient duty cycle requirements.

The excellent results obtained in this coil test have significant implications for superconducting electric machines and also for other high-performance applications of superconducting coils. The processes necessary to supply multifilamentay Nb_3Sn conductor in usable form have all been developed. We have demonstrated that the excellent properties that Airco has achieved with this wire can be retained when the wire is fabricated into a coil of cross section suitable for use in electric machines, using processes suitable for the construction of electric machines. The technical feasibility of applying multifilamentary Nb_3Sn conductors to electric machines and other devices having similar requirements is in large measure proven by these results.

ACKNOWLEDGMENTS

Mr. T. Chapadeau and Mr. R. K. Terbush did a most careful job of winding the coil. Their conscientious efforts were essential to the successful outcome of this experiment.

REFERENCES

1. B. B. Gamble and T. A. Keim, A superconducting generator design for airborne applications, in: "Advances in Cryogenic Engineering," Vol. 25, Plenum Press, New York (1979), p. 127.
2. E. T. Laskaris, Transient thermal analysis of epoxy-impregnated superconducting windings in linearly ramped fields, Trans. ASME, J. Heat Transfer 100:702 (1978).
3. C. R. Spencer, P. A. Sanger, and M. Young, The temperature and magnetic field dependence of superconducting critical current densities of multifilamentary Nb_3Sn and NbTi composite wires, IEEE Trans. Magn. 15:76 (1979).
4. J. W. Ekin, Strain dependence of the critical current and critical field in multifilamentary Nb_3Sn composites, IEEE Trans. Magn. 15:197 (1979).

THE TRIALS AND TRIBULATIONS OF FABRICATING THE PIPE FOR THE "ROPE IN A PIPE" Nb_3Sn SUPERCONDUCTOR

P. Sanger, E. Adam, G. Grabinsky,
E. Gregory, E. Ioriatti, and G. Rothschild

Airco Superconductors, Carteret, New Jersey

INTRODUCTION

A conductor using flowing supercritical helium as a coolant has been adopted for the superconducting magnet being built by the Airco-Westinghouse industrial team for the Large Coil Program (LCP) at Oak Ridge National Laboratory. This conductor utilizes the "rope in a pipe" concept in which a large number of superconducting Nb_3Sn strands are formed into a cable and wrapped in a stainless steel jacket as shown in Fig. 1.

This conductor will operate at 17.6 kA, 4 K, and 8 T and has a square cross section of 20.7 mm x 20.7 mm with rounded corners. Cooling is provided by forcing supercritical helium at 4 K through the interstices of the cable. Other conductors of this type, but smaller, have been tested with favorable results.

This conductor has passed through several configurations before reaching its final state. The design parameter having the most impact on the fabrication process has been the choice of the steel jacket that surrounds the Nb_3Sn cable and contains the supercritical helium coolant. The requirements demanded of this steel and its weld include superior mechanical strength (σ_y > 1034 MPa) with good fracture toughness, good cold-working properties, excellent weldability, and finally, compatability with a 700°C 30-h Nb_3Sn formation heat treatment.

The initial preliminary design called for the use of 304L stainless steel in a 0.89-mm-thick jacket.[1] When the structural analysis was completed, it became obvious that a much higher

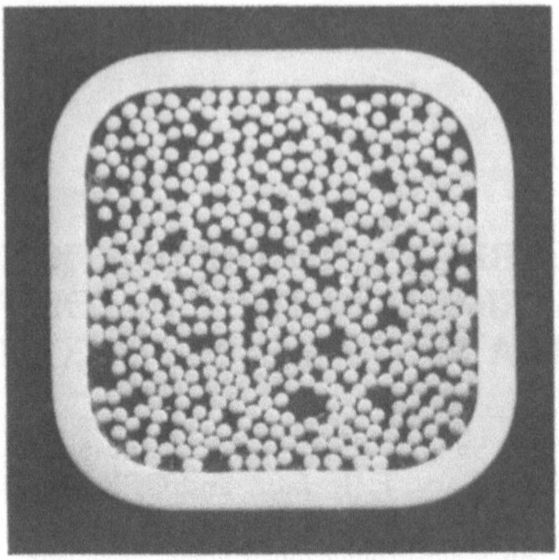

Fig. 1. The Nb_3Sn forced-flow conductor to be used in the Airco-Westinghouse LCP Coil.

strength alloy would be needed. The choice of nitrogen-strengthened 21-6-9 stainless steel with a thicker (1.73-mm) jacket was made.[2]

In late 1979, tests on welds of 21-6-9 steel revealed a dramatic reduction in fracture toughness at 4 K due to the 30-h-700°C heat treatment. Further testing resulted in an abandonment of this alloy as a jacket material, and a search for a new alloy was initiated. This jacket material selection program is discussed in other papers of this conference.[3,4] As a result of these studies, an iron-based superalloy, JBK-75, was selected. This alloy is one of the several alloys based on the A-286-type superalloy and was developed by Sandia Laboratories and Rockwell International to improve the weldability of the basic A-286.[5] The JBK-75 alloy has the particularly convenient property of requiring an aging heat treatment for the same time and temperature as the heat treatment necessary to form the Nb_3Sn superconductor.

JACKET MATERIAL PROCESSING

Although the properties of JBK-75 were suitable for this application, the alloy itself is not commercially available and had primarily been fabricated in bar form with only experimental lots processed to strip form. Utilizing the prior procurement experience of Rockwell International, a material specification was established that required:

1. tight chemistry control on all constituents, particularly C, Si, Mn, and B,
2. vacuum induction melt and vacumm arc remelt,
3. double homogenization heat treatment,
4. supplementary ultrasonic inspection,
5. lower than normal solution heat treatments,
6. long unit lengths, and
7. high degree of cleanliness during processing.

A 9,000-kg lot of this alloy has been processed, and it has met or surpassed the requirements of this application.

CONDUCTOR PROCESSING

The jacket fabrication process is designed to continuously wrap and weld the JBK-75 steel sheet around the cable and to compact the assembly to the final conductor dimensions. The cabled conductor and the flat JBK-75 steel strip enter the jacketing operation with the cable positioned above the strip. The stainless steel strip is progressively roll-formed into a tube enclosing the cable. Fins in the top set of rolls locate the strip symmetrically in the pass as well as restrain the cable from entering the gap between the edges of the strip during forming and welding.

The assembly then enters the welding station where the steel jacket is gas shielded, tungsten arc welded. This step is crucial to the successful operation of the conductor. The jacket is the helium containment vessel of the magnet, and therefore, the 5.3 km of weld must seal 0.4-MPa, 4-K supercritical helium from vacuum with less than a 5×10^{-8} cm^3/s helium leak rate. Also, the cable must not be damaged during the welding process.

The final step in the jacketing process is the compaction and squaring of the conductor. The welded, round jacket containing the cable enters a series of rolls that reduce the cross section of the conductor by 13% and produce a 20.7-mm-square conductor. The various steps of this process are shown schematically in Fig. 2. During this process, the weld integrity is tested utilizing a multifrequency-eddy-current technique. The final check is made by checking for helium leaks at 77 K after the reeling of the conductor.

Extensive development has already gone into this process based on the behavior of types 304L and 21-6-9 stainless steel, and it was initially hoped that JBK-75 would be similar enough that only minor changes would be necessary. It did not turn out to be the case; the differences are discussed below.

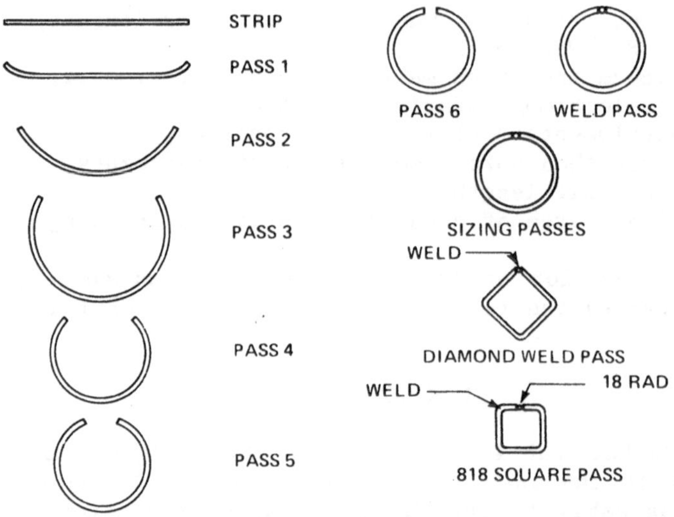

Fig. 2. Sequence of forming stages in producing the seam-welded steel jacket for the Airco-Westinghouse LCP Conductor.

Jacket Forming

The first set of problems to be tackled occurred in the forming of the 80-mm-wide strip. In all tube roll-forming operations, a substantial amount of sliding between the roll and the formed strip is inherent in the process. This particular application also requires a high degree of cleanliness in the final product, necessitating the use of unlubricated roll surfaces. Types 304L and 21-6-9 steel tolerated this unlubricated friction very well, whereas JBK-75 steel did not, and a severely galled and unacceptable surface resulted. The elimination of this condition consisted of many refinements to the process: (1) the strip itself was used in a 5% cold-worked condition, which increased its Rockwell hardness from 73 to 100, (2) all the rolls on the system were hard chrome plated, (3) more use was made of side roll forming to reduce the sliding as the strip entered the vertical stands, and (4) very small amounts of paraffin were applied to the rolls where the galling was the most severe.

The next complication arose from the conductor design demand that the jacket weld always occur on the top flat side of the conductor and not in the corner. For types 304L and 21-6-9 steel, this condition was assured by a diamond pass immediately ahead of the final squaring pass. This preforming had the added feature of preventing the jacket from being trapped in the corners where each of the four rolls in the squaring pass meet. This entrapment can produce sharp grooves in the corners of the jacket of the final square conductor.

This technique was not successful with the JBK-75 alloy and, in fact, enhanced the small existing residual twist in the weld. Removal of the diamond pass practically eliminated any weld twist but resulted in sharp corner grooves in the final conductor. Through experimentation, it was found that adjustment of the side rolls in the preceding sizing stations could control the small amount of residual twist in the jacket weld. The corner groove defects could be eliminated by reducing the diameter of the jacket 5% prior to the squaring operation. This discovery required conversion of the sizing section (originally designed to take 0.1% reduction while rounding the tube) into a roll section capable of the 5% reduction. Besides a modified roll design, changes to the drive train were required to increase the speed of the reduction section with respect to the forming sections. Increased speed is necessitated by the fact that the reduction in diameter directly translates into a jacket length increase that must be accommodated by a speed increase to prevent accumulation of conductor between the sections of the tube mill.

Jacket Welding

The most important function of the conductor jacket is to maintain total containment of the supercritical helium coolant, preventing loss of vacuum to the LCP facility. Since helium leaks at 4 K are rarely small, one leak or failure in the 5.3 km of weld is sufficient to prevent the operation of the coil in the LCP system. To achieve this degree of integrity, gas tungsten arc welding (GTAW) was selected. Early in the program, long lengths of type 304L and 21-6-9 jacketed conductor had been produced without resorting to out-of-the-ordinary techniques. The welding of the type A-286 and JBK-75 alloys revealed new and more stringent requirements than previously anticipated.

The appearance of dross in the welds of A-286, and to a lesser extent in JBK-75, was a condition that is foreign to the welding of types 304L and 21-6-9 steels. We also refer to this dross as "floaters," since they appear as islands floating on the weld. Scanning electron micrographs of typical floaters in A-286 and JBK-75 are presented in Fig. 3. The composition of typical floaters was determined from energy-dispersive x-ray analysis to be primarily aluminum and titanium, presumably in the oxide form. At this time, it has not been determined that the existence of sporadic floaters detrimentally affects the properties of the weld. However, it was felt that large numbers of floaters could reduce the aged weld strength, since the weld metal would be depleted of the constituents from which it derives its aged strength.

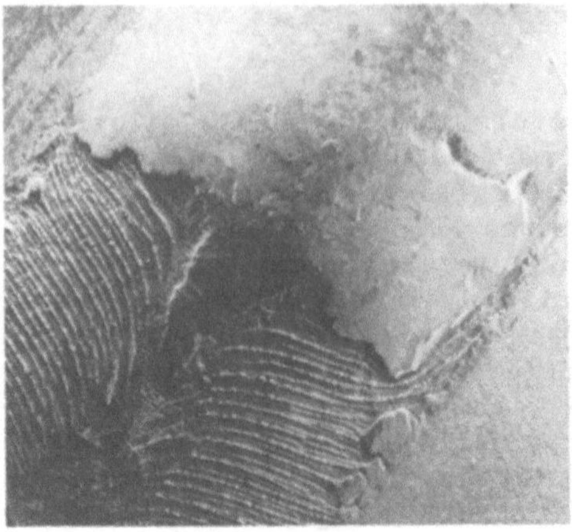

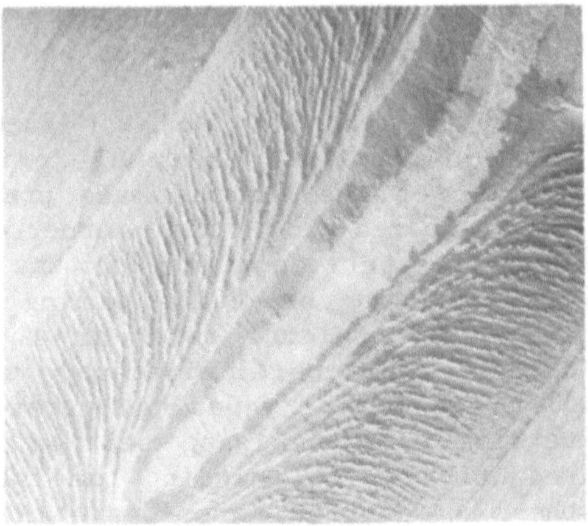

30 X 24 X

Fig. 3. Typical aluminum and titanium oxide floaters on welds of A-286 (left) and JBK-75 (right). (Reduced 12%.)

Three parameters were found to influence the appearance of floaters. Not surprisingly the type, purity, and quality of gas shielding were found to be critical. The use of argon shielding gas was not successful. Helium gas was found to produce a significant improvement. Particular attention was required to avoid contaminating the gas supply while connecting new bottles to the gas system prior to a run. The type and purity of the shielding gas was not sufficient to eliminate weld contamination. It was found that the incoming cable, which is directly underneath the weld, acted as an oxygen sponge. Since purging the cable prior to insertion in the mill was considered impractical, continued purging of the completed conductor was implemented. A gas control system was developed to maintain a constant flow of helium through the continually increasing pressure drop of the conductor length on the rotating take-up table.

Even the maintenance of excellent helium gas shielding did not completely eliminate the floaters. Their existance appears to be coupled with the welding energy input. For energy inputs less than 130 kJ/m, floaters formed on the weld. The floaters disappeared for larger heat inputs. One benefit of using helium gas in the shielding is the potential for higher heat inputs than with argon gas. The welding conditions corresponding to this limit produced a very clean, slightly over-penetrated weld, which completely meets the conductor requirements.

CONCLUSION

The adoption of the iron-based superalloy JBK-75 as the jacket material for the Nb_3Sn forced-flow conductor for the LCP revealed problems significantly different from those of the 304L and 21-6-9 stainless steel jackets. These problems included poor abrasion behavior, different reactions to cold reduction, and critical welding conditions. The number and scope of the changes required to solve the problems only reinforce the importance of material selection as a crucial element in superconducting magnet design and illustrate the fact that many properties of materials impacting a proper selection are not well understood a priori and can only be determined by a trial and error process.

ACKNOWLEDGMENTS

The work presented in this paper presents the constant and enthusiastic energies of the Airco Superconductors technical team and LCP program personnel from both Westinghouse Electric Corporation and Oak Ridge National Laboratory. The authors are glad to have the opportunity to express their appreciation. The development of this conductor is being supported, in part, by DoE funding of Union Carbide Corporation Nuclear Division Subcontract 22X-21083C to Airco Superconductors, a department of Airco, Inc.

REFERENCES

1. P. A. Sanger, E. Adam, E. Gregory, W. Marancik, E. Mayer, G. Rothschild, and M. Young, IEEE Trans. Magn. MAG-15(1): 789 (1979).
2. C. J. Heyne, D. J. Hackworth, S. K. Singh, J. L. Young, and P. A. Sanger, in: "Proceedings of the 8th Symposium on Engineering Problems of Fusion Research," IEEE Publication No. 79CH1441-5NPS, Vol. III, IEEE, New York (1979), p. 1148.
3. R. E. Gold, W. A. Logsdon, G. E. Grotke, and B. Lustman, Evaluation of conductor sheath alloys for a forced-flow Nb_3Sn superconducting magnet coil for the Large Coil Program, in: "Advances in Cryogenic Engineering - Materials," Vol. 28, Plenum Press, New York (1982), p. 759.
4. W. A. Logsdon, G. E. Grotke, R. E. Gold, and B. Lustman, Cryogenic tensile and fracture toughness properties of three candidate structural materials for the Large Coil Program superconducting magnet conductor sheath, in: "Advances in Cryogenic Engineering - Materials," Vol. 28, Plenum Press, New York (1982), p. 771.
5. J. A. Brooks and R. W. Krenzer, Weld. J. 53(6):2425 (1974).

EVALUATION OF CONDUCTOR SHEATH ALLOYS FOR A FORCED-FLOW Nb_3Sn SUPERCONDUCTING MAGNET COIL FOR THE LARGE COIL PROGRAM*

R. E. Gold

Westinghouse Fusion Power Systems, Pittsburgh, Pennsylvania

W. A. Logsdon and G. E. Grotke

Westinghouse Research and Development Center
Pittsburgh, Pennsylvania

B. Lustman

Westinghouse Steam Turbine Generator Division
Pittsburgh, Pennsylvania

INTRODUCTION

The Westinghouse-Airco superconducting toroidal magnet design for the Large Coil Program (LCP) employs forced-flow, supercritical liquid helium to cool the niobium-tin superconducting windings. This requires that the conductor sheath, and in particular the longitudinal seam weld that closes the sheath over the bundled conductor during manufacture, be absolutely leak tight. Normal operating stresses in the sheath are approximately 240 MPa; however, during a fault condition the local stresses can approach 1035 MPa.

During manufacture of the niobium-tin superconductor, after the coil has been fabricated to its final dimensions and configuration, a final reaction-anneal heat treatment of 700°C for 30 h is required to effect the formation of the superconducting compound (Nb_3Sn) by solid-state diffusion. This heat treatment was found to

*Invited paper.

cause significant loss of tensile ductility and fracture toughness properties of the conductor sheath alloy at 4 K; the weld metal regions were found to be particularly susceptible. Closer examination of this problem suggested the loss of ductility was associated with the conversion of delta ferrite to sigma phase and austenite in the weld fusion zone during the reaction-anneal heat treatment. The Nitronic 40 alloy initially selected for the sheath application was found to be unsatisfactory in this regard. Attempts to solve this problem by changes in the welding process or by the use of a more stable filler metal did not show promise. Hence, a testing program was established for the purpose of selecting and qualifying a replacement alloy. The results of that testing program are presented in this paper.

BACKGROUND OF THE PROBLEM

Description of the Application

The Westinghouse superconducting coil design for the DoE Large Coil Program[1] employs forced-flow, supercritical helium to cool the cabled niobium-tin superconducting strands. Each strand is comprised of 2869, 3.5-μm-diameter niobium filaments embedded in a 0.43-mm-diameter bronze (Cu-13 wt.% Sn) core; the bronze core is, in turn, clad with a copper layer resulting in a 0.7-mm-diameter strand. Four hundred eighty-six such strands are present in the final cable. The cable is encased in a 1.63-mm-thick sheath roll-formed to a final 20.8-mm-square configuration from a 78-mm-wide strip by a series of forming and sizing passes. The sheath is sealed by a full penetration autogenous longitudinal gas-tungsten arc weld. After final compaction, the sheathed cable contains about 0.32 void fraction for helium flow.

To prevent sintering of the strands to each other during thermal processing, a 2- to 5-μm layer of Cu_2S is formed on each strand by reaction with an argon-H_2S atmosphere at 400-450°C for approximately six seconds. Also, the cable is wrapped with a 0.025-mm-thick stainless steel strip prior to sheathing primarily to minimize contamination of the autogenous sheath weld by reaction with the strands. After sheathing, the conductor is coiled and heat-treated in an inert atmosphere for about 30 h at 700°C (hereafter called "reaction anneal") to react the niobium filaments with the tin in the bronze core to form the superconducting Nb_3Sn compound.

Because of the long length of autogenous weld required and the necessity for complete leak tightness to pressurized helium, high integrity welds free of cracking or connected porosity must be routinely achievable. Reduction in toughness or ductility of both weld metal and base metal as a result of the reaction anneal must be within acceptable limits. The sheath must resist combined stress loading at 4 K arising from stresses on the coil due to the magnetic toroidal field, differential thermal contraction loads, and internal helium gas pressure. Consequently, sheath stresses are high, both in steady-state and accidental conditions; cyclic loadings are imposed as a result of the variation in stress with magnetic field, internal pressure, and cooldown. The reaction anneal after sheathing requires chemical stability against attack of the sheath by impurities in the inert atmosphere used during the heat treatment, particularly sulfur contamination arising from reaction with the strand insulation.

Initial Sheath Material Selection

Nitronic 40, a nitrogen-strengthened chromium-nickel-manganese austenitic stainless steel, which combines excellent 4 K tensile strength and ductility with exceptionally good weldability characteristics, was originally selected as the conductor sheath material. Tests conducted by Morris[2] and later confirmed by Westinghouse, however, revealed that severe losses in ductility occurred in the welds and, to a lesser extent, in the base metal as a result of the reaction anneal. Subsequent evaluations indicated that these changes were due to metallurgical instabilities, such as sigma formation in the welds (and possibly base metal) and martensite formation during low temperature plastic deformation.

Since the unacceptable deterioration of toughness and ductility occurred as a result of the reaction anneal, it was concluded that either major revisions were necessary to the fabrication scheme or a replacement alloy had to be identified. Revisions to the fabrication sequence that were considered in an attempt to retain the Nitronic 40 alloy included: the use of filler metal instead of autogenous gas-tungsten arc (GTA) welds, the use of high-frequency (solid-state) resistance welding instead of GTA fusion welding, and remelting the weld area after reaction annealing. While cursory evaluations of several of these variations were carried out, their general applicability was rendered moot by the subsequent finding that the Nitronic 40 base metal properties were also borderline following the processing and reaction anneal described.

This latter finding led to the decision that a replacement alloy selection was required for the conductor sheath application. An experimental program was initiated to screen and, ultimately, to qualify an alternate sheath alloy. The experimental program that was carried out pursuant to this goal is described below.

EXPERIMENTAL PROGRAM

To permit rapid screening and evaluation of the candidate alloys, a consistent set of processing and testing procedures was adopted. The processing schedule adopted was established to approximately simulate that expected during the conductor manufacturing operation. Hence, material for testing was solution annealed, GTA welded (where applicable), cold reduced 10% in thickness, and "reaction annealed" 30 h at 700°C. This sequence, described during this program as "Fully Processed," served as the reference test condition for both base metal and weld metal specimens.

All welding operations were performed using a fully automatic gas-tungsten arc (GTA) process at a weld speed of approximately 13 cm/min. Welds were full penetration butt welds produced while the weld blanks were positioned in a specially designed fixture. This fixture permits the use of flowing argon behind (beneath) the specimens as well as for the inert cover gas. All welds were made autogenously (i.e., without the use of a filler metal). Welds were of a uniform and consistently high quality. All alloys that were evaluated demonstrated excellent basic weldability; there was no evidence of hot or cold cracking or any other type of weld defect. Subsequent to welding, specimens were cold rolled directly without additional preparation. The final heat treatment, used to simulate the Airco reaction anneal process, was carried out at 700°C for 30 h. A hydrogen atmosphere was used for these heat treatments to minimize surface contamination.

Acceptance Criteria

To provide a measure of achievement for the testing program, minimum acceptance criteria were established. These are provided in Table 1. The yield strength and fracture toughness requirements were based on stress analysis of the conductor sheath application during fault conditions--viz., local reversion of the superconductive material to the normal state. The tensile elongation criterion further ensures the ability of the sheath to adjust to unanticipated operational conditions.

Candidate Sheath Alloys

Candidate structural alloys were selected for inclusion in the screening stage of this evaluation based on several considerations. They had to be austenitic (hopefully to 4 K) to avoid problems associated with ferromagnetic behavior in large magnet structures, and they had to offer some promise of meeting the 1034-MPa yield strength required at 4 K. Finally, they had to be readily available (preferably within four to eight weeks from the start of the test program).

Structural alloys that generally satisfied these requirements were recommended by personnel within Westinghouse and by others associated with similar superconducting magnet programs. Table 2 lists the composition of the structural alloys that were identified for inclusion in the testing program. Nitronic 40, the original

Table 1. Large Coil Program Conductor Sheath Material Performance Requirements

Material Property at 4 K	Design Requirement	Minimum Acceptance Criteria
Tensile Yield Strength	1014 MPa	1034 MPa
Fracture Toughness	28 MPa $\sqrt{m}$	55 MPa $\sqrt{m}$
Total Tensile Elongation	---	5%

(unsuccessful) selection for the conductor sheath, is included for general comparison. The compositions indicated in Table 2 are not nominal; in each case they represent analyses reported by the appropriate vendor or supplier.

The list of candidate alloys contains alloys representative of a fairly wide range of "austenitic" steels or superalloys. The initial selection, Nitronic 40, is a solid-solution alloy capable of extremely high strength at cryogenic temperatures. It also possesses excellent weldability characteristics. Unfortunately, these appear due, at least in part, to the formation of greater than 3-5% delta ferrite in the weld fusion zone. Subsequent transformation of this delta ferrite to sigma phase at the 700°C reaction annealing temperature is a major reason for the unsuitability of this alloy for the conductor sheath application.

Incoloy 800 H is a widely used iron-base superalloy having excellent weldability. Primarily solid-solution strengthened, its compositional range (particularly titanium and aluminum) provides a small increment of precipitation hardening as well.

Kromarc 58 is a solid-solution-strengthened, fully austenitic stainless steel. Previous examinations of the cryogenic tensile and fracture toughness behavior of this alloy in both the solution-treated and quenched, and solution-treated, quenched and 30% cold-worked conditions gave extremely promising results. Therefore, Kromarc 58, although only a developmental alloy that had not been produced commercially in thin strip form, was considered a candidate for the conductor sheath.

A-286 is a widely used superalloy that derives its strength from both solid-solution effects and gamma-prime [$Ni_3(Al,Ti)$] precipitation hardening. A-286 has a questionable reputation regarding its basic weldability because of a propensity for hot cracking in heavy section welds. This propensity can apparently be controlled or avoided, however, by keeping grain size quite small and controlling such impurities as manganese, silicon, and boron. The

Table 2. Chemical Compositions of Candidate LCP Conductor Sheath Alloys

Alloy	Composition (Weight Percent)														
	Fe	C	Mn	P	S	Si	Ni	Cr	Mo	V	Al	B	Cu	N	Ti	Other
Nitronic 40	Bal.	0.021	8.54	0.022	0.002		7.19	20.48	0.17				0.20	0.33		0.13 Co
Incoloy 800 H	45.64	0.090	1.01		0.002	0.21	31.08	20.62			0.38		0.56		0.41	
Kromarc 58	Bal.	0.017	12.60	0.001	0.007		20.00	16.70	2.61	0.23		0.0120		0.15		0.012 Zr
A-286	Bal.	0.054	0.12	0.006	0.006	0.18	24.55	14.06	(1.2-*1.3)	0.26	0.10	0.0057			2.13	
JBK-75	Bal.	0.018	0.03	0.002	0.003	0.06	30.40	14.70	1.20	0.32	0.17	0.0011		0.003	2.20	<0.001 O_2
Inconel 706	36.38	0.050	0.15	0.100	0.005	0.15	42.01	16.00			0.29	0.0030	0.10		1.84	3.02(Cb+Ta)
304 LN	Bal.	0.024	1.78	0.034	0.006	0.39	9.14	18.30						0.12		
Hastelloy C	0.72	<0.002	0.13	<0.005	<0.002	0.02	Bal.	15.60	15.02						0.24	<0.10 Co

*Not reported by vendor/supplier
Value in () is "nominal" for A-286.

alloy JBK-75 was developed by Sandia Laboratories as a special chemical modification of A-286. A major advantage for both of these alloys for the current application is the fact that the gamma-prime precipitation heat treatment is normally at ~720°C for 16 h. This heat treatment is quite similar to the 30 h at 700°C reaction anneal required for the LCP superconductor.

Inconel 706 is a very high strength nickel-base superalloy. Weldability is usually considered adequate only for material in the solution-annealed condition. In addition to solid-solution strengthening, Inconel 706 is strengthened by precipitation of both gamma prime [γ' - $Ni_3(Al,Ti)$] and gamma double prime [γ'' - $Ni_3(Nb,Ta)$]. Full strengthening requires a two-stage heat treatment. 304 LN is a low-carbon, nitrogen-added version of the popular 18Cr-10Ni austenitic stainless steels. Reliability and availability at a reasonable cost, coupled with a substantial data base, argued for inclusion of this alloy in the initial evaluation phases. Hastelloy C is a nickel-chromium-molybdenum superalloy primarily strengthened by solid-solution hardening. This alloy was primarily designed for corrosion resistance and modest strength at elevated temperatures.

Of the eight alloys identified in Table 2, five were eliminated from consideration very early in the evaluation. These included Inconel 706, Hastelloy C, 304 LN stainless steel, Incoloy 800 H, and of course Nitronic 40 owing to the limitations previously observed. The elimination of Inconel 706 was based on unacceptable ferromagnetic behavior, whereas Hastelloy C and Incoloy 800 H were dropped owing to insufficient strength. The 304 LN stainless alloy was also eliminated because of insufficient strength and because the 304 LN weld metal is not exempt from the same type of structural/metallurgical instability that handicaps Nitronic 40. Hence, the emphasis of the testing program was focused on A-286, JBK-75, and Kromarc 58.

Mechanical Testing Procedures

A major difficulty associated with this study was the developing of reliable mechanical, and particularly fracture toughness, properties of candidate structural alloys in thin sections (0.13- to 0.18-cm thick). While a reasonable volume of mechanical property data are available in the literature on various materials tested in thin sections (although not necessarily at cryogenic temperatures), no thin-section elastic-plastic or fully plastic fracture toughness data exist. Therefore, it was necessary to develop appropriate test specimen geometries and experimental procedures to evaluate thin sections of the candidate structural materials at cryogenic temperatures. A description of both the specimen geometries and the test procedures that were used is provided in detail in a companion paper.[3]

Tensile testing, for both smooth and sharp-notch specimens, observed the standard methods described in the American Society for Testing and Materials (ASTM) Standard Methods of Tension Testing of Metallic Materials (ASTM-E8-79a). The stress concentration factor (K_t) used for the sharp-notched tensile tests was 9.3.

A single-edge notched sheet specimen was selected to develop fracture toughness properties for the various candidate materials. Providing the distance between loading pin centers equals three times the specimen width, the tensile load can be considered to be uniformly distributed across the specimen width at a distance from the crack tip not less than the width,[4] thus eliminating the buckling problem expected with a compact tension specimen. Heretofore, elastic-plastic stress intensity solutions (J solutions) were available only for bend-type specimens (compact tension and bend bars) and center-cracked panels. Consequently, in this investigation, it was necessary to generate the solution for calculating J from single-edge notched specimens. This solution was developed by Ernst and is described in detail in Reference 5. Description of actual test practice used for this evaluation is provided in Reference 3.

RESULTS AND DISCUSSION

Mechanical Property Testing

A detailed reporting of the tensile and fracture toughness test results developed during this evaluation is provided in Reference 3. Only a brief summary of these results is provided here, in Table 3. Tensile and notch-tensile tests were conducted at 4 K, 77 K, and 297 K (room temperature); fracture toughness tests were performed only at 4 K. Since the conclusions and ultimate alloy selection for the conductor sheath application were based on the 4-K properties, only these values are summarized in Table 3.

Note that each of these alloys easily exceeds the minimum acceptance criteria listed in Table 1. Also note that test results are presented for the JBK-75 alloy solution annealed at two different temperatures--899°C and 982°C. Although the 982°C annealed material exhibited slightly better fracture toughness properties at 4 K, the fact that the 899°C annealing treatment is recommended for low temperature service for metallurgical reasons and produces slightly higher tensile strength, led to selection of that heat treatment as the "reference" JBK-75 heat treatment for this application.

Limited fractographic examinations of the candidate alloys following 4-K testing revealed fully ductile, dimpled fracture surfaces for both base metal and weld metal specimens. Under no conditions was brittle or cleavage-type behavior observed.

Table 3. Tensile and Fracture Toughness Properties of Base Metal and GTA Welds of the Candidate Alloys at 4 K

Alloy	Material* Condition	0.2% Y.S. (MPa)	UTS (MPa)	Total Elongation (%)	Sharp-Notch Strength (MPa)	NSR σ_s/σ_{ys}	Fracture Toughness (MPa $\sqrt{m}$)
A-286	Base Metal	1284	1737	33.1	1596	1.24	136.9
	GTA Weld	1215	1550	14.7	1291	1.06	96.2
JBK-75	Base Metal (SA-899°C)	1320	1966	32.7	1653	1.25	131.5
	GTA Weld (SA-899°C)	1240	1634	13.4	1371	1.11	76.8
	Base Metal (SA-982°C)	1248	1832	32.4	1645	1.32	137.2
	GTA Weld (SA-982°C)	1156	1622	12.0	1109	0.96	88.8
Kromarc 58	Base Metal	1351	1622	23.2	1548	1.15	102.1
	GTA Weld	1238	1507	29.3	1528	1.23	145.1

*All material "fully processed" = Solution annealed + GTA welded (where applicable) + cold worked 10% + reaction annealed 30 hrs. at 700°C.

Other Evaluations

In addition to the mechanical property evaluations, some attention was given to the question of possible sulfidation of the structural alloy by the Cu_2S insulating layer during the reaction anneal. To examine this possibility, a series of experiments was carried out. Carefully cleaned sheet specimens of several of the candidate alloys (Incoloy 800 H, A-286, Kromarc 58, JBK-75, and Nitronic 40) were sealed in silica ampules with a 30-cm length of sulfided superconductor strand in an ultrapure argon atmosphere. Following exposure for 30 h at 700°C, the specimens were removed, weighed, and metallographically examined. In addition, 4-K tensile tests were performed on several of the Kromarc 58 and JBK-75 specimens.

The extent of attack varied considerably, not only from alloy to alloy but also from specimen to specimen for a given alloy. Although the sulfidation reaction was not negligible, there was no evidence of a significant effect on mechanical properties due to the reaction products. The presence of titanium or aluminum, as for example in A-286 and JBK-75, restricted the reaction morphology to the formation of a surface layer, whereas internal sulfidation was observed in manganese-containing alloys, such as Kromarc 58 and Nitronic 40. Although this suggests a slight advantage for A-286 or JBK-75, it seems highly unlikely that sulfide formation will have serious impact on the sheath properties.

Finally, because of the importance of weldability to the overall reliability of the Westinghouse-Airco conductor performance, preliminary investigations of the basic weldability behavior of Kromarc 58, JBK-75, and A-286 were carried out. These investigations employed Tigamajig tests to examine for heat-affected-zone (HAZ) cracking and subsize Varestraint tests to explore for the potential for fusion zone cracking or "hot cracking/tearing." The relative crack sensitivities derived from these tests can be summarized as follows:

- JBK-75 is much less sensitive than A-286 to hot cracking in the fusion zone.
- With regard to both HAZ and fusion zone crack sensitivity, JBK-75 and Kromarc 58 are approximately equivalent.
- From the standpoint of fusion zone hot cracking, both JBK-75 and Kromarc 58 are inferior to type 304 stainless steel.

It is important, however, to note that none of the three candidate alloys exhibited a propensity for hot cracking or HAZ cracking that would prohibit their selection for the subject application. Despite the poor reputation of A-286 with regard to weld metal cracking, no evidence was seen of this problem in the fine-grain-size, thin-section strip evaluated.

CONCLUSIONS

Each of the three final candidate alloys A-286, JBK-75, and Kromarc 58 satisfy the acceptance criteria established for the conductor sheath application: a 4-K minimum yield strength of 1034 MPa with a total tensile elongation of at least 5%, and a minimum 4-K fracture toughness value of 55 MPa $\sqrt{m}$. Simple comparison of the tensile and fracture toughness test results argues for selection of Kromarc 58 on the combined basis of better strength, ductility, and fracture toughness in the weld metal at liquid helium temperature. When weldability test results and sulfide attack resistance are considered, however, the ranking of the three candidates is JBK-75, Kromarc 58, and A-286.

Nevertheless, it is important to remember that all three candidates satisfy the minimum requirements. Relative comparisons beyond that level, particularly in view of the extremely limited testing program that was carried out, can lead to specious and unwarranted conclusions.

The extremely tight time schedule for the LCP superconductor manufacture and the absolute reliability required of the conductor sheath suggest greater weight be assigned to such factors as availability and reliability. Kromarc 58 suffers in this regard since it has never been produced as thin strip outside of the Westinghouse Research and Development Center. Preliminary contacts with potential vendors reflected this fact; delivery times promised to be in the nine- to twelve-month time range. Although A-286 enjoys considerably greater commercial status than JBK-75, the latter alloy has been produced in quantity as 2700- to 4100-kg VIM-VAR melts for defense-related applications for Rockwell's Rocky Flats Plant. This experience has been positive; metallurgists at Rocky Flats feel that at least three reliable vendors are currently qualified in the routine production of JBK-75.

Hence, in view of the difficulties anticipated with the procurement of Kromarc 58 strip and the fact that the much tighter composition limits of JBK-75 suggest a greater reliability to go with its intrinsically better weldability compared with A-286, a decision was made to procure approximately 9100 kg of JBK-75 strip for the Westinghouse-Airco conductor sheath application. (At the time of this conference, a preproduction lot of over 2000 kg of the JBK-75 strip has been received from Carpenter Technology. Tensile and fracture toughness tests at 4 K have easily surpassed the acceptance criteria established for this evaluation.)

ACKNOWLEDGMENTS

The authors gratefully acknowledge several of their Westinghouse co-workers who greatly contributed to the successful completion of this evaluation. R. R. Hovan and A. R. Petrush conducted all of the tensile tests, and H. J. Stoutmire capably performed all of the cryogenic fracture toughness tests. G. R. Wagner directed the sulfiding experiments, and L. M. Friedman and W. R. Kuba, respectively, supervised and welded the significant number of weldments tested in this program.

Special thanks are also in order for P. A. Sanger of Airco Superconductors, who supplied important test material and invaluable technical assistance, and to T. R. Leax who assisted in obtaining and processing several of the test materials.

REFERENCES

1. "Superconducting Magnet Coils for the Large Coil Program," Phase 2 Final Report to Union Carbide Corp., Oak Ridge, Tennessee by Westinghouse Electric Corporation, Pittsburgh, Pennsylvania on Contract 22X-31747C (March 31, 1980).
2. J. W. Morris, Jr., University of California, Berkeley, private communication (February 1980).
3. W. A. Logsdon, G. E. Grotke, R. E. Gold, and B. Lustman, Cryogenic tensile and fracture toughness properties of three candidate structural materials for the Large Coil Program superconducting magnet conductor sheath, in: "Advances in Cryogenic Engineering--Materials," Vol. 28, Plenum Press, New York (1982), p. 771.
4. "Plane Strain Crack Toughness Testing of High Strength Metallic Materials," W. F. Brown, Jr. and J. E. Srawley, eds., ASTM STP 410, American Society for Testing and Materials, Philadelphia (1966).
5. H. A. Ernst, Unified solution for J ranging continuously from pure bending to pure tension, unpublished research at Westinghouse R&D Center, Pittsburgh, Pennsylvania (1981).

CRYOGENIC TENSILE AND FRACTURE TOUGHNESS PROPERTIES OF THREE CANDIDATE STRUCTURAL MATERIALS FOR THE LARGE COIL PROGRAM SUPERCONDUCTING MAGNET CONDUCTOR SHEATH*

W. A. Logsdon and G. E. Grotke

Westinghouse Research and Development Center, Pittsburgh, Pennsylvania

R. E. Gold

Westinghouse Fusion Power Systems, Pittsburgh, Pennsylvania

B. Lustman

Westinghouse Steam Turbine Generator Division, Pittsburgh, Pennsylvania

INTRODUCTION

The tensile and fracture toughness properties of three candidate structural materials for the DoE Large Coil Program superconducting magnet conductor sheath were developed at liquid helium temperature. The three candidate materials included A-286, Kromarc 58, and JBK-75. Both base plates and autogenous gas tungsten arc weldments were tested for each material in the fully processed condition (i.e., following solution annealing, 10% cold work, and a reaction annealing heat treatment of 30 h at 700°C, 1292°F). Each material was tested in very thin sections (0.173 cm, 0.068 in) equivalent to that in the actual conductor sheath. The various challenges associated with tensile and especially fracture toughness testing of reasonably high toughness materials in thin sections at liquid helium temperature are described in detail.

The background information regarding the specific design application and the details of the Westinghouse-Airco toroidal

*Invited paper.

magnet for the Large Coil Program are presented in detail in a companion paper in these proceedings[1] and will not be repeated here. It is necessary only to mention that the initial conductor sheath alloy selection, Nitronic 40, proved unacceptable due to metallurgical instabilities that caused severe losses in mechanical properties at the 4-K design temperature. Hence, an experimental program was carried out to identify a replacement alloy. Of the seven candidate alloys identified for this evaluation (see Ref. 1), extended testing was confined to the three most promising--A-286, Kromarc 58, and JBK-75, a modification of the A-286 composition, and test results are provided here only for these three alloys.

These final three candidate conductor sheath materials were tested in the fully processed condition, which is defined as:

- Solution annealed
- Autogenously GTA welded (where applicable)
- Cold rolled, 10%
- Reaction annealed, 30 h at 700°C (1292°F)

This processing schedule was selected because it best simulated the working, welding, and heat-treating experienced by the actual conductor during the manufacturing operation.

ACCEPTANCE CRITERIA

To provide a measure of achievement for the testing program, minimum acceptance criteria were established. These are provided in Table 1. The yield strength and fracture toughness requirements were based on stress analysis of the conductor sheath during fault conditions--viz., local reversion of the superconductive material to the normal state. The tensile elongation criterion further ensures the ability of the sheath to adjust to unanticipated operational conditions.

EXPERIMENTAL PROCEDURES

A major experimental problem dealt with developing the mechanical and, especially, the fracture toughness properties of various candidate structural materials in thin sections (0.13- to 0.18-cm, 0.050- to 0.070-in thick). While a reasonable amount of mechanical property data are available in the literature on various materials tested in thin sections (although not necessarily at cryogenic temperature), no thin-section elastic-plastic or fully plastic fracture toughness data exist. Therefore, it was necessary to develop appropriate test specimen geometries and experimental procedures to evaluate thin sections of the candidate structural materials at cryogenic temperatures.

Table 1. Large Coil Program Conductor Sheath Material Performance Requirements

Material Property at 4 K (-452°F)	Design Requirement	Proposed Acceptance Criteria
Yield Strength	1014 MPa (147 ksi)	>1034 MPa (>150 ksi)
Fracture Toughness	28 MPa$\sqrt{m}$ (25 ksi$\sqrt{in}$)	>55 MPa$\sqrt{m}$ (>50 ksi$\sqrt{in}$)
Total Elongation	---	5%

Tensile Testing

The test procedures set forth in the American Society for Testing and Materials (ASTM) Standard Methods of Tension Testing of Metallic Materials (ASTM E8-79a) were followed throughout this investigation. The thin-section (sheet) specimens utilized to evaluate the tensile properties of the candidate conductor sheath materials are illustrated in Fig. 1. Both types of tensile specimens were machine finished to a thickness of 0.13 cm (0.050 in).

The tensile properties of base plate materials were developed utilizing the constant-width (0.635 cm, 0.25 in), 2.54-cm (1.0-in) gauge-length specimen. This tensile specimen geometry is very similar to the subsize rectangular specimen recommended in ASTM E8-79a with the exception of overall length, (5.22 cm, 2.25 in vs.

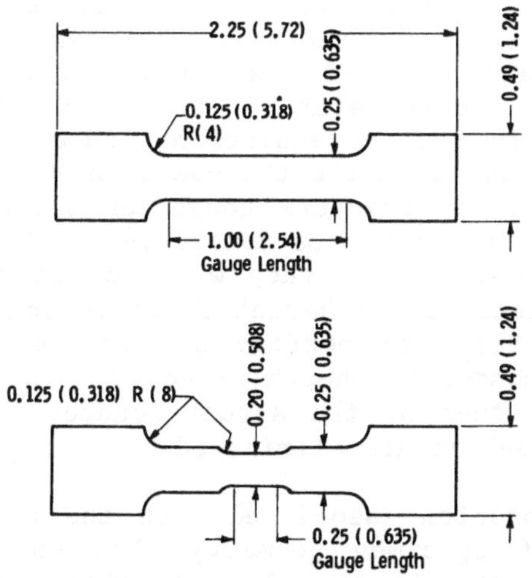

Fig. 1. Sheet tensile specimens. All specimens were 0.050-in (0.127-cm) thick; dimensions are in inches (cm).

10 cm, 4.0 in for the ASTM specimen). This shorter overall specimen length is of minor concern for two basic reasons: First, the standard contains a note that states that the overall specimen length should be as large as the material will permit, and indeed, in this investigation, a 10-cm (4.0-in) long tensile specimen was not practical for several of the candidate materials. Second and more importantly, the specimen geometry recommended in the standard requires extra overall length, all of which occurs in the specimen grip regions, because the specimens are designed to be loaded via a friction grip system and, as such, require as much bearing area as possible. On the other hand, the subject tensile specimens were loaded at the shoulders via a split-grip testing fixture. As such, extra long specimen grip sections were not required.

The reduced gauge length (0.35 cm, 0.25 in) tensile specimen geometry was utilized for testing the various autogenous gas-tungsten arc (GTA) welds. Since all weldment tensile specimens were oriented transverse to the various welds, the shortened specimen gauge length completely spanned the weld fusion zone and was made up entirely of weld metal. As such, this specimen geometry provides a very accurate measure of weldment yield and ultimate strengths. On the other hand, care should be utilized when comparing base versus weld metal total elongation values since they were measured with specimens of different gauge lengths.

Fracture Toughness Testing

The property J_{Ic} characterizes the fracture toughness of materials at or near the onset of crack extension from a preexisting fatigue crack. As long as the prescribed minimum specimen size requirements are satisfied, the value of J_{Ic} is a material property.[2] Because the fracture toughness of the majority of candidate structural materials was anticipated to be rather high, even at liquid helium temperature, it was obvious that the specified minimum specimen size requirements necessary for a valid J_{Ic} test could not be met with the thin-section sizes involved. Therefore, the thin-section fracture toughness values obtained through this investigation could not be considered "valid" per the proposed J_{Ic} testing standard;[3] i.e., they will not be an exact measure of the particular material's inherent fracture toughness. These data can still be utilized to perform an accurate fracture mechanics analysis of the conductor sheath, however, since the test specimen thickness matches that of the actual conductor sheath; i.e., section thickness problems are eliminated.

An initial problem associated with the toughness testing was to select a test specimen geometry that would be adequate for obtaining fracture toughness values in thin sections at cryogenic temperatures. The popular compact tension (CT) specimen was deemed inappropriate because buckling would most likely be introduced upon loading. In addition, a bend bar specimen was obviously not adequate for lack of load-bearing area. Furthermore, center-cracked panels were ruled out because there are potential difficulties in precracking these specimens so that the fatigue cracks at each notch tip are of comparable length. Keep in mind that owing to the large difference between room and liquid helium temperature yield strengths, the final fatigue precracking loads had to be quite low.

The single-edge notched sheet specimen illustrated in Fig. 2 was selected to develop fracture toughness properties for the various candidate materials. Providing the distance between loading pin centers equals three times the specimen width, the tensile load can be considered to be uniformly distributed across the specimen width at a distance from the crack tip not less than the width,[4] thus eliminating the buckling problem expected with the compact tension specimen. The specimen was designed such that after precracking, the specimen a/W ratio equaled 0.5, where a and W are the specimen crack length and width, respectively. The maximum stress intensity for terminal fatigue precracking was limited to 13 MPa$\sqrt{m}$ (12 ksi$\sqrt{in}$). Displacement across the specimen notch was measured via an immersible clip gauge attached to the integral knife edges (see Fig. 2).

To date, elastic-plastic stress intensity solutions (J solutions) have been developed only for bend-type specimens (compact

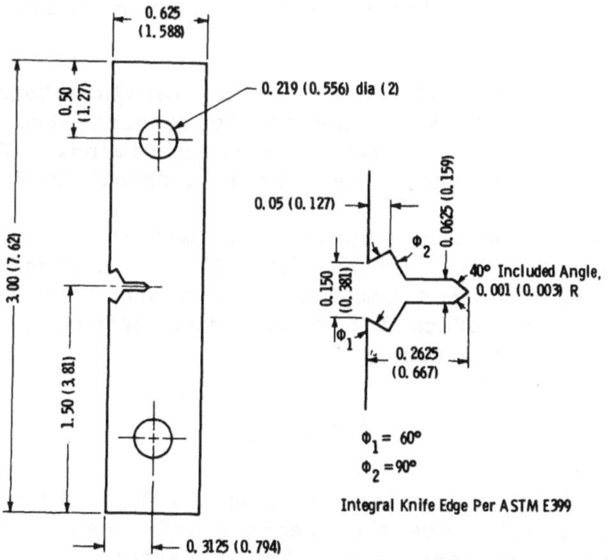

Fig. 2. Single-edge notched sheet specimen. Specimen thickness was 0.068 in (0.173 cm) maximum; dimensions are in inches (cm).

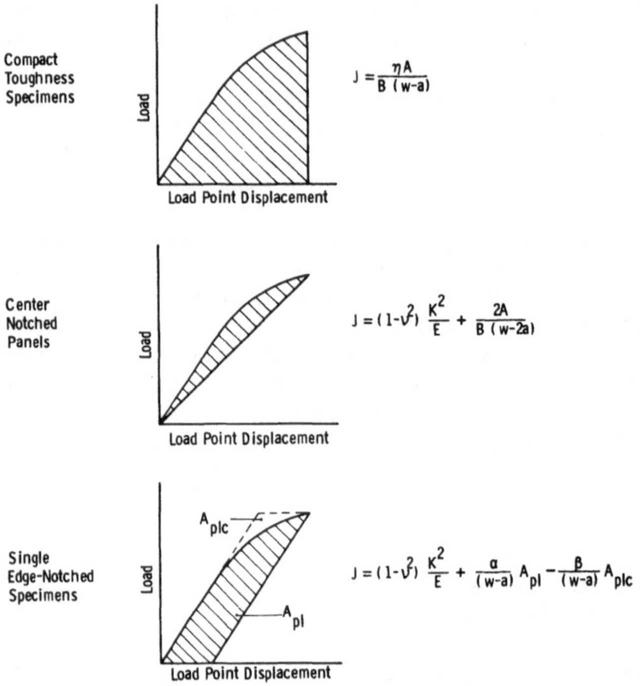

Fig. 3. Description of the graphical evaluation of J from load versus load-point displacement records for various specimen geometries.

tension and bend bars) and center-cracked panels. Consequently, in this investigation, it was necessary to generate the solution for calculating J from single-edge notched specimens. This solution was developed by Ernst and is described in detail in Ref. 5.

A description of the graphical evaluation of J from load vs. load-point displacement records for all three specimen types is presented in Fig. 3. For compact tension and bend bar specimens, the region near the crack tip experiences primarily bending, and the solution for J is given by

$$J = \frac{\eta A}{B(W-a)} \tag{1}$$

where A is the full area under the load vs. load-point displacement record and B, W, and a are the specimen thickness, width, and original crack size, respectively. The η factor, a function of the specimen (a/W), is a dimensionless coefficient that corrects for the tensile component of loading that occurs in a compact specimen. For a three-point-bend bar specimen, η equals 2.0.

The region surrounding the crack tip of a center-cracked panel experiences tensile loading. The solution for J is given by

$$J = (1 - \nu^2)\frac{K^2}{E} + \frac{2A}{B(W - 2a)} \qquad (2)$$

where ν is Poisson's ratio, E is Young's modulus, and K is the linear elastic stress intensity. The first quantity in the above formula is simply equivalent to Griffith's G (elastic energy release rate or crack-driving force). Note that A for the center-cracked panel is simply the area under the load versus load-point displacement record that falls above a line joining the initial and final displacement values.

Because the single-edge notched specimen experiences neither pure bending nor pure tensile loading but a combination of both (in particular with an a/W ratio equivalent to 0.5), the J solution for this specimen falls somewhere between that for a compact tension and a center-cracked panel specimen. The J solution is given by

$$J = (1 - \nu^2)\frac{K^2}{E} + \frac{\alpha}{(W - a)} A_{pl} - \frac{\beta}{(W - a)} A_{plc} \qquad (3)$$

where α and β are dimensionless functions of the specimen's a/W ratio. For an a/W ratio of 0.5, α and β equal 2.0 and 0.293, respectively. The plastic area under the load vs. load-point displacement record is A_{pl}, and the complementary plastic area is A_{plc}.

All of the single-edge notched specimens were loaded until the applied load dropped significantly from that observed at the maximum load point. In addition, each specimen was unloaded several times during the test (approximately ten percent of the applied load) in a fashion similar to that recommended in the single-specimen unloading compliance J test technique.[6] Expanded records (10X) were simultaneously made of each unloading. In every case, crack growth, as evidenced by a change in slope of the expanded unloading curves (i.e., a change in specimen compliance), was always first observed at the unloading immediately following the maximum load point. As a result, the maximum load point was used to calculate J for every single-edge notched specimen tested in this investigation.

RESULTS

The yield strengths and total elongations of the final three candidate conductor sheath materials are illustrated in Figures 4

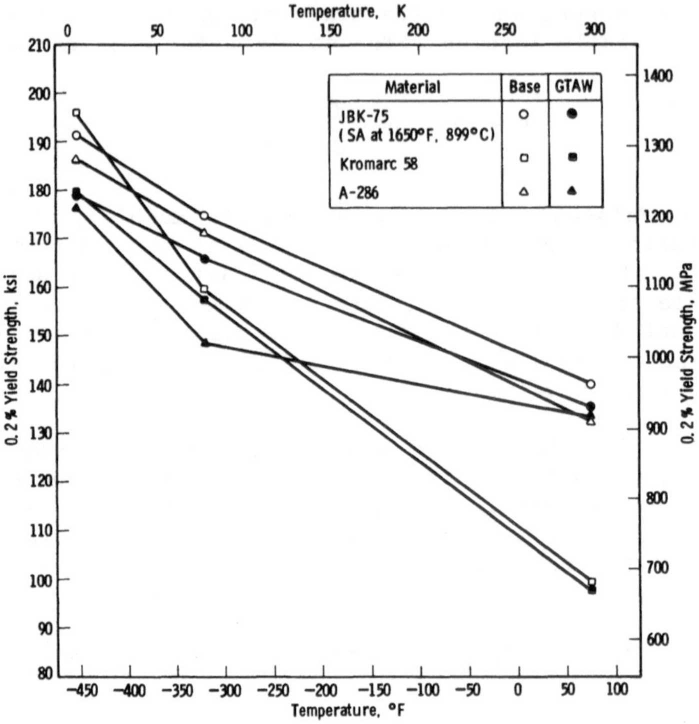

Fig. 4. Comparison of the 0.2% yield strength of fully processed JBK-75, Kromarc 58, and A-286 base plate and autogenous gas tungsten arc welds.

and 5, respectively, and the liquid helium temperature fracture toughness properties are summarized in Table 2. These results are discussed in detail in Ref. 7 and reviewed briefly in Ref. 1 (this volume).

None of the final three candidate structural materials stands out as being clearly superior, although the yield strength, total elongation, and fracture toughness properties of all three materials, both base plates and GTA welds, satisfied the proposed acceptance criteria. As a result, from a tensile and fracture toughness properties standpoint, all three alloys would be adequate structural materials for the conductor sheath application. In view of the difficulties anticipated with the procurement of Kromarc 58 strip, however, combined with the fact that the much tighter chemical composition limits of JBK-75 suggest a greater reliability to go with its intrinsically better weldability as compared with that of A-286, JBK-75 was selected as the LCP superconducting magnet conductor sheath material.

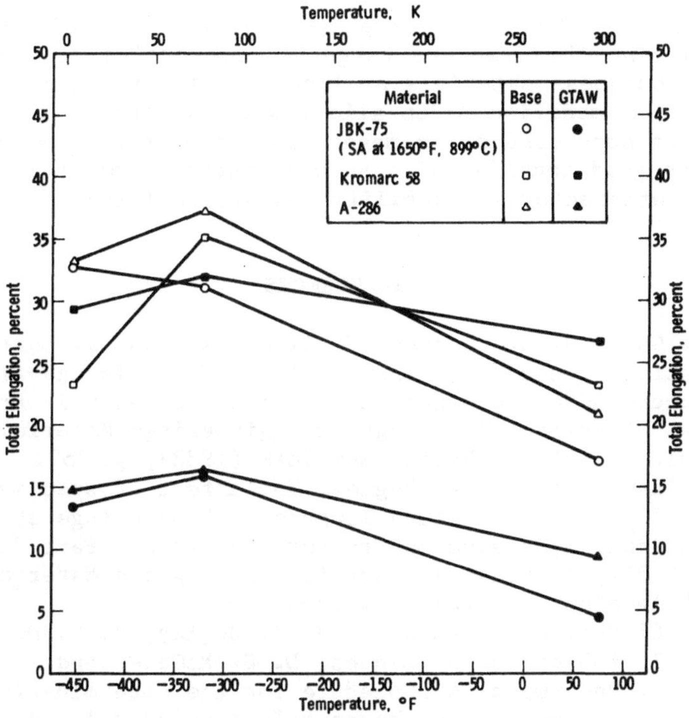

Fig. 5. Comparison of the total elongations of fully processed JBK-75, Kromarc 58, and A-286 base plate and autogenous gas tungsten arc welds.

Table 2. Fracture Toughness of the Final Three Candidate Conductor Sheath Materials in the Fully Processed Condition at 4 K (-452°F)

Material	Material Condition	Fracture Toughness MPa√m	Fracture Toughness ksi√in
A-286	Base	136.9	123.7
	GTAW	96.2	86.9
Kromarc 58	Base	102.1	92.2
	GTAW	145.1	131.1
JBK-75	Base	131.5	118.8
	GTAW	76.8	69.4

CONCLUSIONS

1. The single-edge notch bend bar specimen proved adequate for developing the fracture toughness properties of thin-section, high toughness materials at liquid helium temperature.
2. JBK-75, a modified A-286 alloy, was selected as the Large Coil Program superconducting magnet conductor sheath material owing to its good tensile and fracture toughness properties combined with satisfactory weldability and availability.

REFERENCES

1. R. E. Gold, W. A. Logsdon, G. E. Grotke, and B. Lustman, Evaluation of conductor sheath alloys for a forced-flow Nb_3Sn superconducting magnet coil for the Large Coil Program, in: "Advances in Cryogenic Engineering--Materials," Vol. 28, Plenum Press, New York (1982), p. 759.
2. J. D. Landes and J. A. Begley, The effect of specimen geometry on J_{Ic}, in: "Fracture Toughness, Proceedings of the 1971 National Symposium on Fracture Mechanics, Part II," ASTM STP 514, American Society for Testing and Materials, Philadelphia (1972), pp. 24-39.
3. G. A. Clarke, W. R. Andrews, J. A. Begley, J. K. Donald, G. T. Embley, J. D. Landes, D. E. McCabe, and J. H. Underwood, A procedure for the determination of ductile fracture toughness values using J integral techniques, J. Test. Eval. JTEVA 7(1):49-56 (1979).
4. W. F. Brown, Jr. and J. E. Srawley, "Plane Strain Crack Toughness Testing of High Strength Metallic Materials," ASTM STP 410, American Society for Testing and Materials, Philadelphia (1966).
5. H. A. Ernst, Unified solutions for J ranging continuously from pure bending to pure tension, presented at the Fourteenth National Symposium on Fracture Mechanics, Los Angeles, California (July 1981).
6. G. A. Clarke, W. R. Andrews, P. C. Paris, and D. W. Schmidt, Single specimen tests for J_{Ic} determination, in: "Mechanics of Crack Growth," ASTM STP 590, American Society for Testing and Materials, Philadelphia (1976), pp. 27-42.
7. W. A. Logsdon, R. E. Gold, B. Lustman, and G. E. Grotke, "Material Evaluation and Selection for Superconducting Magnet Sheathing--Large Coil Program," Westinghouse Electric Corporation, Pittsburgh, on Contract 22X-31747C (March 1981).

MECHANICAL PROPERTIES OF THE JAPANESE LCT COIL CONDUCTOR

K. Yoshida, Y. Takahashi, E. Tada, M. Shimada, and A. Tokuchi

Japan Atomic Energy Research Institute, Ibaraki, Japan

N. Tada

Hitachi, Ltd., Ibaraki, Japan

S. Shimamoto

Japan Atomic Energy Research Institute, Ibaraki, Japan

INTRODUCTION

Superconducting fusion magnets suffer high electromagnetic forces not experienced in magnets for other applications. High mechanical strength in a winding is generally inconsistent with stability. However, in fusion magnets, both high stability and high mechanical strength must be attained together with large current and high current density. This is one of the distinctive requirements in fusion magnet design.

Far fewer studies have been carried out on mechanical properties than on other aspects of conductor design. Therefore, the Japan Atomic Energy Research Institute (JAERI) has investigated the mechanical properties of the Japanese LCT conductor, which uses cold-worked oxygen-free copper as stabilizer. The details of the Japanese LCT coil itself and its mechanical design are described in previous papers.[1] This report describes the experimental data on the mechanical characteristics of the conductor and discusses the relation between stress and stability in a fusion magnet.

The purpose of the experimental investigation is to measure the magnetoresistivity, yield strength, and the stress-strain curve at 4 K of individual strands, the stranded cable alone, the formed copper stabilizer alone, and finally the full-size conductor.

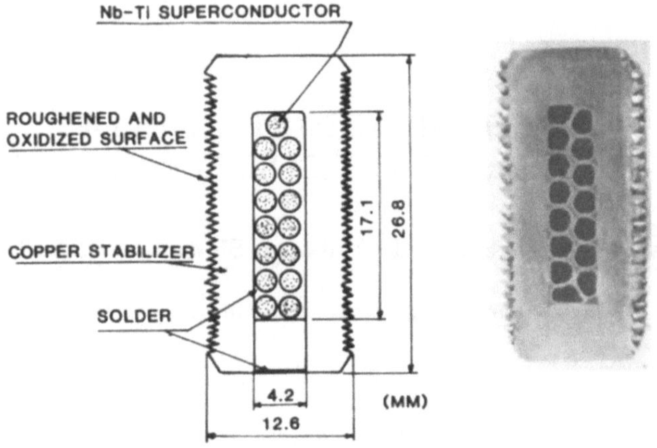

Fig. 1. The cross-sectional view of the LCT 8 T conductor.

Table 1. Characteristics of the Japanese LCT Conductor.

Parameter	8 T Conductor	5 T Conductor
Conductor size (mm)	12.6 x 26.8	12.6 x 21.3
Normal current (kA)	10.22	10.22
Critical current (kA)	19.9	19.9
Yield strength (MPa)	265	265
Heat flux (W/cm^2)	0.85	0.90
Superconductor	Nb - 46.5% Ti	
Cooling surface	Roughened and oxidized surface	
Copper stabilizer		
Resistivity at 4.2 K (Ω•cm)	5.1×10^{-8}	3.9×10^{-8}
Yield strength (MPa)	333	333
Stranded cable		
Cable size (mm)	4.3 x 17.1	3.3 x 9.8
Diameter of strand (mm)	2.3	1.75
Number of strands	15	11
Copper/Nb-Ti ratio	1.0	1.0
Twist pitch of filament (mm)	30	25
Twist pitch of strand (mm)	150	90

THE JAPANESE LCT CONDUCTOR

In selecting the Japanese LCT conductor, shown in Fig. 1, the constraint of pool cooling was adopted at the outset. The winding has two grades of conductor, 8 T and 5 T, designed as shown in Table 1. A feature of the conductor is high heat flux from the roughened and oxidized surface. The oxygen-free, high-purity copper stabilizer is strengthened through cold-work reduction so as to sustain a part of the electromagnetic force. The final reduction rate of copper was chosen on the basis of verification tests described in the following section.

4 K TENSILE TEST FACILITY

High tension at 4 K is required to perform mechanical tests on a high current conductor. For this purpose a specially designed facility of 100 kN (10 ton) capacity was installed at JAERI. Figure 2(a) shows an arrangement for testing one nonstandard sample, such as the LCT conductor. Instead of this single-sample system, a multisample system, which is shown in Fig. 2(b), can be inserted in the cryostat.

In this multisample system, ten samples, which have JIS (Japan Industrial Standard) standard test dimensions, can be simultaneously tested up to 100 kN with one charge of liquid helium. When a number of tests are performed in a short time, the helium consumption is about 10 ℓ/sample. Strain is measured with a strain gauge or a clip-on gauge directly attached to a sample.

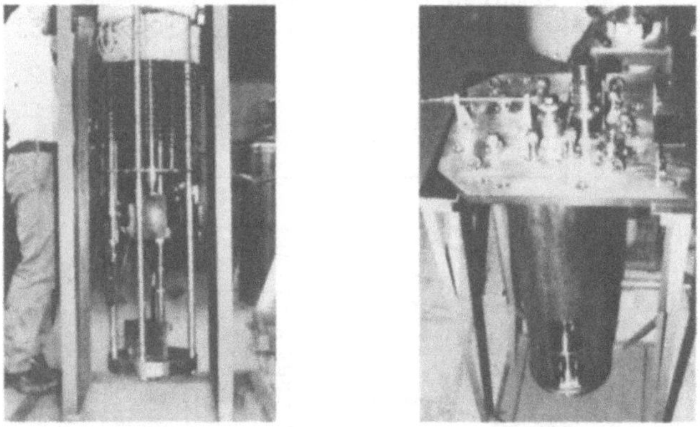

(a) (b)

Fig. 2. 4 K tensile test facility: (a) a single test system; (b) a multisample test system.

TEST RESULTS

The conductor for the Japanese LCT coil consists of a copper stabilizer and a stranded cable composed of strands and solder. The mechanical test results of each conductor component are described in this section.

Copper Stabilizer

Because the copper stabilizer of the LCT coil has to withstand large magnetic forces at 4 K, a hard copper, strengthened by cold reduction, must be used for the stabilizer. When hard copper is used, the following problems arise:

1. the cold work may not be uniformly distributed
2. the resistivity increases
3. the workability for winding decreases.

To find the optimum cold reduction, the mechanical properties, resistivities, and hardness profile inside the conductor were measured as functions of cold reduction.

Tensile tests were carried out at 4 K for various cold reductions to find the optimum cold reduction and the effect of cold reduction. Small samples (0.5-mm diam.) and larger samples (10-mm diam.) cut from the full-size LCT conductor were employed for the testing. Figure 3 shows the yield strength and the magnetic resistivity of both samples at 4 K as functions of cold reduction. The yield strengths of the two samples do not match. The differences are caused by an effect of the cold reduction on the size. The differences increase with increasing cold reduction and reach their maximum at around 12% cold reduction. At 26% cold reduction, the two samples have almost the same yield strength. The results show that cold reduction from 10% to 15% has a large size effect.

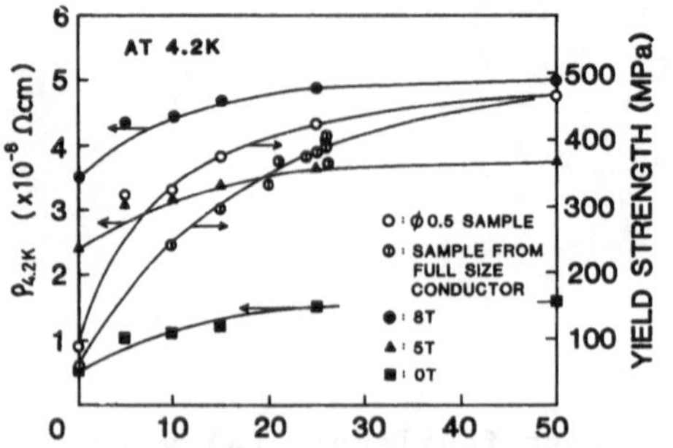

Fig. 3. Yield strength and magnetoresistivity of copper as a function of cold reduction at 4 K.

Table 2. Vickers Hardness Results of 21% Cold Reduction Copper at Room Temperature.

Position	Sample	
	8 T Conductor	5 T Conductor
1,6	97.8	97.1
2,7	93.9	91.4
3,8	90.2	89.6
4,9	91.8	84.2
5,10	92.2	95.4
11,12,13	92.4	92.4
14,19	97.8	95.7
15,20	91.7	93.5
16,21	92.2	92.2
17,22	90.8	92.3
18,23	92.0	94.5
24,25,26,27	95.2	95.2
Average value	93.2	92.8

The Vickers microhardness test was carried out to check the uniformity of the reduction inside the conductor. From the tensile tests, 21% and 26% hard copper were selected for the hardness test. The hardness test results are shown in Table 2 for 21% hard copper. Figure 4 shows the points at which the Vickers hardness was measured. In both hard coppers, edge positions (such as nos. 1 and 2 in Fig. 3) show high Vickers hardness in comparison with center positions (such as nos. 3 and 8). The 26% hard copper has a more uniform profile of the hardness than 21% hard copper, but even in 21% hard copper, the difference of the hardness between the positions is not much. The authors measured Vickers hardness in other samples from 7% to 27% reductions. The results showed explicitly that lower reduction created a nonuniform hardness inside the samples. This effect of the reduction explains clearly the size effect of the yield strength mentioned above.

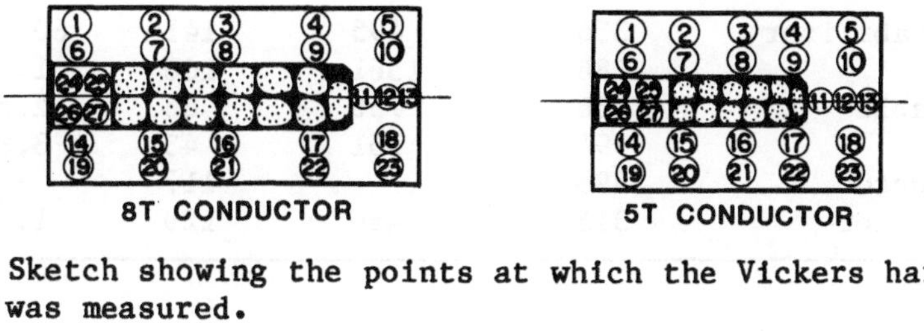

Fig. 4. Sketch showing the points at which the Vickers hardness was measured.

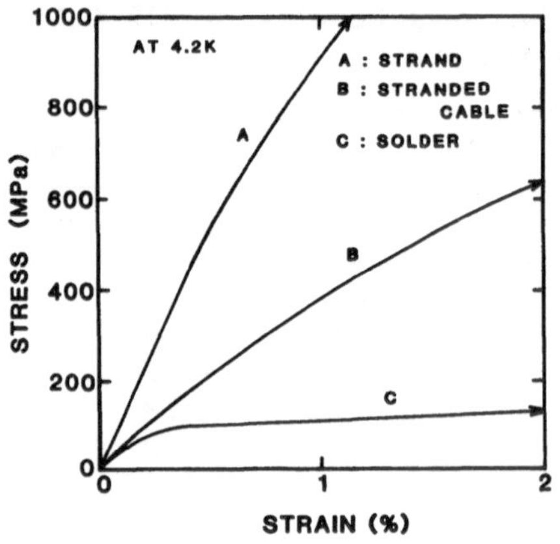

Fig. 5. Stress-strain curves of strand, stranded cable, and solder at 4 K.

The magnetoresistivity of the same tensile samples was measured at 8.5 and 0 T in liquid helium. The results are shown in Fig. 4. The resistivity and the residual resistance ratio of the copper stabilizer of 0% cold reduction are 0.5×10^{-8} ohm·cm and 320 at 0 T, respectively. The resistivity of 21% hard copper is around 5.0×10^{-8} ohm·cm. The stability analysis shows that the Japanese LCT coil is cryostable with this resistivity.

Strand and Solder

A strand consists of 1060 filaments of Nb-46%Ti in a copper matrix. The tensile test results for a single strand are shown in Fig. 5 and Table 3. The ultimate and the yield strengths of a strand are 1561 MPa and 786 MPa at 4 K, respectively. The strain

Table 3. Tensile Test Results of the Conductor Elements at 4 K.

Sample	0.2% Yield Strength, MPa	Ultimate Strength, MPa	Young's Modulus, GPa	Elongation, % (mm)
Copper stabilizer	356	495	141	60.5 (GL=18)
Strand	786	1561	108	1.1 (GL=100)
Strand cable	166	756	66	2.5 (GL=100)
Solder	94	151	43	3.8 (GL=20)
8 T conductor	295	-	117	-
5 T conductor	313	466	129	1.4 (GL=300)

to failure is around 1.1% at 4 K. The Young's modulus calculated by the law of mixtures is 108 GPa.

Low-melting Pb-60%Sn solder was selected to avoid high temperature when the Nb-Ti strands were soldered. Figure 5 and Table 3 show the tensile test results of the solder at 4 K. The yield strength and ultimate strength are 94 MPa and 151 MPa at 4 K, respectively. The bond shear strength of the solder was measured at 4 K. The results show that the bond shear strength between the copper stabilizer and the stranded cable is 48 MPa.

Stranded Cable

The stranded cable of the 8 T conductor consists of 15 soldered strands with a twist pitch of 150 mm. The tensile test results are shown in Figs. 5 and 6. The Young's modulus of the stranded cable is 66 GPa at 4 K, as shown in Table 3. The yield and ultimate strengths and the elongation are 166 MPa, 756 MPa, and 2.5% at 4 K, respectively. The Young's modulus of the stranded cable can be calculated using the law of mixtures as follows:

$$E = (X_1 \times E_1 + X_2 \times E_2)/(X_1 + X_2)$$

where X_1 is the volume of the strand, X_2 is the volume of solder, E_1 is Young's modulus of the strand, and E_2 is Young's modulus of the solder. From the above equation, Young's modulus can be calculated to be 101 GPa. Thus, the law of mixtures cannot be applied to the soldered-twisted strands. That is why a stranded cable has a different Young's modulus from a strand due to a twist effect.

Full-Size Conductor

The tensile test of a full-size conductor, which consists of a stranded cable and a copper stabilizer, was carried out at 4 K. The stress-strain curve of the 8 T conductor is shown in Fig. 6. The Young's modulus and the yield strength are 117 GPa and 295 MPa, respectively, at 4 K, as shown in Table 3. The Young's modulus of the full-size conductor is calculated to be 120 GPa by using the law of mixtures between the stranded cable and the copper stabilizer. The calculated value is in good agreement with the experimental results.

DISCUSSION

On the basis of the mechanical tests, we found that the law of mixtures can be applied to a large-scale conductor consisting of copper stabilizer and stranded cable. Yield strength of the conductor can be calculated by using the law as follows:

$$(\sigma_y)_{con} = f_{co} \times (\sigma_y) + f_{st} \times (\sigma_y)$$

$$f_{co} = A_{co}/A_{tot}, \; f_{st} = A_{st}/A_{tot}$$

$$A_{tot} = A_{co} + A_{st}$$

where σ_y is yield strength, A is cross-sectional area, co is copper stabilizer, st is stranded cable, and con is conductor. Generally, the second term of the right side in this equation is very small, so the yield strength of the conductor can be evaluated from the yield strength of the copper stabilizer. For example, in the case of the Japanese LCT coil, this equation is as follows:

$$(\sigma_y)_{con} = 0.71 \times (\sigma_y)_{co} + 43.4 \quad \text{(MPa)}$$

Therefore, the yield strength of the conductor strongly depends on how much cold reduction is applied to the copper stabilizer.

On the other hand, the cold reduction increases the residual resistivity of the copper stabilizer so that the stability margin decreases. To evaluate stability of the conductor, the stability factor should be considered. The stability factor is heat generation to cooling capacity ratio and can be described as follows:

$$\alpha = \rho I^2/PA\beta q_e$$

where α is stability factor, ρ is magnetoresistivity, I is operating current, P is perimeter, β is cooling efficiency, q_e is effective heat flux, and A is cross-sectional area of the stabilizer. In this equation, magnetoresistivity should be used as the combined resistivity of the copper stabilizer and copper matrix of a strand. The stability factor is in proportion to the resistivity of the copper stabilizer, which strongly depends on the cold reduction.

From the above discussion, both yield strength and stability of the large-scale conductor can be evaluated by cold reduction of copper stabilizer. Mechanical test results reported in this paper clarify the relation between yield strength and cold reduction and between stability and cold reduction.

Figure 7 shows the yield strength and the stability factor as a function of the cold reduction for the Japanese LCT conductor. In Fig. 7, 16% cold reduction satisfies the requirements of strength and stability. Finally, 21% hard copper was selected as the copper stabilizer of the Japanese LCT coil because of its consistency of both stability and strength. In this way, we can get the best reduction of the copper stabilizer when a large superconducting magnet is designed for a fusion reactor.

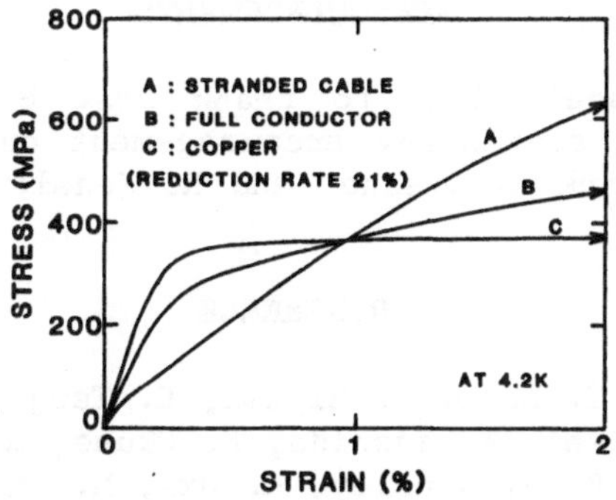

Fig. 6. Stress-strain curves of stranded cable, copper stabilizer, and full-size conductor of 8 T.

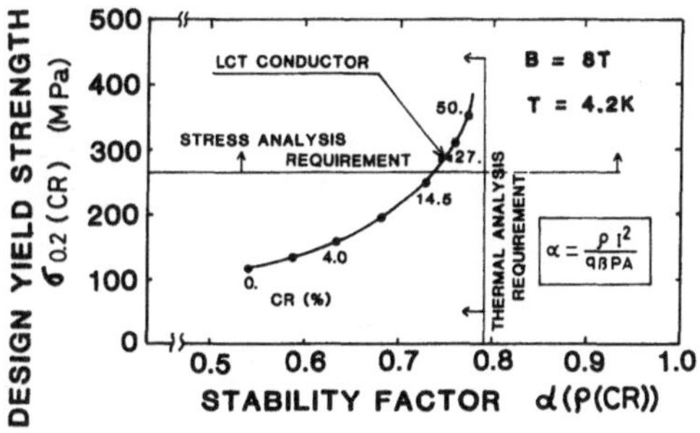

Fig. 7. Relation between yield strength and stability factor of the Japanese LCT coil.

CONCLUSIONS

On the basis of the various mechanical studies, we drew the following conclusions:

1. The law of mixtures can be applied to a large conductor, which consists of stranded cable and copper stabilizer.
2. Cold reduction increases the yield strength of oxygen-free copper. For example, 50% reduction makes the yield strength higher by 7 times and the resistivity higher by 1.5 times at 4 K and 8 T in comparison with that of no reduction.
3. The 21% hard copper was selected as the copper stabilizer for the Japanese LCT coil because of its consistency of both stability and strength. Its Young's modulus and the yield strength are 141 GPa and 365 MPa, respectively, at 4 K.

ACKNOWLEDGMENTS

The authors would like to thank Drs. S. Mori, Y. Iso, and Y. Obata for their continuing encouragement on this project. The authors wish to thank L. Dresner and R. Kensley for their valuable comments.

REFERENCE

1. S. Shimamoto, T. Ando, T. Hiyama, H. Tsuji, Y. Takahashi, E. Tada, M. Nishi, K. Yoshida, K. Okuno, K. Koizumi, T. Kato, K. Yasukochi, R. S. Kensley, K. Oka, M. Shimada, Y. Sanada, Y. Ibaraki, Evolution of the Japanese test coil work for the large coil task," IEEE Trans. Magn. MAG-17:1734 (1981).

THE FABRICATION AND PROPERTIES OF MULTIFILAMENTARY Nb_3Sn SUPERCONDUCTORS BY THE SOLID-LIQUID DIFFUSION METHOD

M. Nagata, S. Okuda, M. Kawashima, M. Yokota, and K. Ohkura

Sumitomo Electric Industries, Ltd., Osaka, Japan

M. Watanabe, Y. Kimura, M. Umeda, H. Yamasaki, and Y. Akiyama

Electrotechnical Laboratory, MITI, Ibaraki, Japan

INTRODUCTION

There are a number of production processes for the fabrication of multifilamentary Nb_3Sn superconductors. Most of them use bronze with an Sn concentration of near 13 wt.%. Many efforts have been made to increase the Sn concentration in order to increase the overall critical current density of superconductors. Essentially, most of the processes use solid-solid diffusion between Nb and solid bronze in the final heat treatment for forming Nb_3Sn. We, however, succeeded in manufacturing a Nb_3Sn superconductor using solid Nb and liquid Sn-Cu alloy.[1]

In this paper we describe the manufacturing process and properties of Nb_3Sn superconductors produced by this solid-liquid diffusion method.

MANUFACTURING PROCESS OF Nb_3Sn SUPERCONDUCTORS

Cross sections of two types of Nb_3Sn superconductors are shown in Fig. 1. The basic production process is as follows:

Niobium composite wires with a single core of Sn-7 wt.% Cu are bundled in a Cu tube and drawn to the final size. The Nb wires are strongly connected only by the area reduction of cold drawing, and annealing is not necessary. But the drawability is limited by the difference of mechanical properties between the Nb and the Sn-Cu core, as shown by the hardness in Fig. 2.

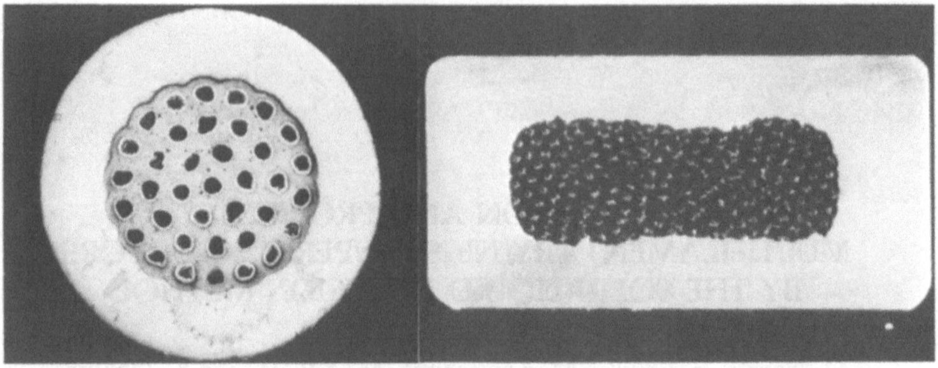

(a) Small-diameter wire (1.64ϕ)

(b) Rectangular wire (2.5 mm (2.5 mm x 5.0 mm)

Fig. 1. Cross sections of multifilamentary Nb_3Sn superconductors made by the solid-liquid diffusion method. The outer shell of each conductor is pure Cu, which acts as a stabilizer, and the matrix is Nb. Nb_3Sn superconductors are formed between the Sn-Cu cores and Nb matrix. After reaction, the cores almost become voids.

We succeeded in producing Nb_3Sn superconductors by choosing an appropriate volume fraction between Nb and Sn-Cu cores and then forming an Nb matrix in the center of the conductor. If the Nb/Sn-Cu ratio becomes larger, the drawability increases, but the overall current density (excluding Cu) decreases.[2] In this configuration, leakage of liquid Sn from the Nb matrix can be prevented, and the bending strain becomes smaller than that of superconductors having a distributed configuration of Nb wires in a Cu matrix.

Finally, the wire is heat-treated for Nb_3Sn formation at around 700°C. The Cu tube outside becomes a stabilizer.

The matrix is Nb with filaments of Sn-7 wt.% Cu; this configuration is the same as an internal bronze process. The Sn-rich Cu alloy cores change to the liquid state at the heat-treatment temperature (approximately 700°C) for formation of Nb_3Sn.

It was commonly believed that at such temperatures A15 Nb_3Sn cannot be formed by diffusion between such an Sn-rich Cu alloy and Nb. Experiments on the diffusion reaction between Nb and liquid Sn-rich Cu alloy revealed that intermetallic compounds of Nb_6Sn_5 and $NbSn_2$ are formed at 700°C. However, when the Sn-Cu core was smaller, the Sn-rich Nb compounds formed in the first stages of heat treatment change to Nb_3Sn within a short period, as shown in

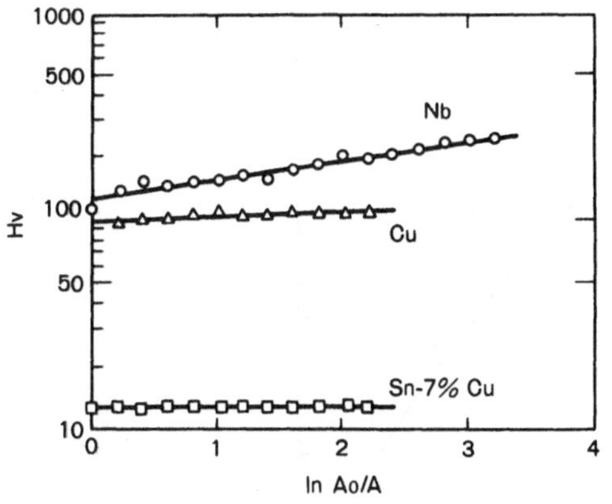

Fig. 2. Change in hardness with area reduction by cold drawing.

Fig. 3. To obtain high performance of Nb_3Sn superconductors, it is essential to decrease the Sn-Cu core diameter to less than 50 µm. The details of this diffusion mechanism will be published elsewhere.[3]

THE PERFORMANCE OF Nb_3Sn SUPERCONDUCTORS

Performance of Short Samples

The newly developed multifilamentary Nb_3Sn superconductors have high current density, as shown in Fig. 4. Critical current density throughout the Nb matrix exceeds 1.5×10^5 A/cm^2 at 10 T.

In the solid-liquid diffusion method, the layer thickness of Nb_3Sn is proportional to the Sn-Cu core diameter, so the overall current density of various sized superconductors is almost constant

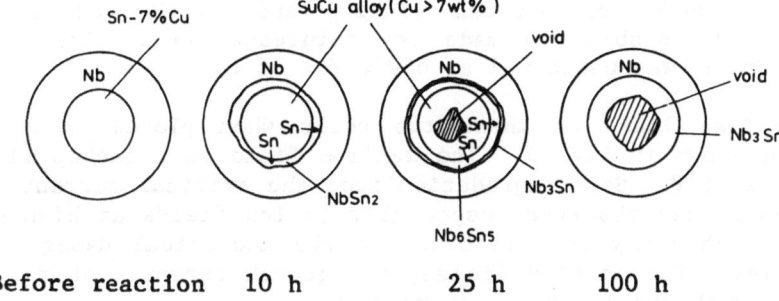

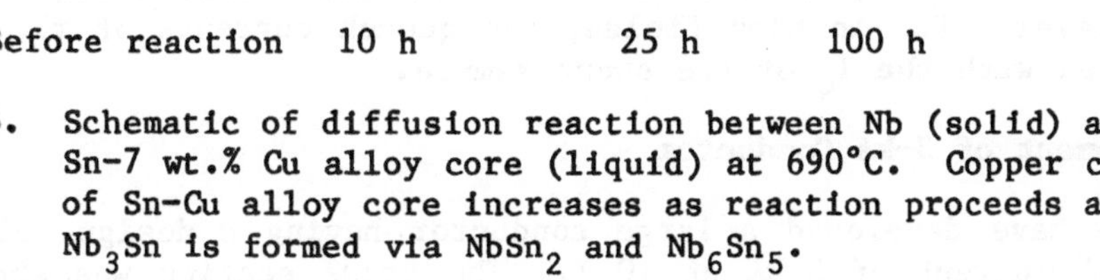

Fig. 3. Schematic of diffusion reaction between Nb (solid) and Sn-7 wt.% Cu alloy core (liquid) at 690°C. Copper content of Sn-Cu alloy core increases as reaction proceeds and Nb_3Sn is formed via $NbSn_2$ and Nb_6Sn_5.

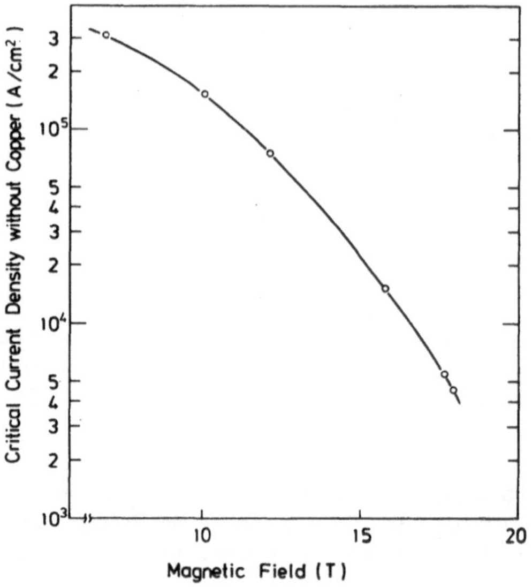

Fig. 4. Overall critical current density (excluding Cu) vs. magnetic field.

after the diffusion reaction has gone as far as possible. The heat treatment for Nb_3Sn formation was done at 690°C for 72 h.

Performance of Small Coils

To investigate the critical current properties of long samples, we made two small coils from 20-m and 100-m lengths of our Nb_3Sn superconductor. The specifications of the two coils are shown in Table 1. Both coils were wound and heat-treated, then impregnated with epoxy. Insulation of the winding was E glass yarn, and insulation between winding and bobbin was a silica ceramic. The bobbin was made of stainless steel. The reaction heat treatment was performed at 690°C for 75 h.

The load lines of these two coils when placed in a backup field are shown in Fig. 5. The maximum field in a backup field of 10 T was 11.6 T. Some degradation from the critical current of the short sample was observed, especially in low fields at high quench currents. This may be attributed to the mechanical damage of the Nb_3Sn layer. But in high fields, the quench currents of the coils coincided with the I_c of the short sample.

Development of 3-kA Conductor

We have developed a large conductor having a design value of critical current of 3 kA at 10 T. The cross section was shown in

Table 1. Nb_3Sn Coil Specifications

Parameter	Coil 1	Coil 2
Outer diameter	40 mm	80 mm
Inner diameter	20 mm	40 mm
Coil length	60 mm	90 mm
Number of turns	197	605
Inductance	0.442 mH	7.75 mH
Conductor length	20 m	110 m
Packing factor	0.67	0.67
Specifications	Wire diam.	1.64 mm
of Nb_3Sn	Sn-Cu core	40 μm
Superconductor	No. of filaments	49
	Cu to SC ratio	2.5

Fig. 1(b). The conductor was produced by the same process as the small conductors, except that the outer diameter of the initial Cu tube was larger.

A critical current measurement was conducted on a 1.5-m-long sample on a bakelite holder, as shown in Fig. 6, in a 10-T backup field. The measured value of critical current was 3500 A at 10 T, and overall critical current density (excluding Cu) was 1.1×10^5 A/cm^2. This current density was somewhat lower than that of the small superconductor shown in Fig. 1(a), because the rectangular shape of the conductor decreases the current density.

CONCLUSION

We have succeeded in manufacturing a Nb_3Sn superconductor using 93 wt.% Sn - 7 wt.% Cu alloy cores surrounded by Nb. In this case, the alloy became liquid at the heat-treatment temperature (690°C). Tin-rich Nb compounds formed initially during heat treatment at 690°C were rapidly transformed into an A15 Nb_3Sn superconductor in 50 - 100 h. The multifilamentary Nb_3Sn superconductors developed have high current density (J_c overall: 1.5×10^5 A/cm^2, 10 T).

This superconductor was made into a small coil and generated 11.6 T with a backup field of 10 T. An effort was made to enlarge the conductor size; a rectangular conductor of 2.5 mm x 5.0 mm could carry 3500 A at 10 T. We are now constructing a 12-T large-scale magnet (inner diam.: 400 mm, outer diam.: 685 mm) with this conductor, using a 7-T backup field.

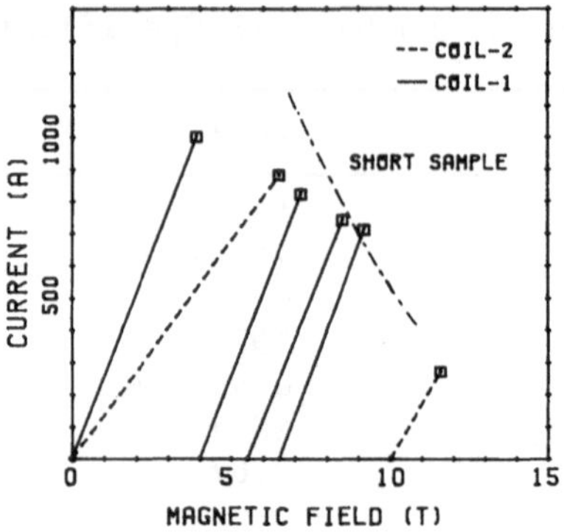

Fig. 5. Load lines of two Nb_3Sn coils.

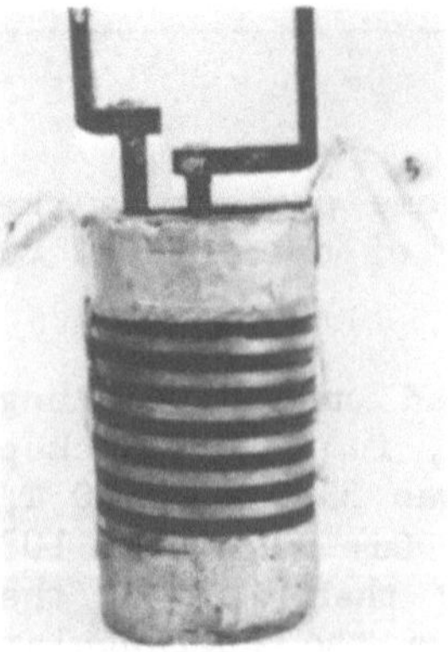

Fig. 6. Sample holder for measuring critical current of rectangular conductor in 10 T. Outer diameter of holder is 60 mm.

ACKNOWLEDGMENTS

We wish to thank Dr. Y. Aiyama and Dr. T. Yasui for their continuous encouragement of this development and Dr. R. Akihama for high field measurement of I_c at MIT. The authors are grateful to Mr. S. Isojima and Mr. H. Takei for technical discussions.

REFERENCES

1. S. Okuda, M. Nagata, M. Yokota, M. Watanabe, and Y. Kimura, Filamentary Nb_3Sn superconductor manufactured by the solid-liquid diffusion method, in: "Filamentary A15 Superconductors," M. Suenaga and A. F. Clark, eds., Plenum Press, New York (1980), p. 81.
2. B. Avitzur, Wire J., August:42 (1970).
3. H. Yamasaki, Y. Kumura, to be published.
4. High field data was obtained by Dr. R. Akihama at Francis Bitter National Magnet Laboratory, MIT.

LONG-LENGTH Nb_3Sn TUBULAR CURRENT-CARRYING ELEMENTS FOR AN AC SUPERCONDUCTING CABLE

V. M. Pan, G. A. Klimenko, Yu. I. Beletsky, and V. S. Flis

Institute of Metal Physics, Kiev, USSR

INTRODUCTION

Investigations carried out in our laboratory[1] have shown that it is possible, in principle, to produce superconducting Nb_3Sn diffusion claddings able to transfer ultrahigh transport currents (700 A/mm in the self-field at 4.2 K) at a low loss level in the superconductor (3-5 $\mu V/cm^2$ at a surface current of 500 rms A/cm, frequency of 50 Hz, and 4.2 K). On the basis of experimental data on critical current and ac loss measurements as a function of conditions under which Nb_3Sn is formed, the technology has been developed to deposit diffusion claddings of Nb_3Sn with high superconducting characteristics on tubular surfaces. The developed technology may be used to produce current-carrying elements of a rigid superconducting cable used for electric power transmission lines, superconducting shields, and so on.

Earlier, only one method of tubular element production with superconducting Nb_3Sn was described. This method had been primarily meant for production of long superconducting cylindrical shields.[2] The idea was as follows: Niobium ribbon clad with Nb_3Sn is soldered with mild tin-lead solder copper tubes. If the ribbon should be soldered to the internal surface of the copper tube, the latter must be cut along the forming line. After the ribbon is soldered, the semicylinders are again connected by means of soldering or a special bandage.

This method of producing tubular components with Nb_3Sn layers suffers essential disadvantages: first of all, normal solder and possible gaps cause additional thermal and electrical losses. Also, one cannot lower ac losses by improving surface quality.

We propose a method to prepare tubular components with Nb_3Sn cladding in the following way: Nb_3Sn layer is formed directly on a ready composite (Cu-Nb) tube by means of a well-known diffusion process.

PRODUCTION OF BIMETALLIC TUBES

Long Cu-Nb tubes 76 and 51 mm in diameter and 2.5- to 3.0-mm thick were prepared according to the scheme: welding by explosion of copper and niobium tubular ingots 0.7-m long, hot pressing with the extrusion more than ten times, and cold rolling.

At explosion welding for the production of the 76-mm-diameter tube, the method of internal plating was used (the Nb plating layer undergoes tension), whereas in the case of the 51-mm-diameter tube, the external plating was applied (the plating layer is compressed). Therefore in the latter case Nb tubular ingots were used (110 mm in diam., 2-mm thick), that were produced by rolling of annealed ribbons and their welding along the forming line by the arc method in a controlled atmosphere. Welding conditions are described.[3] Niobium tubes 100 mm in diameter and 4-mm thick meant for the production of the tube 76 mm in diameter were made by means of hot sewing and extrusion from ingots 120 mm in diameter. To improve the conditions under which Nb and Cu flow simultaneously at hot pressing, one must consider the thickness ratio of initial tubes. If the Nb initial tube has small thickness, continuity of the Nb layer may not be attained. That is why two Nb initial tubes 2-mm thick were used to prepare the tube 51 mm in diameter. During composition, the welding joints were unfolded by an angle of more than 30°. Heating under pressing was done in inductors. Pressing was followed by annealing in the vacuum furnace at 900°C for 1 h. Cold rolling on the installation for the tube cold rolling was performed according to the route: tube 76 mm in diameter, 96 x 8.5 - 76 x 2.5; tube 51 mm in diameter, 83 x 5.7 - 63 x 4.0 - 51 x 2.5.

DEPOSITION OF Nb_3Sn LAYER ON LONG TUBES

To deposit the covering layer on a niobium surface of composite (Cu-Nb) tubes, the tubes must be compiled into complexes, that is, into welded constructions that consist of a composite tube with a Nb cylindrical container welded to its Nb end. The container serves as a sink for bronze, which did not take part in the reaction. After tin and copper had been inserted into this tube, the choke with a Nb stopgap was welded to the free edge of the tube. The stopgap lets the inert gas into the complex. Before welding the complex, the tube was thoroughly degreased, and its edges were deprived of copper for a length of 30 mm; this was done in two stages: In the first stage, a large amount of copper was mechanically removed. In the second stage, the rest of copper was etched away by chemical or electrochemical methods. The complex was welded-up in a special installation in a vacuum with the aid of electron beam.

The welding installation (Fig. 1) is a vacuum cylindrical chamber where sluice chambers horizontally arranged make it possible to weld tubes up to 120 mm in diameter and 14-m long. Evacuation is accomplished through carbon sorption pumps cooled with liquid nitrogen and through a cryosorption panel cooled with liquid helium. Dynamic vacuum during welding is 10^{-7} torr. Sluice chambers are 150 mm in diameter. The electron gun is firmly fixed on the welding chamber. The electronic welding device may work in both the continuous and impulse modes. The power of the electron beam varies in the range 0 to 13 kW. Accelerating voltage may be from 0 to 30 kV. The electron gun is supplied with electromagnetic deviating and focusing systems. The extreme angle of the beam deviation is 11°. To weld ring joints in the tubes, one must rotate the tubes during welding. For this purpose, the sluice chamber contains a vacuum inlet for rotation. To change the rate the rotation driver is used with a smooth regulation.

The complex was thermally treated in a specially designed furnace (Fig. 2). Its container serves as a heater and is a stainless steel tube 150 mm in diameter with a wall thickness of 1.5 mm. The full length of the sectioned heater is 10 m (one section is 2.5-m long). Each section has its own power supply and autonomic regulation. Full length of the furnace, including 0.7-m-long end sleeves (one of them contains the rotation inlet), is 11.4 m. This furnace design can anneal items up to 7.5-mm long and

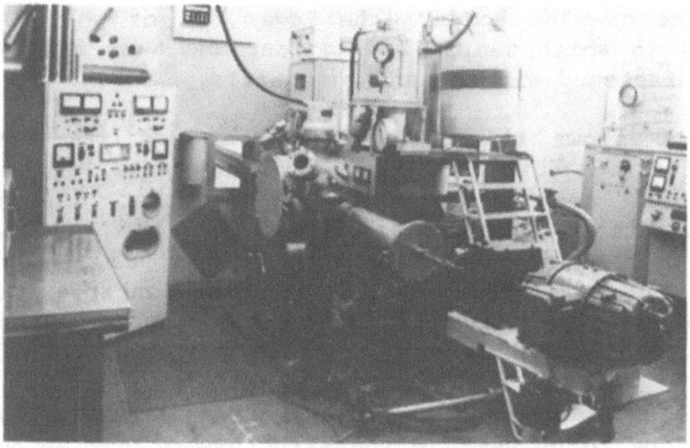

Fig. 1. Vacuum chamber performed technological welding on long composite tubes.

Fig. 2. Furnace for the thermotreatment of long composite tubes.

ensure uniform temperatures throughout the entire furnace length as precisely as ±15°C. The furnace is placed on the holder, which permits it to tilt from a horizontal into an oblique (30°) position. The furnace must be tilted to let melted bronze Sn+30%Cu flow into the container.

EXPERIMENTAL RESULTS

Critical current in samples cut from composite tubes was measured in parallel and perpendicular external magnetic fields up to 6 T created by superconducting solenoids possessing internal holes of 20 mm and 40 mm, respectively. The method of measurement is described. Measurements in the perpendicular field were done also with the coaxial holder, the lower end of which had massive copper shoes to which samples 38-mm long and 4-mm wide were either soldered or fastened with pressing contacts.

Figure 3 shows the critical current and its density as a function of time duration of diffusion annealing for samples treated at 850°C and 900°C. It may be seen that, depending on the annealing duration, the $I_c(\tau)$ curves give the critical current maximum in both cases--annealing at 900°C and 850°C. For samples annealed at 850°C, the maximum is reached at longer exposure and is somewhat blurred in the $I_c(\tau)$ curve as compared with that for samples annealed at 900°C.

Such changes of critical current with the annealing duration are perhaps due to changes in the structure of the Nb_3Sn diffusion layer. As the annealing duration increases, the grains of the Nb_3Sn phase increase from 0.7-2 μm (annealing at 900°C for 5 h) to

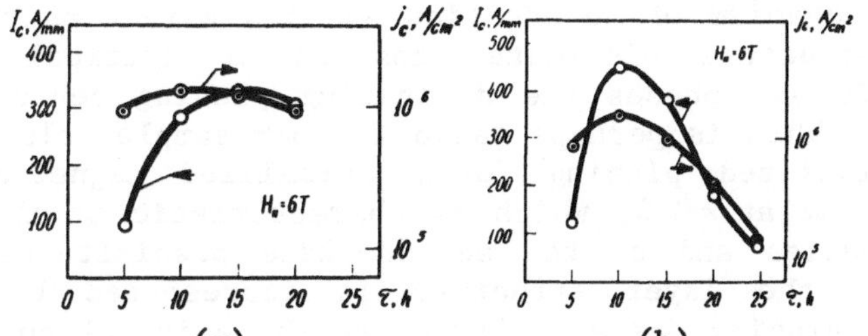

Fig. 3. The critical current and the critical current density vs. annealing time: (a) T = 850°C; (b) T = 900°C.

1.5–6 μm (annealing at 900°C for 24 h). It may be seen from Fig. 4 that the average grain size varies weakly with the annealing duration increase. Only separate grains grow and, with long enough annealing, penetrate through the whole depth of Nb_3Sn layer and thus form weak places that limit the critical current.

The appearance of the Nb_3Sn diffusion layer with relatively small grains may be connected with the fact that the bath melt is alloyed with Cu. Copper additions to Sn are known to help in formation of fine-grain structure, which in turn, leads to higher density of the critical current.[4,5] Thus it has been shown[5] that the addition of 10–15 wt.% Cu to the tin bath decreased the average size of Nb_3Sn grains at annealing 1 h at 950°C from 4.6 μm (for pure Sn) to 2.6–2.2 μm.

Such an effect of Cu on Nb_3Sn grain size is perhaps due to the fact that Cu is a surface-active addition that lowers the surface energy (σ_{LS}, in the interface melt-solid Nb and thus favours formation of Nb_3Sn nuclei. Besides copper being a surface-active addition, it occupies boundaries of Nb_3Sn grains as disperse precipitation of the phase rich in Cu. These precipitates hinder further growth of Nb_3Sn grains.

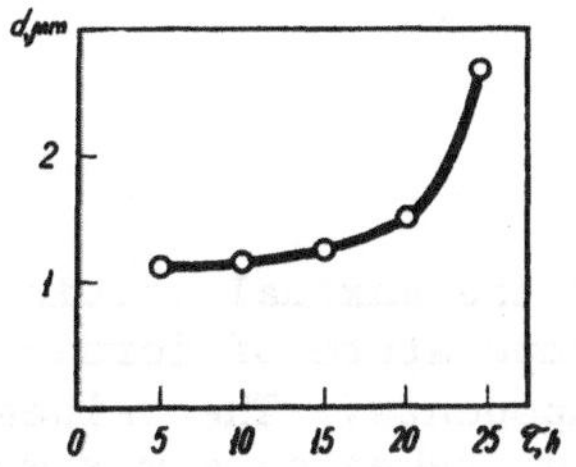

Fig. 4. Variation of the mean value of the grain size with the annealing time (T = 900°C).

This mechanism of Cu addition and its effect on the formation of Nb_3Sn structure elucidates changes of critical currents in samples. In A15 phases the dominating pinning centers are grain boundaries. This is perhaps valid for our samples also because the curves "normalized pinning force--normalized magnetic induction" have a maximum at ~0.3, which is characteristic of the pinning on grain boundaries and on the second-phase precipitates. In other words, when the layer structure is represented by fine Nb_3Sn grains, the pinning force is large and the critical current density remains practically unchanged (the smaller value of the critical current density for samples annealed during 5 h may be connected with the fact that during calculation we considered full layer thickness, including also the low effective subsurface layer of coarse crystals, which is about one-third of the overall thickness). In this case the increase in the layer thickness leads to higher integral critical current. With the growth of annealing duration, (τ = 15–24 h) the compiling recrystallization occurs, resulting in coarser grains. This, in its turn, leads to lower critical current density, which cannot be compensated by the increase in the diffusion layer thickness, and hence the critical current begins to drop (Fig. 3).

The reason for a wider maximum $I_c(\tau)$ for the annealing at 850°C lies perhaps in the fact that when the temperature of annealing is decreased, the processes of compiling recrystallization are sharply hindered, grains enlarge more slowly, and the critical current density for samples annealed at 850°C varies less sharply with time than for those annealed at 900°C (Fig. 3).

The maximum critical current (per width unit) versus the annealing temperature shows that the highest critical current corresponds to annealing at 900°C (Fig. 5).

This is explained, on the one hand, by the velocity with which the diffusion layer is growing, and on the other, by the rate of processes of compiling recrystallization. These processes lead to the most favourable result for the annealing at 900°C, 10 h. Sharp decrease in the critical current in samples annealed at 950°C may

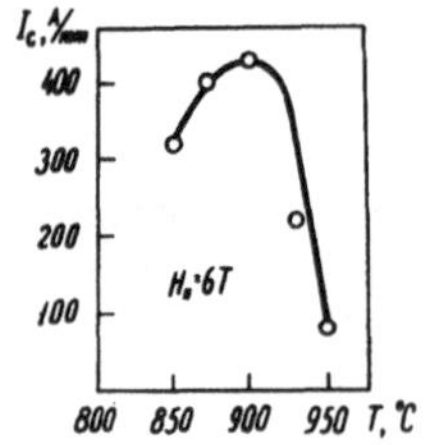

Fig. 5. The dependence of the maximal critical current on the temperature of formation of Nb_3Sn components. The values of critical current taken on a maximum of curves $I_c(\tau)$ at $T_{ann.}$ = const.

apparently be connected with the fact that at the annealing temperature increase the rate of grain growth accelerates faster than the nucleation rate. Therefore at the temperature increase, coarser grains are formed. This is supported[5] (the grain size in Nb_3Sn layer obtained at 900°C, 1 h is 0.9 μm, while that for 950°C is 2.2 μm).

Losses in the superconductor placed into an alternating magnetic field were measured on specially prepared samples (40-mm long, 4-mm wide, 1-mm thick) treated in the regimes similar to those used for composite tubes. All of their sides were covered with Nb_3Sn. These samples were collected into packets of 5 samples each. Measurements were done by the calorimetric and magnetic method before and after polishing of the Nb_3Sn surface. It has been shown that polishing of the surface decreases the loss level by 1.5-2 times (up to 3-5 $\mu W/cm^2$) as compared with that in initial samples.

CONCLUSION

At present, two sets of coaxial conductors are manufactured 7.50-mm long (external tube is 76 mm in diameter; internal tube is 51 mm in diameter). They are meant for a test installation where current tests will be carried out. A general view of the tubes is shown in Fig. 6.

Krzhizhanovsky's electric power engineering institute has performed comparative tests of samples prepared by both the method of superconducting Nb_3Sn ribbon soldered to a copper tube and by our technology at 900°C for 10 h, which showed that the use of our technology (proposed for the production of tubular current-carrying components of rigid cables in electrical power transmission lines

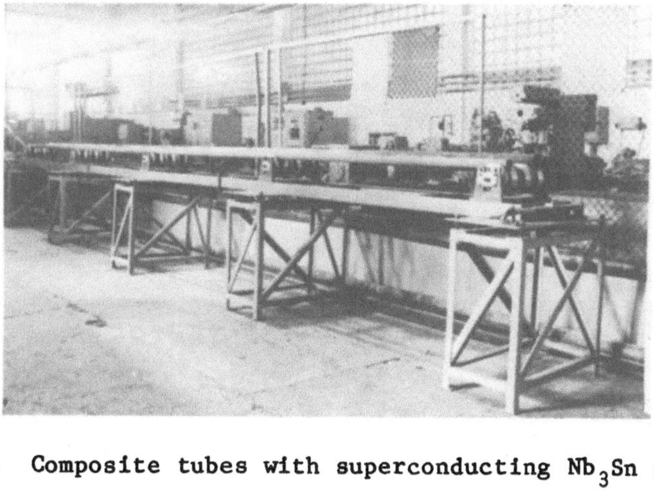

Fig. 6. Composite tubes with superconducting Nb_3Sn coating.

for the temperature range 6-9 K) ensures a current-carrying ability at least 5 times higher than that for the products prepared by the technology proposed.[2]

REFERENCES

1. V. M. Pan, V. S. Fils, Yu. I. Beletski, V. I. Latysheva, and G. A. Klimenko, Superconducting properties of composite rigid tubular elements with Nb_3Sn diffusion layers, in: "Advances in Cryogenic Engineering--Materials," Vol. 26, Plenum Press, New York (1980), pp. 410-414.
2. A. C. Newton, F. Martin, S. I. Lorant, and W. I. Toner, Rev. Sci. Instrum. 44(2):224-245 (1973).
3. M. M. Nerodenko, S. M. Gurevich, E. A. Asnis, A. I. Solodkov, M. D. Rabkina, V. N. Gridnev, and N. P. Kushnareva, Raspredeleniye kisloroda i azota v svarnykh shvakh splavov niobiya, Svar. Proizvod. 3:28-30 (1980).
4. G. Caslaw, Enhancement of the critical current density in niobium-tin, Cryogenics, 11(1):57-59 (1971).
5. E. A. Bondarenko, I. V. Morosov, V. Ya. Pachomov, A. S. Petukhova, V. P. Posyado, V. I. Sokolov, N. L. Speranskaya, and L. P. Revyakina, Issledovaniye diffusionnykh sloyev v sisteme niobii-olovo i niobii-olovo-med', in: "Struktura i Svoistva Sverkhprovodyaschikh Materialov," Nauka, Moscow (1974), pp. 63-38.
6. V. M. Dzugutov, Ustanovka dlya issledovaniya histerezisnykh poter v sverkhprovodnikakh v shirikoy oblasti magnitnykh poley, in: "Kriogenika i Sverkhprovodimost," Energiya, Moscow (1973), pp. 24-28.
7. V. V. Lavrova, L. M. Fisher, V. A. Yudin, Ustanovka dlya izmereniya poter' v sverkhprovodnikakh v peremennom magnitnom pole, in: "Voprosy Krioelektrotechniki i Nizkotemperaturnogo Experimenta," Naukova Dumka, Kiev (1976), pp. 6-11.

FABRICATION AND PERFORMANCE OF AN ALUMINUM-STABILIZED COMPOSITE SUPERCONDUCTOR

K. T. Hartwig, P. Zolliker,* D. Yu,† S. W. Van Sciver, and A. Khalil

Applied Superconductivity Research Center, University of Wisconsin, Madison, Wisconsin

INTRODUCTION

In a number of cases,[1,2] practical superconductors have been fabricated with aluminum as the chief stabilizer. Aluminum was chosen because it is lightweight, can be readily purified to a high degree for modest cost,[3] has a low magnetoresistance at high fields,[4] and is quite radiation transparent.[5] Applications for such superconductors include thin-walled magnets for physics experiments, small lightweight magnets for aerospace applications, and large magnets that require massive amounts of conductor.

Aluminum is not widely used as a stabilizer because it has a low strength[6] and because it is difficult to join. Although the pure metal is inherently weak, conductors utilizing it can be bolstered with supplemental strength members.[1,7] Strength aside, aluminum can be readily joined to itself or other metals by a variety of soldering,[8,9] brazing, or welding techniques. The major joining problems stem from a rapidly forming tenacious surface oxide layer.

The work reported here is part of an effort to develop a manufacturing scheme for long lengths of large aluminum-stabilized superconductors. Our focus has been on developing an assembly line that uses a solder bath. A major concern is that the assembly procedure and choice of solder must be compatible with all conductor

*International Association for the Exchange of Students for Technical Experience (IAESTE) visitor from the Swiss Federal Institute of Technology (ETH).

†Visiting Scholar, the General Research Institute for Nonferrous Metals, The Ministry of Metallurgical Industry, Beijing, China.

materials. A triangular-shaped prototype aluminum-stabilized composite conductor has been successfully manufactured. The fabrication method and preliminary performance characteristics are reported.

MATERIALS

The triangular conductor is composed of two sector-shaped pieces of 260 RRR (RRR $\equiv \rho_{300K}/\rho_{4.2K}$) 99.994% aluminum joined by 89Sn-11Zn solder to a 0.086-cm-diameter wire of multifilamentary NbTi in Cu.[10] The superconducting wire was sized for convenience, whereas the aluminum purity (RRR) was selected for appropriate conductor stability. The fabrication characteristics of this aluminum are similar to those of aluminum with a RRR greater than 1000, the probable choice for large energy storage magnet applications.

A solder composition of 89 wt.% Sn with 11 wt.% Zn was chosen for several reasons:[11] its melting range of 198 to 218°C is low, it wets aluminum, it has good corrosion resistance even in moist air, it is a moderately strong solder able to develop joints with shear strengths in excess of 35 MPa (5000 psi), and it is nonsuperconducting yet highly conductive at 4.2 K. A significant disadvantage of 89Sn-11Zn is that the liquid can embrittle aluminum.[12]

CONDUCTOR FABRICATION

The 99.994% aluminum was extruded at 500-550°C into a long wedge, which in cross section has a 0.33-cm radius, a 30° sectional angle, and a semicircular slot in one side. When two pieces of wedge are pressed together, they form a 60° wedge with a hole about the center of mass.

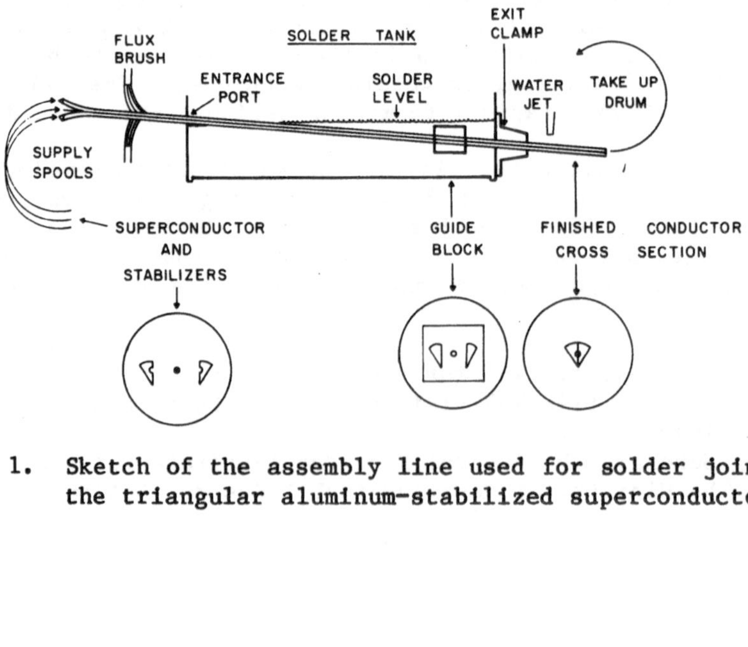

Fig. 1. Sketch of the assembly line used for solder joining the triangular aluminum-stabilized superconductor.

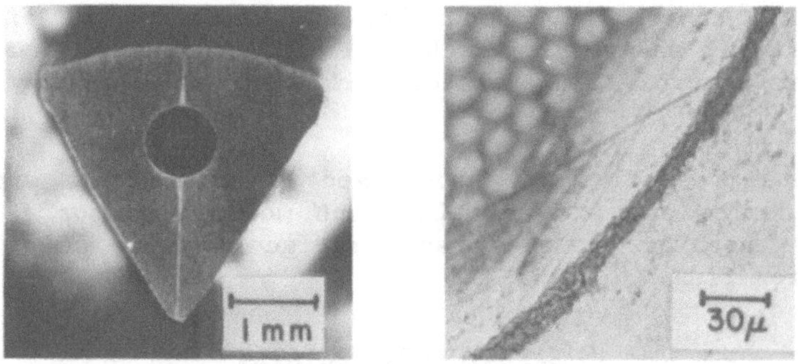

(a) (b)

Fig. 2. Photomicrographs of the conductor cross section (a) and solder-bond region (b).

A 0.086-cm diameter multifilamentary NbTi-in-Cu superconducting wire was joined with two pieces of aluminum wedge as follows: First, the superconducting wire was polished with steel wool. The wire and two aluminum wedges were then degreased with trichloroethylene and acetone, coated with Alcoa aluminum flux #69, and fed into a tilted solder tank containing molten 89Sn-11Zn at 265-270°C. By co-drawing at a rate of 2.5 cm/s, these three components were led through a guide block and joined together in an exit clamping die. The aperture of the clamping die was adjusted to avoid deformation in the clamping process. Finally, cold water was sprayed on the exiting conductor. The fabrication line is sketched in Fig. 1. The fabricated conductor (see Fig. 2) is 90% high-purity aluminum, 6% OFHC copper, 3% NbTi, and about 1% solder by volume.

The scheme mentioned above results in the successful preparation of moderately long conductors limited only by the length of available subcomponents. During an earlier phase of the project, however, we attempted to join superconducting wire to wedges that had been electroplated with 60Pb-40Sn solder or 100Sn. Many of these attempts were unsuccessful. Although the French have been successful,[2] our experience is that it is difficult to prepare long lengths of aluminum wire with a quality surface layer of electroplated solder. In addition, the electroplated surface layer is subject to damage if heated above about 175°C because of substrate metal outgassing and consequent surface blistering. In order to electroplate aluminum with solder or pure tin, the aluminum is often first plated with copper. The copper-aluminum interface is prone to deterioration at elevated temperatures because of the formation of brittle intermetallics, such as Al_2Cu. We observed layer separation in several samples of joined electroplated conductors and decided to pursue flux-bath soldering as an alternative. The flux-bath soldering work was further encouraged by the success of

Morpurgo and co-workers,[1] even though this method of joining also carries with it an array of technical difficulties, such as careful temperature control, flux related pollution, and cleaning off spent flux.

We are presently very encouraged by experiments with ultrasonic soldering and are adapting such methods to our fabrication line. Ultrasonics promotes rapid surface wetting and eliminates the need for flux.

EXPERIMENTAL DETAILS

The bond shear strength, current transfer length, and overall conductor stability were measured to evaluate the bonding quality between the stabilizer and the superconductor wire in the wedge conductor.

Bond Shear Strength

Samples for the bond shear strength test were prepared from long conductors and tested as follows: Aluminum was stripped off one end of each sample leaving about 2 cm of exposed superconducting wire. The sample lengths, not including the exposed superconducting wire, were from 0.15 to 1.20 cm and were limited by the maximum tensile load of about 360 N (80 lbf) that the superconducting wire can bear. Using the sample holder shown in Fig. 3, the strength of the solder bond between the aluminum stabilizer and the superconducting wire was tested in shear at both room temperature and 77 K at a deformation rate of 4.2×10^{-3} cm/s.

Several samples underwent a process of thermal cycling. They were dipped into a liquid nitrogen bath for at least one minute, and then taken back to ambient circumstances. After ten such cycles, the samples were tested at room temperature.

Fig. 3. Drawing of fixture used to measure solder joint shear strength.

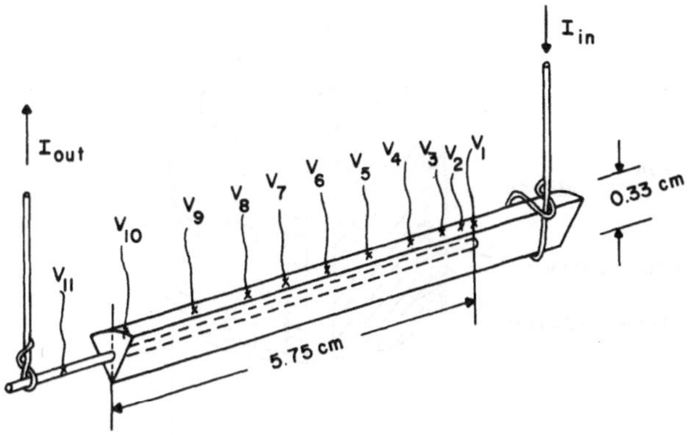

Fig. 4. Schematic of instrumented sample for current transfer length and hence boundary resistance measurement.

Current Transfer Length

Samples 4- to 6-cm long were prepared with a bare NbTi wire at one end and aluminum only at the other end (see Fig. 4). Voltage taps, with a common reference at the superconducting wire, were soldered on the sample surface with an In-Sn alloy solder. The distances between voltage taps were calculated from the electrical potential measurement at room temperature or 77 K. By measuring the electrical potential profile along the sample, a characteristic current transfer length, and hence a bond resistance was determined.[10]

Stability Measurements

The transport current characteristics were measured in a background field and geometry similar to the actual energy storage magnet winding configuration. The test section for the present study consisted of a 1.1-m length of wedge (one-sixth of the cross section of the prototype round conductor)[7] noninductively wound and epoxied into a grooved coil form, 73.0-mm o.d., so that only one surface of the conductor was exposed to helium (see Fig. 5). The conductor was instrumented with voltage taps for normal zone detection. The outer surface was exposed to vertically oriented cooling channels formed by 10-mm-wide, 0.5-mm-thick G-10 strips covering 33% of the conductor-helium interface and surrounded with a G-10 insulating sheet. The cooling channels had an overall length of 70 mm. The sample was measured in a 100-mm-i.d. superconducting magnet with a maximum field strength of 7.6 T at 4.2 K and 9.5 T at 2 K. The stability of the test section was investigated by slowly ramping the sample transport current in a constant background magnetic field, while monitoring the voltage across a 50-mm segment of the conductor.

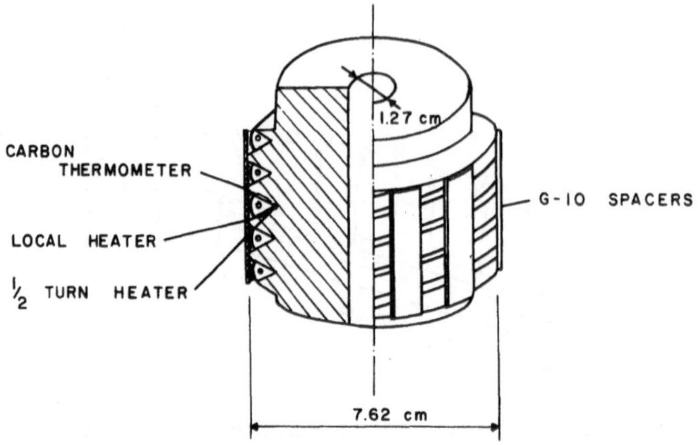

Fig. 5. Drawing of coil form and wound conductor for stability experiments.

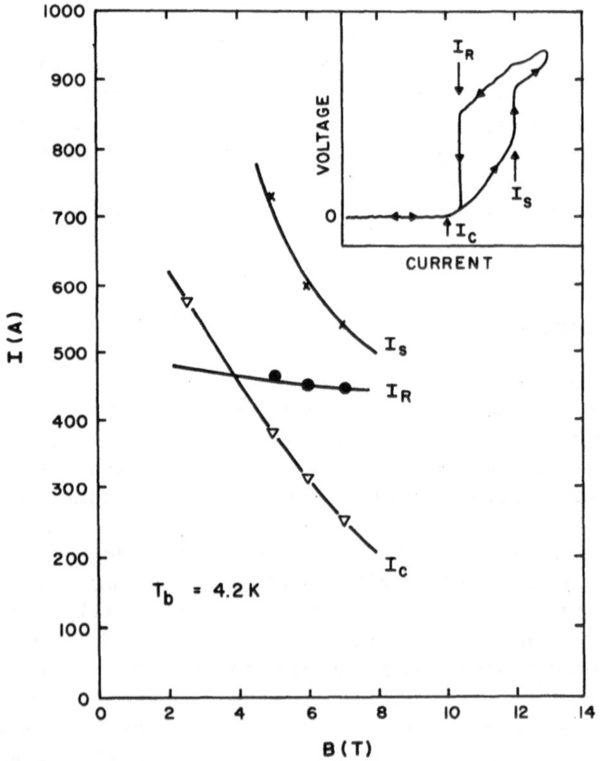

Fig. 6. Critical current (I_c), stable current (I_s), and recovery current (I_r) plotted as a function of magnetic field strength at 4.2 K. Insert shows typical V vs. I plot.

RESULTS AND DISCUSSION

Bond Shear Strength

A shear strength of 31-43 MPa was measured at room temperature for both uncycled and thermally cycled joints. This strength level agreed with that of a larger flat sample of bonded copper prepared with Sn-Zn solder. The results are also consistent with earlier work[13] for Sn-Pb solder joints. The bond shear strength in liquid nitrogen was 54-91 MPa, almost double that for a room temperature joint. It is evident that the bond shear strength does not depend heavily on the solder composition, the stabilizer material, or the low temperature thermal history of the bond after joining. Rather, the solder joining process is the major factor controlling bond quality.

Current Transfer Length

Current transfer lengths of 0.25-0.37 cm were measured on four samples from which the electrical boundary resistances of 3-8 $n\Omega cm^2$ were deduced.[10] Our results are in agreement with Goodrich and Ekin's measurement of 4.5 $n\Omega cm^2$ for a eutectic Pb-Sn joint[14] between multifilamentary NbTi-in-Cu conductors. We conclude that the primary boundary resistance comes from the solder bonding interfaces.[10]

Stability

Voltage-current plots of the conductor are shown in Fig. 6. The critical current, I_c, marks the beginning of the current sharing region above which there is a nonzero voltage across the conductor. We arbitrarily define the actual current at the point where the resistive voltage exceeds 5 μV/cm. In the current sharing region, the V-I plots contain no hysteresis, and one can easily return to the fully superconducting state by reducing the current below I_c. The stable current, I_s, is the end current at which the conductor goes fully normal. The only way to return to the superconducting state once I_s is exceeded is by reduction of the transport current to the recovery value, I_r. In the present experiments, it was necessary to return to I_r in less than 60 s to avoid thermal runaway.

The critical current, stable current, and recovery current versus applied magnetic field are plotted in Fig. 6. The conductor is fully stable, having a recovery current above I_c. However, by extrapolation of these results to low fields, it appears that the conductor will be only marginally stable below 4 T and 4.2 K.

Both I_s and I_r were determined exclusively by the heat transfer limits to the helium. Based on measurements of the heat generated at I_s, we calculated the maximum stable heat flux, q_s, to be 0.43 ±0.02 W/cm^2. We observed no noticeable hysteresis in the heat transfer characteristics, since the conductor recovered when the heat transfer rate decreased below q_s. It is probable that this lack of hysteresis occurred because of the natural circulation of the helium.

The present results showed satisfactory performance of the aluminum-stabilized NbTi conductor under current sharing and recovery processes. We observed no hysteresis due to poor electrical bonds between the superconductor and stabilizer, as has occurred in some previous aluminum conductors.[15] In the present experiments, all hysteretic phenomena can be understood in terms of helium heat transfer.

CONCLUSIONS

1. The reported continuous soldering process is found to be capable of producing long lengths of aluminum-stabilized superconductor with good bonding between the stabilizer and the superconducting matrix and with full stability above 4 T and 4.2 K for a current up to 450 A.

2. The bonding quality (bond shear strength and electrical boundary resistance) does not strongly depend on the solder used or the materials joined. Rather bond quality is controlled by the soldering process.

ACKNOWLEDGMENTS

The authors are grateful to Mr. Richard Schumacher for assistance with conductor fabrication and experimental testing and to Mr. John Egan for assistance with preparing photomicrographs. This work is supported by the U.S. Department of Energy and the Wisconsin Electric Utility Research Foundation.

REFERENCES

1. M. Morpurgo and G. Pozzo, Cryogenics 17(2):87 (1977).
2. P. Genevey and J. LeBars, in: "Proceedings of 6th International Conference on Magnet Technology," ALFA Publishing Co., Bratislava, Czechoslovakia (1977), p. 1033.
3. C. N. Cochran, R. K. Dawless, and J. B. Whitchurch, "1 GWh Diurnal Load-Leveling Superconducting Magnetic Energy Storage System Reference Design, Appendix B, Cost Study High Purity Aluminum Production," LASL Report No. LA-7885-MS, Vol. III, Alcoa (1978).
4. F. R. Fickett, Phys. Rev. B 3(6):1941 (1971).

5. H. Desportes, J. LeBars, and G. Magaux, in: "Advances in Cryogenic Engineering," Vol. 25, Plenum Press, New York (1980), p. 175.
6. K. T. Hartwig and F. J. Worzala, Mater. Sci. Eng. 48(1):41 (1981).
7. K. T. Hartwig, in: "Advances in Cryogenic Engineering - Materials," Vol. 26, Plenum Press, New York (1980), p. 494.
8. "Soldering Alcoa Aluminum," Aluminum Company of America, Pittsburgh (1972).
9. "Aluminum Soldering Handbook," Third Edition, The Aluminum Association, Inc., New York (Dec. 1976).
10. D. Yu, Y. M. Eyssa, and P. Zolliker, Electrical boundary resistance in an aluminum-stabilized superconductor, in: "Advances in Cryogenic Engineering - Materials," Plenum Press, New York (1982), p. 701.
11. Manuscript in preparation.
12. H. Ichinose, Trans. Jap. Inst. Met. 9:35 (1968).
13. K. T. Hartwig and S. W. Van Sciver, in: "Proceedings of 8th Symposium on Engineering Problems of Fusion Research," IEEE Pub. No. 79CH1441-5NPS, Vol. 4, IEEE, New York (1979), p. 1794.
14. L. F. Goodrich and J. W. Ekin, IEEE Trans. Magn. MAG-17:69 (1981).
15. S. T. Wang, R. P. Smith, S. H. Kim, and J. J. Peerson, in: "Advances in Cryogenic Engineering," Vol. 23, Plenum Press, New York (1978), p. 70.

DEVELOPMENT OF AN INTERNALLY STRENGTHENED Nb_3Sn CONDUCTOR

C. R. Spencer, E. Adam, and E. Gregory

Airco Superconductors, Carteret, New Jersey

INTRODUCTION

In large-bore, high-field superconducting magnets, the windings are subjected to high tensile stress levels under normal operating conditions. Since most of these magnets will probably be constructed using Nb_3Sn superconductor, it is necessary that superconductors capable of withstanding stress levels in excess of 105 MPa be developed.

Nb_3Sn composite conductors formed by a standard double-extrusion bronze process[1] are delicate and cannot withstand bend strains larger than 0.6% without irreversible degradation of the critical current density. It is desirable that the Nb_3Sn not be respooled after the reaction stage, and if respooling is necessary, it is recommended that the number of respooling steps be minimized to prevent damage to the fragile Nb_3Sn. Equally important is the fact that reacted Nb_3Sn composite conductors consist of Sn bronze, Nb, and Nb_3Sn with Cu added for electrical stability. These composites cannot be expected to have room-temperature yield strengths exceeding 90 MPa, and so techniques of strengthening the conductor must be developed.

A process in which hardened Cu is soldered to the Nb_3Sn core has been described elsewhere.[2] One inherent problem with soldering to Nb_3Sn conductors is that the soldering must be done after the high-temperature reaction heat treatment, and respooling can lead to bend strain damage. A second problem faced by a fabricator while soldering cold-worked Cu to a conductor is that the time of residence at solder temperatures can lead to loss of strength in the hardened Cu. One final consideration is that the solder bond itself may not have the mechanical integrity needed in high-stress applications.

INTERNALLY STRENGTHENED CONDUCTOR

A Nb_3Sn conductor has been designed using the compacted monolith concept.[1,3] This conductor is composed of seven strands of Nb_3Sn around a strand of strengthener and is compacted within a Cu tube. The strengthener chosen must be of an alloy that has acceptable strength and magnetic properties. The compacted monolith technique allows one to change the proportions of superconductor, Cu, and strengthener to achieve the superconducting and strength properties desired.

Nb_3Sn STRANDS

The Nb_3Sn strands chosen for this design consist of 62 vol.% Cu and 38 vol.% Sn bronze, Nb, Nb_3Sn, and Ta. Each strand contains 9481 filaments, each approximately 3 μm in diameter in the final conductor configuration. This wire is cold-drawn to 1.52 mm by a standard production process. Seven Nb_3Sn strands are positioned around a strand of A-286 iron-based superalloy and sheathed with an oxygen-free Cu tube.

STAINLESS STEEL

Although a variety of stainless steels have acceptable magnetic properties, many of these stainless steels lose some of their strength when subjected for long times to temperatures above 650°C, which are necessary to form Nb_3Sn. The iron-based superalloy A-286

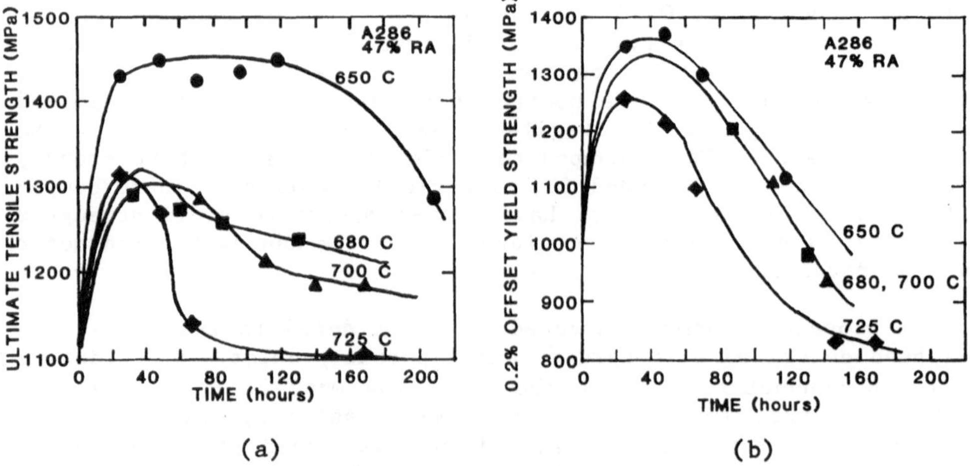

Fig. 1. (a) Ultimate tensile strength of an A-286 wire given a 47% reduction in area and then heat-treated. (b) 0.2% offset yield strength of A-286 wire given a 47% reduction in area and then heat-treated.

has low magnetization and very desirable age-hardening characteristics. Cold-worked A-286 can be expected to have acceptable yield strengths after long-term heat treatments above 650°C. It has been chosen as the sheath used in the Westinghouse-Airco Nb_3Sn Large Coil Project Conductor. These properties have made A-286 an ideal candidate for internally strengthening Nb_3Sn composites. However, since A-286 work-hardens quite rapidly and must be annealed at 1000°C, it is not compatible for large reductions in composites with Nb_3Sn.

Samples of A-286 superalloy were given a 47% reduction in area (RA) and heat-treated for varying times and temperatures to simulate the reduction and reaction conditions in the compacted monolith. The samples were tensile-tested; the ultimate tensile and 0.2% offset yield strengths are plotted in Figure 1. Both strengths reach maximum values at approximately 30 h for the temperatures tested. For longer times, the ultimate and yield strengths drop monotonically with a minimum UTS of 1100 MPa and a minimum 0.2% offset yield strength of about 800 MPa, which appears to be acceptable for use in the conductors under consideration.

SHEATHING

Unit lengths of compacted superconductor as long as 500 m may be required in some applications. Because an elongation of nearly 100% occurs during the compaction process, continuous lengths of wire in excess of 250 m must be sheathed with Cu tube. To accomplish this, a tube mill has been modified such that parallel strands (or cables) may be sheathed with Cu tube, which is formed and TIG welded around the wire. Previous experience[1,3] has proven that the weld penetrates the tube without causing damage to the wires inside. Shielding gas of nitrogen protects the tube and wires from oxidation. This forming and welding process has been reliably used to sheath over 1000 m of High Field Test Facility and Elmo Bumpy Torus conductor with 15.9-mm o.d. oxygen-free Cu tube.

FORMING

The sheathed composite is reduced in-line downstream of the tube mill through two round tungsten carbide drawing dies for a total area reduction of 44%. Final compaction occurs during a third round draw, a turkshead pass, and a final rectangular draw. The total reduction in area of the actual starting material is 47%.

RESULTS

Several designs have been tested for ease of fabrication and strength. Initial samples were made of seven 316 stainless steel strands in a Cu tube around which 10 strands of Nb_3Sn superconducting wire were positioned, and the array was pulled into a Cu tube. The assembled conductor (Fig. 2) was reduced by drawing and turksheading. A similar conductor was made by replacing the 7 stainless steel strands with a single 316 stainless steel strand (Fig. 3). Conductors shown in Figures 2 and 3 can be improved by placing 7 Nb_3Sn strands around a single A-286 stainless steel strand (Fig. 4). Each Nb_3Sn strand contains 25 to 65% Cu, which easily deforms during the rectangular forming processes. A summary of the tested designs is shown in Table 1. Tensile specimens of each design were prepared by heat-treating to simulate the Nb_3Sn reaction conditions and then tensile-tested at room temperature. The results (shown in Table 2) can be predicted by the rule of mixtures.

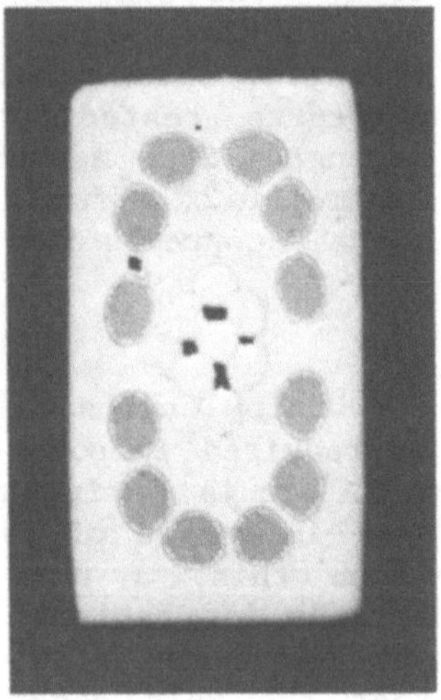

Fig. 2. Twelve Nb_3Sn strands around a Cu-sheathed cable of seven strands of 316 stainless steel wires, compacted (sample 1A) to 2.6 mm x 4.6 mm.

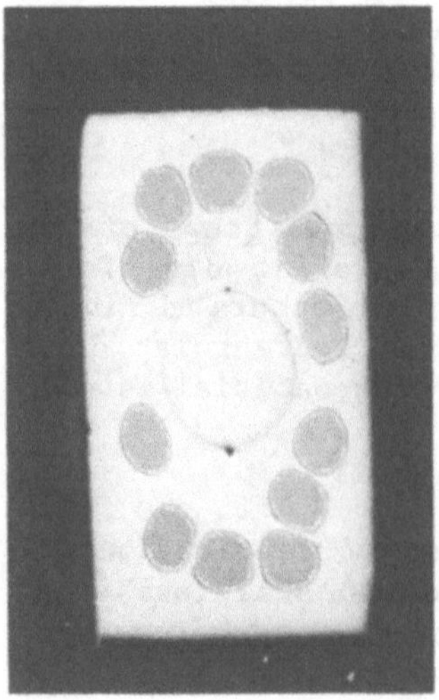

Fig. 3. Twelve Nb_3Sn strands around a Cu-sheathed strand of 316 stainless steel, compacted (sample 1B) to 2.6 mm x 4.6 mm.

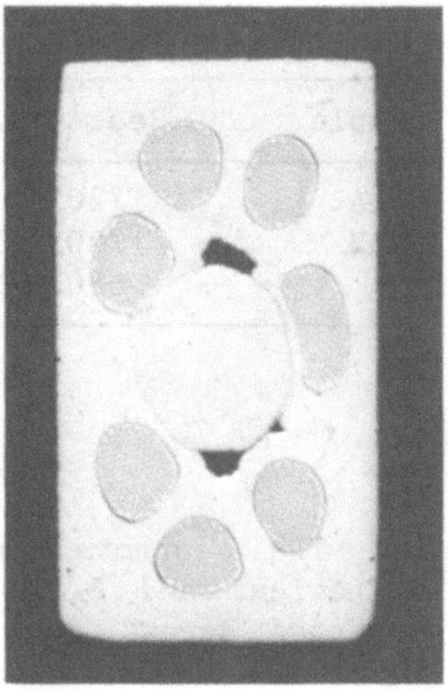

Fig. 4. Seven Nb_3Sn strands around a strand of A-286 stainless steel, compacted (sample 6) to 2.8 mm x 5.0 mm.

Critical current densities as high as 631 A/mm^2 measured in the superconducting cores at 10 T have been achieved with this conductor. These compacted conductors allow the conductor designer to easily vary the percentages of Cu, superconductor, and strengthener by adjusting such variables as Cu tube wall thickness, strand Cu content, and the diameters and number of superconducting and strengthening strands.

CONCLUSION

A process for strengthening superconductors internally has been developed. This process eliminates the soldering and handling steps currently used to strengthen the reacted Nb_3Sn conductor externally. The compacted monolith concept affords much versatility to the superconductor designer because the basic stabilizer, superconductor, and strengthener modules may be apportioned and arranged to meet the magnet designers' needs.

Table. 1. Internally Strengthened Conductor

Sample	No. Nb_3Sn Strands	% Cu on Strand	Core Material	% of Conductor		
				Nb_3Sn Core	S. Steel	Cu
1A	12	25	316ss(7)	15	6.7	78.3
1B	12	25	316ss	14.6	10.1	75.3
6	7	62	A-286ss	17	12	71

Table 2. Tensile and 0.2% Offset Yield Strengths

Sample	Heat Treatment	TS (MPa)	0.2% Yield (MPa)
1A	720°C/55 h	184	76
1B	720°C/55 h	239	72
6	680°C/256 h	305	182

ACKNOWLEDGMENTS

The authors gratefully acknowledge the assistance of F. Lewicki and B. G. Radcliffe in the preparation of the samples reported here. Part of this work was done under Kernforschungszentrum Karlsruhe Purchase Order No. 130/D1/524611.

REFERENCES

1. C. Spencer et al., in: "Proceedings of the 8th Symposium on Engineering Problems of Fusion Research," IEEE Publication No. 79CH1441-5NPS, IEEE, New York (1979), p. 265.
2. P. Singh, E. Adam, E. Gregory, W. G. Marancik, F. T. Ormand, M. Young, and C. Spencer, Prototype multifilamentary Nb_3Sn superconductor for the High Field Test Facility at Lawrence Livermore Laboratory, in: "Advances in Cryogenic Engineering - Materials," Vol. 26, Plenum Press, New York (1980), p. 464.
3. C. Spencer et al., IEEE Trans. Magn. MAG-17:1006 (1981).
4. R. E. Gold, W. A. Logsdon, G. E. Grotke, and B. Lustman, Evaluation of conductor sheath alloys for a forced-flow Nb_3Sn superconducting magnet coil for the Large Coil Program, in: "Advances in Cryogenic Engineering - Materials," Vol. 28, Plenum Press, New York (1982), p. 759.
5. W. A. Logsdon, G. E. Grotke, R. E. Gold, and B. Lustman, Cryogenic tensile and fracture toughness properties of three candidate structural materials for the Large Coil Program superconducting magnet conductor sheath, in: "Advances in Cryogenic Engineering - Materials," Vol. 28, Plenum Press, New York (1982), p. 771.

CUPROUS SULFIDE AS A FILM INSULATION FOR SUPERCONDUCTORS*

G. R. Wagner, P. D. Vecchio, and J. H. Uphoff

Westinghouse Electric Corporation, Research and Development Center
Pittsburgh, Pennsylvania

INTRODUCTION

The test coil being designed and built by Westinghouse for the Large Coil Program (LCP) utilizes a conductor of forced-flow design having 486 strands of multifilamentary Nb_3Sn compacted in a stainless steel conduit.[1] The impetus for the present work stemmed from the need for some form of "insulation" on those strands to prevent sintering during reaction and to reduce ac losses. During the course of the LCP design several substances were considered as candidates for strand insulation but were rejected for various reasons.[2] However, cuprous sulfide was found to perform well as a protective layer with no deleterious effects on the properties of the conductor.[3] The sulfide layer survived the cabling, compacting, and heat-treating processes and effected the required low ac losses. As a result, a good deal of interest has arisen concerning the possible use of cuprous sulfide in other magnet systems. In this paper we describe the continuous process that was developed for applying cuprous sulfide to long lengths of conductor, discuss the properties of the films, and present some encouraging data on test coils that have been wound with sulfide coated conductors.

CUPROUS SULFIDE

Background

Cuprous sulfide is a p-type semiconducting compound that may be formed readily by heating copper in H_2S or sulphur vapor. The

*Supported in part by the Union Carbide Corporation, Contract No. 22X31747C.

copper-sulphur phase diagram has been reviewed by Roseboom[4] and Cook.[5] Three stable phases of cuprous sulfide exist at room temperature. Orthorhombic chalcocite, Cu_2S, inverts to a hexagonal form at 103°C. It may also dissolve enough sulphur, if available, to approach a composition $Cu_{1.8}S$, which is a separate phase called digenite. An intermediate composition, $Cu_{1.96}S$, called djurleite, is a low symmetry phase that inverts to a metastable tetragonal form and a stable cubic form called high digenite near 93°C. The djurleite phase is more stable than chalcocite but forms only in the presence of Cu^{2+} ions. None of the high temperature forms can be quenched to room temperature.

A film of cuprous sulfide formed on copper by reaction with H_2S at elevated temperatures would, in the absence of oxygen and sulphur, form Cu_2S of the high digenite (cubic phase) for $T > 435°C$ and hexagonal chalcocite for $103 < T < 435°C$. In either case, the Cu_2S would convert to orthorhombic chalcocite when cooled to room temperature.

The thermal, electrical, and optical properties of cuprous sulfide have been studied by several authors[6] because of the interest in CdS-Cu_2S solar cells. Stoichiometric Cu_2S has a room temperature resistivity of 100 Ω·cm. However, the resistivity decreases by five orders of magnitude for $Cu_{1.8}S$. It is also known that the electrical resistivity of stoichiometric Cu_2S decreases sharply upon exposure to oxygen.

Experimental Coating

Cuprous sulfide layers about 2- to 4-μm thick were formed on copper wires (0.7-mm diam.) in H_2S vapor at 500°C in about 20 to 30 s. These layers showed reasonably good adhesion, which means they did not scrape off easily with the thumbnail. Since the layer was not insulating at room temperature, quantitative measurements of scrape abrasion could not be performed with meaningful results.

Cuprous sulfide was also formed on copper by heating in sulphur vapor rather than H_2S. Coatings produced in this way were inferior to most of those developed using H_2S. They tended to be grainy and did not adhere well. The layer usually crazed upon bending and flaked off easily. The layer thickness was difficult to control because it depended strongly on the sulphur vapor pressure, which was not easily controlled. The compound formed in this way is probably Cu_xS with $x < 2$ because of the excess sulphur vapor present. In fact, some combination of the phases digenite, djurleite, and chalcocite is likely and may explain the poor mechanical properties.

A three-zone, tube furnace 90-cm long with a 25-mm i.d., 120-cm-long quartz tube was used to coat wires in continuous lengths using H_2S. The quartz tube was sealed at the ends with Teflon cylinders having holes through which the wire passed. The holes are a sliding fit for the wires to seal out air and moisture. Argon and H_2S are admitted through separate openings at the wire exit end of the quartz, flow through the furnace, and exit at a common opening at the wire entrance end. This mixture passes through a cold trap, an oil bubbler, and exits into the top of the hood. The bubbler prevents air from entering the furnace and the cold trap removes oil vapor from the system. Flow meters monitor the flow of argon and H_2S. The argon used is ultrahigh purity and is passed through a titanium gettering system to remove residual oxygen before entering the furnace. It is important to keep the system oxygen free.

Before entering the furnace, the wire was cleaned by passing it through an acid bath, a rinse of water to remove the acid, and a sponge wiper and then blown dry. The acid used was a commercial copper polishing etch, which passivates the copper surface to inhibit oxidation. This cleaning was an important step to achieve uniform layers with reliable mechanical properties. A capstan drive controlled the speed of the wire and a take-up spool was driven by a constant torque motor.

Determining the parameters that produce the "best" cuprous sulfide layer in the shortest time was a task carried out rather empirically with a few guideposts provided by the copper-sulphur phase diagram, the mechanical constraints of our wire pulling-furnace combination, and the chemistry of the reaction. Since the gas-solid reaction rate increases with temperature, the time required to form the Cu_2S layer was minimized by operating at the maximum temperature allowed by all other constraints. One of these constraints was that the presence of sulphur vapor tends to form $Cu_{1.8}S$ and thus the operating temperature should be low enough to preclude any appreciable dissociation of the H_2S. Using the equilibrium constant, K_p, for the reaction[7] $2H_2S \rightleftarrows 2H_2 + S_2$ at a constant pressure of one atmosphere, the calculated temperature dependence indicated less than 5% dissociation below 800°C. Experimentally, we found that for temperatures greater than 800° C in at least one of the furnace zones, excess sulphur formed as a solid phase on the cold, gas-exit end of the furnace. Corresponding with this, the cuprous sulfide coating resembled that formed in the experiments with sulphur vapor. Thus, temperatures in this extreme were avoided.

The furnace temperature profile used for coating LCP wire at 3.5 m/min (11.5 ft/min) is shown in Fig. 1. Also shown is the temperature of the wire as a function of position in the furnace

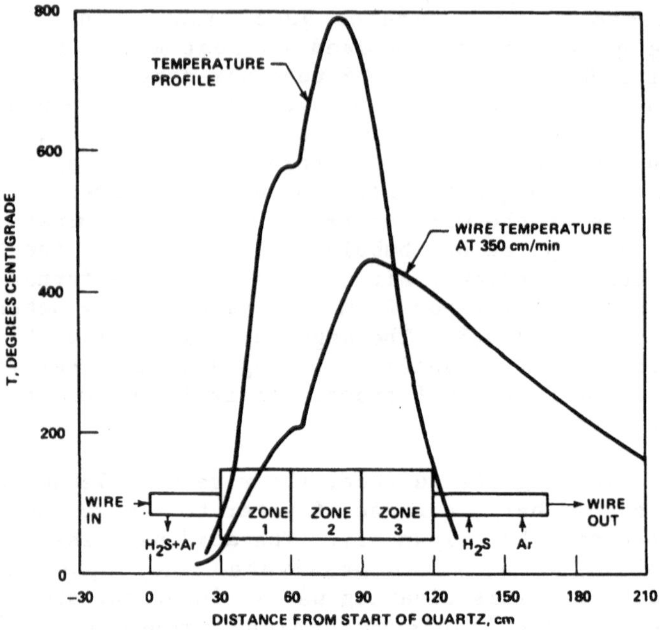

Fig. 1. Furnace temperature profile for typical Cu_2S coating.

at this speed and a schematic of the furnace showing the entrance and exit ports for the gases. With this setup, a flow rate of about 300 cm^3/min of H_2S through the furnace yielded a layer thickness in the range 5 to 7 μm. Thus, about 90% H_2S exited the furnace unreacted.

The apparatus just described was used to apply cuprous sulfide to approximately 18,000 m of wire including the wire used to wind coil #2 described below. This apparatus has subsequently been replaced with one that uses a 3-m long furnace and is capable of running 8 wires in parallel, each at a speed of 7.5 m/min for a total capacity of 60 m/min. The wire used for coil #1 described below was sulfided in this furnace.

Film Testing

Mechanical testing was not quantitative at all. Some films, such as all of those formed in sulphur vapor, were obviously not good because a light wipe with fingers would remove some or all of the film. Films were also considered unacceptable if they could be scraped off with a thumbnail with reasonable pressure. Eventually, we obtained films that passed the thumbnail test and did not flake off when the wire was bent into a hairpin with a radius producing 20% strain. For some films, the wire could be bent into a hairpin, straightened, and the cycle repeated until breakage of the wire occurred, without the film coming off. Thickness of the film is important in determining mechanical strength. For best results, the film should not be greater than 10 μm and preferably less than 5 μm.

Electrical tests were conducted with twisted pairs of sulfide coated copper wires. The volt-amp characteristics of 2 wires twisted about each other 3 times over a length of about 1 cm were determined. The free ends of the wires were anchored with brass screws to a Micarta plate and immersed in liquid helium. Voltage and current leads were soldered to the wires (after removing the Cu_2S) at the screws. The other free ends were passed through Teflon tubes in a flange at the top of the helium Dewar and over a pulley with weights attached to the wire ends. In this way, the only electrical connections between the two wires were at the areas of contact in the twists. A sufficient amount of weight (∿1 kg) was used to pull the twists tight.

The volt-amp results for such a test using "good" films are shown in Fig. 2. These data are characteristic of a back biased metal semiconductor junction. For $V < 0.4$ V the current depends linearly on voltage with a resistance of about 2 Ω for this particular case. Of course, the resistance depends upon the contact area and other parameters. Near 0.4 V, breakdown occurs, and the current rises rapidly with voltage. The same characteristics are

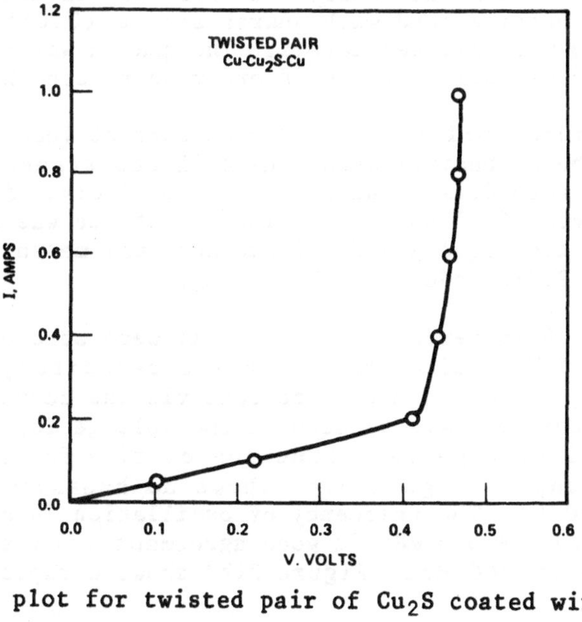

Fig. 2. V-I plot for twisted pair of Cu_2S coated wires at 4.2 K.

observed when the voltage polarity is reversed because there are two junctions in series opposition. We could not determine the maximum reverse current density allowed before damage occurred to the film in this test because the area of contact between wires was not known. A maximum of 1 A was used as a measuring current, but we found some films to be damaged at lower currents.

COIL TESTS

Two solenoids were wound with sulfide coated wires and heat-treated at 700°C for 30 h in two different furnaces. Coil #1 was heated in a controlled atmosphere of flowing ultrapure argon in a clean stainless steel tank. Coil #2 was heated in a furnace that is used routinely for heat-treating a variety of metals with standard-grade flowing argon. In each case, the sulfide layer remained intact and no sintering occurred. The electrical properties of the coils were studied before and after reaction and are described below, as are the winding parameters.

Coil #1: Copper Wire

Coil #1 was wound with copper wire (0.7-mm diam.) that had been coated with sulfide at a rate of 7 m/min with a thickness that varied from 2 to 5 μm over the length of the wire. The coil consisted of 840 turns (218-m total length) wound as 8 layers on a stainless steel former having a 7.5-cm i.d. and a 7.5-cm winding length. The final o.d. was 8.9 cm. A stainless steel band was clamped around the finished coil. The wire was insulated from the former and the outside band with quartz fabric (0.015-cm thick), which was also used between layers. The quartz was heated in air to 600°C for 45 min before using in order to remove the binder.

The finished coil was fitted with current leads and voltage taps. Before being heat-treated, the coil resistance and inductance at 4.2 K were 0.12 Ω and 36 mH, respectively. After 30 h at 700°C, they were 0.07 Ω and 4 mH. The resistance was measured by the four-probe technique and the inductance was measured at 240 Hz with a General Radio Model 1688 Digibridge.

The pulse characteristics of the coil were studied after heat treatment at 4.2 K by discharging a 60-μF capacitor, charged to various voltage levels, into the coil via the current leads and recording the coil voltage as seen at the voltage taps. Figure 3 shows the coil voltage as a function of time for two values of capacitor voltage. Figure 3(a) shows no breakdown with peak voltage at 400 V. The frequency of oscillation is about 280 Hz, corresponding to L = 5.5 mH, in good agreement with the measured value of 4 mH at 240 Hz. Figure 3(b) shows a rapid decrease in

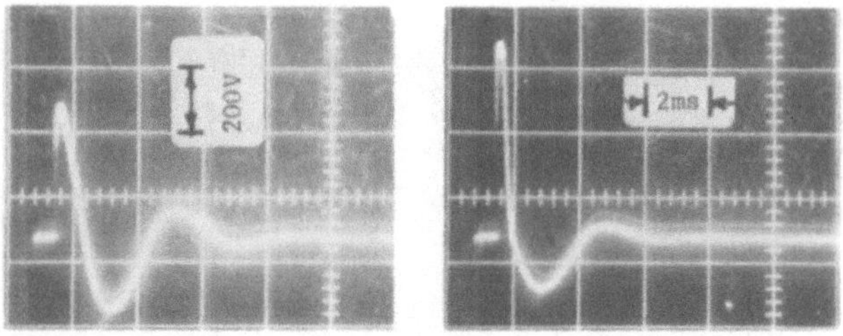

(a) (b)

Fig. 3. Coil #1 voltage vs time after reaction for two values of V_c (capacitor voltage).

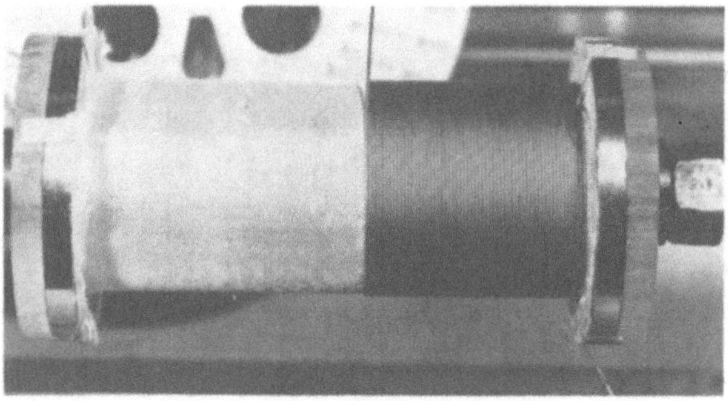

Fig. 4. Coil #2 during winding. Winding length is 10 cm.

coil voltage from the peak of 600 V, indicating turn-to-turn conduction. Threshold for breakdown was found to be 480 V, corresponding to a turn-to-turn voltage of approximately 0.6 V; the voltage was assumed to be uniformly distributed across the coil. This is in good agreement with the dc characteristics shown in Fig. 2. Repeated cycles with capacitor voltages up to 800 V did not damage the coil. The coil behaved as expected, with current flowing turn-to-turn when the junction breakdown voltage was exceeded, but without damage to the sulfide layer.

Coil #2: Nb_3Sn LCP Wire

Coil #2 was wound in the same manner as coil #1 with quartz fabric between layers and next to the former and outside band. The former has an i.d. of 3.8 cm and a winding length of 10.2 cm. The conductor was supplied by Airco Superconductors and is described elsewhere.[8] It is the Nb_3Sn conductor manufactured for the Westinghouse LCP coil. The conductor was coated with sulfide

Fig. 5. Coil #2 on hanger after reaction at 700°C for 30 h.

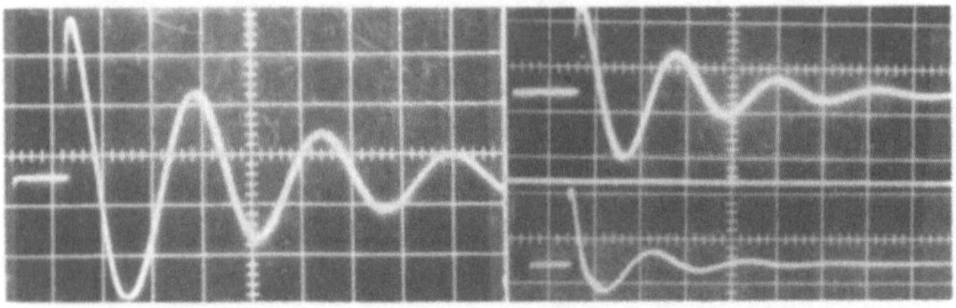

(a) (b)

Fig. 6. (a) Coil #2 voltage vs time before reaction for V_c = 600 V; (b) Coil #2 voltage vs time after reaction for V_c = 400 V. Scales are the same as in Fig. 3.

at a rate of 3 m/min with a thickness that varied from 2 to 6 μm over its length. A total length of 111 m was used to wind 6 layers with 138 turns/layer. Figure 4 shows the coil during winding.

Figure 5 shows the finished coil mounted on a hanger that is fitted with 500-A vapor-cooled leads. The coil ends were clamped and soldered to copper blocks for current contact, and the voltage taps were soldered below the current leads. Before reaction, the coil inductance at 4.2 K and 240 Hz was 9 mH, and after, 6 mH.

Figure 6(a) shows the coil voltage as a function of time at 4.2 K before reaction. Again, a capacitance of 60 μF was discharged into the coil. No breakdown was observed with as much as 640 V across the coil. The coil was superconducting, as expected,

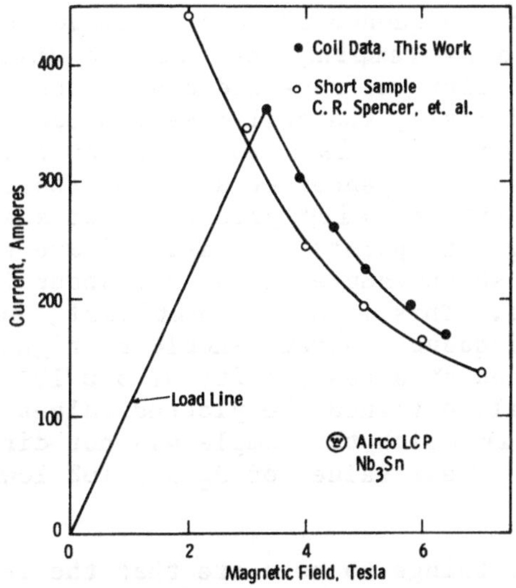

Fig. 7. Coil #2 quench current vs magnetic field and predicted load line. Open circles are short-sample critical current data from Ref. 8.

and had a reasonable Q with a ringing frequency of 200 Hz, corresponding to an inductance of 10.6 mH, in good agreement with the value measured with the Digibridge.

The two traces in Fig. 6 (b) were taken after reaction with the same values of capacitor voltage. The upper trace shows no breakdown, whereas the lower shows a more rapid decrease in coil voltage after the initial peak of 320 V, followed by a recovery to the same oscillatory behavior as in the upper trace. The level of 320 V was the threshold for breakdown and the behavior in the lower trace was observed in about half the pulses. For voltages greater than about 350 V, breakdown occurred consistently in every case. With the voltage assumed to be uniformly distributed across the coil, 320 V corresponds to 0.39 V between turns. This value again agrees well with the behavior noted in Fig. 2.

The frequency of oscillation after reaction is 250 Hz, corresponding to L = 6.8 mH, in good agreement with the Digibridge value.

Coil #2 was designed to produce $9.5(10^{-3})$T/A at its center.[9] Its physical size was chosen to fit easily into an available NbTi magnet which has a 13-cm bore and produces a maximum field of 5 T, which was used to provide a uniform background field for measuring the quench characteristics of coil #2. Figure 7 shows these data.

The solid circles are quench current vs field for the coil. Each point was obtained by ramping the coil to quench at a rate of 50 A/min. Every point was repeated and no training was observed. In zero background field, the coil itself produced 3.3 T at 360 A. The calculated load line is also shown for zero bias. The open circles are the data of Spencer et al.[8] for short-sample critical current measurements on wire produced from a similar billet and reacted similarly. The points at 2 and 3 T are extrapolated from their data. The short-sample values are about 20% lower than the coil quench values. This is not significant, but should be expected because the quench current should be higher than the critical current measured at a resistivity of 5×10^{-12} $\Omega \cdot cm$. In addition, Spencer et al. obtained the plotted values using a variable temperature rig in which the sample was not directly immersed in the helium bath. Their values of J_c are 10% low owing to heating in the sample.

The important things to note are that the coil achieved short sample I_c-H values and produced magnetic fields according to the expected load line. All of the ampere turns were there, and the cuprous sulfide layer provided adequate insulation between turns.

SUMMARY

Two solenoids that were wound with cuprous sulfide coated wires and heat-treated at 700°C for 30 h have demonstrated that the film is effective in providing turn-to-turn insulation for less than about 0.5 V between turns. The sulfide layer provides a metal-semiconductor junction that becomes conducting at approximately 0.5 V. Repeated cycling of the coil voltage in excess of this value has produced no damage to the sulfide layer. The junction provides self-protection to the coil as long as the maximum allowable current density in the cuprous sulfide is not exceeded.

The superconducting solenoid wound with cuprous sulfide insulated wire achieved short-sample critical current with no training up to 6.4 T and the expected ampere turns.

In addition to the work described here, it should be mentioned that sulfide coated wire has been used with great success in another magnet program under U.S. Air Force sponsorship. The wire was used to form a six-around-one cable, which was then glass insulated and wound into a fully potted coil. The dc and pulse characteristics of this coil are described in these proceedings.[10]

REFERENCES

1. C. J. Heyne, D. T. Hackworth, S. K. Singh, and J. L. Young, in: "Proceedings of the 8th Symposium on Engineering Problems of Fusion Research," No. 79CH1441-5NPS, IEEE, New York (1979), p. 1148.
2. Superconducting Magnet Coils for the Large Coil Program, Phase 2 Final Report, Westinghouse Electric Corporation, Research and Development Center, Pittsburgh, Pennsylvania (1980).
3. G. R. Wagner, in: "Proceedings of the 8th Symposium on Engineering Problems of Fusion Research," No. 79CH1441-5NPS, IEEE, New York, (1979), p. 1446.
4. E. H. Roseboom, Jr., Econ. Geol. 61:461 (1966).
5. William R. Cook, Jr., Ph.D. Thesis, Case Western Reserve University, Cleveland, Ohio (1971).
6. For a review, see A. G. Stanley, in: "Applied Solid State Science," Vol. 5, R. Wolfe, ed., Academic Press, New York (1975), p. 251.
7. "JANAF Thermochemical Tables," 2nd Ed., NSRDS-NBS37, U.S. National Bureau of Standards, Washington, D.C. (1971).
8. C. R. Spencer, P. A. Sanger, and M. Young, IEEE Trans. Magn. MAG-15:76 (1979).
9. D. Bruce Montgomery, "The Magnetic and Mechanical Aspects of Resistive and Superconducting Systems," Robert E. Krieger Co., Huntington, New York (1980).
10. B. B. Gamble, T. E. Laskaris, and T. A. Keim, Transient performance of a prereacted, fully impregnated multifilamentary Nb_3Sn Coil, in: "Advances in Cryogenic Engineering--Materials," Vol. 28, Plenum Press, New York (1982), p. 739.

WELDABLE STRUCTURAL STEELS FOR ROTORS OF CRYOGENERATORS

K. A. Yushchenko, V. I. Belotzerkovetz, O. G. Kvasnevskii, and A. V. Shavel

E. O. Paton Institute of Electrowelding, Ukr.SSR Academy of Sciences, Kiev, USSR

INTRODUCTION

During the next ten to fifteen years, there will probably be a transition to use of high power cryoelectrical machines. The first industrial cryogenerators are already being designed in some countries. The initial publications indicate that by 1984 or 1985 the first cryogenerators (with 300 MVA of power) will be placed on a test bed and put under pressure. This will be the intermediate step between extensive laboratory investigations and the actual use of superconductivity in cryoelectrical machines. Therefore, the choice and optimization of structural materials for rotors of cryogenerators is urgent. Owing to service conditions and safety considerations, the materials have especially strict requirements:

- high strength at 293 K and sufficient brittle fracture resistance at 4.2 K;
- fatigue loading resistance;
- structural and property stability in magnetic fields;
- resistivity to crack growth rate and large critical crack sizes preceding their metastable development;
- good weldability, even for thick specimens, by all welding methods;
- possibility of producing large bars by ordinary metallurgical manufacturing methods;
- low magnetic susceptibility;
- the least possible values of Young's modulus and thermal expansion coefficient;
- low susceptibility to stress concentration.

There are some other requirements (for example, cost) that are also important and should be considered when choosing a structural material. The analysis of these requirements and the specifics of operation indicate that special steels and alloys must be developed for rotors of cryoelectrical machines.

In the USSR, scientists at the E. O. Paton Institute of Electrowelding and the Physical and Technical Institute of Low Temperatures, Ukr. Academy of Sciences, have developed a group of steels and alloys for cryogenic applications that are being used. These materials possess an austenitic structure with a high degree of stability at 293 K and are characterized by various strength levels. All have good weldability.

AUSTENITIC STEELS

Let us analyze these stable austenitic 03Kh13M10G19AM2 and 03Kh20N16AG6 steels. Their chemical composition is given in Table 1; mechanical properties of the steels and weldments are given in Table 2.

Table 1. Chemical Composition of Cryogenic Steels

Steel	%C	%Mn	%Si	%Cr	%Ni	%N	%Mo	%S	%P
03Kh13N10G19AM2	<0.03	19–20.5	<0.4	13–14.5	8.5–10	0.25–0.4	1.8–2.4	<0.02	<0.03
03Kh20N16AG6	<0.03	5.0–7.0	<0.04	18.0–20.0	15.0–17.0	0.12–0.22	-	<0.02	<0.02

Table 2. Mechanical Properties of 03Kh20N16AG6 and 03Kh13N10G19AM2 Steels and Weldments

Steel, Weldment	Test Temperature, K	σ_{UTS}, MPa	$\sigma_{0.2}$, MPa	δ, %	Ψ, %	$\sigma_{NTS}/\sigma_{UTS}$	a_v, MJ/m^2
03Kh20N16AG6	293	370	724	58	67	1.46	3.2
	4.2	1200	1684	26	46	1.42	0.87
Weldment 03Kh20N16AG6	293	350	695	62	65	1.5	1.7
Flux ANK-45 w-01Kh19N15G6M2AV2	4.2	1160	1600	38	48	1.45	0.52
03Kh13N10G19AM2	293	450	750	46	67	1.35	2.7
	4.2	950	1420	23	38	0.84	0.9
Weldment 03Kh13N10G19AM2	293	380	690	52	65	1.4	1.8
Flux ANK-45 w-01Kh19N15G6M2AV2	4.2	930	1370	33	34	1.35	0.61

For both steels, automatic gas-shielded, submerged, and electroslag welding may be used. Steel 03Kh20N16AG6 is the most completely characterized; its properties have been investigated in conditions of static and fatigue loading and also in magnetic fields.

It should be noted that owing to their high nitrogen content, both steels in the as-quenched condition at 4.2 K possess yield strengths of more than 900 MPa and ultimate strengths of more than 1300 MPa. The weldments are practically equal in strength to the base metal. The steels are not very sensitive to stress concentrators and their strength does not decrease while testing notched specimens. As theoretical and experimental investigations show, a magnetic field of about 4 T at 4.2 K does not lead to any significant change of the metal structure or increase the characteristics of the yield strength.

The steels possess exceptionally high brittle fracture resistance. The stress intensity factor is more than 160 $MPa \cdot m^{1/2}$ for the 03Kh20N16AG6 steel (>140 $MPa \cdot m^{1/2}$ for its weld) and 110 $MPa \cdot m^{1/2}$ for the 03Kh13N9G20AM2 steel (90 $MPa \cdot m^{1/2}$ for its weld), as defined by the J-integral method.

Investigations of fatigue strength showed that the fatigue life limit on the basis of 10^7 cycles for 03Kh20N16AG6 steel during fatigue bending is more than 600 MPa. The specific strength of both steels at cryogenic temperature is also exceptionally high (Table 2). Comparatively low nickel content, complete structural stability of steels, good adaptability to manufacture and low cost place these materials into the category of materials that could be used for various elements of the cryogenerator rotor.

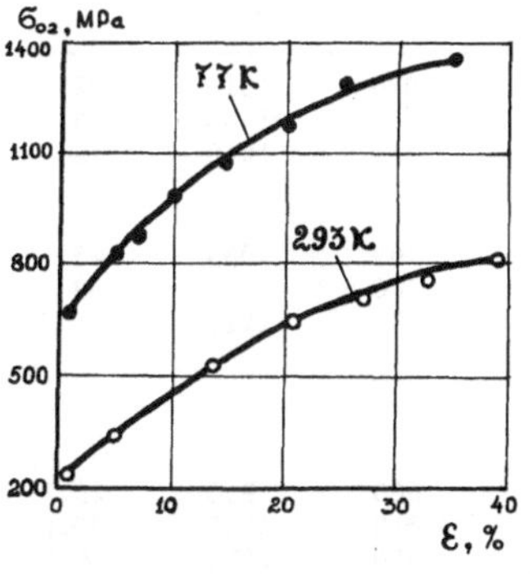

Fig. 1. The influence of the degree of preliminary deformation on the yield strength of 03Kh20N16AG6.

The essential drawback of the steels is their comparatively low strength (yield strength = ∿400 MPa and strength limit = ∿700 MPa) in the as-quenched condition at room temperature. As shown in our latest investigations, the strength of steels containing nitrogen may be significantly increased by strain hardening. After 10-15% deformation of the stable austenitic steel at 293 K, the yield strength may be increased up to 600 MPa and the ultimate strength up to 900 MPa (Fig. 1) without any significant decrease in ductile qualities or brittle fracture resistance.

NICKEL ALLOYS

The principal advantage of the nickel alloys, strengthened by the formation of γ'-phase, is: in the aged condition they possess high strength. As a rule, the high nickel alloys are considered difficult to weld because of their increased sensitivity to crack formation during the welding process in the weld and HAZ. In the

Table 3. Chemical Composition of Nickel Alloys

Alloy	%C	%Si	%Mn	%Cr	%W	%Mo	%Al	%Nb	%Fe	%Ni	Others
KhN60MVYu	<0.03	<0.3	0.6-3.5	19-22	4-7	8-12	2.8-3.4	-	<5	rest	B 0.01
KhN55MYuB	<0.08	<0.5	0.4 0.8	18-20	-	8.5-10.0	1.2-1.8	1.5-2.5	10-15	rest	-
KhN30MYuB	<0.03	<0.3	0.5	15-17	-	4-6	-	4.5-5.5	rest	28-32	-

Table 4. Mechanical Properties of Nickel Alloys

Alloy	Heat Treatment	Test Temp., K	Properties Exponents					
			σ_{UTS}, MPa	$\sigma_{0.2}$, MPa	δ, %	Ψ, %	a_v, MJ/m^2	$\sigma_{NTS}/\sigma_{UTS}$
KhN60MVYu	Quenching 1150°C, cooling in air	293	1020	1310	32	34	0.57	1.36
		77	1150	1480	25	27	0.44	1.33
		4.2	1315	1610	18	24	0.42	1.32
KhN55MYuB	Quenching 1100°C, cooling in water, aging at 700°C for 16 h	293	560	1060	45	48	0.86	1.32
		77	740	1270	39	40	0.72	1.32
		4.2	890	1420	31	29	0.63	1.30
KhN30MYuB	Quenching 1100°C, cooling in water, aging at 700°C for 15 h	293	680	950	25.2	48.3	0.45	1.2
		77	810	1250	30.0	36.4	0.38	1.2
		4.2	910	1410	24.2	18	0.35	1.1

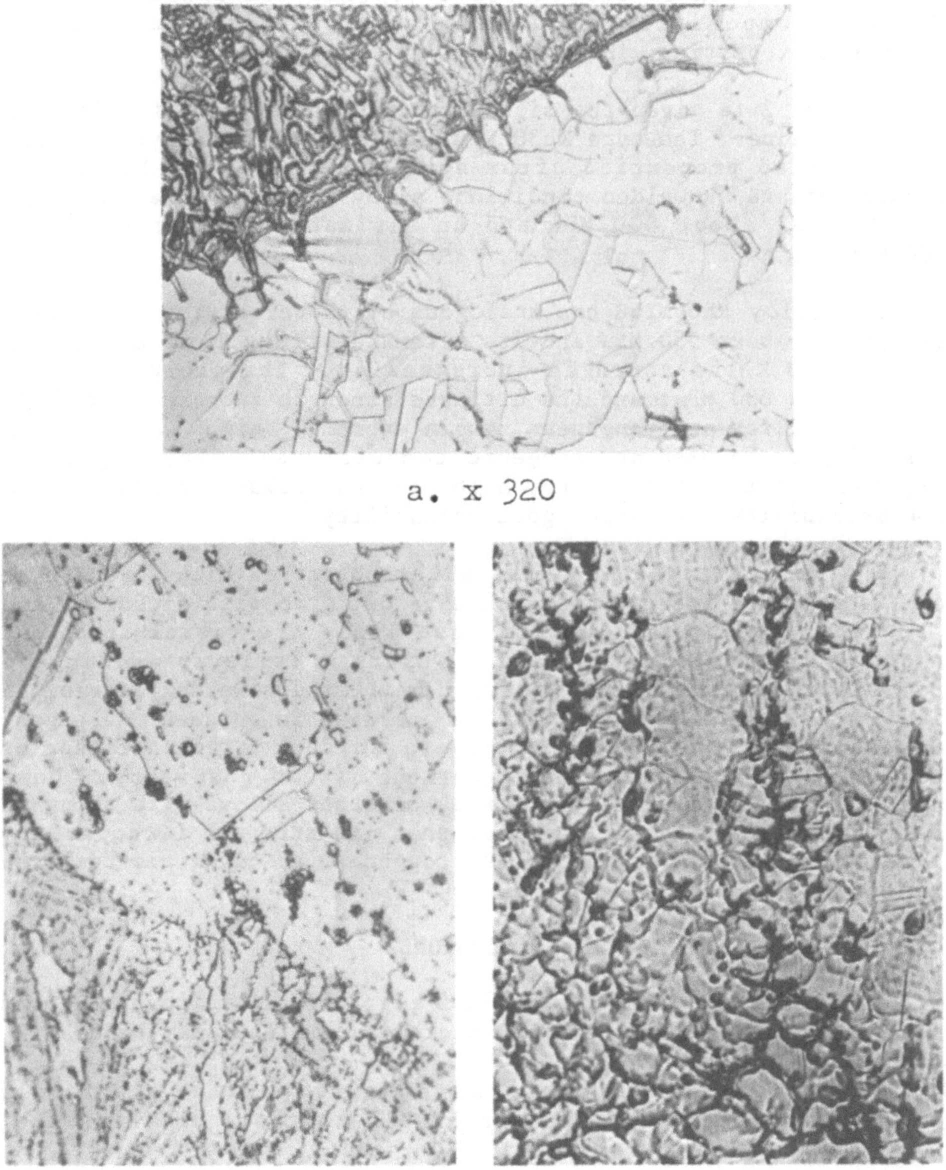

a. x 320

b. x 200

c. x 320

Fig. 2. Microstructure of nickel alloys near the fusion line. (a) KhN60MVYu, (b) KhN30MYuB, (c) KhN55MYuB. (Reduced 5%.)

USSR, various weldable alloys with high and moderate nickel content have been developed for cryogenic temperatures. The chemical composition of some of them is given in Table 3; mechanical properties, in Table 4.

In Fig. 2 is given the microstructure of the HAZ of alloys at the fusion line. The type Kh30MVYu and Kh55MYuB alloys acquire the indicated properties after aging at 973 K for 15 h. In the as-quenched and as-welded condition, alloy KhN60MVYu has a yield strength of about 800 MPa and an ultimate strength of about 1100 MPa.

The alloy Kh65MYuB has analogous characteristics. Its yield strength is about 550 MPa and its ultimate strength is about 900 MPa. At 20 K the strength increases 40%; the yield strength increases to 800 MPa; and the ultimate strength increases to 1400 MPa. Ductility and toughness remain extremely high, which allows this alloy to be used at cryogenic temperatures. The degree of γ'-phase precipitation is not high, and the alloy, in addition to good weldability, possesses good workability.

The more alloyed Kh60MVYu alloy possesses high strength at room temperature following quenching. Owing to significant Mo and W content, thick specimens of this alloy are resistant to hot crack formation during welding. During multilayer welding, a sufficient quantity of γ'-phase in multipass welds is precipitated, which leads to a significant increase in weldment strength. In some cases, the weldments have a uniform strength factor of about 0.9, and subsequent heat treatment is not necessary. Comparison of the mechanical properties of other alloys with those of the Kh60MVYu alloy (Table 4) shows that its strength is superior; it maintains a high ductility and toughness at cryogenic temperatures.

CONCLUSIONS

1. Depending on the required strength level, nitrogen-alloyed stable austenitic steels or nickel-base alloys may be used.

2. At cryogenic temperatures, the properties of steels containing nitrogen are preferable to those of the nickel alloys.

SHIELDED METAL-ARC AND FLUX-CORED METAL-ARC STAINLESS STEEL WELDMENTS: MAGNET CASES FOR 4-K SERVICE

E. N. C. Dalder

Lawrence Livermore National Laboratory, Livermore, California

O. W. Seth

Chicago Bridge and Iron Company, Houston, Texas

T. A. Whipple

National Bureau of Standards, Boulder, Colorado

INTRODUCTION

Load-bearing structures of superconducting magnet systems involve the manufacture of defect-free fusion welds in nitrogen-strengthened austenitic stainless steels in thicknesses to 150 mm. These welds must be capable of fracture-safe operation at 4 K at stresses close to yield and must resist failure caused by growth of fatigue cracks to critical sizes and rapid propagation to fracture. Many factors are involved in making satisfactory welds for this purpose; they have been examined and discussed in Refs. 1 and 2, for example.

This paper presents an evaluation of shielded metal-arc (SMA) and flux-cored metal-arc (FCMA) welding consumables designed to deposit ferrite-free 316L stainless steel weld metal in the flat position at high deposition rates. The significance of the results to the design and manufacture of the Mirror Fusion Test Facility (MFTF-B) superconducting magnet set[3] are discussed.

MATERIALS

Welds were made in 75-mm-thick type 304LN austenitic stainless steel (Fe-19Cr-9Ni-2Mn-0.02C-0.14N) using one of three manufacturers' 6.4-mm-diameter type E316L-15 SMA electrodes or using one manufacturer's 2.4-mm-diameter type 316L FCMA electrodes. Compositions of weld rods had been balanced to ensure a ferrite-free weld.

PROCEDURE

Welds were made in the flat position, using a single-vee joint with a 30° included angle, a 9.5-mm root opening, and a 9.5-mm-thick backup bar. Restraint was produced by welding the assembly to a 2.7- to 15.2-cm-thick carbon-steel plate after completion of root passes. Approximately 250 mm of weld per joint was sent to Lawrence Livermore National Laboratory (LLNL) for evaluation, which consisted of chemical analyses near the weld face and root, room-temperature side-bend tests and metallographic examination, 77-K Charpy V-notch impact tests with the notch oriented to cause crack propagation either along the direction of welding or from top to bottom of the weld, 4-K weld-metal tension tests, and elastic-plastic fracture-toughness tests.[4] Tension test specimens were removed from near the face or root of the weld and pulled in the longitudinal direction (parallel to the direction of welding). The notch in the elastic-plastic fracture-toughness specimens was oriented to produce cracking parallel to the welding direction.

Side-bend testing was conducted per paragraph QW163 of Ref. 5. Impact testing at 77 K involved immersion of specimens in liquid nitrogen for 20 to 30 min, removal of a specimen, insertion in the holder of a 326-J impact machine, and breaking within 5 s after removal from the liquid-nitrogen bath. Experience has shown that negligible temperature rise occurs under these conditions. Energy absorption, lateral expansion at the root of the notch, and amount of shear fracture were measured. Tensile tests at 4 K were run at a crosshead speed of 1.27 mm/min in a relatively "hard" tensile-testing machine with specimens immersed in liquid helium for 30 min. Specimen temperature was measured with C-resistance thermometers, and extension up to 2% strain with 2 to 3 strain gages mounted on the specimen gage length, 180° or 120° apart around the circumference of the specimen. Elastic-plastic fracture-toughness tests at 4 K were conducted at the National Bureau of Standards in Boulder according to procedures described in Ref. 4.

Metallographic specimens were prepared for examination by standard polishing techniques and etched with a 10% oxalic acid solution at 1.5 V, 6 A, for 15 to 45 s. Ferrite numbers (FN) were measured on polished and etched metallographic specimens using a magnetic inductive type instrument.

Table 1. Compositions and Ferrite Numbers of Type 316L Welds

Vendor	Weld-Process	Location of Analysis	Composition, Weight Percent									Cr* Equiv.	Ni** Equiv.	Ferrite No.	
			C	Mn	P	S	Si	Cr	Ni	Mo	N_2			Calc.	Meas.
A	S.M.A.	Top	0.034	2.62	0.013	0.010	0.25	17.94	13.51	2.14	0.029	20.4	16.7	-1	0
		Bottom	0.031	2.70	0.024	0.010	0.26	18.03	13.33	2.07	0.035	20.5	16.6	-1/2	0.5
B	S.M.A.	Top	0.029	1.90	0.018	0.012	0.37	17.98	13.08	2.20	0.056	20.7	16.6	+1/2	1
		Bottom	0.027	1.99	0.019	0.012	0.39	18.23	12.99	2.13	0.070	21.0	16.9	0	1.5
C	S.M.A.	Top	0.032	1.81	0.034	0.010	0.36	18.65	14.34	2.21	0.052	21.4	17.8	0	0
		Bottom	0.034	1.79	0.034	0.009	0.34	18.91	13.91	2.10	0.052	21.5	17.4	+1	0
A	F.C.M.A.	Top	0.019	2.86	0.012	0.020	0.33	17.31	13.72	2.23	0.065	20.0	17.7	-1	0
		Bottom	0.020	2.79	0.012	0.017	0.33	17.23	13.59	2.14	0.059	19.9	17.4	-4.5	0
		Weld Pad	0.023	2.51	0.013	0.017	0.39	18.29	13.29	2.30	–	–	–	–	2.0

Notes:

*Cr. Equiv. = % Cr + % Mo + 1.5% Si + 0.5% Cb

**Ni. Equiv. = % Ni + 30(% C + % N_2) + 0.5% Mn

RESULTS

Chemical compositions and FNs, both measured and calculated, are presented in Table 1. The method of calculating FNs is that described by Szumachowski and Reid.[6] With exception of a discrepancy between the calculated FN of -4.5/-5.0 for the FCMA weld with a measured FN of 2.0, agreement between calculated and measured FN values was generally good, i.e., calculated FNs had a range of -1 to 0, and the measured FNs were 0 to 1.5.

Side-bend tests passed requirements of paragraph QW163 of Ref. 5 (Table 2). The ductility of the FCMA weld bend barely meets these requirements: two 3-mm-long cracks were noted.

Tension-test results at 4 K are presented in Table 2. The SMA weld made from Vendor A's products had an average yield strength (σ_y) of 701 MPa, just above the 689 MPa value needed for MFTF-B to ensure that minimum base-metal and weld-metal σ_y values were equal for load sharing. Elongation and reduction in area were satisfactory. The weld made from Vendor B's products showed a somewhat lower average σ_y, 673 MPa, and the weld made from Vendor C's products had a higher average σ_y, 773 MPa. Both sets of ductility parameters were satisfactory. The average σ_y of the FCMA weld was low, 643 MPa, and there was about a factor of 2 variation in elongation and reduction in area values between the two specimens tested. Such variations are often an indication of welding-related inhomogeneities.[2]

Table 2. Toughness and Tensile Properties of Type 316L Welds

Weld Process and Vendor	Notch Orientation in CVN Test	77 K CVN				4 K Tensile Properties**			
		Energy-Absorption*	Lateral Expansion*	4 K K_{1c}(J)**	Ultimate Strength	0.2% Yield Strength	Elong. in 25.4 mm	Red. in Area	Room Temp.* Side Bend
S.M.A.:A	L†	61 J	0.84 mm	168.7 MPa$\sqrt{m}$	1335 MPa	701 MPa	50.5%	39.6%	Passed, 1 1.6 mm crack
	ST††	52	0.75						
S.M.A.:B	L	47	0.57	91.3	1328	673	40.0	26.3	Passed, 2 1.6 mm crack
	ST	47	0.70						
S.M.A.:C	L	55	0.76	165.4	1256	773	35.5	28.0	Passed, 1 3.2 mm crack
	ST	56	0.64						
F.C.M.A.:A	L	48	0.62	152.2	1170	643	28.3	23.1	Passed, 2 3.2 mm crack
	ST	65	0.61						
A.S.M.E. B. & P.V. Code, min.		20.4	0.38						None over 3.2 mm
LLNL MFTF-B requirement min.				132.0	1034	690	20.0		

*Average of 3 samples
**Average of 2 samples
***Worst of 2 samples
† L longitudinal
†† ST short transverse

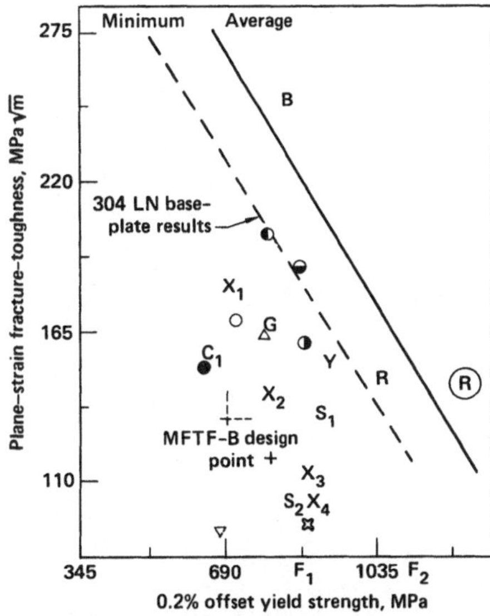

Coating	Symbol	Process	Filler	Vendor
-15	○	SMA	316L	A
-15	▽	SMA	316L	B
-15	△	SMA	316L	C
	●	FCMA	316L	A
-15	+	SMA	316L	
-15	¤	SMA	316L	
-15	◒	SMA	316L	
-15	◐	SMA	316L	
-15	◑	SMA	316L	
-16	X_1 X_4	SMA	316L	
-16	Y	SMA	316L	
	S_1, S_2	SAW	316L	
-15	C_1	SMA	308L	
	F_1, F_2	FCMA	308L	
	G	GMA	308L	
	R	SAW	20-16-6	
	Ⓡ	GMA	Cr Ni Mn	
-15	B	SMA	17Cr-13Ni-4Mn-2Mo	

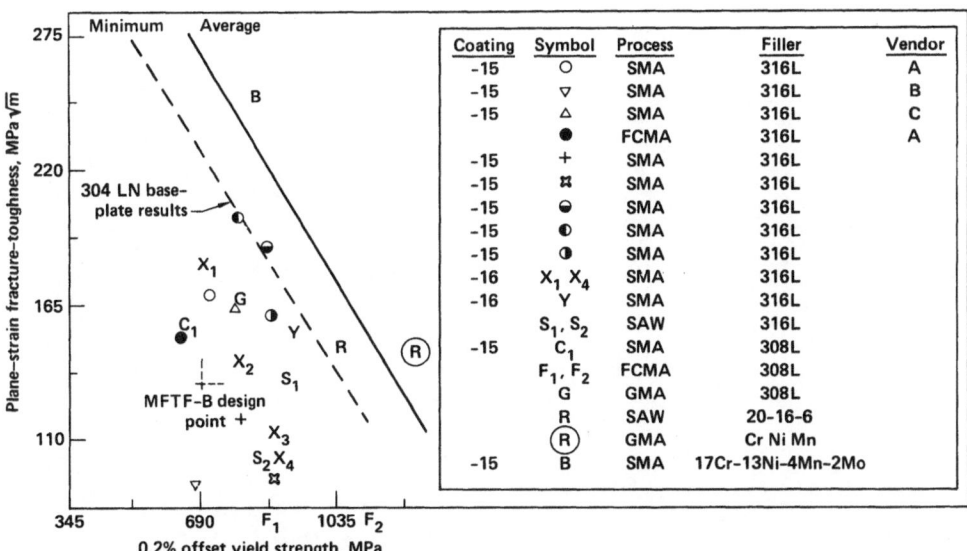

Fig. 1. Plane-strain fracture toughness versus yield strength of type 316L weld metals at 4.2 K.

In Table 2 results of both 77-K impact tests and 4-K elastic-plastic fracture-toughness tests are presented. Welds made from all products showed excellent 77-K energy absorption and lateral-expansion values compared with the 20.4-J (min.) and 0.38-mm (min.) values of Ref. 7. SMA welds from the products of Vendors B and C showed isotropy of energy absorption between the longitudinal and short-transverse directions. Both SMA and FCMA welds made with Vendor A's products showed superior energy absorption in the short-transverse direction. Little or no shear fracture was seen on all impact specimens.

Elastic-plastic fracture-toughness test results, when converted to plane-strain fracture-toughness (K_{Ic}) values, tell a different story. Welds made from Vendor A's products yielded excellent results, an average K_{Ic} for the FCMA weld of 152.2 MPa$\sqrt{m}$ and an average K_{Ic} for the SMA weld of 168.7 MPa$\sqrt{m}$. An average K_{Ic} of 165.4 MPa $\sqrt{m}$ was obtained from a weld made from Vendor C's SMA weld rod. A weld made from vendor B's SMA weld rod yielded an average K_{Ic} value of 91.3 MPa $\sqrt{m}$. A minimum K_{Ic} of 132.0 MPa $\sqrt{m}$, obtained by conversion from elastic-plastic fracture-toughness test results, is required of welds and base metals used in MFTF-B. The results of comparing σ_y:K_{Ic} trends at 4 K are plotted in Fig. 1, which also shows the minimum and average performance trends established for type 304LN stainless steel and the design point of 690 MPa σ_y:132 MPa$\sqrt{m}$ K_{Ic} for the MFTF-B magnet cases. Acceptable weld metals lie both to the right and above the design point.

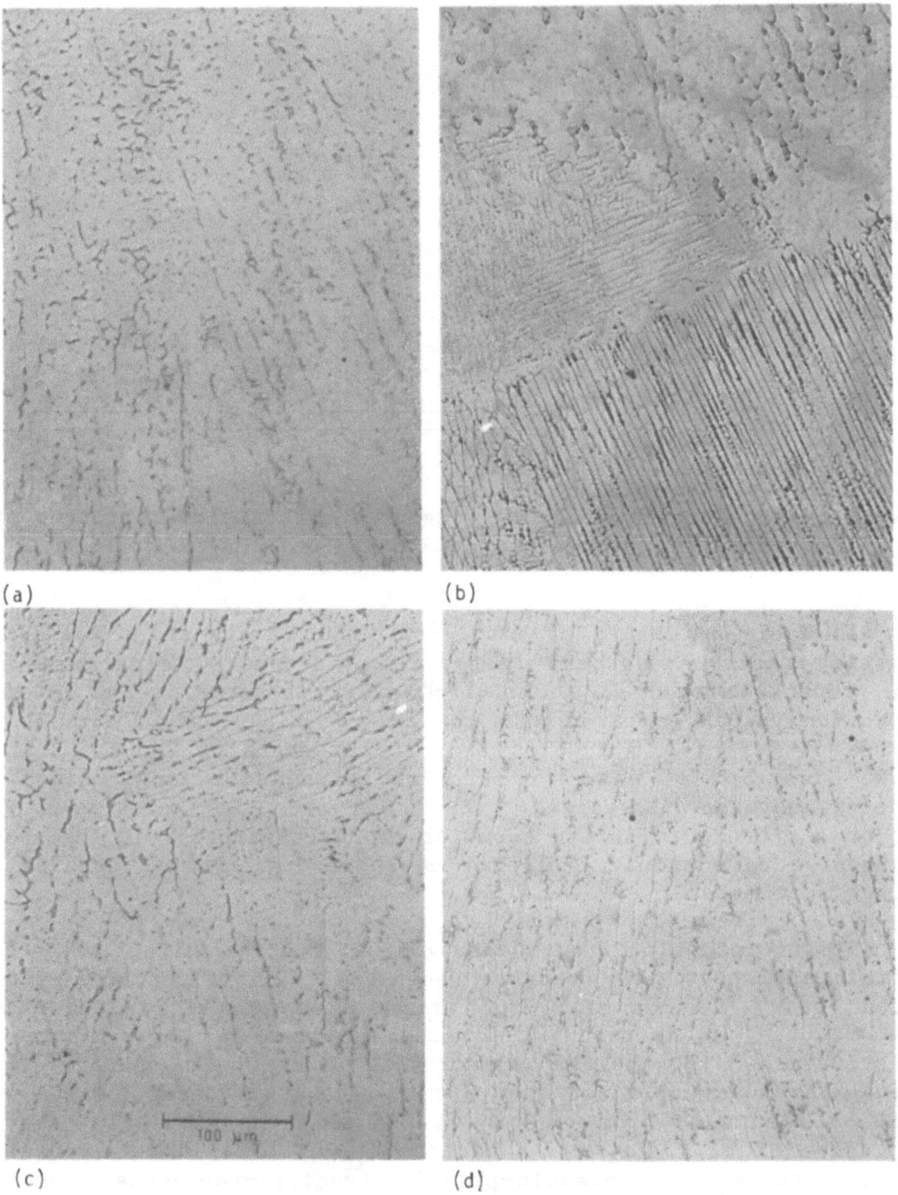

Fig. 2. Microstructure of type 316L weld metal deposited by (a) SMA Vendor A; (b) SMA Vendor B; (c) SMA Vendor C; (d) FCMA Vendor A. All X250. (Reduced 28%.)

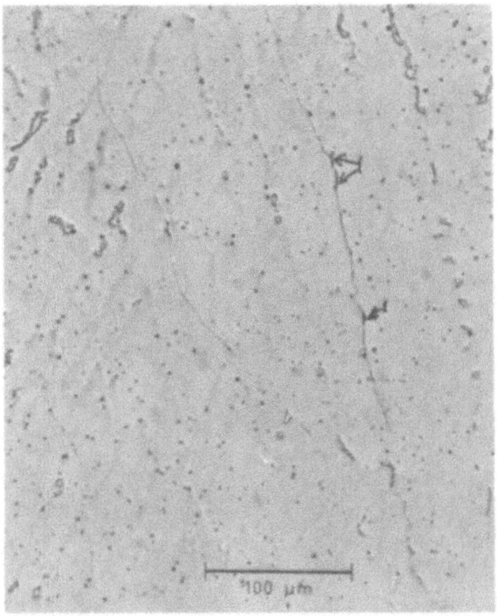

Fig. 3. Microstructure of type 316L weld metal deposited by SMA Vendor B. Note grain-boundary precipitates (1).

Figures 2 and 3 show the microstructures of each weld. No evidence of microfissuring was found on examination to magnifications of 1000. Figure 2 shows evidence of a divorced island of ferrite at interdendritic boundaries, which supports measured FNs (Table 1). Note the precipitates along some of these boundaries (Fig. 3). Such precipitates have been shown by Witherell[8] to be associated with a susceptibility for microfissuring.

Figure 4 shows microstructures of the FCMA weld. Figure 4a shows many scattered defect regions located at interpass boundaries. At higher magnifications (Fig. 4b), all defect regions fall either along interpass boundaries or at intersections of the defect region with the fusion line. Pores, 0.15-0.25 mm in diameter, are scattered throughout the defect regions. Dimensions of the defect regions are about 4.8 mm (transverse to weld) by 7.9 mm (along weld axis). Regions of partially-dissolved metallic flux constituents are also present.

Figure 5 shows scanning-electron microscopy (SEM) and energy-dispersive x-ray results from the defect regions. Both metallic and nonmetallic inclusions are present. Metallic inclusions are Cr (Fig. 5a), Ni (Fig. 5b), and alloys containing various amounts of Cr, Fe, Mn, and Ni. Nonmetallic inclusions are rich in Ca. Extensive SEM studies of fracture surfaces of the fracture toughness

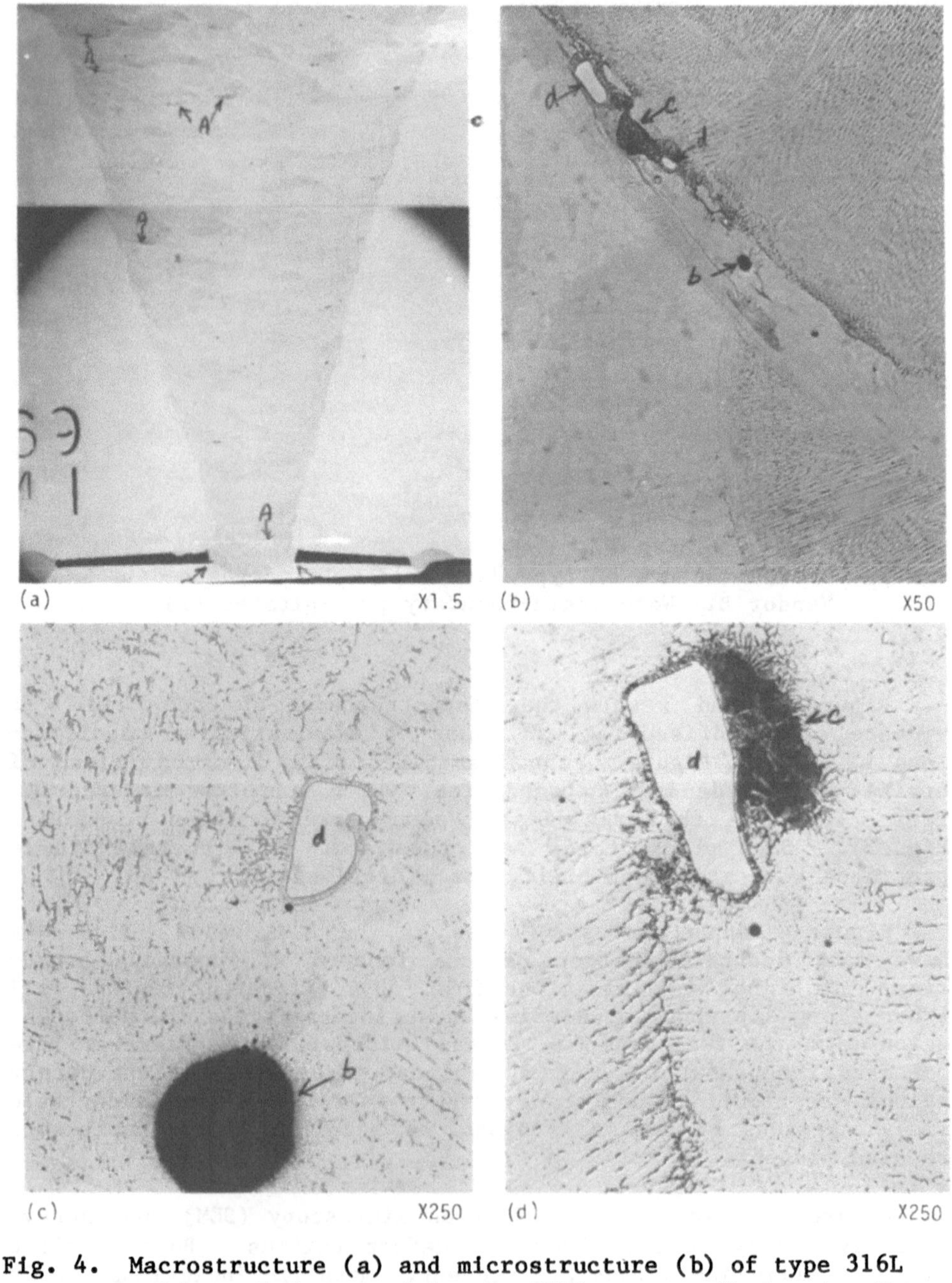

Fig. 4. Macrostructure (a) and microstructure (b) of type 316L weld metal deposited by the FCMA process Vendor A. Note defect regions at interpass boundaries (a), porosity (b), Cr-rich ferrite (c), and Cr-rich inclusions (d). (Reduced 28%.)

specimens show all fractures were predominantly ductile rupture, with fracture emanating from inclusions (Fig. 6a). All SMA welds showed separations along isolated stringers of ferrite (Fig. 6b). All welds showed discontinuous microfissures that were partially closed prior to testing (Fig. 6c). The SMA welds made with Vendors B and C's products showed continuous microfissures that were completely open prior to testing (Fig. 6d), as well as a tendency to separate along primary and secondary dendrite interfaces (Fig. 7a). Defect regions in the FCMA weld separated under load, forming straight cracks (Fig. 7b).

DISCUSSION

Presented in Fig. 1 is 4-K $\sigma_y:K_{Ic}$ information on austenitic stainless steel weld deposits.[1,2,8,9] Superposed on this figure are results of this investigation: minimum and average trend lines for type 304LN stainless steel, and the MFTF-B magnet-case design point. Most welds that exceed MFTF-B design-point requirements are type 316L, deposited in the flat position by the SMA process with compositions balanced to yield welds with no ferrite. Coating type does not make an appreciable difference in performance. Other weld compositions that exceed MFTF-B design-point requirements include type 316/SMA, type 308L/GMA, a Soviet 20Cr-16Ni-6Mn alloy deposited by either the SA or GMA process, and an Austrian-manufactured 17Cr-13Ni-4Mn-2Mo SMA product. Two of three SMA welds evaluated herein also exceeded the MFTF-B design-point requirements.

Consider which welds failed to meet MFTF-B design-point requirements and possible reasons why. Of the FCMA weld deposits, the two with type 308L welds had poor fracture toughness attributed to high nitrogen levels (above 0.12%),[10] which is supported by the results of Ref. 11. The type 316L FCMA weld had a slightly low σ_y but adequate K_{Ic}. One SA and four SMA welds made with type 316L products had low K_{Ic} values (probably due to too much ferrite)[2,6] for all but the SMA weld made with Vendor B's product. This weld had a rather low FN of 1/1.5, but also had the highest nitrogen level (0.056-0.070%) of the three SMA welds tested herein. Excessively high nitrogen levels are known to lower cryogenic toughness in stainless-steel welds.[11]

The presence of microfissures, whether discontinuous or continuous, did not affect the 4-K K_{Ic} or 77-K impact performance, as has been noted before.[2,8,12] Regions of undissolved FCMA flux constituents, porosity, and Cr-rich ferrite in the FCMA weld, all of which contributed to opening of continuous cracks (Fig. 7b) in cryogenic testing, did not seriously degrade either the 4-K K_{Ic} or 77-K impact performance. Microfissures, which become visible only after plastic deformation at cryogenic temperatures, were oriented normal to the cracking plane in impact and K_{Ic} specimens. These microfissures act as obstacles to crack growth by diverting the

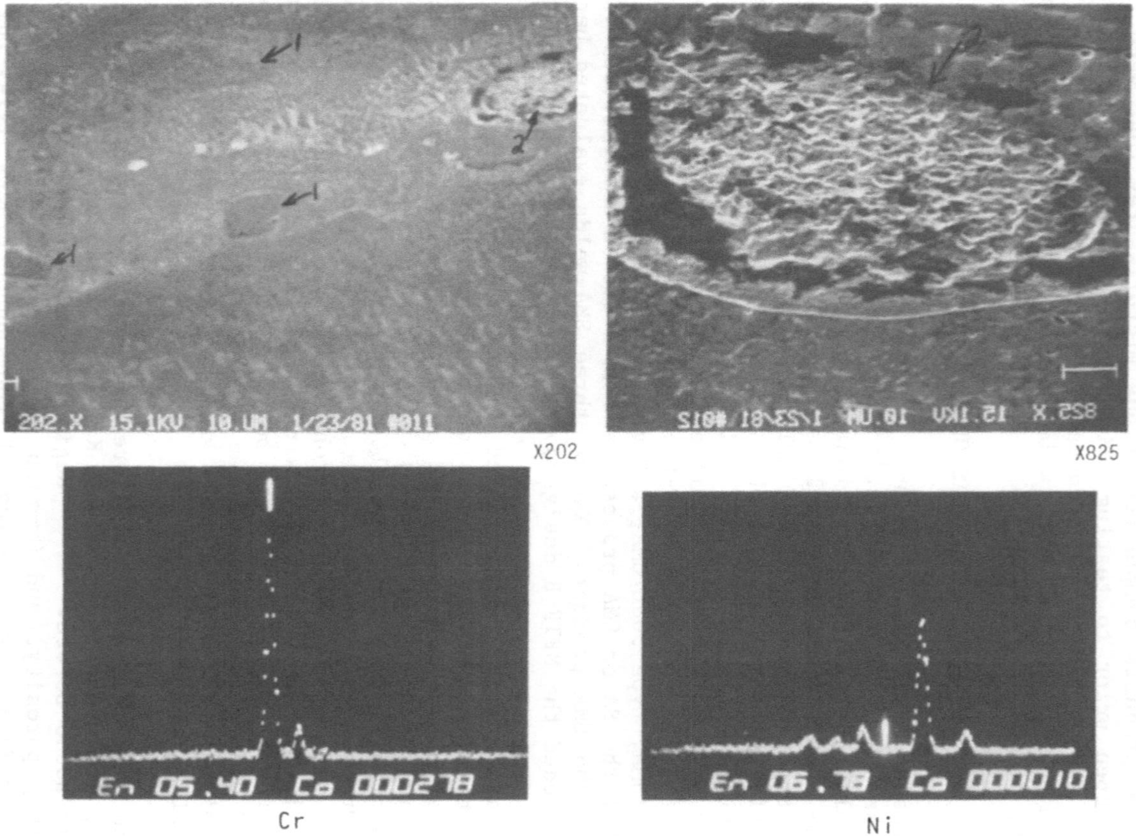

Fig. 5. Results of energy dispersive x-ray analyses on various defect regions in a FCMA Vendor A weld. Note regions of high Cr (1), high Ni (2). (Reduced 28%.)

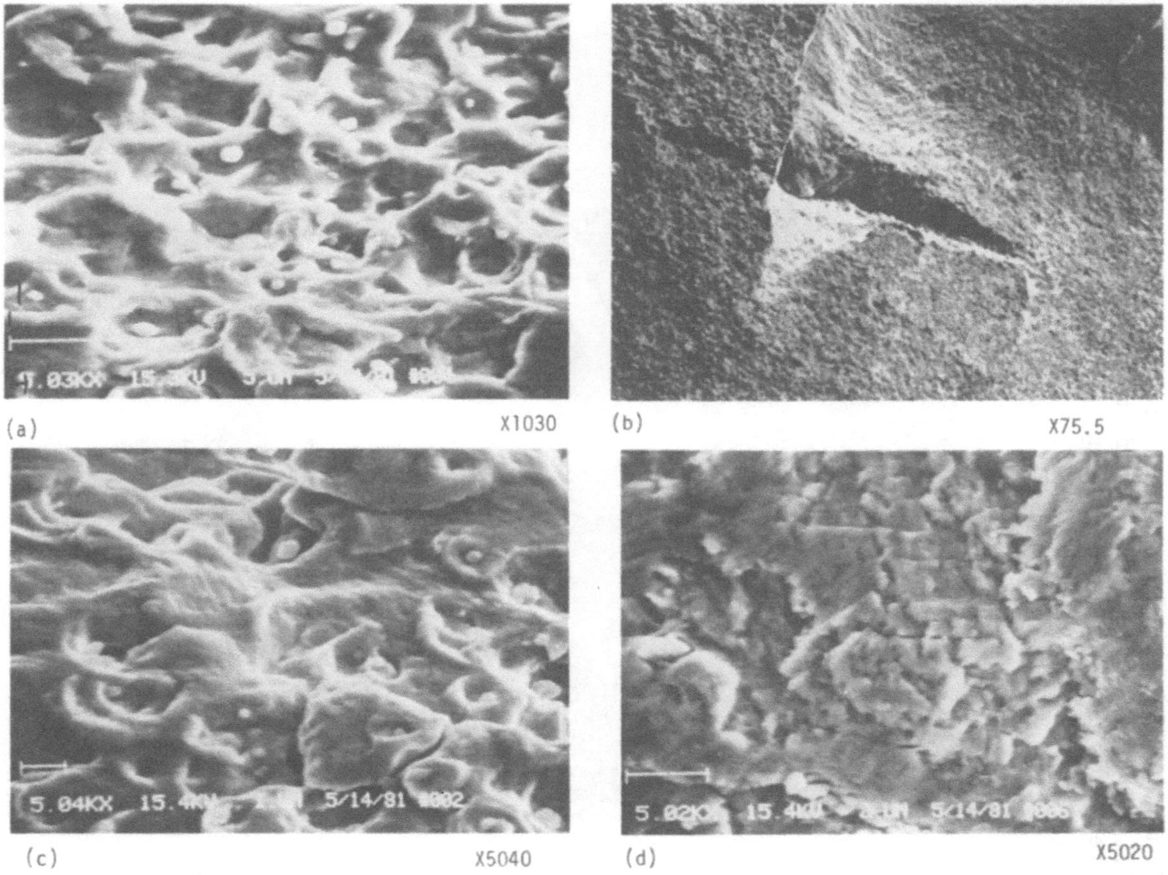

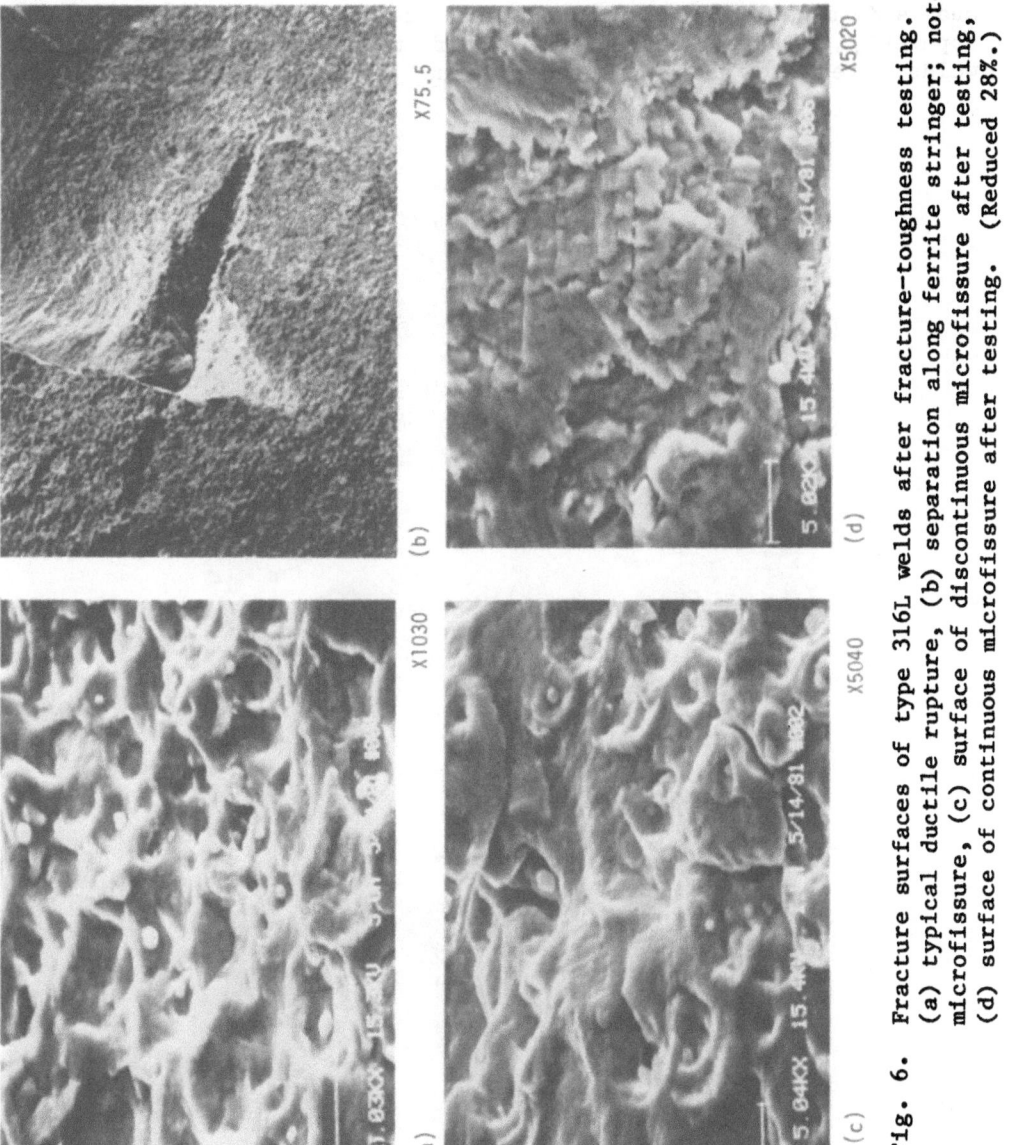

Fig. 6. Fracture surfaces of type 316L welds after fracture-toughness testing. (a) typical ductile rupture, (b) separation along ferrite stringer; note microfissure, (c) surface of discontinuous microfissure after testing, (d) surface of continuous microfissure after testing. (Reduced 28%.)

crack from its original path. Separations along primary and secondary interdendritic boundaries that were shown by SMA welds made with both Vendor B and C's weld rod do not correlate with K_{Ic} performance, as the latter weld showed an 82% higher K_{Ic} than the former weld. Also, no definite correlation between microfissuring and 4-K tensile ductility, as had been seen previously,[2] was apparent, perhaps because levels of microfissuring in all four welds were below some threshold value. Poor K_{Ic} of the SMA weld

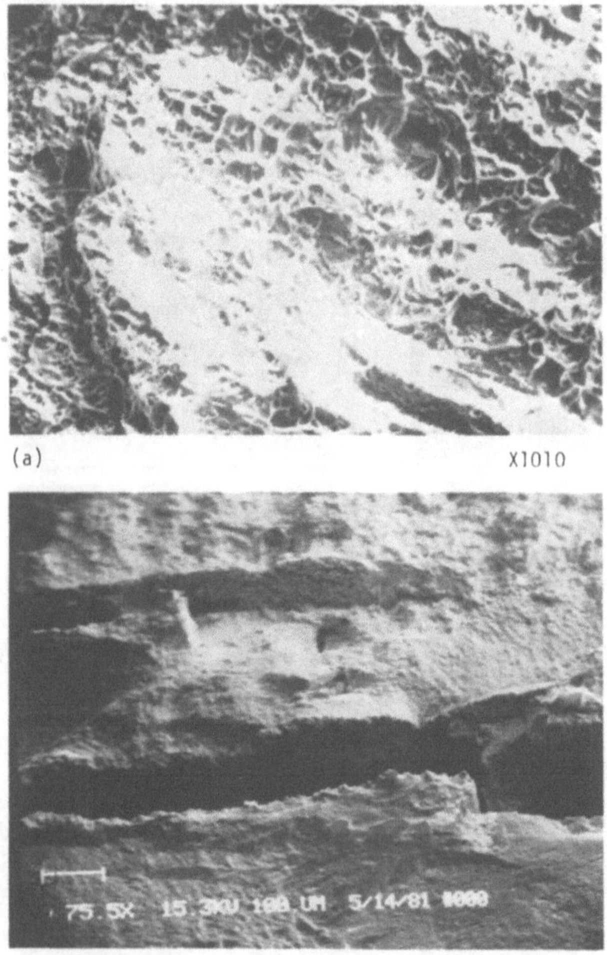

Fig. 7. Fracture surfaces of type 316L welds after fracture-toughness testing. (a) separation along primary and secondary dendrite interfaces in an SMA weld, (b) separation along defect regions in a FCMA weld. (Reduced 28%.)

made with Vendor B's weld rod may be explained on the basis of a combination of the highest ferrite (FN: 1/1.5) and nitrogen levels (about 0.056/0.070%) of the three SMA welds tested herein.

CONCLUSIONS

1. Of three highly restrained SMA welds made in the flat position with large-diameter type E316L-15 welding electrodes supplied by different vendors, two welds showed adequate combinations of 4-K tensile and fracture toughness properties to be considered for use in manufacture of MFTF-B superconducting magnet cases.

2. The poor 4-K performance of the third SMA weld is explainable, in part, by a combination of higher ferrite level and nitrogen concentration than in the other two SMA welds.

3. Despite the presence of undissolved metallic flux constituents and porosity in a type 316L FCMA weld, its 4-K K_{Ic} would be adequate for use on MFTF-B. However, its 4-K σ_y was slightly below MFTF-B requirements.

ACKNOWLEDGMENTS

Excellent mechanical testing services of R. Brady and metallographic services of R. Kershaw and D. Diaz are gratefully acknowledged. Editorial assistance of J. Bantley is gratefully acknowledged. Work performed under the auspices of the U.S. Department of Energy by Lawrence Livermore National Laboratory under Contract No. W-7405-Eng-48.

REFERENCES

1. E. Dalder, Welding of austenitic stainless steels for liquid helium service, in: "Proceedings of the 1980 Superconducting MHD Magnet Design Conference," Massachusetts Institute of Technology, Cambridge, Massachusetts (1980).
2. T. A. Whipple, H. I. McHenry, and D. T. Read, Weld. J. 60(14):72s.
3. R. E. Bulmer and J. Van Sant, "Tandem Mirror Magnet System for the Mirror Fusion Test Facility," UCRL-84309.
4. D. T. Read, The computer-aided J-integral test facility at NBS, in: "Materials Studies for Magnetic Fusion Energy Applications at Low Temperatures--II," NBSIR 80-1620, National Bureau of Standards, Boulder, Colorado (1980), p. 290.
5. Welding and brazing qualifications, Section IX, in: "ASME Boiler and Pressure Vessel Code," ASME, New York (1980).
6. E. Szumachowski and H. Reid, Weld. J. 57(11):325s.
7. Section VIII, Divisions 1 and 2, in: "ASME Boiler and Pressure Vessel Code," ASME, New York (1977).

8. C. Witherell, Weld. J. 59(11):326s (1980).
9. D. T. Read, H. I. McHenry, P. Steinmayer, and R. Thomas, Jr., Weld. J. 59(4):104s (1980).
10. F. Schneider, Jr., General Dynamics-Convair, San Diego, California, personal communication (1979).
11. E. Szumachowski and H. Reid, Weld. J. 58(2):340s (1979).
12. T. Gooch and J. Honeycomb, Weld. J. 59(8):233s (1980).

USE OF HEAVY SECTION AUSTENITIC WELDS FOR 4-K SERVICE*

A. Nyilas and H. Krauth

Institut für Technische Physik, Kernforschungszentrum Karlsruhe
Karlsruhe, Federal Republic of Germany

INTRODUCTION

Both the magnetically confined fusion reactors of the future and the large demonstration fusion devices will have to use superconducting magnets. Very large forces are involved in these magnets owing to their size and the high magnetic fields. The 4-K magnet structure must safely withstand the stresses arising under all operational conditions. For this purpose, nitrogen-strengthened fully austenitic stainless steels show an optimum combination of strength, stiffness, and toughness. The use of these materials in large complex constructions necessitates material joining, which gives rise to varying material properties along the joined section. Thus the weakest points in the material behaviour of such large-scale constructions are their weldments.

This report covers mainly the weldment properties associated with the Euratom coil for the Large Coil Task development.[1] The selected structural material for the casing of this D-shaped coil is the fully austenitic chromium-nickel stainless steel German Mat. No. 1.4429, which is similar to AISI 316LN. The construction of this case requires about 50-mm-thick austenitic weldments with an overall weld length of about 100 m. The structural integrity of this component at 4 K demands a quantitative assessment of the weld metal to guarantee this discontinuous zone and a maximum margin of safety against ductile and brittle failure.

Different fusion welding processes, e.g., SMA (Shielded Metal Arc), GMA (Gas Metal Arc), and EB (Electron Beam) welding can be used to join the selected structural material depending on the

*Invited paper.

joint thickness, the component dimensions, and the specified construction. EB welding of a smaller D-shaped coil case using the same material and a sound ~20-mm single-pass weld penetration was reported by Friesinger et al.[2] The possibility of joining the LCP case using single-pass EB welding can be excluded because of the size and the complexity of the construction, although the EB technique is capable of being used to manufacture ~50-mm joint thickness. Early investigations to join this material by the common GMA process revealed poor wetting behaviour of the base metal by the liquid weld metal, which resulted in large-scale lack of fusion. Further technological investigations on the various types of GMA power sources, the proper selection of welding rod material, and the gas mixture were indispensable for the development of a suitable GMA welding process to achieve sound welds. The known practice of manual welding process with the disadvantage of low deposition rate seemed to be the only acceptable way at this stage of development. The SMA welding of fully austenitic materials, especially of heavy-walled components, however, still gives rise to technological difficulties. This is evident from the production of numerous types of internal defects and microstructural changes. Therefore, much effort has been put into the improvement of the weld quality by proper selection of the covered consumables.

The engineering application of fracture mechanics to the weld zone is imperative for a safe design. Therefore, the fracture toughness values of several weldments welded by different consumables were determined at 4 K by the application of the J-integral test with compact tension (CT) specimens. A commercially available consumable stick electrode for the welding process was chosen by considering the toughness and yield strength level at 4 K together with the size, shape, type, and location of internal weld defects.

Table 1. Composition of Deposited Weld Metals and the Calculated Chromium and Nickel Equivalents

Mat. des.	Commercial name	Joint type	Chemistry, wt%++										
			C	Si	Mn	P	S	Cr	Ni	Mo	N	Cr_{eq}	Ni_{eq}
base	Nirosta 4429	-	0,032	0,41	1,26	0,016	0,016	16,7	13,7	2,68	0.16	20.0	20.0
A1 A2 A3	Thermanit 19/15 Thermanit 19/15 Thermanit 19/15	Double U Double U T-joint	0,039	0,31	5,36	0,020	0,005	19,3	15,4	2,99	0.17	22.8	24,4
B1 B2	Novonit 4455B Novonit 4455B	Double U SMA-dep.+	0,057	0.33	6,74	0,012	0,022	18,0	17,0	2,93	0,19	21.4	27,8
C1	Thermanit 18/17	Double U	0,037	0,25	3,10	0,018	0,013	17,5	16,1	3,99	-	21,9	18,8
D1	Grinox 23	Double U	0,075	0,84	4,48	0.017	0,007	23,3	17,9	0.41	-	25.0	22.4

+ SMA multipass weld deposit on 316 LN base plate

++ Fe in balance

WELD MATERIALS

The weldments presented in this paper are all SMA welded and were produced as part of the LCP materials testing programme. Table 1 summarizes the weld zone characteristics of the tested different commercially available consumable electrodes. The covering of the weld rods was in all cases basic coating. Before use the electrodes were dried at 300°C to remove moisture from the coating. The joints were prepared by machining the base metal out of a ~50-mm-thick plate material. The 45-50-mm-thick joints were produced in the flat position by multipass welding. The overall length of each joint was ~250 mm.

WELD METALLURGY

Fully austenitic welds are readily attainable by an exact balance of the minor and alloying elements of the weld chemistry with the purpose of avoiding an intersection of the δ-ferrite phase during the solidification period of the weldments. The δ-ferrite, however, is accepted as a beneficial microstructural constituent because of its ability to reduce the tendency towards fissuring in austenitic multipass welds. According to WRC, a 3 FN (Ferrite Number) minimum for welding consumables is recommended in any multipass weld to prevent the fissuring, as shown by Irving.[3] The disadvantage of δ-ferrite in the microstructure derives from the low toughness properties of δ-ferrite-bearing austenitics at 4 K, as determined by Whipple et al.[4] for 316L weld metal. Therefore, the allowable ferrite quantity is restricted by the decrease in toughness properties when these weldments are subjected to 4-K service. A compromise acceptable toughness level and fissuring-free weld beads are often impossible. The problem will be more serious with the heavy weldments for complex joints, since the main cause of fissuring lies in the presence of internal stresses. The actual role of δ-ferrite in avoiding cracking can be seen in its higher solution potential for low-melting-point sulphides and oxides compared with that of the austenitic phase and its lower resistance against plastic deformation at elevated temperatures.

The detrimental effect of low-melting-point sulphide and oxide grain boundary films has been shown by Brooks.[5] Recently Takalo et al.[6] derived an empirical relationship between the hot cracking tendency of austenitic welds and the impurities sulfur and phosphorus. Investigations with several types of austenitic welds with differing chromium and nickel equivalents revealed a marked decrease of fissuring susceptibility on lowering the phosphorus and sulfur content. Besides the phosphorus and sulfur content in the melt, the high oxygen concentration, as present in the low-carbon series of austenitics, gives rise to a serious problem. In practice, the carbon content in the alloy is lowered at the price of increasing the oxygen in the metal. This behaviour is explainable by the carbon-oxygen reaction. Because of the high oxygen-dissolving potential, oxygen pickup during the welding can be expected if no effective deoxidizing agents are involved in the process. Oxidation occurs whenever the protection of the weld pool by the shielding atmosphere is not entirely effective, or the exact control of emissive oxides resulting from the presence of alkaline carbonates in the covering has not been thoroughly balanced.

High oxygen content in the molten stainless steel phase is deleterious to weld quality. The formation of refractory nature oxides causes the lack of fusion defects in multipass welds. Furthermore, dissolved oxygen and sulfur, both strong surface active elements, adsorb at the liquid metal interface (Gibbs Adsorption). Also, segregation can occur during the solidification period, giving rise to low-melting-point films at austenitic grain boundaries. To prevent the formation of refractory-type oxides, the coating should contain energetic agents, e.g., alkaline earth and alkaline fluorides. The negative effect of these agents is, however, poor arc stability. The operator can control these arc irregularities manually.

Thus reducing the sulfur, phosphorus, and oxygen content to a low level, one is able to weld fully austenitic joints with little risk of fissuring. Moderate manganese content in weld metal favours sound welds.[7] In addition, manganese increases the nitrogen solubility of 316LN and compensates the solubility decrease of the nitrogen in the weld pool, which results from the temperature rise during the welding process. All the above facts have been considered in the choice of a commercially available consumable electrode to weld 316LN.

WELD RESULTS

To determine the weldability of 316LN, tests were conducted with commercially available electrodes (A, B, C, and D in Table 1). The test results show a correlation between manganese content of the weld metal and the defects found. The tests A1 and A2 with double U-joints show negligible lack-of-fusion defects and low fissuring tendency. Hot cracking has been found in one case with about a 1-mm crack length at the weld centerline. Poor arc stability of these A series presumably resulted from the coating. The welded T-joint with the same electrode (A3) showed similar behaviour with minor lack-of-fusion defects at the root section. In this case, it has been proposed to improve the weld quality at the root section by GTA (Gas Tungsten Arc) welding prior to first pass. In addition, the T-joint should be prepared by overlaying a single pass on the flat side to gain enough space for a later GTA root pass.

The weld test B1 revealed large lack-of-fusion defects mostly at the base-metal interface. Fissuring tendency was low and the arc stability was better than in the preceding test. In this case the higher manganese content, compared with the A series, was not able to prevent the formation of refractory-type oxides in the weld pool, owing to the higher emissive oxide content of the covering necessary for better arc stability.

The weldment C1, with low manganese and silicon in the melt, produced the poorest weld quality; there were numerous lack-of-fusion defects and cracks, revealing a tendency towards refractory oxide-type defects. The consumable electrode D1 demonstrated the best results in the weld quality: no fissuring and no lack-of-fusion defects.

The foregoing results emphasize the weldability of 316LN heavy joints by the selection of ~5% manganese bearing consumable and applicable covering. The additional serious problem is to preserve the weld quality during the course of the entire fabrication.

WELD MATERIAL PROPERTIES

Uniaxial tensile tests with round, standard specimens were conducted to determine the yield strength, ultimate strength, reduction of area, and percent elongation of the base metal and its weldments. The tensile data were obtained in different laboratories (see Acknowledgment). The base metal was tested in the as-received condition. The specimens were machined out of the plate material in the long transverse and in the longitudinal rolling directions. The welded specimens were oriented along the transverse direction of the weld joint. The midportion of the specimens included the weld zone, and the elongation values were obtained by placing the extensometer in this region. The mean of these measured values with their standard deviations are shown in the Table 2 for various temperatures.

Charpy V-notch tests were conducted at 4 K using the DIN-standard method. For the base metal, impact values were obtained using samples from different locations of the 50-mm-thick plate material. The mean of these measurements with their standard deviations show energy absorption values of 122 ±21 J at 4 K, 131 ±22 J at 77 K, and 225 ±40 J at ambient. The determination of Charpy impact data for the A-type weldments was done by machining standard V-notch specimens out of two different 50-mm-thick joint configurations. For these reasons, a double U-joint and a T-joint were welded with the A-type consumables. The notch positions were oriented in different weld seam locations to establish a maximum of information about the weldment. Table 3 shows the energy absorption values for both the tested joints. The varying impact energy

Table 2. Mechanical Properties of the Base Metal and Its Weldments

Mat. design.	Temp. K	Yield strength MPa	Tensile strength MPa	Elongation (2.5 cm gauge length)%	Red. of area %	Young's Modulus GPa
316 LN base	4.2	1195±46	1363±58	35±7	65±4	214±13
	77	838±180	1177±261	54±5	64±3	196±23
	293	429±53	675±19	48±2	74±6	185±2
SMA weld A-type	4.2	1248±9	1519±33	21±6	28±7	189
	293	476±12	716±28	40	63	-
SMA weld B-type	4.2	1291	1420	7	10	203
	77	1021	1209	8	9	165

Table 3. Charpy Impact Energy 4-K Data of A-Type Weldments

Notch position	Double U-joint (Two identical tests): Impact direction perpendicular to multipass at the weld and parent metal interface	Impact direction parallel to multipass in weld metal
Top weld seam	25 ; 25 J	29 ; 29 J
Mid weld seam	41 ; 57 J	29 ; 29 J
Bottom weld seam	67 ; 59 J	40 ; 37 J

T-joint (Impact direction perpendicular to multipass)

Notch position	right weld seam	mid weld seam	left weld seam
Upper weld and parent metal interface	35 J	23 J	48 J
Middle of the weld metal	57 J	36 J	17 J
Lower weld and parent metal interface	44 J	45 J	26 J

data indicate a qualitative difference in the tested zones. Measured Vickers hardnesses in the vicinity of the fractured surface revealed no significant correlation between hardness and toughness. The scatter of the values should be seen in the nature of this special test method: Uncontrolled rise of temperature during the impact and machined notch radii variation can affect the energy absorption of the individual test very strongly.

Fracture toughness tests at 4 K were the main investigation of the materials testing program. The base metal (~316LN) value for K_{Ic} was recently found to be 208 MPa$\sqrt{m}$ ±7% at 4 K.[8] The method adopted for these investigations was the J-integral multispecimen test with Merkle-Corten correction coefficients. The nominal size

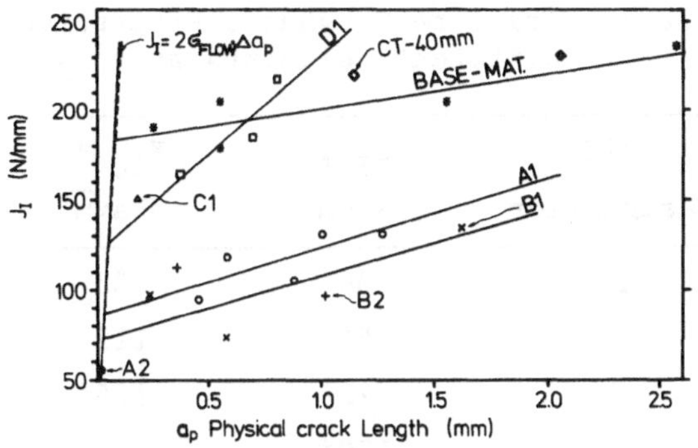

Fig. 1. J_I integral values of austenitic materials at 4 K.

of the standard CT specimens was fixed at 23-mm thickness. To reveal the size independency of these measurements, 40-mm-thick standard CT specimens were checked at 4 K. The J_I values of these tests confirmed the regression line of the base metal within the experimental accuracy, as given in Fig. 1.

The fracture toughness tests of the weldments were conducted in the elastic plastic regime similar to base metal tests with 23-mm-thick CT specimens. The crack extension lines in all cases were parallel to the weld centerline in the welding direction and covered the entire range of the root section. All specimens were machined from the middle of the joints. The tested joints are tabulated with their critical J values and the calculated K_{Ic} values in Table 4. Figure 1 gives the J_I data of each individual test vs. the physical crack extension. For the tested joint C1, only one measurement out of five was successful, because of instable crack extension of the specimens due to large lack-of-fusion zones. Similarly, one of the B1 tests, after the crack extension line was marked, showed a spontaneous instable cracking at 4 K with crack branching, because of lack-of-fusion defects at the weld-base metal interface. The D1 weldments toughness results were the best values at 4 K; also, no defects could be detected. In this case, it is assumed that the absence of the nitrogen increased the K_{Ic} value to 164 MPa$\sqrt{m}$. It is further assumed that this increase of toughness is counterbalanced by the decrease of yield point at 4 K. But up to now, no measurements were performed to verify this.

The compromise for the choice of the consumable electrode was the use of A-type weldment for the joining of the 316LN-type base metal. Additional tests were performed to confirm the above results, and therefore, the joint A2 was tested using 3 specimens. Two CT specimens with a nominal 25-mm thickness were tested independently at the National Bureau of Standards using the J-integral

Table 4. Fracture Toughness Properties of Different Weldments at 4 K.

Weld Designation	J_{Ic}, N/mm	K_{Ic},* MPa$\sqrt{m}$
A1	86	133
A2	54†	111
B1	72	125
C1	150‡	-
D1	125§	164

*Calculated according to $K_{Ic}^2 = J_{Ic} \cdot E/(1-\nu^2)$ with $\nu = 0.28$.
†Average of two tests determined at NBS-USA.
‡J_I value.
§Assumed Young's modulus E = 200 GPa in case of D1 weld metal.

compliance method, which gave a mean value of 54 N/mm with a K_{Ic} of 111 MPa$\sqrt{m}$. The third CT specimen with a thickness of 23 mm was tested at Kernforschungszentrum Karlsruhe by the ASTM E-399 procedure and gave a K_Q value of 135 MPa$\sqrt{m}$. The test result is not completely valid according to the thickness criterion, but the J_{Ic} of the same type weldment in Table 4 reveals data comparable to the LEFM test. The reason for the discrepancy between the compliance method and the multispecimen method is still not clear. One possibility could be in the difficulty of the exact determination of the Young's modulus for weld zones, which is essential for the compliance procedure.

NONDESTRUCTIVE TESTING OF WELD AUSTENITICS

The question as to how safe the selected weldments are can be answered with the question of how reliable is the detectability. During the course of the materials testing, much effort has been put into gaining experience of the nondestructive testing (NDT) capability of the ~50-mm-thick weld metal. Large lack-of-fusion defects can be readily inspected by x-raying. From the practical viewpoint, however, the ultrasonic technique seemed the most promising NDT inspection for smaller flaws and cracks.

It is well known that the austenitic weldments with their coarse-grain microstructure cannot be inspected with the same ultrasonic techniques used for ferritic weldments. The high attenuation in austenitic weldments reduces the testing capability drastically. The test procedure cannot be improved by simply lowering the testing frequency of the common pulse-echo transducers because of the decrease in resolving power. To test the ultrasonic attenuation, the T-joint (A3) given in Table 1 was surface machined to achieve a well-defined coupling condition between the probe and testpiece. Metallographic examination of both outer ends of the T-joint revealed defects at the root section. This prepared joint was subjected to ultrasonic testing with a commercially available 1 MHz transducer by the common pulse-echo procedure. The testing was carried out perpendicularly from the top bead of the weldment with longitudinal waves. The discontinuities at 22-mm weld depth caused an acoustic mismatch of 46-48 dB.

Further testing was conducted with the transmitter-receiver technique using two separated ultrasonic transducers. This technique has a higher sensitivity and lower microstructural scattering echoes than the common pulse-echo technique. At the testing frequency of 2 MHz and at angles of incidence of 45°, 60°, and 70°, the weld zones were subjected to the detection procedure. Ultrasonic attenuation in the range of 48-64 dB was achieved and the discontinuities could be readily found. To determine the sensitivity function of the crossing point of the two transducers, 2-mm holes as test reflectors were drilled in the weldment at different weld seam locations. This function could be used in flaw size determination, but doubling the flaw size changed the echo height only by a few percent, which was too insensitive. The focal diameter of the crossing point was estimated at around 10 mm, and reduction of this size will increase the sensitivity of the size determination. Future work with focussing transducers acting as lenses will reduce the focal area, thus increasing the inspection capability.

APPLICATION TO STRUCTURAL DESIGN

The concepts of fracture mechanics are being strictly applied to the weldments. In this field, the stress intensity factor, K_I, is accepted as a single parameter for the mode I-type load, allowing a correlation between flaw size and design stress. According to the design recommendation[9] for 4-K magnet structures, a maximum loading of 67% of yield strength and 50% of ultimate strength is regarded as safe against ductile failure. Against brittle failure, 60% of the measured K_{Ic} is assumed to give a safe design stress limit. Inserting these safety margins in each individual case results in values of 832 MPa for yield strength and 760 MPa for ultimate strength (Table 2 data). After adoption of a shape factor of 1.2 and a penny-shaped crack of 4-mm diameter, the equation $\sigma = K_{Ic}/(2 \cdot \sqrt{\pi \cdot a})$ yields an allowable stress limit of 820 MPa with K_{Ic} of 130 MPa$\sqrt{m}$. However, for weldments, the residual stresses must also be considered. In this case, the peak of the residual stresses is assumed to be the ambient yield strength of about

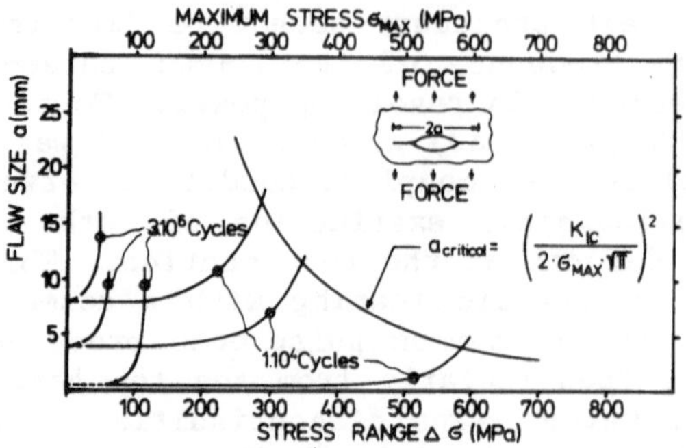

Fig. 2. Fatigue crack growth of the 316LN weld material under cyclic load.

400 MPa, which reduces the calculated stress limit to ~420 MPa. This value is regarded as a safe design stress limit.

During service at 4 K, the weldments will be subjected to cycling load. Therefore the crack growth under alternating loads must be estimated. Calculations have been performed by numerical integration of the Paris law with the measured[10] constants m = 3.36 and $C = 1.43 \cdot 10^{-9}$ mm/cycle for 4-K 316LN weld metal. The results for two cycle numbers and different initial flaw sizes are shown in Fig. 2, demonstrating the drastic decrease of the allowable design stress limit.

CONCLUSIONS

1. Weldments were produced with commercially available electrodes with standard chemical composition of both metal and covering. The results at 4 K are an increase of yield strength and ultimate strength of ~4% and ~10%, respectively, over those of the parent metal. However, a ~35% decrease in fracture toughness at 4 K was measured.

2. Optimization of yield strength and toughness level seems to be possible by proper balancing of the manganese and nickel in the weld chemistry.

3. Much work is still needed in the field of NDT and in the determination of residual stresses for safe application of fracture mechanics for large components.

4. Welding processes with high deposition rates (GMA) must be further developed and optimized to reduce the time and costs involved in producing the weldments.

ACKNOWLEDGMENTS

The authors would like to thank H. I. McHenry for carrying out the toughness tests at NBS, Boulder, Colorado. For assistance in ultrasonic testing, the authors are grateful to E. Neumann and his staff of the Bundesanstalt für Materialprüfung, Berlin. Also, the authors are grateful to K. Henninger for carrying out the J_I measurements at 4 K. The tensile and the impact testing were performed under the supervision of Siemens AG, as part of the LCT design effort.

This work is based on a contract between Kernforschungszentrum Karlsruhe and Max-Planck-Institut für Plasmaphysik, Garching, concerning the cooperation in the field of superconductivity. It was supported by EURATOM.

REFERENCES

1. H. Krauth, S. Förster, G. Messemer, A. Nyilas, and H. Zehlein, The mechanical design of the Euratom test coil for the Large Coil Task, Fusion Technol. 1:539 (1980).
2. G. Friesinger, K. P. Jüngst, A. Nyilas, and G. Kuhnen, Electron beam weldability of nitrogen-strengthened austenitic structural materials for fusion magnets, Fusion Technol. 2:1281 (1980).
3. R. R. Irving, How to avoid hot cracking in stainless welds, Iron Age (Jan. 3, 1977), p. 60.
4. T. A. Whipple, H. I. McHenry, and D. T. Read, Fracture behaviour of ferrite-free stainless steel welds in liquid helium, Weld. J. 60:72s (1981).
5. J. A. Brooks, Weldability of high N, high Mn austenitic stainless steel, Weld. J. 54:189s (1975).
6. T. Takalo, N. Suutala, and T. Moisio, Austenitic solidification mode in austenitic stainless steel welds, Metall. Trans. A 10A:1173 (1979).
7. C. D. Lundin, C.-P. D. Chou, and C. J. Sullivan, Hot cracking resistance of austenitic stainless steel weld metals, Weld. J. 59:226s (1980).
8. H. Krauth and A. Nyilas, Fracture toughness of nitrogen strengthened austenitic steels at 4 K, in: "Fracture and Fatigue," J. C. Radon, ed., Pergamon Press, New York (1980).
9. C. D. Henning and E. N. C. Dalder, Structural materials for fusion magnets, in: "Transactions of the 5th International Conference on Structural Mechanics in Reactor Technology," Vol. N, North Holland, Amsterdam and Elsevier North-Holland, New York (1979), p. 2.2/1.
10. Phase II detailed design analysis report for LCP, prepared by GE under UCC contract 22X-31745 C (not published).

WELDING OF AUSTENITIC STAINLESS STEEL FOR 4-K SERVICE*

F. R. Schneider, Jr.

Manufacturing Technology Department, General Dynamics/Convair Division
San Diego, California

SUBJECT

High deposition rate welding processes for superconducting magnets operating at 4.2 K.

INTRODUCTION

Fabrication of superconducting magnets requires welding thick sections of austenitic stainless steel. Full penetration joints in thicknesses of two (2) to five (5) inches are common with current fabrication for 4-K service. Magnets in this thickness range result in tons of deposited weld metal with a requirement of good mechanical properties at 4 K.

This strenuous welding requirement indicates the need for qualified high deposition rate welding processes. The common welding process of shielded metal arc welding has a deposition rate of approximately two (2) pounds per hour. An additional concern is the many starts and stops with this process.

OBJECTIVE

To develop and implement high deposition rate welding processes for fabrication of superconducting magnets at 4-K service.

APPROACH

Initial welding processes explored were gas metal arc welding (GMAW) and flux core arc welding (FCAW). The GMAW, requiring less

*Invited paper.

initial development, was first utilized by Convair on the "Pre-prototype Magnet System" (PMS) in the second quarter of 1979. The materials were 316L base and 316L filler alloy.

Initial production welding on the "Large Coil Program" (LCP) began in the first quarter of 1980 utilizing the GMAW process. Final development of the FCAW process was nearing completion at that time. The materials were 304L base metal and 308L filler alloy.

The FCAW process was first implemented into production on the LCP during the first quarter of 1980. FCAW is now our first choice process on the LCP, with position normally being the deciding factor. Ease of welding, operator skill level, interpass cleaning, soundness of weld, good wetting action, deposition rate, as well as properties, are the main attributes responsible for its acceptance.

During this period of time, the "Mirror Fusion Test Facility A" (MFTF-A) was designed and fabricated. Base material was 304LN and 316L filler alloy. The yin-yang magnet had a weld metal deposit weight of over 20,000 lbs. The SMAW process was utilized. This pointed out the desirability of an approved high deposition rate welding process.

With the initiation of the MFTF-B program, an effort was made to qualify the FCAW process with 316L filler alloy for 304LN base

Fig. 1. The PMS helium vessel inner cylinder weldment after machining.

Fig. 2. The LCP three-inch side plate to adapter butt joint being mechanized GMA welded.

Fig. 3. The LCP post to key 4-in-thick full penetration tee joint root passes being welded with the GMAW process. This was the first production application of the FCAW process, which was used for the fill passes.

materials. This effort has been successful and has been approved by Lawrence Livermore National Laboratories for use on the MFTF-B program. The estimated deposited weld metal requirement for this program is over 57,000 lbs. Approximately 35,000 lbs. appears suitable for FCAW process consideration.

The Elmo Bumpy Torus (EBT) program will be utilizing the GMAW, GTAW, and SMAW processes, as appropriate. The base material is 304L and the filler alloy is 316L. The weld-joint thicknesses on this program range from 1/2 to 1-1/2 in.

Initial work with the submerged arc welding process resulted in highly promising values. Additional work in this area is continuing with emphasis on improving the yield strength at 4 K.

A yield strength of 94.9 ksi was obtained. The goal is for 110 ksi minimum with 120 ksi desired. The K_{Ic} average was 149.3, well above the minimum goal of 120 ksi $\sqrt{in.}$

Fig. 4. The LCP bobbin assembly being raised into the vertical position. Root passes of side plates to inner ring were made in the horizontal position. Fill passes were completed with the FCAW process in this position, which provided for down-hand welding of the full penetration corner joints.

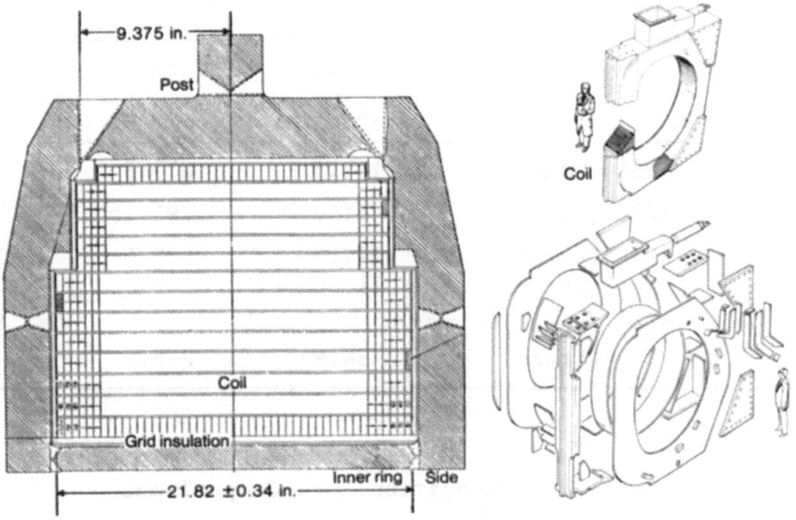

Fig. 5. A cross section of the LCP superconductor case indicating the major weld joints and an exploded view of the details.

Fig. 6. One of the yin-yang magnet cases for the MFTF-A program being loaded for shipment to LLNL where the superconductor pack was assembled prior to close out welding of these two major subassemblies.

Table 1. Average Properties at 4 K

Process	Filler Alloy	Yield (ksi)	Tensile (ksi)	K_{Ic} (ksi $\sqrt{in}$)
GMAW	308L (Trans)	112.5	217.3	121.5
FCAW	308L (Trans)	81.0	191.25	122.3
FCAW	316L (Long)	122.3	200.6	183.1
	(Trans)	112.0	185.8	
SAW	316L (Long)			149.3
	(Trans)	94.9	215.6	

Table 2. Deposition Rate Comparisons

Weld Process	Arc Deposition Rate (lbs./h)	Utilization Factor (%)	Realized Deposition Rate (lbs./h)
GTAW	0.5 - 2	25	0.13 - 0.5
SMAW	8.0 - 9	25	2.0 - 2.3
GMAW	12.0 - 16	60	7.0 - 9.6
FCAW	16.0 - 18	55	8.8 - 9.9
SAW	18.0 - 20	55	9.9 - 11.0

Fig. 7. A set of typical tensile and K_{Ic} fracture toughness specimens.

CONCLUSION

High deposition rate welding processes are attainable for the stringent requirements of 4-K service. Flux core arc welding has proven both desirable and satisfactory in production on superconducting magnets at Convair. The FCAW transverse yield strengths of over 110 ksi and K_{Ic} values of 183 ksi $\sqrt{in}$ at 4 K have been developed. Submerged arc welding appears to be suitable for designs with a yield strength requirement of 90 ksi at 4 K. It is anticipated that we will achieve over 100 ksi yield within the next year. Both of these processes are position limited to flat or horizontal using a fillet technique.

There is a need for welding processes with a higher deposition rate than GTAW or SMAW for out of position welding. It would appear that the GMAW process with small diameter wire (0.030 to 0.063 in) would be the most promising, yielding medium deposition rates. Our work with GMAW to date has been with 1/16- and 3/32-in diameter wire with the goal of maximum deposition rates.

To achieve high quality at minimum cost, the welding engineer must have the freedom to select the appropriate welding processes for efficient fabrication. Therefore, there is the need to qualify more than one process.

ACKNOWLEDGMENTS

Particular credit and appreciation is due the following people for their cooperation and assistance in performing portions of the development work: J. A. Horvath, W. A. Roden, and C. Leal of Convair, E. N. C. Dalder of Lawrence Livermore National Laboratory, J. E. Sims of Chicago Bridge and Iron Company, D. J. Kotecki of Teledyne McKay, D. Beard of Department of Energy, and R. P. Reed of the National Bureau of Standards.

FERRITIC WELDMENT OF GRAIN-REFINED FERRITIC STEELS FOR CRYOGENIC SERVICE

H. J. Kim, C. K. Syn, and J. W. Morris, Jr.

Lawrence Berkeley Laboratory, University of California, Berkeley, California

INTRODUCTION

Previous research in this laboratory led to the development of thermal cycling techniques that establish an extremely fine effective grain size in 12Ni[1-4] and 9Ni[4,5] cryogenic steels. In the grain-refined condition these alloys offer a good combination of strength and toughness at 4 K and have possible applications in the structures of cryogenic devices such as high field superconducting magnets. But the structures of these devices are usually welded, and candidate structural alloys must, hence, retain good properties in the welded condition. To weld the grain-refined steels one must overcome the dual problems of brittleness in the weld deposit and brittleness due to grain coarsening in the heat-affected zone (HAZ).

The problem of welding grain-refined Fe-12Ni-0.25Ti for 4-K service was first approached in this laboratory[6] by using high-nickel filler metals such as are often specified for ferritic steel weldments at 77 K. This approach led to an undesirable brittleness in the fusion zone and a low yield strength in the weld metal. Similar problems were encountered by Ishikawa and Maruyama[7] in the welding of 13Ni cryogenic steel.

A more promising approach was developed in joint research between the Japanese steel companies Nippon Kokan, K.K. and Kobe Steel;[8-10] they showed that quench-and-tempered 9Ni steel may be welded for 77-K service with a matching ferritic filler if a multipass GTAW technique is employed. The GTAW process provides a relatively clean weld deposit and permits a largely independent control of the heat input and the metal deposition rate. In controlled GTAW welding, subsequent weld passes may be made to impose thermal cycles on the solidified material, hence refining the structures of the weld deposit and preserving a refined microstructure in the heat-affected zone. This process is now used commercially in Japan in the manufacture of vessels for inert cryogens and has been successfully employed in a prototype spherical LNG tank.[11] Exploratory work at the NASA Lewis Research Center[12] showed that GTAW could also be used for autogenous welding of Fe-12Ni-0.5Ti and that it yields weldments with good fracture toughness at 77 K. Continuing this line of research, we have recently developed a 14Ni filler metal and multipass GTAW welding process for grain-refined Fe-12Ni-0.25Ti.[13] The resulting welded plates show high strength and good impact toughness in liquid helium in both the weld metal and heat-affected zone and show promising properties in fracture toughness tests.

This paper reports the initial results of similar studies on ferritic GTA weldments in grain-refined 9Ni steel. The microstructural problem in welded 9Ni steel is superficially more formidable than that in the 12Ni alloy since the 9Ni steel base plate only has good 4-K properties when it is given in a final intercritical temper to introduce a distribution of stable high-temperature austenite phase into the grain-refined structure. The austenite provides an additional grain refinement and getters carbon and carbides from the matrix. Its retention near the fusion line and in the heat-affected zone may be important to the 4-K toughness of welded plates.

ALLOY PREPARATION AND EXPERIMENTAL PROCEDURE

The 9Ni steel used in this work was a commercial grade provided by Nippon Kokan, K.K. Its composition is given in Table 1. The material was received in the form of 15-mm-thick plates in the quench-and-temper (QT) condition. The plates were solution-annealed at 900°C for 2 h to remove the effects of prior deformation and thermal treatment. They were then given the 2BT heat treatment[5] diagrammed in Fig. 1. This treatment consists of a four-step alternating thermal cycle to refine the effective grain size followed by an intercritical temper to introduce an appropriate distribution of precipitated austenite. A transmission electron micrograph of the 2BT microstructure is shown in Fig. 2.

The weld filler metal was cast in this laboratory in 4.5-kg ingots of nominal composition Fe-14Ni-0.2Ti-0.003B. The ingot composition is given in Table 1. After homogenizing at 1200°C for 24 h, the ingots were hot-rolled and swaged into wire of 1.6-mm diameter. This wire was used as the filler for manual GTA welding.

The plates to be welded were machined into one of two joint configurations, a 60° single V or a 45° single bevel. The plates

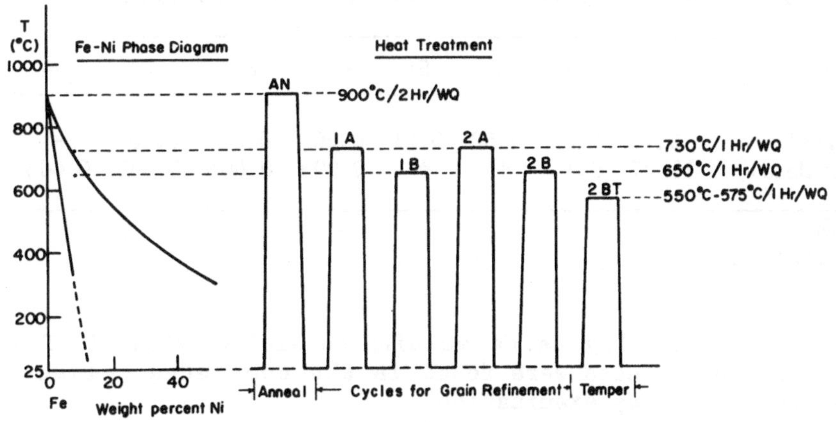

Fig. 1. The 2BT heat treatment of 9Ni steel plotted alongside the Ni-rich segment of the Fe-Ni phase diagram.

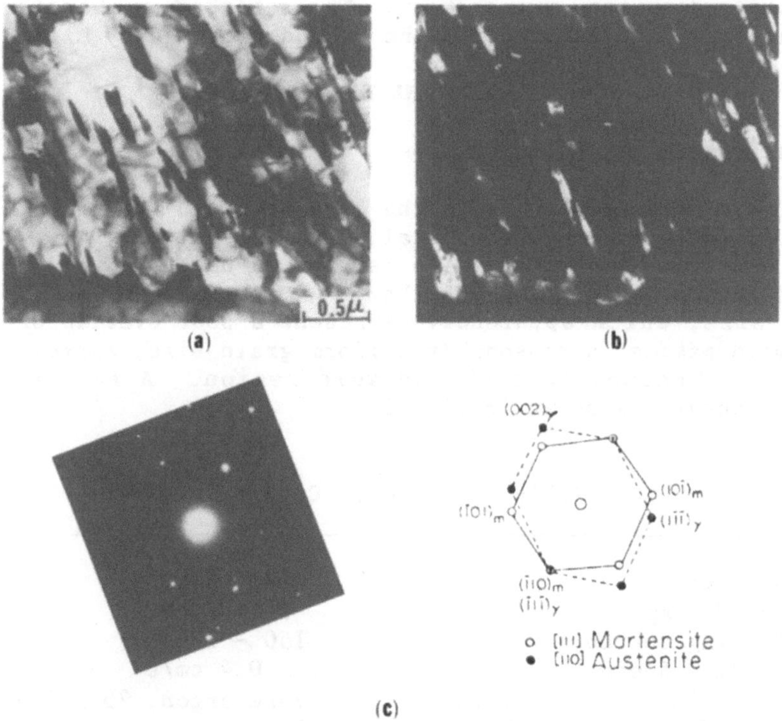

Fig. 2. TEM micrographs of retained austenite in 2BT 9Ni steel. (a) bright field, (b) dark field taken from $(002)_\gamma$ diffraction spot in (c).

Table 1. Chemical Compositions* of Base and Filler Metal

	Ni	Ti	Mn	Si	C	P	S	B	Fe
9Ni Steel	9.18	0.06	0.60	0.21	0.06	0.01	0.002	-	Bal.
Filler Metal	13.87	0.16	†	†	0.003	0.001	0.002	0.003	Bal.

* Wt.%.
† Negligible.

were welded manually using welding conditions shown in Table 2. No significant defects were found in the completed weldments by nondestructive x-ray examination.

Specimens for mechanical testing were cut from the welded plates, as described below. Specimens for x-ray diffraction analysis of the residual austenite content were sliced parallel to the fusion boundary (see insert in Fig. 4), ground, and chemically polished. The distribution of austenite was found by sequential grinding and surface preparation. The retained austenite volume fraction was calculated using the method proposed by Miller.[14]

RESULTS AND DISCUSSIONS

Microstructure

Grain Refinement. As in the case of grain-refined 12Ni steel welded with the same filler metal, the multipass GTA-welded 9Ni steels were efficiently grain refined in both the weld metal and heat-affected zone. Although there were some islands of larger grain size, which apparently represent a poor overlap of sequential weld passes, a reasonably uniform grain size, approximately 5 μm, was obtained through the weld region. A full-thickness microstructure is shown in Fig. 3.

Table 2. Welding Conditions

Heat Input	7 - 8 kJ/cm
Arc Voltage	14 - 18 V
Welding Current	150 - 180 A
Welding Speed	0.4 cm/s
Shielding Gas Flow Rate	Pure argon, 25 ft^3/hr.
Root Gap	1.6 mm
Interpass Temperature	50°C - 250°C

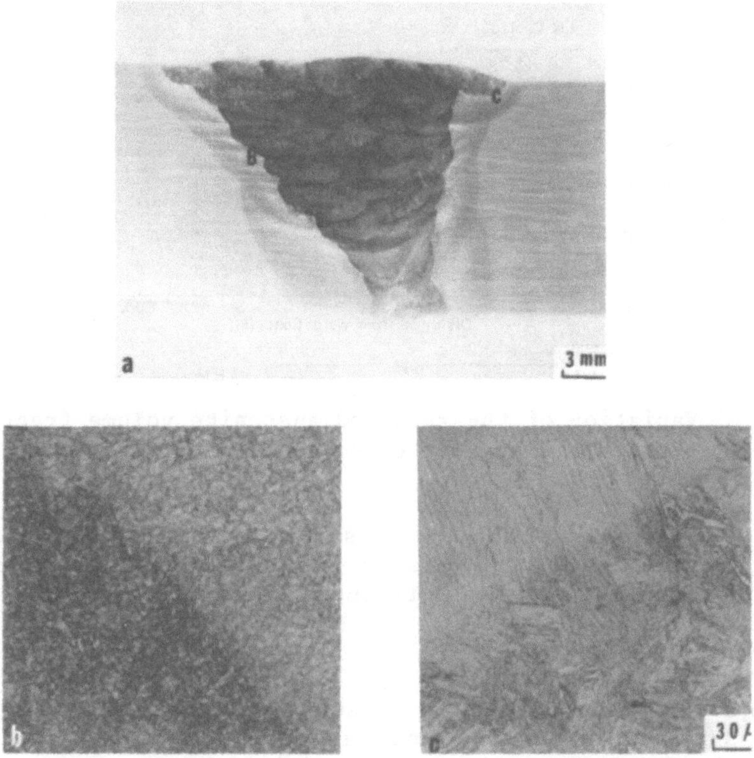

Fig. 3. Macro- and microstructures of 9Ni welded joint: (a) macrostructure of welded joint, (b) microstructure of well-refined region marked B in (a), (c) grain-coarsened last pass region marked C in (a).

Retained Austenite. The volume fraction of retained austenite in the welded 9Ni steel is plotted as a function of distance from the fusion boundary in Fig. 4. In agreement with results obtained earlier from the welding of 12Ni[13] and 9Ni[16] steel, there is no retained austenite within the weld metal. The volume fraction of austenite decreases gradually through the heat-affected zone from a value of approximately 15% in the base metal. These results are in some disagreement with the earlier observations of Tamura et al.[17] on the welding of quench-and-tempered 9Ni steel with ferritic filler metal. They reported an increase in retained austentie in the heat-affected zone to approximately 10% from a lower value of 4% within the base metal. They also found significant austenite retention within the weld metal and argued that the austenite fraction is well correlated to cryogenic impact toughness.[17,18] In the present work, the

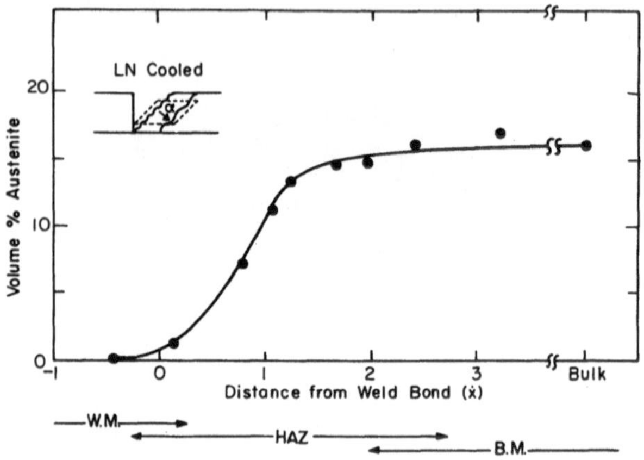

Fig. 4. Variation of the retained austenite volume fraction through the heat-affected zone in welded 9Ni steel.

austenite volume fraction decreases monotonically through the heat-affected zone and has no obvious correlation with the variation of impact toughness through the weld region.

Cryogenic Toughness

Impact Toughness. Charpy impact tests were conducted at 77 K and 4.2 K to give a general indication of the toughness of the welded plate. The results are listed in Table 3 and are compared with those obtained for 12Ni steel welded with the same filler metal and similar procedure. The variation of impact toughness through the heat-affected zone at 4 K is plotted in Fig. 5. The Charpy impact energy of the welded 9Ni steel is relatively constant through the heat-affected zone and reaches a maximum within the weld metal. The sharp maximum in impact toughness within the

Table 3. Charpy V-notch Impact Values of 9Ni and 12Ni Welded Joints

Welded Joint	Impact Energy, J (ft-lb)		
	Weld Metal	HAZ	Base Metal
9Ni joint at 77 K	184(136)	169(125)	156(115)
at 4.2 K	153(113)	149(110)	146(108)
12Ni joint at 77 K	182(135)	217(160)	156(115)
at 4.2 K	173(128)	179(132)	136(100)

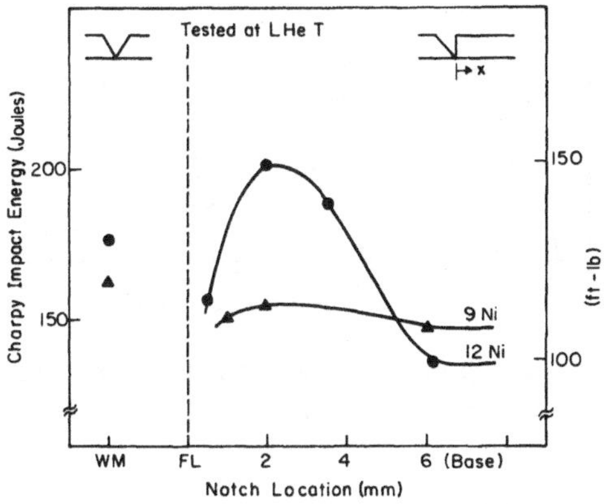

Fig. 5. The variation of Charpy impact energy with notch location for samples impacted in liquid helium.

the heat-affected zone, which was found in 12Ni steel, is not present. There is, moreover, no obvious correlation between the impact toughness of the welded plate and the retained austenite level.

All of the broken specimens in both 9Ni and 12Ni fractured in a completely ductile mode. The relevant scanning electron fractographs taken from the 9Ni welded specimens are presented in Fig. 6. The impact ductile-brittle transition temperature of the ferritic weldment is, therefore, below liquid helium temperature. There was no evidence of significant fusion zone brittleness in these tests. Although there was a slight relative drop in Charpy

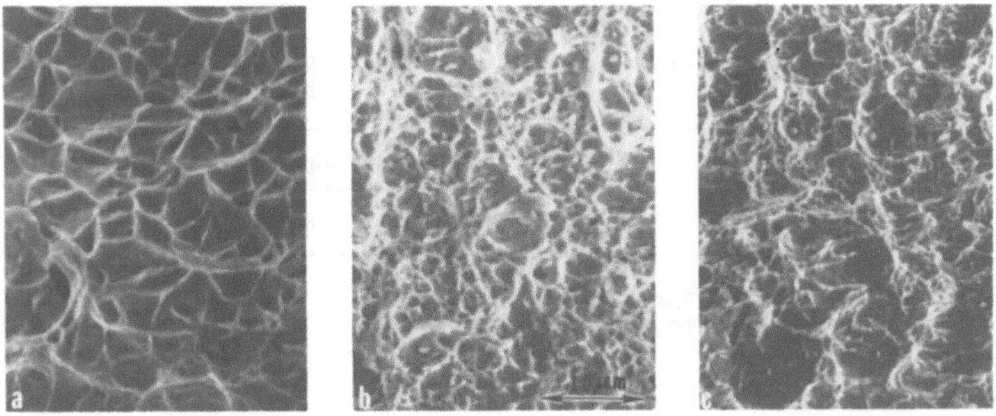

Fig. 6. Scanning electron fractographs of 9Ni Charpy specimens broken at 4.2 K (a) weld metal (b) HAZ (c) base metal.

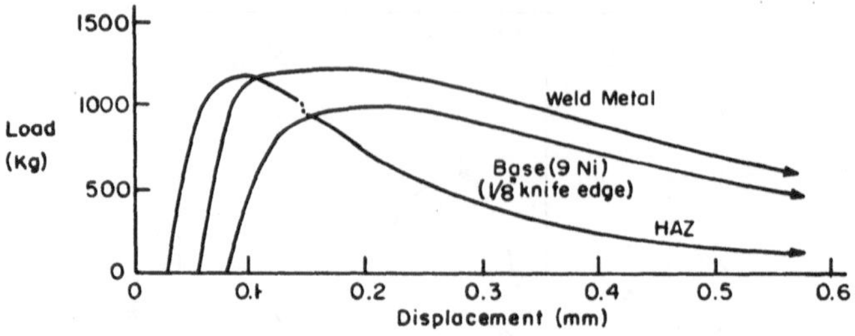

Fig. 7. Load-crack opening displacement curves of three-point bend specimens at 77 K.

impact energy near the fusion line, the fusion zone fracture was ductile, and cracks initiated from the heat-affected zone did not follow the fusion boundary.

Cryogenic Fracture Toughness. The cryogenic fracture toughness tests conducted to date on ferritic welded 9Ni plate have been at 77 K. Tests at 4 K are now in progress. The completed test series employed three-point bend test specimens of 1-cm thickness prepared according to ASTM specifications[19] and tested on a MTS machine equipped with a liquid nitrogen cryostat. The load-crack opening displacement (COD) curves are plotted in Fig. 7. The specimens from the base plate and weldment showed no evidence of unstable crack propagation at 77 K. The HAZ specimen showed only a slight indication after the peak load had been passed. Fractographic examination of the surfaces showed that the fracture mode was completely ductile rupture in all cases.

Since the specimens tested were well away from plane strain conditions, valid K_{Ic} values were not obtained. The fracture

Table 4. Fracture Toughness of 9Ni Welded Joint at 77 K

	Fracture Toughness*	
	MPa $\sqrt{m}$	ksi $\sqrt{in}$
9Ni Base Metal	220	200
HAZ (0.5 mm)	247	225
HAZ (2.0 mm)	212	193
Weld Metal	328	298

* All values were calculated using the Equivalent Energy method.[20]

toughness was estimated using the equivalent energy method;[20] the result of the calculation is shown in Table 4. Fracture toughness of the weld metal is near 330 MPa $\sqrt{m}$ (300 ksi $\sqrt{in}$ at 77 K. High fracture toughness was preserved in the heat-affected zone.

Fracture toughness tests of the ferritic welded 9Ni plates at 4 K are now in progress. The 4-K toughness of the weld metal itself should, however, be nearly the same as that obtained in tests on 12Ni steel welded with the same filler metal and welding procedures.[13] Tests of the fracture toughness of the ferritic weldment in 12Ni have been complicated by the difficulty of obtaining good grain refinement in the final weld passes. The tests nonetheless demonstrate that the grain-refined weld material remains ductile and tough near 4 K.

CONCLUSIONS

The results presented above show that it is possible to weld grain-refined 9Ni steel with ferritic weld filler metal so as to retain good toughness at cryogenic temperatures. The results of this work may permit the utilization of retreated commercial grade 9Ni steel in structural applications within helium-cooled cryogenic devices where high strength and good toughness are required.

ACKNOWLEDGMENTS

The 9Ni steel employed in this research was provided by Nippon Kokan, K.K. The research was supported by the Director, Office of Energy Research, Office of Development and Technology, Magnet Systems Division of the U.S. Department of Energy under Contract No. W-7405-ENG-48.

REFERENCES

1. S. Jin, J. W. Morris, Jr., and V. F. Zackay, in: "Advances in Cryogenic Engineering," Vol. 19, Plenum Press, New York (1974), p. 379.
2. S. Jin, J. W. Morris, Jr., and V. F. Zackay, Metall. Trans. 6A:141 (1975).
3. S. Jin, S. K. Hwang, and J. W. Morris, Jr., Metall. Trans. 6A:1969 (1975).
4. J. W. Morris, Jr., S. Jin, and C. K. Syn, in: "Proceedings of First JIM International Symposium," Kobe, Japan (1976), p. 393.
5. C. K. Syn, S. Jin, and J. W. Morris, Jr., Metall. Trans. 7A:1827 (1976).
6. D. E. Williams, M.S. Thesis, University of California, Berkeley; LBL Report #10639, Lawrence Berkeley Laboratory, Berkeley, California (1979).

7. K. Ishikawa and N. Maruyama, Cryogenics 18:585 (1980).
8. I. Watanabe, T. Takamura, J. Tanaka, and F. Koshiga, 1980 AWS 61st Annual Meeting, Session 22.
9. K. Ikeda, T. Godai, T. Sugiyama, M. Aoki, and Y. Nishikawa, 1980 AWS 61st Annual Meeting, Session 22.
10. Report No. RDPO-7902, Kobe Steel, Ltd., Japan (1979).
11. I. Watanabe, Central Research Laboratory, Nippon Kokan, K.K., private communication.
12. J. H. Devletion, J. R. Stephens, and W. R. Witzke, Weld. J. 56:97-S (1977).
13. H. J. Kim and J. W. Morris, Jr., LBL Report #13046, Lawrence Berkeley Laboratory, Berkeley, California (1981).
14. R. L. Miller, Trans. ASM 57:892 (1964).
15. S. Jin, W. A. Horwood, J. W. Morris, Jr., and V. F. Zackay, in: "Advances in Cryogenic Engineering," Vol. 19, Plenum Press, New York (1974), p. 373.
16. K. W. Mahin, Ph.D. Thesis, University of California, Berkeley; LBL Report #10922, Lawrence Berkeley Laboratory, Berkeley, California (1980).
17. H. Tamura, T. Onzawa, S. Vematsu, and K. Maekawa, J. Jpn. Weld. Soc. 48:931 (1979).
18. H. Tamura, T. Onzawa, and S. Vematsu, J. Jpn. Weld. Soc. 49:855 (1980).
19. ASTM Standard E399-78a, in: "1980 Annual Book of ASTM Standards," Part 10, ASTM, Philadelphia (1980).
20. F. J. Witt and T. R. Mager, Nucl. Eng. Des. 17:91 (1971).

DEVELOPMENT OF FORGING AND HEAT TREATING PRACTICES FOR AMS 5737 FOR USE AT LIQUID HELIUM TEMPERATURES

E. N. C. Dalder

Lawrence Livermore National Laboratory, Livermore, California

M. Greenlee

Viking Metallurgical Corporation, Verdi, Nevada

INTRODUCTION

To achieve a combination of high yield strength (σ_y), plane-strain fracture toughness (K_{Ic}), and resistance to galling when turned against austenitic stainless steels in highly loaded threaded turnbuckles in the MFTF-B (Mirror Fusion Test Facility),[1] AMS 5737 (Fe-15Cr-25Ni-1Mo-V-Ti-Al-B), a heat-treatable Fe-base superalloy that is slightly ferromagnetic[2] under high magnetic fields at 4 K was chosen for large (~340 kg), forged turnbuckles. This report describes the forging and heat-treatment optimization program that resulted in good σ_y and K_{Ic} values over the 4-300 K range of service temperatures and the verification tests run on a preproduction forging and actual production parts.

PROCEDURE

Because of a tendency towards notch brittleness if incorrectly thermomechanically processed,[3,4] a preproduction forged block was manufactured by a combination of double upsetting and cross drawing a piece of 406-mm-diameter billet conforming to AMS 5737H (Table 1) to a 229-mm-diameter x 254-mm-long piece that is representative of the rough-forged production part. Maximum forging temperature was limited to 1394 K to avoid melting of Fe_2Ti.[5] A finishing temperature and postforging solution treatment of 1172 K was selected to yield a fine grain size in the finished part, a prerequisite to good low temperature toughness. Since the rate of grain growth

Table 1. Chemical Composition of AMS 5737

Heat no.	Forging nos.	Composition, weight percent C	Mm	P	S	Si	Cu	Mo	Aℓ	Co	Ni	Cr	V	Ti	B
9-6673	1-2, 4-12	0.05	0.40	<0.01	0.002	0.52	+	1.44	0.26	0.32	26.8	14.2	0.22	2.22	0.007
9-7261	16, 17	0.04	0.18	0.011	0.004	0.40	0.10	1.29	0.20	0.21	24.6	14.8	0.24	2.28	0.008
9-7086	15	0.05	0.38	0.013	0.003	0.56	+	1.32	0.16	0.28	24.9	14.2	0.25	2.29	0.006
AMS5737H		*0.08	2.00	0.025	0.025	1.00	+	1.50	0.35	+	27.00	16.00	0.50	2.35	0.01
								1.00			24.00	13.50	0.10	1.90	0.003

+ Not reported or specified.
*Maxima unless a range is quoted

during solution treatment is proportional to the square of the grain diameter, it was necessary to determine the minimum time required at 1172 K to insure complete dissolution of alloy additions. This was accomplished by cutting an 89-mm-thick transverse slice from the preproduction forging, quartering the slice, and heat-treating the quartered pieces for 1, 2, 4, and 24 h at 1172 K, followed by oil quenching. Standard round, smooth, and notched (K_t = 3) tensile specimens and metallographic specimens were machined from the first three slices, and a metallographic specimen was machined from the fourth slice to check for eta phase formation.[4] A K_t of 3 was chosen to match that anticipated to be created in the threaded portion of the turnbuckle under load.

After optimization of the postforging solution-treatment time, smooth and notched (K_t = 3) tensile specimens and elastic-plastic fracture toughness specimens were removed from the remainder of the preproduction forging. Tensile specimens were removed from near the surface of the forging, oriented in both the longitudinal and tangential directions.

Fracture-toughness specimens were oriented so that the crack grew in the longitudinal direction in a radial plane (C-L designation per Ref. 6). Tensile specimens, both smooth and notched (K_t = 3), oriented in both the longitudinal and tangential directions, were also removed from prolongations on selected production forgings. All specimens were given a standard aging treatment of 718 K, 16 h, air cool.

Room temperature tensile testing was done in a 45.5 ton electrohydraulic tensile testing machine per Ref. 7. Tensile testing at 4 K was done in a "hard" machine (Instron, 4.55 ton) per Ref. 8. Single-specimen elastic-plastic fracture-toughness tests at 300 K and 4 K were done in LLNL's facility per Ref. 9. Metallographic specimens were prepared by standard methods and examined as described in Ref. 8.

Table 2. Mechanical Properties of AMS 5737 Forgings

Orient.	Time at 1172K+ (hour)	Test temperature (K)	Ultimate strength (MPa)	0.2% yield strength (MPa)	Elong in 25.4 mm (%)	Red in area (%)	Notch UTS+++ / Smooth UTS	Plane strain frac tough
Optimization of time at 1172K								
Longitudinal++	1	4	1399	921	22.0	18.6	1.40	***
	2		1217	790	23.5	33.7	1.36	
	4		1233	806	29.0	36.9	1.31	
Properties from verification – forging								
Longitudinal*	2	4	1506	1047	25.5	22.7	1.31	132.7 $MPa\sqrt{m}$
Tangential++	2		1581	1128	19.0	17.4	1.18	
Longitudinal*	2	300	1098	823	21.5	30.7	1.40	136.3
Tangential++	2		1093	801	20.0	25.3	1.38	
Properties of specimens cut from production forgings								
Longitudinal*	2	300	1044-1063 / 1053	807-809 / 808	20.6-21.0 / 20.8	26.7-52.5 / 29.6	1.38-1.44 / 1.41	
Tangential**	2		992-1099 / 1038	766-821 / 792	19.0-21.0 / 20.4	27.7-30.8 / 30.0	1.34-1.44 / 1.39	
AMS 5737H (min values)			965	631	12.0	15.0		

Notes:

+1172K (2 hrs), oil quench, 718K (16 hrs), air cool

++Single specimen

+++K_t = 3.0

*Average of 2 specimens

**Average of 5 specimens

***Converted from J-integral test results

RESULTS

Chemical analyses of 3 heats of AMS 5737 used in manufacture of the 1 preproduction and 14 production forgings are given in Table 1. The requirements of AMS 5737H were met. Table 2 summarizes tensile and fracture toughness test results, the latter quantity being reported as K_{Ic}, converted by the method discussed in Ref. 10.

Considering first the results of the study to optimize solution treatment time at 1172 K, presented in Table 2 and Fig. 1, a time of 2 h at 1172 K was chosen as a compromise between maximizing strength and notch toughness (short times) and maximizing ductility (longer times). A metallographic specimen exposed for 24 h at 1172 K (not shown) contained massive eta precipitates,[3,4] indicative of poor notch toughness.

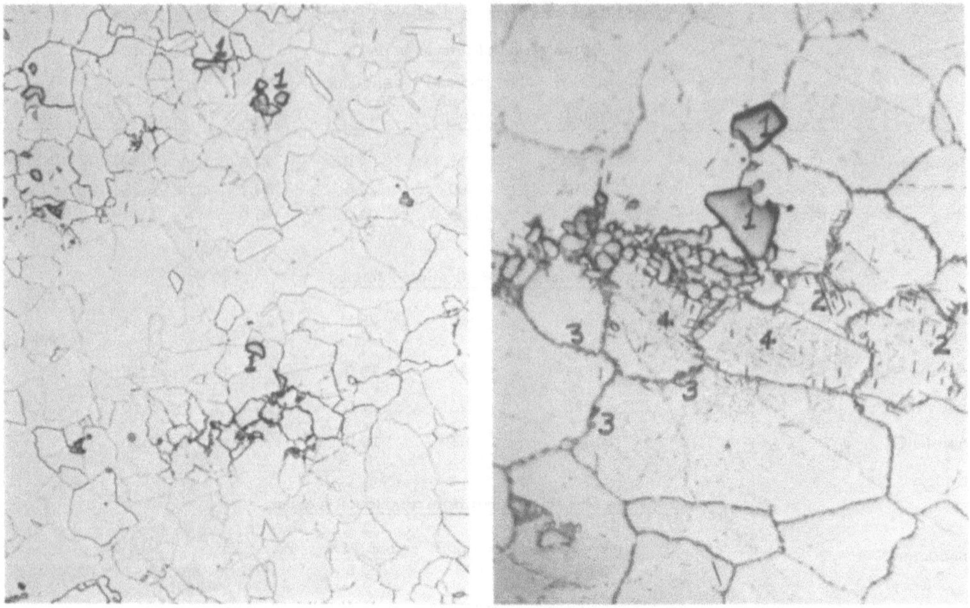

(a) X250 (b) X1050

Fig. 1. Transverse section through sample 2N after a postforging heat treatment of 1172 K for 2 h, oil quench, plus 718 K for 16 h, air cool, and a tensile test at 4 K. (Reduced 28%.) Note MC-type carbonitrides (1), Widman Statien-type of phase (2), a grain-boundary phase (3), a possible "G" phase and an unidentified intragranular precipitate (4).

Considering next the 4-K and 300-K results on the preproduction forging given the now "standard" solution treatment, 2 h at 1172 K, several points are apparent:

1. Tensile properties at 300 K exceed the requirements of AMS 5737H.

2. Tensile properties at 4 K show higher ultimate tensile strength, σ_u (24%), yield strength, σ_y (33%), and elongation to fracture, e_f (9%), and lower reductions in area, RA (48%), and notch tensile strengths ratio, NTR (4%), based upon longitudinal properties, than the results obtained from the solution-treatment optimization study.

3. The 4-K K_{Ic} (average) of 132.7 MPa$\sqrt{m}$ is excellent and is based on the average of two specimens, one with a K_{Ic} value of 108.6 MPa$\sqrt{m}$, the other with a K_{Ic} of 156.9 MPa$\sqrt{m}$. The K_{Ic} at 300 K is 136.3 MPa$\sqrt{m}$.

4. Tangential tensile properties at 4 K exhibit higher σ_u and σ_y and lower e_f, RA, and NTR values than the 4-K longitudinal properties.

The 300-K tensile properties, σ_u, σ_y, and e_f, of longitudinal and tangential specimens removed from production forgings surpassed the requirements of AMS 5737H. Both these properties and the RA and NTR values were comparable to results obtained from the tests on the preproduction forging.

DISCUSSION

Examination of microstructures taken from 4-K notched tensile specimens oriented in the tangential (Figs. 2a, 3a) and longitudinal (Figs. 2b, 3b) directions indicates that although both specimens have the same duplex grain structure, the former specimen shows no preferential banding of second phases, such as M(C,N), eta, and "G" phase, whereas the latter specimen shows extensive

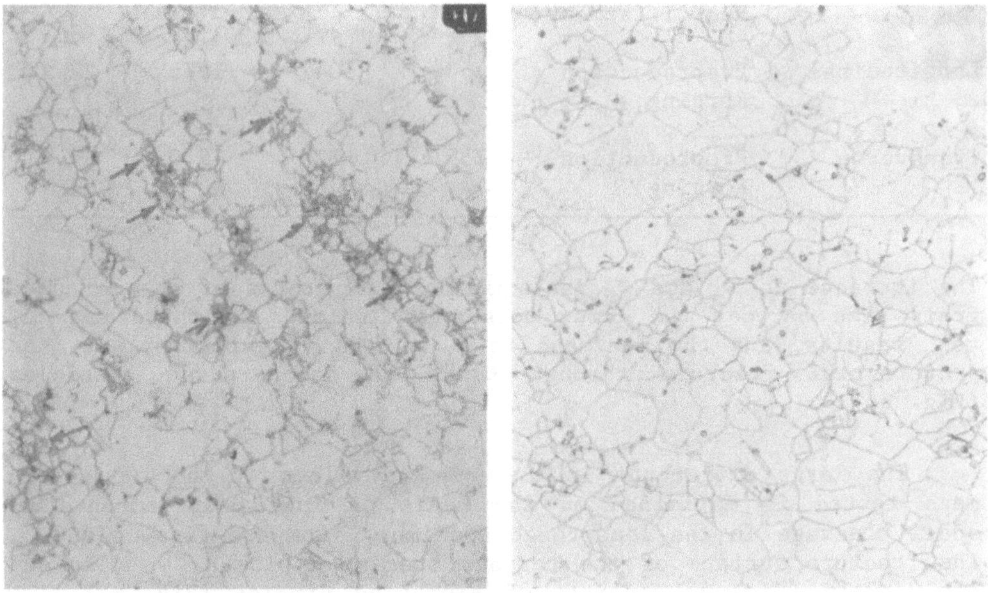

(a) Tangential; X100 (b) Longitudinal; X100

Fig. 2. Microstructure of notched tensile specimens tested at 4 K. Note heavy banding of second phases and fine-grained regions (arrows) in the tangential specimen and absence of same in the longitudinal specimen. (Reduced 28%.)

banding occurring at an angle of about 45° to the tensile axis (horizontal in photomicrographs); compare Figs. 2a and 2b. At higher magnifications (Fig. 3b), the heavy banding of second phases is seen to be closely associated with the primary fracture. The decrease in NTR between the longitudinal (NTR = 1.31) and tangential (NTR = 1.18) orientations at 4 K is explainable on the basis of preferential fracture through bands of low-toughness second phases.

A similar explanation is proposed as the reason for increased 4-K e_f and RA on going from smooth longitudinal tensile specimens tested to optimized solution-treatment time to smooth longitudinal specimens to smooth tangential specimens tested after removal from the preproduction forging, as shown below:

Orientation	Occurrence	σ_u, MPa	σ_y, MPa	e_f, %	RA, %
Longitudinal	Optimized Heat Treatment	1217	790	23.5	33.7
Longitudinal	Preproduction Forging	1506	1047	25.5	22.7
Transverse	Preproduction Forging	1581	1128	19.0	17.4

The increase in σ_u and σ_y is explainable in terms of an increased grain size between the first versus the second and third sets of test results, and the decrease in e_f and RA is explainable by the preferential fracture through bands of second-phase particles (Fig. 3b).

The large difference in the 4-K K_{Ic} values of the two specimens tested is explained on the basis of increasing amounts of quasi cleavage in the less-tough specimen. Compare Figs. 4 and 5. The fracture surface of the tougher specimen, with a K_{Ic} value of 156.9 MPa$\sqrt{m}$ (Fig. 4) consists of elongated dimples (Fig. 4a) with one isolated area of quasi cleavage (Fig. 4b). The fracture surface of the less-tough specimen, with a K_{Ic} of 108.6 MPa$\sqrt{m}$, consists of mixed areas of elongated dimples and quasi cleavage (Figs. 5a, 5b), with an occasional grain-boundary separation (Fig. 5a).

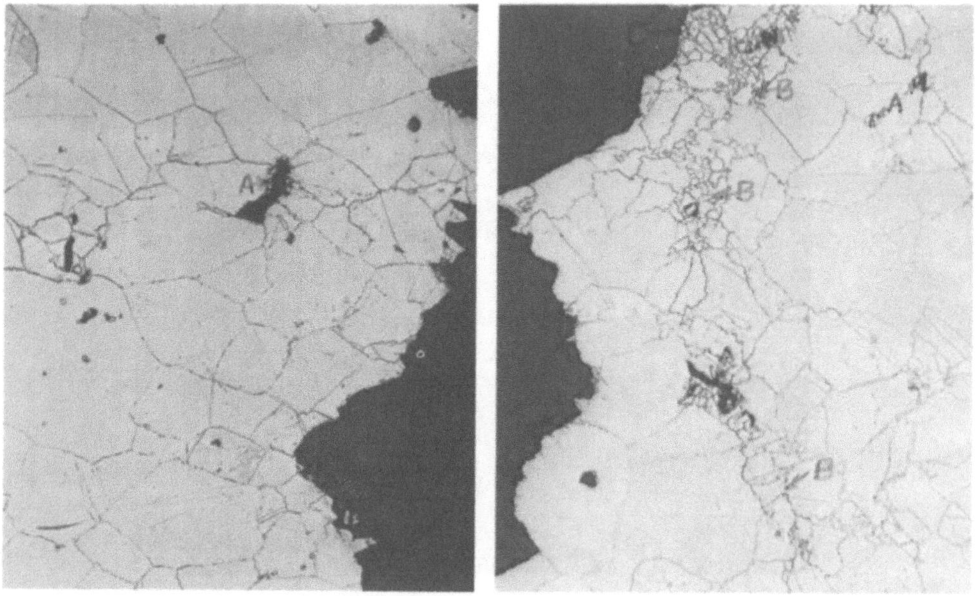

(a) Tangential; X250 (b) Longitudinal; X250

Fig. 3. Microstructure of fractures in notched tensile specimens tested at 4 K. Note secondary cracking associated with MC-type carbonitrides (A), heavy banding of second phases and fine-grained regions (B) in tangential specimen, and association of primary fracture (C) with (B). (Reduced 28%.)

CONCLUSION

A double-upset and cross-draw forging sequence, with finish forging in the 1172-1200 K range, followed by a heat treatment of 1172 K for 2 h, oil-quench, 718 K for 16 h, air cool, produced useful combinations of smooth and notched (K_t = 3) tensile and K_{Ic} performance in large AMS 5737 forgings over the range 4-300 K. Differences in mechanical properties were explained on the basis of microstructure differences arising from the forging practice.

ACKNOWLEDGMENT

Work was performed under the auspices of the U.S. Department of Energy by the Lawrence Livermore National Laboratory under Contract Number W-7405-ENG-48.

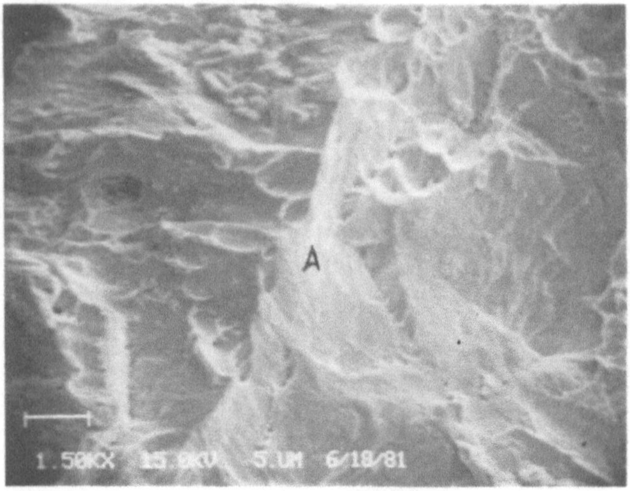

(a) X1500

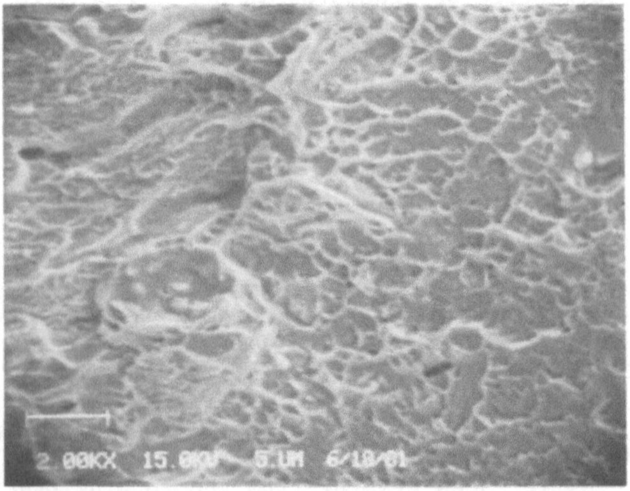

(b) X2000

Fig. 4. Fracture surface of 4-K K_{Ic} specimen 3B1, showing elongated dimples, quasi cleavage, and a grain-boundary film (A). (Reduced 28%.)

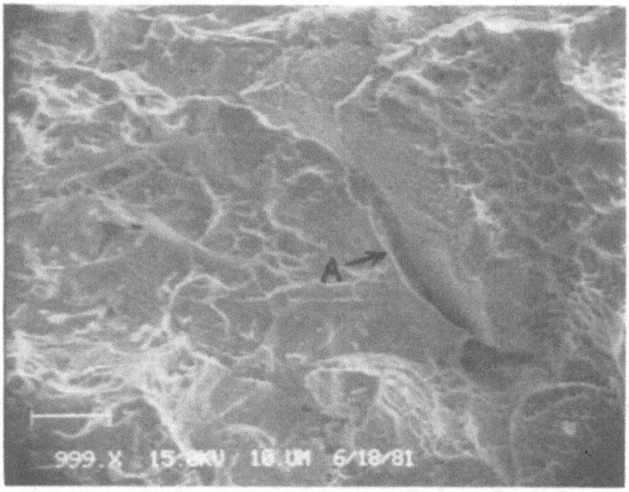

(a) X999

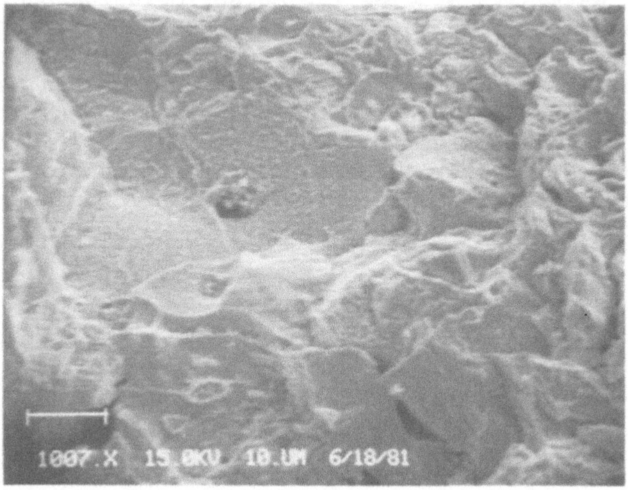

(b) X1027

Fig. 5. Fracture surface of 4-K K_{Ic} specimen 3C1, showing quasi cleavage, a grain-boundary crack (A), and elongated dimples. (Reduced 28%.)

REFERENCES

1. C. D. Henning et al., UCRL-59955, Lawrence Livermore National Laboratory, Livermore, California (1980).
2. K. Efferson et al., ORNL-4150, Oak Ridge National Laboratory, Oak Ridge, Tennessee (1967), p. 126.
3. C. Sullivan and M. J. Donachie, Met. Eng. Q. 1 (1971).
4. D. Muzyka, in: "The Super Alloys," C. Sims and W. Hagel, eds., John Wiley & Sons, New York (1972), p. 113.
5. A. Sabroff et al., "Forging Materials and Practices," Reinhold (1968), p. 366.
6. "Standard Method of Test for Plane Strain Fracture Toughness of Metallic Materials," ASTM E-399-78, Part 10, American Society for Testing and Materials, Philadelphia (1978), p. 512.
7. "Standard Method of Test for Tensile Properties of Metallic Materials," ASTM E8-78, Part 10, American Society for Testing and Materials, Philadelphia (1978).
8. E. N. C. Dalder, O. W. Seth, and T. A. Whipple, Shielded metal-arc and flux-cored metal-arc stainless steel weldments: magnet cases for 4-K service, in: "Advances in Cryogenic Engineering--Materials," Vol. 28, Plenum Press, New York (1982). p. 839.
9. D. T. Read, The computer aided J-integral test facility at NBS, in: "Materials Studies for Magnetic Fusion Energy Applications at Low Temperatures--III," NBSIR 80-1627, National Bureau of Standards, Boulder, Colorado (1980), p. 215.
10. R. L. Tobler and R. P. Reed, Interstitial carbon and nitrogen effects on the cryogenic fatigue crack growth of AISI 304 type stainless steels, Refs. 9 and 13, in: "Materials Studies for Magnetic Fusion Energy Applications at Low Temperatures--IV," NBSIR 81-1645, National Bureau of Standards, Boulder, Colorado (1981), p. 101.

PRELIMINARY RESULTS ON THE DEVELOPMENT OF VACUUM-BRAZED JOINTS FOR CRYOGENIC WIND TUNNEL AEROFOIL MODELS

D. A. Wigley

Mechanical Engineering Department, University of Southampton, Southampton, United Kingdom

P. G. Sandefur, Jr. and P. L. Lawing

NASA Langley Research Center, Hampton, Virginia

INTRODUCTION

As the speed of aeroplanes has increased into the transonic regime, it has not been possible to test aerodynamic models in conventional wind tunnels at Reynolds numbers equivalent to those achieved in flight. The cryogenic wind tunnel, in which liquid nitrogen is injected into the tunnel circuit and evaporates to give a gas temperature of approximately 80 K, has removed this restriction, but brought with it the complexities of operation at cryogenic temperatures. The proceedings of AGARD Lecture Series No. 111[1] covers in detail most aspects of cryogenic wind tunnel operation and, in particular, the papers by Wigley[2] and Kilgore[3] deal with the properties of materials and construction of aerofoil models, respectively. Furthermore, the factors to be considered in choosing materials for the construction of the pilot models for the National Transonic Facility (NTF), currently nearing completion at the NASA Langley Research Center, are discussed in detail in a National Bureau of Standards (NBS) report by Tobler.[4]

The necessity of ensuring that two- and three-dimensional models can operate safely in cryogenic wind tunnels has triggered a reappraisal of the techniques used to construct such models. Of particular interest is a method of achieving the necessary connections between orifices drilled into the aerofoil surfaces to measure the pressure distribution over the aerofoil surfaces and the measuring equipment generally located remote from the model. Current practice is to machine a recess into one surface of the aerofoil to expose the bottom of the pressure tappings, braze fine-diameter tubes to connect with each individual hole, and then lead a bundle of such tubes out via the wing root or support sting. The aerodynamic profile of the model is then restored by welding a cover plate, usually fitted with its own pressure orifices, over the recess.

This whole operation is labour intensive and therefore expensive; an alternative approach has been developed by Lawing, Sandefur, and Wood.[5] In essence, this involves machining a network of open channels into the surfaces of two flat plates, one of which is shown in Fig. 1a. A bonding agent is then applied to one surface, and the two plates are clamped together and heated under vacuum long enough to achieve a metallurgically sound joint, thus forming a network of passages within a monolithic block. The bond between the two plates has to be strong and tough, especially for cryogenic use, and the technique clean and precise enough to prevent either obstruction of the passages and orifices or cross leakage between adjacent passages. Initial trials carried out at the NASA Langley Research Center demonstrated that diffusion-assisted brazed joints could be formed in 17-4 PH, 15-5 PH, AISI type 347 and Nitronic 40 stainless steels using electrodeposited copper as the bonding agent. Subsequent work has concentrated on 15-5 PH and Nitronic 40 using thin foils of pure copper and Nicrobraz LM, a commercially available nickel-based alloy containing boron and silicon melting point depressants. This paper summarizes the work carried out to understand and evaluate the metallurgical characteristics of these bonds. Further details of this work are given in a report by Wigley.[6]

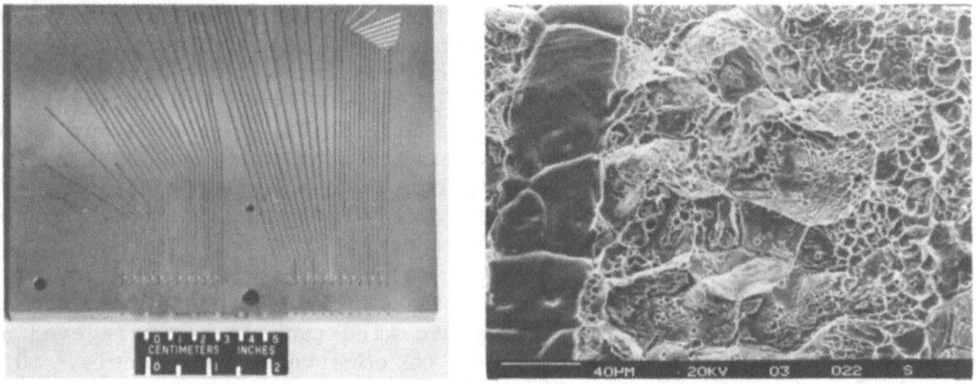

(a) (b)

Fig. 1. a) Open channels machined into surface of flat plate.
b) Stereoscan view of 15-5 PH fracture surface.

15-5 PH STAINLESS STEEL

This material is known to have poor toughness at the low operating temperatures of a cryogenic wind tunnel, but its widespread use at the NASA Langley Research Center for the construction of aerofoil models for conventional tunnels made it the natural point of reference for the evaluation of alternative materials and bonding systems. Examination of one of the early specimens, which was electroplated with copper, revealed a pattern of light and dark areas on the fracture surface that corresponded with that formed by the original grinding operation. This indicated that the surface was too rough for the formation of a void-free bond with the limited amount of copper available. Figure 1b gives a higher magnification view of part of the fractured surface taken on a scanning electron microscope, and this shows clearly the smooth nature of the unwetted (dark) regions. The lighter region indicates that the fracture is basically intergranular in nature, with fine scale "ductile-dimple" features on the exposed faces formed by the ductile failure of the copper interlayer. Sections taken perpendicular to an unfractured bond and then polished and etched for metallographic examination indicated not only the basic nature of the bond, but also a high degree of voidage, as revealed by the continued weeping of etchant from the bond line. These voids correspond to the unwetted areas seen on the fracture surface of Fig. 1b, and they would probably cause unacceptably high cross leakage between adjacent pressure channels in a model. Subsequent specimens were therefore prepared by using thin foils of copper, rather than by electroplating.

The next series of tests were carried out using 0.001-in and 0.002-in copper foils at temperatures of 1150 and 1200°C and for holding periods of 30, 60, and 120 min, respectively. The essential features of the types of bond obtained are illustrated in Figs. 2a through 2d. Figure 2a shows the bond formed using 0.001-in copper foil and holding at 1200°C for 30 min. This is, in essence, a 100% brazed joint with a continuous layer of copper between the two 15-5 PH surfaces; furthermore, there is no indication of staining or weepage, and the bond is virtually void-free. The grain boundaries in contact with the bond have been attacked by the molten copper to give the rounded, triangular nature visible in Fig. 2a. Evidence that copper has diffused rapidly down the grain boundaries can be seen from the precipitates visible within the boundaries of grains a few diameters away from the joint. In contrast, the bond line in the sample with a 0.001-in copper foil held at 1200°C for 120 min can be seen from Fig. 2b to be very flat in nature and almost indistinguishable from the parent metal. Almost all of the copper has diffused away from the bond line leaving only isolated islands of visible copper between the diffusion bonds formed between the two 15-5 plates. This loss of copper has two significant effects: it causes void formation in the bond line and it increases the intergranular nature of the failure when samples are fractured across the bond line.

The joint formed by holding at 1200°C for 60 min can be seen from Fig. 2c to consist of both copper-brazed and diffusion-bonded regions and to be more undulating in nature than in either the 30-min or 120-min samples. It is believed that recrystallization and grain growth are initiated from the bond zone and that in these 60-min specimens undiffused copper is redistributed around these grain boundaries. The highly intergranular nature of the fracture surface, created by breaking a specimen across the bond line, is illustrated vividly in Fig. 2d. The fine structure visible on the surface of the grain is caused by the ductile nature of the micromechanism of failure in the brazed and diffusion-bonded region.

One final metallurgical feature shown by all of these samples is the very large grain size produced by prolonged holding at the elevated temperatures needed to form a bond. Since high toughness

(a) (b)

(c) (d)

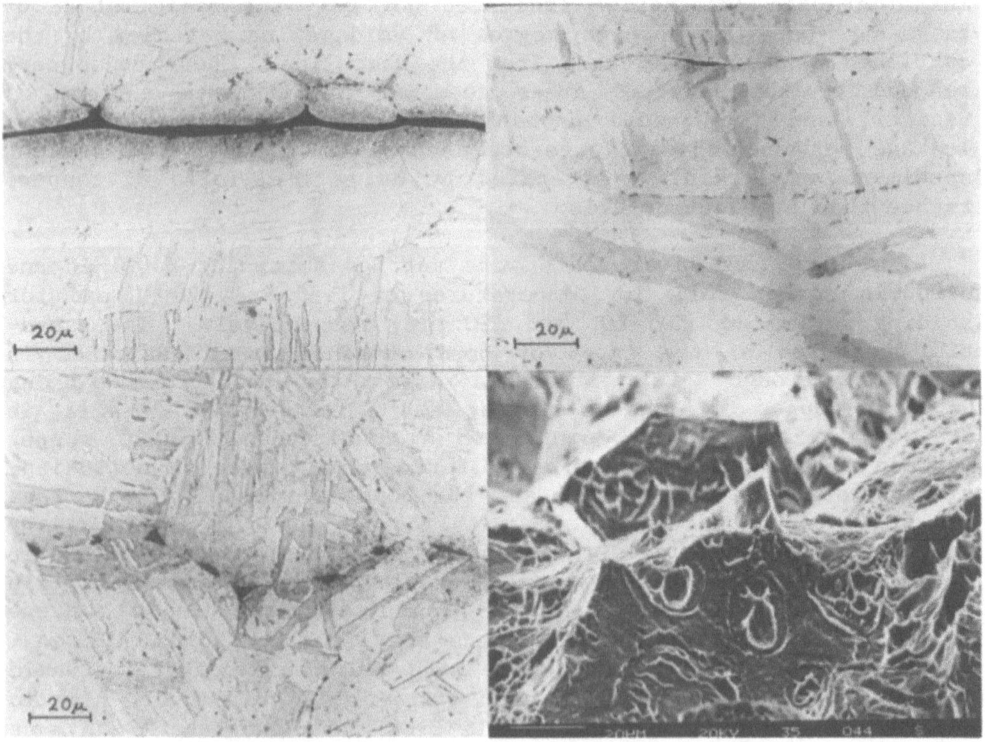

Fig. 2. Bonds in 15-5 PH stainless steel formed at 1200°C for a) 30, b) 120, and c) 60 min. d) Stereoscan view of intergranular fracture in the 60-min sample.

demands a fine grain size, this characteristic reduces even further the already marginal toughness shown by 15-5 PH stainless steel at low temperatures. For cryogenic use, the H1150M heat treatment, corresponding to a tensile strength of 793 MPa (115 ksi) is considered mandatory; the majority of our samples showed parent metal strengths of the order of 1380 MPa (200 ksi), and thus they would have to undergo a heat-treatment cycle before they could be considered suitable for low temperature applications.

Only small samples were available for evaluating mechanical properties of the bonded joints, and so three-point-bend tests were carried out on samples 2-mm thick, 10-mm wide, and 20-mm long with the bond located through-thickness at centrespan. When tested at room temperature all specimens, regardless of the type of bond, showed a definite yield at approximately the same stress as that at which yield occurred in the parent metal. Failure followed, however, after negligible plastic deformation in all specimens except one batch that had been heat-treated to the H1075 condition (1000 MPa, 145 ksi tensile stress), which gave deformation equivalent to approximately 5% in a tensile test. At liquid-nitrogen temperatures, all specimens failed in an apparently brittle manner at stresses in the range 60-95% of the yield stress of the parent metal, a further indication of the generally poor cryogenic properties of both the parent metal and the bonds formed by using thin copper foil.

The final tests on 15-5 PH stainless were carried out using 0.003-in thick foils of Nicrobraz LM, a commercially available, nickel-based alloy containing boron and silicon melting point depressants, which enables brazed bonds to be formed at temperatures as low as 1000°C. Figure 3a shows a view of the bond line and adjacent grains of the parent metal, and it indicates clearly that there is some form of diffusion zone on either side of the bond line. It was initially thought that the second-phase particles visible in this zone and in the grain boundaries of the parent metal were borides and silicides formed as the two melting point depressants diffused out of the LM foil into the 15-5 parent metal. Microprobe analysis showed, however, that they were iron-chromium sigma phase, an embrittling phase known to be formed in stainless steels during prolonged holding at high temperatures and whose formation is particularly encouraged by high silicon contents.

It might, therefore, have been expected that the presence of the sigma phase would produce an embrittled joint, but further three-point-bend tests at room temperature showed that the joint was, in fact, more ductile than any formed using copper foils. Not only was the deformation prior to failure greater than before, but fracture took place by ductile localized necking within the nickel layer. As yet, however, insufficient samples have become available to determine the low temperature properties of these bonds.

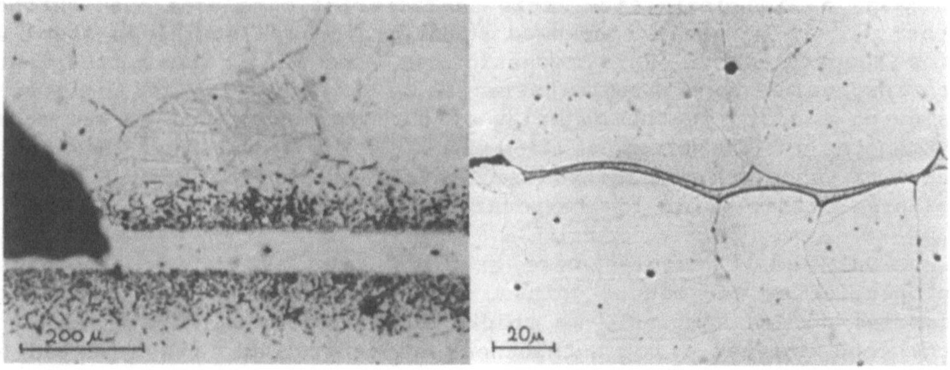

(a) (b)

Fig. 3. a) 15-5 PH/Nicrobraz LM bond, 1180°C, 30 min.
b) Nitronic 40/0.001-in copper bond, 1200°C, 30 min.

NITRONIC 40

This alloy is an austenitic stainless steel containing approximately 9% manganese and 6% nickel, which is nitrogen strengthened to give properties generally superior to those of the 300-series stainless steels. It has been chosen for the construction of the pilot model to be used in the NTF cryogenic tunnel because of its optimum combination of strength and toughness at low temperatures. The ability to form bonded joints in Nitronic 40 in the manner discussed earlier would, consequently, be a very significant advantage for the construction of the models that will follow the pilot model in the NTF tunnel.

Initial samples were coated with 0.0003-in-thick electrodeposited copper layers and bonds formed by holding in a vacuum oven for 30 min at a temperature of 1180°C, whereas subsequent samples utilized 0.001-in-thick copper foil and a holding temperature of 1200°C. Figure 3b shows a polished and etched metallurgical section taken through the joint formed using the copper foil. In the section illustrated, the copper is virtually continuous and the joint is of a brazed nature, but other sections showed that there were also diffusion-bonded regions present. Figure 4a is a stereoscan photograph of the tensile face of a small sample deformed through a 60° angle in a three-point-bend test at room temperature. Prior to testing, the surface had been polished to a one-micron diamond finish to show the slip lines formed when a metal is stressed beyond its yield point. Within a grain, the lines produced by a single slip system are parallel to each other, and their direction changes at the boundaries owing to the differing orientations of the slip planes in the various grains.

Close examination of Fig. 4a shows that slip lines are visible not only in the parent metal, but also within the copper-rich phase in the bond line.

> Thus, when tested at room temperature, the bond formed by diffusion-assisted brazing of Nitronic 40 using copper as the filler is strong and ductile enough to permit yield and very significant plastic deformation of the Nitronic 40 itself.

A similar three-point-bend test carried out in liquid nitrogen showed that the bond strength still exceeded that of the parent metal; evidence for this is given by the stereoscan photograph of Fig. 4b. This shows the edge of one of the fracture halves with the fracture surface oriented across the picture. The ductile nature of the failure process is indicated by the heavy deformation visible at the surface. The vertical face on the left-hand side of the photograph was the tensile face in the bend test, and the slip lines visible on the Nitronic 40 grains, together with their generally deformed nature, bear witness to the yield and plastic flow that occurred prior to failure. This deformation was, however, significantly smaller at 77 K than at 300 K.

Further three-point-bend tests were carried out on samples of Nitronic 40. Figure 5a shows a schematic indication of the location of the bond in the different types of specimen. The specimens described already are type A, with the through-thickness bond located at midspan, whereas in type B the bond is located through thickness, but parallel to the span. Type C has the bond located

(a) (b)

Fig. 4. a) Slip lines in Nitronic 40 tested at 300 K in three-point bend.
b) Fracture surface of Nitronic 40 tested at 77 K.

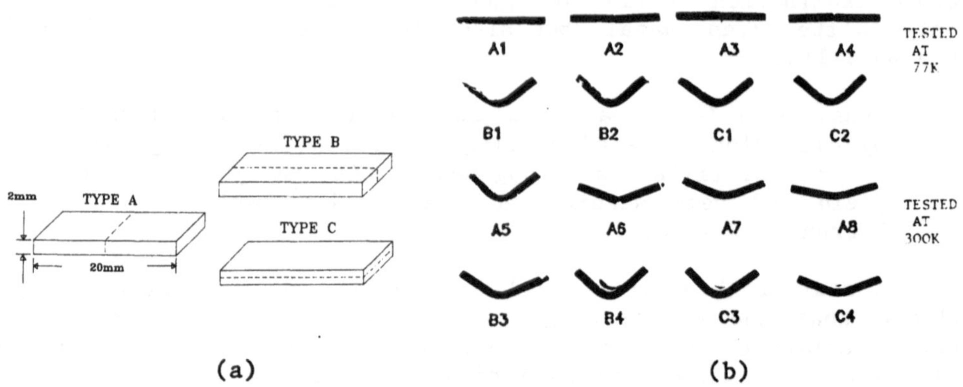

Fig. 5. a) Location of bond plane in bend tests.
b) Nitronic 40 tested at 300 and 77 K.

in the plane of the sample and at its neutral axis, a configuration used to detect delamination in a weak-bonded system. The results of tests at room temperature and in liquid nitrogen are illustrated pictorially in Fig. 5b where the as-tested samples are viewed edge-on. No failures occurred in any of the room temperature tests. The varying degrees of deformation were chosen simply for experimental convenience. At 77 K all type A samples yielded in the parent metal before fracturing across the bond, some after a significant amount of plastic deformation. No fracture or debonding occurred in either type B or C samples indicating that a mode I, tensile crack-opening stress system is necessary for failure at liquid nitrogen temperatures.

The results obtained from these tests are summarized in Table 1. It should be noted that the values obtained from three-point-bend tests rarely correlate exactly with those from tensile tests, especially with small samples of the size used in these experiments. Nevertheless, the results are extremely encouraging. At room temperature, all three types of samples showed yield strengths similar to that of the parent metal, and even the most stressed A types showed maximum strengths averaging 90% of that of the parent metal. At 77 K, two of the A type samples yielded at the same stress as the B and C types, one even work-hardening to a stress 150% of its yield value before failing.

As yet, relatively few samples of Nitronic 40 have been produced, but initial trends indicate that the thickness of the copper layer should be kept to the minimum necessary for complete surface coverage. For example, Fig. 6a illustrates the effect of liquid copper attack on the grain boundaries of a sample that had been misaligned, thus permitting copper to accumulate locally in a thick

Table 1. Nitronic 40 Three-Point-Bend Test Results

Test Temp. (K)	Specimen I.D.	Yield Stress (MPa),	(ksi)	Maximum Stress (MPa),	(ksi)	Defln. (mm)	Comment
300	A8	538	78	897	130	2	
300	A7	490	71	972	141	3	
300	A6	448	65	634	92	3	
300	A5	538	78	1021	148	6	
300	B4	565	82	1083	157	6	
300	B3	538	78	1035	150	4	
300	C4	(420)	(61)	(738)	(107)	3	50% bond
300	C3	538	78	1048	152	6	
300	Average	522	75				
300	Quoted	400	58				
77	A1		-	1180	171	-	Snapped
77	A2	1269	184	1882	273	1.2	Yielded
77	A3		-	(669)	(97)	0.4	50% Bond
77	A4	1310	190	1330	193	0.4	Yielded
77	B1	1365	198	2345	340	6	
77	B2	1365	198	2345	340	7	
77	C1	1227	178	2317	336	6	
77	C2	1145	166	1145	312	7	
77	Average	1280	186				
77	Quoted	1034	150				

Note: Specimens: 2-mm thick x 10-mm wide x 20-mm long.
Test span: 16 mm.

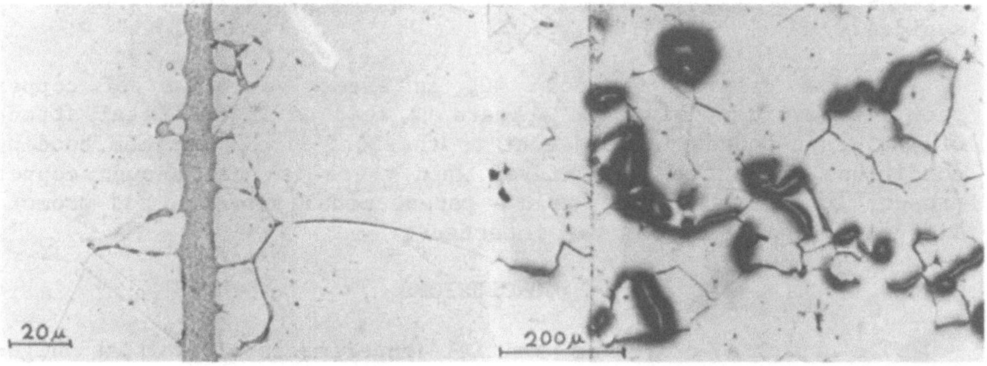

(a) (b)

Fig. 6. Grain boundary attack by liquid copper in Nitronic 40, a) light b) severe.

layer. An even more striking indication of potential pitfalls is shown by Fig. 6b, which was obtained from a small development sample in which a high bonding temperature was achieved by Joule heating using an alternating electric current. The specimen was held at temperature for less than 5 min, but this was sufficient to drive molten copper up to 5 mm each side of the bond line, literally forcing the grains apart along their boundaries. Obviously, more development work is still required before this process is fully characterized.

MECHANISM OF BOND FORMATION

As a result of this work, it is possible to understand in general terms the basic mechanisms involved in bond formation. At temperatures in excess of approximately 1000°C and in the moderate vacuum involved, it appears that the oxide film on the surface of stainless steel is no longer self-repairing and that when copper melts at 1083°C it is able to "wet" the metal surface. Heat treatments involving short times and relatively low temperatures produce essentially a brazed joint between the two surfaces, and if enough copper is present, the bond is void-free. At longer times and higher temperatures, diffusion of copper away from the bond line permits the asperities on the two surfaces to come into contact and allows the formation of diffusion bonds. If, however, too much copper diffuses from the bond line, voids are left behind that could lead to the possibility of cross leakage between adjacent pressure passages and also degradation of the surface finish of aerofoil models where the joint intersects the surface. In 15-5 PH stainless steel, it appears that the optimized bond is essentially of a brazed nature. Enough filler metal must be present to fill all the crevices and so prevent void formation, and in this respect the lower fluidity of the nickel-based LM brazing alloy is advantageous if the deleterious effects of sigma-phase formation can be avoided.

In the case of Nitronic 40, an excessive amount of copper should be avoided, since it appears to lead to liquid metal attack of the grain boundaries adjacent to the bond line. Optimum bonding conditions appear to be achieved when there is just enough copper present to remove oxide from the parent metal surfaces and promote diffusion bonding across the interface.

CONCLUSIONS

The results of these initial experiments show that high-strength, void-free bonds can be formed by vacuum brazing of stainless steels using copper and nickel-based filler metals. In Nitronic 40, brazed joints have been formed with strengths in excess of the yield strength of the parent metal and, even at liquid nitrogen temperatures, the excellent mechanical properties of the parent metal are only slightly degraded.

The poor toughness of 15-5 PH stainless steel at cryogenic temperatures is lowered even further by the presence of diffusion-brazed bonds, and it is highly unlikely that the technique would be used for any critical areas of aerofoil models intended for low-temperature service. Nevertheless, the potential advantages of this simplified method of construction still have attractions for use at ambient temperatures.

ACKNOWLEDGMENTS

The authors wish to acknowledge the interest, support, and assistance of NASA-Langley, Hampton Technical Center, Hampton, Virginia and the Mechanical Engineering Department, The University, Southampton, U.K.

REFERENCES

1. Cryogenic Wind Tunnels, in: "Proceedings of the AGARD Lecture Series No. 111," May 1980, AGARD LS 111, ISBN 92-835-1360-6 (1980).
2. D. A. Wigley, The physical properties of metals and non-metals and The effect of low temperatures on the strength and toughness of materials, papers 4 & 6, AGARD LS 111 (1980).
3. R. A. Kilgore, Model design & instrumentation experiences with continuous-flow cryogenic tunnels, paper 9, AGARD LS 111 (1980).
4. R. L. Tobler, "Materials for Cryogenic Wind Tunnel Testing," NBSIR 79-1624 National Bureau of Standards, Boulder, Colorado (1980).
5. P. L. Lawing, P. G. Sandefur, Jr., and W. H. Wood, "A Construction Technique for Wind Tunnel Models," NASA Tech. Brief LAR-12710 (Fall 1980).
6. D. A. Wigley, "The Structure and Properties of Diffusion Assisted Bonded Joints in 17-4PH, 347, 15-5PH and Nitronic 40 Stainless Steels," NASA CR 165 745 (1981).

COMPOSITE ALUMINUM-FIBERGLASS EPOXY PRESSURE VESSELS FOR TRANSPORTATION OF LNG AT INTERMEDIATE TEMPERATURE

S. G. Ladkany

The Johns Hopkins University, Baltimore, Maryland

INTRODUCTION

The design of large, 6-m diameter, composite aluminum-fiberglass epoxy pressure vessels for the transportation of liquefied natural gas (LNG) at intermediate temperature is presented. The pressure vessels are designed to have an operating pressure range of up to 6.21 MPa (900 psi) and pressure-to-burst ratio close to two. The cylindrical pressure vessels are circumferentially reinforced with layers of high strength fiberglass epoxy or pultruded glass polyester overwrap. The vessels are prestressed at ambient temperature, using a sizing technique (autofrettage). The large pressure vessels are to be used in the economical storage or transport of LNG at temperature and pressure conditions between the critical conditions, 191 K, 4.69 MPa (-116°F, 680 psi) and atmospheric conditions, 106 K, 0.1 MPa (-268°F, 14.7 psi). The pressure vessels are designed for ultimate failure modes in the circumferential direction to prevent the possibility of an axial separation of a portion of the vessel at failure.

TRANSPORTATION OF LNG AT INTERMEDIATE TEMPERATURE

Pressure vessels for the transportation of LNG at high pressure and at low temperature levels of 172 K (-150°F) and 144 K (-200°F) were fabricated in the mid-1960s and were installed in a Liberty ship, the SigAlpha, for use in the transport of North Sea gas to Europe.[1] It has been recently suggested[2] that LNG could be transported quite economically at the intermediate temperature and pressure conditions of 178 K and 3.11 MPa (-140°F and 450 psi). The method contemplates loading the refrigerated and pressurized gas into specially designed tankers for shipment and unloading at the port of arrival, by simply: "throttling it adiabatically to atmospheric pressure, sending the high-enthalpy vapor directly by pipeline to the consumer or other market and transferring the low-enthalpy liquid to conventional storage. This split results in 51% liquid and 49% vapor." Such a method may save up to: "$66 million per year for 50¢/$10^3 ft^3$ gas in a 1 x 10^6 scfd liquefaction plant, if a 100% load factor is assumed."[2]

Explosion hazards of LNG or LPG carriers however, increase dramatically with the volume of liquefied gas spilled in an accident. The cryogenic liquid flash evaporates upon contact with water, forming an air-gas cloud with potentially considerable detonation magnitude.[3] Economic and safety considerations thus warrant the transportation of LNG in carriers containing a number of very strong pressure vessels of the free-standing, self-supporting type that possess high structural integrity and large factors of safety. Aluminum vessels, reinforced with fiber composites are ideal for such usage.

FILAMENT REINFORCED METAL COMPOSITE PRESSURE VESSELS

Research and development programs to evaluate filament-wound composite pressure vessels with internal metal liners have been conducted for over twenty years at NASA-Lewis Research Center. Materials, construction processes, and performance of such pressure vessels under ambient and cryogenic conditions have been evaluated for aerospace vehicle applications.[4,5,6] Commercial applications of metal composite pressure vessels are now widespread[4] and the technology for the design and construction of thick-walled cylindrical[5] and spherical[6] vessels has been developed. Steel, titanium, and aluminum liners have been tested along with overwrap composites of S glass, Kevlar, Thornell, graphite, and other fibers. Such vessels enjoy the characteristics of high strength and high reliability; they are 25-40% lighter in weight and also safer than comparable all-metal, high performance pressure vessels.

Stressing of a filament-wound pressure vessel results in some matrix cracks, thus requiring an impermeable liner. Two kinds of metal liners have traditionally been used: the thin-metal type and the load-sharing type. Thin metal liners are totally surrounded by and bonded to the overwraps that cover them. The bonding is necessary for the structural support of the liner and to prevent the separation of the composite shell and the liner at cryogenic temperature due to the differential thermal contraction between metal and composite. The liners are in the postyield state of stress under normal operating pressure conditions and thus have a limited cyclic fatigue life, which is on the order of 2000 cycles for an aluminum liner.[7] Such an aluminum-composite vessel containing liquid nitrogen was successfully tested at the Beech Aircraft Corporation.[8]

Load-sharing metal liners do not require a bond between the metal shell and the composite overwrap. Because the liners can resist axial tension, the overwrap does not necessarily have to include the cap, thus avoiding the difficult and costly helical and axial winding processes, which are not well developed for large pressure vessels. The circumferential overwrapping, however, of a large cylinder with a uniaxial fiber composite is a relatively straightforward procedure; similarly, the on-site construction techniques of large diameter (6 m) profile pultruded tanks of glass-polyester, have been prototyped and tested.[9] This "profile pultrusion" technique allows the variety of reinforcements that make up the profile, such as fiberglass roving, continuous strand mat, and woven fabrics, to be fed in together to a powered storage cannister located on the tank foundation. The construction is achieved by: "back feeding the profile into an erector unit which places it in the form of a continuous helix." The simultaneous feeding of the bonding resin along with the hardener to the tank profile produces: "an exothermic reaction in the bonding resin, thereby continuously bonding the tank wall together as it is being erected."[9] Interest in the structural and cryogenic applications of pultruded glass polyester is growing; it has been proposed for the design of large cryogenic structures, such as the supporting struts of energy storage magnets.[10] Recently, the nonlinear stresses and displacements of the fibers and matrix in radially stressed pultruded polyester circular structures have also been studied.[11] Although pultruded glass polyester does not have the great strength and homogeneity of glass epoxy, its relatively low cost and ease of manufacturing make it a likely candidate for metal-composite pressure vessels.

The materials to be used for the glass-epoxy overwrap are S2 glass and NASA Resin 2 epoxy or the Lincoln (LRF-092) Resin formula, which has also proven successful in cryogenic applications. Kevlar-49-epoxy, Thronel (300 and 755)-epoxy, and graphite-epoxy overwraps are not used for the designs since the economic considerations overwhelmingly favor S glass, because weight reduction is not critical.

For the large prestressed aluminum composite tanks under consideration, aluminum alloys 5083-H 113 and 6061-T6 are adequate, since they have been used successfully for cryogenic applications and are commercially available and readily welded. Aluminum alloy 6061-T6 is used in this paper for the vessel design since it is resistant to corrosion and of high strength; it has been proposed in the design of large helium Dewar systems.[12]

VESSEL DESIGN

Designs are presented for two LNG vessels, 6 m in diameter and 12 m in length; one is overwrapped with epoxy glass and the other with pultruded glass polyester. The cylindrical portions of the vessels are 6 m in length and the caps are hemispherical, 3 m in radius. The hemispherical caps are welded to the cylinders, and the joints are reinforced with rings (frames) that insure the strain compatibility between cylinders and caps at ultimate pressure, close to burst. The aluminum cylinders are circumferentially overwrapped with the glass composite and stiffened against buckling (under the compressive prestress at zero internal pressure) by circumferential frames that are placed at 2.16-m intervals for a critical circumferential buckling stress of -172 MPa. These stiffening frames are also used for structurally supporting and fastening the free-standing vessels during transportation and operation. The aluminum liners are 47-mm thick, the glass-epoxy overwrap is 17-mm thick, and the pultruded polyester glass overwrap is 51-mm thick.

FAILURE MODES

The cylindrical vessels are designed for failure in the circumferential direction. Experiments have shown that the failure mode is controlled by the ratio of the axial stress to the circumferential stress at burst pressure. Instability occurs at a stress ratio of one, whereas a circumferential mode of failure will take place if the stress ratio is less than one.[5] A maximum burst ratio of axial stress to circumferential stress of 0.9 σ_y is assumed for the vessel designs shown.

During the normal operating pressure conditions, both the aluminum liner and the composite overwrap are in the elastic stress range; however, during the sizing (autofrettage) pressure cycle or when an accidental rise in pressure causes the aluminum to yield, the liner reaches the strain hardening, elastoplastic range, while the overwrap remains in the elastic range up to burst. Thus the circumferential failure mode is achieved when the overwrap reaches the ultimate strain level, at which it will fail, and the liner in turn strains circumferentially to a rapid failure. The axial stress in the liner is kept below yield at all times.

Because the metal liner in a metal composite vessel is relatively thin compared with the required wall thickness of an equivalent all-metal tank, the leak-before-burst failure mode is more likely to be achieved; furthermore, the thinner vessel walls and the existence of the composite overwrap allow the selection of a tough metal, which improves the probability of a leak-before-burst failure.[4]

THE PRESTRESSING CYCLE—AUTOFRETTAGE

Prior to operation, the aluminum composite vessel is pressurized to a level that drives the aluminum liner into the postyield, strain-hardening range. When the pressure is relieved, the predetermined residual stress is overcome by the tension in the composite, which in turn creates an equilibrating compression in the liner. This process, called "autofrettage" by NASA, puts a state of compressive prestress in the metal liner and a tensile prestress in the overwrap, in a manner analogous to the prestressing operation of reinforced concrete beams. This compressive prestress in the aluminum expands the useful elastic range of the liner by a factor of two.

The vessel designs presented have the properties of rather thin shells since their radius-to-thickness ratio, R/t_{ℓ}, is larger than 50. Thus, the cylindrical shells have to be stiffened against elastic buckling by circumferential stiffening frames placed at the intervals shown in Table 1. A wide flange section having flange plates of 0.193 m x 20 mm and a web plate of 0.47 m x 13 mm is adequate for the vessel designs presented.

Since pressurized vessels are, in effect, energy storage devices, with the energy stored being equal to the product of the internal pressure, ρ, and the volume of the vessel, V, the energy stored, E, is:

$$E = V\rho \tag{1}$$

The mass of the vessel, M, is subject to the virial theorem:[13]

$$M = p_d E/\sigma_T \tag{2}$$

in which p_d is the density and σ_T is the average tensile stress of the vessel. Keeping p_d and σ_T constant and changing the radii, the overall lengths, and thicknesses, a family of equivalent constant mass cylinders could be designed that may contain the same volume of pressurized LNG. Three equivalent pressurized tank designs are shown in Table 1. Since the average stress in and the densities of the glass epoxy and the pultruded glass polyester are not the same, the masses of any two equivalent tanks having dissimilar overwraps are different. It is important to note that the prestressing operation does not in any way change the ultimate burst pressure of a metal-composite vessel, it simply increases the internal pressure level at which the elastoplastic bifurcation starts in the liner.

The prestressing action easily overcomes the differential radial displacements between the aluminum liner and the composite overwrap at cryogenic temperatures. The effect of the cryogenic cooling thus is to slightly reduce the magnitude of the prestress in the liner and in the overwrap.

Table 1. Equivalent Constant Mass and Pressure Aluminum Composite Cylinders

R (m)	L (m)	t_ℓ (mm)	$(t_w)_e$ (mm)	$(t_w)_p$ (mm)	L_s (m)
2	13.5	31	11.5	34	1.44
3	6	47	17	51	2.16
4	3.38	63	23	68	2.88

R: radius
L: total length of cylinder
t_ℓ: thickness of liner
$(t_w)_e$: thickness of epoxy overwrap
$(t_w)_p$: thickness of polyester overwrap
L_s: maximum (critical) distance between two stiffening frames for a buckling stress of -172 MPa

DESIGN PROCEDURE AND RESULTS

The analysis and design of a large aluminum composite pressure vessel is highly nonlinear, thus, iterative by nature. It requires the establishment, at every iteration cycle, of stress equilibrium, strain compatibility, and the corresponding yield stress level generated by the biaxial stress conditions existing in the shell. The circumferentially wound composite overwraps offer no resistance to the axial deformation of the liner and thus develop negligible axial stresses. Ignoring the radial stresses, which is standard for thin shell analysis, and remembering that the overwrap remains elastic up to burst, while the aluminum liner becomes orthotropic in the elastoplastic, postyield state, the simultaneous equations that must be solved are obtained from the general theory of elasticity:

$$\left(\frac{u}{r} = \varepsilon_{\theta\ell} = \varepsilon_{\theta w}\right)_{\text{interface}} \tag{3}$$

$$\varepsilon_{x\ell} = \varepsilon_{xw} \tag{4}$$

In the elastic range:

$$\varepsilon_{\theta\ell} = \frac{1}{\Sigma_\ell}\left(\sigma_{\theta\ell} - \mu_\ell \sigma_{x\ell}\right) + \alpha_\ell T \tag{5}$$

$$\varepsilon_{\theta w} = \frac{1}{\Sigma_w}\left(\sigma_{\theta w} - \mu_w \sigma_{xw}\right) + \alpha_{\theta w} T \tag{6}$$

$$\varepsilon_{x\ell} = \frac{1}{\Sigma_\ell}\left(-\mu_\ell \sigma_{\theta\ell} + \sigma_{x\ell}\right) + \alpha_\ell T \tag{7}$$

In the elastoplastic range:

$$\Delta\varepsilon_{\theta\ell} = \left(\frac{\Delta\sigma_{\theta\ell}}{\Sigma_t} - \frac{\mu_{x\theta}}{\Sigma_\ell}\Delta\sigma_{x\ell}\right) + \alpha'_{\theta\ell}T \quad (8)$$

$$\Delta\varepsilon_{x\ell} = \left(-\frac{\mu_{\theta x}}{\Sigma_t}\Delta\sigma_{\theta\ell} + \frac{\Delta\sigma_{x\ell}}{\Sigma_\ell}\right) + \alpha'_{x\ell}T \quad (9)$$

The Von Mises yield condition for the biaxial state of stress in the liner is:

$$\sigma_{y\ell}^2 = \sigma_{\theta\ell}^2 + \sigma_{x\ell}^2 - \sigma_{\theta\ell}\sigma_{x\ell} \quad (10)$$

The symbols used are:

- u: radial displacement
- ε: strain
- μ: Poisson's ratio
- T: temperature
- Σ, Σ_t: Young's modulus and tangent modulus
- σ: stress
- α, α': coefficient of thermal expansion
- Δ: incremental change

The subscripts are:

- ℓ: liner
- y: yield
- θ: circumferential direction
- w: overwrap
- x: axial direction

Details of the characteristic behavior of the two aluminum composite vessels under various pressure conditions are shown in Table 2.

EXTERNAL FOAM INSULATION

The high pressure may prohibit using the usual type of internal vessel insulation; thus externally applied unreinforced polyurethane foam is used with this design. Two foam types are recommended (Stepan Bx250A and General Electric Polyurethane) because of their proven excellent thermal performance and resistance to structural and fatigue damage under thermal stress cycling (over 2400 cycles).[14]

CONCLUSIONS

Designs for high pressure aluminum composite LNG vessels are presented. The prestressed tanks exhibit high performance and safety characteristics. They are designed for a circumferential, leak-before-burst mode of failure.

Table 2. Pressure-Stress-Strain Summary of Aluminum-Composite Vessels

Temp./Operation	ρ (MPa)	$\sigma_{\theta\ell}$ (MPa)	$\sigma_{\theta w}$ (MPa)	$\sigma_{x\ell}=\sigma_{cap}$ (MPa)	$\varepsilon_{\theta\ell}$ (%)
		S Glass Epoxy Overwrap			
300 K/Yield	5.94	307	210	191.8	0.358
300 K/Sizing	6.78	319	561.9	272	0.958
300 K/Prestress	--	-110	265.6	--	--
178 K/Contraction	--	29.4	-80.4	--	--
178 K/Normal	6.21	205	538.2	200.1	0.507
178 K/Yield	9.14	361.6	631.7	295.4	0.581
178 K/Failure	10.55	376.8	828	340.2	1.117
		Glass Polyester Overwrap			
300 K/Yield	6.01	307	74	194	0.358
300 K/Sizing	8.58	320	210.7	269	1.01
300 K/Prestress	--	-119	107.5	--	--
178 K/Contraction	--	29	-26.5	--	--
178 K/Normal	6.21	229.3	156.1	200.1	0.543
178 K/Yield	8.74	361.6	191	281.5	0.581
178 K/Failure	10.59	380.9	276	341.5	1.025

E(aluminum) = 72 GPa

E(glass polyester) = 20 GPa

$E_{tangent}$(aluminum) = 2.0 GPa

E(glass epoxy) = 58.6 GPa

σ_{yield}(aluminum, 178 K) = 345 MPa

$\sigma_{ultimate}$(glass polyester = 276 MPa

σ_{yield}(aluminum, 300 K) = 276 MPa

$\sigma_{ultimate}$(glass epoxy) = 828 MPa

REFERENCES

1. R. E. Petsinger, Gas Mag. (Jan. 1978).
2. C. P. Bennett, Marine transportation of LNG at intermediate temperature, in: "Advances in Cryogenic Engineering," Vol. 25, Plenum Press, New York (1980).
3. M. M. Kamel and A. Khahil, Explosion hazards of LNG and LPG carriers during transport, in: "Advances in Cryogenic Engineering," Vol. 25, Plenum Press, New York (1980).

4. E. E. Morris, W. P. Patterson, R. E. Landes, and R. Gordon, Composite pressure vessels for aerospace and commercial applications, in: "Composites in Pressure Vessels and Piping," PVP-PB-021, ASME, New York (Sept. 1977).
5. B. H. Jones, Design and analysis of circumferentially reinforced prestressed pressure vessels, in: "Composites in Pressure Vessels and Piping," PVP-PB-021, ASME, New York (Sept. 1977).
6. F. P. Grestle, Jr. and M. Moss, Thick-walled spherical composite pressure vessels, in: "Composites in Pressure Vessels and Piping," PVP-PB-021, ASME, New York (Sept. 1977).
7. S. S. Manson, "Thermal Stress and Low Cycle Fatigue," McGraw-Hill, New York (1966).
8. R. L. Condor and N. L. Newhouse, Cyclic pressure test of a filament wound vessel containing liquid nitrogen, paper HB-4, presented at the International Cryogenic Materials Conference, Madison, Wisconsin (Aug. 21-24, 1979).
9. W. B. Goldsworthy, Modern manufacturing methods for pressure vessels and piping, in: "Composites in Pressure Vessels and Piping," PVP-PB-021, ASME, New York (Sept. 1977).
10. S. G. Ladkany, Laminated fiberglass composites for cryogenic structures in underground superconductive energy storage magnets, in: "Nonmetallic Materials and Composites at Low Temperatures," Plenum Press, New York (1979).
11. S. G. Ladkany, Nonlinear stresses and displacements of the fibers and matrix in radially loaded circular composite ring, in: "Fundamentals and Applications of Nonmetallic Materials at Low Temperatures," Plenum Press, New York (in press).
12. S. G. Ladkany, Underground rippled Dewar system for 10,000 MWh superconductive energy storage magnets, in: "Proceedings of the Seventh International Cryogenic Engineering Conference," IPC Science and Technology Press, Guildford, Surrey, England (1978).
13. W. C. Young, R. W. Boom, and S. G. Ladkany, Structural design for large superconductive magnets, presented at the World Electrotechnical Congress, Moscow, USSR (June 21-25, 1977).
14. S. L. Sharpe and R. G. Helenbrook, Durability of foam insulation for LH_2 fuel tanks of future subsonic transports, in: "Nonmetallic Materials and Composites at Low Temperatures," Plenum Press, New York (1979).

2. R. Morris, [illegible] Wettergren, [illegible] and [illegible] Composite [illegible] vessels for aerospace and commercial applications, in: "Composites in Pressure Vessels and Piping," [illegible]

3. [illegible] Design [illegible] (Sept. 1977).

4. [illegible] and M. [illegible] "[illegible]-filled composite pressure vessels," in: "Composites in Pressure Vessels and Piping," [illegible] New York [illegible]

5. S. Hanson [illegible] and [illegible] [illegible]

6. L. [illegible] and [illegible] [illegible] liquid nitrogen [illegible]

[illegible]

[illegible] Materials [illegible]

[illegible]

[illegible] Press, [illegible] (1977).

[illegible] Plenum Press, New York (1979).

AUTHOR INDEX

MATERIALS INDEX

SUBJECT INDEX